WITHDRAWN BY THE
UNIVERSITY OF MICHIGAN LIBRARY

REARRANGEMENTS IN GROUND AND EXCITED STATES

Volume 1

This is Volume 42 of
ORGANIC CHEMISTRY
A series of monographs
Editor: HARRY H. WASSERMAN

A complete list of the books in this series appears at the end of the volume.

REARRANGEMENTS IN GROUND AND EXCITED STATES

edited by
Paul de Mayo
Photochemistry Unit
Department of Chemistry
The University of Western Ontario
London, Ontario, Canada

1

1980

ACADEMIC PRESS
A Subsidiary of Harcourt Brace Jovanovich, Publishers
New York   London   Toronto   Sydney   San Francisco

COPYRIGHT © 1980, BY ACADEMIC PRESS, INC.
ALL RIGHTS RESERVED.
NO PART OF THIS PUBLICATION MAY BE REPRODUCED OR
TRANSMITTED IN ANY FORM OR BY ANY MEANS, ELECTRONIC
OR MECHANICAL, INCLUDING PHOTOCOPY, RECORDING, OR ANY
INFORMATION STORAGE AND RETRIEVAL SYSTEM, WITHOUT
PERMISSION IN WRITING FROM THE PUBLISHER.

ACADEMIC PRESS, INC.
111 Fifth Avenue, New York, New York 10003

United Kingdom Edition published by
ACADEMIC PRESS, INC. (LONDON) LTD.
24/28 Oval Road, London NW1 7DX

Library of Congress Cataloging in Publication Data

Main entry under title:

Rearrangements in ground and excited states.

   (Organic chemistry series ;)
   Includes bibliographical references and index.
   1. Rearrangements (Chemistry)--Addresses, essays,
lectures. I. Mayo, Paul de. II. Series: Organic
chemistry series (New York) ;
QD281.R35R42   547.1'39   79-51675
ISBN 0-12-481301-1 (v. 1)

PRINTED IN THE UNITED STATES OF AMERICA

80 81 82 83   9 8 7 6 5 4 3 2 1

CONTENTS

List of Contributors — vii
Foreword — ix
Contents of Other Volumes — xiii

Essay 1 **Rearrangements of Carbocations**
MARTIN SAUNDERS, JAYARAMAN CHANDRASEKHAR, and
PAUL VON RAGUÉ SCHLEYER

    I. Introduction — 1
    II. Directly Observable Carbocations — 3
    III. Multiple Rearrangement Reactions — 22
    IV. Theoretical Studies — 34
    V. Conclusions — 47
    References — 48

Essay 2 **Gas-Phase Ion Rearrangements**
RICHARD D. BOWEN and DUDLEY H. WILLIAMS

    I. Introduction — 55
    II. Importance of Metastable Peaks — 57
    III. Potential Energy Profile Approach — 62
    IV. Types of Potential Energy Profiles — 80
    V. Conclusions — 90
    References — 90

Essay 3 **Rearrangements of Carbenes and Nitrenes**
W. M. JONES

    I. General Introduction — 95
    II. 1,2 Rearrangements of Carbenes and Nitrenes — 97
    III. Type 1 Carbene–Carbene and Carbene–Nitrene Rearrangements — 119
    IV. Type II Carbene–Carbene Rearrangements — 149
    References — 153

Essay 4	**Free-Radical Rearrangements**
	A. L. J. BECKWITH and K. U. INGOLD

	I.	Introduction	162
	II.	Measurement of Rates of Unimolecular Radical Reactions	165
	III.	Rearrangement by Transfer of a Carbon-Centered Group	168
	IV.	Ring-Closure Reactions	182
	V.	Ring-Opening Reactions	220
	VI.	Rearrangement by Transfer of a Heteroatom-Centered Group	241
	VII.	Isomerization by Transfer of a Halogen Atom	248
	VIII.	Isomerization by Hydrogen Atom Transfer	251
	IX.	Rotation	270
	X.	Inversion	276
		References	283

Essay 5	**Hypothetical Biradical Pathways in Thermal Unimolecular Rearrangements**
	JEROME A. BERSON

	I.	Introduction	311
	II.	1,1 Biradicals (Carbenes)	313
	III.	1,3 Biradicals	324
	IV.	1,4 Biradicals	353
	V.	Biradicals in [1,3] and [1,5] Sigmatropic Rearrangements: Some Comments on the Principle of Least Motion	372
		References	383

Essay 6	**Rearrangements in Carbanions**
	D. H. HUNTER, J. B. STOTHERS, and E. W. WARNHOFF

	I.	Migration of Saturated Groups	392
	II.	Migration of Unsaturated Carbon	400
	III.	Migrations in Doubly Bonded Oxygen	410
		References	465

Index   471

LIST OF CONTRIBUTORS

Numbers in parentheses indicate the pages on which the authors' contributions begin.

A. L. J. Beckwith (161) Department of Organic Chemistry, The University of Adelaide, Adelaide, South Australia

Jerome A. Berson (311) Department of Chemistry, Yale University, New Haven, Connecticut 06520

Richard D. Bowen (55) University Chemistry Laboratory, Cambridge, CB2 1EW, England

Jayaraman Chandrasekhar (1) Institut für Organische Chemie, Universität Erlangen-Nürnberg, 8520 Erlangen, Federal Republic of Germany

D. H. Hunter (391) Department of Chemistry, University of Western Ontario, London, Ontario N6A 5B7, Canada

K. U. Ingold (161) Division of Chemistry, National Research Council of Canada, Ottawa K1A OR6, Canada

W. M. Jones (95) Department of Chemistry, University of Florida, Gainesville, Florida 32611

Martin Saunders (1) Department of Chemistry, Yale University, New Haven, Connecticut 06520

J. B. Stothers (391) Department of Chemistry, University of Western Ontario, London, Ontario N6A 5B7, Canada

Paul von Ragué Schleyer (1) Institut für Organische Chemie, Universität Erlangen-Nürnberg, 8520 Erlangen, Federal Republic of Germany

E. W. Warnhoff (391) Department of Chemistry, University of Western Ontario, London, Ontario N6A 5B7, Canada

Dudley H. Williams (55) University Chemistry Laboratory, Cambridge CB2 1EW, England

FOREWORD

This volume had an elder sibling, but that was by another sire. Nonetheless a family resemblance remains. In both cases the term "rearrangement" has been interpreted to mean what the editor wanted it to, at, following the precedent set by Humpty Dumpty, whatever cost, provided the interest justified it. The subjects of these separate essays, while not covering the entire span of possibilities, do indeed enclose much territory. When compared with the previously mentioned work on rearrangements ("Molecular Rearrangements," Part I, 1963) one may see, in part, the remarkable and vigorous development of chemistry over the past fifteen years.

One can discern at least three major surges of creative activity. First, the flood from the lower reaches of the periodic table now makes it impossible for the educated chemist to insist (while looking askance at boron) that, with the exception of sulfur, the essentials of chemistry are contained in the first row. Second, the period of the previous volume coincided with, and so missed, the photochemical renascence, the latter being such that virtually all laboratories are now equipped with the pale blue lights which have provided rearrangements in abundance. The principles guiding these excited state transformations are beginning to emerge, and chemistry is no longer a single surface subject. Indeed, even the ubiquitous interest in energy hypersurfaces dates from the birth of the earlier volumes. Further, some aspects of photochemistry were the stimulus for, perhaps, the most fruitful concept of this period: the emergence, or, better, recognition of the view that the orbitals of reacting molecules guided their transformation by placing barriers on certain pathways, hitherto apparently as permitted as others, in order to maintain symmetry or maximum overlap: the orbital correlation diagram has, in this period, already found its way into the undergraduate text. The third supernova is, therefore, the descent of orbital theory into the marketplace. In short, it has become respectable, indeed barely highbrow, to be a theoretician; and their mixing in polite society at meetings with experimentalists has become a commonplace and a stimulation. All these changes—imagine the state of the art of the subjects of these essays fifteen years ago—in a subject deemed, by some, dead.

As previously, the essays are intended to be critical, stimulating, personal, and creative: they are not intended to be comprehensive, though adequate referencing should make them a starting point for research.

Some subjects will be found absent. In some of these too little change has occurred in fifteen years to justify a new treatment. In other instances adequate treatment elsewhere rendered an essay unnecessary. And it must be admitted some lacunae are caused by the willingness of the spirit of some potential authors, but, alas, their weakness of the flesh. . . .

The arrangement is roughly the following: first, ground state transformations; a bridge passage of theory, with a codetta of those interesting transformations that start on the ground state surface and finished in the excited state; and, finally, transformations on upper surfaces. But the separation is, in part, a formalism: silicon chemistry includes photochemistry, and cis–trans isomerism may be thermal—a sign that chemistry, once a whole and then fragmented, is gradually becoming unified again.

PAUL DE MAYO

"Contrariwise", *continued Tweedledee, "if it was so, it might be; and if it were so, it would be: but as it isn't, it ain't. That's logic."*

L. Carroll

. . . Western culture has been profoundly affected by the findings, the methodology, the attitudes, and the outlook of scientists. It is not wholly the fault of scientists that what has passed into the general culture is grossly distorted in two critical ways. One is the mistaken identification of science with rationality. The other is the exaggerated dichotomy between science and nonscience. *

G. Vickers

* Reprinted with permission of MIT Press, Cambridge, Massachusetts.

CONTENTS OF OTHER VOLUMES

VOLUME 2

Rearrangements: A Theoretical Approach
  Nicolaos D. Epiotis, Sason Shaik, and William Zandar

Rearrangements Involving Boron
  Andrew Pelter

Molecular Rearrangements of Organosilicon Compounds
  A. G. Brook and A. R. Bassindale

The Polytopal Rearrangement at Phosphorus
  F. H. Westheimer

Rearrangements in Coordination Complexes
  W. G. Jackson and A. M. Sargeson

Fluxional Molecules: Reversible Thermal Intramolecular Rearrangements of Metal Carbonyls
  F. A. Cotton and B. E. Hanson

Index

VOLUME 3

Chemical Generation of Excited States
  N. J. Turro and V. Ramamurthy

Cis–Trans Isomerism of Olefins
  J. Saltiel and J. L. Charlton

Photochemical Rearrangements in Trienes
  W. G. Dauben, E. L. McInnis, and D. M. Michno

The Di-π-Methane (Zimmerman) Rearrangement
  Howard E. Zimmerman

Photochemical Rearrangements of Enones
  David I. Schuster

Photochemical Rearrangements of Conjugated Cyclic Dienones
  Kurt Schaffner and Martin Demuth

Rearrangements of the Benzene Ring
  D. Bryce-Smith and A. Gilbert

Photorearrangements via Biradicals of Simple Carbonyl Compounds
  Peter J. Wagner

Photochemical Rearrangements Involving Three-Membered Rings
  Michel Nastasi and Jacques Streith

Photochemical Rearrangements of Five-Membered Heterocycles
  Albert Padwa

Photochemical Rearrangements of Coordination Compounds
  Franco Scandola

Index

ESSAY 1 | **REARRANGEMENTS OF CARBOCATIONS**

MARTIN SAUNDERS,
JAYARAMAN CHANDRASEKHAR,
AND PAUL VON RAGUÉ SCHLEYER

I.	Introduction	1
II.	Directly Observable Carbocations	3
	A. Rapid Degenerate 1,2-Shifts	5
	B. Deuterium Perturbation Method for the Nonclassical Ion Problem	8
	C. Hydride Shifts to More Distant Carbons	11
	D. Solvents and Solvent Effects	14
III.	Multiple Rearrangement Reactions	22
	A. Monte-Carlo Methods of Analysis	23
	B. Use of Graphs to Elucidate Rearrangement Mechanisms	24
IV.	Theoretical Studies	34
	A. Migratory Aptitudes	34
	B. Effect of Ring Size on the Ease of 1,2-Shifts	41
	C. Rearrangements in Cycloalkenyl Cations	44
V.	Conclusions	47
	References	48

I. INTRODUCTION

Rearrangements are characteristic of carbonium ion reactions. Carbanions, free radicals, and other intermediates are far less prone to such transformations. Carbocations are electron-deficient species; they do not possess enough electrons to satisfy the octet rule for each atom and simultaneously to provide two electrons for each bond. Electron distributions that differ significantly from classical bonding models result. Often a repositioning of the nuclei occurs in order to increase the number of bonding contacts. In effect, carbocations have a lower inherent structural integrity than more electron-rich species; carbocation potential energy surfaces are flatter, and the tendency to rearrange and to form bridged or otherwise delocalized structures is great.

The first work in which carbonium ion rearrangements were observed was carried out in nucleophilic solvents where cation lifetimes were immeasurably short (*1*). Under these conditions, the rates of individual rearrangement steps could not be determined. The structures of the carbocation intermediates themselves could be inferred only indirectly from, for example, the rate- or product-determining transition states in a solvolysis. Carbocation systems capable of undergoing degenerate rearrangement to the same chemical structure are illustrative. If isotopic or other labels in the products observed are not distributed statistically, it is generally assumed that classical carbocation intermediates that do not have sufficient time to equilibrate before quenching are involved. When the statistical mixture of products is obtained, either fully equilibrated sets of classical cations or symmetrically bridged species (the so-called nonclassical ions) are considered to be responsible. Kinetic data have often been used to attempt to differentiate between these possibilities. However, solvolysis rates give information only about the *transition states* between the starting materials and the ions. The transition states are not always closely related in structure and energy to the ions themselves. Quantitative prediction of solvolysis data is not yet possible. As a result, the interpretation of solvolytic data in order to draw conclusions about the structures of the ions often involves uncertainty and engenders controversy.

Since the appearance of Part I of this volume's predecessor (*2*), several major developments have occurred that have increased our understanding of carbonium ions and their rearrangements. First, carbocations as stable species in solution in nonbasic media can now be investigated directly (*3*). Many new techniques have been developed for this purpose or have been applied to carbocation problems.

Second, the chemistry of gas-phase ions has come of age (*4*). Although not much structural information can be obtained directly, an increasing body of invaluable quantitative data on the energies and reactions of ions in the gas phase is being assembled, which, *inter alia*, facilitates the interpretation of processes occurring in solution. We can now begin to ascertain the degree to which solvent affects the properties of ions (*5*).

Third, theoretical calculations have become a valuable source of detailed information not available otherwise (*6*). The structures of only a few stable carbocations and of no rearrangement transition states are known experimentally. Ground and transition state structures can, however, be calculated theoretically with a sufficiently high degree of accuracy, at least for some interpretative purposes.

Fourth, it is now appreciated that typical solvolysis reactions, especially of primary and secondary substrates, do not involve "free" carbocation intermediates but rather highly solvated "cationoid" species with partially positive character (*7*). Nucleophilic solvent assistance can play a

1. REARRANGEMENTS OF CARBOCATIONS

dominant role in such reactions. Often there is competition between such solvent assistance (which leads to unrearranged products) and neighboring group participation (often leading to rearrangement) (8). Early experiments on the "ethyl cation" in solution provide an example (9). The failure to find large amounts of label scrambling would now be attributed to the inability to form an unencumbered ethyl cation under the conditions employed, rather than to any difficulty in forming a hydrogen-bridged transition state or intermediate. Rather than attempt a comprehensive literature survey of examples of carbonium ion rearrangement reactions, this review will concentrate on the carbocations themselves and the rearrangements they undergo. It should be appreciated that many rearrangements involve *static* carbocation intermediates. For example, conversion of **1** to **2** via **3** is considered to be a "carbonium ion rearrangement," although the allyl cation intermediate **3** is a static species not undergoing any transformation.

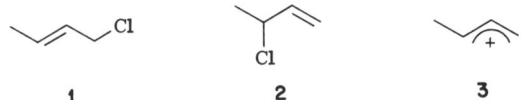

     1   **2**   **3**

A number of reviews subsequent to that referred to above (2) have discussed carbonium ions and their rearrangements (*10–15*). Besides those appearing in advanced texts (*10*), we mention, in particular, the five-volume series *Carbonium Ions* (*11*), a monograph of the same title (*12*), and others (*13,14*). One (*14a*) discusses lucidly, and in great detail, the orbital symmetry rules governing sigmatropic shifts in carbocations, and another debates the nonclassical ion problem (*15*).

This review is intended to be interpretative and is restricted largely to smaller systems. The application of newer experimental techniques based largely on NMR spectroscopic investigations of stable carbonium ions and the results of theoretical studies are emphasized. Sections discussing the analysis of complex multistep rearrangements by graph and by computer methods are complementary. The rapidly developing area of gas-phase rearrangements is discussed by Williams in Essay 2 of this volume.

II. DIRECTLY OBSERVABLE CARBOCATIONS

A great number of carbocations are directly observable in stable ion media (*3–3b*) or in the gas phase (*4, 16*). Stable solutions of most simple acyclic, monocyclic, and polycyclic tertiary cations, many ions with stabilizing groups (substituted allyl, benzyl, dienyl, *anti*-7-norbornenyl, 2-norbornyl, 2-bicyclo[2.1.1]hexyl, etc.), and cyclopropylcarbinyl and a few simple secondary cations (isopropyl, *sec*-butyl, and cyclopentyl) have been prepared.

It has recently been found that the medium-ring secondary cations (C_8–C_{11}) can be observed directly (*17, 18*). They are apparently stabilized and protected against ring contraction to the smaller tertiary cations by transannular hydride bridging. Such ring contraction does occur at higher temperatures.

Although it is often convenient to formulate carbonium ion rearrangements by invoking primary carbocations, the available evidence suggests that most, if not all, such species are incapable of independent existence (*19*) unless stabilizing substituents are present. That is, simple primary cations do not seem to be local minima on potential energy surfaces (*20*). Protonated cyclopropanes or other bridged ions are probably involved instead (*21*). However, in a simple case, the isopropyl cation has been found to undergo hydrogen scrambling about three or four times faster than carbon scrambling (*22*), a result that can readily be explained using an *n*-propyl cation or equivalent intermediate.

In other cases, it may not be possible to observe a carbocation species directly, because the barrier to rearrangement to an isomeric ion of lower energy is too easily overcome. Thus, although there is much evidence for the intermediacy of the cyclohexyl cation **5** in solvolysis experiments, rearrangement of this ion to the 1-methylcyclopentyl cation **6** occurs too rapidly under stable ion conditions [or in the gas phase (*23*)] to permit detection.

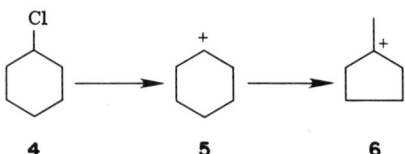

As a consequence of such rapid rearrangements, only a handful of secondary and a few stabilized primary ions are directly observable in stable ion media. Seemingly simple ions, such as the methyl and ethyl cations, cannot be prepared in currently available stable ion media, since the equilibrium appears to be unfavorable to their formation. It is doubtful if the methyl and ethyl cations can have more than a transitory existence in solution.

Indirect studies of species that cannot be prepared as stable ions are often possible. A number of stable carbonium ions undergo 1,2-shifts uphill in energy (i.e., tertiary to secondary or secondary to primary rearrangements). Such transformations are reversible. They can be observed because the complete cycle leads to carbon or hydrogen scram-

1. REARRANGEMENTS OF CARBOCATIONS

bling. Such reactions can be followed through the use of isotopic tracers, but it is more convenient to study them in stable ion solutions at temperatures (0°–20°C) where the reactions are so fast that NMR lineshape changes can be observed (*3b, 24–25*).

From lineshape analysis, the rates of the overall reactions can be obtained. Since the downhill part of the reaction is expected to be rapid and to require negligible activation, the barrier for the uphill process is expected to *approximate* the energy difference between the more stable and the less stable ion. Such energy differences do not relate directly to the 1,2-shift step itself. Reactions that do not proceed uphill (degenerate rearrangements) are necessary to obtain such information.

Line broadening in the proton spectrum of methylcyclopentyl cation (*21, 25*) can be accounted for by the scheme shown in Eq. (1). The activation energy E_a is found to be 15.4 kcal/mol, which is an approximate value for the difference in energy between the tertiary and secondary ions here. The exact value must be reduced by the barrier to the *downhill* step (possibly 2–3 kcal/mol).

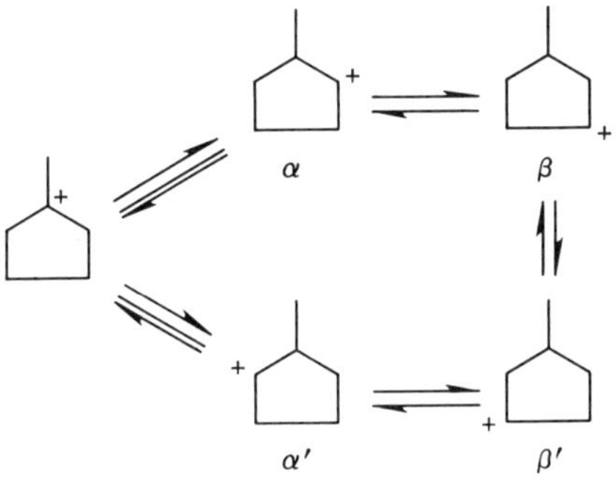

$$(1)$$

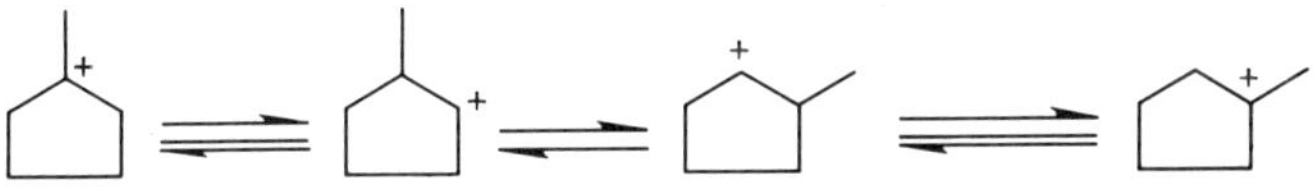

$$(2)$$

The methide shift in this system [Eq. (2)] can be studied through the observation of the *carbon* spectrum, which is unaffected by the hydride shifts described above (*25a*). The rate of the process involving methide shift was found to be one-sixteenth the rate of the hydride shift process. This implies that the transition state for methide shift is 1.5 kcal/mol higher in energy than that for the hydride shift.

A. Rapid Degenerate 1,2-Shifts

If the product of a rearrangement reaction is chemically indistinguishable from starting material (degenerate rearrangement), the barrier may simply be the energy required to form a symmetrically bridged transition state. Alternatively, this bridged species could be a local minimum or even the more *stable* form. In the latter case, there would be *no* barrier. Many simple acyclic and monocyclic tertiary and secondary cations, e.g., **7–10**, capable of undergoing degenerate 1,2-shifts of hydride and methide, give averaged proton and carbon peaks in their NMR spectra with the usual methods of measurement, even at low temperatures. If a rate process is more rapid than a limit determined by the NMR chemical shift difference in frequency [approximately (hertz)^2] between the nuclei that are being interchanged, single sharp lines will appear in the spectrum for groups of nuclei that are interchanged by this process at positions which are the weighted averages of the frequencies being interchanged.

However, it is possible to measure the rearrangement rates for some of these ions by observing the CMR spectra at low temperatures ($-140°C$) using a superconducting spectrometer to increase the frequency separation. The rates are of the order of 10^7 sec^{-1}, and the barriers range from 3.1 to 4.4 kcal/mol (26). Therefore, these ions possess double-minimum energy surfaces with barriers for rearrangement.

Other degenerate rearrangements have been found to have higher barriers. The transition states for these rearrangements seem very likely to be similar to those in the reactions described above. The higher barriers must then be due to lower energies in the ions themselves. In the following two cases [Eqs. (3) and (4)], this stabilization is probably a result of nonclassical delocalization:

1. REARRANGEMENTS OF CARBOCATIONS

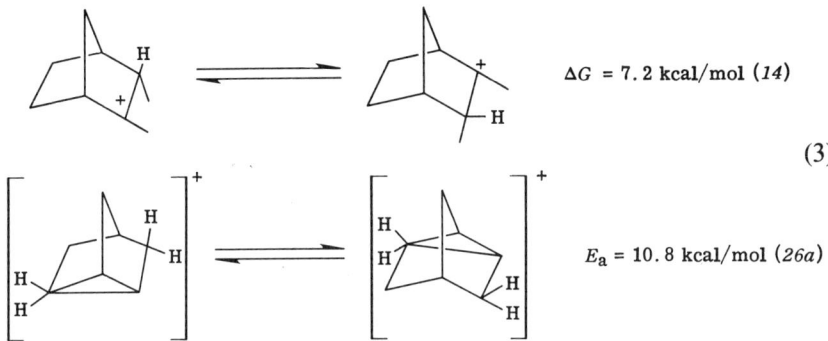

$\Delta G = 7.2$ kcal/mol (*14*)

$E_a = 10.8$ kcal/mol (*26a*)

(3)

Failure to observe line broadening at low temperature for ions capable of undergoing degenerate rearrangements may, however, still be consistent with a similar energy surface with a somewhat lower barrier, in the range from 3 kcal/mol to zero. If, however, the barrier *vanishes* and the double-minimum surface goes to a single-minimum surface, the stable structure is now the *bridged* ion. The NMR lineshape does not distinguish between these two situations. Therefore, since neither the interpretation of solvolytic rates nor the observation of NMR lineshapes settles the question, there has been room for much speculative argument, based mainly on whether certain models are adequate for the prediction of solvolysis rate constants.

In cases in which the bridged ions are very much more stable than the nonbridged ions, the rates of solvolysis reactions often provide compelling evidence. Reaction rates are typically accelerated by very large factors compared with the solvolysis of model substances. A case in point is the solvolysis of *anti*-7-norbornenyl tosylate (**11b**) compared with 7-norbornyl tosylate (**11a**) (*27*).

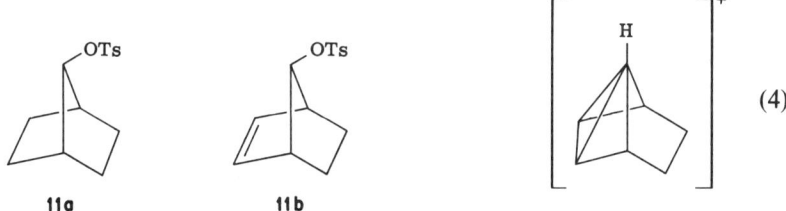

11a   **11b**

(4)

A rate enhancement of 10^{11} is convincing evidence for the involvement of the double bond, not only in the ionization transition state, but also in the structure of the resulting ion. However, when the rate enhancements are smaller, one can dispute the suitability of the model systems and how the results might be corrected for various steric and structural effects. Such discussions have not led to universally accepted conclusions (*15*).

B. Deuterium Perturbation Method for the Nonclassical Ion Problem

The introduction of deuterium into ions undergoing rapid degenerate 1,2-shifts breaks the symmetry and leads to revealing results (*26, 28–31*). Equilibrium constants between the structures rapidly interchanging are no longer unity, since the structures and hence the energies are made slightly different by the isotope. As a consequence, pairs of nuclei that gave averaged, single peaks in the nonisotopic substance now give separate lines. The observed splitting between these lines δ is related to the chemical shift difference Δ that would be seen if there were *no* exchange and the isotopic equilibrium constant K. The equation for this splitting is

$$\delta = \Delta(K - 1)/(K + 1)$$

In the following case, the observed CMR splitting between carbon atoms 1 and 2 in **12** and **13** was found to be 81.8 ppm at $-142°C$ as a result of an equilibrium deuterium isotope effect, $K = 1.91$ (*31*):

$$D_3C\ H\ CH_3 \quad \rightleftharpoons \quad D_3C\ H\ CH_3$$

12 **13**

The largest effects seen so far are in the ^{13}C spectra of carbonium ions where the atoms, alternating between being positively charged and neutral, are characterized by particularly large frequency differences. The β-deuterium isotope effects are due to hyperconjugative effects on the force constants for stretching and bending CH bonds adjacent to a carbonium center; these are also a function of the dihedral angle (*31, 32*). Splittings in simple monocyclic systems are in the range of 25 ppm per deuterium on methyl and 50 ppm per deuterium on methylene. The cases in which the heights of the barriers on the double-minimum energy surface have been found to be 3–4 kcal/mol show such splitting. However, large splitting has also been observed in several cases in which the height of the barrier is lower and it is not possible to measure the rate using the NMR lineshape method (*32a*) [Eq. (5)].

$$D\diagup\!\!\!\!\diagdown\overset{+}{\diagup}\!\!\!\!\diagdown \quad \rightleftharpoons \quad D\diagup\!\!\!\!\diagdown\underset{+}{\diagup}\!\!\!\!\diagdown \qquad (5)$$

When deuterium is used to break the symmetry of substances known to be symmetric, nuclei originally identical due to symmetry are also split in frequency. Cyclohexenyl (**14**) and cyclopentenyl (**15**) cations, with *no* rapid rearrangement processes occurring and with a single energy minimum, show very small splittings in the ^{13}C spectrum upon the introduction of deuterium ($\delta \cong 0.5$ ppm) (*33*).

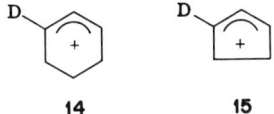

14   15

In the tricyclononyl cation (**17**) reported by Coates (*34*), the proton and carbon spectra establish that the ion *must* have a nonclassical structure (**16**) (rapidly equilibrating C_s or C_{2V} alternatives are ruled out since they must lead to more extensive degeneracy than is observed). The introduction of deuterium results in a spectrum in which the splitting is only 0.1 ppm or less (*32a*).

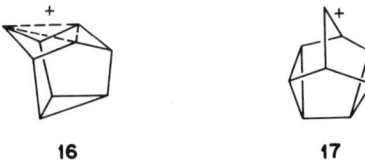

16   17

Thus, in species where rapid, degenerate, equilibria are occurring, deuterium substitution results in characteristically large splittings. On the other hand, when a system possesses *true* symmetry rather than time-averaged molecular symmetry, the corresponding splittings are more than an order of magnitude smaller.

This method has been applied to two controversial cases. In the bicyclo[2.1.1]hexyl cation **18**, the introduction of deuterium as shown resulted in a splitting in the ^{13}C spectrum of 1.2 ppm, a value far smaller than that to be expected for a rapidly equilibrating system (*35*). It was concluded that the ion is bridged (nonclassical) (**19**).

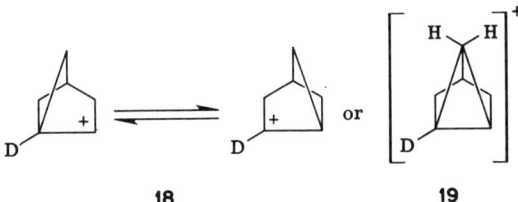

18   19

There is a complication in the norbornyl cation **20** (*36*). Besides the Wagner–Meerwein rearrangement, a further rapid process, the 6,2-hydride shift [Eq. (6)], has a barrier of only 5.9 kcal/mol (*36a*) and results in a certain amount of unavoidable line broadening.

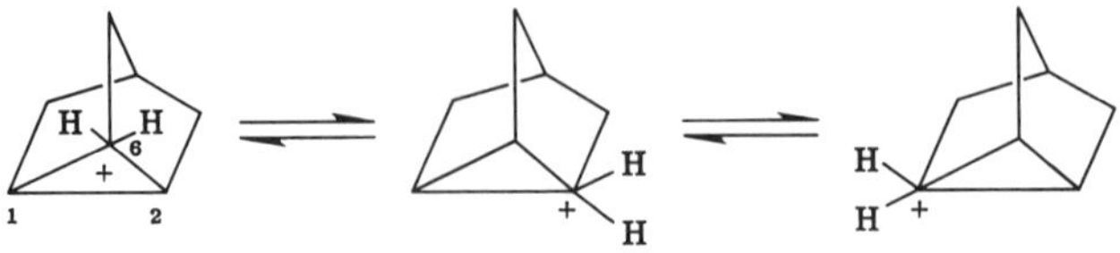

Even in the ion with no deuterium, the downfield carbon line (due to the equilibrating carbons 1,2, and 6) is found to be 2 ppm wide (*32a*). Nevertheless, no *additional* isotopic splitting or broadening can be discerned in either the 2-monodeuterio or the 3,3-dideuterio ions, and therefore the isotopic splitting δ can be no more than 2 ppm. This result can be compared with those for methylene-deuterated dimethylcyclopentyl (**21**) and dimethylnorbornyl (**22**) cations, models for rapidly equilibrating classical ions. The <2 ppm splitting found for the norbornyl cation **23** is quite different from the values measured in the cases of equilibrating, classical ions and is in the range expected for a symmetric, nonclassical ion.

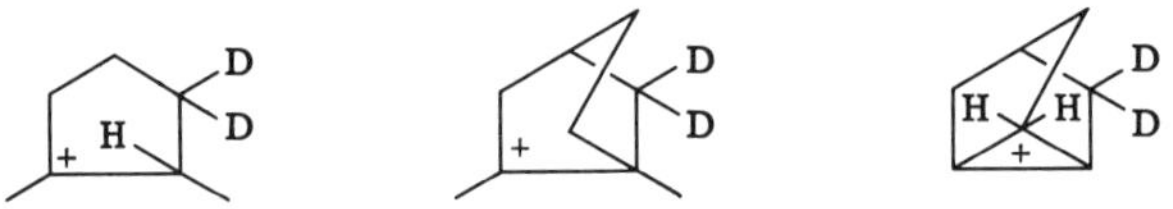

The 1,2-dimethylnorbornyl cation **22** shows a splitting (δ = 24 ppm) on deuteration of one of the methylene groups, which is reduced by about a factor of 4 from the splitting observed in the analogous open-system 1,2-dimethylcyclopentyl cation **21** (δ = 104 ppm). This may indicate partial delocalization in this ion as it tends toward the bridged structure due to reduction in the barrier between the equilibrating ions. The smaller this barrier, the smaller the difference between these structures. The equilibrium isotope effect K would become closer to unity and the chemical shift difference between interchanging carbons Δ would decrease, both changes reducing δ. Thus, this case may represent an *intermediate* situation, with a low barrier between two structures that are partly delocalized. The NMR splitting is reduced as a result.

Recently, this new method has been applied to the question of whether the 1,6-dimethylcyclodecyl cation is classical (**24**) or bridged (**25**). The ^{13}C spectrum showed a *single* peak for carbons 1 and 6 resulting either from a very rapid 1,6-hydride shift or from the true symmetry of the bridged, nonclassical, ion. It was found that the ion prepared with one deuteromethyl group showed a splitting of only 0.5 ppm, clearly supporting the bridged structure (**25**) (*18*).

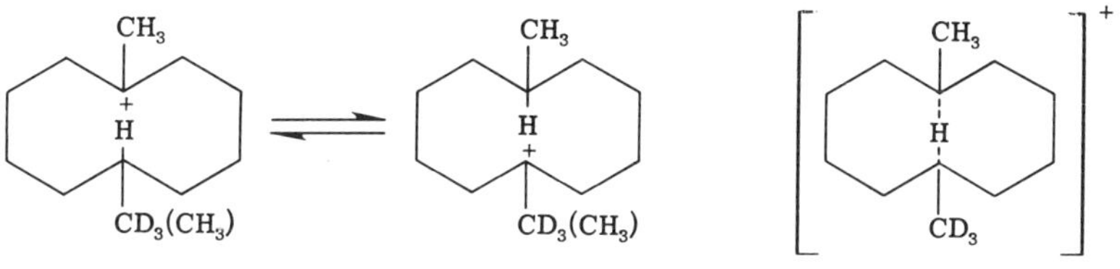

24 **25**

C. Hydride Shifts to More Distant Carbons

Since 1,2-shifts in carbonium ions are so common and rapid, the possibility of similar 1,3-, 1,4-, and 1,5-shifts aroused early speculation. However, a sequence of 1,2-shifts can often yield the same result as a 1,3- or more distant shift, and the unambiguous demonstration of mechanism can be difficult. Transannular hydride rearrangements in medium-sized rings were the first well-established class of such migrations (*37*).

1. 1,3-Hydride shifts

The original evidence obtained from the use of isotopic labeling in solvolysis studies (*38*) was disputed (*39*), and the first unambiguous evidence was obtained through the study of the stable cation **26**, which was designed to undergo a *degenerate* 1,3-hydride shift (*40, 41*).

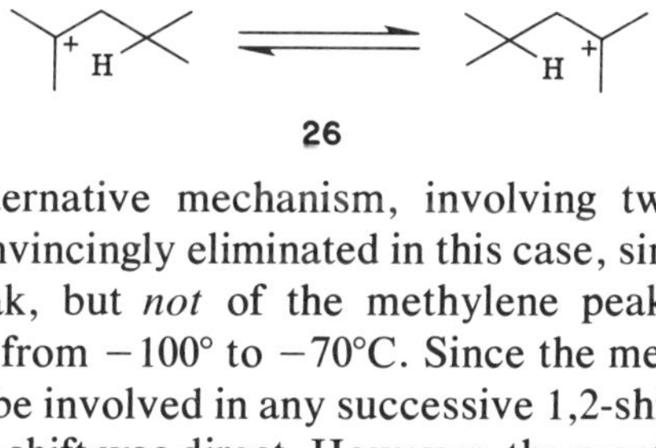

26

The nagging alternative mechanism, involving two successive 1,2-shifts, could be convincingly eliminated in this case, since line broadening of the methyl peak, but *not* of the methylene peak, occurred in the temperature range from $-100°$ to $-70°C$. Since the methylene hydrogens would necessarily be involved in any successive 1,2-shift mechanism, this proved that the 1,3-shift was direct. However, the reaction may still occur in successive steps involving closure to a corner-protonated cyclopropane, a corner-to-corner proton shift, and then opening to the product ion [Eq. (7)].

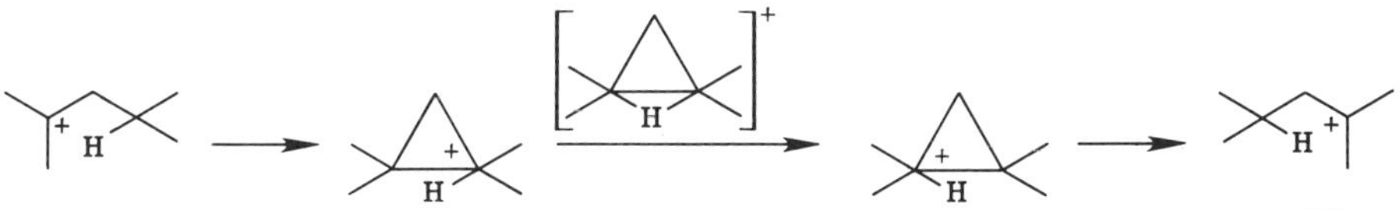

(7)

This possibility seems unlikely on the basis of estimates of the energies of substituted protonated cyclopropanes (21); their involvement would require more energy than the experimentally found 8.5 kcal/mol activation energy. Therefore, at least in this ion, a direct 1,3-hydride shift appears to occur. This is actually a question of the geometry of the transition state, which now can be explored by theoretical calculations. A similar degenerate rearrangement was explored in 1,3-dimethylcyclohexyl cation 27; the barrier was found by NMR line shape methods to be 10.5 kcal/mol (22).

27

Incorporating the system into a ring constrains the transition state and raises the activation energy somewhat. Perhaps this implies that a transition state in which the hydride is co-planar with the three carbons involved (28) is preferred in the open system. This geometry cannot be achieved in the cyclohexyl system.

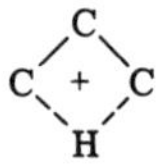

28

2. 1,4-Hydride shifts

The same strategy can be used to explore 1,4-hydride shifts. Up to $-70°C$, no line broadening was observed in the proton spectrum of the 2,5-dimethyl-2-hexyl cation 29 (41). Above this temperature, isomerization to other tertiary ions occurred. However, preparation of ions labeled with two deuteromethyl groups on one end led to complete scrambling (both methyl peaks were reduced 50%). It was possible to measure the rate, using the magnetization transfer technique, by irradiating one methyl peak and observing that the intensity of the other peak was decreased. A value of approximately 12–13 kcal/mol was estimated for the barrier.

29

The possibility of a protonated cyclobutane (42), analogous to protonated cyclopropane, seems unlikely, and a direct hydride transfer mechanism appears to be appropriate. The distinction is a rather subtle one depending on the degree of carbon–carbon bonding involved (see below).

When such a 1,4 system was incorporated into a ring, the same process

1. REARRANGEMENTS OF CARBOCATIONS

could be studied; NMR line broadening could be observed in the proton spectrum of the 1,4-dimethylcyclohexyl cations **30**. Line broadening could be observed here, but not in the open system. Only the 1,2- and

30

1,3-dimethylcyclohexyl cations were in equilibrium with the starting ions at temperatures ($-60°$ to $-40°C$) where line broadening could be seen. In the open system, many other isomers formed rapidly at temperatures above $-70°C$. Magnetization transfer was used to ascertain whether the process seen in the 1,4-dimethylcyclohexyl cation could have occurred via a series of successive 1,2-shifts, and, under conditions where strong magnetization transfer was observed between the peaks due to the different methyls, the peaks due to the methine and methylene groups showed no such effects.

This experiment seems to rule out the successive 1,2-shift mechanism. If so, the 1,4-shift barrier is about 13 kcal/mol. The geometry of this system does not permit a transition state in which hydrogen is collinear with the two carbons, so this can be ruled out. If appreciable bonding between the carbons were required, a considerable amount of additional strain due to approach to the geometry of the highly strained bicyclo[2.2.0]hexyl system would result. Since the barrier seems to be only a few kilocalories per mole above the barrier in the open compound, it may be inferred that the transition state does not require a close approach between these carbons. A bent transition state, but not one requiring extensive cyclic three-center bonding, seems required for this case.

In more general terms, there appears to be a spectrum of transition states for these hydride shifts ranging from the collinear to forms so bent that considerable interaction between the carbon atoms occurs. These alternatives have been investigated theoretically for the simplest system, $C_2H_7^+$ (H_3C—H—CH_3^+) (6, 43, 43a). The results agree. They imply that no great energy penalty is paid by changing the CHC angle over a considerable range. Another way of saying the same thing is that the force constant for bending the bond angles of the dicoordinated (hypervalent) hydrogen is thus indicated to be small.

3. 1,5-Hydride shifts

Stimulated by reports of transannular hydride transfers in medium-sized rings (37, 44) and by a suggestion of Arigoni concerning a hydride shift mechanism for a remote oxidation that occurred in the course of an acid-induced reaction (45), a possible 1,5-hydride shift in the stable 2,6-dimethylheptyl cation **31** was examined. At the lowest temperatures available, the stable ion exhibited only a *single averaged peak* for the four methyl groups. This implies that a 1,5-hydride shift occurs with a barrier of less than 6 kcal/mol or that a symmetric bridged structure (**32**) may be present. Here the carbon framework permits the transition state for hydride transfer to be practically linear; the energy of the transition state compared to the energy of the starting ions apparently decreases to at least 2.5 kcal/mol below that for the 1,3-shift. As already mentioned, similar observations were recently made in medium-sized ring cations (17).

31   32

The rate of hydride transfer between isobutane and *t*-butyl cation, which is the intermolecular counterpart of the rearrangements discussed above, has been measured. An E_a of 3.6 kcal/mol was obtained, and $\Delta S^\ddagger$ was found to be -27 e.u. (46). The transition state for this transfer is presumed to be linear.

D. Solvents and Solvent Effects

Since much of the experimental work discussed in this chapter concerns stable ion solutions, it is necessary to consider the solvents used and their possible effects on the properties and energies of the ions. This leads to the broader question of solvation of carbocations, especially in more nucleophilic media. In what way, and to what extent, do carbocations and their rearrangements differ in the gas phase, in nonnucleophilic media, and in typical solvolysis solvents?

Many tertiary aliphatic, acyclic, and cyclic cations can be prepared by the reaction of halide, alcohol, or alkane precursors with excess SbF_5 in liquid SO_2 or HF with or without admixture of HSO_3F. However, secondary isopropyl, *sec*-butyl, and cyclopentyl ions cannot be prepared using these solvents. In order to make stable solutions of these ions, it is necessary to use SO_2ClF or SO_2F_2 or their mixtures. The media containing SbF_5 and these two solvents are, at present, the best available for keeping the widest variety of carbocations in stable solution.

These media are often collectively referred to as "super acid" ("magic acid" is a mixture of SbF_5 and HSO_3F and is an inferior solvent for preparing ions that are difficult to obtain) (3). Although it has been in use

1. REARRANGEMENTS OF CARBOCATIONS

since 1927 (47), it is felt that "super acid" is a poorly descriptive term, since it is not just the acidic (either protonating or Lewis acid) properties of the solvent that are important in ensuring the stability of the carbonium ions, but the lack of either basic or nucleophilic reactivity. Naturally, in order to form solutions of these substances, which are actually salts, a medium with fairly high dielectric constant is necessary as well. The extreme lack of basicity of SbF_5 was illustrated by the observation that t-butyl cation d_6 (two CD_3 groups) suffered only a small amount of deuterium scrambling ($\sim 10\%$) after several hours at 100°C (48). When HSO_3F was added, the exchange process was faster due to the somewhat greater basicity of this solvent.

Solutions of t-amyl cation in pure SbF_5 can be heated above 150°C without decomposition. On cooling, the NMR spectrum is found to be unchanged. Traces of water or other nucleophiles do not affect the properties of these media significantly since the excess SbF_5 present is such a powerful Lewis acid that it complexes such nucleophiles very effectively. The solutions described above, shall be referred to collectively, as stable ion media.

What are the effects on the structures and energies of carbocations in going from the gas phase into stable ion media or into media commonly used for solvolysis reactions, e.g., acetic acid or aqueous acetone? Since solvation energies are very large (~ 80 kcal/mol for Cl^- in H_2O and a similar value for the t-butyl cation) (47, 49), differences of solvation among different ions in different media might well be appreciable. This expectation was supported by early calculations based on electrostatic models and assumed ionic radii for various carbocations (50). The estimated solvation energies were indicated to depend on the solvent, the number of carbon atoms, and the position of the carbocation center. If so, changes in equilibria and in related energies from one solvent to another would be substantial. In extreme cases, detailed carbocation structures might change on altering the medium. Thus, observations on carbocations in the gas phase, in stable ion media, and during solvolysis reactions might pertain to quite different species, and it might be expected that data from such sources would not be comparable.

Nevertheless, the picture that emerged after sufficient experimental data had been collected is much less complicated. The extent of solvation is found in many instances "not to vary significantly among different carbonium ions" (51). For example, several resonance-stabilized carbocations have nearly identical heats of solution in water (in comparison with the gas phase) (52).

The differences in the energies of isomeric tertiary and secondary carbocations (t-butyl versus 2-butyl) in the gas phase are around 16 kcal/mol (*16*). This value is similar to that obtained in solution from scrambling reactions of tertiary ions proceeding via secondary ions (*53*). The direct measurement of the heat of isomerization of the *sec*-butyl to the *t*-butyl cation is 14.5 ± 0.5 kcal/mol in stable ion media (*54*). The uncertainties of measurement in these experiments are comparable to the differences in the reported values. Finally, when solvolysis rate data, *corrected for nucleophilic solvent assistance,* are compared with directly measured heats of ionization in stable ion media, a linear correlation with a slope near unity is found (*55*). This important finding underscores the long-assumed relevance of solvolysis data to carbocation stability.

The methyl scrambling process in the *t*-amyl cation [Eq. (8)] was measured in SbF_5/SO_2, SbF_5/liquid HF, SbF_5/SO_2ClF, and a SO_2ClF/HSO_3F mixture (*24, 48, 56*). The rates were found to be the same within experimental error. Thus, there is not appreciable change in solvation of the ion and of the rearrangement transition state despite the differences in dielectric constants and in hydrogen-bonding abilities of the solvents.

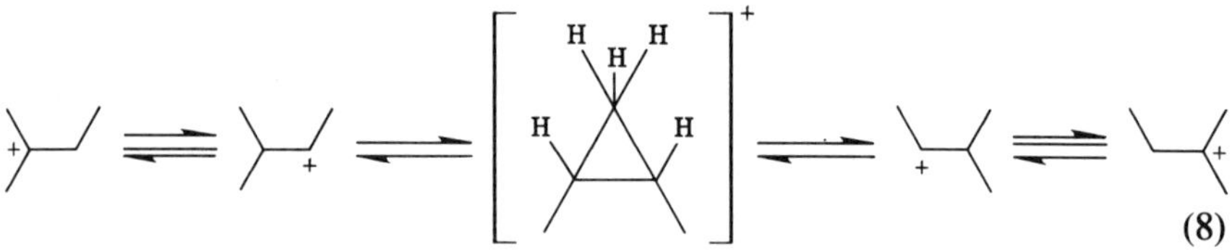

$$\tag{8}$$

The activation entropies for a number of scrambling processes of carbonium ions in stable ion media have been reported (*21*). All are close to zero. If a substantial change in solvation occurred on going to the transition state, one would expect that solvent would be released or gained and large entropy changes should result. None are noticeable.

Solvent effects on a variety of rearranging systems have been studied. Methyl scrambling in the heptamethylbenzenonium ion (**33** ⇌ **35**, etc.), demonstrated to occur by 1,2 shifts (*25a*), has the same activation energy in 9.3 N H_2SO_4, in concentrated HCl, and even in HCOOH, a commonly used solvolysis solvent (*56*). This means that the relative solvation energies of the ground state (**33**) and the transition state (**34**) are not altered in solvents of widely differing character.

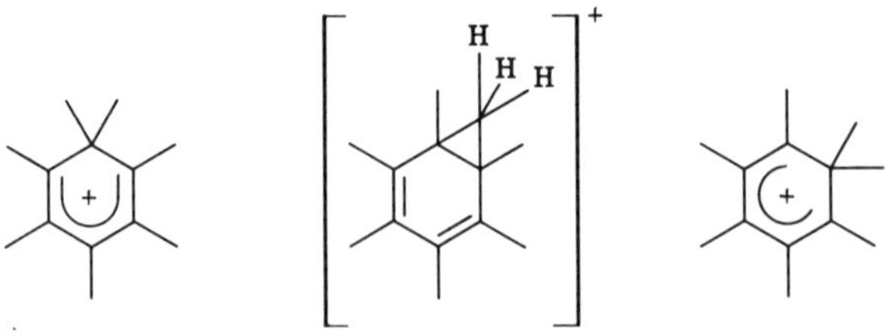

33 **34** **35**

The effect of various concentrations of aqueous sulfuric acid on the reversible isomerization between allylic systems, e.g., **36** ⇌ **37** and **38** ⇌ **39** (where R = i-C$_3$H$_7$, t-C$_4$H$_9$) was investigated; practically no changes in the equilibrium constants were found (*57*).

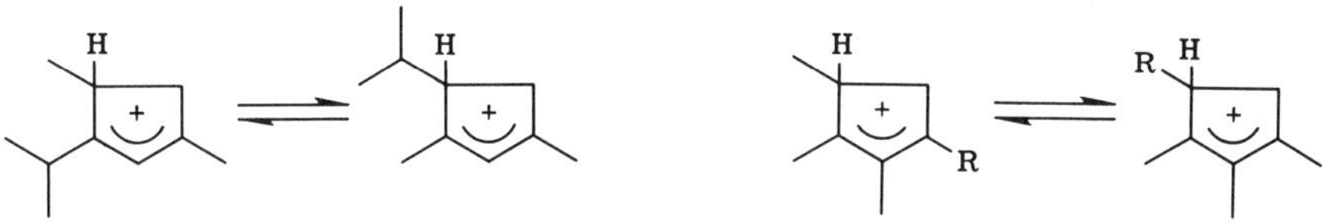

36   **37**   **38**   **39**

A much more extensive study, in which 11 different solvents were employed, involved equilibrating systems **40** ⇌ **41** and **42** ⇌ **43** (*58*). The equilibrium constants K for both reactions, summarized in Table 1, show a notable lack of solvent dependence (+ *ca*. 1.5%). It is even more remarkable that the rate constants of the forward reactions k_f vary only by factors of 2–3 despite the wide range of solvents used. Cation **40** isomerizes to **41** via an intermediate, identified as **45** by further mechanistic study (*59*). This means that at least two transition states are involved, presumably **44** and **46**, for the required methyl shifts. The results (Table 1) show that stable ions **40** and **41**, as well as the transition state between them (e.g., **44** or **46**), are solvated to nearly the same relative extent despite their different structures and charge distributions.

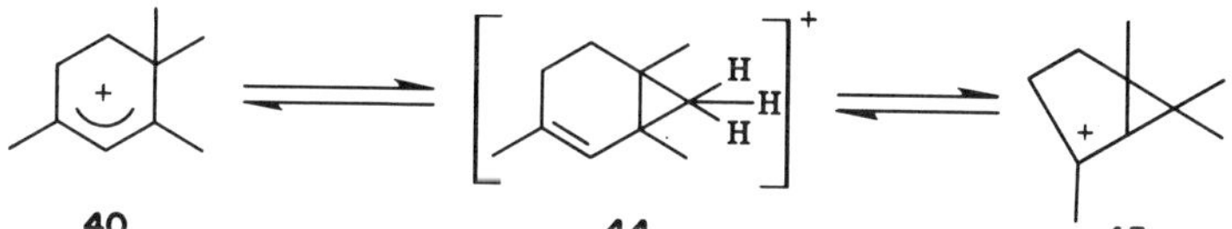

40   **44**   **45**

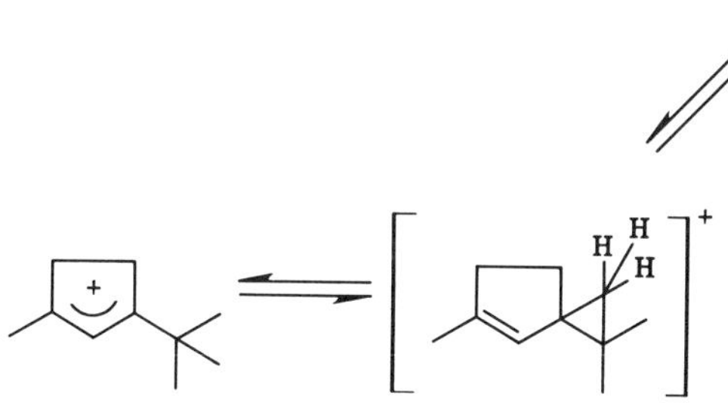

41   **46**

42   **43**

TABLE 1

Solvent Effects on Kinetics and Equilibria of Two Carbonium Ion Rearrangements[a]

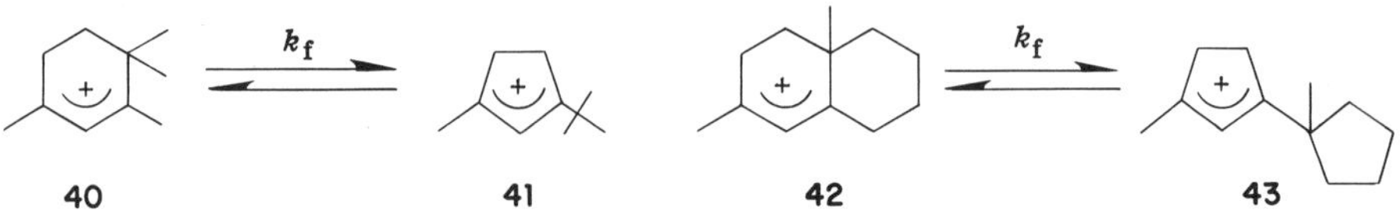

	40 ⇌ 41, 25°C		42 ⇌ 43, 0°C	
Acid solvent	K	$k_f \times 10^4$	K	$k_f \times 10^4$
ClSO$_3$H	13.4	3.1	0.082	6.6
FSO$_3$H	13.8	3.1	0.080	6.8
ClSO$_3$H + 20% Cl$_2$S$_2$O$_5$	12.8	2.8	0.075	5.0
CF$_3$SO$_3$H	14.0	2.8	0.083	4.4
H$_2$S$_2$O$_7$ (20% fuming H$_2$SO$_4$)	11.0	2.0	0.075	3.1
1 : 1 FSO$_3$H–SbF$_5$	12.4	1.6	0.080	3.8
100.6% H$_2$SO$_4$	11.1	1.65	0.063	3.3
97.0% H$_2$SO$_4$	10.3	1.53	0.059	2.8
94.2% H$_2$SO$_4$	10.4	1.43	0.065	3.1
89.4% H$_2$SO$_4$	11.0	1.64	0.069	2.4
83.3% H$_2$SO$_4$	11.4	2.18	—	2.8

[a] Sorensen (58).

It is certainly *not* to be expected that all ions will be solvated equally (49). The cases we have considered until now have dealt with *isomers* (or their transition states). At least for such species, electrostatic (general) solvation, as distinct from specific solvation by or interaction with the solvent, appears to be rather independent of structure. However, all these ions and transition states are highly delocalized. A detailed analysis of the rotational barriers of allyl cations (59a) indicated that these are considerably smaller in solution than in the gas phase. The charge in the perpendicular transition states is much less delocalized and benefits preferentially from electrostatic solvation. A general study of allyl rotational barriers as a function of solvent is needed.

That carbocations of different size or molecular weight can be solvated to various extents is illustrated by results on protonated alkylbenzenes 47 (60–62):

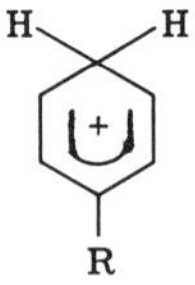

47

The fact that the stability order found in solution is the opposite of that found in the gas phase is conclusive evidence for Schubert's suggestion that the "Baker–Nathan" order (Me > Et > i-Pr > t-Bu) is due to steric hinderance to solvation rather than to "hyperconjugation" or other inherent electronic effects (*63*).

Extensive, recent work in the gas phase demonstrates that the stability of ions depends generally on their size. This is especially true of smaller systems, which tend to be stabilized to a greater extent by substituents. The charge, even in such classical cations as ammonium ions, is highly delocalized. In effect, an ion in the gas phase must provide its own "solvation" by distributing its charge; the more atoms available, the more effective this polarization stabilization can be. In solution, electrostatic solvation takes over this function, and the results show little or no "size" dependence. For this reason, it is important to employ systems of comparable size in interpreting gas-phase data or to correct for such size effects.

The relative stabilities of the 1-adamantyl and t-butyl cations are illustrative. Using the corresponding bromides as the ion precursors, in the gas phase the 1-adamantyl cation **48** is considerably more stable (*64a*); in contrast, in solvolysis, t-butyl bromide is 10^3 *more* reactive (*ca.* 4 kcal/mol) (*65*). In stable ion media the heat of ionization of 1-adamantyl chloride is 3.8 ± 1.1 kcal/mol less exothermic than t-butyl chloride (*66*), a result agreeing well with the solvolysis data. The 1-adamantyl cation is more stable only in the gas phase, where its much larger size permits more effective intramolecular charge distribution than in the t-butyl cation **49**.

The rearrangement reactions considered in this essay are *isomerizations*, in which the number of atoms is constant. Such rearrangements must involve transition states that differ significantly from ground states in structure, bonding, and charge distribution. It has often been suggested that nonclassical (bridged) carbocations or transition states should have lower solvation energies than their classical counterparts (*67*). This is based on the assumption that the charge in the classical structure is more localized and is subject to greater electrostatic (Born equation) stabilization. This expectation was supported in a series of papers by Jorgensen (*67–70*) reporting results of molecular orbital calculations on solvent–carbocation interactions.

As a model "solvent," Jorgensen employed HCl for computational simplicity; an extensive series of carbocations, both classical and nonclassical, was examined. The results afford a "first look," and a fascinating one indeed, at the structural and energetic details of carbocation–solvent interactions. Jorgensen's conclusions are noteworthy (*69a*), but it should be kept in mind that they will be subject to some change when further work of this type is carried out.

Bridged and allylic ions, in which the charge is formally distributed to more than one carbon center and the LUMO's are high in energy, interact more weakly with HCl than do classical cations. In some nonclassical ions, no coordination of Cl with carbon is indicated at all; solvation takes place through C—H···Cl—H interactions. Such interactions show little preference among various C—H bonds and are similar in magnitude in classical and in nonclassical cations. In the ethyl cation, multiple solvation (up to four additional HCl molecules) did not change the preference for the classical form indicated by attachment of the first HCl. Jorgensen concluded:

> The computational evidence . . . suggest that the relative energies of isomeric ions in the gas phase and in solution of moderate solvating ability like HCl are not necessarily the same. . . . In solutions that lack good electron-donating species such as superacid the effects of differential solvation should be minimized. However, the relative viability of nonclassical ions compared to classical isomers under conditions such as acetolysis appears diminished (*69a*).

These conclusions do not seem to be consistent with the experimental evidence cited above. Differences in solvation energies between carbocations and their isomerization transition states have not as yet been detected. One can find fault with Jorgensen's use of ethyl to model solvation differences between classical and nonclassical species. The interaction of the first HCl with classical $C_2H_5^+$ is so strong that the species resulting may be better regarded as protonated ethyl chloride than as a solvated ethyl cation. Such "specific" solvation will naturally be stronger when covalent bond or partial covalent bond formation is possible, but an encumbered "cationoid" species, rather than a true carbocation, in the sense of the present discussion, results. Although indirect, some evidence appears to favor some of Jorgensen's conclusions. The energy of the norbornyl cation has been established in the gas phase (*64, 71*). A variety of comparisons with other secondary cations, or involving comparison of tertiary–secondary energy differences, indicate an "extra" stabilization of 10–12 kcal/mol, consistent with the bridged structure **50** being more stable than the classical **51** (*15, 55, 64, 71*).

In stable ion media *or under solvolysis conditions*, a significantly smaller "extra" stabilization of 6–8 kcal/mol is indicated by many lines of evidence (*15, 55, 72*). This reduction in "extra" stabilization is consistent with preferential solvation, of modest magnitude, of the classical over the bridged form. However, the acetolysis-derived data do not seem different

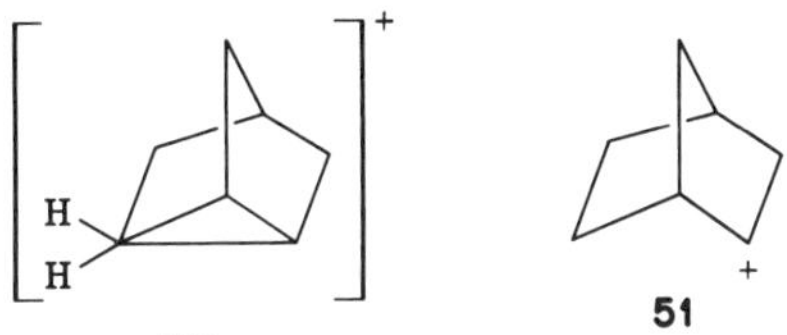

1. REARRANGEMENTS OF CARBOCATIONS

50 **51**

from those in SbF_5/SO_2ClF; nor is this distinction supported by other work (55).

In considering possible solvation energy effects in solvolytic media, it is worth discussing another case (73). The relative rates of 3,2-hydride versus 6,2-hydride shift during solvolysis of norbornyl tosylate in acetic acid were measured through the use of isotopic tracers. The 6,2-shift [Eq. (9)] was found to be about 200 times faster. This ratio was compared with the ratio between the 6,2-shift rate under stable ion conditions (36a) and the rate of the 3,2-shift (26a). Since the ratio of these two rates is $10^{8.8}$, it was implied that this was a striking example of "the importance of the environment in carbonium ion processes." The writers do not agree. Correcting for the measured temperature dependencies of these rates is necessary. When the rates of the 6,2-shift (measured at $-115°C$) and the 3,2-shift (measured at $-50°C$) are extrapolated to $+50°C$ (the solvolysis temperature), the ratio is 4000. Considering the uncertainties in $\Delta H^{\ddagger}$ for these reactions and the long extrapolation, these ratios of 200 and 4000 are indistinguishable within experimental error.

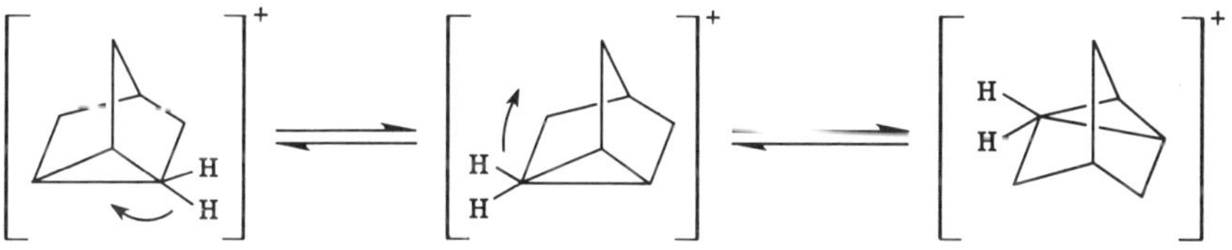

6, 2 shift   3, 2 shift   (9)

It is concluded that *differential* solvation effects on carbocation rearrangements probably are not of large magnitude. Although ions are stabilized a great deal by solvation, the interaction forces can be long range or not specific, so that many isomeric ions and transition states may be stabilized to comparable extents. At the other extreme, solvation of less stable ions may be structure specific. In these instances, solvent-bond cationoid species may be involved rather than solvated carbocations, but such reactions belong in a different category. Rotational barriers in allyl cations or equilibria including allyl and classical cycloalkyl carbocation isomers would appear to be ideally suited for critical experimental tests of differential solvation effects.

III. MULTIPLE REARRANGEMENT REACTIONS

In numerous reactions proceeding through carbonium ion intermediates, the structural relation of the products to the starting materials indicates that very many rearrangement steps have occurred. Either an ion has been formed and has undergone sequential transformation to other ions before being quenched to yield the product, or the rearrangement has proceeded via a number of equivalent nonionic intermediates. These may result from addition of nucleophiles to partly rearranged ions or by elimination from these ions to give olefins, which can then be reprotonated to continue the sequence. With a certain amount of patience and imagination, it usually is possible to find a route proceeding by known and accepted carbonium ion rearrangement steps which can yield the product. Such complex rearrangements are often the subject of examination questions for graduate students.

Under stable ion conditions there is no quenching step. Therefore, when a carbocation can rearrange, this rearrangement may be repeated without limit. As we have seen, many such rearrangements occur millions of times per second or faster. However, known rearrangement rates span a very large range and much slower steps, which may not be seen to any measureable extent under solvolytic conditions, also occur. These can often be detected in solutions of carbocations in stable ion media through the observation of isotopic scrambling. Depending on the rate of the rearrangement processes that are occurring, observation of changes in the PMR or CMR spectrum may indicate rapid rearrangements. Even if much is known about each of the elementary reaction steps, it may still be difficult to deduce overall reaction mechanisms for isomerizations or for redistributions of isotopic labels. In either slow or fast rearrangement reactions, nuclei move from one position to another within the ion. The problem of scrambling of isotopic labels turns out to be intellectually *identical* to the problem of accounting for lineshapes of NMR spectra. In both cases, one must determine the probability of a given nucleus going from one position in an ion to another, when a particular overall rearrangement process occurs.

How can a series of elementary rearrangement steps be connected with an overall reaction result? The rates of individual steps and the energies of the intermediates may be used to determine whether the observed result can be fit or to predict a new result. Alternatively, information about individual steps or the ratio of rates may be deduced from an observed overall result. In many simple cases, which involve a few steps, it is necessary only to write down carefully all of the possible rearrangements and to evaluate their probabilities at each stage. However, even for rela-

1. REARRANGEMENTS OF CARBOCATIONS

tively simple systems, this method quite often is not applicable because there are too many possibilities. In some cases, relatively simple-looking problems are extremely difficult to solve by inspection because the rearrangement can proceed via a variable or even an unlimited number of steps before eventually yielding the product.

In the case of mechanism (2) (Section II) it is straightforward to compute the probability of interchange of α and β protons when the reaction occurs. First, we note that the initial step, the uphill rearrangement from the tertiary ion to the secondary ion, does not itself ensure hydrogen scrambling. Going back to the tertiary ion returns all the atoms to their original positions. However, when a methide shift occurs followed by a downhill hydride shift to a rearranged tertiary ion, simply labeling the hydrogens and following them through the reaction path indicates that two of the four α hydrogens have become β and, naturally, two of the β hydrogens have become α. The probability of hydrogen exchange as a result of this reaction process is therefore one-half.

Mechanism (8), which is also a very simple process, cannot be treated in this way at all. The difficulty is that, *a priori*, one secondary-to-secondary hydride shift is as probable as another. Therefore rearrangements among α, β, β', and α' ions can occur in many ways, and one cannot even predict the number of steps before downhill rearrangement to the tertiary ion ends the process. Depending on the exact path, different numbers of α- and β-hydrogens will have interchanged. This random element in the process is the source of the difficulty.

A. Monte-Carlo Methods of Analysis

Computers can be applied to simulate such networks of rearrangements and to achieve the desired analysis as accurately as one wants. Saunders and Budiansky (25a) used a computer in this manner to study carbonium ion rearrangements. Such simulations are usually called Monte-Carlo procedures since, at each stage, choices of the path to be followed are made with predetermined probability factors using a pseudorandom number generator in order to establish the pathway. By the use of the program described here, any acyclic or monocyclic ion can be represented and analyzed.

The representation of an ion structure is in the form of a connection matrix between the carbon atoms, linearized for ease of handling by the computer. The hydrogen substituents can, if desired, be distinctly labeled. Their positions are recalled by the program in the form of a table of their attachments to the carbon atoms. The fate of any carbon and any hydrogen can be followed through all subsequent rearrangements; thus, the program can be used to explore either isotopic scrambling reactions or lineshape changes in the NMR spectra when fast reactions occur. The program is able to perform 1,2-hydride and methide shifts and can use Wagner–Meerwein shifts of carbon atoms in order to ring-contract and to ring-expand with predetermined probabilities. In addition, protonated cyclopropane rearrangements are also treated using simple general rules concerning which protonated cyclopropanes are formed and the probability of corner-to-corner migration in these protonated cyclopropanes. Many scrambling processes in simple cations require such protonated cyclopropane steps in order to occur at all (21).

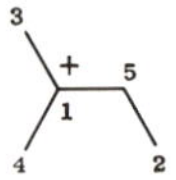

$$\begin{vmatrix} & 0 & 1 & 1 & 1 \\ 0 & & 0 & 0 & 1 \\ 1 & 0 & & 0 & 0 \\ 1 & 0 & 0 & & 0 \\ 1 & 1 & 0 & 0 & \end{vmatrix}$$

0 1 1 1 0 0 1 0 0 0

t-Amyl cation   Connection matrix   Linear form for computer handling

B. Use of Graphs to Elucidate Rearrangement Mechanisms

As indicated, many carbocation rearrangements are enormously complex, and the usual assumption that a minimal number of steps separates reactant from product is often belied by the observations. The point is well illustrated by the numerous degenerate rearrangements. Thus, the cyclopentyl cation **52**, despite its C_{2v} or C_2 symmetry, exhibits only a single proton and a single carbon NMR signal down to very low temperatures owing to rapid 1,2-hydrogen rearrangement (36a, 75). Even carbon scrambling occurs, at least in the gas phase (see Essay 2) (76). Such degenerate carbocation rearrangements have been reviewed extensively (78).

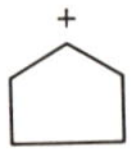

52 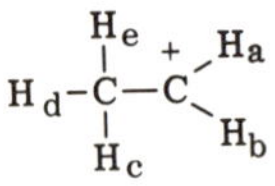   **53** 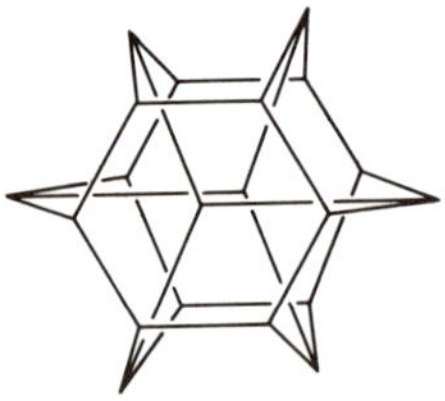   **54**

Balaban first used graph theory to attack degenerate carbocation isomerizations systematically (79). Considering the classical ethyl cation

1. REARRANGEMENTS OF CARBOCATIONS

53, he pointed out the 20 possible permutations obtainable by interchanging hydrogens (labeled H_a, H_b, etc., in **53** for illustration). A 20-vertex graph (**54**) represents all 1,2-hydride shift possibilities. It is interesting that similar graphs have been developed to illustrate the five-coordinate permutational possibilities for the reorganization of ligands around a central atom such as phosphorus (*79*).

Independently, others (*80*) applied graph theory in quite another way to analyze complex nondegenerate rearrangements in which many different pathways were possible. In the presence of strong Lewis acids such as $AlBr_3$, tetrahydrodicyclopentadiene (**55**) isomerizes to adamantane (**56**)

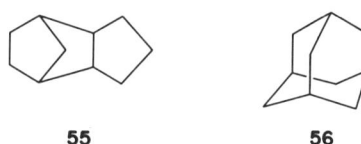

55   **56**

(*81*). Under such conditions, reversible hydride abstraction occurs; carbocations as reactive intermediates can be generated at virtually any position on a carbon skeleton. Hence, such reactions tend to be controlled by the thermodynamic stability of the hydrocarbon products, since virtually every favorable rearrangement can take place. In fact, a large number of diamond and cage hydrocarbons have been synthesized by this isomerization method (*82*).

The mechanisms of such reactions, or merely the pathways traversed in going from one carbon skeleton to another, can be dreadfully complex. It has been shown (*80*) how such cases can be treated systematically by means of a graph of rearrangement possibilities. Figure 1 represents the latest version of the adamantane graph (*83*). The steps in constructing this and similar graphs are the following:

1. Only those $C_{10}H_{16}$ tricycloalkane isomers are considered that have a reasonable chance thermodynamically of being isomerization intermediates. This excludes isomers with three- or four-membered rings or those with obviously strained structures. Methyltricyclononanes, alkyltricyclooctanes, etc., are excluded on similar grounds. A set of tricyclodecanes results (Fig. 1).

2. As a further refinement (*83*), empirical force field (*84, 84a*) calculations are employed to estimate the stability of each isomer quantitatively. The calculated heats of formation are given within each circle of Fig. 1.

3. Assuming that a cation can be generated at any carbon atom, all interconversions between members of the set involving 1,2-carbon shifts were considered. Computer programs are available for this purpose (*85*) if the problem is too cumbersome to be solved by hand. The resulting graph

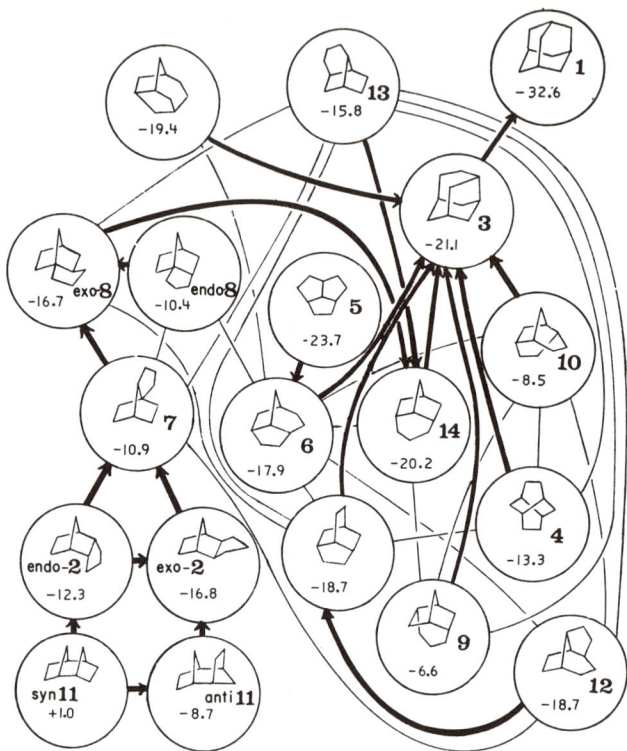

Fig. 1 Tricyclodecane graph (*83*). The most likely rearrangement pathways are shown by darkened circles and arrows. Calculated heats of formation (*84, 84a*) are given under each structure. [Reproduced from *J. Am. Chem. Soc.* **95**, 5769 (1973). Copyright by the American Chemical Society.]

(Fig. 1) (*83*) shows the many pathways whereby tricyclodecanes can be interconverted; there are 2897 ways to go from **55** to **56** (*80*)!

4. Unlikely pathways are now discarded (*83*). These may involve (a) high-energy bridgehead cation intermediates (e.g., with a 1-norbornyl cation part structure (**57**), (b) steps indicated by the empirical force field

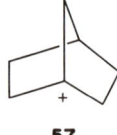

57

calculations to be endothermic by more than a few kilocalories per mole (unless these represent the only ways of reaching the indicated products), or (c) unfavorable dihedral angle relationships (*83, 85–87*) between the

1. REARRANGEMENTS OF CARBOCATIONS

"vacant" carbocation orbital and the migrating C—C bond. As discussed in greater detail in Section IV, B, the rates of 1,2-shifts in rigid polycyclic systems depend strongly on these dihedral angles (*88–90*); angles greater than *ca.* $\theta = 30°$ (**58**) are assumed effectively to preclude migration. A

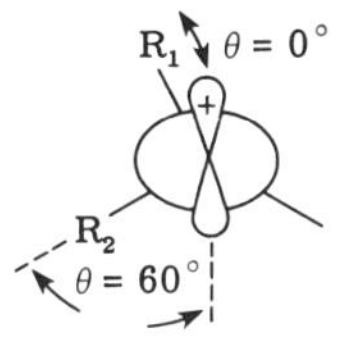

R_1 should migrate preferentially to R_2

58

simplified graph, indicated by the darkened lines in Fig. 1, results. The arrows suggest the most favorable pathways from any given tricyclodecane isomer to adamantane (**56**). The available chemical evidence (and quite a bit has been gathered) is in agreement with these deductions although a new pathway (in Fig. 1 from **7** to **12**) has recently been discovered (*80, 83, 91*).

As Fig. 2 emphasizes, the inclusion of only one more carbon atom into the system results in a fearful increase in complexity (*87*). The real situation is even worse. Figure 2 includes only tricycloundecane isomers, but not the end products of their isomerization, the two methyladamantanes **59** and **60**! Nevertheless, such graphs greatly simplify inherently complicated problems and compress a great deal of information into manageable form.

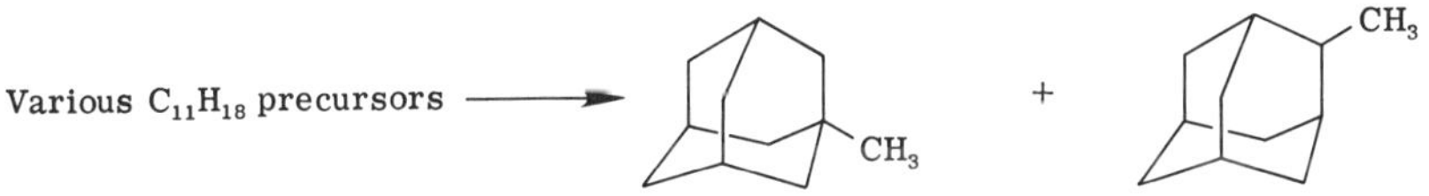

Various $C_{11}H_{18}$ precursors ⟶

(major) **59**   +   (minor) **60**

The graphic treatment of such rearrangements has now become "standard operating procedure" and has been used, for example, to analyze the $C_{11}H_{16}$ tetracycloundecane rearrangements leading to the two ethanonoradamantanes **61** and **62** (*86*):

Various $C_{11}H_{16}$ precursors ⟶

61   +   **62**

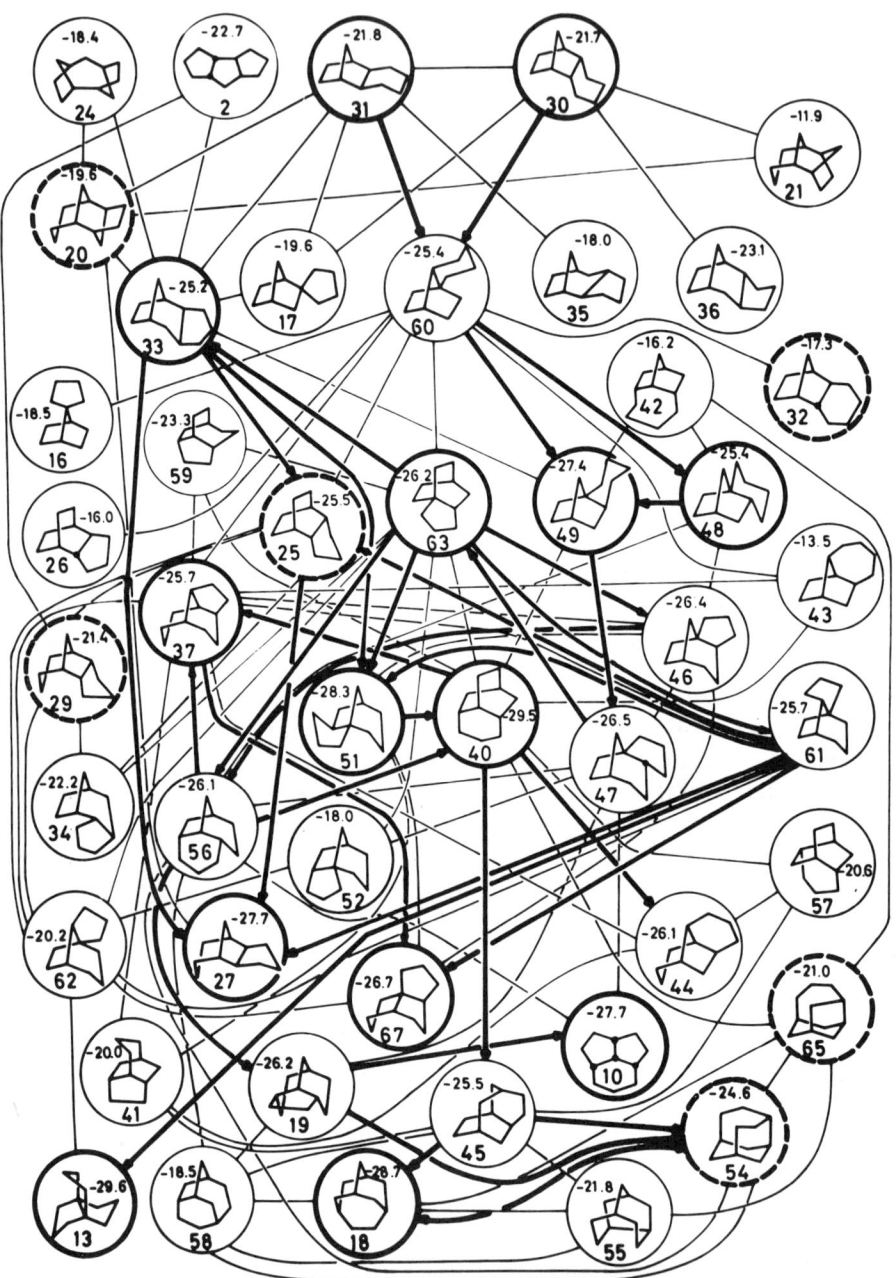

Fig. 2 Main portion of the tricycloundecane graph. Structure numbers refer to the original paper (87). Negative numbers are calculated heats of formation (84). The darkened lines show likely pathways; the darkened circles designate intermediates that actually have been identified experimentally. Compounds in dashed-line circles have been shown not to be intermediates. [Reproduced from *J. Am. Chem. Soc.* **99**, 5362 (1977). Copyright by the American Chemical Society.]

1. REARRANGEMENTS OF CARBOCATIONS

The transformation of $C_{14}H_{20}$ pentacyclotetradecane isomers, e.g., **63**, into diamantane (**67**) carries such analyses one step further (Fig. 3) (*85*).

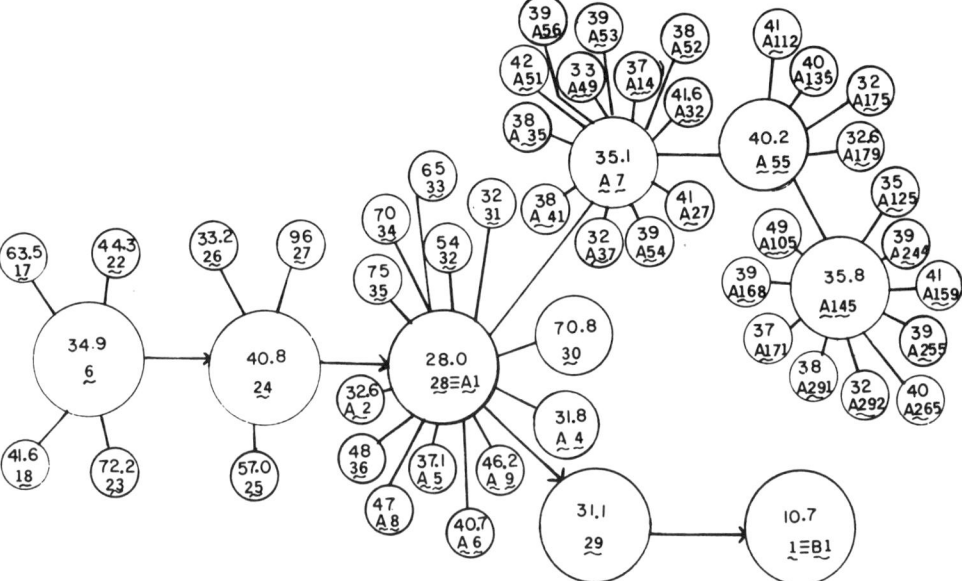

Fig. 3 Simplified pentacyclotetradecane graph. Each circle represents a $C_{14}H_{20}$ isomer. Structures **63–70** are shown in the text; the remainder can be found in the original paper (*85*). Calculated strain energies (kilocalorie per mole) are shown within each circle. Those given to three significant figures were calculated using the EAS force field (*84*); the others were calculated by a more approximate method. [Reproduced from *J. Am. Chem. Soc.* **97**, 743 (1975). Copyright by the American Chemical Society.]

There must be at least 40,000 pentacyclotetradecane isomers; even after most of these are rejected on the grounds of strain, there are still too many to treat by graphs similar to those in Figs. 1 and 2. (Subsets of the $C_{14}H_{20}$ rearrangement surface are amenable to such analysis, however.) Figure 3 outlines the simplified procedure employed (*85*). Each circle represents a pentacyclotetradecane isomer; calculated heats of formation also are given. All possible 1,2-carbon shifts from **63** were considered; the thermodynamics of the five most favorable were suggested by force field calculations (*84, 84a*). The least unfavorable step of these five leads to **64**. That all possible steps from **67** are endothermic agrees with experimental findings: **63** is isolated as a rearrangement intermediate after partial isomerization of other pentacyclotetradecane isomers (*85*). Continuation of the analysis procedure leads, in turn, to **65**, to **66**, and, finally, to diamantane (**67**) (Fig. 3).

Figure 3 also shows the exploration of a rearrangement tributary (also shown by **68**, **69**, and **70**). This proves to be a "dead end" as far as

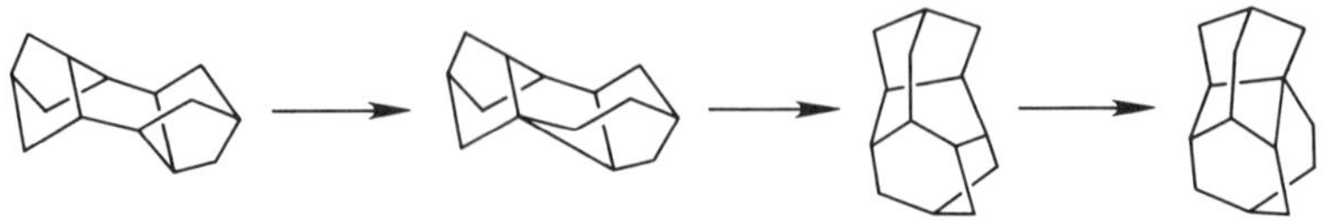

63 → 64 → 65 → 66

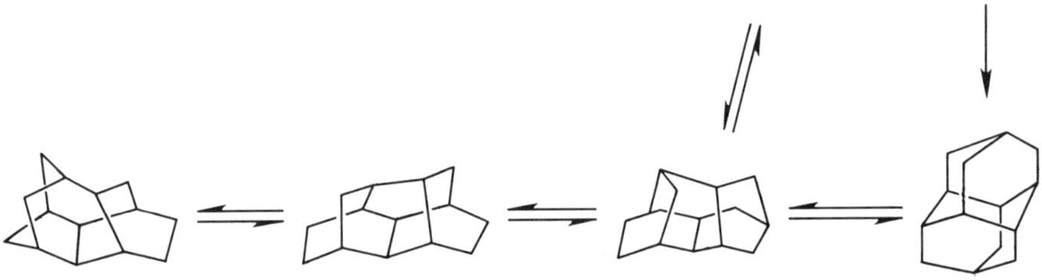

70 ⇌ 69 ⇌ 68 ⇌ 67

diamantane is concerned. If such a tributary were followed during an analysis, "doubling back" (e.g., to **65** in Fig. 3) would be required until a branching point were reached that would lead to the most stable product.

Readers will appreciate that the accuracy of such analyses reflects the approximations employed. Systematic approaches certainly are preferable to "mechanistic guessing" and have proved their worth in all cases examined. For example, the simple four-step pathway from **63** to diamantane (**67**) seems obvious once it has been pointed out, but it does not take long for a chemist, armed only with a set of molecular models, to realize just how difficult finding *any* mechanism for this transformation, let alone the best one, would be.

Others (*14, 92–94*) have employed graph theory in another way—to analyze complex rearrangements in norbornane systems. Five rearrangement mechanisms are recognized. Using the norbornane numbering scheme, these mechanisms are (1) Wagner–Meerwein skeletal rearrangement of C-6 from C-1 to C-2, (2) hydride shift from *endo*-C-6 to *endo*-C-2 (such rearrangements are not established for other groups), (3) hydride, methide, or other exo migrations from C-3 to C-2; (4) similar endo 3,2 shifts, and (5) a "double Wagner–Meerwein" skeletal rearrangement initiated by migration of C-7 from C-1 to C-2 and involving the bicyclo[3.1.1]hexyl cation. All these transformations, written in classical cation form for convenience, are shown below [Eq. (10)].

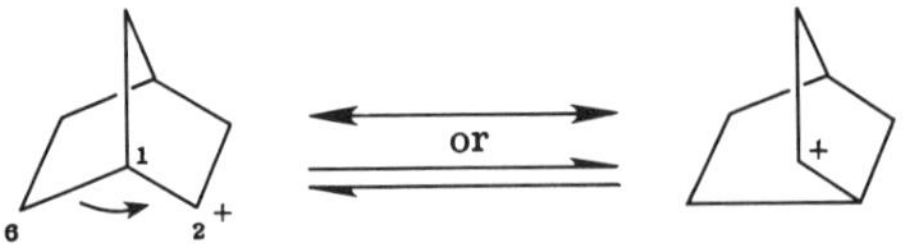

Wagner–Meerwein rearrangement (WM)

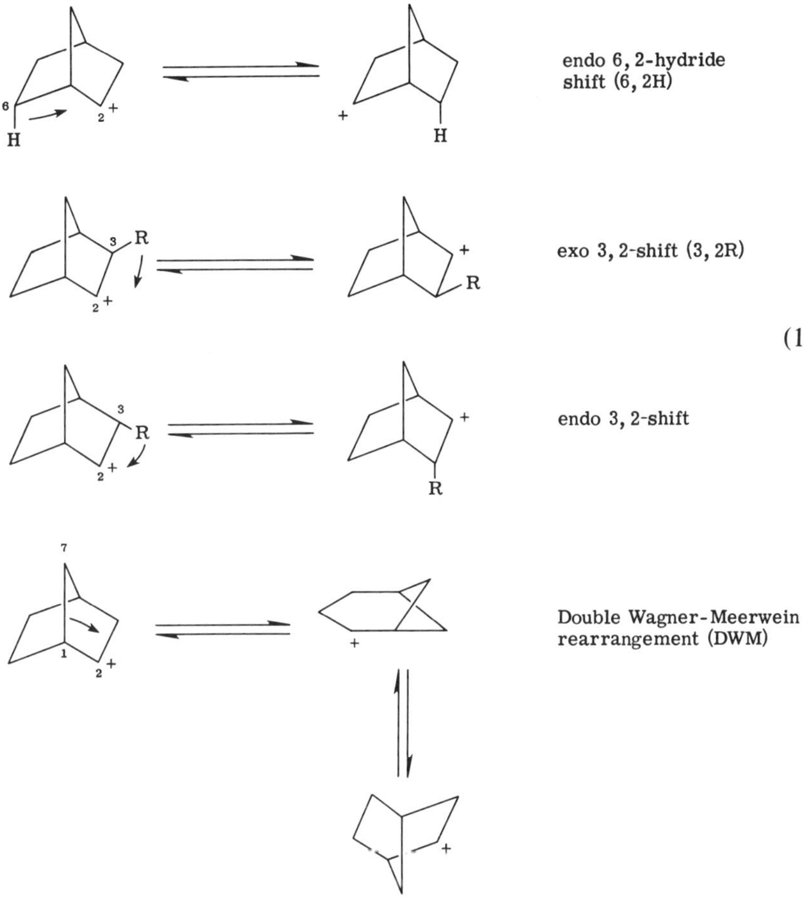

1. REARRANGEMENTS OF CARBOCATIONS

endo 6,2-hydride shift (6,2H)

exo 3,2-shift (3,2R)

(10)

endo 3,2-shift

Double Wagner-Meerwein rearrangement (DWM)

In the parent 2-norbornyl cation, all five of these processes are degenerate and lead from one enantiomer to the other and to different patterns of exchange of carbon and hydrogen atoms. Norbornyl rearrangements have been studied by following the loss of optical activity, by isotopic labeling, and by dynamic NMR experiments (*14–15, 36, 95*). When substituents are present, the situation becomes even more complicated. It has been calculated that there are 18,480 distinguishable norbornyl cations with seven hydrogen atoms, three methyl substituents, one hydroxyl substituent, and one labeled skeletal carbon atom (*93*).

Graphic analysis, aided by a computer program (*93*), has revealed the simplest mechanisms necessary to explain a given result (*92, 94*). For example the conversion of labeled camphors to 3,4-dimethylacetophenone [Eq. (11)] results in isotopic distributions that can be explained *without* invoking endo 3,2-hydroxyl shifts. The rearrangement is complicated enough, and involves Wagner–Meerwein, double Wagner–Meerwein, 6,2-hydride, 3,2-hydride, 3,2-methide, and exo 3,2-hydroxyl shifts (*92*)! Figure 4 presents the graphic analysis (*94*) of the sequential rearrangements of fenchyl (trimethylnorbornyl) cations shown below (*14*). The letters within the ellipses indicate the positions of attachment of the methyl groups; the key structures are shown explicitly. The mechanisms involved are indicated by the abbreviations between each double-headed arrow; some of these lead only to enantiomers.

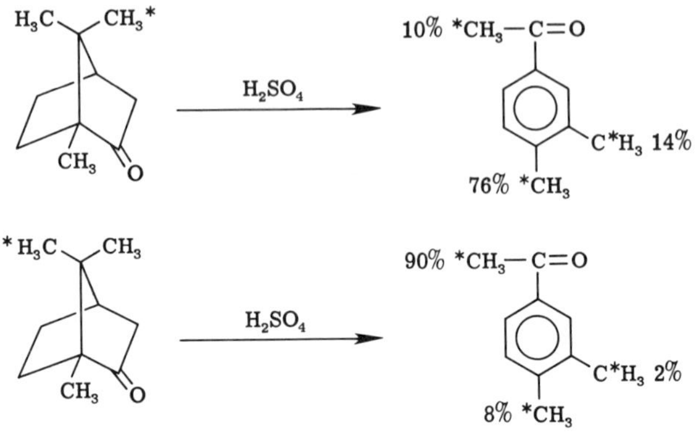

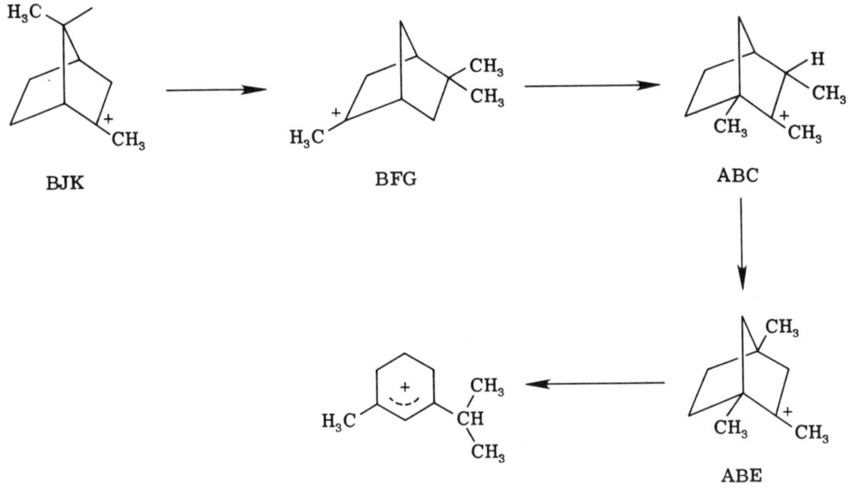

Sorensen (*14*) further adapted this program to the elucidation of rearrangements among dimethyl-2-norbornyl cations. The complete map (not distinguishing enantiomers) is shown in Fig. 5. Circles indicate tertiary and squares secondary ions; the letters again indicate positions of methyl attachment. Mechanisms are coded by numbers (1, WM; 2, exo 3,2R; 3,

1. REARRANGEMENTS OF CARBOCATIONS

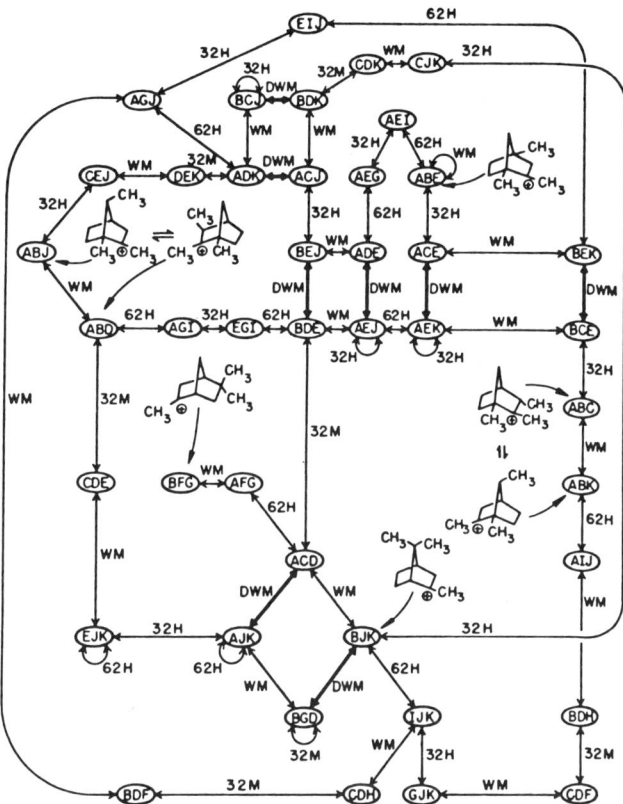

Fig. 4 An abbreviated graph showing rearrangements among some fenchyl (trimethylnorbornyl) cations (94). Structure are coded by letters within ellipses, and rearrangements by abbreviations. [Reproduced from *J. Am. Chem. Soc.* **96**, 2524 (1974). Copyright by the American Chemical Society.]

endo 6,2H; 4, DWM; 5, endo 3,2R). Sorensen estimated the barriers involved for each possible step. He concluded that the pathways of lowest energy may not involve the smallest number of steps. For example, two routes from BH are shown in Fig. 5 by hatched, "railroad tie" lines; one leads to AB in 3 steps, and the second to BE in 12. Experimentally BE is observed before AB is formed!

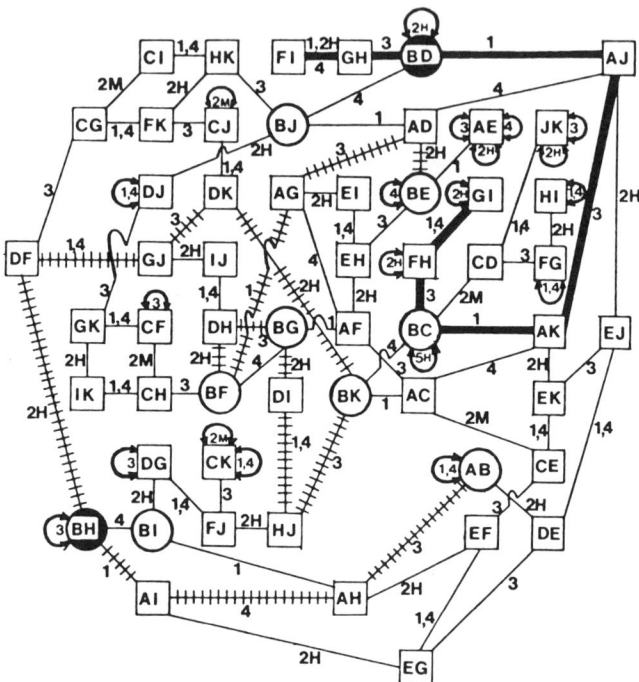

Fig. 5 Sorensen's dimethyl-2-norbornyl graph (*14*). See text for an explanation of the symbols. (Reproduced by permission of the publisher.)

Such graphic treatments are also useful in designing experiments (e.g., in indicating the best place to incorporate isotopic labels) to distinguish among mechanistic alternatives. Many other systems are amenable to such analyses (*93*), e.g., the homoadamantyl cation, which can undergo both degenerate Wagner–Meerwein and 1,2-hydride shifts (*96*).

IV. THEORETICAL STUDIES

A. Migratory Aptitudes

Two factors determine the ease of the rearrangement process (*12*): (a) the propensity of R_1 to migrate and (b) the ability of the rest of the molecule to sustain the migration. Several attempts have been made to

1. REARRANGEMENTS OF CARBOCATIONS

quantify the first effect by establishing a scale of migratory aptitudes for various groups (*97–108*). A general method is to generate a carbonium ion

$$R_3-\overset{R_1}{\underset{R_2}{C}}-\overset{+}{\underset{R_5}{C}}\overset{R_4}{\diagdown} \longrightarrow \overset{R_3}{\underset{R_2}{\diagup}}\overset{+}{C}-\underset{R_5}{C}-R_4 \overset{R_1}{\diagdown} \quad (12)$$

at a position adjacent to two or more substituents (R_1, R_2, or R_3), any of which can migrate, and to observe the ratio of products formed after competitive migration. This method of intramolecular comparison often has led to conflicting results owing to the complicating effect of the second factor mentioned above. The groups, R_2 and R_3, which do not migrate affect the rearrangement rates significantly: The measured migratory aptitudes are influenced by the conformational preference of the various substituents, by strain effects, and by the relative stability of the ions formed. It is difficult to devise experiments to obtain estimates of the inherent migratory ability of various groups in which such complications are insignificant.

Theoretical methods are particularly suited for resolving problems of this kind. Calculations have been performed (*98, 99*) on the simple model system **71** ⇌ **72** ⇌ **73**. Since this is a degenerate process, the energy

$$\underset{\textbf{71}}{\overset{R\quad H}{H\cdots\overset{+}{C}-C}}\overset{\Delta E}{\rightleftharpoons} \underset{\textbf{72}}{\overset{R}{H_2C\overset{+}{\triangle}CH_2}} \rightleftharpoons \underset{\textbf{73}}{\overset{H\quad R}{\overset{+}{C}-C-H}}$$

difference between **71** and **72** (which is a corner-protonated cyclopropane) provides an estimate of the barrier of R migration free from steric complications. Because of competition from nucleophilic solvent participation, such cases are not generally amenable to experimental study. Although a given theoretical method (e.g., MINDO/3 or minimal basis set *ab initio*) may have an inherent bias toward the classical or toward the bridged structure, it is hoped that effective cancellation of errors will occur if the energy differences are considered only for a series of substituents. The MINDO/3 energy differences between **71** and **72** for a wide variety of R groups (Table 2) show an interesting trend. The calculated migratory aptitude for the alkyl groups follows the order methyl > ethyl > *i*-propyl > *t*-butyl. *Ab initio* calculations using the STO-3G basis set suggest the same trend: methyl > ethyl > *t*-butyl.

Experimentally, methyl and ethyl groups do not exhibit a large difference in their migratory aptitudes. Estimates of ethyl/methyl migration rate ratios have varied from 0.5 to 55 (*100–112*). The more recent values are around 25.

TABLE 2

MINDO/3 and STO-3G Energy Differences[a] between 71 and 72 for Various R Groups

R	$\Delta E^{b,c}$	R	ΔE^b
Methyl	6.9 (−1.0)	Cyclobutyl	4.3
Ethyl	3.8 (−2.2)	Bicyclo[1.1.1]pentyl	8.2
i-Propyl	−4.3	COOCH$_3$	−1.2
t-Butyl	−7.6 (−5.3)	CH$_2$OH	3.1
Cyclopropyl	20.4		
Methylcyclopropyl	14.3		

[a] Expressed as kilocalories per mole. Ref (98).
[b] A positive ΔE indicates that **72** is more stable than **71**.
[c] STO-3G values in parentheses.

The predicted aptitude of the *t*-butyl group relative to methyl is quite surprising. Usually, **72** is considered to be a π complex between ethylene and the alkyl cation R⁺. The greater stability of *t*-butyl cation relative to methyl cation has previously been assumed to imply that **75** is more stable than **74**. The *t*-butyl group was therefore expected to migrate more rapidly.

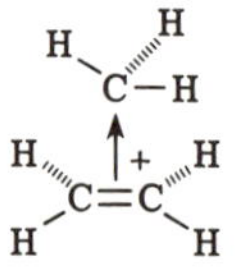

74 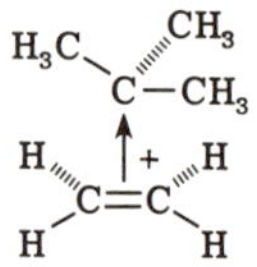   **75**

The MINDO/3 structures of **74** and **75** suggest why this is not so. To maximize the stabilizing interaction between R⁺ and ethylene, the cationic center has to be pyramidally distorted significantly. The vacant orbital on R⁺ thus gains s character, leading to greater overlap with the HOMO on ethylene. The structure **74** achieves the desired pyramidal geometry easily, whereas steric crowding of the methyl groups encumbers **75**. Therefore, the *t*-butyl group is indicated to have a low migratory aptitude in carbocations.

Large rate enhancements have nonetheless been observed with *t*-butyl migration. For example, in the pinacol–pinacolone rearrangement, *t*-butyl migrates 4000 times faster than methyl (*102*). However, this process is strongly dominated by steric effects; *t*-butyl migration is faster only because it leads to a greater reduction of strain in the parent ion. This complication is typical.

The reduction in the solvolysis rate due to *gem*-dimethyl substitution at the 6 position in the 2-norbornyl system was attributed to steric inhibition of bridging (*113*). This is analogous to the interpretation suggested by the theoretical results.

Not all tertiary groups are expected to have low inherent migratory aptitudes. The reluctance of these groups to pyramidalize significantly in the bridged structure **72** may be overcome by simply locking them into suitable ring systems. The bicyclo[1.1.1]pentyl group represents an extreme example. This group should migrate easily, despite its tertiary character, if the previous analysis is correct. Indeed, the calculated energy difference between **71** and **72** (Table 2) suggests a migratory aptitude for bicyclo[1.1.1]pentyl similar to that of a methyl group. Thus, the result obtained for *t*-butyl is not an artifact of any deficiency in the method of calculation used for tertiary systems.

On extending this analysis, small rings are expected to migrate even better. The MINDO/3 energies for **71** and **72** with cyclopropyl and cyclobutyl groups support this conclusion. In fact, the cyclopropyl group is calculated to have an additional stabilizing interaction present in the corresponding bridged structure. Two idealized conformations (**76a** and **76b**) may be considered for this structure.

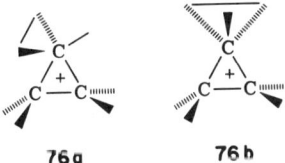

In **76a**, the fragments are held together by the interaction between the vacant orbital on a cyclopropyl cation moiety and the π MO on ethylene. In **76b**, in which the cyclopropyl group has been rotated 90°, another stabilizing interaction is present: One of the Walsh orbitals of the three-membered ring has the right symmetry to interact with the π orbital on ethylene (Fig. 6).

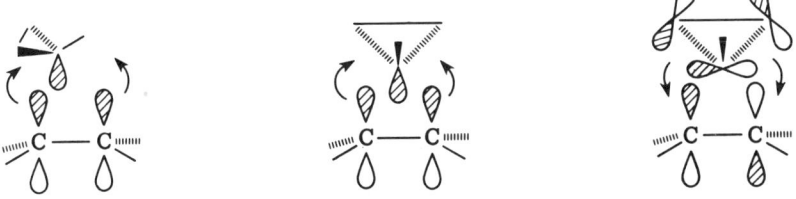

Fig. 6 Stabilizing orbital interactions in **76a** and **76b**.

The structure **76b** may be recognized as a complex between CH⁺ and two ethylene molecules (**76c**). It represents one of the simplest examples of a class of compounds that may be treated as a complex between CH⁺ and four electron donors. Ions **77–81** provide further illustration. The species **77–80** have been made under stable ion conditions (*114–116*). Compound **81** has been suggested to be involved in the rearrangement of the homocubyl cation (*68*).

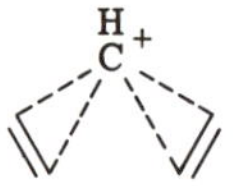

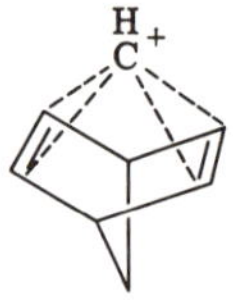

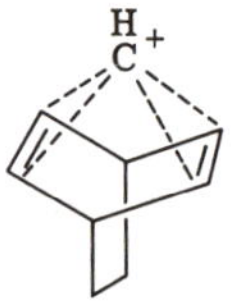

In spite of the favorable electronic interactions in **76c**, the most stable $C_5H_9^+$ isomer is the cyclopentyl cation **82**. The barrier to the conversion of **76c** to this ion along the symmetry-allowed C_s reaction pathway is calculated by MINDO/3 method to be only 9 kcal/mol. This is consistent with the solvolysis study of 2-cyclopropylethyl esters in which cyclopentyl products were obtained (*117*). In the corresponding neopentyl system, cyclopropyl migration was the only process observed (*118*). The formation of a tertiary carbenium ion **83** is obviously the driving force in this

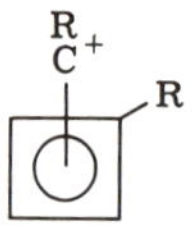

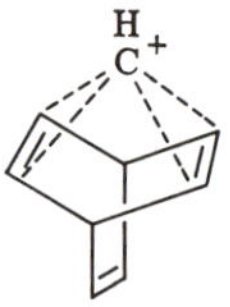

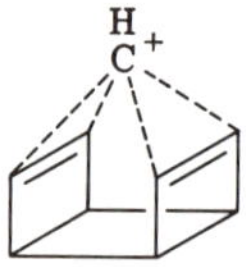

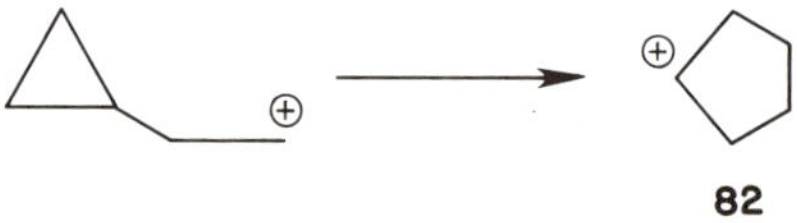

case. The 1-methylcyclopropyl group was found to retard the solvolysis rate (*118*). This is consistent with the expected adverse methyl steric effect on migratory aptitude (compare Table 2).

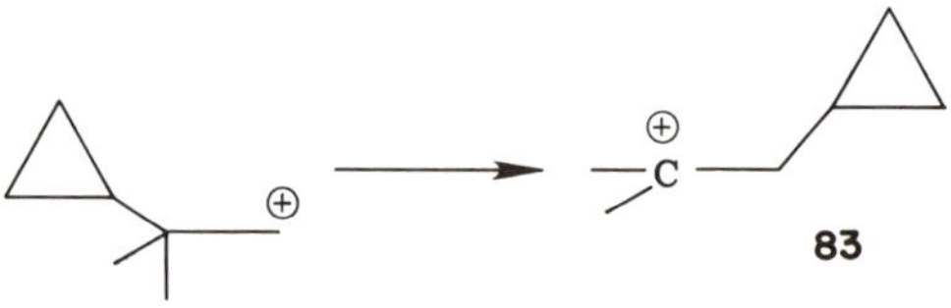

1. REARRANGEMENTS OF CARBOCATIONS

The effects of substituents other than alkyl groups on migratory aptitudes have been examined using *ab initio* techniques. The energies of classical and bridged structures of 3-substituted propyl cations (**84–88**) have been calculated using the STO-3G basis set with model geometries (*98*).

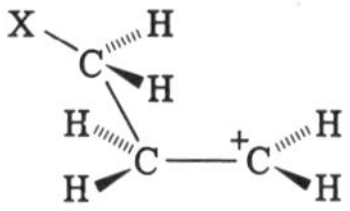

84 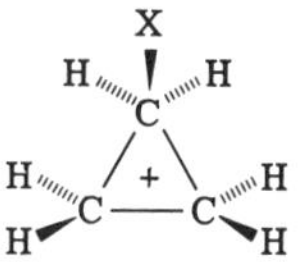   **85** 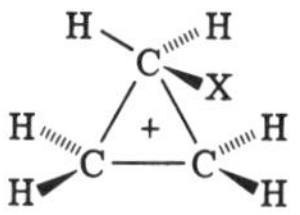   **86**

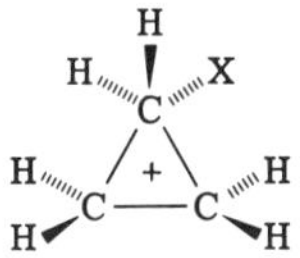

87 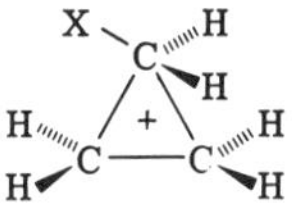   **88**

The energy differences between structures **84** and **85** for a large range of groups, X, vary remarkably little—only 10 kcal/mol (Table 3). Typical ranges of the order of 30–90 kcal/mol are found for such "first-row substituent sweeps" in studies of cationic species (*119–121*). Electronic factors apparently do not play a major role in influencing the migration tendencies of CH_2X groups.

However, the energy of the bridged form is strongly dependent on the

TABLE 3

STO-3G Relative Energies[a] of Conformations 85–88 with Respect to 84

Substituent	Conformation			
	85	86	87	88
Li	−2.6	−12.9	−29.9	−37.2
BeH	1.6	−3.1	−11.4	−15.0
BH_2[b]	1.3	−0.4	−4.6	−5.6
CH_3	2.2	3.6	6.9	11.6
NH_2[a]	−1.6	1.6	8.2	11.2
OH[b]	−1.3	1.5	7.7	11.7
F	0.9	4.4	12.0	16.4
CN	4.4	6.6	11.2	14.0
C≡CH	0.4	2.5	6.9	9.4
H	1.0	1.1	1.0	1.1

[a] Expressed as kilocalories per mole (*98*).
[b] Data for the most stable substituent conformation in each case.

conformation. Only σ-donating, π-accepting substituents prefer **88** or **85**. All other substituents studied prefer to move away from the plane of the three-membered ring. The rotational barriers are surprisingly large.

There are experimental systems in which substituent groups are rigidly held in fixed conformations. For example, the bridged, 6-substituted 2-norbornyl cation **89** resembles **87**. Consequently, all σ-withdrawing and/or π-donating substituents [e.g., OCH_3 (*122*), COOR (*123*), and CH_3 (*124*)] are expected to reduce the tendency to bridge in this system. A large number of 6-substituents have recently been studied by Grob and his coworkers (*124*).

89

Interestingly, the bridged form of the 3-hydroxypropyl cation is considered to be at least 10 kcal/mol in energy above the classical structure on the basis of mass spectroscopic fragmentation studies (*125*). Since no conformational rigidity is present in the parent ion, this conclusion does not agree with either the MINDO/3 (Table 2) or the STO-3G results (Table 3).

Strongly electron-withdrawing groups might well be expected not to be able either to bridge or to migrate effectively. Surprisingly, large migratory aptitudes for electronegative groups have been observed in certain systems (*126–137*). Examples of the migration of keto, ester, thio ester, amidate, phosphonate, and phosphinyl groups are known. In a few cases these groups migrate as easily as phenyl.

MINDO/3 calculations on carbomethoxy-substituted ethyl cations indicate no large intrinsic aptitude for the migration of such an electron-withdrawing group (Table 2). Other factors are probably involved in the rapid rearrangements observed. For example, in the dienone–phenol rearrangement shown in Eq. (13), path a is followed only because path b

1. REARRANGEMENTS OF CARBOCATIONS

results in an ion destabilized by the ester group attached to a charged center (*137*). The strong dependence of the rate of carboethoxy migration on the group left behind supports this interpretation. Migration is 135 times faster when R is phenyl instead of methyl. Also, when R is a second carboethoxy group, no migration is observed; only fragmentation occurs.

The 1,2-shift of any group is expected to be less facile in a tertiary carbenium ion than in secondary or primary ions. MINDO/3 calculations indicate a significant reduction in the hydride shift rate along the cation series ethyl, 2-butyl, 2,3-dimethyl-2-butyl. Accurate measurement of the low barrier to 1,2-hydride shifts in the last ion has been made recently (*26*). Although reproducing the trend correctly, MINDO/3 overestimates the magnitude of the variation (Table 4).

TABLE 4

MINDO/3 Energy Differences between Classical and Bridged Structures and Experimental Barriers to 1,2-Hydride Shifts[a]

System	MINDO/3	Experimental
Ethyl	-8.0^b	—
2-Butyl	-5.1^c	$\leq 2.4^d$
2,3-Dimethyl-2-butyl	$+1.2^e$	3.1^d

[a] Expressed as kilocalories per mole.
[b] Ref (*98a*).
[c] Ref (*43*).
[d] Ref (*26*).
[e] Ref (*98*).

B. Effect of Ring Size on the Ease of 1,2-Shifts

The barriers to 1,2-shifts are often higher in rings than in acyclic systems. For example, the methyl shift in the rearrangement of the 1,3- to the 1,2-dimethylcyclopentyl cation requires about 1.5 kcal/mol more energy than in model acyclic cation systems (*13, 89*). This greater activation barrier has been rationalized on the basis of orbital orientation. To have a low barrier, the empty p orbital on the cationic center should have a zero dihedral angle with the C—X bond orbital (X being the migrating group). It is more difficult to achieve this orientation in cyclopentyl cations (compare **90** and **91**).

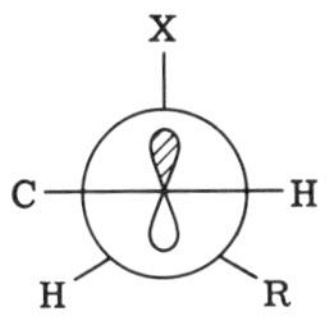

90 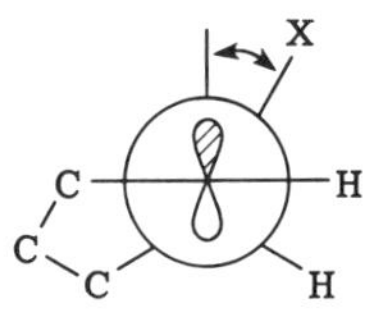   **91**

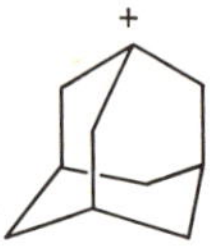

92 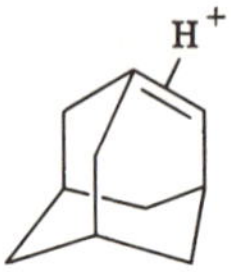 **93**

The orbital orientation effect is seen most dramatically in the 1-adamantyl cation **92** (*82*). Structure **93** permits very little stabilization in the usual bridged or π-complex transition state, since the adamantane fragment is a "perpendicular olefin." A lower limit of 30 kcal/mol has been suggested for the hydride shift barrier on the basis of unsuccessful attempts to observe deuterium scrambling (*138*).

However, the influence of dihedral angle variation does not suggest a large difference in 1,2-shift barriers in smaller rings, relative to cyclopentyl cation. Nevertheless STO-3G calculations on ethyl cations deformed to model small rings (**94** and **95**) indicate appreciable effects (*98a*); **95** becomes less stable relative to **94** as θ decreases (Table 5).

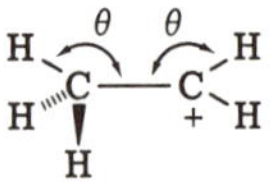

94 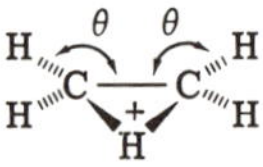 **95**

MINDO/3 and *ab initio* calculations on cyclopropyl, cyclobutyl, and cyclopentyl cations (*98a*) support this conclusion (Table 6). The experimental barrier to 1,2-hydride shift in the cyclopentyl cation is low. The cyclobutyl cation undergoes a variety of rearrangements, but the simple 1,2-shift rendering CH_2 and CH protons equivalent was not observed until recently. The barrier has been estimated to be greater than 10 kcal/mol (*139*).

TABLE 5

RHF/STO-3G Energy Differences between 94 and 95[a]

Θ (degrees)	ΔE[b]	Increase
Optimum[c]	11.2	0
109.5[d]	13.1	0.9
90.0	23.1	11.9
70.0	61.1	49.9

[a] Expressed as kilocalories per mole. Ref (*98a*).
[b] Structure **94** is always more stable.
[c] Optimized value of Θ.
[d] All other structural parameters as in the fully optimized structure.

1. REARRANGEMENTS OF CARBOCATIONS

TABLE 6

Energy Differences between Classical and Bridged Structures[a]

Cation	MINDO/3	STO-3G	Experimental barrier to 1,2 shift
Cyclopentyl	−0.7	14.5	3.1[b]
Cyclobutyl	2.2	19.3	>10[c]
Cyclopropyl	7.3	33.6	Not obsd.[d]

[a] Expressed as kilocalories per mole. Ref (98a).
[b] In the tertiary ion. Ref (26).
[c] Ref. (139).
[d] Ref. (139a).

The cyclopropyl cation rearranges to the allyl cation, making studies under stable ion conditions difficult (14a). However, the failure of specific attempts to observe 1,2-shifts during the solvolysis of cyclopropyl derivatives provides indirect evidence for large 1,2-barriers (124).

The ring size effect is also found in bicyclic systems. Thus, the barrier of 7 kcal/mol to 3,2-hydride shift in the 2,3-dimethyl-2-norbornyl cation is ca. 3 kcal/mol higher than that in the 2,3-dimethyl-2-butyl cation (140). However, the much larger barrier in the parent 2-norbornyl cation (> 10.8 kcal/mol for the exo 3,2-shift) (36a) cannot entirely be due to this ring size effect. The assumption of extra stabilization of the secondary 2-norbornyl cation initial state, owing to C—C bridging, best explains this barrier. But in the 2-bicyclo[2.1.1]hexyl cation, less prone to C—C bridging, much of the high barrier to 3,2-hydride shift (> 13 kcal/mol) (141) must be attributed to the larger angle strain effect. It would be of interest to examine the tertiary cation.

The 2-norbornyl cation endo 3,2-hydride shift barrier is higher than the exo 3,2-barrier. Torsional effects present in the idealized transition states **96** and **97** have been implicated in the remarkable stereospecificity of the hydride shift (142) (in this reference, previous explanations are critically examined and rejected).

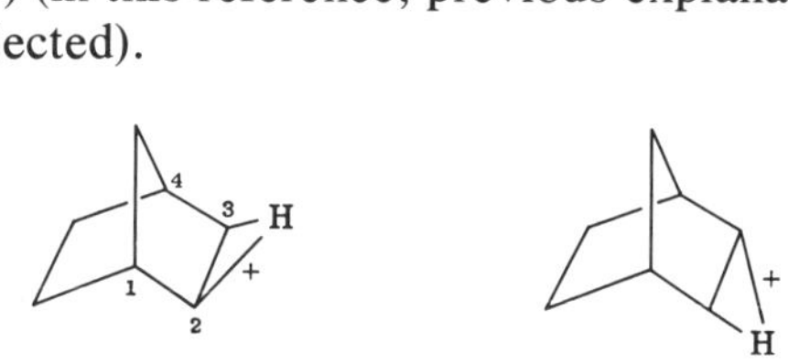

96   **97**

In **96**, the torsional arrangements around C-1—C-2 and C-3—C-4 bonds are nicely skewed. In contrast, in the endo conformer **97**, the arrangements around the same bonds are nearly eclipsed, which is an energetically unfavorable situation. Because of such torsional effects, **96** may be preferred by as much as several kilocalories per mole.

An alternative, orbital, explanation has also been advanced (*143*). The HOMO of norbornene is said not to be a symmetric, pure π MO. Rather, it is said to extend to a nonequivalent degree in the exo and in the endo directions. This is because of mixing of the unperturbed C-2—C-3 π and σ orbitals via mutual interaction with the methano bridge orbital (Fig. 7).

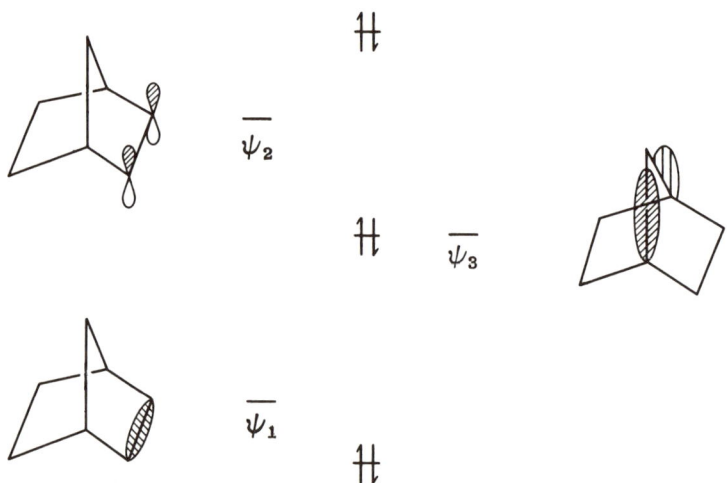

Fig. 7 Three-orbital interaction in norbornene. The HOMO is $a\psi_1 + b\psi_2 - c\psi_3$, where $b \gg a \simeq c > 0$, based on peturbation theory.

This results in the HOMO having increased electron density on the exo side. Hence, an exo proton bridge is more stable. However, recent theoretical reexaminations indicate that norbornene has a symmetrical π MO (*143a*). The reason for the *exo*-3,2-migrational preference must be sought elsewhere.

C. Rearrangements in Cycloalkenyl Cations

The ease and stereochemistry of sigmatropic shifts in protonated, cyclic conjugated systems generally follow orbital symmetry rules (*14, 144*). Thus, 1,2-shifts in benzenium ions are rapid (*145*). Both hydride and methyl shifts have been observed. The energy difference between **98** and **99** is overestimated by *ab initio* methods (4-31G basis set) (*146*) (Table 7). Use of larger basis sets and inclusion of electron correlation are likely to reduce the calculated barrier (*43*). In protonated fluorobenzene (**100**), fluorine shift (**101**) appears to be energetically unfavorable. However, a

1. REARRANGEMENTS OF CARBOCATIONS

TABLE 7

Calculated and Experimental Barriers[a] to Rearrangements in Cycloalkenyl Cations

Ion	Idealized transition state or intermediate	Calculated[b] energy difference	Experimental barrier
Benzenium (98)	H-Bridged form (99)	21	10
1-Fluorobenzenium (100)	F-Bridged Form (101)	40	—
2-Fluorobenzenium (102)	H-Bridged Form (102)	32	—
Heptamethylbenzenium	—	—	15, 18
Bicyclo[3.1.0]hexenyl (104)	Cyclopentadienylcarbinyl		
	Bisected (105)	20	15
	Perpendicular (106)	32	—
Cyclobutenyl (107)	H-Bridged form (108)	71[c]	—
	Cyclopropenylcarbinyl		
	Bisected (111)	27	—
	Perpendicular (112)	65	—

[a] Expressed as kilocalories per mole.
[b] *Ab initio* with the 4-31G basis.
[c] *Ab initio* with the STO-3G basis.

hydride shift (**102**) is calculated to be nearly as facile as in the parent ion, since it leads to a stabilized benzenium ion (**103**).

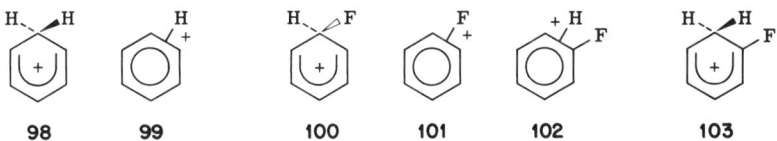

In the isomeric $C_6H_7^+$ bicyclo[3.1.0]hex-3-en-2-yl cation **104**, an interesting circumambulatory rearrangement is observed (*147*). The [1,4] sigmatropic shift mechanism shown below [Eq. (14)] is predicted to occur with inversion of configuration at the migrating carbon atom. This has been confirmed by experiment and by *ab initio* calculations on **104–106** (*148*) (Table 7).

In cyclobutenyl cations **107**, 1,2-shifts do not occur, in agreement with orbital symmetry arguments. The transition state **108** has antiaromatic cyclobutadienoid character and is high in energy. A direct shift across the ring (**109**) also is calculated to be energetically unfavorable (*98*).

A redistribution of —CD_3 labels in the pentamethyl derivatives has revealed the occurrence of a different rearrangement (*150*). In ion **110** scrambling of —CD_3 and —CH_3 groups was observed among positions 1,2 and 3, but not to position 4 [Eq. (15)]. This rules out the occurrence of 1,2-methyl shifts. Two mechanisms have been proposed to account for

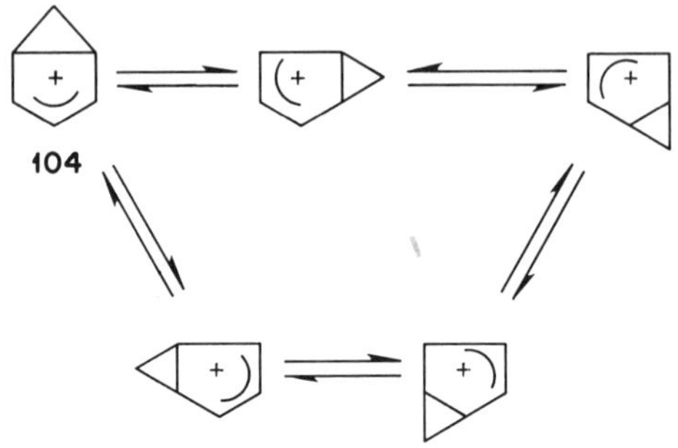

(14)

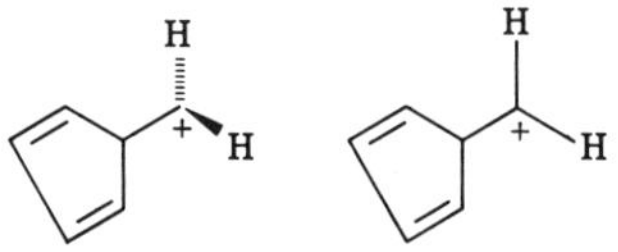

104

105   106

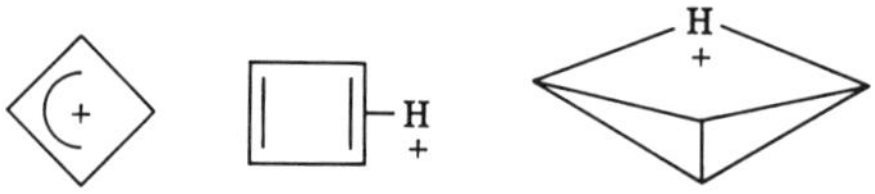

107   108   109

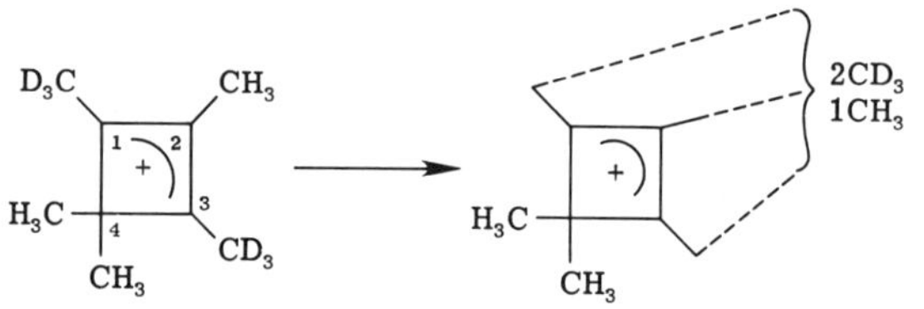

(15)

110

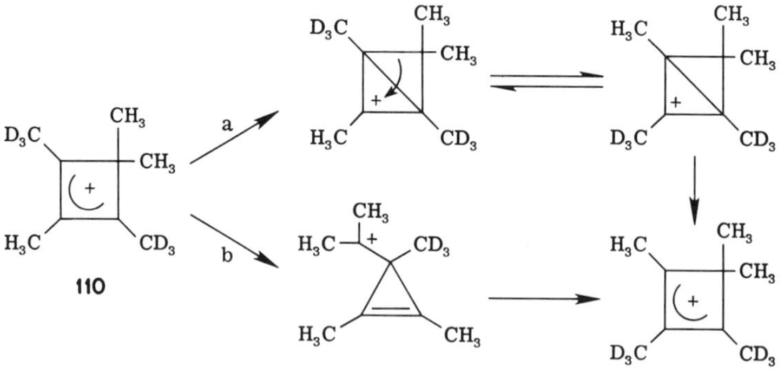

(16)

110

1. REARRANGEMENTS OF CARBOCATIONS 47

the observed partial scrambling [Eq. (16)]. A precedent for path b is known from solvolysis work: Cyclopropenylcarbinyl tosylates yield cyclobutenyl products (151). The transition state structure for the two paths, however, may well be similar. The alkyl shift in path a is indicated to be a viable process on the basis of *ab initio* calculations (Table 7) (152). Qualitative arguments predict inversion of configuration at the migrating carbon in this circumambulatory rearrangement. The opposite stereochemistry is indicated by the *ab initio* calculations. The idealized transition state for retention, namely, the bisected conformation of cyclopropenylcarbinyl cation **111**, is more stable than the perpendicular conformation **112**. This is because of the well-known interaction between a

111   112

cyclopropyl ring Walsh orbital and an acceptor orbital of the right symmetry. This reversal of predicted stereochemistry is thus an example of *subjacent orbital control* (153).

V. CONCLUSIONS

A change of emphasis has occurred in the last two decades. Products from carbonium ion rearrangements and the rates of processes (e.g., solvolyses) passing through ionic intermediates no longer are the main source of information concerning carbocations. Carbocations can now be studied directly in stable ion media, in the gas phase, and by means of theoretical calculations. The energies of the cations and their interconversion barriers provide a much more detailed understanding of the basic processes carbocations undergo. Comprehensive analyses of rearrangement possibilities, aided by graphic treatments and computer programs, bring order to problems that otherwise seem hopelessly complex.

We still do not understand all the factors that influence carbocation rearrangements. It is seldom possible in realistically chosen cases to predict what will happen before a reaction is carried out experimentally. Future studies will reveal more details of carbocation potential energy surfaces, particularly rearrangement transition states. Theoretical calculations appear to be best suited for this purpose. Although much has been learned already, considerable developmental work is necessary before the larger systems of experimental interest can be probed reliably and cheaply.

Carbocation rearrangements have great synthetic potential. As the preparations of adamantane and of cage hydrocarbons show, many C—C bonds can be made and broken to give completely different skeletal patterns during a single reaction. Nevertheless, carbocation rearrangements are not yet considered to be generally effective synthetic procedures. Such reactions often are not controllable and do not lead to single products. Reaction conditions, which can control carbocation lifetimes and their nature, are particularly important. A more detailed understanding of rearrangement processes and carbocation behavior will make possible the design of new syntheses that use the structural flexibility of carbocations to advantage.

REFERENCES

1. C. D. Nenitzescu, in "Carbonium Ions" Volume I (G. A. Olah and P. von R. Schleyer, ed.), Vol. 1, Chapter 1, p. 1. Wiley (Interscience), New York, 1968.
2. P. de Mayo, ed., "Molecular Rearrangements," Part I. Wiley (Interscience), New York, 1963.
3. For an historical account of the development of this area, see G. A. Olah, "Carbocation and Electrophilic Reactions." Wiley, New York, 1974.
3a. H. Meerwein, K. Bodenbenner, P. Borner, F. Kunert, and H. Wunderlich, *Justus Liebigs Ann.* **632**, 38 (1960); D. M. Brouwer, S. McLean, and E. L. Mackor, *Chem. Discuss. Faraday Soc.* **39**, 121 (1965).
3b. M. Saunders and E. L. Hagen, *J. Am. Chem. Soc.* **90**, 2436 (1968).
4. See, e.g., J. L. Franklin, ed., "Ion-Molecule Reactions," Vols. 1 and 2. Plenum, New York, 1972; P. Ausloos, ed., "Interactions between Ions and Molecules." Plenum, New York, 1975; S. G. Lias and P. Ausloos, "Ion-Molecule Reactions." Am. Chem. Soc., Washington, D.C., 1975; M. T. Bowers, ed., "Gas-phase Ion Chemistry." Academic Press, New York, 1979; D. H. Williams, this volume.
5. P. Kebarle, *J. Am. Chem. Soc.* **99**, 360 (1977); *Annu. Rev. Phys. Chem.* **28**, 445 (1977).
6. L. Radom, D. Poppinger, and R. C. Haddon, in "Carbonium Ions" (G. A. Olah and P. von R. Schleyer, ed.), Vol. V, Chapter 38, p. 2303. Wiley (Interscience), New York, 1976; W. J. Hehre, in "Modern Theoretical Chemistry" (H. F. Schaefer, III, ed.), Vol. 4. Plenum, New York, 1977.
7. T. W. Bentley and P. von R. Schleyer, *Adv. Phys. Org. Chem.* **14**, 1 (1977).
8. F. L. Schadt, III, C. J. Lancelot, and P. von R. Schleyer, *J. Am. Chem. Soc.* **100**, 228 (1978).
9. J. D. Roberts and J. A. Yancey, *J. Am. Chem. Soc.* **14**, 5943 (1952).
10. S. P. McManus and C. U. Pittmann, Jr., in "Organic Reactive Intermediates" (S. P. McManus, ed.), p. 194. Academic Press, New York, 1973; N. S. Isaacs, "Reactive Intermediates in Organic Chemistry." Wiley, New York, 1974; B. Capon and S. P. McManus, "Neighboring Group Participation." Plenum, New York, 1976; T. H. Lowry and K. S. Richardson, "Mechanism and Theory in Organic Chemistry." Harper, New York, 1976; F. A. Carey and R. J. Sundberg, "Advanced Organic Chemistry," Parts A and B. Plenum, New York, 1977; J. March, "Advanced Organic Chemistry," 2nd ed. McGraw-Hill, New York, 1977; W. J. LeNoble, "Highlights of Organic Chemistry." Dekker, New York, 1974; V. A. Koptyug, ed., "Contemporary Problems in Carbonium Ion Chemistry." Science Publishers, Novosibirsk, USSR, 1975.

1. REARRANGEMENTS OF CARBOCATIONS 49

11. G. A. Olah and P. von R. Schleyer, eds., "Carbonium Ions," Vols. I-V. Wiley (Interscience), New York, 1968–1976.
12. D. Bethell and V. Gold, "Carbonium Ions, An Introduction." Academic Press, New York, 1967.
13. D. M. Brouwer and H. Hogeveen, *Prog. Phys. Org. Chem.* **9**, 179 (1972).
14. T. S. Sorensen, *Acc. Chem. Res.* **9**, 257 (1976).
14a. T. S. Sorensen and A. Rauk, in "Pericyclic Reactions." (A. P. Marchand and R. E. Lehr, ed.) Vol. II, chapter 1, p. 1. Academic Press, New York, 1977.
15. H. C. Brown (with comments by P. von R. Schleyer), "The Nonclassical Ion Problem." Plenum, New York, 1977.
16. H. M. Rosenstock, K. Draxl, B. W. Steiner, and J. T. Herron, *J. Phys. Chem. Ref. Data* **6**, Suppl. 1 (1977).
17. R. P. Kirchen, T. S. Sorensen, and K. Wagstaff, *J. Am. Chem. Soc.* **100**, 6761 (1978); R. P. Kirchen and T. S. Sorensen, *J. Chem. Soc., Chem. Commun.* p. 769 (1978).
18. T. S. Sorensen, private communication.
19. D. H. Williams, B. J. Stapleton, and R. D. Bowen, *Tetrahedron Lett.*, 2919 (1978).
20. See, e.g., L. Radom, J. A. Pople, V. Buss, and P. von R. Schleyer, *J. Am. Chem. Soc.* **94**, 311 (1972); P. C. Hariharan, L. Radom, J. A. Pople, and P. von R. Schleyer, *ibid.* **96**, 599 (1974); P. K. Bischof and M. J. S. Dewar, *ibid.* **97**, 228 (1975).
21. M. Saunders, P. Vogel, E. L. Hagen, and J. Rosenfeld, *Acc. Chem. Res.* **6**, 53 (1973).
22. A. P. W. Hewett, Ph.D. Thesis, Yale University, New Haven, Connecticut (1975).
23. W. J. Hehre, private communication. C. Wesdemiotis, R. Wolfschütz, and H. Schwarz, to be published.
24. M. Saunders and E. L. Hagen, *J. Am. Chem. Soc.* **90**, 6881 (1968).
24a. M. Saunders, E. L. Hagen, and J. Rosenfeld, *J. Am. Chem. Soc.* **90**, 6882 (1968).
25. M. Saunders and J. Rosenfeld, *J. Am. Chem. Soc.* **91**, 7756 (1969).
25a. M. Saunders and S. P. Budiansky, *Tetrahedron* **35**, 929 (1979).
26. M. Saunders and M. R. Kates, *J. Am. Chem. Soc.* **100**, 7082 (1978).
26a. M. Saunders, P. von R. Schleyer, and G. A. Olah, *J. Am. Chem. Soc.* **86**, 5680 (1964).
27. S. Winstein, M. Shatavsky, C. Norton, and R. B. Woodward, *J. Am. Chem. Soc.* **77**, 4183 (1955).
28. M. Saunders, M. H. Jaffe, and P. Vogel, *J. Am. Chem. Soc.* **93**, 2558 (1971).
29. M. Saunders and P. Vogel, *J. Am. Chem. Soc.* **93**, 2559 (1971).
30. M. Saunders and P. Vogel, *J. Am. Chem. Soc.* **93**, 2561 (1971).
31. M. Saunders, L. Telkowski, and M. R. Kates, *J. Am. Chem. Soc.* **99**, 8070 (1977).
32. D. E. Sunko, I. Szele, and W. J. Hehre, *J. Am. Chem. Soc.* **99**, 5000 (1977).
32a. M. R. Kates, Ph.D. Thesis, Yale University, New Haven, Connecticut (1978).
33. M. Saunders and M. R. Kates, *J. Am. Chem. Soc.* **99**, 8071 (1977).
34. R. M. Coates and E. R. Fretz, *J. Am. Chem. Soc.* **97**, 2538 (1975); **99**, 297 (1977). H. C. Brown and M. Rauindranathan, *ibid.* **99**, 299 (1977).
35. M. Saunders, M. R. Kates, K. B. Wiberg, and W. Pratt, *J. Am. Chem. Soc.* **99**, 8072 (1977).
36. P. von R. Schleyer, W. E. Watts, R. C. Forst, Jr., M. B. Comisarow, and G. A. Olah, *J. Am. Chem. Soc.* **86** 5679 (1964); F. R. Jensen and B. H. Beck, *Tetrahedron Lett.* p. 4287 (1966); G. A. Olah, G. Liang, G. D. Mateescu, and J. L. Riemenschneider, *J. Am. Chem. Soc.* **95**, 8698 (1973).
36a. G. A. Olah, A. M. White, J. R. De Member, A. Commeyras, and C. Y. Lui, *J. Am. Chem. Soc.* **92**, 4627 (1970).

37. V. Prelog and J. G. Traynham, in "Molecular Rearrangement" (P. de Mayo, ed.), Vol. 1, p. 593. Wiley (Interscience), New York, 1963.
38. O. A. Reutov and T. N. Shatkina, *Tetrahedron* **18**, 237 (1962).
39. G. J. Karabatsos and C. E. Orzech, Jr., *J. Am. Chem. Soc.* **84**, 2838 (1962).
40. D. M. Brouwer and J. A. van Doorn, *Recl. Trav. Chim. Pays-Bas* **88**, 573 (1969).
41. M. Saunders and J. J. Stofko, Jr., *J. Am. Chem. Soc.* **95**, 252 (1973).
42. T. Pakkanen and J. L. Whitten, *J. Am. Chem. Soc.* **98**, 6336 (1976).
43. H.-J. Kohler and H. Lischka, *J. Am. Chem. Soc.* **101**, 3479 (1979).
43a. P. K. Bischof and M. J. S. Dewar, *J. Am. Chem. Soc.* **97**, 2278 (1975); J. A. Pople, R. Krishnan, and P. von R. Schleyer, unpublished calculations.
44. W. Parker and C. I. F. Watt, *J. Chem. Soc., Perkin Trans. 2* p. 1642 (1975); L. Stehelin, J. Lhomme, and G. Ourrison, *J. Am. Chem. Soc.* **93**, 1650 (1971); L. Stehelin, L. Kannelias, and G. Ourisson, *J. Org. Chem.* **38**, 847 and 851 (1973).
45. See Q. Branca and D. Arigoni, *Chimia* **23**, 189 (1969); Q. Branca, Dissertation, Federal Institute of Technology, Zurich (1970).
46. S. Brownstein and J. Bornais, *Can. J. Chem.* **49**, 7 (1971).
47. N. F. Hall and J. B. Conant, *J. Am. Chem. Soc.* **49**, 3047 (1927); see R. J. Gillespie, *Acc. Chem. Res.* **1**, 201 (1968).
48. E. L. Hagen, Ph.D. Thesis, Yale University, New Haven, Connecticut (1965).
49. R. M. Noyes, *J. Am. Chem. Soc.* **84**, 513 (1962); R. H. Stokes, *ibid.* **86**, 979 and 982 (1964); R. W. Taft, J. F. Wolf, J. L. Beauchamp, G. Scorrano, and E. M. Arnett, *ibid.* **100**, 1240 (1978).
50. Cf. J. L. Franklin, *Trans. Faraday Soc.* **49**, 443 (1952); G. A. Olah and P. von R. Schleyer, eds., "Carbonium Ions," Vol. I, p. 107. Wiley (Interscience), New York, 1968; V. Gold, *J. Chem. Soc. Faraday Trans. 1* **68**, 1611 (1972); D. Bethell and V. Gold, "Carbonium Ions. An Introduction." p. 139 f. Academic Press, New York, 1967.
51. J. L. Fry, J. M. Harris, R. C. Bingham, and P. von R. Schleyer, *J. Am. Chem. Soc.* **92**, 2540 (1970).
52. J. F. Wolf, P. G. Harch, and R. W. Taft, *J. Am. Chem. Soc.* **97**, 2904 (1975).
53. M. Saunders and E. L. Hagen, *J. Am. Chem. Soc.* **90**, 2436 (1968).
54. E. W. Bittner, E. M. Arnett, and M. Saunders, *J. Am. Chem. Soc.* **98**, 3734 (1976).
55. E. M. Arnett, C. Petro, and P. von R. Schleyer, *J. Am. Chem. Soc.* **101**, 522 (1979).
56. M. Saunders, in "Magnetic Resonance in Biological Systems" (A. Ehrenberg, B. G. Malmström, and T. Vanngard, eds.), p. 85. Pergamon, Oxford, 1967.
57. T. S. Sorensen, *J. Am. Chem. Soc.* **89**, 3782 and 3794 (1967).
58. T. S. Sorensen, *J. Am. Chem. Soc.* **91**, 6398 (1969).
59. T. S. Sorensen and T. Ranganayakulu, *J. Am. Chem. Soc.* **92**, 6539 (1970).
59a. H. Mayr, W. Förner, and P. v. R. Schleyer, *J. Am. Chem. Soc.* **101**, 6032 (1979).
60. W. J. Hehre, R. T. McIver, Jr., J. A. Pople, and P. von R. Schleyer, *J. Am. Chem. Soc.* **96**, 7162 (1974).
61. L. Radom, *Aust. J. Chem.* **27**, 231 (1974).
62. E. M. Arnett and J.-L. M. Abboud, *J. Am. Chem. Soc.* **97**, 3865 (1975).
63. W. M. Schubert and W. A. Sweeney, *J. Org. Chem.* **21**, 119 (1956).
64. R. H. Staley, R. D. Wieting, and J. L. Beauchamp, *J. Am. Chem. Soc.* **99**, 5964 (1977).
64a. H. Schwarz, private communication.
65. P. von R. Schleyer and R. D. Nicholas, *J. Am. Chem. Soc.* **83**, 2700 (1961).
66. E. M. Arnett and C. Petro, *J. Am. Chem. Soc.* **100**, 5408 (1978).
67. E.g., W. L. Jorgensen, *J. Am. Chem. Soc.* **99**, 280 (1977).
68. W. L. Jorgensen, *J. Am. Chem. Soc.* **99**, 4272 (1978).
69. W. L. Jorgensen and J. L. Munroe, *Tetrahedron Lett.* p. 581 (1977).

1. REARRANGEMENTS OF CARBOCATIONS

69a. W. L. Jorgensen, *J. Am. Chem. Soc.* **100**, 1049 (1978).
69b. W. L. Jorgensen, *J. Am. Chem. Soc.* **100**, 1057 (1978).
70. W. L. Jorgensen and J. E. Munroe, *J. Am. Chem. Soc.* **100**, 1511 (1978).
71. J. J. Solomon and F. H. Field, *J. Am. Chem. Soc.* **98**, 1567 (1976); P. Kebarle, *Annu. Rev. Phys. Chem.* **28**, 445 (1977); P. P. S. Saluja and P. Kebarle, *J. Am. Chem. Soc.* **101**, 1084 (1979). Earlier data [E. Kaplan, P. Cross, and R. Prinstein, *J. Am. Chem. Soc.* **92**, 1445 (1970)] has been reevaluted in these references; J. L. Beauchamp and F. A. Houle, *ibid.* **101**, 4067 (1979).
72. E. M. Arnett, N. Pienta, and C. Petro, *J. Am. Chem. Soc.* (submitted for publication); P. von R. Schleyer, N. L. Allinger, E. M. Arnett, C. Petro, and N. Pienta, *ibid.* (submitted for publication).
73. C. J. Collins and C. E. Harding, *J. Am. Chem. Soc.* **91**, 7194 (1969).
74. M. Saunders and S. P. Budiansky, *Tetrahedron* **35**, 929 (1979).
75. G. A. Olah and J. Lukas, *J. Am. Chem. Soc.* **90**, 933 (1968).
76. H. Schwarz, M. Saunders, J. Chandrasekhar, and P. von R. Schleyer, unpublished observations.
77. A. T. Balaban, D. Farcasiu, and R. Banca, *Rev. Roum. Chim.* **11**, 1205 (1966); cf. A. T. Balaban, *ibid.* **15**, 1960 (1970).
78. R. E. Leone, J. C. Barborak, and P. von R. Schleyer, in "Carbonium Ions" (G. A. Olah and P. von R. Schleyer, eds.), Vol. IV, Chapter 33, p. 1837. Wiley (Interscience), New York, 1973; *Angew. Chem., Int. Ed. Engl.* **9**, 860 (1970); L. A. Telkowski and M. Saunders, in "Dynamic Nuclear Magnetic Resonance Spectroscopy" (L. M. Jackman and F. A. Cotton, eds.), Chapter 13, p. 523. Academic Press, New York, 1975.
79. See M. Gielen, in "Chemical Applications of Graph Theory" (B. T. Balaban, ed.), Chapter 9, p. 261. Academic Press, New York, 1976.
80. H. W. Whitlock, Jr. and W. Siefkin, *J. Am. Chem. Soc.* **90**, 4929 (1968).
81. P. von R. Schleyer, *J. Am. Chem. Soc.* **79**, 3292 (1957); P. von R. Schleyer and M. M. Donaldson, *ibid.* **82**, 4645 (1960).
82. Reviews: E. M. Engler and P. von Schleyer, *Org. Chem., Ser. One* **5**, 239 (1973); M. A. McKervey, *Chem. Soc. Rev.* **3**, 479 (1974); R. C. Fort, Jr., "Adamantane: The Chemistry of Diamond Molecules." Dekker, New York, 1977; S. Hala, *Chem. Listy* **71**, 18 (1977).
83. E. M. Engler, M. Farcasiu, A. Sevin, J. M. Cense, and P. von R. Schleyer, *J. Am. Chem. Soc.* **95**, 5769 (1973).
84. E. M. Engler, J. D. Andose, and P. von R. Schleyer, *J. Am. Chem. Soc.* **95**, 8005 (1973).
84a. N. L. Allinger, M. T. Tribble, M. A. Miller, and D. H. Wertz, *J. Am. Chem. Soc.* **93**, 1637 (1971).
85. T. M. Gund, P. von R. Schleyer, P. H. Gund, and W. T. Wipke, *J. Am. Chem. Soc.* **97**, 743 (1975).
86. S. A. Godleski, P. von R. Schleyer, E. Osawa, Y. Inamoto, and U. Fujikura, *J. Org. Chem.* **41**, 2596 (1976).
87. E. Ōsawa, K. Aigami, N. Takaishi, Y. Inamoto, U. Fujikura, Z. Majerski, P. von R. Schleyer, E. M. Engler, and M. Farcasiu, *J. Am. Chem. Soc.* **99**, 5361 (1977).
88. L. K. M. Lam, D. J. Raber, J. L. Fry, M. A. McKervey, J. R. Alford, B. D. Cuddy, V. G. Keizer, H. W. Geluk, and J. L. M. A. Schlatmann, *J. Am. Chem. Soc.* **92**, 5246 (1970); Z. Majerski, P. von R. Schleyer, and A. P. Wolf, *ibid.* p. 5731.
89. D. M. Brouwer and J. Hogeveen, *Recl. Trav. Chim. Pays-Bas* **89**, 211 (1970).
90. A Nickon and R. C. Weglein, *J. Am. Chem. Soc.* **97**, 1271 (1975).
91. M. Farcasiu, E. W. Hagaman, E. Wenkert, and P. von R. Schleyer, *Pap.* ORGN 187, 173rd Natl. Meet. Am. Chem. Soc., New Orleans, LA., March, 1977; unpublished

observations. P. v. R. Schleyer, P. Grubmüller, W. F. Maier, O. Vostrowsky, L. Skatterøl, and K. J. Holm, *Tetrahedron Lett.*, (1980).
92. C. J. Collins and C. K. Johnson, *J. Am. Chem. Soc.* **95**, 4766 (1973).
93. C. K. Johnson and C. J. Collins, *J. Am. Chem. Soc.* **96**, 2514 (1974).
94. C. J. Collins, C. K. Johnson, and V. F. Raaen, *J. Am. Chem. Soc.* **96**, 2524 (1974).
95. J. A. Berson, *in* "Molecular Rearrangements" (P. de Mayo, ed.), Part I, Chapter 3, p. 111. Wiley (Interscience), New York, 1963.
96. J. E. Nordlander, F. Y.-H. Wu, S. P. Jindal, and J. B. Hamilton, *J. Am. Chem. Soc.* **91**, 3962 (1969); P. von R. Schleyer, E. Funke, and S. H. Liggero, *ibid.* p. 3965; J. E. Nordlander, J. B. Hamilton, Jr., F. Y.-H. Wu, S. P. Jindal, and R. R. Gruetzmacher, *ibid.* **98**, 6658 (1976).
97. C. D. Gutsche and D. Redmore, "Carbocyclic Expansion Reactions." Academic Press, New York, 1968.
98. J. Chandrasekhar and P. von R. Schleyer, unpublished calculations.
98a. Chandrasekhar and P. v. R. Schleyer, *Tetrahedron Lett.*, p. 4057; G. Wenke, Dissertation, Technischen Universität München, 1979.
99. Y. E. Rhodes, G. Wenke, and D. Lenior, unpublished calculations.
100. D. J. Cram and J. D. Knight, *J. Am. Chem. Soc.* **74**, 5839 (1952).
101. R. H. Burnell, *J. Chem. Soc.* p. 1307 (1958).
102. M. Stiles and R. P. Mayer, *J. Am. Chem. Soc.* **81**, 1497 (1959).
103. E. N. McElrath, R. M. Fritz, C. Brown, C. Y. Legall, and R. B. Duke, *J. Org. Chem.* **25**, 2195 (1960).
104. R. L. Heidke and W. H. Saunders, Jr., *J. Am. Chem. Soc.* **88**, 5816 (1966); J. R. Owen and W. H. Saunders, *ibid.* p. 5809.
105. For an interesting discussion of C to O migration, see E. Hedeya and S. Winstein, *J. Am. Chem. Soc.* **89**, 1661 (1967). Also see ref. 99.
106. J. E. Dubois and P. Bauer, *J. Am. Chem. Soc.* **90**, 4510 and 4512 (1968).
107. B. Miller, *in* "Mechanisms of Molecular Migration" (B. S. Thyagarajan, ed.), Vol. 1. p. 247. Wiley (Interscience), New York, 1968.
108. T. Yvernant and M. Mazet, *Bull. Soc. Chim. Fr.* p. 638 (1969).
109. M. Mazet, *Bull. Soc. Chim. Fr.* p. 4309 (1969).
110. R. B. Carlin and K. P. Sivaramakrishnan, *J. Org. Chem.* **35**, 3368 (1970).
111. J. W. Pilkington and A. J. Waring, *Tetrahedron Lett.* p. 4345 (1973).
112. G. I. Borodkin, M. M. Shakirov, V. G. Shubin, and V. A. Koptyug, *Zh. Org. Khim.* **14**, 989 (1978); A. N. Dezina, V. I. Mamatuk, B. G. Deredeijev, and V. A. Koptyug, *ibid.* **12**, 610 (1976); D. V. Korchagina, B. G. Deredeijev, V. G. Shubin, and V. A. Koptyug, *ibid.* p. 384; V. G. Shubin and V. A. Koptyug, *Proc. Siberian Dep., Akad. Sci. USSR* p. 131 (1976).
113. P. von R. Schleyer, M. M. Donaldson, and W. E. Watts, *J. Am. Chem. Soc.* **87**, 375 (1965); D. E. McGreer, *Can. J. Chem.* **40**, 1554 (1962).
114. H. Hogeveen and P. W. Kwant, *Acc. Chem. Res.* **8**, 413 (1975).
115. H. Hart and M. Kuzuya, *J. Am. Chem. Soc.* **94**, 8958 (1972); **96**, 6436 (1974).
116. R. M. Coates and E. R. Fretz, *Tetrahedron Lett.* p. 1955 (1977).
117. M. J. S. Dewar and J. M. Harris, *J. Am. Chem. Soc.* **92**, 6557 (1970).
118. Y. E. Rhodes and T. Takino, *J. Am. Chem. Soc.* **92**, 5270 (1970).
119. Y. Apeloig, P. von R. Schleyer, and J. A. Pople, *J. Am. Chem. Soc.* **99**, 1291 (1977).
120. J. D. Dill, P. von R. Schleyer, and J. A. Pople, *J. Am. Chem. Soc.* **99**, 1 (1977).
121. Y. Apeloig, P. von R. Schleyer, and J. A. Pople, *J. Am. Chem. Soc.* **99**, 5901 (1977).
122. P. von R. Schleyer, P. T. Stang, and D. J. Raber, *J. Am. Chem. Soc.* **92**, 4725 (1970).
123. G. W. Oxer and D. Wege, *Tetrahedron Lett.* p. 457 (1971).
124. W. Fischer, C. A. Grob, and G. V. Sprecher, *Tetrahedron Lett.* p. 473 (1979); W.

Fischer, C. A. Grob, G. V. Sprecher, and A. Waldner, *ibid.* p. 1901, 1905 (1979); A. K. Coulter and R. A. Reith, *J. Org. Chem.* **44**, 3529 (1979).
125. R. D. Bowen, J. R. Kalman, and D. H. Williams, *J. Am. Chem. Soc.* **99**, 5481 (1977).
126. M. Sprecher and D. Kost, *Tetrahedron Lett.* p. 703 (1969).
127. J. Wemple, *J. Am. Chem. Soc.* **92**, 6694 (1970).
128. R. M. Acheson, *Acc. Chem. Res.* **4**, 177 (1971).
129. R. N. McDonald, *in* "Mechanisms of Molecular Migrations" (B. S. Thyagarajan, ed.), Wiley (Interscience), New York, 1971.
130. P. F. Cann, D. Howell, and S. Warren, *J. Chem. Soc., Chem. Commun.* p. 1148 (1971).
131. L. A. Paquette, T. Kakahano, and J. F. Kelly, *J. Org. Chem.* **36**, 435 (1971).
132. V. Tortorella, L. Toscano, C. Vetuschi, and A. Romeo, *J. Chem. Soc.* p. 2422 (1971).
133. J. Kagan, D. A. Agdeppa, Jr., and S. P. Singh, *Helv. Chim. Acta* **55**, 2252 (1972).
134. D. Howells and S. Warren, *J. Chem. Soc., Perkin Trans.* 2 p. 1645 (1973).
135. J. N. Marx, J. C. Argyle, and L. R. Norman, *J. Am. Chem. Soc.* **96**, 2121 (1974).
136. P. Brownbridge, P. K. G. Hodgson, R. Shepherd, and S. Warren, *J. Chem. Soc., Perkin Trans. 1* p. 2024 (1976).
137. J. N. Marx and E. J. Bombach, *Tetrahedron Lett.* p. 2391 (1977).
138. P. Vogel, M. Saunders, W. Thielecke, and P. von R. Schleyer, *Tetrahedron Lett.* p. 1429 (1971).
139. J. S. Staral and J. D. Roberts, *J. Am. Chem. Soc.* **100**, 8018 (1978); J. S. Staral, I. Yavari, J. D. Roberts, G. K. S. Prakash, D. J. Donovan, and G. A. Olah, *ibid.* p. 8016.
139a. T.-M. Su, Ph.D. Thesis, Princeton University, 1970.
140. A. J. Jones, E. Huang, R. Haseltine, and T. S. Sorensen, *J. Am. Chem. Soc.* **97**, 1133 (1975).
141. G. Seybold, P. Vogel, M. Saunders, and K. B. Wiberg, *J. Am. Chem. Soc.* **95**, 2045 (1973).
142. P. von R. Schleyer, *J. Am. Chem. Soc.* **89**, 699 (1967).
143. S. Inagaki, H. Fujimoto, and K. Fukui, *J. Am. Chem. Soc.* **98**, 4054 (1976).
143a. P. v. R. Schleyer and W. F. Maier, unpublished calculations; R. Gleiter, private communication.
144. R. B. Woodward and R. Hoffmann, *Angew. Chem., Int Ed. Engl.* **8**, 781 (1969).
145. W. von E. Doering, M. Saunders, H. G. Boyton, H. W. Earhart, E. F. Wadley, W. R. Edwards, and G. Laber, *Tetrahedron* **4**, 178 (1958); B. B. Derendyaev, V. I. Mamatyuk, and V. A. Koptyug, *Tetrahedron Lett.* p. 5 (1969); G. A. Olah, R. H. Schlosberg, D. P. Kelly, and G. D. Mateescu, *J. Am. Chem. Soc.* **92**, 2546 (1970).
146. W. J. Hehre and J. A. Pople, *J. Am. Chem. Soc.* **94**, 6901 (1972); W. J. Hehre and P. C. Hiberty, *ibid.* **96**, 7163 (1974).
147. R. F. Childs and S. Winstein, *J. Am. Chem. Soc.* **90**, 7146 (1968); P. Vogel, M. Saunders, N. M. Hasty, and J. A. Berson, *ibid.* **93**, 1551 (1971).
148. W. J. Hehre, *J. Am. Chem. Soc.* **94**, 8908 (1972); **96**, 5207 (1974); R. C. Haddon, *J. Org. Chem.* **44**, 3608 (1979).
149. D. M. Brouwer and J. A. van Doorn, *Recl. Trav. Chim. Pays-Bas* **89**, 333 (1970); D. M. Brouwer and H. Hogeveen, *ibid.* p. 211.
150. V. A. Koptyug, I. A. Shleider, and I. S. Isaev, *Zh. Org. Khim.* **7**, 852 (1971); V. A. Koptyug, I. A. Shleider, I. S. Vasilyena, and A. I. Rezvukhin, *ibid.* p. 108989; I. A. Shleider, I. S. Isaev, and V. A. Koptyug, *ibid.* **8**, 1337 (1972).
151. R. Breslow, J. Lockhart, and A. Small, *J. Am. Chem. Soc.* **84**, 2793 (1962).
152. A. P. Devaquet and W. J. Hehre, *J. Am. Chem. Soc.* **96**, 3644 (1974).
153. J. A. Berson, *Acc. Chem. Res.* **5**, 406 (1972); J. A. Berson and L. Salem, *J. Am. Chem. Soc.* **95**, 8917 (1972).

ESSAY 2

GAS-PHASE ION REARRANGEMENTS

RICHARD D. BOWEN
AND DUDLEY H. WILLIAMS

I.	Introduction	55
	A. Historical Development	55
	B. Metastable Ions	56
	C. Gas-Phase versus Solution Experiments	56
II.	Importance of Metastable Peaks	57
	A. Slow Reactions	57
	B. Channeling	58
	C. Kinetic Energy Release	59
III.	Potential Energy Profile Approach	62
	A. Concept of the Potential Energy Profile	62
	B. Metastable Abundance Data	63
	C. Isotope Labeling	64
	D. Energetics	67
	E. Kinetic Energy Release	70
	F. Collisional Activation and Ion Cyclotron Resonance	77
	G. Predictive Capacity	80
IV.	Types of Potential Energy Profiles	80
	A. Complete Equilibration of Isomers before Unimolecular Dissociation	80
	B. No Equilibration of Isomers before Unimolecular Dissociation	82
	C. Rate-Determining Isomerization before Unimolecular Dissociation	85
V.	Conclusions	90
	References	90

I. INTRODUCTION

A. Historical Development

Mass spectrometry has been used as an analytical and structural probe in a wide variety of ways; pioneering examples of its application in the chemical field include the discovery of isotopes (*1*) and the mass defect (*2*). After the expansion of the petrochemical industry, it became a routine tool for the organic chemist. Interest initially centered on the fragmentation patterns of organic molecules, the interest resulting in the publication of several text (3–7). Later, it became apparent that further information concerning the compound of interest and, in particular, individual ions derived from it could be obtained from a consideration of the peaks termed "metastable" in the mass spectrum. In fact, it has emerged that metastable peaks contain a wealth of useful information concerning the slow reactions of ions in the gas phase (8–10). A careful investigation of the reactions that give rise to these peaks constitutes a powerful and convenient method for examining the slow dissociations of ions in the gas phase.

B. Metastable Ions

An ion that leaves the source of the mass spectrometer intact but dissociates before arriving at the detector is called a metastable ion (8). The dissociation of metastable ions may result in the appearance of a metastable peak. Thus, for example, metastable ions formed in the second field-free region of a mass spectrometer, in which the ions enter the magnetic analyzer (MA) after having been transmitted through the electrostatic analyzer (ESA), are evidenced by a metastable peak in the normal mass spectrum. Moreover, decompositions occurring in the first field-free region of such an instrument may be detected, free from any possible interference from daughter ion peaks in the normal mass spectrum, by simple modification of the instrumental conditions (11, 12). Similar considerations apply to mass spectrometers having different geometries. Conceptually, these metastable peaks form a "direct picture" of what happens when dissociation takes place.

C. Gas-Phase versus Solution Experiments

When the reactions of ions are being studied, there are distinct advantages in conducting the experiments in the gas phase. First, any complications caused by the influence of solvent molecules are avoided. In solution experiments collisions lead to rapid interconversion of the internal energy of the system among the various rotational, vibrational, and translational modes that are available. A Maxwell–Boltzmann energy distribution is set up, and any energy released in the course of the reaction is quickly absorbed into this common energy "pool." Thus, any component of a reverse activation energy which is partitioned as translation (i.e., kinetic energy release associated with the reaction) cannot be distinguished from that component which goes into internal energy of the products. This information, which is lost in solution experiments, may be acquired in the gas phase, provided that the pressure in the reaction vessel

2. GAS-PHASE ION REARRANGEMENTS

is sufficiently low to eliminate collisions; this condition is readily satisfied. Any kinetic energy release accompanying dissociation is evidenced by a broadening of the metastable peak for the process concerned (Section II,C) and may thus be easily observed. Second, the dissociation of an ion is readily detected by the very appearance of this metastable peak. Furthermore, when two or more decomposition channels are observed, their relative abundances may be deduced from the sizes of the corresponding metastable peaks. Therefore, these data not only are easily accessible, but can also be rapidly analyzed and interpreted. Third, the nature of the experiment permits a very tight control over the chemistry; this is discussed in detail in the next section.

II. IMPORTANCE OF METASTABLE PEAKS

A. Slow Reactions

The usual origin of metastable peaks is the slow, unimolecular dissociation of ions having a well-defined range of internal energies just above the threshold for reaction. The fact that metastable ions dissociate while traveling between the source and collector fixes a definite time scale for their decomposition. Although there are many parameters that influence the time taken for an ion to pass through the instrument, metastable ions may generally be considered to have a lifetime of $ca.$ 10^{-5} sec. This corresponds to a unimolecular rate constant in the range 10^4-10^6 sec^{-1}, which is slow compared to bond vibrational frequencies ($ca.$ 10^{14} sec^{-1}). Consequently, metastable ions have well-defined lifetimes and dissociate after some 10^8-10^{10} vibrations; it is thus possible to sample essentially all energetically accessible reactant configurations before decomposition. Furthermore, because the rates at which ions decompose increase very rapidly above the threshold for reaction, the dissociations of metastable ions normally occur with excess energies in the transition states, which are small and comparable to those found in solution experiments (9). Although definite examples of the intervention of isolated electronic states are known (13), this behavior appears to be relatively uncommon and is often discounted in the slow, unimolecular reactions of ions. For example, the slow dissociations of 11 members of the $C_nH_{2n+1}^+$ ($n = 2-12$) homologous series of ions may be rationalized or correctly predicted without invoking the intervention of isolated states (14).

Since metastable ions usually decompose with small excess energies in the transition states (9), the dominant parameter in determining the occurrence or nonoccurrence of a reaction is the activation energy for the process concerned (15). This is illustrated elegantly by the decomposition of suitably labeled ions, which frequently reveal the operation of primary deuterium isotope effects. These isotope effects span the range known for solution reactions, and, for several small ions, are sometimes much larger. In the case of ionized methanes, where the threshold for D· loss is calculated to be only *ca.* 2 kcal/mol above that for H· loss, exclusive H· loss is observed from CH_4^+, CH_3D^+, $CH_2D_2^+$, and CHD_3^+; CD_4^+ alone eliminates D· (*16*). Investigations (*17–19*) of labeled ethane and propane have also uncovered large isotope effects; for instance, $CD_3CH_2^+$ eliminates H_2 and D_2 in the ratio of *ca.* 150:1, and loss of H· from $CH_3CD_3^+$ is over 600 times more abundant than that of D·.

B. Channeling

When an ion of a given molecular formula is observed to undergo two or more different decompositions, useful information can usually be deduced from the relative abundance of the various decay channels. On the assumption that the reactions are in competition, the observation that two isomeric ions, when generated from different precursors, undergo the same dissociations in similar ratios leads to the conclusion that these ions are able to interconvert before decomposition (*15*). In other words, the activation energies for dissociation of each ion must be greater than those for rearrangement to the other structure. Minor changes in the extent to which the common reactions take place may be due to slight variations in the internal energies of ions produced from different precursors (*20*). However, when two ions are found to undergo completely different dissociations, it may be concluded that these ions do not interconvert before slow, unimolecular decompositions, i.e., that the energy barriers for rearrangement of either ion to the other are greater than those for dissociation.

Two early and important examples of the use of this "metastable abundance criterion" to obtain information about the structures of ions in the gas phase concern the ions $C_6H_{13}^+$ and $C_2H_5O^+$. For $C_6H_{13}^+$, irrespective of whether the precursor contains an incipient primary, secondary, or tertiary hexyl cation, C_2H_4 and C_3H_6 losses are observed in the ratio of *ca.* 1:1.6. This indicates that interconversion of all energetically accessible reactant configurations precedes dissociation (*15*). In contrast, $C_2H_5O^+$ ions may be classified into two distinct groups (see also Section IV,B). The members of one class, generated from compounds having an ether moiety (CH_3OCH_2Y, where Y is a variable group, e.g., Cl, CH_3), presumably have the structure $CH_3O^+=CH_2$ and are found to lose almost exclusively CH_4. Other ions, formed from precursors of the general formula CH_3CHYOH or CH_2YCH_2OH, probably form an interconverting mixture

2. GAS-PHASE ION REARRANGEMENTS

of $CH_3CH=OH^+$ and $\overline{CH_2CH_2O}^+H$. The ions in this class undergo two slow dissociations, CH_4 and C_2H_2 loss, in a ratio of *ca.* 1:1.9 *(21)*, which precludes interconversion with $CH_3O^+=CH_2$ before decomposition.

C. Kinetic Energy Release

Useful conclusions concerning the unimolecular reactions of ions may be deduced not only from the abundances of metastable peaks but also from their shapes, which are, in effect, "pictures" of what happens when dissociation occurs.

Consider, first, an ion, of unit charge and mass m_1, which decomposes in the second field-free region of a mass spectrometer in which ions are transmitted by the ESA before entering the MA (similar arguments may be applied to instruments of different geometry). The dissociation produces a fragment ion of mass m_2, together with a neutral (see Fig. 1; closed circles denote the product ions formed, whereas the neutrals are represented by open circles). In the absence of repulsion between the neutral and the fragment ion, a relatively slow separation occurs, and the molecules drift apart. The resulting metastable peak is narrow and gaussian (Fig. 2a). Suppose, however, that the fragment ion and neutral repel one another; this leads to a relatively discrete kinetic energy release accompanying the dissociation, the ion and neutral receiving equal and opposite momenta in addition to that which they would have possessed in the absence of kinetic energy release. Thus, decompositions occurring from orientation A (Fig. 1) give rise to fragment ions with slightly more momentum than that which would have been expected in the absence of kinetic energy release. These ions are "pushed along the beam" (x direction). Consequently, they are deflected slightly less readily by the MA and appear at an m/e value marginally higher than the normal value $[m_2^2/m_1$ *(22)*$]$ (Fig. 2b). Conversely, ions that dissociate in configuration B give rise to fragment ions with slightly reduced momentum because these fragment ions are "pushed back along the beam." As a result, they are

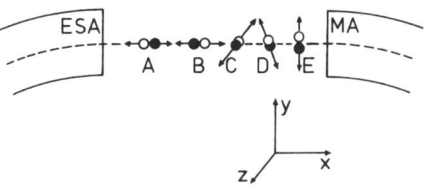

Fig. 1 Dissociation of metastable ions in the second field-free region of a mass spectrometer in which ions are transmitted by the ESA before entering the MA. Reproduced from *Angew. Chem., Int. Ed. Engl.* **18**, 451 (1979). Copyright by Verlag Chemie GMBH.

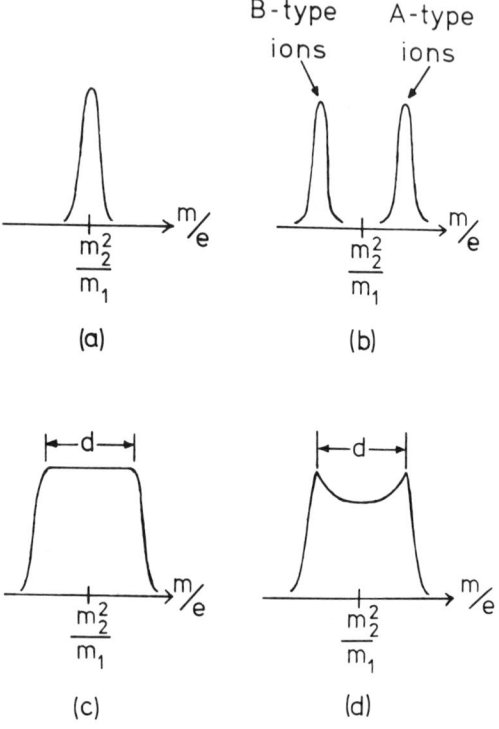

Fig. 2 Metastable peaks corresponding to dissociation of ions shown in Fig. 1. (a) Without a relatively large and specific kinetic energy release; (b) with a relatively large and specific kinetic energy release, extreme components only; (c) with a relatively large and specific kinetic energy release, overall peak; (d) with a relatively large and specific kinetic energy release, together with discrimination against ions deflected in the z direction (C-type ions). Reproduced from *Angew. Chem., Int. Ed. Engl.* **18**, 451 (1979). Copyright by Verlag Chemie GMBH.

deflected somewhat more readily by the MA and are displaced to an m/e value marginally lower than m_2^2/m_1. Dissociations from all intermediate configurations (e.g., C, D, and E) also occur; this situation gives rise to an "exploding sphere" of ions. The net result is a flat-topped metastable peak (Fig. 2c). A mathematical treatment of the problem leads to a simple relationship between the kinetic energy release (T, electronvolts) and the width (d, atomic mass units) of the flat-topped metastable peak (*23, 24*). In some cases, there may be discrimination against fragment ions formed from decompositions occurring in orientations such as C. This discrimination originates because ions produced in this way are deflected so extensively that they are not able to pass through the slits aligned in length

2. GAS-PHASE ION REARRANGEMENTS

along the z direction. This effect, which is due to the finite length of the slits (24), results in a "dish-topped" metastable peak (Fig. 2d). An example of this situation is given in Fig. 3, where the metastable peak for CH_4 loss from protonated acetaldehyde is shown as recorded on two instruments having different geometries.

The fundamental point is that a flat-topped or dish-topped metastable peak indicates that the dissociation is accompanied by a relatively specific kinetic energy release. In contrast, a gaussian peak reveals that decomposition occurs without the release of a discrete amount of kinetic energy. Nevertheless, an average kinetic energy release may be computed from the width at half-height of a gaussian metastable peak (25). The average kinetic energy release is very sensitive to the excess energy present in the transition state for the dissociation step (9); as the amount of excess energy increases, the average kinetic energy release becomes greater. This is evidenced by broad, gaussian metastable peaks and may be used to detect slow, rate-determining isomerizations of ions before relatively fast dissociations (26).

Occasionally, a metastable peak is observed which is not simply gaussian, flat-topped, or dish-topped, but composite. This indicates that two or more competing reaction channels exist for the process concerned. Several examples of this behavior are known in systems in which $C_3H_3^+$ is formed in the course of dissociation (27–30). The most plausible general explanation is that two structures (presumably cyclopropenyl and propargyl) for the $C_3H_3^+$ product ion are accessible with two mechanisms for decomposition, each of which is characterized by a different kinetic energy release. In the case of H_2 loss from $C_3H_5^+$, this explanation is established by appearance potential measurements (30), which reveal that the transition state energies for the two competing channels are approxi-

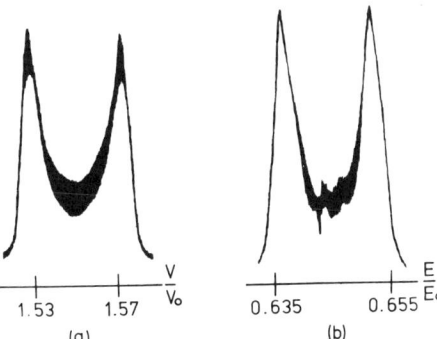

Fig. 3 Metastable peak for the reaction $CH_3CH{=}OH^+ \rightarrow H^+C{=}O + CH_4$. (a) Recorded using an AEI MS 902 instrument; (b) recorded using a VG Micromass ZAB 2F instrument.

mately 291 kcal/mol. Since the kinetic energy releases that accompany H_2 loss are 17 and 3 kcal/mol, it follows that the heats of formation of the $C_3H_3^+$ product ions must be at the most 274 and 288 kcal/mol. These results, together with known heats of formation for $C_3H_3^+$ [256 and 281 kcal/mol, respectively, for cyclopropenyl and propargyl cations (*31*)], show that the channel for H_2 loss that releases 17 kcal/mol of kinetic energy must give rise to cyclopropenyl cation since no other $C_3H_3^+$ structure is energetically accessible (*30*). It may therefore be deduced that the alternative decay route, which releases only 3 kcal/mol of kinetic energy, results in the formation of the higher-energy propargyl cation, which is now energetically accessible (*30*).

III. POTENTIAL ENERGY PROFILE APPROACH

A. *Concept of the Potential Energy Profile*

The potential energy profile is a model that can be used to rationalize the slow reactions and rearrangements of ions. The basic idea is to analyze all possible isomerization and dissociation pathways in terms of the energy required to form the appropriate intermediates, reacting configurations, and product combinations. Certain assumptions are then made concerning the factors that determine the occurrence, or nonoccurrence, of a given decomposition pathway. These are the following:

1. All possible decay channels are available to each ion. This corresponds to competition between all available dissociation routes and the absence of any isolated electronic states that might interfere by preventing this competition or by giving rise to fast dissociations. In fact, although there are definite systems in which complications caused by isolated electronic states arise (*13*), these instances appear to be rare and need not be considered in most cases.

2. The decomposition channels having the lowest overall activation energies are greatly favored. That is, energy is considered to be the dominant parameter, and probability or "entropy" effects are only of secondary importance. This is equivalent to the view that there is enough time for all energetically accessible reactants, intermediates, and reacting configurations to be explored. Any product combination, therefore, that can be reached via such energetically accessible species may be formed. This assumption also seems plausible and is usually valid; however, there are a few cases in which "entropy" effects are important and must be considered. These examples seem to involve rearrangements in which the probability of attaining the correct geometry of the relevant activated complex is so small that, even after *ca.* 10^9 vibrations, there is still not

2. GAS-PHASE ION REARRANGEMENTS

enough time for these configurations to be explored. A notable instance of this effect is observed in the slow dissociations of $C_3H_6^+$. Here H· loss is the major slow decomposition, even though H_2 loss requires less energy (32). This behavior, which corresponds to a discrimination against any process that must occur via a very improbable activated complex, appears to be fairly common when H_2 losses are involved. In a recent study (33) of the homologous series of unsaturated carbonium ions $C_nH_{2n-1}^+$, loss of H_2 was often found to be a minor reaction in the first field-free region. However, when the same ions decomposed in the second field-free region, H_2 loss was often the major process. This is consistent with the operation of a systematic discrimination against H_2 loss. As ions of longer lifetimes are sampled (second, as opposed to first, field-free region metastable ions), this discrimination becomes less pronounced because there is more time for the improbable activated complex, which is required for H_2 loss, to be sampled. Hence, H_2 loss becomes more important for longer-lived ions.

When the assumptions outlined above hold true, it is frequently possible to explain, or, in certain cases, to predict correctly, the behavior of organic ions in the gas phase. It is important to note that the total heat of formation of the products is not always the deciding factor for the selection of decomposition channels. This is because high-energy intermediates and/or reacting configurations may be involved in routes to energetically favorable products. Thus, for example, elimination of C_2H_4 from carbonium ions is frequently preempted, even though it may lead to thermodynamically stable products, because it must proceed via a relatively high-energy primary cation (14, 33, 34).

$$R\overset{\frown}{-CH_2}\overset{+}{-CH_2} \longrightarrow R^+ + CH_2=CH_2 \qquad (1)$$

There are several sources of experimental data that can be used in the construction of potential energy profiles. Some are discussed briefly below.

B. Metastable Abundance Data

When two (or more) isomeric ions are observed to undergo the same slow dissociations in very similar ratios, it may be deduced that these ions interconvert before decomposition (15). This leads to the conclusion that the activation energies for rearrangement of either of the isomeric ions are less than those for dissociation. Conversely, when two isomeric ions (or groups of isomeric ions) are found to undergo different reactions, it may be deduced that the activation energy for rearrangement of either structure to the other is greater than that needed to promote decomposition.

This relative competition between two or more decay routes (the so-called metastable abundance criterion) is a powerful method of discovering whether isomeric ions interconvert by rearrangement. For example, isomeric carbonium ions [$C_nH_{2n+1}^+$ (14, 15, 35), $C_nH_{2n-1}^+$ (33, 36), and $C_nH_{2n-3}^+$ (34, 36)] are usually found to undergo the same slow reactions irrespective of the precursor used to generate the ion. This shows that, in general, rearrangement processes have lower activation energies than decomposition pathways for these ions. A specific case of historical importance, $C_6H_{13}^+$ (15), was discussed above. In other systems, ions of a given chemical formula may be assigned to groups the members of which interconvert with each other but not with members of the other groups.

An early example of this type of classification is the $C_2H_5O^+$ system (21) discussed earlier (Section II,B; see also Section IV,B). Others are $C_3H_8N^+$ (37, 38), $C_3H_7O^+$ (26), $C_4H_9O^+$ (39), and $C_3H_5O^+$ (40). However, the relative competition between decay channels depends on the internal energies of the decomposing ions (20), and minor differences in the abundances of competing dissociations may reflect slight differences in the internal energies of ions formed from different precursors.

The "metastable abundance criterion" may also be used to detect rate-determining rearrangements of ions; this application is discussed in Section IV,C.

C. Isotope Labeling

Whereas metastable abundance data yield information as to which ions interconvert before decomposition, labeling results indicate which atoms (or groups of atoms) become equivalent before dissociation occurs. Thus, if two isomeric ions are able to interconvert via a series of hydride shifts, all the hydrogen atoms that are able to participate in this series of hydride shifts may be rendered equivalent. In some cases, these rearrangements may involve all the atoms of a given element in the ion. An example is furnished by extensive labeling studies on the isomeric $C_4H_9^+$ ions. The dominant slow dissociation of these ions is always CH_4 elimination, accompanied by a small and almost constant percentage of C_2H_4 loss (35). Moreover, ^{2}H- and ^{13}C-labeling studies show that the constituent atoms in the expelled CH_4 neutral are selected at random from all those in the ion (35). This can be explained by considering the possible mechanisms whereby the isomeric butyl ions may interconvert (Scheme 1). The 1,2-hydride or 1,2-methyl shifts required in these rearrangements are symmetry-allowed (41) and are known to be facile processes both from solution NMR experiments (42) and from calculations (43). Accurate heats of formation of the four "classical" isomers of $C_4H_9^+$ are known (44), as is the energy required to cause CH_4 elimination from $C_4H_9^+$ (45).

2. GAS-PHASE ION REARRANGEMENTS

These energy data lead to the conclusion that, at internal energies sufficient to cause dissociation [208 kcal/mol (45)], even the thermodynamically least stable n-butyl cation [heat of formation 201 kcal/mol (44)] is readily accessible. Rapid and reversible interconversion of all the four "classical" butyl cations is therefore expected to precede slow elimination of CH_4. The most plausible mechanisms whereby these rearrangements may occur involve either hydride or methyl shifts; their rapid occurrence must lead to statistical distribution of all hydrogen and carbon atoms in the ion, and the decomposition of labeled ions ought to involve statistical selection of the hydrogen and carbon atoms.

$$CH_3CH_2CH_2\overset{+}{C}H_2 \underset{\text{shift}}{\overset{1,2\text{-H}}{\rightleftharpoons}} CH_3CH_2\overset{+}{C}HCH_3 \underset{\text{shift}}{\overset{1,2\text{-}CH_3}{\rightleftharpoons}} \overset{+}{C}H_2CHCH_3$$
$$|$$
$$CH_3$$

$$1,2\text{-H} \updownarrow \text{shift}$$

$$CH_3\overset{+}{C}CH_3$$
$$|$$
$$CH_3$$

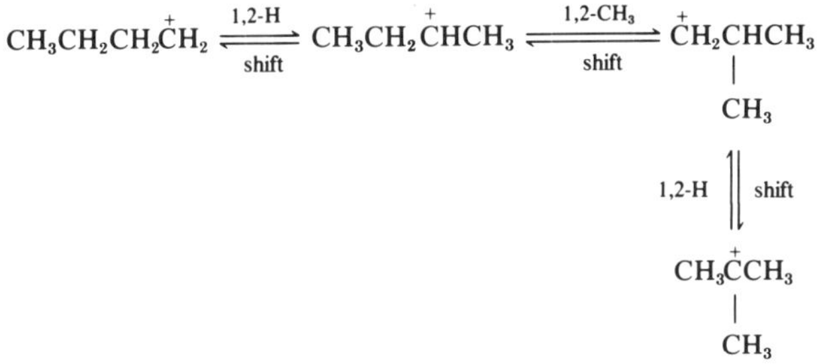

Scheme 1

This statistical loss of atoms from labeled species is a fairly common phenomenon; other examples include $CH_3\cdot$ loss from $C_4H_8^{\ddot{+}}$ molecular ions (46, 47) and C_2H_4 loss from $C_7H_{11}^+$ ions (48). This behavior is sometimes referred to as "randomization" or "scrambling." Such expressions should be used with caution because they may appear to imply that the chemistry of the system is not understood when, in fact, it may be rationalized in terms of a simple mechanism such as that in Scheme 1. Moreover, it is important to grasp that it is the equilibration of isomeric species that causes loss of identity of atoms and not the presence of a suitably symmetric intermediate on the reaction pathway. Such is evident from the $C_4H_9^+$ system discussed above; statistical loss of CH_4 is observed from this ion (35) even though no single structure for $C_4H_9^+$ exists in which all the hydrogen and carbon atoms are equivalent.

Another important aspect of labeling experiments is the uncovering of isotope effects in the slow decomposition of ions in the gas phase. This is especially useful in cases in which it gives rise to a primary deuterium isotope effect. These isotope effects are occasionally very large, especially in cases involving hydrogen radical loss from the molecular ions of small hydrocarbons (16–19). These reactions are probably simple bond cleavages; however, isotope effects may also be observed in rearrangements in which a hydrogen atom is transferred in the rate-determining step. An example of this situation is hydrogen transfer from carbon to oxygen in $C_3H_7O^+$ ions, generated as protonated propionaldehyde, which eliminate H_2O via the mechanism given in Scheme 2. The results of ^2H-labeling studies on this system reveal that the hydrogen atom originally bound to oxygen is predominantly retained in the H_2O molecule that is lost (49). The other hydrogen atom is selected at random from the six carbon-bound hydrogens, and an isotope effect of roughly 2:1 operates in favor of hydrogen rather than deuterium transfer to oxygen. Thus, for instance, $CD_3CH_2CH=OH^+$ and $CH_3CD_2CD=OH^+$ undergo essentially the same slow dissociations, H_2O, HOD, and D_2O being eliminated in the ratios 64:34:2 and 61:36:3, respectively (49). This is conclusive evidence that all hydrogen atoms attached to carbon become equivalent via a series of reversible 1,2-hydride shifts and that hydrogen transfer from carbon to oxygen is the rate determining step (RDS) in H_2O loss.

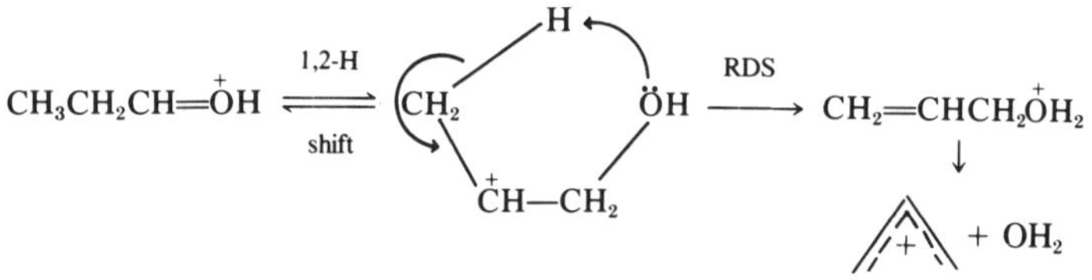

$$CH_3CH_2CH=\overset{+}{O}H \underset{\text{shift}}{\overset{1,2\text{-H}}{\rightleftarrows}} \begin{pmatrix} CH_2 \\ \overset{+}{C}H-CH_2 \end{pmatrix} \overset{H}{\underset{}{\overset{}{\rightarrow}}} \overset{..}{O}H \overset{RDS}{\longrightarrow} CH_2=CHCH_2\overset{+}{O}H_2 \downarrow \triangle^+ + OH_2$$

Scheme 2

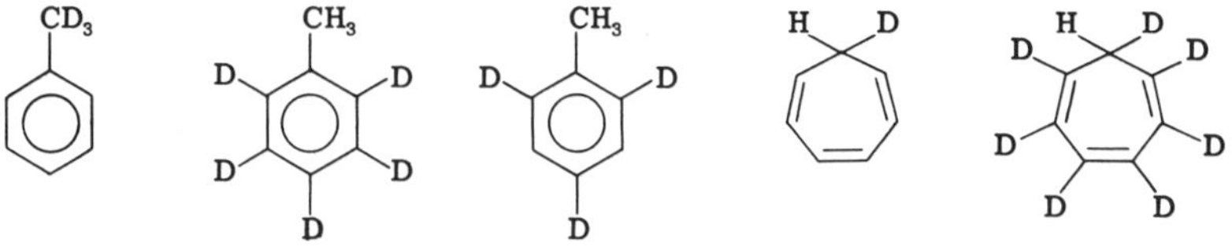

Another interesting example of the need to consider isotope effects in conjunction with partial or complete loss of identity of atoms in labeled species stems from work on H· loss from $C_7H_8^{+\cdot}$ isomers (50, 51). For the three labeled toluenes **1–3**, the loss of H· and D· can be accommodated by a model in which all hydrogen and deuterium atoms occupy the same sites to a statistical extent, thus giving rise to random loss of H· and D·, together with an isotope effect of 2.8:1 favoring H· loss (50). Exactly the same behavior is observed for the molecular ions of the labeled cycloheptatrienes **4** and **5** (50). This constitutes strong evidence that the toluene and cycloheptatriene molecular ions are able to interconvert at lower internal energies than those required to promote H· loss. Very similar results were obtained in another study of the ionized toluene isomers (51), although here a slightly larger isotope effect of 3.5:1 in favor of H· loss was observed. The slight discrepancy between the two studies is presumably due to a population of ions of marginally longer lifetimes [and hence lower average internal energies (20)] in the latter work (51).

D. Energetics

The use of energy data in constructing potential energy profiles for the unimolecular reactions of ions is in many ways self-evident. The success of the approach depends on the availability of reasonably reliable heats of formation for all the species that are likely to be involved in isomerization or decomposition pathways. The lack of accurate thermochemical data is therefore a potentially troublesome problem.

Sophisticated instrumentation is required to determine high-precision heats of formation for organic ions in the gas phase. Energy measurements using conventional mass spectrometers are prone to systematic, in addition to random, errors (52, 53). Reliable heats of formation are available for relatively few organic ions. These data are obtained from photoionization studies (54, 55), sometimes accompanied by mass analysis, thus giving accurate appearance potential measurements (55), from appearance potential determinations using monoenergetic electron beams (31, 44, 56–59), or from proton affinity results (60–63). A comparison of the heats of formation obtained for some protonated aldehydes and ketones is given in Table 1. In general, the agreement between the values determined by the three techniques is good, although there are some nontrivial differences (e.g., for protonated acetone). The proton affinity method is most promising for several reasons: (a) The measurements are based on equilibrium studies. Therefore, provided that sufficient care is taken to ensure that the equilibrium is actually achieved, the thermodynamically most stable conformation of the ion in question must be

TABLE 1

Heats of Formation Determined by Photoionization (PI), Proton Affinity (PA), and Appearance Potential (AP) Measurements

Ion	Heat of formation (kcal/mol)		
	PI (55)	PA (63)	AP (59)
$CH_2=\overset{+}{O}H$	170	164	169
$CH_3CH=\overset{+}{O}H$	140	142	139
$CH_3CH_2CH=\overset{+}{O}H$	134	130	132
$(CH_3)_2C=\overset{+}{O}H$	128	121	120
$CH_3CH_2CH_2CH=\overset{+}{O}H$	—	123	—
$(CH_3)_2CHCH=\overset{+}{O}H$	—	121	—
$CH_3CH_2(CH_3)C=\overset{+}{O}H$	—	112	115

generated. There are thus no possible complications analogous to those inherent in simply ionizing a molecule, a process that may give rise to a conformation of the ion of interest which is not the lowest in energy. (b) A whole series of related species may be analyzed; the resulting order of proton affinities may then be deduced with great accuracy even though the absolute values may not be known to better than ±1 kcal/mol. In this way the heats of formation of homologous ions may be accurately correlated; this is not possible with data obtained for individual molecules by photoionization or appearance potential measurements. (c) Proton affinity measurements are now becoming more extensive in scope, and many useful heats of formation are now available (63). The technique is also extremely versatile; for instance, the heat of formation of 2-propenyl cation is known from the proton affinity of propyne (62).

When experimental values for the heats of formation of cations are not available, recourse is often had to calculations or estimation methods. The latter are usually based on the group equivalent method for estimating the heats of formation of neutrals (64, 65); this procedure has been used with satisfactory results for many years. Nevertheless, it should be emphasized that experimental results of reasonable accuracy are always preferable to calculations or estimates and should be used whenever possible.

A more serious problem arises in cases in which there are as yet no experimental methods for obtaining the heats of formation of the relevant species. In such cases, calculations or estimation methods must be employed, at least at present. This difficulty in obtaining heats of formation of certain species is especially pronounced for open-chain carbonium ions; these ions (e.g., **7** and **8**) are plausible intermediates in the decomposition of ions such as protonated propionaldehyde (**6**) (Scheme 3). However, attempts to obtain the heat of formation of **8** via measurement of the appearance potential for Br˙ loss from ionized 3-bromopropan-1-ol (**9**) are likely to fail because, at threshold, participation of the oxygen lone pair is probable, and thus protonated oxetane (**10**) is formed rather than **8** (66). Although calculations of the heat of formation of species such as **8** are possible (67), a simple estimation procedure would clearly be helpful. Such a procedure (66), based on an isodesmic substitution (67), is illustrated in Scheme 4. The appropriate butyl cation is considered, and the energy required to replace the terminal —CH_3 group with —OH is assumed to be the same as that known for the analogous change in the corresponding neutrals (butan-1-ol and *n*-butane) (66). This yields a value [169 kcal/mol (66)] for the heat of formation of **8** which assumes that no stabilization of **8**, by orbital overlap of the oxygen lone pair with the cationic center, or destabilization, by inductive withdrawal of electrons

2. GAS-PHASE ION REARRANGEMENTS

by the electronegative oxygen atom, occurs. Lack of orbital overlap is precisely the situation of interest; however, inductive withdrawal of electrons by oxygen ought to occur, thus resulting in destabilization of the cation **8**. Calculations suggest that the magnitude of this effect depends on the conformation of the ion (67). For electronegative substituents attached to the β- and γ-carbon atoms (the carbon at the cationic site being designated α), maximal values of 10 and 3 kcal/mol are found for the inductive destabilization. Use of these corrections gives an estimated value of 172 kcal/mol for the heat of formation of **8** (66). It must be understood that this procedure is designed to give approximate values only; however, despite the fact that these estimates may be in error by several kilocalories per mol, the values are considerably better than none at all. Moreover, application of the estimation method in numerous systems results in analyses that are at least self-consistent (33, 34, 69–73), and, in some cases (33, 34, 69, 73), genuine predictions can be correctly made concerning the decompositions of previously uninvestigated ions.

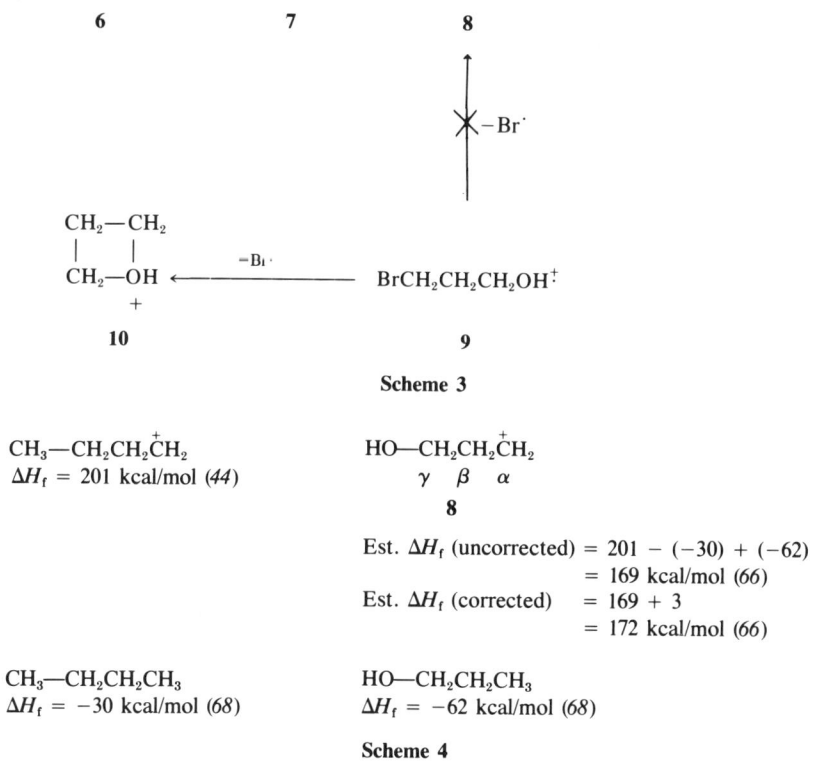

Scheme 3

Scheme 4

E. Kinetic Energy Release

It has already been emphasized that the metastable peak for a given slow reaction constitutes a "direct picture" of what happens when dissociation occurs. Broadly speaking, there are three main classes of peak shapes, each of which yields information concerning the shape of the potential energy profile for the final step.

1. The case of narrow and approximately gaussian peaks.

These shapes indicate that the products drift apart in the course of dissociation and correspond to the general potential energy profile shown in Fig. 4. This situation corresponds to no reverse activation energy; in other words, no intermediate or transition state is as high in energy as the products. Dissociation of the reacting configuration is therefore continuously endothermic and occurs with little excess energy in the transition state for the final step (which is, in this case, entirely productlike). An excellent example of this behavior, already described, is C_2H_4 loss from protonated propionaldehyde (Scheme 3) (69). The potential energy profile in Fig. 5 may be deduced using known (68, 70) or estimated (66) heats of formation; it is assumed that 1,2-hydride shifts may occur essentially without activation energy apart from those associated with the inherent exo- or endothermicities of the reaction concerned. Figure 5 shows that protonated propionaldehyde may attain the reacting configuration **8** without passing through any intermediates or transition states with energies greater than that of **8**. Dissociation of **8**, via simple bond cleavage assisted by the lone pair on oxygen, can now occur without reverse activation energy. This view is derived from two pieces of experimental evidence: (a) appearance potential measurements are consistent with the reaction occurring with no significant reverse activation energy (69); (b) the

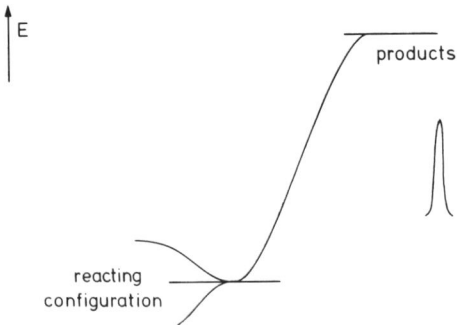

Fig. 4 Potential energy profile corresponding to narrow gaussian metastable peaks.

2. GAS-PHASE ION REARRANGEMENTS

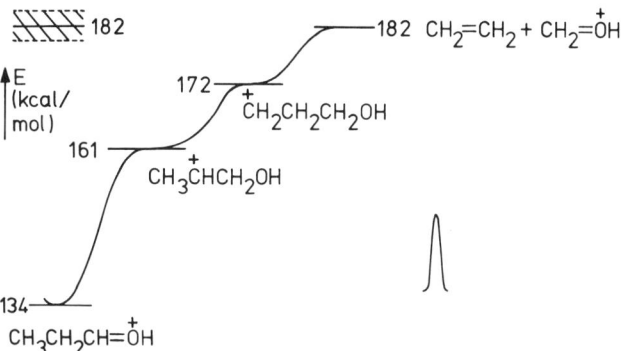

Fig. 5 Potential energy profile, schematic metastable peak, and measured transition state energy for C_2H_4 loss starting from $CH_3CH_2CH=OH^+$. Reproduced from *J. Am. Chem. Soc.* **99**, 3192 (1977). Copyright by the American Chemical Society.

metastable peak for the process is narrow and gaussian (see Fig. 5) and corresponds to an average (25) kinetic energy release of only 0.8 kcal/mol (74).

2. The case of flat-topped or dished metastable peaks.

This shape of peak shows that the products are formed in such a way that they repel one another; dissociation is therefore accompanied by a relatively large and specific kinetic energy release. It may therefore be concluded that there is a reverse activation energy for the final step and that this reverse activation energy contains a relatively specific translational component. This in turn indicates that a transition state is involved in the final step (Fig. 6). Once the reaction progresses beyond this transition state, the products begin to form on a potential energy profile that is repulsive, and kinetic energy release may consequently be observed. An example of this situation is furnished by the nitrogen analogue (**11**) of protonated propionaldehyde, which may be considered to eliminate C_2H_4 via a mechanism (Scheme 5) parallel to that postulated for the oxygen analogues. The potential energy profile shown in Fig. 7 can be constructed by the use of known (68) or estimated (66, 69) heats of formation (69). It is at once evident from Fig. 7 that, in contrast to the oxygen analogues, the postulated reacting configuration (**13**) is now higher in energy than the products of dissociation. As a result, **13** should now approximate a transition state for the overall reaction of C_2H_4 loss from **11**, dissociating exothermically to products. Consequently, kinetic energy release accompanying the reaction is expected; this is observed to be the case (37), the metastable peak for the process being flat-topped (see Fig. 7) and corre-

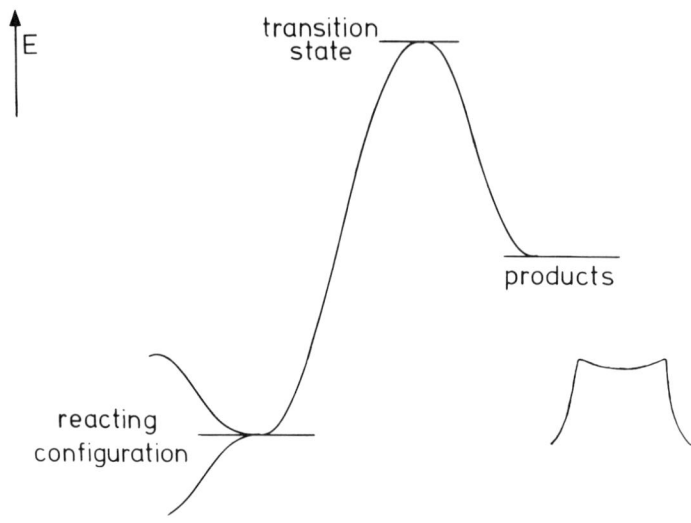

Fig. 6 Potential energy profile corresponding to possible occurrence of flat-topped (or dish-topped) metastable peaks.

sponding to a kinetic energy release of 9 kcal/mol. Other experimental evidence in favor of the potential energy profile depicted in Fig. 7 stems from energy measurements, which yield a transition state energy of approximately 224 kcal/mol, in satisfactory agreement with the estimated heat of formation of **13**.

$$CH_3CH_2CH=\overset{+}{N}H_2 \rightleftharpoons CH_3\overset{+}{C}HCH_2NH_2 \rightleftharpoons \overset{+}{C}H_2CH_2CH_2\overset{\frown}{N}H_2 \rightarrow CH_2=CH_2 + CH_2=\overset{+}{N}H_2$$
$$\textbf{11} \qquad\qquad \textbf{12} \qquad\qquad \textbf{13}$$

Scheme 5

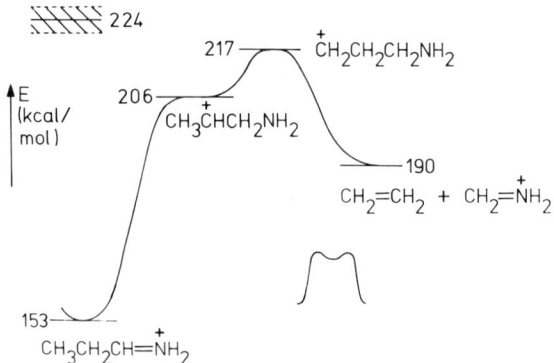

Fig. 7 Potential energy profile, schematic metastable peak, and measured transition state energy for C_2H_4 loss starting from $CH_3CH_2CH=NH_2^+$. Reproduced from *J. Am. Chem. Soc.* **99**, 3192 (1977). Copyright by the American Chemical Society.

2. GAS-PHASE ION REARRANGEMENTS

There are many other instances of the correlation between the occurrence of a reverse activation energy for a dissociation channel and kinetic energy release accompanying the reaction. An important general class is that of symmetry-forbidden (41) 1,2 eliminations. These are found to give rise to flat-topped metastable peaks, whereas the symmetry-allowed (41) 1,1 eliminations are found to give rise to gaussian peaks (75, 76).

It should be noted, however, that flat-topped or dished metastable peaks correspond to the release of potential energy concurrent with the dissociation step. When potential energy is released before the final step, a different metastable peak shape results.

3. The case of broad gaussian peaks.

Such a shape of peak reveals that there is a relatively large excess energy in the transition state for the final step; part of this excess energy is partitioned as translation and is therefore evidenced by kinetic energy release (26). The corresponding potential energy profile is depicted in Fig. 8. As with case 2 above, a transition state is involved, although this transition state (y, Fig. 8) does not now collapse directly and exothermically to products. Instead, it rearranges to a distinct reacting configuration (z), which then undergoes dissociation to products. Consequently, potential energy is released before decomposition rather than concurrent with dissociation. Looking at the reaction in reverse, the products of dissociation attract one another and can associate to form the reacting configuration (z) with little or no activation energy. However, in order to progress

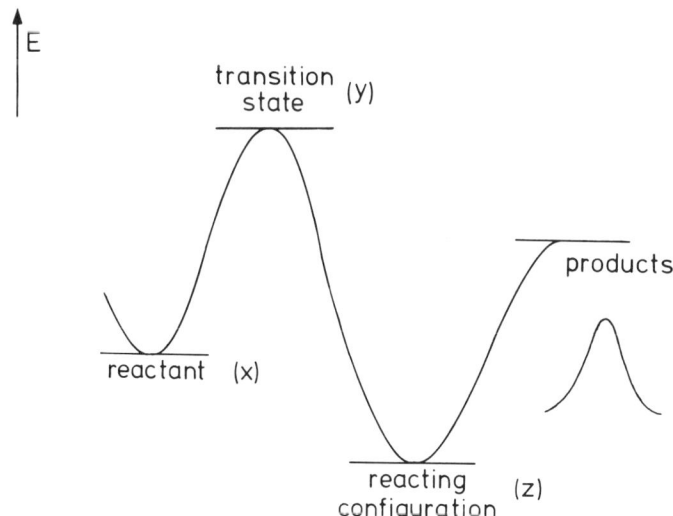

Fig. 8 Potential energy profile corresponding to broad gaussian metastable peaks.

from this reacting configuration to the reactant (*x*), a transition state (*y*) of higher energy than the products must be surmounted. Thus, dissociation of the reacting configuration (*z*) is continuously endothermic (or nearly so) and requires less energy than rearrangement back to the reactant (*x*). Several instances of this kind of behavior have been reported recently (*26, 71, 74, 77*). For example, CO loss from ions of general formula R⁺CO (where R is a carbonium ion) may occur via direct cleavage or after rearrangement of the incipient carbonium ion. The most plausible mechanisms whereby dissociation and isomerization may occur in the simplest case (R = C_3H_7) are given in Scheme 6. Elimination of CO from $(CH_3)_2CH$—⁺CO (**14**) via simple bond cleavage may take place, probably

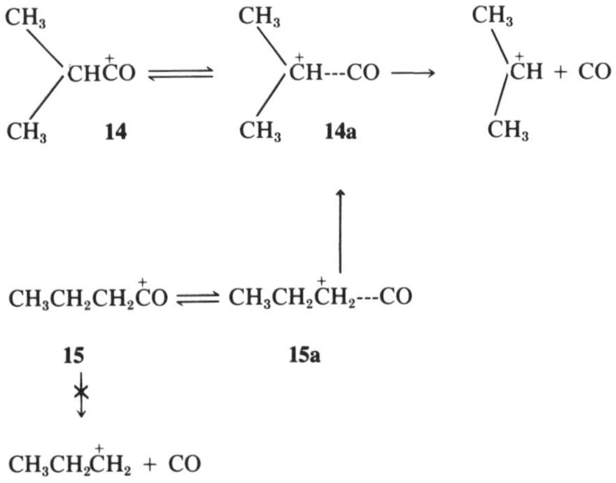

$$(CH_3)_2\overset{+}{C}HCO \;\rightleftharpoons\; (CH_3)_2\overset{+}{C}H\text{---}CO \;\longrightarrow\; (CH_3)_2\overset{+}{C}H + CO$$
$$\textbf{14} \qquad\qquad \textbf{14a}$$

$$\uparrow$$

$$CH_3CH_2CH_2\overset{+}{C}O \;\rightleftharpoons\; CH_3CH_2\overset{+}{C}H_2\text{---}CO$$
$$\textbf{15} \qquad\qquad \textbf{15a}$$

$$\updownarrow$$

$$CH_3CH_2\overset{+}{C}H_2 + CO$$

Scheme 6

without reverse activation energy, to give energetically favorable products. This is consistent with the observation that CO loss from **14** gives rise to a very narrow gaussian metastable peak (Fig. 9) (*77*). In contrast, CO loss from the straight-chain isomer **15** is evidenced by a broad gaussian metastable peak (Fig. 9). This result precludes direct cleavage of **15** to form the $CH_3CH_2CH_2^+$ cation and CO, which would be expected to produce a narrow gaussian peak. The most likely explanation is that, as CO loss begins to occur from **15**, a 1,2-hydride shift takes place in the incipient $C_3H_7^+$ ion (**15a** → **14a**). This process leads to the release of *ca*. 16 kcal/mol of potential energy (i.e., the difference in stability of isomeric primary and secondary carbonium ions) (*57*). The presence of this extra 16 kcal/mol of energy in **14a** is more than sufficient to cause dissociation to $(CH_3)_2CH^+$ and CO; isomerization of **15a** to **14a** is therefore the rate-determining step in CO loss from **15**. After the slow isomerization, relatively fast dissociation of **14a** occurs, with excess energy present in the transition state,

2. GAS-PHASE ION REARRANGEMENTS

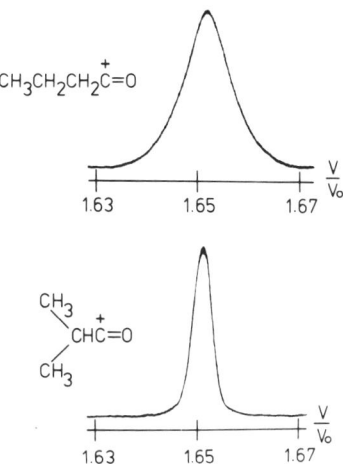

Fig. 9 Metastable peaks for CO loss from $CH_3CH_2CH_2{}^+CO$ and $(CH_3)_2CH^+CO$ ions dissociating in the first field-free region of an AEI MS 902 instrument. Reproduced from *Tetrahedron Lett.* p. 2919 (1978). Copyright by Pergamon Press.

which results in a broadening of the metastable peak for CO loss (77). This behavior, in which rearrangement of incipient primary carbonium ions to more stable isomers precedes decomposition, is general for higher homologues (e.g., the isomeric ions $C_4H_9{}^+CO$). Indeed, a consideration of the kinetic energy release that accompanies CO loss from such ions constitutes an effective probe for the occurrence of isomerization before dissociation.

Finally, in connection with kinetic energy release, the observation of a composite metastable peak indicates that two different channels for dissociation are in operation, each having closely similar energy requirements. As has already been mentioned (Section II,C), this behavior is frequently encountered when $C_3H_3{}^+$ is a product ion. Another interesting example is C_2H_4 loss from $C_7H_{11}{}^+$; this process is observed to give rise to a broad composite metastable peak (Fig. 10) (34). Two possible mechanisms for this reaction are given in Scheme 7; known or estimated heats of formation are given in brackets. The corresponding potential energy diagram is shown in Fig. 11. Elimination of C_2H_4 from **16** may occur via a symmetry-allowed (41) pericyclic process, possibly without significant reverse activation energy; however, it results in the formation of relatively high-energy products. In contrast, C_2H_4 loss from **17** leads to products of considerably lower energy, but the heat of formation of the primary carbonium ion **17** is rather high. Both these channels are likely to require a total energy of *ca.* 230 kcal/mol; **16** → products, being endothermic, is expected to result in a narrow metastable peak, whereas **17**

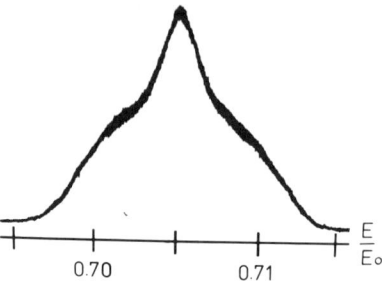

Fig. 10 Composite metastable peak for C_2H_4 loss from $C_7H_{11}^+$ ions dissociating in the second field-free region of a VG Micromass ZAB 2F instrument. Reproduced from *Org. Mass Spectrom.* **13**, 330 (1978). Copyright by Heyden and Son.

→ products is exothermic and might release kinetic energy, which would be evidenced by a broader metastable peak. The composite overall peak for C_2H_4 loss from $C_7H_{11}^+$ may therefore be explained. The central narrow component is considered to arise from decomposition of **16**, and the broad component is ascribed to dissociation of **17** (*34*). Further support for this explanation stems from a successful prediction, made on the basis of the behavior of $C_7H_{11}^+$, of the observed reactions of the higher homologue $C_8H_{13}^+$ (Scheme 8). The process homologous to **17** → products (**19** → products) can now proceed via the energetically more favorable secondary carbonium ion (**19**) in a reaction pathway that is now approximately

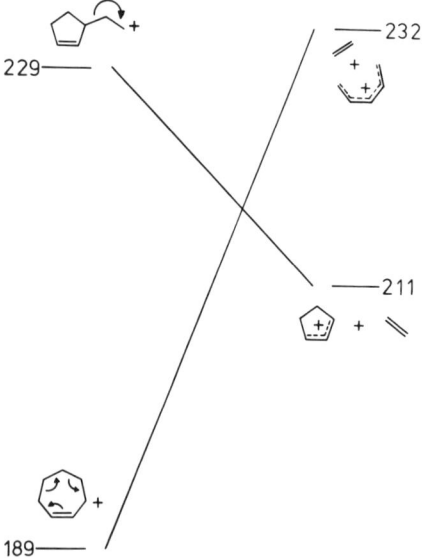

Fig. 11 Potential energy diagram for C_2H_4 loss from $C_7H_{11}^+$. Reproduced from *Org. Mass Spectrom.* **13**, 330 (1978). Copyright by Heyden and Son.

2. GAS-PHASE ION REARRANGEMENTS

[189] (34)   [220] (58)   [12] (68)
             ($\Sigma \Delta H_f = 232$)
16

[229] (66)   [199] (58)   [12] (68)
             ($\Sigma \Delta H_f = 211$)
17

Scheme 7

thermoneutral. Loss of C_3H_6 via the process homologous to **16** → products (**18** → products) is therefore preempted because it requires *ca.* 20 kcal/mol more energy. Elimination of C_2H_4 from $C_8H_{13}^+$ is also unfavorable because of energy considerations (*34*). Consequently, $C_8H_{13}^+$ ought to lose C_3H_6 and should do so via only one channel (i.e., **19** → products), which is almost thermoneutral. Experimentally, the main slow dissociation of $C_8H_{13}^+$ (94%) is C_3H_6 loss; the metastable peak for this process is not composite but is gaussian and only slightly broadened [average (*25*) kinetic energy release 2.6 kcal/mol] (*34*).

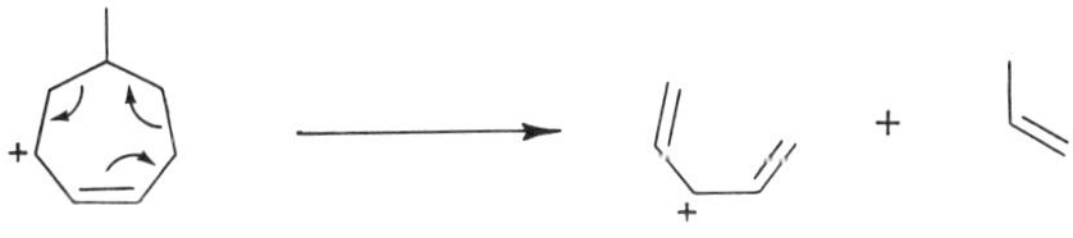

[184] (34)   [220] (58)   [5] (68)
             ($\Sigma \Delta H_f = 225$)
18

[206] (66)   [199] (58)   [5] (68)
             ($\Sigma \Delta H_f = 204$)
19

Scheme 8

F. Collisional Activation and Ion Cyclotron Resonance

Both collisional activation (CA) and ion cyclotron resonance (ICR) are techniques that permit the detection of any ion that exists in a significant well in the potential energy profile. For $C_2H_5O^+$, for instance, CA studies indicate that only two structures, presumably **20** and **21**, exist in relatively deep potential energy wells (78). This conclusion is in agreement with that reached by considering the slow dissociations of $C_2H_5O^+$ (21, 79) and with an ICR analysis (80).

$$CH_3CH=\overset{+}{O}H \qquad\qquad CH_3\overset{+}{O}=CH_2$$

$$\textbf{20} \qquad\qquad\qquad \textbf{21}$$

Of the two techniques, CA is the more commonly employed and is now a fairly routine method for investigating the structures of ions in the gas phase (81). The basis of the method is to allow low-energy ions, which have not undergone dissociation after some 10^8 vibrations, to collide with an "inert" gas. These collisions result in the excitation of the nondecomposing ions, which now acquire enough internal energy to dissociate relatively rapidly. In the case of two ions for which interconversion requires a significant activation energy, there is usually not enough time for much isomerization to occur after collision. Therefore, if two ions are found to exhibit the same CA spectra, it follows that they must have isomerized before collision. This means that the energy barriers toward interconversion of the ions must be less than those for decomposition. Conversely, the observation of different CA spectra indicates little or no interconversion of the ions before collision and subsequent fast dissociation, i.e., that the energy barriers toward interconversion of the two ions are at least comparable to those for dissociation. A difference in CA spectra for two ions, however, does not necessarily preclude interconversion before slow dissociation because the range of internal energies sampled by CA spectroscopy is much larger than that appropriate for ions undergoing dissociation in metastable transitions. Metastable ions are such that they have a well-defined range of internal energies just above the threshold for reaction; in contrast, CA spectroscopy also examines low-energy, nondecomposing ions (Fig. 12). Consequently, CA spectra do not depend significantly on the internal energy of the ions (unless, perhaps, these ions lie in a very shallow potential energy well) (82), whereas the relative abundance of reactions undergone by metastable ions frequently does (20). Thus, the activation energies for interconversion of isomeric ions may be less than those for dissociation, but nevertheless large enough to cause significant differences in the CA spectra.

An example of this situation is found in the long-standing "benzyl versus tropylium" problem. Several $C_7H_7^+$ species with different CA spectra are observed (83, 84). ²H-Labeling results reveal that the $C_7H_5D_2^+$ ions of nominal structure **22** eliminate CD_2 in CA spectroscopy

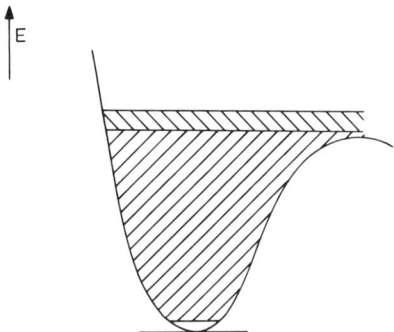

Fig. 12 Potential energy profile showing range of internal energies of ions dissociating in normal metastable transitions (▧) and ions dissociating under collisional activation (▨).

(*84*), thus proving that such low-energy ions do not collapse to tropylium structures. Moreover, ICR experiments show that two distinct $C_7H_7^+$ populations (presumably benzyl and tropylium), with different reactivities, exist (*85*). These results (*83–85*) are strong evidence that in the gas phase both benzyl and tropylium cations exist in significant potential energy wells.

22

Notwithstanding this, studies on higher-energy ions (metastable and source dissociations) produced from [2,6-^{13}C]toluene (**23**) (*86*), a tropylium salt **24** (*87*), and doubly labeled cycloheptatriene (*88*) suggest that isomerization of benzyl to tropylium (or interconversion of the two) occurs. Thus, for instance, the molecular ion of **23** fragments to yield some unlabeled $C_5H_5^+$ and some $^{13}C_2H_2^+$. This is not consistent either with dissociation of ionized **23** without rearrangement, or with simple collapse to a tropylium structure, which then dissociates without further isomerization because the molecular ion of **23** cannot lose $^{13}C_2H_2$ or C_5H_5 without prior rearrangement.

23   **24**

G. Predictive Capacity

Some of the important methods that are used in the construction of potential energy profiles have been discussed above. In most cases, the data obtained by these methods are combined to produce a potential energy profile, which is then used to explain the observed slow dissociations of an ion or a group of ions. However, in favorable cases, a potential energy profile may be constructed for ions that are as yet uninvestigated, and, for such systems, predictions may be made concerning the slow reactions. These predictions may refer to the observed decomposition routes, the channeling ratios, the energy needed to cause dissociation, the results of isotope labeling experiments, the shapes of the metastable peaks for dissociation, or a combination of these factors. The test of all scientific theories lies in their ability to formulate predictions in cases for which no results are known. In this respect, the success of the potential energy profile approach, outlined above, is both detailed and general in scope.

IV. TYPES OF POTENTIAL ENERGY PROFILES

When the methods discussed above are applied, three basic kinds of potential energy profiles occur.

A. Complete Equilibration of Isomers before Unimolecular Dissociation

This behavior corresponds to the general form of potential energy profile depicted in Fig. 13. The activation energies for interconversion of two isomeric ions A^+ and B^+ are less than those for decomposition, and

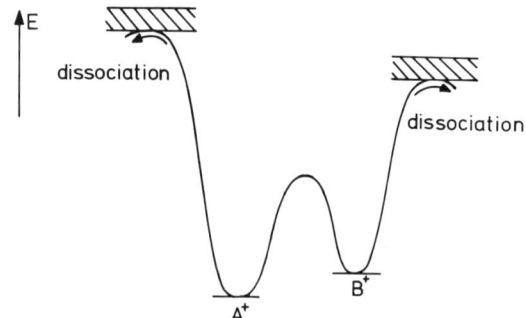

Fig. 13 Potential energy profile corresponding to interconversion of isomers before dissociation. Reproduced from *Acc. Chem. Res.* **10**, 280 (1977). Copyright by the American Chemical Society.

2. GAS-PHASE ION REARRANGEMENTS

rapid interconversion of A^+ and B^+ precedes dissociation, which occurs from the same structure (or structures) irrespective of origin. The occurrence of such a potential energy profile can be deduced from four experimental criteria:

1. The ions must undergo the same slow dissociations in similar ratios [although, as noted, minor differences in channeling ratios may arise because of slightly different ion internal energies (20)].

2. The kinetic energy release associated with each decay route must be the same, irrespective of whether the ions are generated as A^+ or B^+. Moreover, if a composite metastable peak is found for a dissociation of A^+, a peak of identical shape must be observed for the corresponding dissociation starting from B^+.

3. The energy required for each decomposition pathway must be independent of whether the ion is formed as A^+ or B^+. This criterion may be more difficult to apply when conventional mass spectrometers are used because of the difficulties of making appearance potential measurements with such instruments (52, 53). However, it should be possible to detect large differences even with standard machines.

4. Labeling studies should reveal that any atoms in sites which are rendered equivalent by reversible isomerization of A^+ and B^+ participate statistically in the eventual decomposition channels.

An example of this type of potential energy profile is that for isomerization and dissociation of $C_4H_9^+$ (Fig. 14) (89). The evidence in favor of dissociation occurring after equilibration of all accessible reacting configurations is as follows:

1. Irrespective of whether the precursor used to generate $C_4H_9^+$ con-

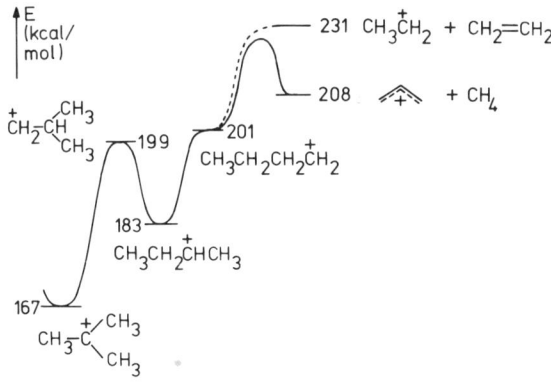

Fig. 14 Potential energy profile for interconversion and decomposition of isomeric $C_4H_9^+$ ions. Reproduced from *Acc. Chem. Res.* **10**, 280 (1977). Copyright by the American Chemical Society.

tains an incipient primary, secondary, or tertiary cation, the main slow dissociation is elimination of CH_4; however, a small and almost constant fraction of the metastable ion current from m/e 57 is due to C_2H_4 loss (35).

2. The metastable peak shape for CH_4 loss from $C_4H_9^+$ does not depend on the means whereby the ion was generated.

3. Energy measurements reveal that the transition state energy for CH_4 loss is the same irrespective of the source of $C_4H_9^+$ (14, 45).

4. The results of ^{13}C- and ^2H-labeling experiments show that loss of CH_4 involves random selection of the necessary atoms (35). This is consistent with the potential energy diagram of Fig. 14 insofar as equilibration of n-, sec-, iso- and t-butyl cations, via a series of 1,2-hydride and 1,2-methyl shifts, would lead to random distribution of the label.

Thus, extensive data can be cited in support of the potential energy profile shown in Fig. 14. This type of potential energy profile is common; other examples have already been cited in Sections III,B and C (14, 15, 32–36, 46–48, 50, 51). Perhaps the most extensive application of all four criteria to deduce interconversion of isomers before decomposition is a definitive study of $C_3H_6^{+\cdot}$ (32).

B. No Equilibration of Isomers before Unimolecular Dissociation

This behavior corresponds to the general form of potential energy profile shown in Fig. 15. Here the activation energies for interconversion of two isomeric ions, A^+ and B^+, are greater than those for decomposition. As a result, A^+ and B^+ undergo decomposition over separate potential energy profiles. Four general criteria may be used to deduce that this type of potential energy profile is appropriate:

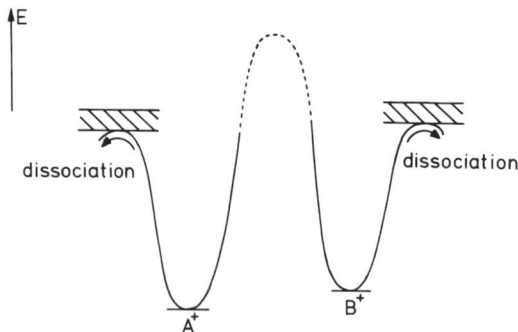

Fig. 15 Potential energy profile corresponding to no interconversion of isomers before dissociation. Reproduced from *Acc. Chem. Res.* **10**, 280 (1977). Copyright by the American Chemical Society.

2. GAS-PHASE ION REARRANGEMENTS

1. The ions ought to dissociate via different channels; furthermore, the extent to which any common processes occur is expected to vary with the original structure of the ion under investigation.

2. It is highly unlikely that the kinetic energy release associated with any common decay routes will be the same.

3. Any common decomposition pathways are likely to proceed via transition states of different energies. This is because, although the same (or isomeric) neutrals are expelled, the reactions are in fact different starting from A^+ and B^+.

4. The behavior of labeled analogues of A^+ and B^+ ought to reveal that different isomerization pathways operate before decomposition of A^+ and B^+, and the extent to which they lose their identity will be, in general, different. In particular, when isotope effects are observed, these are unlikely to be the same for dissociation of A^+ and B^+.

The occurrence of this type of potential energy profile is quite common; indeed, the use of mass spectrometry as a structural probe depends on ions behaving differently when formed from isomeric precursors. Thus, the potential energy profile of isomeric $C_2H_5O^+$ ions (Fig. 16) can be

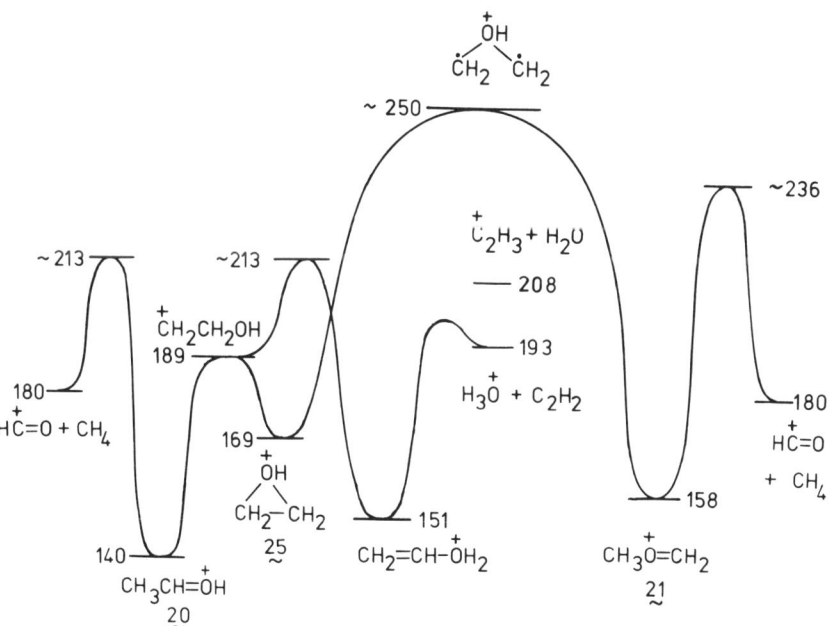

Fig. 16 Potential energy profile for interconversion and dissociation of isomeric $C_2H_5O^+$ ions. Reproduced from *J. Am. Chem. Soc.* **99**, 7509 (1977). Copyright by the American Chemical Society.

constructed (*90*) using known (*55, 56, 59, 63, 68, 91, 92*) or estimated (*66, 90*) heats of formation. It is supported by the following experimental data:

1. Ions generated from precursors having the ether group YCH_2OCH_3, as stated (see Sections II,B and III,B) (**21**), undergo almost exclusively CH_4 loss (*21*), whereas from YCH_2CH_2OH or CH_3CHYOH elimination of CH_4 and C_2H_2 occurs in a constant ratio (*21*).

2. Although both classes of ions referred to above eliminate CH_4 in slow dissociations, there are clearly two independent pathways in operation. Loss of CH_4 from $CH_3O^+{=}CH_2$ gives rise to a gaussian, but fairly broad, metastable peak, thus indicating that a range of kinetic energy is released when decomposition occurs. In contrast, CH_4 loss from $CH_3CH{=}OH^+$ (**20**) is evidenced by a broad, dished metastable peak, which shows that a relatively large and specific kinetic energy release accompanies dissociation (*21*).

3. Appearance potential measurements give approximate transition state energies of 236 and 213 kcal/mol for CH_4 loss starting from $CH_3O^+{=}CH_2$ and $CH_3CH{=}OH^+$, respectively (*90*). These values, although only approximate, are substantially different.

4. Elimination of CH_4 from $CH_3O^+{=}CH_2$ involves statistical selection of any four of the five hydrogen atoms (*26*). In contrast, CH_4 loss from $CH_3CH{=}OH^+$ involves specifically the hydrogen atom originally bound to oxygen together with any three of the four carbon-bound hydrogen atoms (*93, 94*).

Further evidence can be cited in favor of the details of the potential energy profile given in Fig. 16 (*90*). For instance, rapid interconversion of **20** and **25** is expected to occur before slow decomposition via the mechanism of Scheme 9. This should render all hydrogen atoms bound to carbon equivalent while permitting the hydrogen attached to oxygen to retain its identity. Moreover, the carbon atoms in **20** are expected to become equivalent via this mechanism. Consequently, when CH_4 loss occurs, presumably via a 1,2 elimination in **20**, the origin of the constituent atoms may be assigned as follows: either carbon atom may be selected together

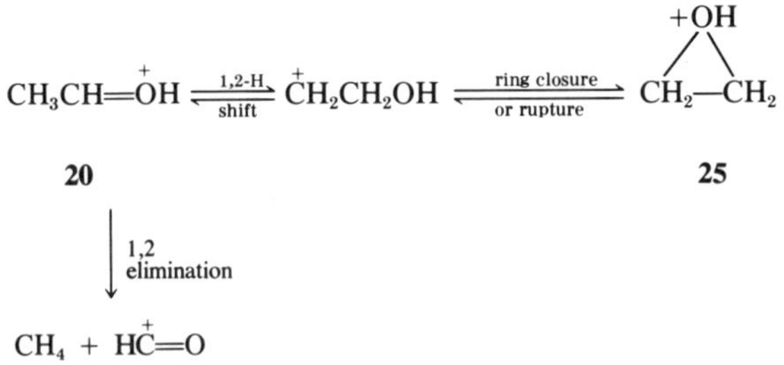

$$CH_3CH{=}\overset{+}{O}H \underset{\text{shift}}{\overset{\text{1,2-H}}{\rightleftharpoons}} \overset{+}{C}H_2CH_2OH \underset{\text{or rupture}}{\overset{\text{ring closure}}{\rightleftharpoons}} \overset{+OH}{\overset{/\ \backslash}{CH_2{-}CH_2}}$$

20 $\qquad\qquad\qquad\qquad\qquad\qquad\qquad$ **25**

$$\downarrow \text{1,2 elimination}$$

$$CH_4 + H\overset{+}{C}{=}O$$

Scheme 9

2. GAS-PHASE ION REARRANGEMENTS

with the hydrogen atom bound to oxygen and any three of the four attached to carbon. This is the experimental result (92–95).

Other examples of this general type of potential energy profile are numerous; some have been mentioned in Section III,B (26, 37–40). In particular, the $C_3H_7O^+$ system has attracted considerable attention (20, 26, 96–99).

C. Rate-Determining Isomerization before Unimolecular Dissociation

This corresponds to the general type of potential energy profile given in Fig. 17. The energy required to cause dissociation of one ion, B^+, is less than that required to promote isomerization to A^+. Hence, B^+ dissociates in preference to rearranging to A^+. However, the lowest-energy decay route for A^+ is isomerization to B^+, which is now formed with considerably more internal energy than is required to bring about decomposition. These high-energy ions therefore dissociate relatively rapidly rather than return to A^+. The overall result is that ions of structure A^+ undergo slow, rate-determining isomerization to B^+ followed by fast dissociation.

The occurrence of this general form of potential energy profile can be detected as follows:

1. Starting from B^+, the relative abundance of any possible competing channels for decomposition is critically dependent on the activation energy for the processes concerned (15). Energy is the dominant parameter in determining the occurrence or nonoccurrence of a given decay route. Thus, the processes of lowest activation energy are expected to be the major slow dissociations, irrespective, to a first approximation, of the

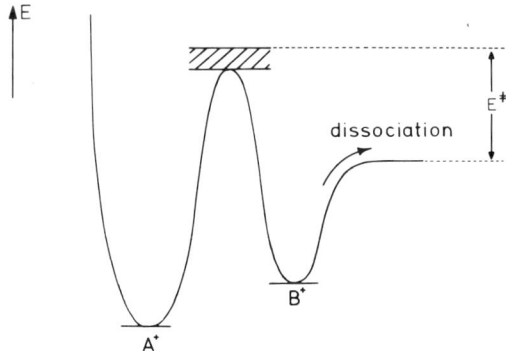

Fig. 17 Potential energy profile corresponding to rate-determining isomerization of ion A^+ to ion B^+ before dissociation. Reproduced from *Acc. Chem. Res.* **10**, 280 (1977). Copyright by the American Chemical Society.

probability of attaining the correct geometry of the relevant activated complex. However, starting from A^+, after the rate-determining isomerization, the resultant B^+ ions have internal energies considerably in excess of that normally associated with slow dissociation of these ions. Consequently, there is no longer enough time for all intermediates and reacting configurations to be explored. The probability of attaining the correct geometry of the transition state for each dissociation channel is now important and may become more important than energetic factors. Thus, there will be a systematic discrimination against any process with a rigid geometric requirement (e.g., rearrangement processes), whereas routes with little or no geometric requirements (e.g., simple bond cleavage reactions) will be favored.

2. The kinetic energy released in any common decomposition channels must be greater starting from A^+ than from B^+. This is because part of the excess energy ($E^{\neq}$) present in the transition states when rearranged ions of structure B^+ dissociate is partitioned as translation. Therefore, the metastable peak widths should be greater starting from A^+.

3. The energy needed to cause each dissociation of B^+ depends, in general, on the process concerned. However, the corresponding energy required to promote decomposition of A^+ depends on the height of the energy barrier for isomerization to B^+. Therefore, the activation energies required for each decay route of A^+ must (a) be the same and (b) be greater than that for any decomposition channel of B^+.

4. Labeling studies may reveal a greater degree of specificity in the decomposition of A^+ because, if statistical distribution of certain atoms were possible only after isomerization to B^+, the rearrangement processes required to effect this might be unable to compete. However, starting from B^+, there may be enough time for these rearrangements to lead to statistical distribution of the relevant atoms before dissociation.

Although the possibility of ions reacting over this kind of potential energy profile was noted (37, 40), only relatively recently was the first definite example, in the $C_3H_7O^+$ system, reported (26). The second proved example, also in the $C_3H_7O^+$ system, is discussed below.

1. Both protonated acetone (**26**) and protonated propionaldehyde (**6**) are observed to undergo two slow dissociations, loss of H_2O and C_2H_4. For protonated acetone, C_2H_4 loss is the major process; however, H_2O elimination dominates starting from protonated propionaldehyde (74). Moreover, as the internal energy of the decomposing ions is reduced by increasing their average lifetimes before dissociation, C_2H_4 loss remains the major reaction of **26** whereas H_2O loss becomes increasingly dominant starting from **6** (20). These data suggest that rate-determining rearrangement of **26** to **6** occurs before decomposition. This is because H_2O loss

2. GAS-PHASE ION REARRANGEMENTS

from **6** (and hence from **26** after isomerization to **6**) must occur via a hydrogen transfer from carbon to oxygen, which proceeds through a five-membered ring transition state (Scheme 10) and which therefore should have a highly stringent geometry requirement. Since C_2H_4 loss from **6** may occur after only 1,2-hydride shifts, which have much less demanding geometry requirements, followed by σ-bond cleavage of **8**, this process is favored at high internal energies (rearranged ions of structure **6**). However, H_2O loss has the lower activation energy (74) starting from **6**; therefore, it is the major slow dissociation of **6** and becomes increasingly more dominant as ions of lower average internal energy are sampled. In contrast, reduction of the average internal energy of decomposing ions of original structure **26** has little effect on the competition between H_2O and C_2H_4 loss (20). This is consistent with the postulated rate-determining isomerization because such an isomerization always gives rise to ions of structure **6**, which have considerably more internal energy than that needed to cause H_2O or C_2H_4 loss.

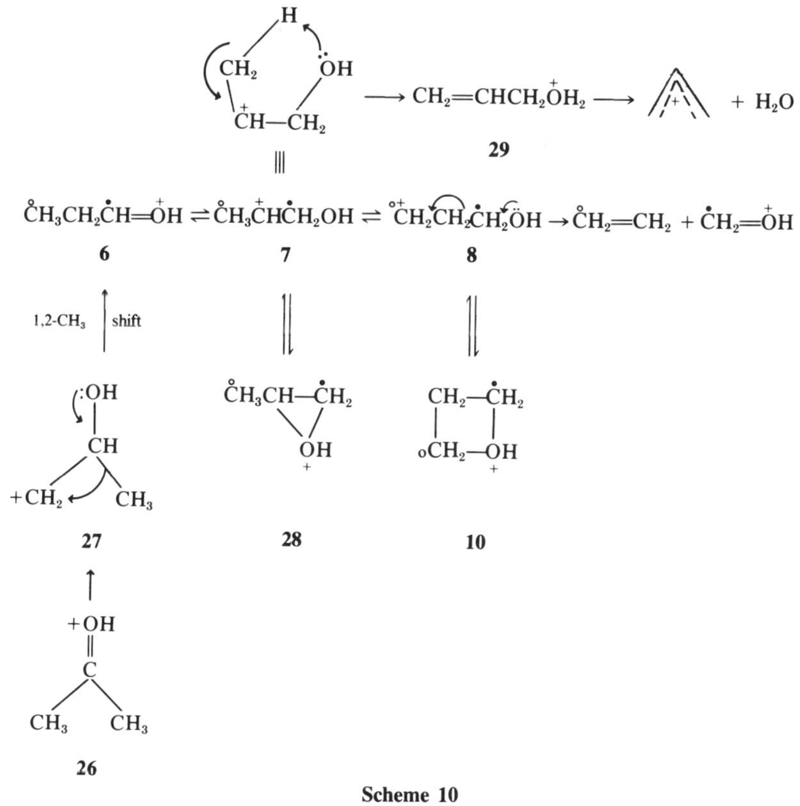

Scheme 10

2. The metastable peaks for H_2O and C_2H_4 elimination from **6** and **26** are gaussian in each case. Nevertheless, the average (25) kinetic energies released when **26** decomposes (1.2 and 0.9 kcal/mol for H_2O and C_2H_4 loss, respectively) are greater than the corresponding values starting from **6** [1.0 and 0.8 kcal/mol, respectively (74)].

3. The energies needed to cause H_2O and C_2H_4 losses from **6** are different, H_2O loss having the lower activation energy. In contrast, the energy required to decompose **26** via H_2O or C_2H_4 elimination is the same within experimental error and is significantly greater than that needed to dissociate **6** (74). These data are summarized in Fig. 18, from which it is evident that the potential energy profile is indeed of the general form depicted in Fig. 17.

4. A more detailed analysis of the mechanism whereby **26** may rearrange to **6** reveals that there are two possible routes. Both of these go through the high-energy primary carbonium ion **27** near the transition state for the process (Scheme 10). In the former, a direct 1,2-methyl shift leads to **6**; in the latter, ring closure of **27** to protonated methyloxirane (**28**) followed by ring opening to **7** and a 1,2-hydride shift gives **6**. The former mechanism does not involve breakage of the original C—O bond in **26**. Hence, ^{13}C-labeled protonated acetone (**30**) might show retention of the original carbonyl carbon atom in the $CH_2\!\!=\!\!OH^+$ fragment ion because the steps **6** → **7** → **8** → products ought to occur *rapidly* once rate-determining isomerization has taken place. This is found to be the case; fast dissociations (source reactions) of **30** involve complete retention of the original carbonyl carbon atom in the $CH_2\!\!=\!\!OH^+$ fragment (97, 100).

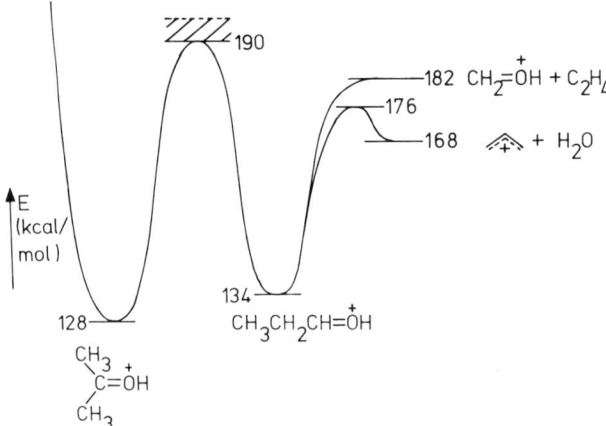

Fig. 18 Potential energy profile for isomerization and decomposition of $(CH_3)_2C\!\!=\!\!OH^+$ and $CH_3CH_2CH\!\!=\!\!OH^+$.

2. GAS-PHASE ION REARRANGEMENTS

Partial (73) or complete (97) retention is also observed in slower dissociations (metastable ions). These data rule out the alternative mechanism for isomerization (**27** → **28** → **7** → **6**) because this mechanism does involve breaking the original C—O bond in **26**. Moreover, they also indicate that after the rate-determining rearrangement (**26** → **6**) has occurred, there is not enough time for other rearrangements (e.g., **8** ⇌ **10**) which would lead to loss of identity of the carbonyl carbon atom to a significant extent. In contrast, starting from **6**, these rearrangement processes, which are low-energy reactions (see Fig. 19 for the estimated potential energy profile),

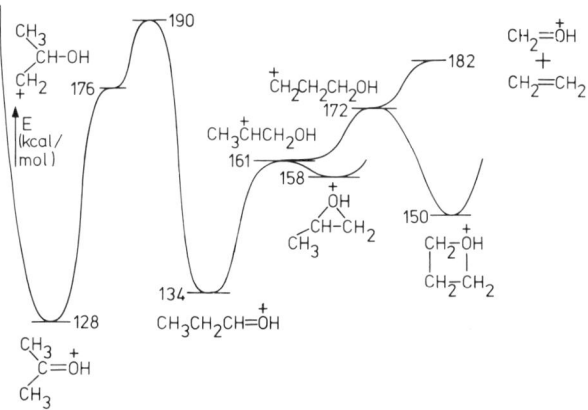

Fig. 19 Detailed potential energy profile for isomerization and decomposition, via C_2H_4 loss, of $(CH_3)_2C=OH^+$, $CH_3CH_2CH=OH^+$, and related ions. Reproduced from *J. Am. Chem. Soc.* **99**, 5481 (1977). Copyright by the American Chemical Society.

ought to be rapid and reversible before dissociation. Thus, **6** is expected to interconvert with **7, 8, 28,** and **10** before eventually eliminating C_2H_4 in slow reactions. Ion **27** is not, however, accessible to **6** via **28**. Hence, equilibration of **6** and **10** (via **7** and **8**) renders the α- and γ-carbon atoms of **6** (denoted by closed and open circles, respectively, in Scheme 10) equivalent but permits the β-carbon atom to retain its identity. When dissociation eventually occurs, the carbon atom bound to oxygen in **8** is retained in the resultant $CH_2=OH^+$ ion. Hence, ^{13}C-labeling studies on **31** and **32** ought to show that the ratio of C_2H_4 loss to $C^{13}CH_4$ loss is 50:50 and 0:100,

$$\underset{30}{\overset{+OH}{\underset{CH_3}{\overset{\parallel}{\underset{}{{}^{13}C}}}\diagdown CH_3}} \qquad \underset{31}{CH_3CH_2{}^{13}CH=OH} \qquad \underset{32}{CH_3{}^{13}CH_2CH=\overset{+}{O}H}$$

respectively. This is found to be the case, within an experimental error of 1% (73).

Thus, there is considerable experimental evidence in the form of channeling data, kinetic energy release measurements, transition state energy measurements, and isotope labeling results which suggests that the rate-determining isomerization **26** → **6** precedes decomposition. The overall case is extremely strong.

Other examples of these rate-determining isomerizations have recently been established. One general case is discussed in Section III,E; another important system in which they occur is $C_4H_9O^+$ (71). In fact, it would seem probable that these rate-determining isomerizations are quite common (101) and are a general cause of broad, gaussian metastable peaks.

V. CONCLUSIONS

The importance of metastable ions in investigating the slow, unimolecular rearrangements and dissociations of ions has been discussed. These reactions can be considered to occur over a suitable potential energy profile, and some of the major experimental methods for constructing such potential energy profiles have been examined. Three general types of potential energy profiles occur; these correspond to complete interconversion before dissociation, no interconversion before dissociation, and rate-determining isomerization before dissociation.

REFERENCES

1. J. J. Thomson, *Philos. Mag.* [6] **21**, 225 (1911).
2. F. W. Aston, *Philos. Mag.* [6] **45**, 934 (1923).
3. J. H. Beynon, "Mass Spectrometry and Its Applications to Organic Chemistry." Elsevier, Amsterdam, 1960.
4. K. Biemann, "Mass Spectrometry." McGraw-Hill, New York, 1962.
5. H. Budzikiewicz, C. Djerassi, and D. H. Williams, "Interpretation of Mass Spectra of Organic Compounds." Holden-Day, San Francisco, California, 1964.
6. H. Budzikiewicz, C. Djerassi, and D. H. Williams, "Mass Spectrometry of Organic Compounds." Holden-Day, San Francisco, California, 1967.
7. J. H. Beynon, R. A. Saunders, and A. E. Williams, "The Mass Spectra of Organic Molecules." Elsevier, Amsterdam, 1968.
8. D. H. Williams and I. Howe, "Principles of Organic Mass Spectrometry." McGraw-Hill, New York, 1972.
9. R. G. Cooks, J. H. Beynon, R. M. Caprioli, and G. R. Lester, "Metastable Ions." Elsevier, Amsterdam, 1973.
10. R. D. Bowen, D. H. Williams, and H. Schwarz, *Angew. Chem., Int. Ed. Engl.* **18**, 451 (1979).

2. GAS-PHASE ION REARRANGEMENTS

11. M. Barber and R. M. Elliott, *12th Annu. Conf. Mass Spectrom. Allied Top.* Committee E.14 (1964).
12. K. R. Jennings, *J. Chem. Phys.* **43**, 4176 (1965).
13. I. G. Simm, C. J. Danby, and J. H. D. Eland, *J. Chem. Soc. Chem. Commun.* p. 832 (1973).
14. R. D. Bowen and D. H. Williams, *J. Chem. Soc., Perkin Trans. 2* p. 1479 (1976).
15. H. M. Rosenstock, V. H. Dibeler, and F. N. Harllee, *J. Chem. Phys.* **40**, 591 (1964).
16. L. P. Hills, M. L. Vestal, and J. H. Futrell, *J. Chem. Phys.* **54**, 3834 (1971).
17. U. Lohle and C. Ottinger, *J. Chem. Phys.* **51**, 3097 (1969).
18. M. L. Vestal and J. H. Futrell, *J. Chem. Phys.* **52**, 978 (1970).
19. C. Lifshitz and L. Sternberg, *Int. J. Mass Spectrom. Ion Phys.* **2**, 303 (1969).
20. A. N. H. Yeo and D. H. Williams, *J. Am. Chem. Soc.* **93**, 395 (1971).
21. T. W. Shannon and F. W. McLafferty, *J. Am. Chem. Soc.* **88**, 5021 (1966).
22. J. A. Hipple, R. E. Fox, and E. U. Condon, *Phys. Rev.* **69**, 347 (1946).
23. J. H. Beynon, R. A. Saunders, and A. E. Williams, *Z. Naturforsch., Teil A* **20**, 180 (1965).
24. J. H. Beynon and A. E. Fontaine, *Z. Naturforsch., Teil A* **22**, 334 (1967), and references cited therein.
25. D. T. Terwilliger, J. H. Beynon, and R. G. Cooks, *Proc. R. Soc. London, Ser. A.* **341**, 135 (1974).
26. G. Hvistendahl and D. H. Williams, *J. Am. Chem. Soc.* **97**, 3097 (1975).
27. P. Goldberg, J. A. Hopkinson, A. Mathias, and A. E. Williams, *Org. Mass Spectrom.* **3**, 1009 (1970).
28. J. L. Holmes, A. D. Osborne, and G. M. Weese, *Org. Mass Spectrom.* **10**, 867 (1975).
29. D. K. Sen-Sharma, K. R. Jennings, and J. H. Beynon, *Org. Mass Spectrom.* **11**, 319 (1976).
30. G. Hvistendahl and D. H. Williams, *J. Chem. Soc., Perkin Trans. 2* p. 881 (1975).
31. F. P. Lossing, *Can. J. Chem.* **50**, 3973 (1972).
32. J. L. Holmes and J. K. Terlouw, *Org. Mass Spectrom.* **10**, 787 (1975).
33. B. J. Stapleton, R. D. Bowen, and D. H. Williams, *Tetrahedron* **34**, 259 (1978).
34. R. D. Bowen, B. J. Stapleton, and D. H. Williams, *Org. Mass Spectrom.* **13**, 330 (1978).
35. B. Davis, D. H. Williams, and A. N. H. Yeo, *J. Chem. Soc. B* p. 81 (1970).
36. M. A. Shaw, R. Westwood, and D. H. Williams, *J. Chem. Soc. B* p. 1773 (1970).
37. N. A. Uccella, I. Howe, and D. H. Williams, *J. Chem. Soc. B* p. 1933 (1971).
38. G. Cum, G. Sindona, and N. A. Uccella, *Ann. Chim. (Rome)* **64**, 169 (1974).
39. T. J. Mead and D. H. Williams, *J. Chem. Soc., Perkin Trans. 2* p. 876 (1972).
40. T. J. Mead and D. H. Williams, *J. Chem. Soc. B* p. 1654 (1971).
41. R. B. Woodward and R. Hoffmann, *Angew. Chem., Int. Ed. Engl.* **8**, 781 (1969).
42. G. A. Olah and P. von R. Schleyer, eds., "Carbonium Ions." Wiley, New York (see esp. Vol. 4, Chapter 33).
43. See, for example, L. Radom, J. A. Pople, V. Buss, and P. von R. Schleyer, *J. Am. Chem. Soc.* **94**, 311 (1972).
44. F. P. Lossing and G. P. Semeluk, *Can. J. Chem.* **48**, 955 (1970).
45. J. L. Holmes, A. D. Osborne, and G. M. Weese, *Org. Mass Spectrom.* **10**, 867 (1975).
46. W. A. Bryce and P. Kebarle, *Can. J. Chem.* **34**, 1249 (1956).
47. G. G. Meisels, J. Y. Park, and B. G. Giessner, *J. Am. Chem. Soc.* **91**, 1555 (1969).
48. F. Bohlmann, M. Brehm, and H. Schwarz, *Org. Mass. Spectrom.* **11**, 783 (1976).
49. R. D. Bowen, D. H. Williams, G. Hvistendahl, and J. R. Kalman, *Org. Mass Spectrom.* **13**, 721 (1978).

50. I. Howe and F. W. McLafferty, *J. Am. Chem. Soc.* **93,** 99 (1971).
51. J. H. Beynon, J. E. Corn, W. E. Baitinger, R. M. Caprioli, and R. A. Benkeser, *Org. Mass Spectrom.* **3,** 1371 (1970).
52. J. H. Beynon, R. G. Cooks, K. R. Jennings, and A. J. Ferrer-Correia, *Int. J. Mass Spectrom. Ion Phys.* **18,** 87 (1975).
53. H. M. Rosenstock, *Int. J. Mass Spectrom. Ion Phys.* **20,** 139 (1976).
54. K. Watanabe, T. Nakayama, and J. Mottl, *J. Quant. Spectrosc. & Radiat. Transfer* **2,** 369 (1962).
55. K. M. A. Refaey and W. A. Chupka, *J. Chem. Phys.* **48,** 5205 (1968).
56. F. P. Lossing, *Can. J. Chem.* **49,** 357 (1971).
57. F. P. Lossing and A. Maccoll, *Can. J. Chem.* **54,** 990 (1976).
58. F. P. Lossing and J. C. Traeger, *Int. J. Mass Spectrom. Ion Phys.* **19,** 9 (1975).
59. F. P. Lossing, *J. Am. Chem. Soc.* **99,** 7526 (1977).
60. M. A. Haney and J. L. Franklin, *J. Phys. Chem.* **73,** 4328 (1969).
61. J. L. Beauchamp and M. C. Caserio, *J. Am. Chem. Soc.* **94,** 2638 (1972).
62. D. H. Aue, W. R. Davidson, and M. T. Bowers, *J. Am. Chem. Soc.* **98,** 6700 (1976).
63. J. F. Wolf, R. H. Staley, I. Koppel, M. Taagepera, R. T. McIver, Jr., J. L. Beauchamp, and R. W. Taft, *J. Am. Chem. Soc.* **99,** 5417 (1977).
64. J. L. Franklin, *Ind. Eng. Chem.* **41,** 1070 (1949).
65. J. L. Franklin, *J. Chem. Phys.* **21,** 2029 (1953).
66. R. D. Bowen and D. H. Williams, *Org. Mass Spectrom.* **12,** 475 (1977).
67. L. Radom, J. A. Pople, and P. von R. Schleyer, *J. Am. Chem. Soc.* **94,** 5935 (1972).
68. J. L. Franklin, J. G. Dillard, H. M. Rosenstock, J. T. Herron, K. Draxl, and F. H. Field, "Ionization Potentials, Appearance Potentials, and Heats of Formation of Gaseous Positive Ions." Natl. Bur. Stand., Washington, D.C., 1969.
69. D. H. Williams and R. D. Bowen, *J. Am. Chem. Soc.* **99,** 3192 (1977).
70. R. D. Bowen and D. H. Williams, *Org. Mass Spectrom.* **12,** 453 (1977).
71. R. D. Bowen and D. H. Williams, *J. Am. Chem. Soc.* **99,** 6822 (1977).
72. R. D. Bowen and D. H. Williams, *J. Chem. Soc., Perkin Trans. 2,* p. 68 (1978).
73. R. D. Bowen, J. R. Kalman, and D. H. Williams, *J. Am. Chem. Soc.* **99,** 5481 (1977).
74. G. Hvistendahl, R. D. Bowen, and D. H. Williams, *J. Chem. Soc., Chem. Commun.* p. 244 (1976).
75. D. H. Williams and G. Hvistendahl, *J. Am. Chem. Soc.* **96,** 6753 (1974).
76. D. H. Williams and G. Hvistendahl, *J. Am. Chem. Soc.* **96,** 6755 (1974).
77. D. H. Williams, B. J. Stapleton, and R. D. Bowen, *Tetrahedron Lett.* p. 2919 (1978).
78. F. W. McLafferty, R. Kornfeld, W. F. Haddon, K. Levsen, I. Sakai, P. F. Bente, III, S.-C. Tsai, and H. D. R. Schuddemage, *J. Am. Chem. Soc.* **95,** 3886 (1973).
79. B. G. Keyes and A. G. Harrison, *Org. Mass Spectrom.* **9,** 221 (1974).
80. J. L. Beauchamp and R. C. Dunbar, *J. Am. Chem. Soc.* **92,** 1477 (1970).
81. K. Levsen and H. Schwarz, *Angew. Chem., Int. Ed. Engl.* **15,** 509 (1976).
82. K. Levsen and F. W. McLafferty, *J. Am. Chem. Soc.* **96,** 139 (1974).
83. J. Winkler and F. W. McLafferty, *J. Am. Chem. Soc.* **95,** 7533 (1973).
84. F. W. McLafferty and J. Winkler, *J. Am. Chem. Soc.* **96,** 5182 (1974).
85. R. C. Dunbar, *J. Am. Chem. Soc.* **97,** 1382 (1975).
86. K. L. Rinehart, Jr., A. C. Buchholz, G. E. van Lear, and H. L. Cantrill, *J. Am. Chem. Soc.* **90,** 2983 (1968).
87. A. Siegel, *J. Am. Chem. Soc.* **96,** 1251 (1974).
88. R. A. Davidson and P. S. Skell, *J. Am. Chem. Soc.* **95,** 6843 (1973).
89. D. H. Williams, *Acc. Chem. Res.* **10,** 280 (1977).
90. R. D. Bowen, D. H. Williams, and G. Hvistendahl, *J. Am. Chem. Soc.* **99,** 7509 (1977).

2. GAS-PHASE ION REARRANGEMENTS

91. D. H. Aue, H. M. Webb, and M. T. Bowers, *J. Am. Chem. Soc.* **97,** 4137 (1975).
92. M. A. Haney and J. L. Franklin, *Trans. Faraday Soc.* **65,** 1794 (1969).
93. D. van Raalte and A. G. Harrison, *Can. J. Chem.* **41,** 3118 (1963).
94. B. G. Keyes and A. G. Harrison, *Org. Mass Spectrom.* **9,** 221 (1974).
95. A. G. Harrison and B. G. Keyes, *J. Am. Chem. Soc.* **90,** 5046 (1968).
96. C. W. Tsang and A. G. Harrison, *Org. Mass Spectrom.* **3,** 647 (1970).
97. C. W. Tsang and A. G. Harrison, *Org. Mass Spectrom.* **5,** 877 (1971).
98. F. W. McLafferty and I. Sakai, *Org. Mass Spectrom.* **7,** 971 (1973).
99. C. W. Tsang and A. G. Harrison, *Org. Mass Spectrom.* **7,** 1377 (1973).
100. A. S. Siegel, *Org. Mass Spectrom.* **3,** 1417 (1970).
101. We thank the Science Research Council (U.K.) and Sidney Sussex College, Cambridge (a Research Fellowship to RDB) for financial support.

ESSAY 3 | REARRANGEMENTS OF CARBENES AND NITRENES

W. M. JONES

I.	General Introduction	95
II.	1,2 Rearrangements of Carbenes and Nitrenes	97
	A. Introduction	97
	B. Theory and Mechanism	102
	C. Stereochemistry	105
	D. Migration of the Vinyl Group: Foiled Methylenes	108
	E. 1,2 Rearrangements of Vinylidenes	113
	F. Retro 1,2 Rearrangements	114
III.	Type I Carbene–Carbene and Carbene–Nitrene Rearrangements	119
	A. Introduction	119
	B. Theory	122
	C. Carbene–Carbene Rearrangements in Solution	126
	D. Carbene–Carbene, Carbene–Nitrene, and Nitrene–Nitrene Rearrangements in the Gas Phase	137
IV.	Type II Carbene–Carbene Rearrangements	149
	References	153

I. GENERAL INTRODUCTION

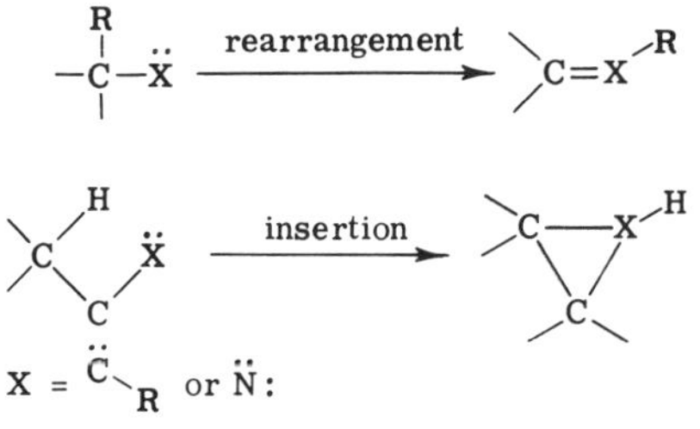

Any definition of carbene and nitrene rearrangements must be arbitrary at best. As a result, it is no surprise that tradition has been more prominent than logic in classifying some carbene and nitrene reactions as rearrangements while excluding others. For instance, insertion of a carbene or nitrene into an α,β-sigma bond is called a 1,2 rearrangement, whereas insertion into a more remote bond is termed an intramolecular insertion. Similarly, intramolecular addition to a double bond may be designated a ring closure unless the new ring is a cyclopropene or azirine that subsequently reopens under the conditions of the reaction. If so, the reaction becomes a carbene–carbene or carbene–nitrene rearrangement.

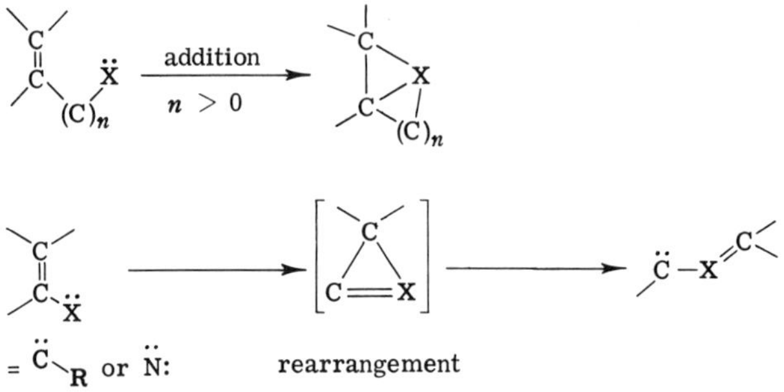

$$\underset{(C)_n}{\overset{\diagup}{C}}\ddot{X} \xrightarrow[n>0]{\text{addition}} \underset{(C)_n}{\diagup C\diagdown X}$$

$$\underset{\ddot{X}}{\overset{\diagup}{C}} \longrightarrow \left[\diagup C=X \right] \longrightarrow \ddot{C}-X$$

$X = \ddot{C}_R$ or $\ddot{N}$:   rearrangement

Nonetheless, these are the two reactions of carbenes and nitrenes that seem to be most frequently thought of as rearrangements and will, therefore, receive most of our attention in the following pages. In addition, the rearrangement of vinylcyclopropylidenes to cyclopentenylidenes will be discussed. The ring opening of cyclopropylidenes to allenes, a type of reaction that is sometimes included under the heading of rearrangements

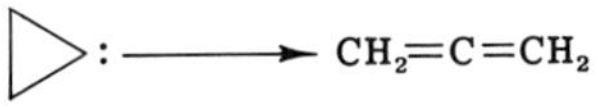

$$\triangleright: \longrightarrow CH_2{=}C{=}CH_2$$

(1), is not discussed in this chapter. However, for the interested reader a review of this topic (through 1974) has appeared (2), and recent developments, including one particularly important theoretical study (3), have also been described (4–9). A few related reactions (along with leading references) that are also omitted from our discussion but that could legitimately be called rearrangements are given in the following equations:

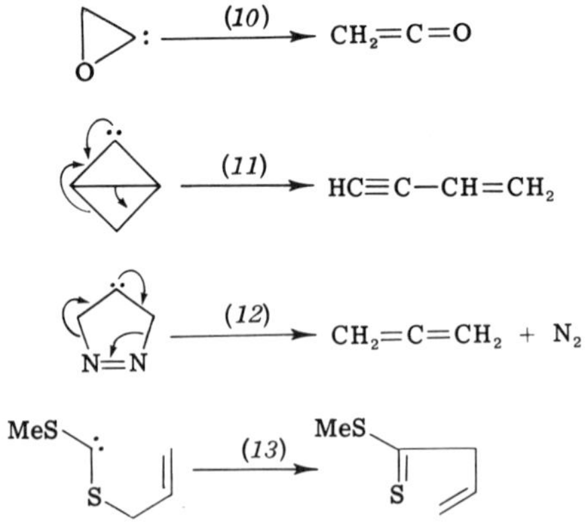

$$\underset{O}{\triangleright}: \xrightarrow{(10)} CH_2{=}C{=}O$$

$$\xrightarrow{(11)} HC{\equiv}C-CH{=}CH_2$$

$$\xrightarrow[N{=}N]{(12)} CH_2{=}C{=}CH_2 + N_2$$

$$\underset{S}{\overset{MeS}{\diagdown}}{:} \Vert \xrightarrow{(13)} \underset{S}{\overset{MeS}{\diagdown}}{\diagup}$$

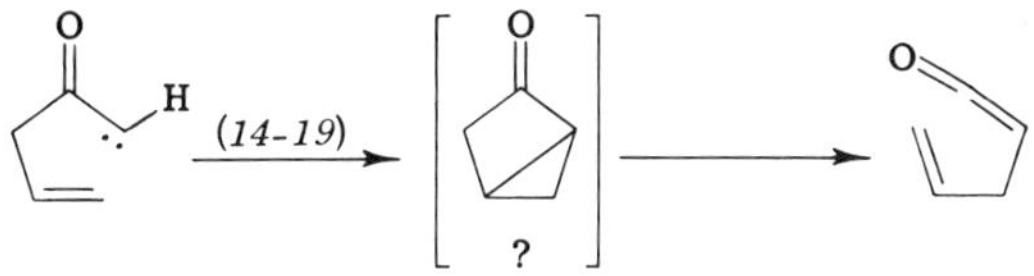

3. REARRANGEMENTS OF CARBENES AND NITRENES

[Scheme: cyclopentanone-CH-H (14-19) → bicyclic intermediate (?) → cyclopentenone-like product]

Finally, the two (to this author's knowledge) examples of 1,4 rearrangements (20, 21) also are not detailed. One of these, for which a radical fragmentation–recombination mechanism has been demonstrated, has found important use in the synthesis of small paracyclophanes (20).*

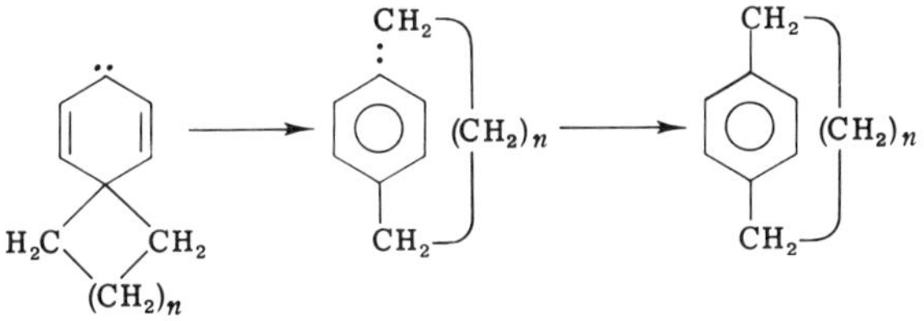

[Scheme: paracyclophane precursor → intermediates with $(CH_2)_n$ bridges]

II. 1,2 REARRANGEMENTS OF CARBENES AND NITRENES

A. Introduction

Groups that have been reported to undergo 1,2 rearrangement to carbenes from saturated carbon (1) include hydrogen, alkyl, benzyl, vinyl, aryl, RS, RO, C_2F_5, F, and Cl. In contrast, rearrangement from vinyl carbon is rare and, in fact, to this author's knowledge, only one case (1) is known (22); normally, vinyl carbenes close to cyclopropenes.

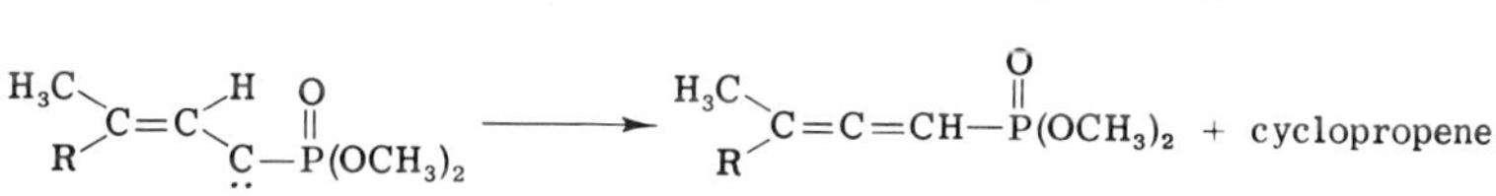

[Scheme: $H_3C,R\text{-}C{=}C\text{-}H,C(=O)P(OCH_3)_2 \longrightarrow H_3C,R\text{-}C{=}C{=}CH\text{-}P(=O)(OCH_3)_2$ + cyclopropene]

    1   4–7%

Migration from carbonyl is one of the most common 1,2 rearrangements of carbenes. The best known of this type of reaction, the Wolff rearrangement, has been thoroughly reviewed (23). The reaction is believed to normally proceed via the free carbene, whether it be generated from decomposition (thermal or photolytic) of diazocarbonyl compounds or from nitrogen-free precursors (2, 20, 24–26). However, in $HCCl_3$, photoinduced rearrangement may proceed by concerted rearrangement of a photoexcited singlet diazoketone (27, 28).

---

* Coverage of the three primary topics emphasizes the recent literature (through 1977) and is not intended to be comprehensive for periods covered by other referenced reviews. Particularly pertinent references that appeared in early 1978 have also been included.

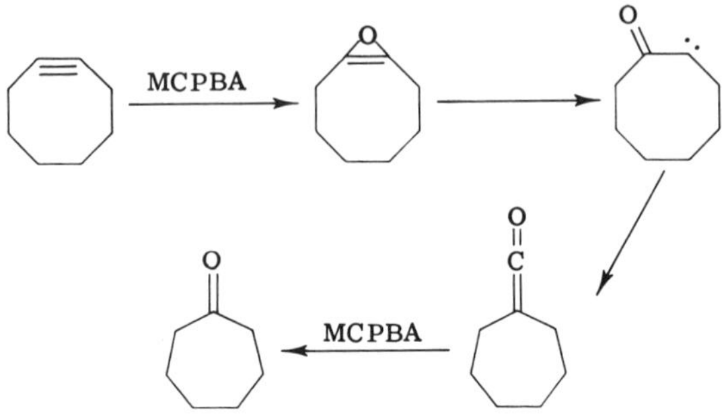

Rearrangements from RSO$_2$ (23), RSO (23), triazoles (23), silicon (29–33), and phosphorus (34, 35) have also been reported. Rearrangement from RPO is the subject of a recent review (34).

1,2 Rearrangements occur at temperatures as low as −196°C (36) and

$$PhC(=N_2)-CH_3 \xrightarrow[-196°C]{h\nu} PhCH=CH_2$$
$$42\%$$

under even the most adverse steric circumstances (20, 37). However,

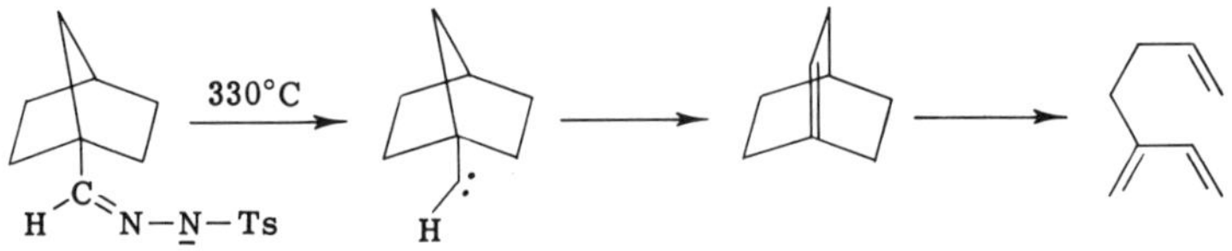

despite their facility, they are sensitive to the nature of the migrating group, and the following general migratory aptitude trends have been observed:
1. From saturated carbon:
    RS > H > Ph (38) > alkyl > RO > R$_2$N (1)
    C$_2$F$_5$ > F (1)
    Cl > H > F(1, 39, 40)
2. From C=O, generated by thermolysis of CO—CN$_2$ (23):
    H > C$_6$H$_5$ > CH$_3$ > R$_2$N > RO
3. From C=O, generated by photolysis of CO—CN$_2$(23):
    H > CH$_3$ > C$_6$H$_5$ > R$_2$N > RO

To date, no unifying theory to explain these trends has appeared. However, this is no surprise, since there are no less than four different types of groups: (1) alkyl and hydrogen, (2) aryl, (3) alkoxy and amino, and (4) thioalkyl.

We will briefly discuss some aspects of each of these:
1. Of the four types, only migration of hydrogen and alkyl has been

3. REARRANGEMENTS OF CARBENES AND NITRENES

examined by the use of any of the quantitative MO methods. This is discussed in some detail in Section II,B. Qualitatively, it may be worth noting that hydrogen migrates faster than alkyl or aryl; although this is no rationalization, it is consistent with intermolecular insertions for which it is well established (*1*) that C—H insertion is much faster than C—C insertion.

2. Aryl migration to the singlet, which may be viewed as a special type of C—C insertion, is faster than alkyl migration (*41–43*), although under some conditions this preference may be blurred by competitive rearrangement to the triplet (*44*) (*vide infra*). Substituent effects (*45, 46*) on the migratory aptitude of aryl groups show a better Hammett correlation ($\rho = -0.28$) with σ^+ than with σ, which points to a transition state with some involvement of, and electron deficiency in, the π system, albeit less than in carbonium ion rearrangements (*45*). The rearrangement of the aryl group has been treated theoretically by Zimmerman's MO following method (*46*).

3. Despite the presence of nonbonded electrons, alkoxy and amino groups migrate slower than hydrogen, aryl, or alkyl. In fact, they normally do not rearrange at all (*47–50*). To at least some extent this reluctance to migrate may only be apparent since an alkoxy group has been found to accelerate the rearrangement of hydrogen, alkyl, and even alkoxy (*47, 48*). Stabilization of an electron-deficient rearrangement origin has been proposed to rationalize this effect (*47*).

4. Thioalkoxy groups migrate faster than any other studied to date (*49, 50*). An interesting demonstration of this tendency is found in the recently reported comparison of carbenes **2** and **3** (*50*). This high migratory ap-

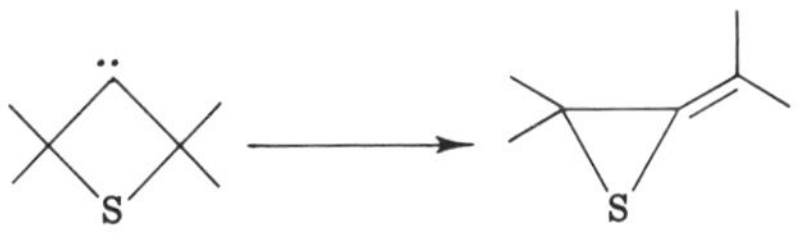

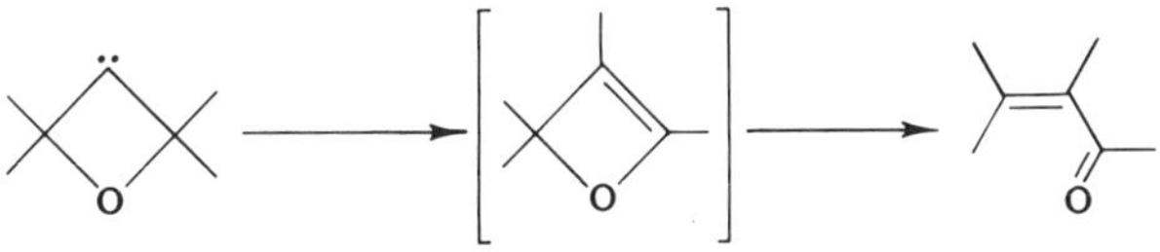

titude has been ascribed to a "greater participative ability of sulfur," which may, at least in part, be due to zwitterionic stabilization of an intermediate by the vacant 3d orbitals of sulfur (*49*).

The multiplicity of the carbene also appears to affect 1,2 rearrangements. In view of the reluctance of alkyl groups and of hydrogen to migrate to radical centers (51, 52), it is not surprising that the triplet is less inclined to rearrange than is the singlet (43, 44, 53–56) and, with few exceptions (55), these groups may not rearrange at all. Drawing further on analogy with free radicals, an aryl group might be expected to migrate to a triplet carbene center more readily than an alkyl group or hydrogen. In fact, it has been reported that conditions that should promote triplet carbene formation increase the Ph/CH$_3$ migration ratio in **4**. These include (Table 1) photosensitization, a heavy-atom solvent, and tetraphenylethylene catalysis.

$$\text{Ph}-\underset{\underset{\text{H}_3\text{C}}{|}}{\overset{\text{H}_3\text{C}}{\underset{|}{\text{C}}}}-\overset{\text{N}_2}{\overset{\|}{\text{CH}}} \longrightarrow \underset{\text{H}_3\text{C}}{\overset{\text{H}_3\text{C}}{\diagdown}}\text{C}=\text{CHPh} + \underset{\text{H}_3\text{C}}{\overset{\text{Ph}}{\diagdown}}\text{C}=\text{CHCH}_3$$

4

In contrast to carbenes, relatively few groups have been found to migrate from saturated carbon to nitrene centers. The most frequently noted are H, aryl, and alkyl, although at least one example of CO_2CH_3 has been reported (57). Migration from C=O to a nitrene center has been frequently postulated for the Lossen, Curtius, and Hofmann rearrangements. However, in a review of nitrene chemistry published in the late 1960's, the relevant arguments were critically evaluated (58), and it was concluded that in *all cases,* including the photoinduced Curtius rearrangement, the evidence to that date pointed to the occurrence of concerted rearrangements rather than to the participation of free nitrenes. More recent experimental (59, 60) and theoretical (61) studies do not refute this conclusion. A comprehensive review can be found in Smith (62).

Migratory aptitudes of groups migrating to nitrenes have been reviewed (63) through the late sixties and, to this author's knowledge, little on this

TABLE 1

Effect of Conditions on 1,2 Rearrangements of 4

Condition	Ph/CH$_3$ rearrangement
Heat in benzene or hexane	10.1
Heat in 1-bromonaphthalene	20.4
Direct photolysis	5.0
Sensitized photolysis	18.0–22.5
Benzene or hexane; catalyzed by tetraphenylethylene	15.0–22.4

3. REARRANGEMENTS OF CARBENES AND NITRENES

subject has appeared since (64). We shall briefly recapitulate this excellent review:

1. With a single, early exception (65), in thermal reactions of azides of type **5** migratory aptitudes generally parallel those of carbenes, that is, H > aryl > Me > F or acyl (66–71).

$$R_2-\underset{R_3}{\overset{R_1}{\underset{|}{\overset{|}{C}}}}-N_3 \xrightarrow{\Delta} R_2-\underset{R_3}{\overset{R_1}{\underset{|}{\overset{|}{C}}}}-\ddot{N}: \longrightarrow \underset{R}{\overset{R}{>}}C=N-R$$

5

2. Thermal reactions of azides such as **8** show migratory aptitudes consistent with an electron-deficient transition state (72). However, in this case the aryl migration may assist in the loss of nitrogen.

3. Photolysis of azides of types **6–8** show *no* preference among mi-

$$H_3C-\underset{C_6H_5}{\overset{CH_3}{\underset{|}{\overset{|}{C}}}}-N_3 \qquad H_5C_6-\underset{C_6H_5}{\overset{CH_3}{\underset{|}{\overset{|}{C}}}}-N_3 \qquad Ar_1-\underset{Ar_3}{\overset{Ar_2}{\underset{|}{\overset{|}{C}}}}-N_3$$

6   **7**   **8**

grations of alkyl, phenyl, or para-substituted phenyl groups (69, 71). Sensitized photolysis of **8** shows the same chemistry as direct photolysis, both qualitatively and quantitatively (71, 73). Furthermore, convincing arguments for rearrangement in a singlet nitrene from the direct photolysis of the triarylmethylazide were put forward (71). From this it would appear that (a) aryl and alkyl groups have the same migratory aptitudes when rearranging to a photogenerated singlet; (b) aryl substituents do not affect migratory aptitudes when rearranging to a photogenerated singlet or a triplet nitrene; and (c) it is not known how the migratory aptitudes of aryl and alkyl groups compare when rearranging to a triplet.

As in the case of vinylcarbenes, vinylnitrenes normally close to azirines faster than they undergo 1,2 rearrangement. However, in isolated cases (74–78) the latter rearrangement has been observed. For example, thermolysis of **9** in the gas phase gives, in addition to 50–60% azirine, small amounts (5–6%) of **10** (74, 75). Electron-withdrawing groups on the dou-

$$CH_2=C\underset{R}{\overset{N_3}{<}} \xrightarrow[\text{vapor}]{\Delta} CH_2=C\underset{R}{\overset{\ddot{N}:}{<}} \longrightarrow \underset{H_2C\text{———}C-R}{\overset{N}{\overset{/\!\!\backslash}{\|}}} + CH_2=C=N-R$$

    9   50–60%   5–6% **10**

ble bond appear to promote rearrangement, presumably by retarding the competing ring closure (77).

B. Theory and Mechanism

Two rearrangement modes have been considered for the early stages of concerted* 1,2 rearrangements of carbenes to alkenes: (a) a mode that finds analogy in carbonium ion rearrangements (usually assumed to be rearrangement of H_A in **11**) in which the migrating group carries its electrons into the vacant orbital of the singlet state and (b) rearrangement into the filled sp^2 orbital (migration of H_B in **12**), a reaction that could be viewed as analogous to a rearrangement of carbanions (see Essay 6).

11   **12**

Since the earliest considerations of the mechanism of this rearrangement (*47, 80*), the former pathway has been preferred (*2*). With the exception of one least-motion treatment (*81*) (which gives a perpendicular excited alkene), this initial preference has found support in all theoretical treatments that have been applied to this problem. These have ranged from the very simple Woodward–Hoffmann (*2*), HOMO–LUMO (*2*), and Zimmerman MO following (*46*) methods, all of which simply predict this mode to be allowed, to more sophisticated approaches such as MINDO/2 (*82*), MINDO/3 (*83*), MNDO (*83*), and least-motion LCAO–MO–SCF (*84–86*). In all of the latter treatments, only migration of hydrogen has been considered. In addition, with one exception (*87*) (least-motion triplet to perpendicular alkene), the rearrangement of the singlet state has been predicted to be preferred.

Although all of the MO treatments agree on the preferred rearrangement of hydrogen into the vacant orbital of the carbene, most have not addressed the question of whether rearrangement of H_B in **11** (with the necessary conformational change) could compete with H_A. Dewar (*82*) alluded to this possibility when pointing out that migration of H_B (in **12**) into the filled carbene orbital could be studied (using MINDO/2) only if retention of the dihedral angle 1234 were forced to remain fixed. Otherwise, rotation around the CC bond would occur, bringing H_B into line with the vacant carbene orbital. In Dewar's study, however, migration of H_A in **11** showed zero activation energy and, as a result, migration of H_B could have been competitive *only* if rotation also had zero activation energy. In contrast, a recent study (*83*) applied MNDO to the ethylidene–ethylene

---

* In at least one instance a diradical mechanism has been proposed to explain 1,2 rearrangement products of simple carbenes (*79*).

3. REARRANGEMENTS OF CARBENES AND NITRENES

problem and found a substantial activation energy (21.9 kcal/mol) for rearrangement of both H_A and H_B. Since this was much higher than would be expected for conformational changes, it is not surprising that the activation energies for rearrangement of the two hydrogen atoms were found to be identical. MINDO/3 (*83*) gave much lower activation energies (0.7 kcal/mol) but, again, they were identical for H_A and H_B.

Dewar and Bodor (*82*) also applied the MINDO/2 method to the rearrangement of cyclohexanylidene and concluded that there should be an "overwhelming preference" for migration of H_a in **13**. This led to the prediction of a high degree of stereospecificity "so long as the conformational integrity of the ring is maintained." Kyba (*83*) used MNDO to study migration in **13**. As in the case of ethylidene, he found substantial

13

and nearly identical activation energies for migration of both hydrogens (23.9 and 23.6 kcal/mol, respectively). Intuitively this may be surprising. However, from a consideration of molecular models, it is not unreasonable if a three-membered ring can be taken as a viable model for the transition state. Thus, rearrangement of H_a (in **13**) would lead to **14**, whereas rearrangement of H_e (either through a twist boat followed by migration or via concomitant torsional change and migration) would lead to **15**. Since these two should be of nearly equal energy and since the

14   **15**

conformational changes required to generate **15** should be less than the calculated activation energy, the activation energy to form each should also be nearly equal. The weaknesses in this argument are the assumptions that the transition state looks like a three-membered ring and that the activation energy for rearrangement is high relative to that for conformational changes. The former assumption is bothersome because in a highly exothermic reaction hydrogen migration might not be expected to have proceeded this far; in fact, a recent study of primary and secondary deuterium isotope effects points to an early transition state (*88*). Concerning the second assumption, it has been suggested that a "limited torsional process" (a conformational change concomitant with the rearrangement) may require less energy than conversion to the twist boat. As a result rearrangement may be nonstereospecific even if its activation energy is lower than would be expected for the conformational change. In any event, both MNDO and MINDO/3 (83) predict little if any stereospecificity in the rearrangement of cyclohexanylidine, and experimental results (see the next section) appear to support this.

As mentioned above, all theoretical studies of 1,2 rearrangements agree that, in the early stages of the reaction, hydrogen migrates into the vacant orbital of the carbene. However, in only one instance (85) has a description of the details of the later stages of the rearrangement been given. Even though we do not know whether this picture will stand the test of time, a clear statement of at least one possibility is of value and will therefore be briefly described.

By the use of a nonempirical LCAO–MO–SCF calculation, the reaction profile for the rearrangement of methylcarbene to ethylene was examined (85). From a charge distribution analysis, the following electron redistribution process was proposed. (a) In the early stages of the rearrangement, the migrating hydrogen resembles a hydride ion as in a typical carbonium ion rearrangement. (b) "After the transition state [which is somewhat earlier than half-reaction] the hydrogen continues its motion as a pseudo proton toward the lone pair which, because of its spatial environment, does not contribute to the π bond but is utilized for the formation of the new C—H bond" (85).

Supporting evidence was obtained by computing electron density contours of the occupied MO's and "following" them from reactant to transition state to product. By means of this technique it was found that the MO corresponding to the bonding electron pair of the migrating hydrogen correlates with the π bond of ethylene and that the nonbonded carbene pair correlates with the new C—H bond.

Although qualitative rules, such as Woodward–Hoffmann, HOMO–LUMO, and MO following, would predict the rearrangement of nitrenes to imines to be allowed, to this author's knowledge this reaction has not been treated by any of the more quantitative MO methods. However, it has been pointed out (86) that, since isoelectronic systems tend to behave similarly, the mechanism of this rearrangement might be expected to be the same as that for carbenes. Translated into the picture described above for carbenes, this would predict the following:

C. Stereochemistry

In the last section it was noted that virtually all of the theoretical treatments of 1,2 rearrangements of carbenes predict preferential migration of hydrogen into the vacant p orbital of the singlet carbene. Recent calculations (83), however, further predict that, in a simple acyclic system such as **11**, this preference would not lead to different rates of migration of

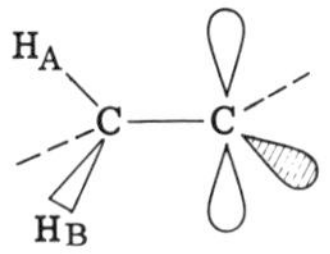

11

H_A and H_B because conformational changes that would bring H_B into the vacant orbital would require less energy than the activation energy for the rearrangement. Migration of H_A and H_B would therefore give identical (actually enantiomeric) transition states. To test experimentally for preferential migration into the vacant or filled orbital of a carbene would therefore require a molecule in which conformational change were slow relative to the rearrangement. Intuitively, and from some of the earlier calculations (82), a properly substituted cyclohexanylidene might be expected to meet this requirement. Thus, in **13**, H_a is favorably aligned with

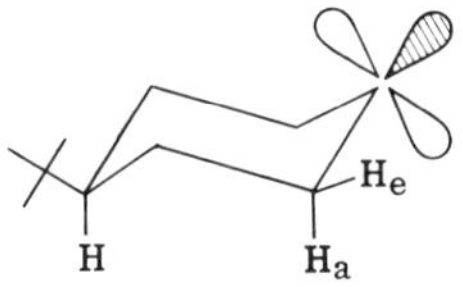

13

the vacant carbene orbital, as is H_e with the filled orbital. Furthermore, to reverse these alignments would require a conformational change that would bring the chair well along toward a boat conformation. However, as discussed in the last section, this is probably a relatively low-energy process (*ca.* 3 kcal/mol), and even less energy may be involved since a "limited torsional" process will generate transition states of nearly identical energy for migration of either H_a or H_e into the vacant orbital. From this it was concluded that, experimentally, by the use of a cyclohexanylidene one cannot distinguish between the two possibilities. In fact, the stereochemistry of hydrogen rearrangement in a number of cyclohexanylidene systems (**16–20**) has been studied. Quite clearly, in no case is there an overwhelming preference for migration of either H_a or H_e. Perhaps it should be noted, however, that in those cases in which both migrating groups are hydrogen, II_a docs show some preference. This is reasonable if H prefers to migrate into the *vacant* carbene orbital since any conformational change required to bring the rearranging hydrogen into the preferred orbital can only slow the rearrangement. It is also interesting that migration of C_6H_5 occurs to a greater extent in **20** than in **19**. This is the first case of conformational differences leading to preferential rearrangement of a group other than hydrogen. This difference, however, cannot be due to the suggested (*91*) contribution of **21** because the

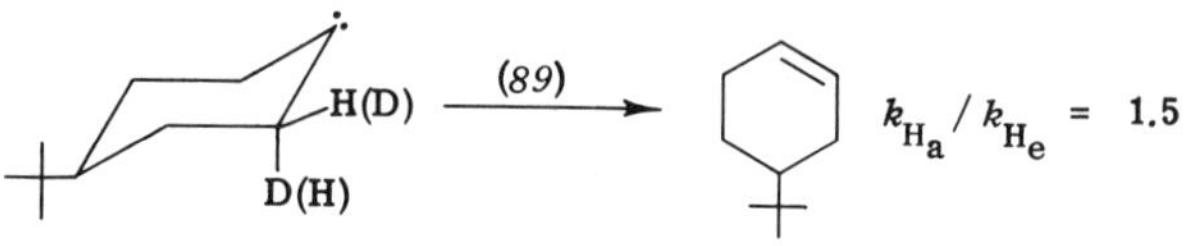

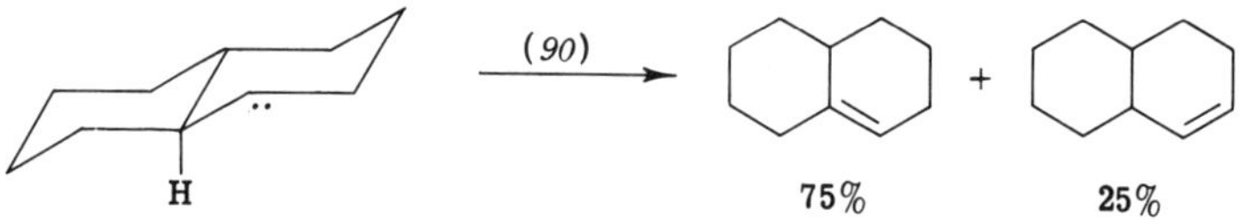

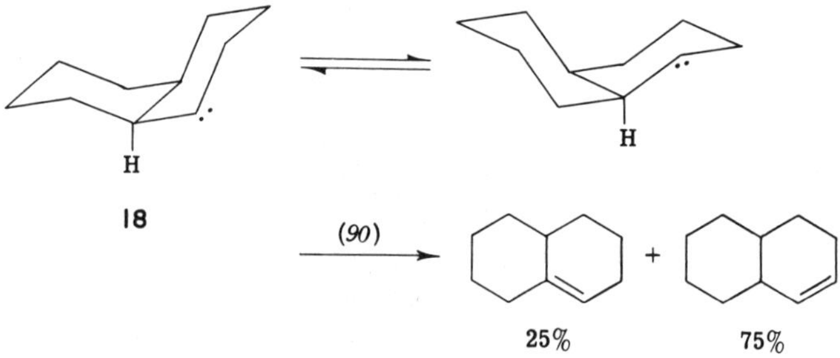

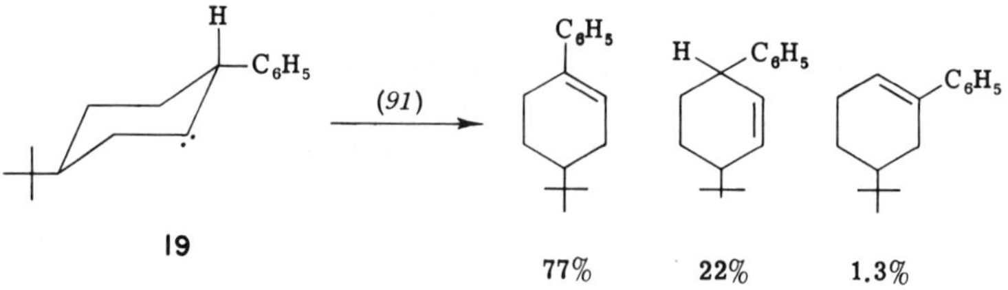

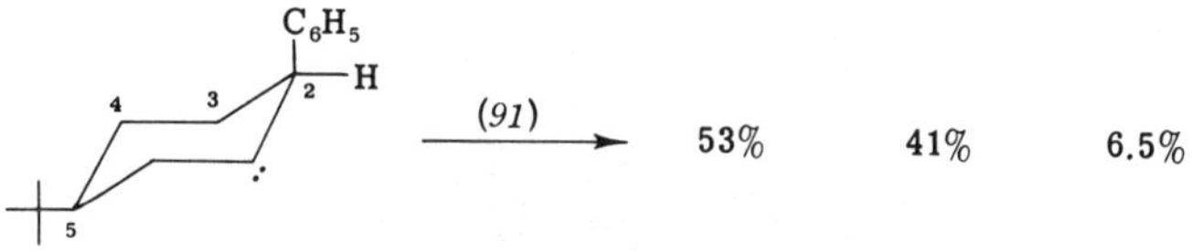

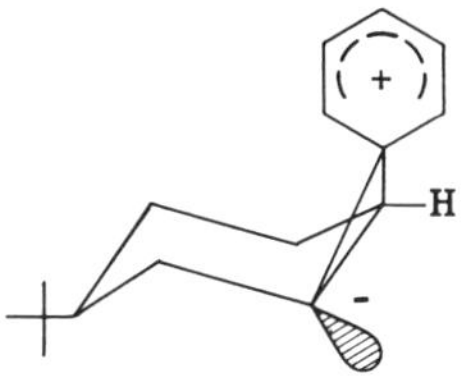

3. REARRANGEMENTS OF CARBENES AND NITRENES

21

migration of C_6H_5 in **19** would give essentially the same intermediate. It is also interesting that H_2 in **20** migrates faster than H_{6a} and H_{6e}. In this case conjugation effects are presumably dominant.

Rigidity retards conformational changes. This suggests that a properly designed polycyclic system might meet the requirements to test for stereoelectronic preference in 1,2 rearrangements. Indeed, the stereochemistry of rearrangement in at least one such system has been studied. Brexanylidene (**22**) is a tricyclic system in which the dihedral angle between H_x and the vacant p orbital of the carbene is believed to be less than for H_n, and, of course, conversely, the dihedral angle between H_n and the filled orbital is less than for H_x. Furthermore, from models, it is

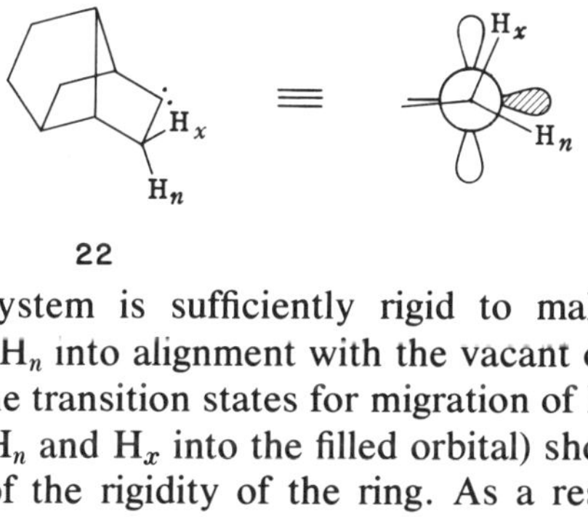

22

known that the system is sufficiently rigid to make conformational changes that bring H_n into alignment with the vacant orbital difficult. To be more precise, the transition states for migration of H_x and H_n into the vacant orbital (or H_n and H_x into the filled orbital) should have different energies because of the rigidity of the ring. As a result, H_x would be expected to migrate significantly faster than H_n if H prefers to migrate to the vacant orbital, and vice versa if hydrogen prefers to migrate to the filled orbital. In fact, when **22** was generated thermally (from the tosylhydrazone salt at 170°–200°C), k_{H_x}/k_{H_n} was found to be 138 (*92*).

Unfortunately, even this is not as clear-cut as it may appear to be because, as pointed out (*93, 94*), preferential rearrangement in bicycloheptanylidene systems may be affected by factors other than orbital alignment. For instance, it is well known (*95*) that H_x rearranges faster than H_n in carbonium ion **23** despite the fact that the dihedral angle between the *p* orbital and the two hydrogens is about the same. To determine whether this is also true for carbenes, rearrangements of **24** (*96*), **25** (*93, 94*), and **26** (*93, 94*) were studied. Indeed, in each case H_x

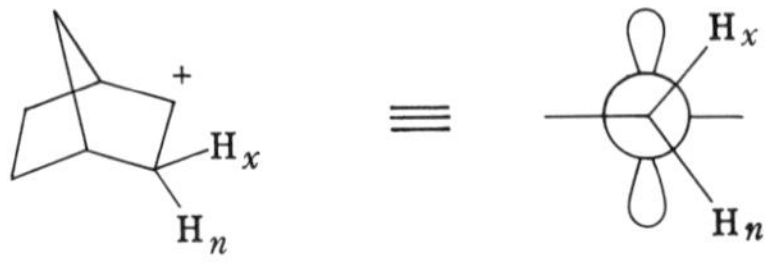

showed a preference. However, the difference in migration preferences in these systems [which has been attributed to a torsional influence (96)] is clearly less than that observed by Nickon and leaves the latter's results as the most compelling evidence to date for migration of hydrogen into the vacant orbital of the carbene.

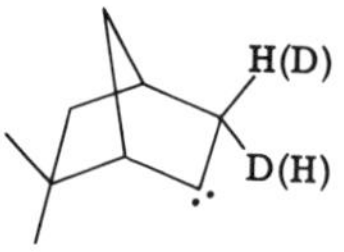

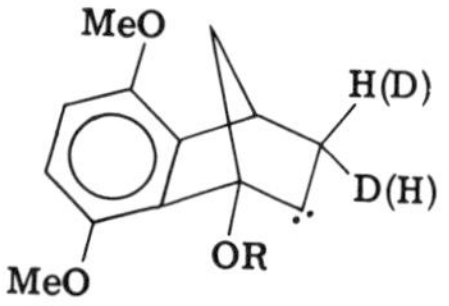

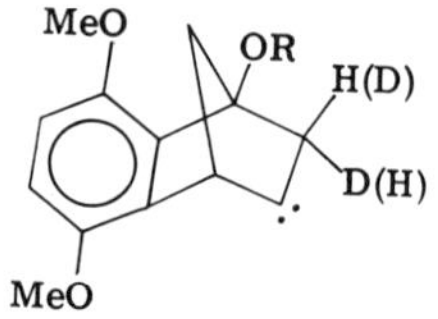

Thus, to briefly summarize, *all* experimental results to date in both cyclic and acyclic (2, 47) systems are consistent with preferential initial migration into the vacant orbital of a carbene. None suggest preferential rearrangement into the filled carbene orbital. In flexible molecules conformational changes appear to be faster than rearrangement.

D. Migration of the Vinyl Group: Foiled Methylenes

In 1968 Gleiter and Hoffmann (97) noted that allylcarbenes **27** and **28** could interact with their double bonds, as in the early stages of intermolecular addition, but, due to strain, could not complete the reaction.

3. REARRANGEMENTS OF CARBENES AND NITRENES

27   28

Since earlier calculations had predicted zero activation energy for the addition process, this led the authors to conclude that the singlet states of these "foiled" carbenes should experience a special "nonclassical" stabilization analogous to carbonium ions. They further predicted that this stabilization would manifest itself in (a) large σ–p splittings (favoring singlet ground states), (b) unsymmetric structures in which the carbene carbon is bent (up to 20°) toward the double bond, and (c) an increase in the nucleophilicity of the carbene. These predictions stimulated a flurry of activity aimed at the generation of a variety of "nonclassical" or "foiled" carbenes. The chemistry of these carbenes as it relates to the "nonclassical" question is discussed in this section.

Although the predictions of Gleiter and Hoffmann primarily forecast anomalous intermolecular chemistry of "foiled" carbenes, in fact, with the exception of **27** (*98, 99*) the recorded reactions of this type of intermediate have been limited to intramolecular reactions. To lend perspective to the chemistry of bicyclic allylcarbenes, it should first be noted that unrestricted allylcarbenes are unexceptional in that typical intramolecular reactions are observed [as, for example, in **29** (*100*)]. It is particularly noteworthy that, in most cases of *thermally generated* allylcarbenes, vinyl migration is facile (*100–104*) and, in those systems in which it can be

distinguished, the vinyl group migrates faster than the alkyl group (*101, 104*) and is even competitive with hydrogen (*100, 102, 103*). Photochemically generated allylcarbenes appear to give more hydrogen rearrangement (*100, 105*).

The chemistry of a number of "foiled" carbenes is summarized in Scheme 1. Also included for reference are the saturated bridged carbenes **30** and **31**. From these results, clearly the most notable generality is that the predominant reaction of thermally generated "foiled" carbenes is vinyl migration. This propensity for rearrangement has been interpreted in terms of special p–π interaction (*98, 107, 110*). However, in view of the

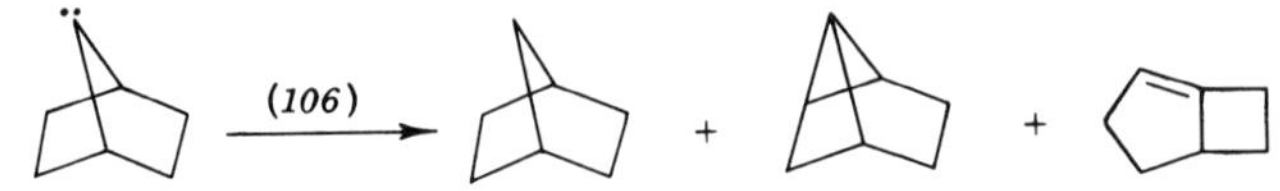

30	(106)	14% + 12% + 74%		
31	(107)	98%		
27	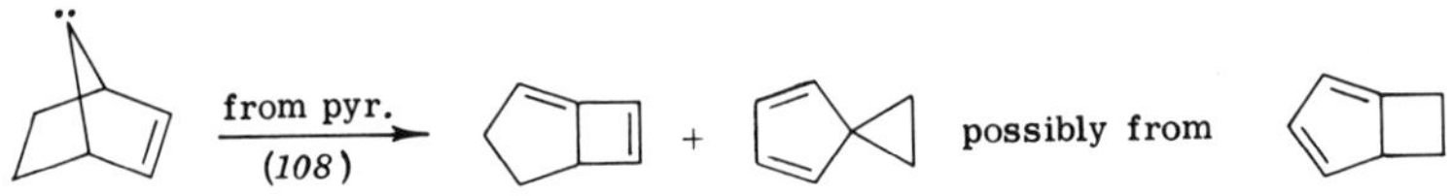 from pyr. (108)	67% (56%) + 6.9% (10%) possibly from		
32	(109)	CH₃ + CH₃		
33	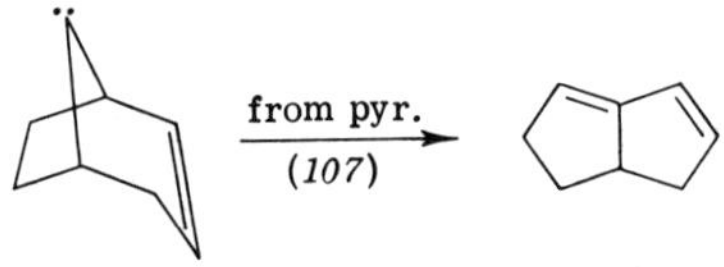 from pyr. (107)	97%		
34	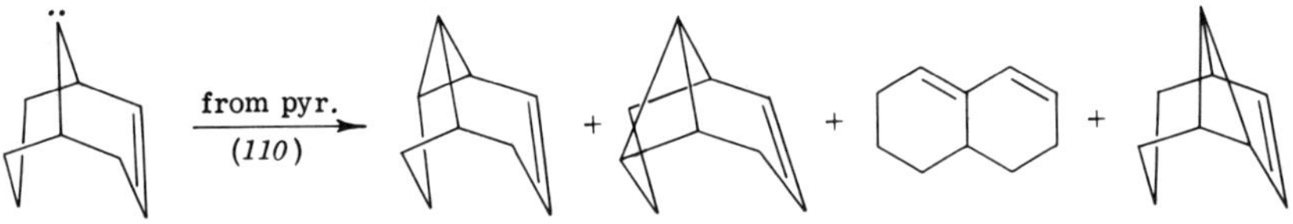 from pyr. (110)	18% **35** + 18% **36** + 31% **37** + 0% **38**		
28	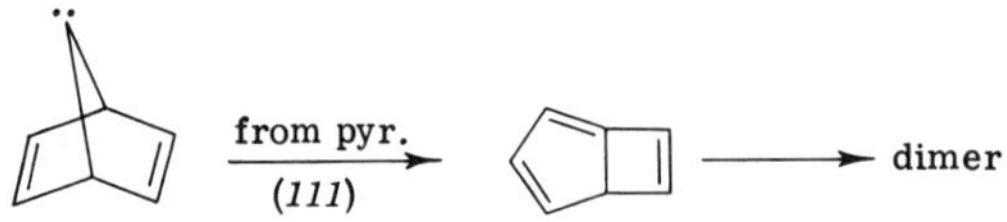 from pyr. (111)	→ dimer		

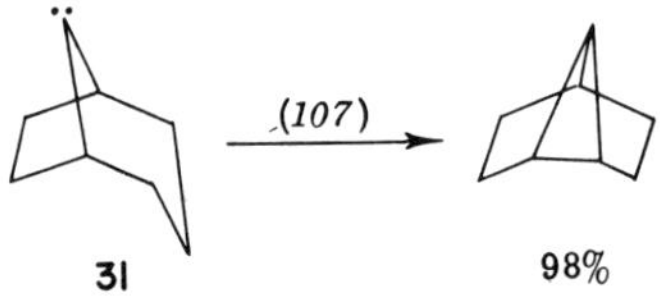

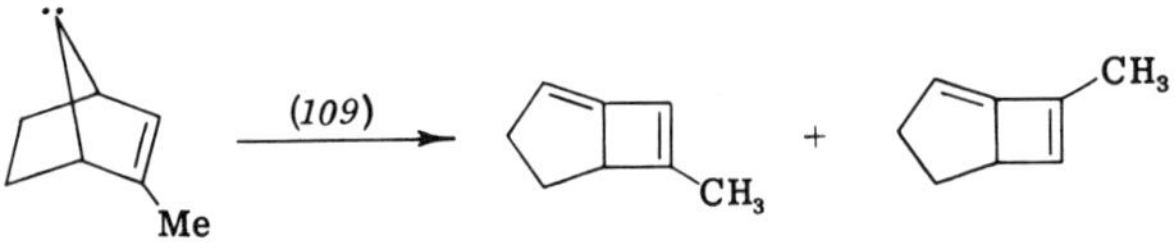

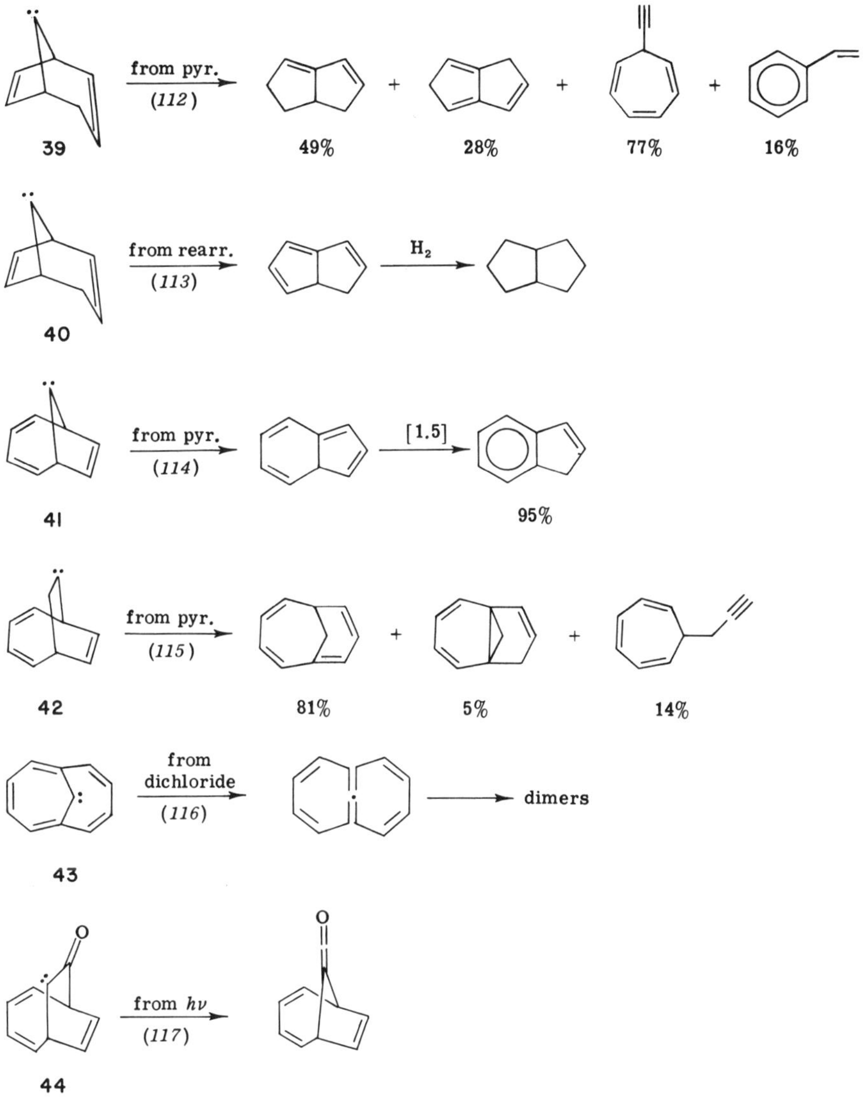

3. REARRANGEMENTS OF CARBENES AND NITRENES

Scheme 1

normal vinyl migratory aptitude, this clearly cannot be taken as strong evidence for a "special" effect.

The absence of insertion products from **33** (*107*) has also been presented as evidence for nonclassical character. Unfortunately, there is no suitable model that permits comparison of vinyl migration with intramolecular C—H insertion. However, perhaps it should be noted that in at least two cases of thermally generated allylcarbenes, in which C—H insertion could have occurred, none was reported (*101, 104*). Two of the more convincing arguments for a ground state p–π interaction are found in the chemistry of **34** and **42**. In the former case the formation of **35** and **36**, to the exclusion of **38**, is difficult to rationalize without invoking some special effect, because there does appear to be some normal preference for allylic C—H insertion (*118, 119*). The latter case is a unique "foiled" carbene in that it is possible to compare directly 1,2-hydrogen rearrangement with vinyl migration. In interesting contrast to **29**, the thermal reaction gave *only* migration of the π system (the dienyl unit). Models show no trivial reason for slow hydrogen migration.

Thus, in sum, "foiled" carbenes are indeed foiled in their attempts to undergo intramolecular addition, and it is quite possible that they are stabilized by a ground state interaction of their vacant "p" orbital with π systems. However, in most cases, it is far from clear that their chemistry is dominated by this stabilization. The intermolecular chemistry of **27** (*98, 99*) is also consistent with some ground state stabilization but, again, attempts to define this more exactly have been disappointing.

The calculations of Gleiter and Hoffmann predicted an even greater stabilization from cyclopropane in carbenes such as **45** than from allylic double bonds. Such carbenes permit direct comparison of exo and endo isomers. In fact, isomer pairs **45** and **47** have been generated by thermolysis of the tosylhydrazone salts (*120, 121*), and, in accord with the

predicted differences in ground state interactions, the chemistry of the two is quite different. Unfortunately, whereas the chemistry of **47** is clearly dominated by simple alkyl migration, that of of **45** is complex. However, it gave no recognizable products from migration of the ethano bridge, and some of the products obtained were consistent with the intervention of the delocalized carbene **46**. Products from **48** were equally complex and difficult to interpret (*112*). In both **45** and **48** the authors

3. REARRANGEMENTS OF CARBENES AND NITRENES

48

argue for interaction between the carbene and the cyclopropane, this, in the latter case, being favored over interaction with the double bond.

Finally, it should be noted that generation of **45** and **47** from pyrolysis of the dry sodium salts gave products that were different from those from decomposition of the same tosylhydrazones with sodium methoxide in diglyme (*122, 123*). The latter probably gave carbonium ion products.

E. 1,2 Rearrangements of Vinylidenes (124)

In its simplest form, the 1,2 rearrangement of vinylidene may be treated as a simple allowed [$\sigma^2 a + \omega^2 s$] pericyclic reaction (*2*). The reaction is

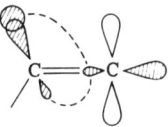

characterized by the ideal orientation of the migrating group for rearrangement into the vacant orbital of the singlet state [which is probably the ground state (*97*)] of the carbene and by the somewhat shorter distance that the group must traverse when compared with alkylidenes. The mechanism of the rearrangement was recently studied using the SCEP (self-consistent electron pairs) MO method (*125*). The rearrangement was calculated to have an activation energy of 8.6 kcal/mol with a transition state in which the migrating hydrogen is nearly symmetrically located between the two carbons. The terminal hydrogen was found to be nearly collinear with the remaining carbon atoms. As in the rearrangement of alkylidenes, the transition state is much farther along the reaction coordinate than the Hammond postulate would forecast for such an exoergic reaction. A theoretical study of the rearrangement of fluorine to the reactive site of a vinylidene has also been reported (*126*).

In contrast to alkylidenes, for which migratory aptitudes can be determined simply by measuring the relative amounts of isomeric olefins, either indirect methods or isotopic labeling are required to gain the same information for vinylidenes. It is therefore no surprise that even though

$$\underset{R_2}{\overset{R_1}{>}}\overset{*}{C}=C: \longrightarrow R_1\overset{*}{C}\equiv CR_2 + R_1C\equiv \overset{*}{C}R_2$$

there are numerous examples (*124, 127–130*) of 1,2 rearrangements of vinylidenes relatively little is known about migratory aptitudes. Nonetheless, some trends have emerged and, although very limited, in general they appear to be similar to those for alkylidenes. For instance, both hydrogen and aryl migration are quite rapid. In fact, as noted in a fine review of the chemistry of unsaturated carbenes (*124*), except in unusual circumstances (such as steric retardation) these rearrangements normally occur to the exclusion of intermolecular chemistry. It has also been found that aryl migration occurs to the exclusion of intramolecular C—H insertion (*131*).

Vinylidenes are also similar to alkylidenes in that alkyl migration is relatively slow. This reluctance to rearrange, coupled with the fact that vinylidenes apparently do not tend to undergo intramolecular insertion [except into a remote C—H (*131*)], has recently been exploited (*132–137*) in a number of useful applications, which include addition to triple bonds to give relatively unencumbered methylenecyclopropenes. Interestingly,

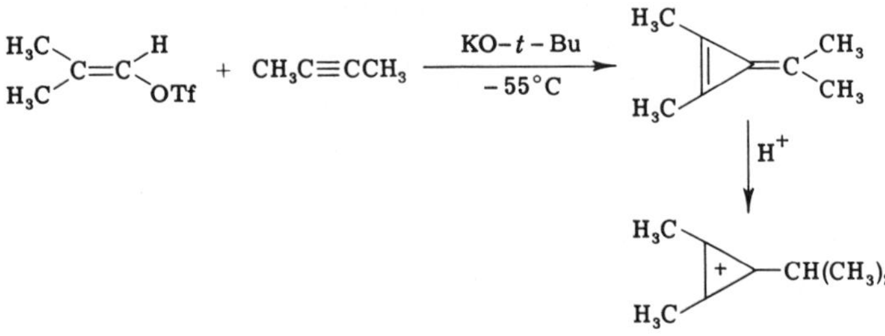

under forcing conditions alkyl migration will occur (*124*), and, in fact, this has found use in the generation of strained cyclic alkynes such as **49** (*138–140*).

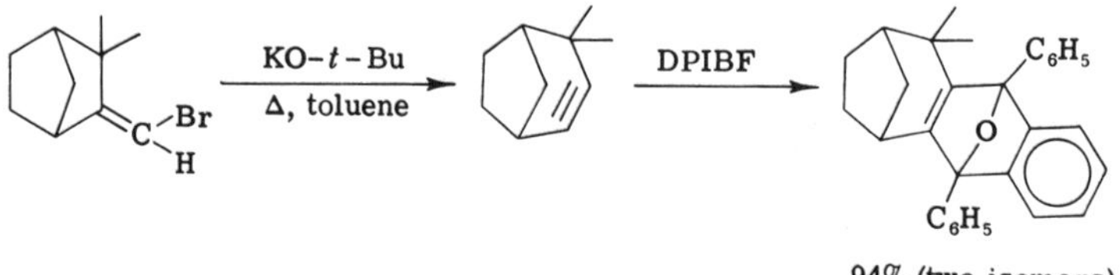

94% (two isomers)

Finally, difluorovinylidene does not rearrange (*141*).

F. Retro 1,2 Rearrangements

In recent years a number of reactions have been recorded that are formally the reverse of 1,2 rearrangements of carbenes. Such rearrangements have been found for both ketones and alkenes.

3. REARRANGEMENTS OF CARBENES AND NITRENES

$$\underset{R'}{\overset{R}{>}}C=O \longrightarrow \underset{R'}{\overset{R}{>}}\ddot{C}-O-R \qquad \underset{R'}{\overset{R}{>}}C=C\underset{}{\overset{}{<}} \longrightarrow \underset{R'}{\overset{R}{>}}\ddot{C}-\underset{|}{\overset{R}{C}}-$$

The rearrangement of ketones to oxycarbenes was discovered by Yates (142) in 1964 when it was noted that photolysis of **50** in alcohol and air gave acetals **51** and lactone **52**, respectively. Since its discovery this

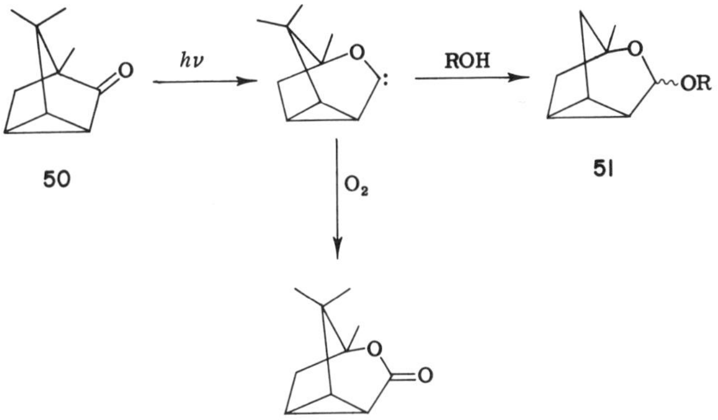

rearrangement has received much attention from a number of research groups and has been the subject of at least three fine reviews (143–145). The salient points of these reviews (with some updating) are as follows:

1. In carbocyclic systems, this rearrangement is a common photoreaction of cyclobutanones. It is often observed in the photolysis of cyclopentanones that are either incorporated into bicyclic or tricyclic systems, such as **53** and **50**, substituted with cyclopropanes, as in **54** and **55**, or constrained in such a way as to retard more common photoreactions (145, 146), as is the case in **56** (147). The rearrangement has not been reported

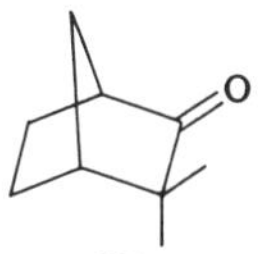

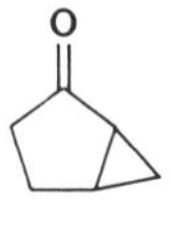

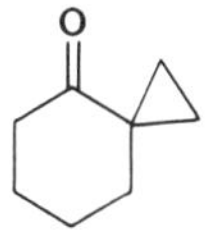

for larger carbocyclic systems, but isolated cases of rearrangements in six-membered heterocyclic ketones have been reported (148, 149). The rearrangement has also been reported for acyclic ketones in which R_3Si is the migrating group (150, 151) (see Essay 8).

$$\underset{}{\overset{O}{\underset{\|}{R_3Si-C-CHR_2}}} \xrightarrow{h\nu} \underset{}{\overset{O}{\underset{\|}{R_3Si-O-C-CHR_2}}}$$

2. In cyclobutanones, the α-carbon that is the more highly substituted with alkyl groups shows a migratory preference (143).

3. Evidence for the carbene as an intermediate includes trapping with alcohol, oxygen, and, in a few cases, double bonds (143, 152) and C=O (153).

4. Depending on the system, the reaction can occur from either the singlet or triplet n,π^* state (145).

5. Both biradical and concerted processes have been proposed for the rearrangement step, and, despite intensive effort, this question remains unresolved. In fact, arguments for each mechanism are sufficiently compelling to lead Yates to suggest that both processes may be operating, with cyclobutanones favoring concerted rearrangement and larger ring ketones favoring Norrish I radicals.

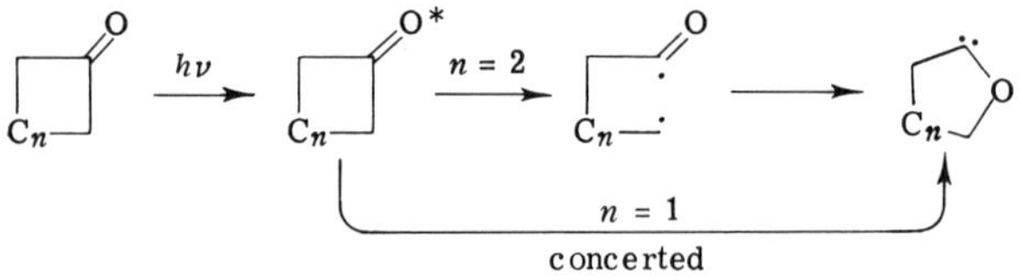

6. Both qualitative MO (154) and extended Huckel calculations (144, 155) predict the observed predisposition of cyclobutanones to rearrange, but neither distinguishes between concerted and diradical mechanisms. However, nonempirical LCAO–SCF–MO calculations address this question and arrive at the interesting prediction that, whereas hydrogen rearranges from the first triplet (of formaldehyde) by a concerted mechanism (156, 157), methyl (of acetaldehyde) rearranges via a diradical (157).

A number of rather simple alkenes have also been found to undergo retro 1,2 rearrangements. This reaction may have been discovered as early as 1962 (158), when the formation of methylcyclopropane from the mercury-sensitized photolysis of 1-butene was reported. Although the authors did not interpret this reaction in terms of a carbene, in 1967 hydrogen migration followed by insertion was suggested as a rational mechanism (159). More recently it was (160) pointed out that ethyl migra-

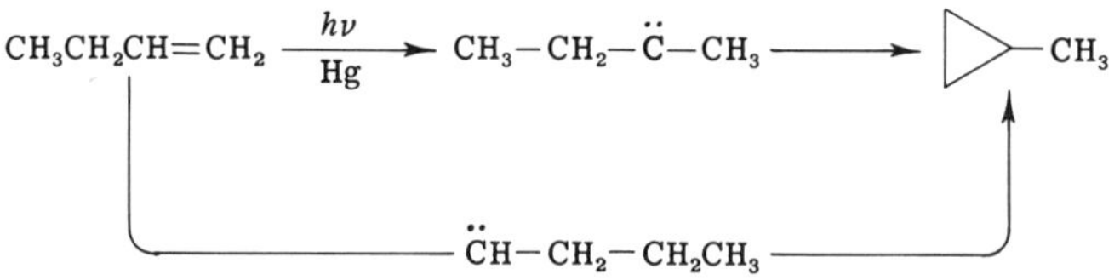

tion would give the same product and, if this reaction does indeed involve a carbene, this may well be the case (vide infra).

Following the probable discovery of retro 1,2 rearrangements of simple

3. REARRANGEMENTS OF CARBENES AND NITRENES

alkenes, the study of these reactions lay essentially dormant [however, see Tschuikov-Roux et al. (*159*)] until the mid 1970's, when no less than six different groups reported reactions that may be interpreted in these terms.

It has been found (*161–164*) that direct photolysis of simple tetrasubstituted alkenes such as **57** and **58** in aprotic solvents gives products expected of a carbene mechanism. In these reactions the photoexcited state

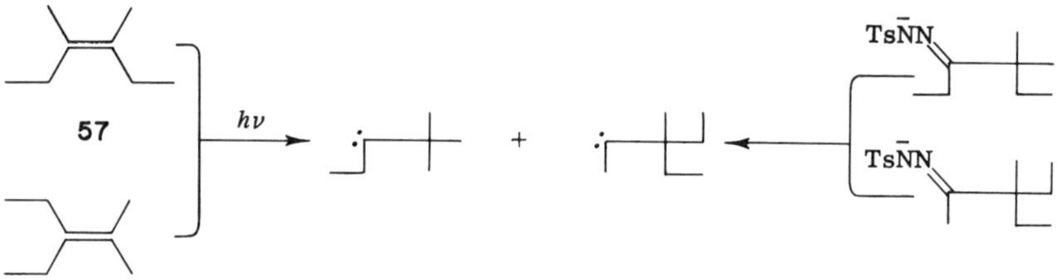

that rearranges is believed to be a π, R (3s) Rydberg state (*165*) that has been represented (following Mulliken) as **59** (*162*). A similar rearrangement has also been found for disubstituted cycloalkenes. For example, photolysis (184.9 nm) of 1,2-dideuterocycloheptene gave **60** and **61**—products and isotope distributions that would be expected from a carbene reaction (*166*). 3-Phenylcycloheptene behaves similarly when sensitized with benzene (*160*).

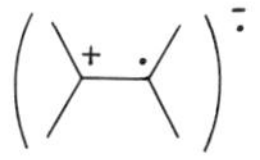

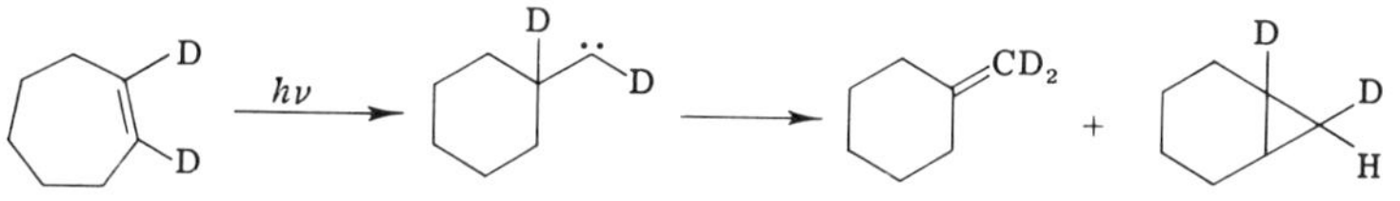

Direct photolysis of hindered diphenylethylenes such as **62** also gives products expected of a carbene intermediate. However, in these cases involvement of a Rydberg state would not be expected. It has been suggested that the reaction may go by a twisted π,π^* state (*167, 168*).

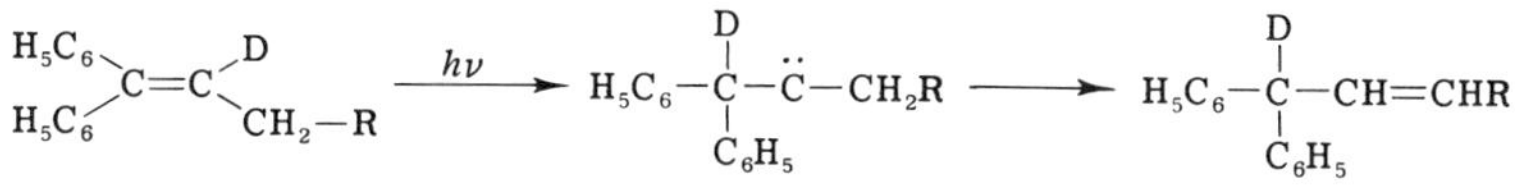

$$H_5C_6\!\!\underset{H_5C_6}{\overset{}{>}}\!\!C\!=\!C\!\!\underset{CH_2-R}{\overset{D}{<}} \xrightarrow{h\nu} H_5C_6-\underset{C_6H_5}{\overset{D}{\underset{|}{C}}}-\overset{..}{C}-CH_2R \longrightarrow H_5C_6-\underset{C_6H_5}{\overset{D}{\underset{|}{C}}}-CH=CHR$$

R = C(C$_6$H$_5$)$_2$CH$_3$

In a most impressive example of a nonphotoinduced retro 1,2 rearrangement, it was shown (*169*) that β elimination of Me₃SiBr from **63** at 110°C gives products consistent with the intermediacy of the arylcarbene **65**. To rationalize this remarkable rearrangement, the intuitively attractive suggestion was made that the twisted bridgehead double bond has dipolar character, as represented by **64**. A trapping experiment provided evidence for the bridgehead double bond.

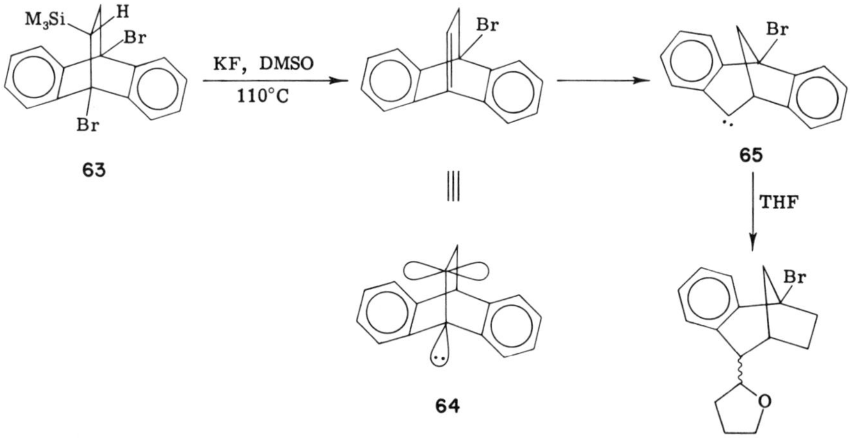

There is some evidence that unstrained alkynes undergo retro 1,2 rearrangements to vinylidenes if exposed to high temperatures. For example, flash vacuum pyrolysis (flash thermolysis) of **66** at 700°C led to ^{13}C scrambling, which can be conveniently accommodated by this mechanism (*170*). 1-Adamantylethyne was also found to undergo scrambling,

$$C_6H_5-{}^{13}C\equiv CH \xrightarrow[\text{flash vac.}]{700°C} \begin{matrix} H \\ H_5C_6 \end{matrix}\overset{13}{C}=\overset{..}{C}: \longrightarrow C_6H_5C\equiv{}^{13}CH$$

66

but higher temperatures were required in this case (*171*). A retro 1,2 rearrangement also provides a reasonable explanation for the formation of **68** from **67** (*171*).

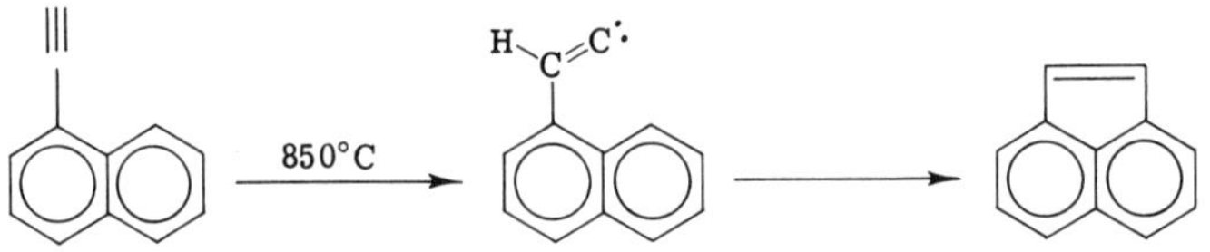

Retro 1,2 rearrangements of alkenes are induced both photochemically and thermally. Furthermore, depending on substitution, the former might

3. REARRANGEMENTS OF CARBENES AND NITRENES

occur via both excited Rydberg and π,π^* states. In view of this variety of states and conditions, any generalizations must be taken as tentative. Nonetheless, a few reaction characteristics appear to be common and may warrant mentioning.

First, most retro 1,2 rearrangements would be expected to be highly endoergic [for example, *ca.* 68 kcal/mol for **69** to **70** (*160*)] and would

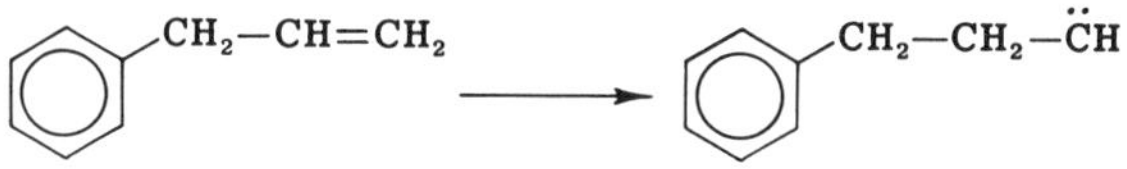

$$\text{Ph-CH}_2\text{-CH=CH}_2 \quad \longrightarrow \quad \text{Ph-CH}_2\text{-CH}_2\text{-}\ddot{\text{C}}\text{H}$$

69 → **70**

therefore be expected to require either photoreactions, highly strained double bonds, or very high temperatures. Second, unlike the retro rearrangements of ketones, there is no evidence that any of the alkene rearrangements go by stepwise mechanisms, although this has not been exhaustively tested. Third, the rearrangement of alkenes may be accelerated (in both thermal and photochemical reactions) by structural characteristics that favor a twisted alkene. However, this is not an absolute requirement. Fourth, in alkenes not substituted with aryl groups, it appears that alkyl migration is preferred to hydrogen migration. The opposite is true for the reverse reaction.

To this author's knowledge, no retro 1,2 rearrangement of imines (or related compounds) to nitrenes has been reported.

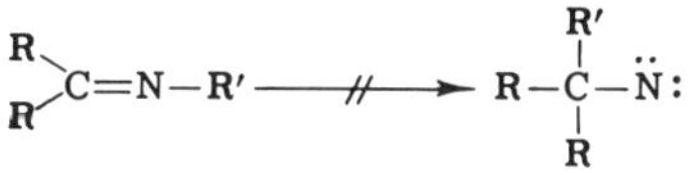

$$R_2C=N-R' \quad \xrightarrow{\quad // \quad} \quad R-\underset{R}{\underset{|}{\overset{R'}{\overset{|}{C}}}}-\ddot{N}:$$

III. TYPE I CARBENE–CARBENE AND CARBENE–NITRENE REARRANGEMENTS

A. Introduction

For convenience, carbene–carbene rearrangements are divided into two groups: type I rearrangements, in which the divalent carbons of the unrearranged and rearranged carbenes are different, and type II rearrangements, in which generation of a carbene induces molecular reorganization to give a carbene the structure of which is different but the divalent carbon of which has retained its integrity. This section is devoted to type I rearrangements.

Although type I carbene–carbene rearrangements and comparable rearrangements involving nitrenes are of relatively recent origin, they have been studied extensively and have been the subject of at least four recent reviews (2, 172–174). With the exception of trivial cases such as **71** (175),

$$R-C\equiv C-\ddot{C}-C\equiv C-R \rightleftarrows R-\ddot{C}-C\equiv C-C\equiv C-R$$

71

$$\begin{array}{c}X\\ \diagup\end{array}C-\ddot{C}\diagdown \longrightarrow \ddot{C}-C\begin{array}{c}\diagup X\\ \diagdown\end{array} \qquad \begin{array}{c}X\\ \diagup\end{array}C-\ddot{N}: \longrightarrow \ddot{C}-N\begin{array}{c}\diagup X\\ \diagdown\end{array}$$

most type I rearrangements reported to date can be represented as simple 1,2 rearrangement of double bonds. In most cases the atom that migrates is carbon. However, rearrangement of oxygen (to a carbene site) is relatively common, and there are isolated examples of reactions that may involve migration of sulfur, nitrogen, or selenium (2, 172, 176).

Type I carbene–carbene and carbene–nitrene rearrangements in which carbon is the migrating group usually involve the interconversion of arylcarbenes and arylnitrenes with their aromatic counterparts [or the corresponding allenes and keteneimines (*vide infra*)]. In general, these rearrangements appear to occur most readily in systems in which the carbene or nitrene is either substituted on a ring with six or more members or is part of a ring that has at least seven members. However, at least one example has been reported in which a carbene substituted on a five-membered carbocyclic ring (**72**) may have expanded to a six-membered ring (177). Attempts to observe similar rearrangements in heterocyclic

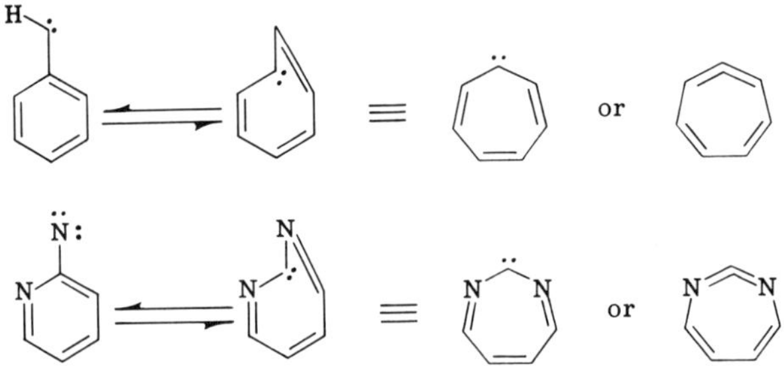

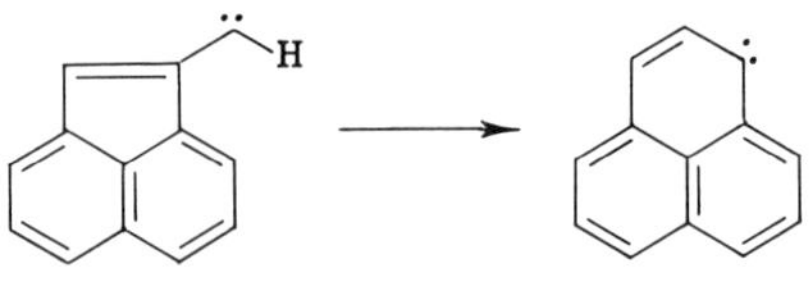

72

five-membered ring systems have failed. The heterocycle either fragments, as in **73** (178), or undergoes normal intermolecular chemistry

3. REARRANGEMENTS OF CARBENES AND NITRENES

73

(*179*). One attempt to expand a carbene substituted on a four-membered ring was reported to have failed (*20*).

The paucity of rearrangements in small rings is certainly due, at least in part, to the strain of the bicyclic intermediates (*vide infra*), such as **74** and **75**, that would be required in the smaller systems. However, this cannot be the whole story since there is evidence for the formation of cyclopropenes that would be at least as strained as those required for carbene–carbene rearrangement in small rings. For example, strong evidence for **76** has been presented (*180*), and, reactions believed to originate from **77** have been observed (*181*); recently evidence for **78** (as a transition state) has been reported (*182*).

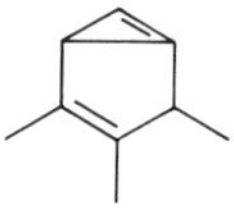

74 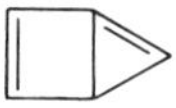   **75**

76    **77**    **78**

However, this cannot be the whole story since there is evidence for the formation of cyclopropenes that would be at least as strained as those required for carbene–carbene rearrangement in small rings. For example, strong evidence for **76** has been presented (*180*), and, reactions believed to originate from **77** have been observed (*181*); recently evidence for **78** (as a transition state) has been reported (*182*).

Finally, mention should be made of one experiment that did not work and one fascinating nitrene–carbene rearrangement that follows a unique mechanism. In the former, the possibility was explored (*183*) that **79** might undergo a carbene–carbene rearrangement via the interesting dipolar intermediate **80**, but no rearrangement was observed.

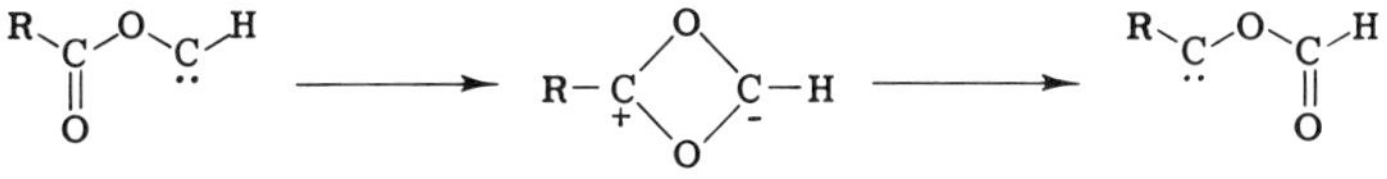

79   **80**

Regarding the latter, the mechanism of isomerization of nitrene **81** to

83 is a problem of long standing (*184*), and, although a nitrene–carbene mechanism via **82** is attractive, in reality the reaction probably goes by fragmentation to nitrogen and atomic carbon followed by recombination.

$$:\ddot{N}-C\equiv N \longrightarrow :C{<}^{N}_{N}\!\!\parallel \longrightarrow :\bar{C}-\overset{+}{N}\equiv N$$

$$\quad\;\; 81 \qquad\qquad\qquad 82 \qquad\qquad\qquad 83$$

$$N_2 + C(^3P)$$

B. Theory

Carbonylcarbenes, thiocarbonylcarbenes, and iminocarbenes probably interconvert via antiaromatic three-membered heterocycles. (Experimental evidence for these intermediates is discussed in the next section.) Most theoretical studies (*2, 172*) of this reaction have focused on the energies of carbonylcarbene and oxirene; MINDO/3, NDDO, *ab initio* SCF–MO, and thermochemical calculations place the carbene at higher energy than the oxirene, whereas extended Huckel and *ab initio* SCF–MO with extended basis set calculations (*185*) predict the opposite ordering. Thermochemical considerations place the iminocarbene about 17 kcal/mol above azirine (*172*).

$$\overset{X}{\underset{}{\parallel}}_{C-\ddot{C}} \rightleftharpoons \overset{\ddot{X}}{\underset{C=C}{\triangle}} \rightleftharpoons \overset{X}{\underset{}{\parallel}}_{\ddot{C}-C}$$

X = O, N, S, Se

MINDO/3 calculations also predict essentially zero activation energy for the ring closure of carbonylcarbenes to oxirenes (*186*). This is reasonable if the oxirene is lower than the carbene since the closure may involve little more than an intramolecular Lewis acid–base reaction between one of the nonbonded electron pairs on the carbonyl oxygen and the vacant *p* orbital of the carbene (*2*). This, of course, would be the case only if the preferred structure of the singlet carbene were one in which the nonbonded pair of the carbene were conjugated with the carbonyl, as is predicted by extended Huckel calculations (*187*).

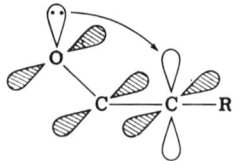

3. REARRANGEMENTS OF CARBENES AND NITRENES

In contrast to the simple mechanism for type I carbene–carbene rearrangements of carbonylcarbenes (and its relatives), the mechanism for the interconversion of arylcarbenes and arylnitrenes with their respective aromatic "carbenes" may be rather complex. A state of the art mechanism for this reaction is summarized for phenylcarbene.

In the first step of this mechanism, the carbene carbon, in a pericyclic reaction, bends out of the plane of the ring as it rotates and bonds to the ortho carbon to give the bicyclo[4.1.0]heptatriene **84**. The reaction proba-

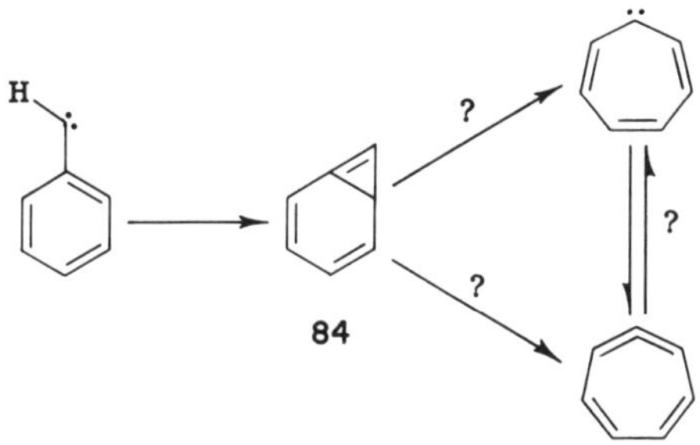

84

bly involves a singlet state. This contention is supported by MINDO/3 calculations (*188*) that predict an activation energy of 6.3 kcal/mol for rearrangement of the singlet. Rearrangement of the triplet (bypassing the bicyclic intermediate) is computed to have an activation energy of 26.5 kcal/mol.

In the first step of the rearrangement of unsymmetric arylcarbenes, such as **85** and **86**, ring closure can occur at nonequivalent ortho positions.

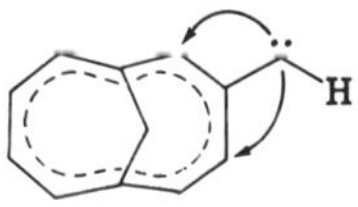

85 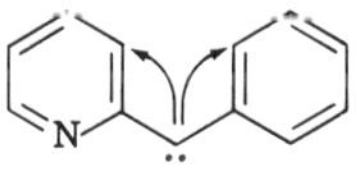 **86**

Two approaches have been used to rationalize the regiochemistry in these types of systems. In one (*173*), the bicycloheptatriene intermediate was likened to the familiar aromatic substitution intermediate **87**, with a conjugated double bond replacing the reactive site. Localization energies

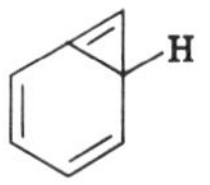

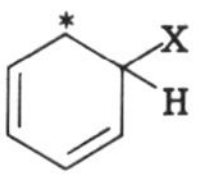

87

based on those calculated for substitution reactions were then successfully used to rationalize the observed (*vide infra*) regiochemistry in two carbocyclic systems. The second approach (*172, 174, 189, 190*), which has been particularly successful in explaining results in heterocyclic systems, such as **86**, that are otherwise difficult to interpret, takes cognizance of the fact that the first step in the rearrangement is a pericyclic reaction that can be represented by HOMO–LUMO notation (**88** and **89**). In this treatment,

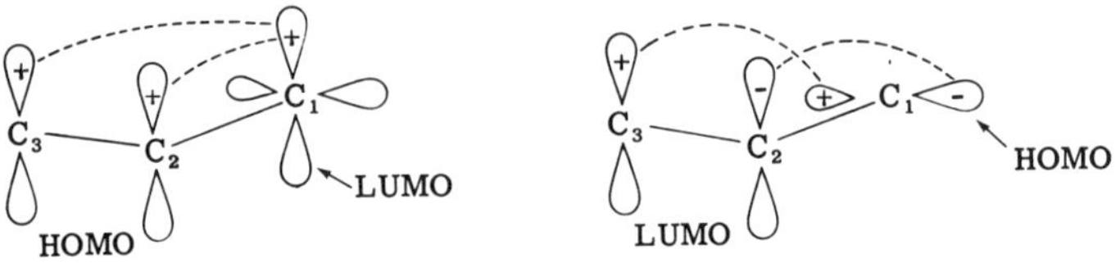

it is assumed that the role of the carbene HOMO (the nonbonded pair) is dominated by its interaction with the ring LUMO at the ipso carbon of **86** (C_2 in **89**), whereas the principal interaction of the carbene LUMO (the vacant "p" orbital) is with the HOMO at the ortho carbon (C_3 in **88**). This leads to a synergic electrophilic–nucleophilic attack by the carbene that has been formulated as **90**. This approach predicts the reaction to be

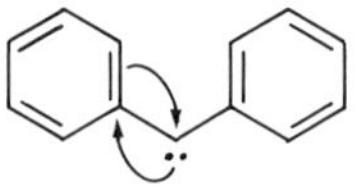

favored by a low-lying LUMO with a large coefficient at the ipso position and a high-lying HOMO with a large coefficient at the ortho position.

Another version of the first step in carbene–carbene rearrangements can be developed from an *ab initio* generalized valence bond (with configuration interaction) treatment of the closure and opening of simple vinylcarbene (*191, 192*). [For a MINDO/3 treatment of the same reaction see Pincock and Boyd (*193*).] This is shown as it would apply to closure of phenylcarbene (**91**). In this mechanism ring closure proceeds by initial

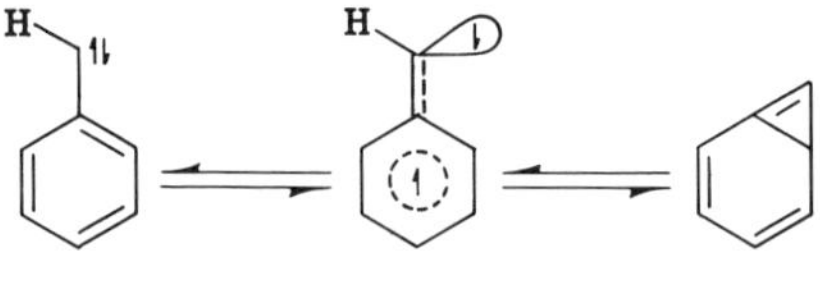

promotion of one nonbonded carbene electron into the benzyl π system (the LUMO) followed by ring closure. The primary drawback of this mechanism is that it is not clear how it could rationalize the regiochemistry of the reactions mentioned above.

Although there is little doubt that the essence of the first step in the rearrangement of arylcarbenes is closure to a bicycloheptatriene, the detailed mechanism of the second step has not yet been resolved. Until recently, it has been assumed to be simple ring opening of the cyclopropene to give cycloheptatrienylidene (**92**). However, both theoretical

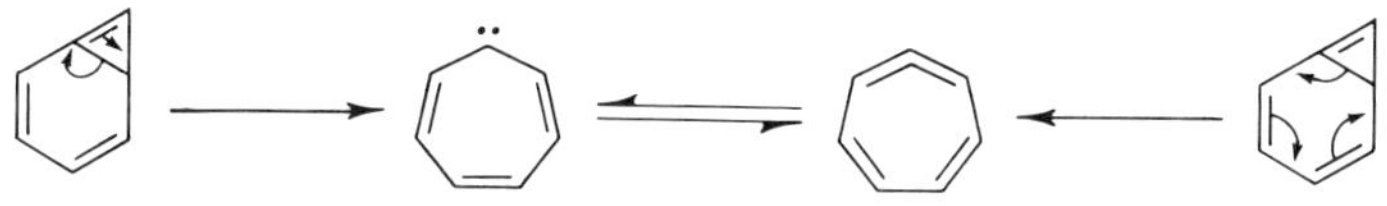

$$\text{84} \longrightarrow \text{92} \rightleftharpoons \text{93} \longleftarrow \text{84}$$

(*188, 194*) and experimental (Section III,C) evidence has mounted for the twisted allene structure **93** as a species that may be of lower energy than the planar carbene. Since bicyclo[4.1.0]heptatriene (**84**) is simply the norcaradiene form of cycloheptatetraene, this suggests the interesting possibility that the second step of the rearrangement is, in reality, direct ring opening by a norcaradiene rearrangement, bypassing the carbene. Although experimental arguments against this have appeared (Section III,C), they are not compelling, and a recent MINDO/3 search (*188, 195*) of the phenylcarbene–cycloheptatrienylidene–cycloheptatetraene energy surface revealed a smooth conversion from bicyclo[4.1.0]heptatriene to cycloheptatetraene with no energy minimum corresponding to cycloheptatrienylidene. This is a problem that clearly deserves further attention.

In the gas phase, pyridylcarbene rearranges to phenylnitrene. In the matrix, phenylnitrene rearranges to the azacycloheptatetraene **99**, and, in solution, phenylnitrene gives an intermediate (probably **96, 97,** or **99**) that reacts with amines to give azepines. The state of the art mechanism (experimental evidence is discussed in Section III,C) for these interconversions is given in Scheme 2.

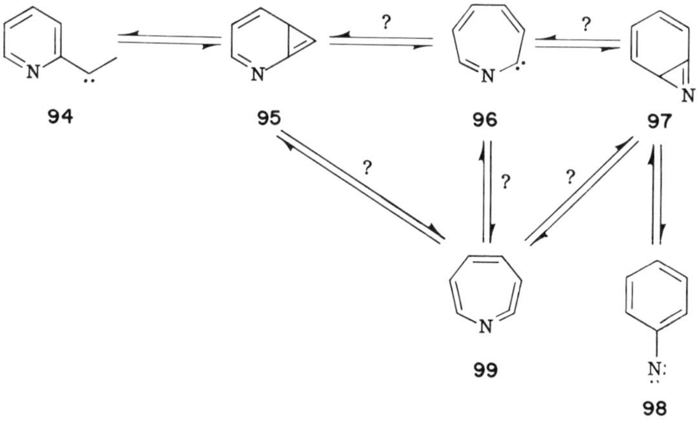

Scheme 2

Theoretical calculations related to these interconversions are sparse. MO calculations (*196*) predict **96** and **99** to be of higher energy than **97**, whereas thermochemical calculations (*172*) predict the opposite, that is, **97** to be of higher energy than **96**. Thermochemical and MO calculations predict **96** to be of higher energy than either **94** or **98** (*172*), but no mention was made of **99**. CNDO/2 and EH calculations (*198*) were used to follow the conversion of **94** to **96**, and, although both methods predict that the nitrogen atom moves out of the molecular plane, which would be required to give **95**, they do not predict the azirine to be a discrete intermediate. These calculations did not track the possible direct conversion of **95** to **99**.

Finally, the relative energies (based on MO calculations) of a number of nitrenes and carbenes with two and three nitrogens have been reported(*172*).

C. Carbene–Carbene Rearrangements in Solution

1. Interconversion of carbonylcarbenes, iminocarbenes, thiocarbonylcarbenes, and selenocarbonylcarbenes

In Section III,B it was pointed out that, although there is no agreement as to their relative energies, there appears to be agreement among theoreticians that the interconversion of carbonylcarbenes occurs via an oxirene intermediate. There is experimental evidence for this. Although attempts to observe directly the oxirene intermediate have failed—even at a temperature as low as 8°K in an argon matrix (*199, 200*)—indirect evidence is compelling. For instance, by tracking with tagged carbon, it was found that the products from many photochemical Wolff rearrangements require intermediates or a transition state (processes actually in competition with the Wolff rearrangement itself) with the symmetry of an

oxirene (*2, 172, 201, 202*). The nature of the products from the thermolysis of **100** leads one to the same conclusion (*203, 204*). In one instance, with **102** as starting material, the oxirene reportedly was trapped (*201*), an observation that argues for the heterocycle as an intermediate. The idea of an oxirene as an intermediate in these reactions is reinforced by the fact that attempts to generate the heterocycle from other sources, such as by the peracid oxidation of alkynes (*2*) and by retro Diels–Alder reactions

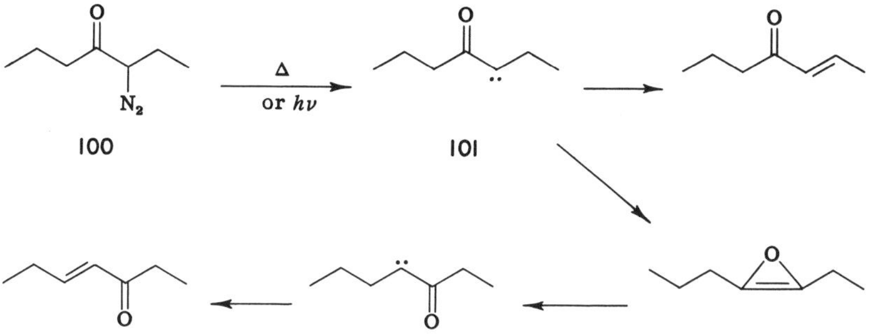

3. REARRANGEMENTS OF CARBENES AND NITRENES

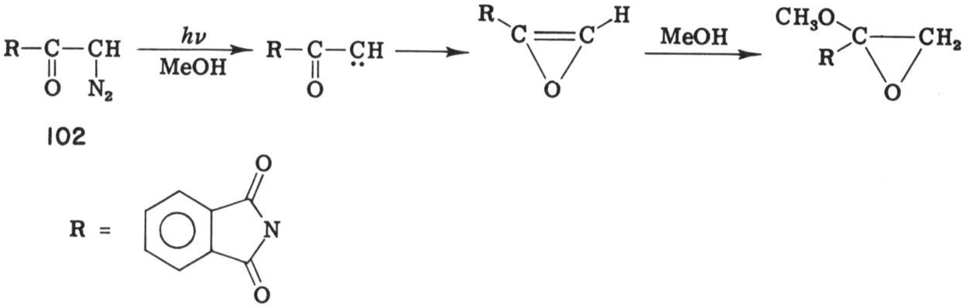

(205, 206), invariably lead to carbonylcarbene products that are usually Wolff rearrangement products.

To date, perhaps the most conspicuous characteristic of carbene–carbene rearrangements of carbonylcarbenes is their near absence when the carbene is generated thermally. In fact, with the single exception of **101**, they have been observed only in the photodecompositions of diazocarbonyl compounds. If, as the evidence seems to indicate (23), thermally induced Wolff rearrangements are not concerted, in most instances the intermediate that undergoes carbene–carbene rearrangement must be different from the carbene that is generated from thermal decomposition of diazocarbonyl compounds. This has led to the suggestion (207) that carbene–carbene rearrangements occur from a vibrationally excited ground state of the singlet carbene, a state that normally is not available in the thermal reaction. This suggestion finds support in the fact that photolysis in the gas phase appears to favor the rearrangement (208).

Carbene–carbene rearrangements of carbonylcarbenes also appear to be retarded by ring strain (172, 209–211) and by phenyl substitution on the rearranging carbene (212).

Just as carbonylcarbenes may undergo carbene–carbene rearrangements via oxirenes, imino-, thiocarbonyl-, and selenocarbonylcarbenes might be expected to rearrange through their respective heterocycles. Although such rearrangements have not been observed in solution, evidence has appeared for high-temperature gas-phase rearrangements of iminocarbenes such as **103** and **104** (*213–217*) and the even more remark-

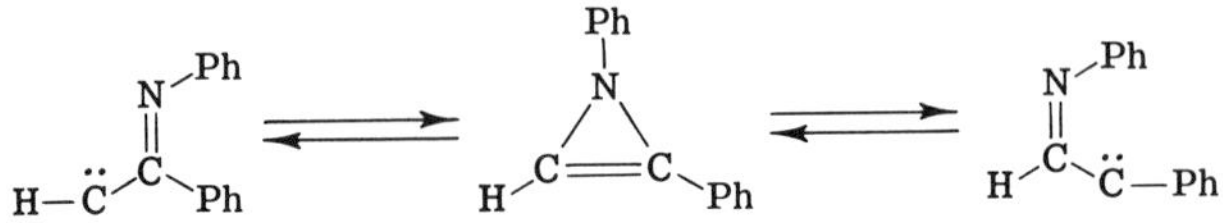

able automerization of **105** (*218*). In the case of the sulfur and selenium analogues, evidence for **106** has been obtained (*219, 220*), and **107** and **108** have been observed directly in an argon matrix at 8°K (*176, 221*).

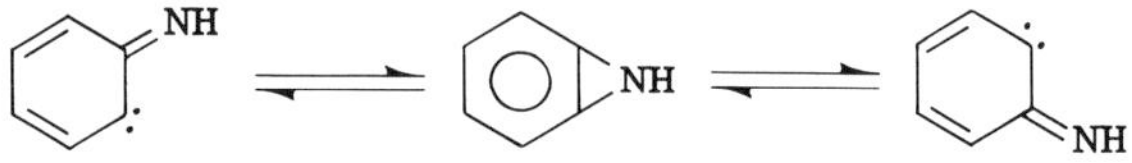

2. Interconversion of aromatic carbenes, arylcarbenes, and arylnitrenes

Despite its low reactivity with solvent, conditions have not been found in which the rearrangement of cycloheptatrienylidene to phenylcarbene competes successfully with dimerization (*222*). However, relatively modest perturbations* that retard dimerization permit the rearrangement to occur (*222*). Thus, 2,7-diphenylcycloheptatrienylidene (**109**) (from the tosylhydrazone salt) rearranges when generated photolytically at 10°C or thermally at 148°C. Even 2,7-dimethylcycloheptatrienylidene gives *o*-

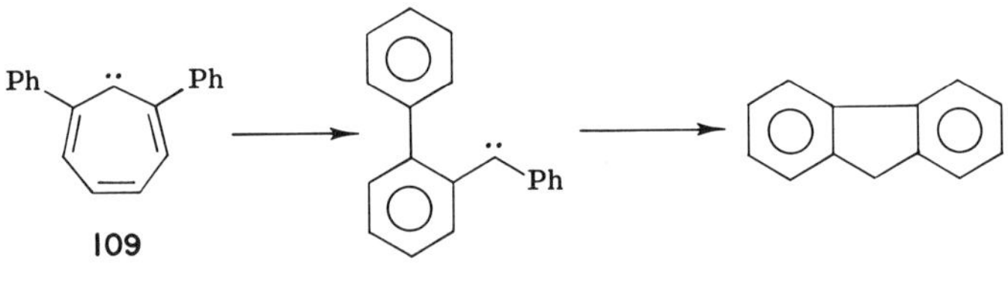

methylstyrene at 150°C, although it gives only tetramethylheptafulvalene at lower temperature. Benzannelation of cycloheptatrienylidene also ac-

---

* Ferrocenylcycloheptatrienylidene (generated from deprotonation of the carbonium ion) has been reported to rearrange under mild conditions (*223*), but carbene involvement in this unusual reaction remains to be confirmed.

3. REARRANGEMENTS OF CARBENES AND NITRENES

celerates the rearrangement (*173*). For example, thermolysis or photolysis, at about −30°C, of the benzotropone tosylhydrazone salt **110** leads to smooth conversion to the corresponding naphthylcarbene. This drama-

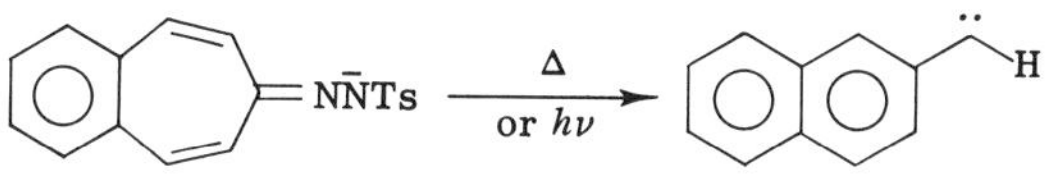

[structure] =N̄NTs →(Δ or hν)→ [naphthyl-CH]
110

tic rate increase over that of the parent has been attributed to the effects that benzannelation would be expected to have on both K and K_2 in Eqs. (1) and (2); not only should it shift the initial equilibrium toward the carbene, but less aromaticity should be lost in the closure of **115** than in that of **112**.

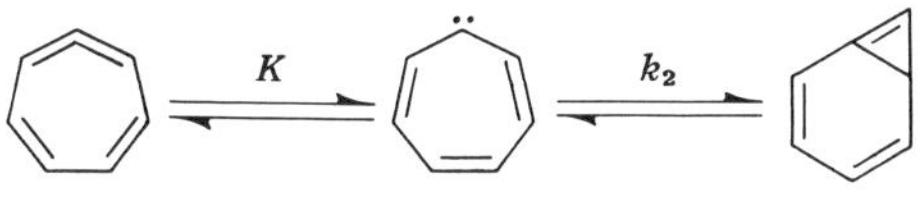

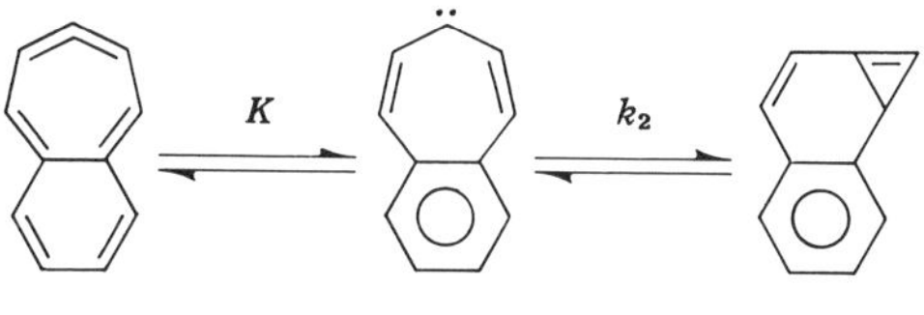

$$\text{111} \underset{}{\overset{K}{\rightleftharpoons}} \text{112} \underset{}{\overset{k_2}{\rightleftharpoons}} \text{113} \tag{1}$$

$$\text{114} \underset{}{\overset{K}{\rightleftharpoons}} \text{115} \underset{}{\overset{k_2}{\rightleftharpoons}} \text{116} \tag{2}$$

However, in Section III,B it was pointed out that MO calculations on the opening of **113** predict conversion directly to the allene, bypassing the carbene (*188*). Are the benzannelation results inconsistent with direct conversion of **111** and **114** to their cyclopropenes? In fact, they are not. Consider Eqs. (3) and (4). If the allenes are of significantly lower energy

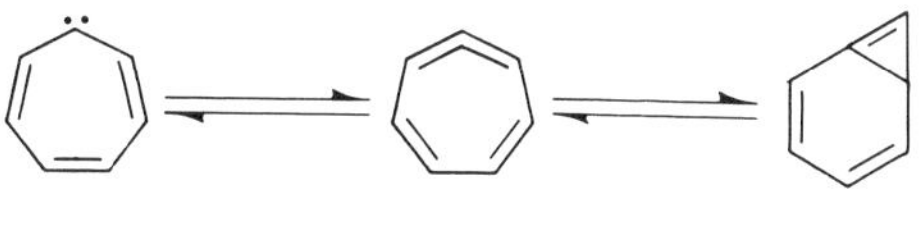

$$\text{112} \rightleftharpoons \text{111} \rightleftharpoons \text{113} \tag{3}$$

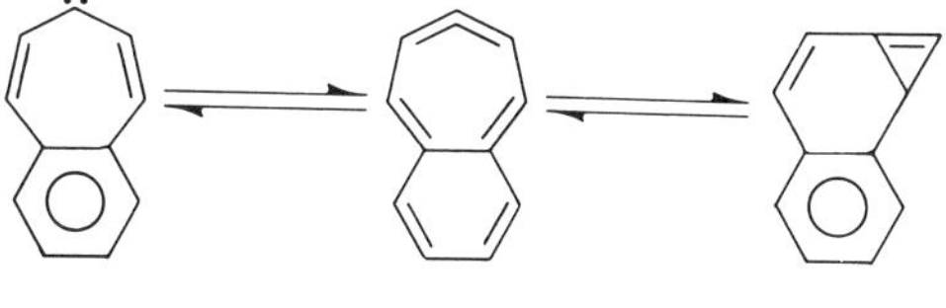

$$\text{115} \rightleftharpoons \text{114} \rightleftharpoons \text{116} \tag{4}$$

than their carbenes, shifting the equilibrium toward the carbene (by benzannelation) should have little effect on the rate of the rearrangement. On the other hand, benzannelation should have a large effect on the rate of ring closure. As a result, a mechanism involving direct closure of the allene predicts the observed results. If, on the other hand, the carbenes are of lower energy than their allene isomers, benzannelation would shift the first equilibrium toward the carbene but would increase the rate of closure to the bicyclo[4.1.0]heptatrienes. Overall, less aromaticity would be lost in closing **115** than **112** (*173*) and, again, acceleration would be predicted. Thus, the benzannelation results do not enable one to distinguish between closure of a carbene and closure of an allene.

It may be worth noting, however, that at least one experiment indirectly favors rearrangement of a carbene. The cycloheptatrienylidene derivative **118** undergoes closure under the same conditions as in the benzannelated systems (*224*). In this case, an allene form would require charge separation, as depicted in **117**.

Unlike cycloheptatrienylidene, which shows unusual intermolecular chemistry [behavior like a nucleophilic species (*225–227*)] that could conceivably be explained by the intermediacy of a single conjugated allene (*228, 229*), the little intermolecular chemistry known of at least two of the annelated cycloheptatrienylidenes is more like that of typical carbenes. For example, **115** adds to cyclohexene to give the normal carbene adduct (*230*), and **118** adds to benzene to give a good yield of the C—H insertion product (*221*). Such observations argue against a single allene intermediate in these systems.

In summary, the information that is available does not allow a clean distinction between the different mechanisms to be made and, in fact, there is no *a priori* reason to assume that the mechanism of rearrangement may not be different in different systems or under different conditions. Nonetheless, all things considered, rearrangement of a carbene remains the simplest mechanism, and this intermediate will therefore be presumed, tentatively, in most cases in the following discussion.

Regardless of the detailed mechanism of ring closure, the facile rearrangement of benzannelated cycloheptatrienylidenes to arylcarbenes in solution has been used to obtain evidence both for the multiplicity of the

3. REARRANGEMENTS OF CARBENES AND NITRENES

rearranged carbene and for the intermediacy of bicyclo[4.1.0]heptatrienes.

By the effect of triplet traps on the stereochemistry of addition to cis-2-butene, it was shown that benzocycloheptatrienylidene **115** rearranges to singlet β-naphthylcarbene **119** (231). This information, combined with the fact that triplet vinylcarbenes do not close to cyclopropenes (232) leaves little doubt that carbene–carbene rearrangements normally involve only singlet states.

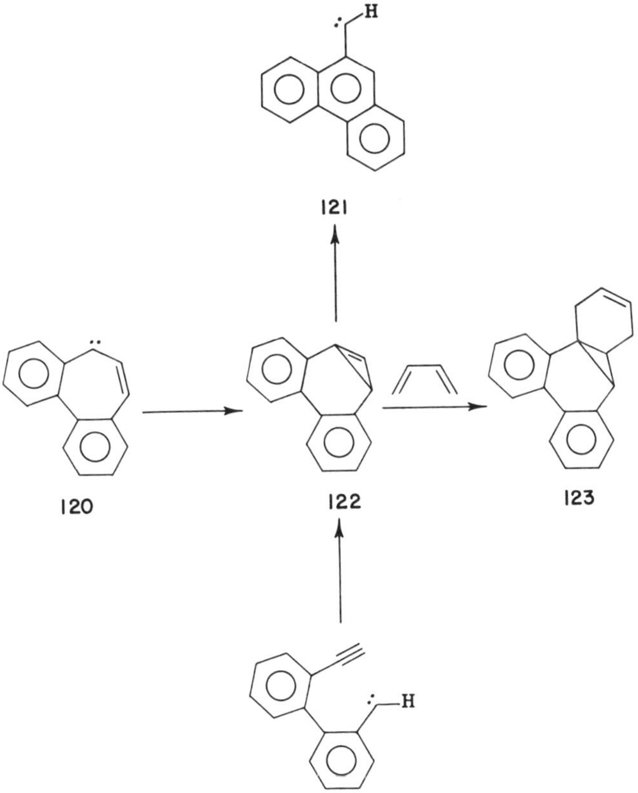

It has been found that benzannelated cycloheptatrienylidenes **115** and **120**, when generated in the presence of carbene traps such as cyclohexene and benzene, give products expected of the rearranged arylcarbene. However, when generated in the presence of cyclopropene traps such as butadiene, cyclopentadiene, and furan, typical cyclopropene adducts such as **123** are obtained (*233*). The conclusion that cyclopropenes are the intermediates that give the Diels–Alder adducts finds support in the fact that **123** was also formed from **124** (*234*). Evidence against the view that these adducts are secondary rearrangement products has been reported (*233*).

Further evidence for cyclopropenes as intermediates in the interconversion of arylcarbenes and aromatic carbenes is found in the nature of the products from dehydrohalogenation of *gem*-dichlorocyclopropanes (*235, 236*). For example, reaction of **125** with base gave a 1:2 mixture of **129** and **132**, products that are reasonably accounted for by the outlined mechanism. Formation of products from both α- and β-naphthylcarbene is particularly important because opening of **126** to **128** as well as **127** is the

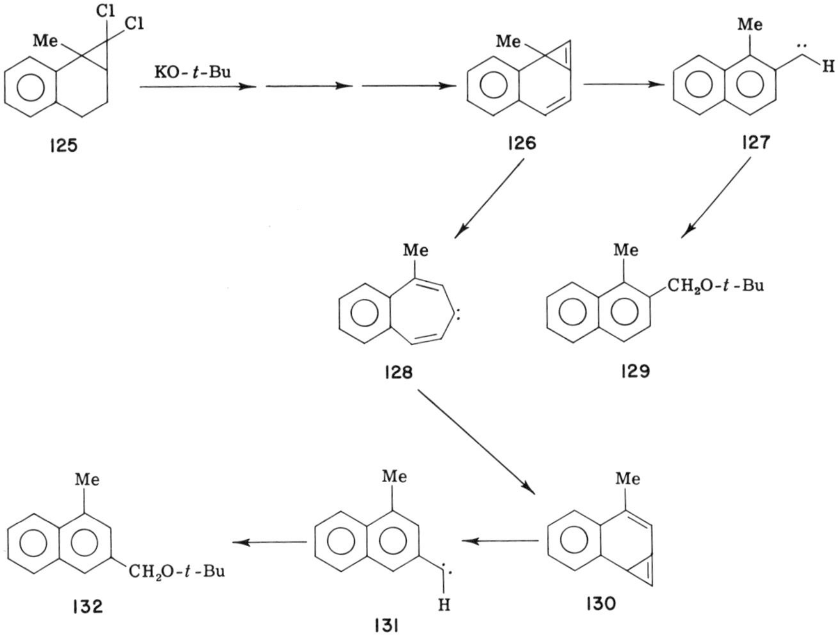

first clear evidence that, under mild conditions, bicyclo[4.1.0]heptatrienes can open to seven-membered rings as well as arylcarbenes.

The only arylcarbenes that have been reported to rearrange to aromatic carbenes under mild conditions in solution are the methanoannulenylcarbenes **133** and **134** (*237–239*). Both rearrange regiospecifically and, as discussed in Section III,B, localization energy arguments satisfactorily

3. REARRANGEMENTS OF CARBENES AND NITRENES

explain the regiochemistry. The same arguments also provide a reasonable explanation for the facility of the rearrangements. Different types of dimeric products from **134** ⇌ **135** and **136** ⇌ **137** were originally (*237*) suggested to be evidence for different kinds of intermediates (carbene and

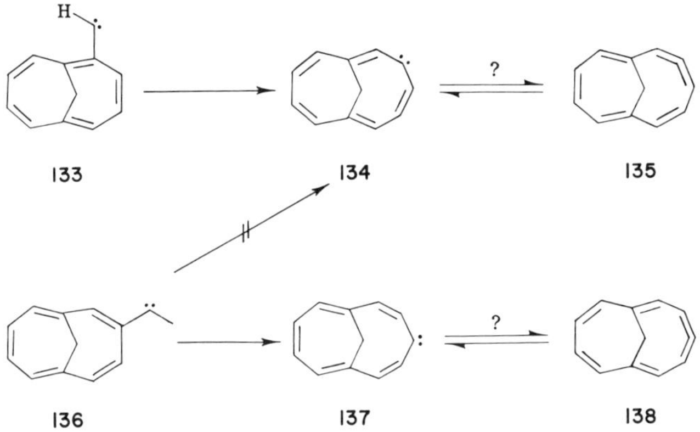

allene). However, this argument has recently been shown to be invalid (*240*).

In solution, intramolecular bond reorganization of phenylcarbene is not competitive with intermolecular reactions. In contrast, in the presence of nucleophilic traps (especially amines) the chemistry of phenylnitrene is dominated by rearrangement.

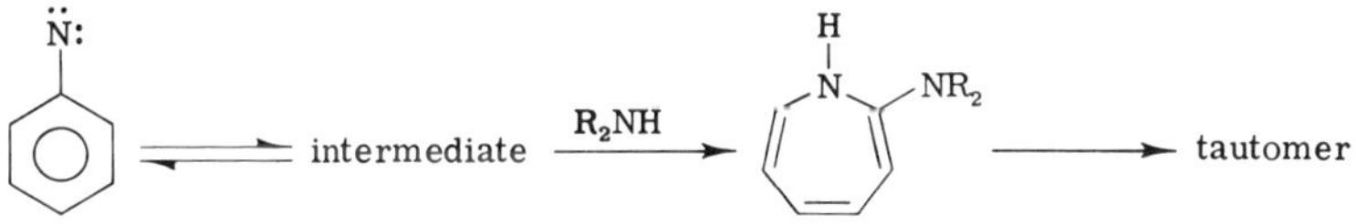

This reaction (*241*) has been reviewed (*242, 243*). These reviews can be summarized and the present position presented as follows:
1. The reaction was discovered by Wolff in 1912.
2. The correct structure of the product was not assigned until 1958 (*244*).
3. Nucleophiles that trap the intermediate include ammonia, primary and secondary amines, H_2S, mercaptans (*245*), triethyl phosphite, and nitrosobenzene.
4. Nitrene sources that give trappable intermediates include azides [by thermolysis or photolysis; the latter is sensitive to the wavelength (*246*)], nitro- (*247*) and nitrosobenzenes (deoxygenated with tervalent phosphorus), *N*-phenyloxaziridines (photolysis), and a novel deoxysilylation (*248*).

5. The rearrangement probably occurs from a singlet nitrene; azepine formation does not occur with triplet sensitizers and is unaffected by triplet quenchers.
6. An azirine (**140**) was first postulated by Michaelis and Schaför as the trappable intermediate in 1913 (*249*).
7. Bond reorganization (from ^{14}C labeling) is inconsistent with the 1-H azirene **143**.

Since the publishing of the two reviews, most of the work on azepine formation from phenylnitrene has focused on the mechanism of the reaction, with primary emphasis on identifying the intermediate that is trapped. As pointed out by Sundberg (*197*) at least four different contenders must be considered: **139, 140, 141** [first suggested by Smith (*242*)],

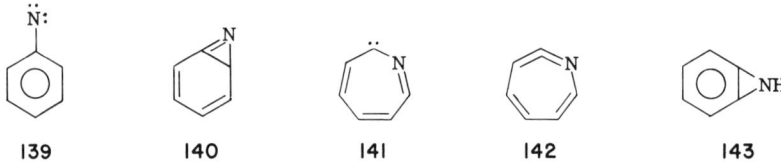

and **142**. These are heterocyclic analogues of the four intermediates believed to be involved in carbene–carbene rearrangements (*vide supra*).

Although at the time of this writing the structure of this intermediate remains unknown, to a large extent as a result of a series of elegant competitive trapping experiments and flash photolysis studies (*197, 250*) a number of important properties can be assigned to it. (a) It must be able to revert to the initially formed intermediate [the nitrene (*vide infra*) (*250–253*)]; (b) it has a lifetime in the millisecond range; and (c) it has no strong absorption above 300 nm.

From these results, it is argued (*197*) that the nitrene is an unlikely choice since it would be expected to be much too short-lived. In addition, on the basis of MO calculations and model systems it should also show strong absorption above 300 nm.

Of the three remaining intermediates there appears to be little question that the azirine **140** is initially formed, and from recent matrix isolation studies (*254*) the keteneimine **142** has probably the lowest energy of the intermediates. However, the question that is still unanswered is, Which intermediate is trapped by the nucleophile? Sundberg points out that the azirine would explain the properties listed above. Furthermore, MO calculations (which erroneously place the azirine at lower energy than either **141** or **142**) predict a long-wavelength absorption (which is not observed) for **141**. For these reasons, he favors the azirine as the intermediate that is trapped.

3. REARRANGEMENTS OF CARBENES AND NITRENES

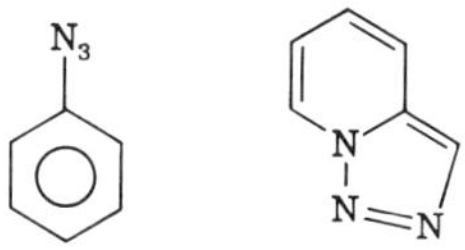

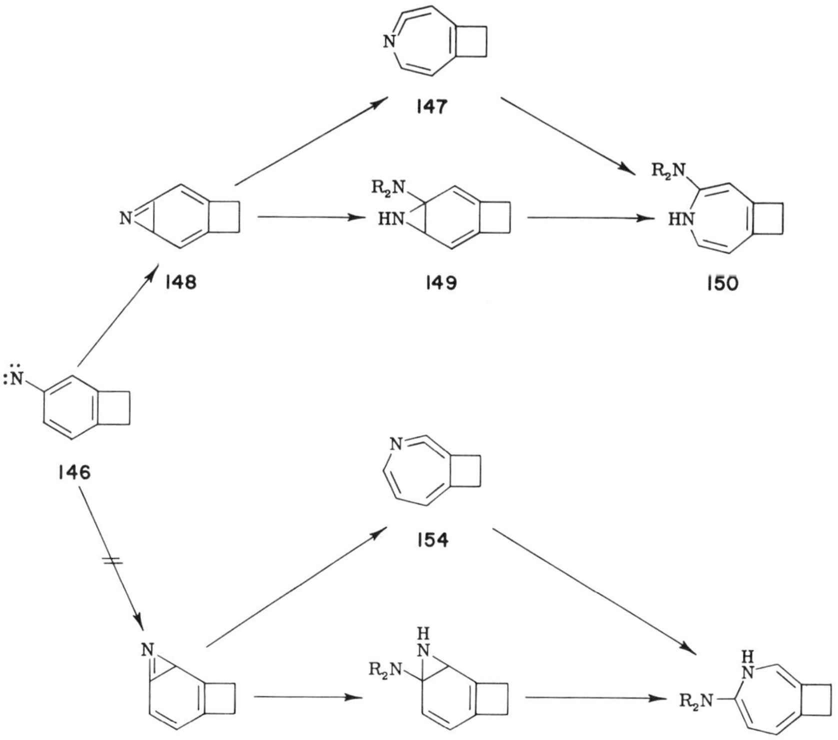

On the other hand, on the basis of the IR spectrum of the species formed from photolysis of either **144** or **145** (argon matrix at 8°K), Chapman (254) concludes that "the intermediate which reacts with nucleophiles is 1-aza-1,2,4,6-cycloheptatetraene [**142**]." Although this is not necessarily true (an earlier intermediate could be that which is trapped), the regiochemistry of closure of arylnitrenes such as **146** also points in this direction (255). Thus, if the azirine is the reactive intermediate, it is not clear why formation or trapping of **148** should be favored over **151**. If, however, the intermediate that is trapped is the keteneimine, the regiochemistry could be readily explained, since models show **147** to be noticeably less strained than **154**. Inhibited opening (to keteneimine) of an initially formed azirine also conveniently explains why the naphthylnitrenes do not undergo this reaction (*244, 256*).

Thus, in summary, the nature of the trappable intermediate remains an open question. If it is not the azirine, the most likely candidate is the ketenimine, which, by analogy with the carbocyclic system, is probably in equilibrium with carbene **141**. However, if this is the case, its properties must be quite different in at least two respects from those of its carbocyclic relative. First, no evidence has been reported of carbene dimers such as **155**, the product that dominates the chemistry of cyclohep-

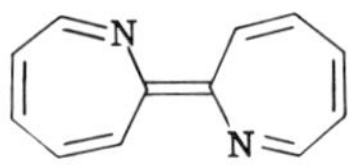

155

tatrienylidene – cycloheptatetraene. Second, the seven-membered ring intermediate must be in rapid equilibrium with its arylnitrene counterpart, even under very mild conditions; in contrast, cycloheptatrienylidene–cycloheptatetraene does not rearrange to phenylcarbene below 300°–350°C.

A discussion of the rearrangement of arylcarbenes and arylnitrenes under mild conditions would not be complete without mention of the rearrangement of **156**, a reaction that is at least 78% complete when the nitrene is generated at 180°C (in benzene) (*218*). This reaction is remark-

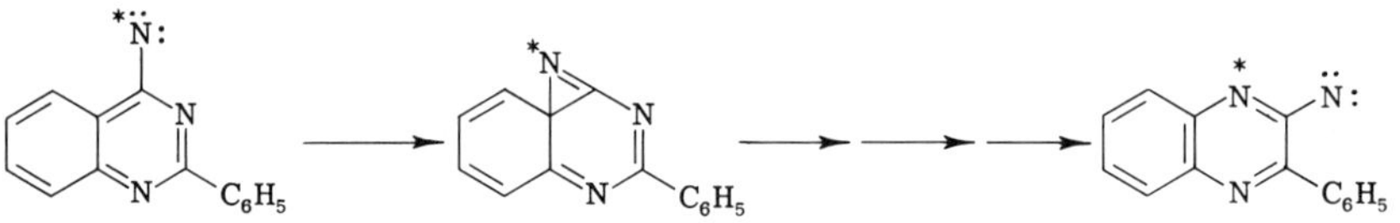

156   **157**   **158**

able because a mechanism analogous to those discussed above would require rearrangement to give the high-energy intermediate **157**. The Wolff type of rearrangement proposed to explain this apparent anomaly is reasonable (*172*).

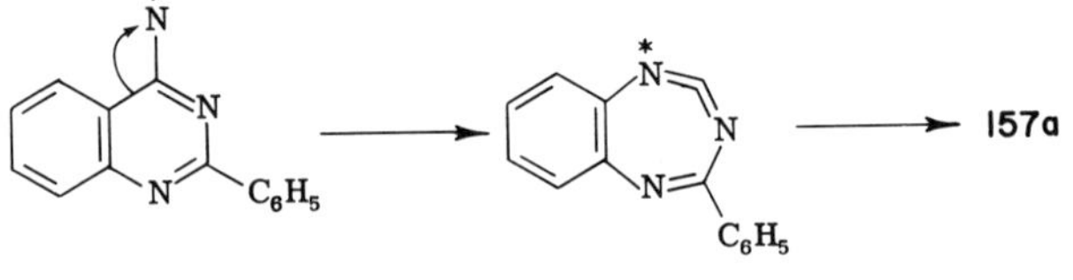

⟶ **157a**

3. REARRANGEMENTS OF CARBENES AND NITRENES

D. Carbene-Carbene, Carbene-Nitrene, and Nitrene-Nitrene Rearrangements in the Gas Phase

1. Introduction

In the previous section it was noted that the interconversions of arylcarbenes and aromatic carbenes (and possibly nitrenes) in solution at modest temperatures are relatively simple reactions involving few intermediates. In addition, they are limited to systems that meet rather severe structural requirements. In contrast, in the gas phase at high temperatures (up to 1000°C) isomerizations and automerizations of arylcarbenes, aromatic carbenes, arylnitrenes, and aromatic nitrenes occur in even the most unlikely systems, and both multiple rearrangements and complex competitive reorganizations are commonplace. Although of relatively recent origin, work in this area has been extensive, and this chemistry has been the subject of a number of recent comprehensive reviews (2, 172, 174). For this reason, no attempt will be made in this chapter to present either a detailed history or a complete coverage of these reorganizations. Instead, in the following paragraphs we will present a state of the art discussion of the three fundamental (and often similar) energy surfaces that describe C_7H_6, C_6H_5N, and $C_4H_4N_2$. Reports of carbene-carbene rearrangements in the gas phase that are neither included in previous reviews nor discussed below have been published recently (257–262).

2. C_7H_6 energy surface

The current version of the C_7H_6 energy surface is summarized in Scheme 3. In this scheme, precursors to most intermediates and all dimeric products have been omitted. It will be noted the the scheme includes no less than five different carbenes as well as a number of other intermediates that probably have heats of formation in the same general range as that of the carbenes. We will attempt to dissect this scheme, placing primary emphasis on evidence for the carbenes and the intermediates involved in their interconversions.

Beginning with phenylcarbene, perhaps the most direct evidence that it rearranges to the seven-membered ring is the fact that, when generated in the gas phase between 250° and 600°C (263–265), it gives the dimer 171.

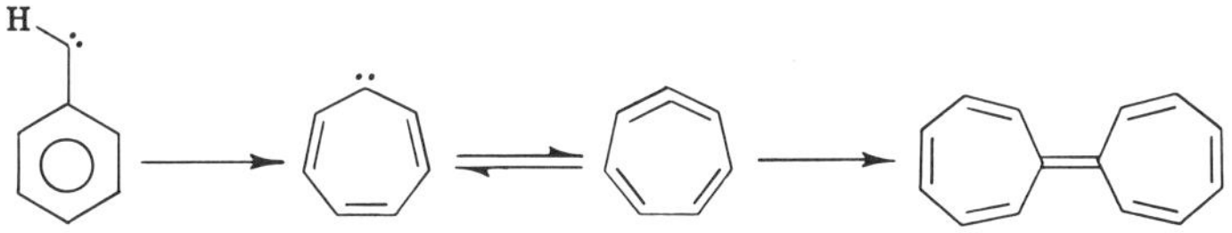

|171|

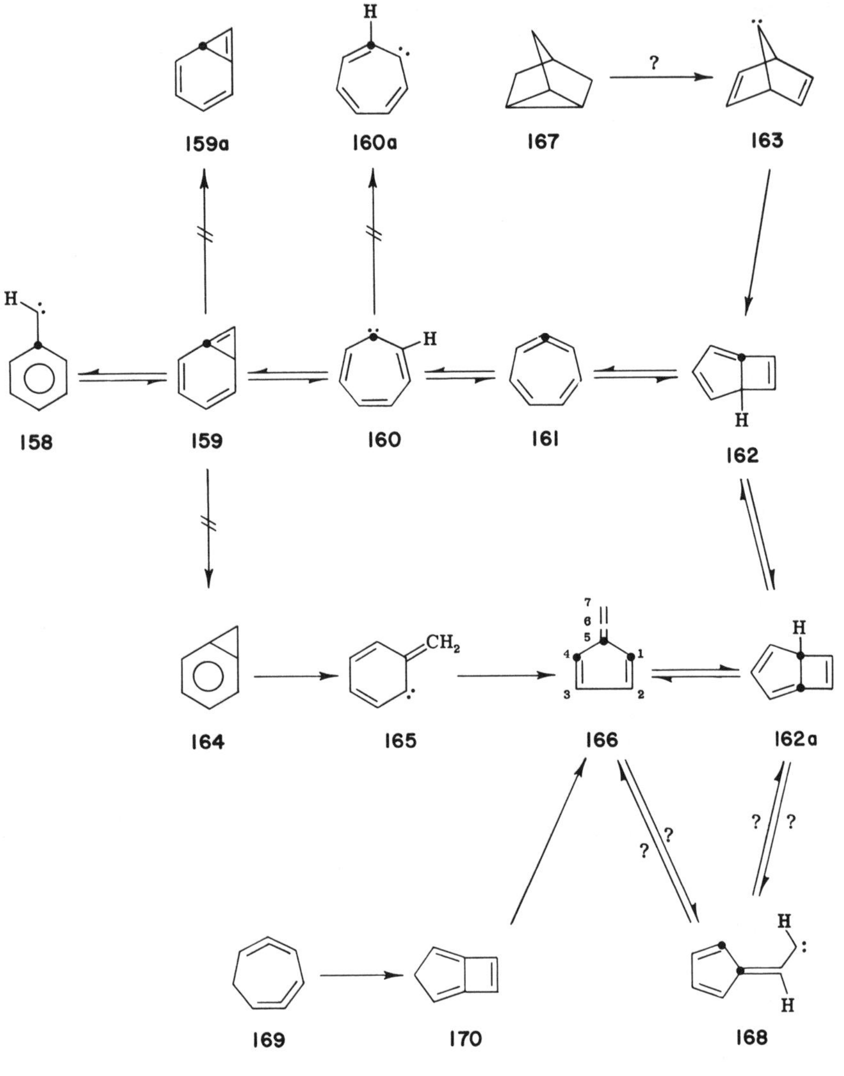

Scheme 3

This product is characteristic of cycloheptatrienylidene–cycloheptatetraene when generated in solution.

Experimental evidence for the cyclopropene intermediate in the solution-phase rearrangement of benzannelated aromatic carbenes was discussed in Section III,C,2. This, along with MINDO/3 (*188*) and

3. REARRANGEMENTS OF CARBENES AND NITRENES

thermochemical calculations (172) make it likely that bicyclo-[4.1.0]]heptatriene (**159**) is an intermediate in the rearrangement of phenylcarbene to cycloheptatrienylidene. However, to this author's knowledge, there is no direct evidence for this intermediate in the gas phase.

Evidence for the reversibility of the phenylcarbene–cycloheptatrienylidene rearrangement is found in the formation of benzocyclobutene and styrene from hot-tube pyrolysis of methyltropone tosylhydrazone salts such as **172**(266). Rearrangement of p-methylphenylcarbene (**173**) to

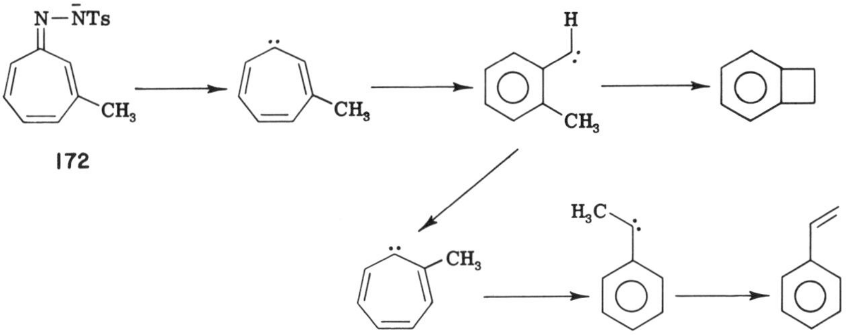

styrene and benzocyclobutene has also been cited as evidence for the reversibility of the phenylcarbene rearrangement. Indeed, such reactions are cleanly rationalized by a series of consecutive interconversions of methylphenylcarbenes and methylcycloheptatrienylidenes(267). This also satisfactorily rationalizes the distribution of the tag in **174** and **175**(268).

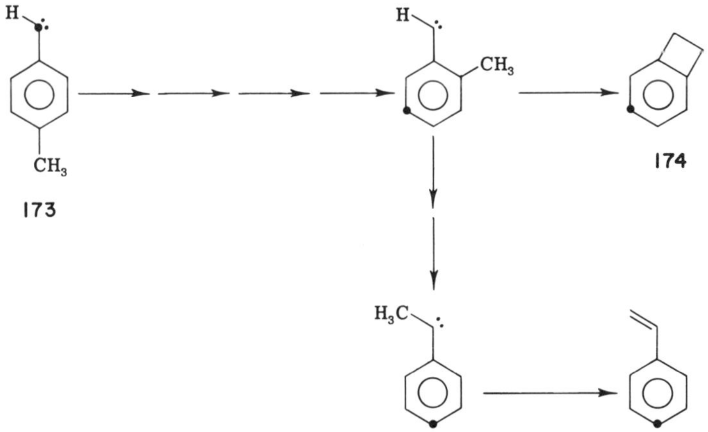

As is the case in solution (Section III,C,2), at high temperature in the gas phase there is no experimental evidence distinguishing between direct opening of **159** to **161** and initial carbene formation followed by isomerization.

Equilibration of **161** and **162** involves electrocyclic opening and closing of a butadiene. This reaction was originally suggested (*2*) as an alternative to the unlikely rearrangement of **159** and **160** to **159a** and **160a**. These rearrangements had been postulated to rationalize complete scrambling of tagged carbon in fulveneallene (*166*) when phenylcarbene is generated at 770°C (*269, 270*). In the scrambling mechanism involving the bicyclo-[3.2.0]heptatriene as an intermediate, 1,5-hydrogen shifts in **162** were postulated as the key steps.* This suggestion has recently found strong support in further studies of the conversion of tagged phenylcarbene to fulveneallene (**166**) (*271*). At 900°C, evenly distributed label was found in fulveneallene. However, at 590°C the label decreases in the order C-5 > C-1 and C-4 > C-2 and C-3 > C-6 > C-7. This is the order predicted by the bicyclo[3.2.0]heptatriene mechanism if total equilibration is not realized.

Evidence that **162** can open to **161** is found in the properties of the intermediate generated by hot-tube pyrolysis of **176** (*111*). Above 200°C

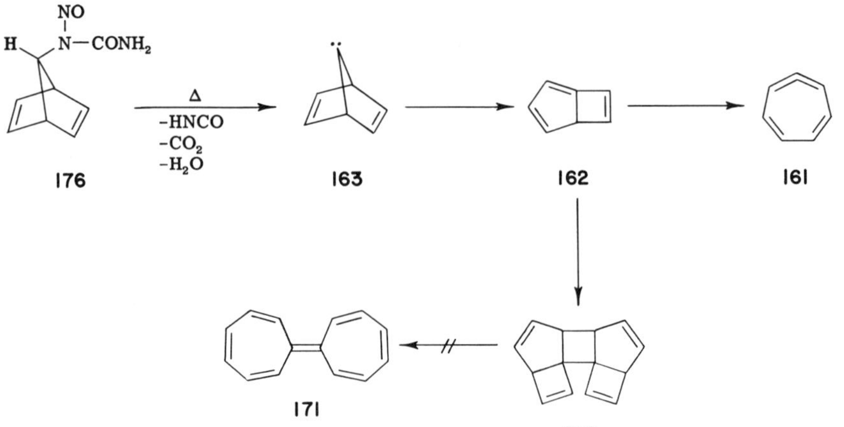

176 loses the elements of HNCO, H_2O, and N_2 to give an intermediate that, by analogy (*272*), is believed to be 7-norbornadienylidene.† At 200°C, the principal hydrocarbons from this intermediate are a mixture of dimers **177**. At higher temperatures, heptafulvene (**171**) appears, and the

---

* Only scrambling to C—1, C—4, and C—5 is shown in **166**. Complete scrambling requires multiple cycles of opening bicyclo[3.2.0]heptatriene to cycloheptatetraene, reclosure, and hydrogen migration.

† This intermediate may also be formed from **167**, but no C_7H_6 or $C_{14}H_{12}$ products were identified when the salt of quadracyclanone tosylhydrazone was pyrolyzed (*273*).

yield of dimers diminishes, disappearing above 300°C. The formation of hydrocarbons **177** is compelling evidence for **162**, and the formation of heptafulvalene is equally strong evidence for **160–161** (**177** does not open to **171** under the reaction conditions). From this it would appear that **162** can open to **161** at relatively modest temperatures (below 300°C). From the evidence available, it appears that reclosure may require considerably more severe conditions.

Pyrolysis of 7-norbornadienyl acetate at 450°C has also been found to give heptafulvalene (274). At first glance this appears similar to the pyrolysis of **176**. However, both deuterium labeling (274) and a similar reaction of 7-phenylnorbornadienyl acetate (275, 276) (in this case giving diphenylcarbene products) rule out 7-norbornadienylidene as an intermediate from **178**. The favored mechanism for entry into the C_7H_6 reactive intermediate manifold from **178** is via the diradical **179**.

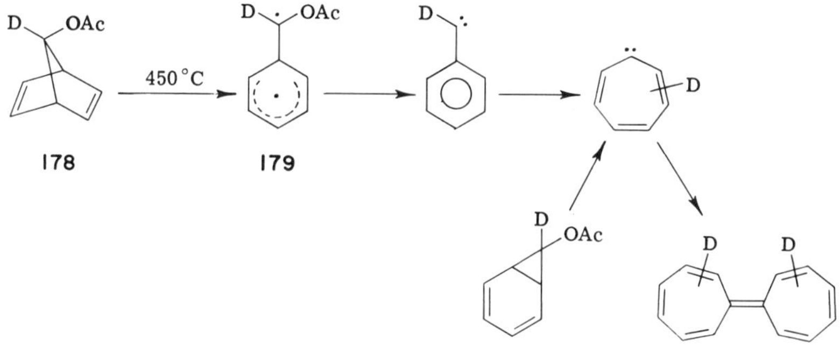

At high temperatures, fulveneallene (**166**) (and its tautomer, cyclopentadienylacetylene) is the energy sink for C_7H_6 (172, 174), and it is believed that, depending on the progenitor, there is more than one mechanism for its formation. When generated from phenylcarbene, it is believed to be formed by the series of conversions shown in Scheme 2. The strongest evidence for this mechanism is the tag distribution discussed above (271). The details of the pathway from **162** to **166** have not been established, although a diradical (or carbene **168**) has been suggested as a reasonable possibility (174).

Evidence for the reversibility of the conversion of **162** to **166** is found in the thermolysis of methylfulveneallene substituted in such a way as to provide the phenylcarbene with an energy sink that is lower than the allene. Thus, pyrolysis of **180** at 1000°C led to complete conversion to styrene and benzocyclobutene (174), products that are typical of methylarylcarbenes. It is interesting that if isomerization of **180** goes via methyl-substituted **168**, this constitutes another example of a thermally induced retro 1,2 rearrangement (see Section II,F).

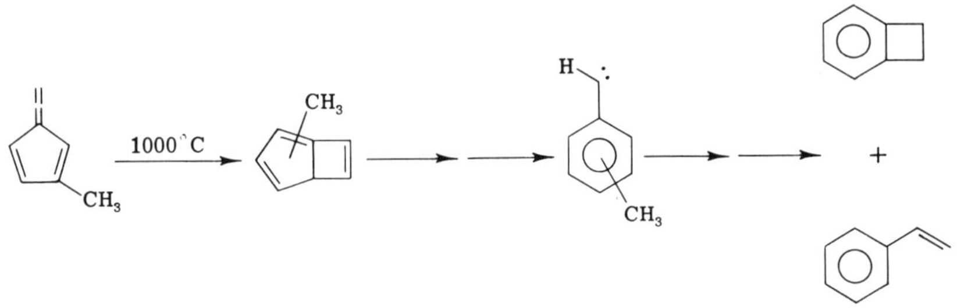

Fulveneallene is also formed from the gas-phase pyrolysis of methylenecyclohexadienylidene precursors such as **181** (*277*) and **182** (*277, 278*) and from *cis*- and *trans* -1,2- diethynylcyclopropanes **183** (*279*). These

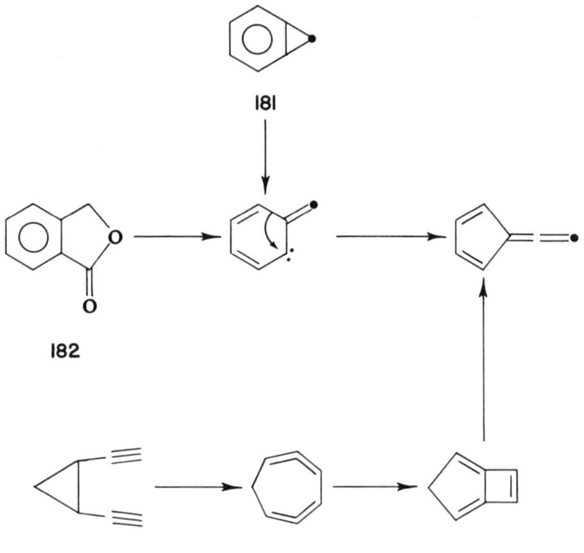

reactions occur at temperatures that are lower than that required for rearrangement of phenylcarbene to fulveneallene and are believed not to involve the cycloheptatrienylidene manifold. For the rearrangement of methylenecyclohexadienylidene to fulveneallene, a simple 1,2-vinyl shift has been proposed (*174*). As predicted by this mechanism, benzocyclopropene labeled at the CH_2 gives fulveneallene with most of its tag (83.5%) at C-7 (*271*).

The mechanism of the rearrangement of diethynylcyclopropane is not clear, but a labeling study is consistent with the participation of cycloheptatetraene (**169**) as an intermediate. It is also noteworthy that pyrolysis of **183** at 460°C (0.02 mm Hg) gives a low yield of heptafulvalene (*279*); this

3. REARRANGEMENTS OF CARBENES AND NITRENES 143

suggests leakage into the cycloheptatrienylidene surface, possibly from "hot" fulveneallene (*174*), but there is no evidence for this.

3. C_6H_5N and related energy surfaces

The chemistry of C_6H_5N at elevated temperatures in the gas phase has been very well reviewed (*172, 174*). The following remarks will therefore only highlight this area, with particular emphasis on where it differs from C_7H_6.

The C_6H_5N energy surface, as it is currently understood, is summarized in Schemes 4, 5, and 6. By analogy with phenylcarbene (Section III,D,2), phenylnitrene (**189**) and 2-, 3-, and 4-pyridylcarbenes (**184, 185**, and **188**,

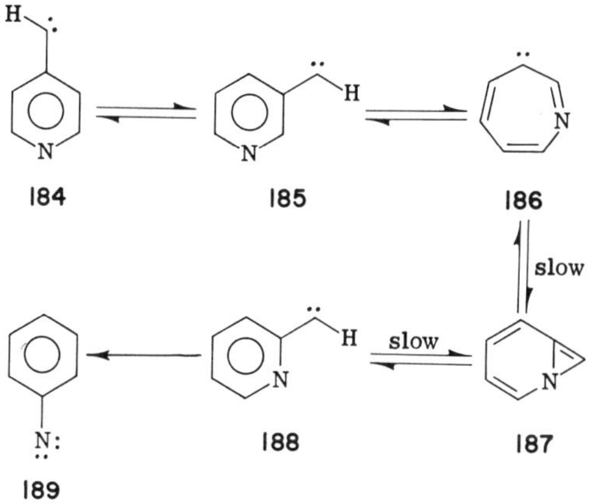

Scheme 4

respectively) might be expected to interconvert by a series of carbene–carbene rearrangements.* In fact, in many respects this is true. For instance, 4- and 3-pyridylcarbenes (**190** and **191**, respectively) readily interconvert, as evidenced by reasonable yields of vinylpyridines **192** and **193** (*280*).

Similarly, 2-pyridylcarbene rearranges to phenylnitrene (*281*). In none of these cases is there reason to assume a mechanism that is different from that normally applied to the automerization of phenylcarbene, although direct evidence for either seven-membered rings or azirines is lacking.

On the other hand, there are at least two important ways in which the phenylnitrene–pyridylcarbene interconversions differ from phenylcarbene automerizations. First, due to the presence of the heteroatom, the intermediates are not of identical energy; 2-pyridylcarbene has been estimated to be about 14 kcal/mol higher in energy than phenylnitrene (*172*).

---

* These are pictured as going through azacycloheptatrienylidene. However, as with the C_7H_6 interconversions, they may actually involve keteneimines instead.

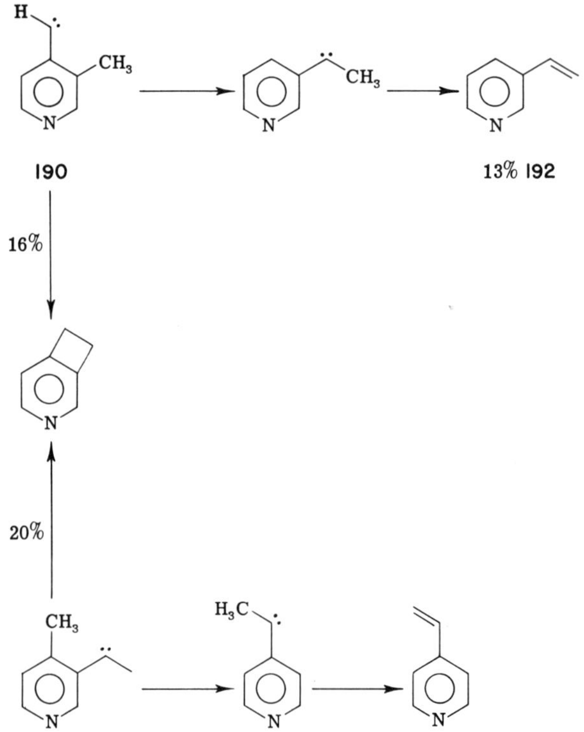

It is therefore no surprise that rearrangement of phenylnitrene to pyridylcarbene does not readily occur; at modest temperatures both phenylnitrene and 2-pyridylcarbene give high yields of the phenylnitrene dimer, azobenzene (282, 283). In fact, even when nitrene dimerization is retarded

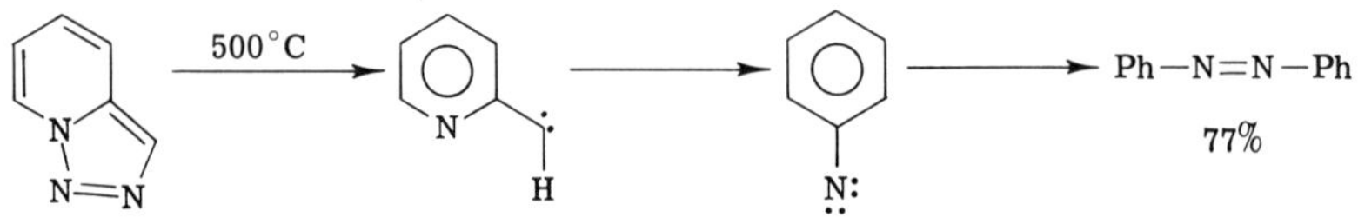

by 2,6 disubstitution and the pyridylcarbene has an internal trap, only 8% of the rearranged product is observed. Again, the primary product is derived from intermolecular chemistry of the nitrene (284).

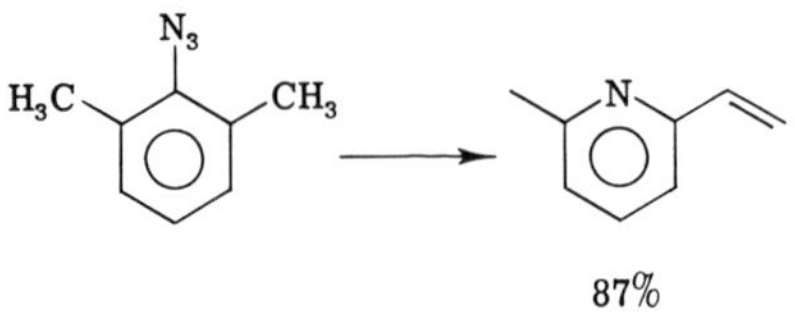

3. REARRANGEMENTS OF CARBENES AND NITRENES

A second important way in which the pyridylcarbenes differ from phenylcarbenes is in the step that connects **186** with **188**. In this step alone, an antiaromatic azirine is required as an intermediate. As a result, the interconversion of 2- and 3-pyridylcarbenes is retarded relative to other rearrangements (*285*). For instance, 2-pyridylcarbenes invariably rearrange to phenylnitrenes, even when an internal trap is present (*280*). Also, a general tendency for 3-, 4-, and 5-pyridylcarbenes to interconvert with relatively little leakage to C-2 and C-6, has been noted (*280*).

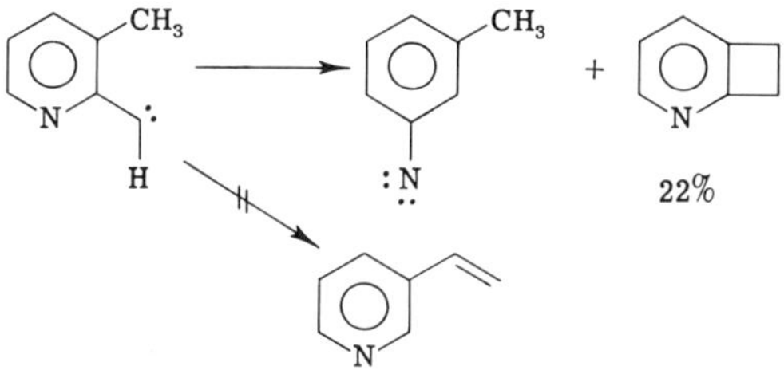

On the other hand, this leakage must occur in some cases despite the high-energy step. For instance, pyrolysis of **194** gives 17% of 2-vinyl-6-methylpyridine (*280*) (which requires this step), and aniline (presumably from phenylnitrene) has been reported as one product from pyrolysis (650°C) of **184** (*286*).

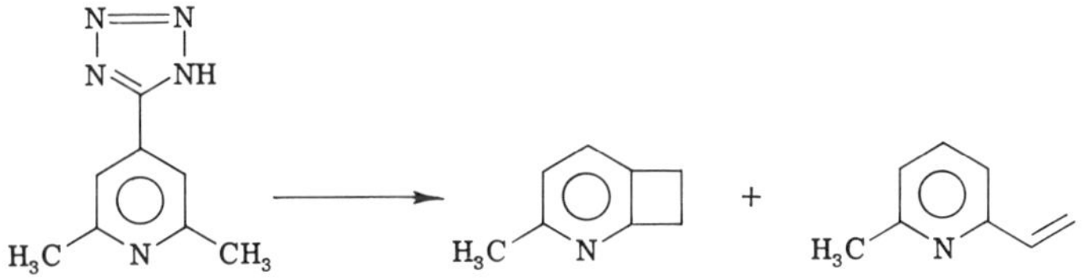

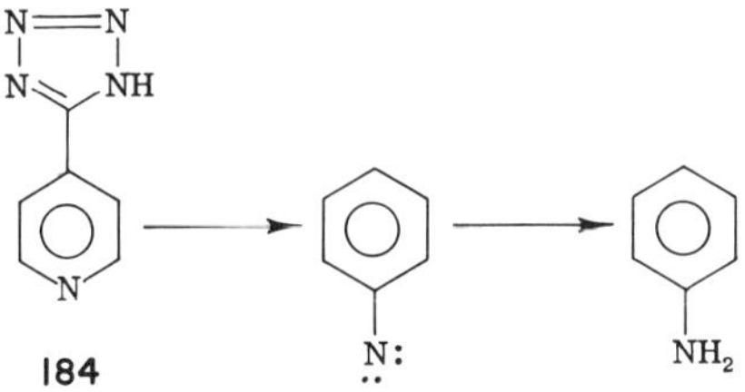

An alternative suggested (*2*) to the azirine intermediate for the interconversion of **185** and **188** is the route via **195**, which has analogy in the C_7H_6 surface. However, to this author's knowledge, there is no evidence for this.

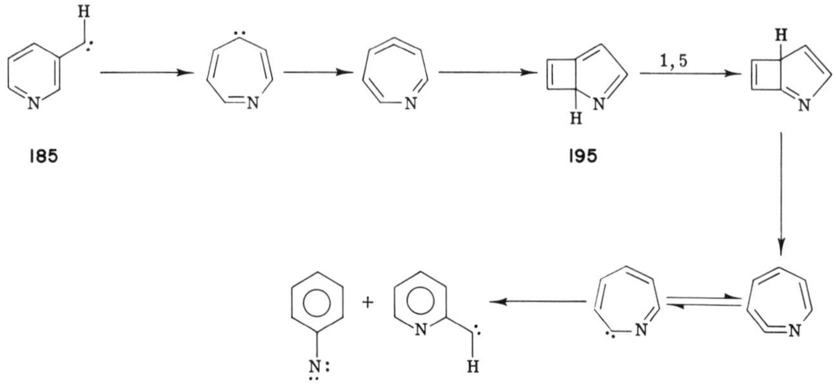

185 195

188

Just as phenylcarbene and methylenecyclohexadienylidene contract to fulveneallene and a mixture of ethynylcyclopentadienes (Scheme 3), so phenylnitrene and iminocyclohexadienylidene (**197**) contract to a mixture of cyanocyclopentadienes, which are believed to originate from **198** (*218*).

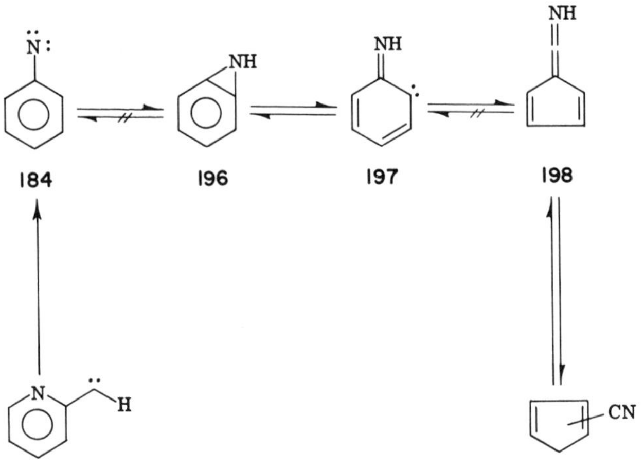

184 196 197 198

188

Scheme 5

However, there is an important difference between the C_7H_6 and C_6H_5N contractions. In the C_6H_5N system the nitrene inserts into the ortho C—H to give **196**. As a result, in contrast to fulveneallene, which requires two different mechanisms depending on the precursor (*vide supra*), cyanocyclopentadienes from all C_6H_5N precursors are believed to derive from iminocyclohexadienylidene (**197**). This contention finds its most compel-

3. REARRANGEMENTS OF CARBENES AND NITRENES

ling support in a study of the tag distribution in cyanocyclopentadienes from labeled **188** (*218*).

It is noteworthy that the distribution also requires that conversion of **184** to **196** be irreversible. Consistent with this is the fact that pyrolysis of **199** also gives cyanocyclopentadienes but no phenylnitrene products (*282, 287*).

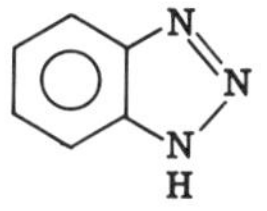

199

Pyridylcarbene precursors show rearrangements to cyanocyclopentadienes at considerably lower temperatures than that required of phenylazide (*288*). This has been attributed to "hot" intermediates activated by energy released during their formation. The phenomenon of "hot" molecules on these energy surfaces and the problems it engenders have been discussed in detail (*174*).

Formation of ethynylpyrroles **202** and **205** from pyrolysis of **200** (*289*) has been taken as evidence for the rearrangement (between 650° and 1000°C) of the heterocyclic analogue of methylenecyclohexadienylidene to fulveneallene. Isolation of both isomers has been rationalized in terms of the bicyclic intermediates **203** and **204** (*174*). These rearrangements are

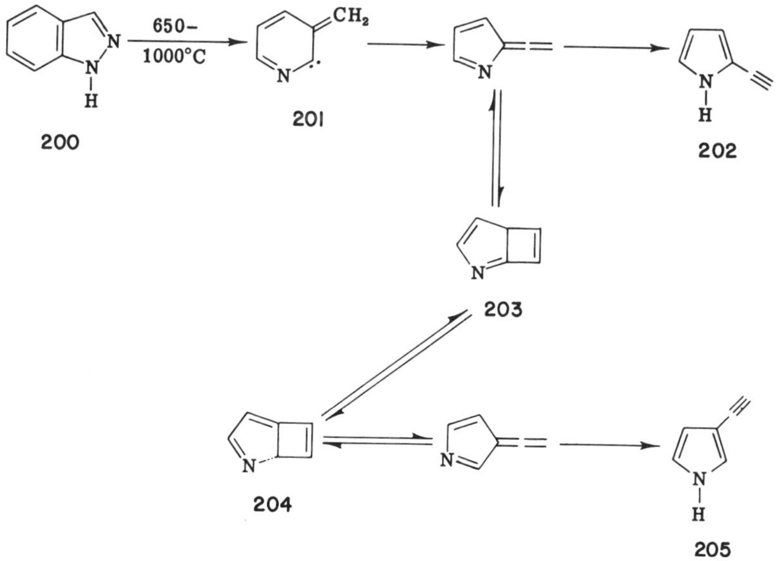

Scheme 6

exactly analogous to those outlined in Scheme 3 for C_7H_6. As in the carbocyclic case, the steps connecting the fulveneallene and the bicyclic heterocycles remain a mystery.

Finally, a number of arylnitrenes and arylcarbenes containing more than one nitrogen have been generated at high temperatures in the gas phase. In most cases, although different mechanisms may be involved, their chemistry generally resembles that of phenylcarbene, pyridyl carbene and phenylnitrene. For example, carbene **206** (*172*) and **208** (*174*) give products that would be expected from rearrangement to nitrenes **207** and **209**. Similarly, nitrenes **210** (*174*) and **212** (*172*) give pyroles and

206 207

208 209

pyrazoles, such as **211** and **214**, which require at least some equilibration of the tagged nitrene nitrogen (*290*).

210 211

212 213 214

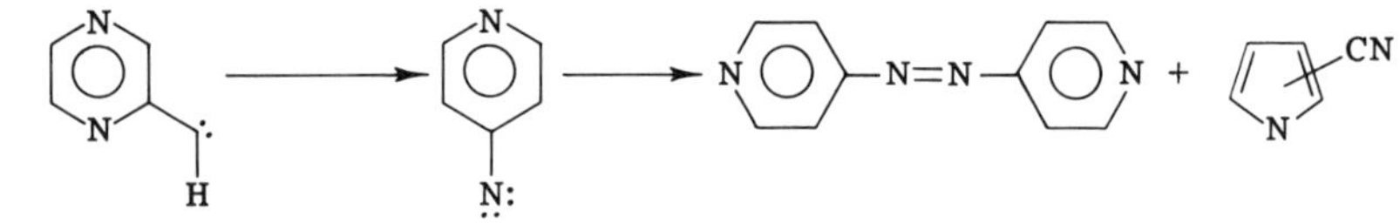

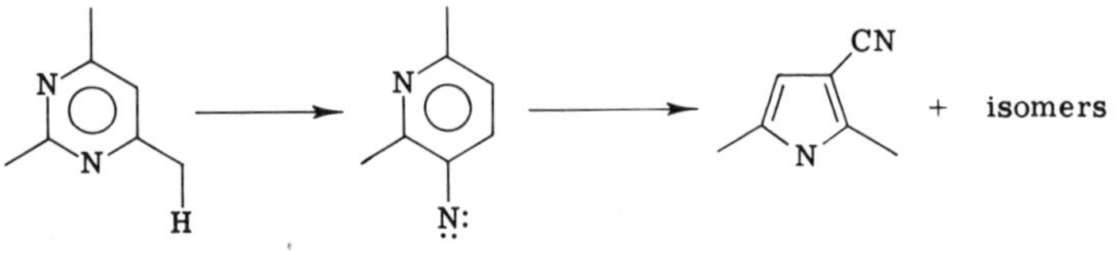

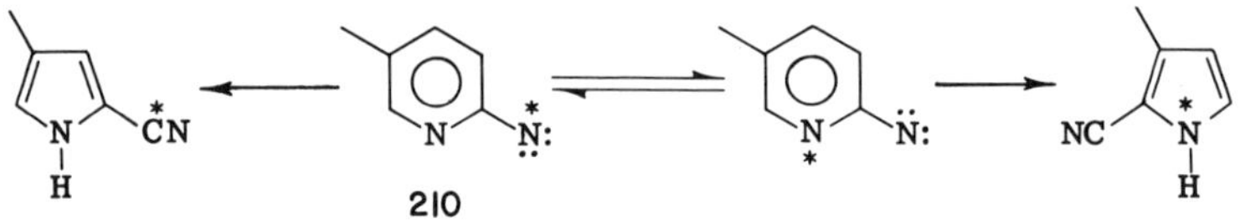

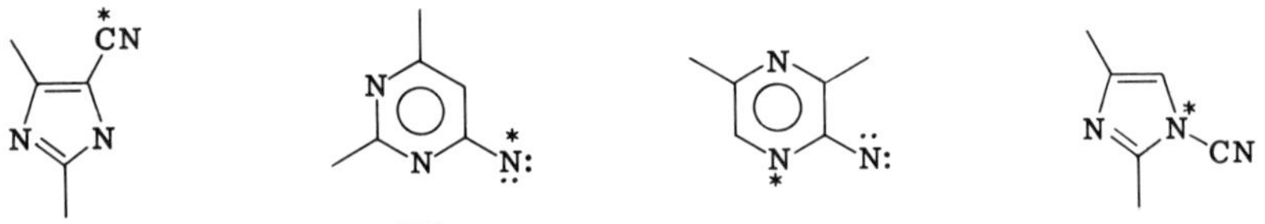

If the nitrene or carbene is positioned between two nitrogens, the normal rearrangement is retarded. For instance, **215** gives no products expected of **216** (*172*), and carbene **217** undergoes predominant rearrangement of the phenylcarbene moiety with relatively little expansion–

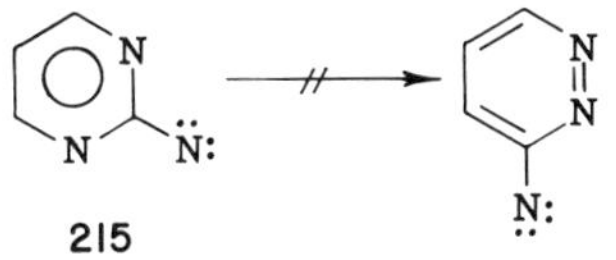

3. REARRANGEMENTS OF CARBENES AND NITRENES

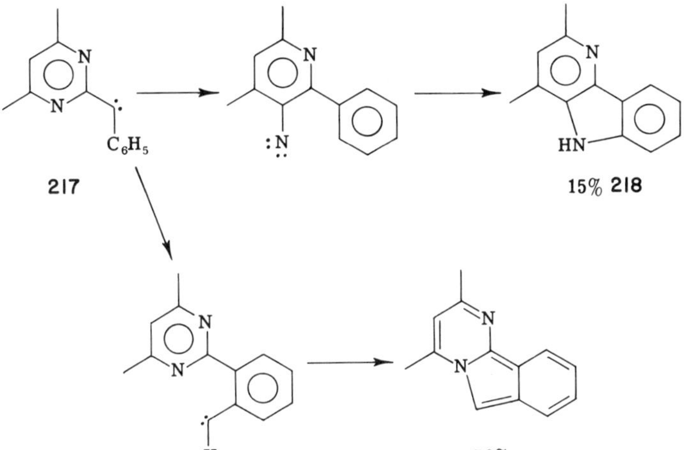

contraction of the heterocyclic ring (*172, 189*). In cases such as these, the antiaromaticity of intermediates such as **219** would clearly create a barrier

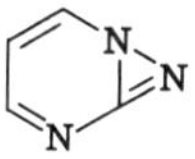

to rearrangement by the normal mechanism. However, the formation of 15% of **218** does require either that this barrier be surmountable or that heterocyclic systems of this type find a different mechanism (probably Wolff) for the rearrangement.

A number of possible mechanisms for contractions of heterocyclic nitrenes such as **210, 212, 213,** and **215** have been discussed in detail (*174*).

IV. TYPE II CARBENE–CARBENE REARRANGEMENTS

In a second type of carbene–carbene rearrangement, generation of a carbene induces molecular reorganization to give a different carbene, but one in which the divalent carbon has retained its integrity. To date, three types of reactions have been recorded that appear to fit this definition: (a) the rearrangement of vinylcyclopropylidenes to cyclopentenylidenes (the Skattebøl rearrangement), (b) the rearrangement of the homovinylcyclopropylidene **220** to **221**, and (c) the rearrangement of 7-norbornadienylidene (**163**) to cycloheptatrienylidene.

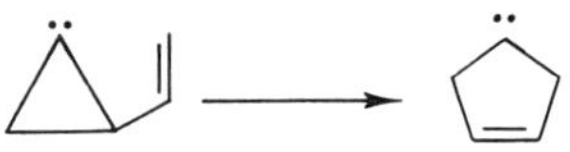

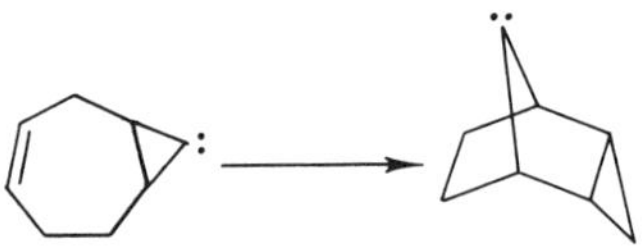

[Structure: vinylcyclopropylidene → cyclopentenylidene]

*[Structure: **220** → **221**]*

*[Structure: **163** → cycloheptatrienylidene]*

Of these three, only the rearrangement of vinylcyclopropylidene to cyclopentenylidene has been studied in any detail. The latter two are limited to single examples (*291, 292*), and the rearrangement of 7-norbornadienylidene was discussed in Section III. The following remarks deal only with the Skattebøl rearrangement.

The rearrangement of vinylcyclopropylidenes to cyclopentenylidenes was discovered by Skattebøl (*293*) in 1962, when he noted that the reaction of **222** with methyllithium gave, in addition to the normal allene product, 86% cyclopentadiene.

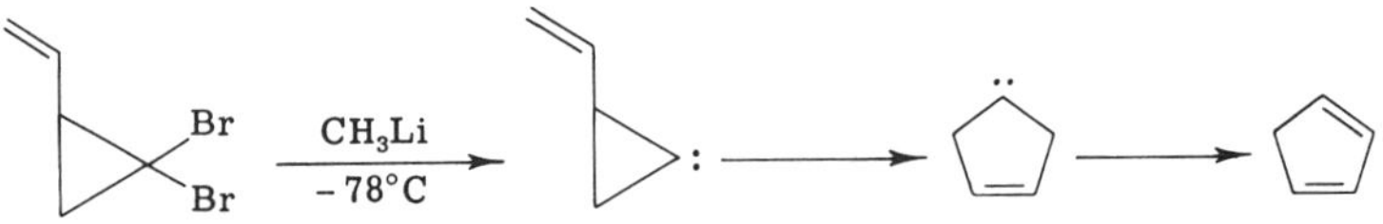

*[Scheme: **222** (gem-dibromide) →(CH$_3$Li, −78°C)→ **223** → cyclopentenylidene → cyclopentadiene, 86%]*

Since its discovery many examples of this type of rearrangement have been reported. These have included rearrangements of cyclopropylidenes (or their carbenoids) generated both from dehalogenation of *gem*-dihalocyclopropanes (*113, 294–298*) and from the reaction of *N*-nitrosourethanes (*299*) and ureas (*300*) with base.

In the rearrangement of vinylcyclopropylidene (**224**) to cyclopentenylidene, either bond a or bond b could, in principle, break (*293, 296*). Both reactions are analogous to simple vinylcyclopropane–cyclopentene

3. REARRANGEMENTS OF CARBENES AND NITRENES

rearrangements. They differ, however, in that the carbene could play a role in the former, but it is not clear how it might be involved in the latter.

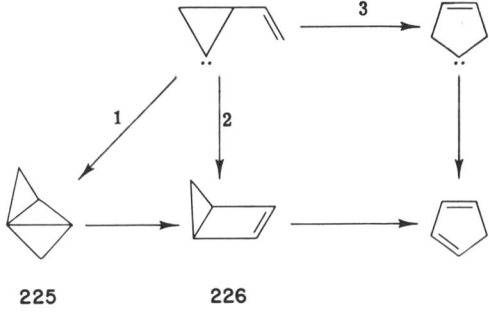

That it is bond a that breaks has been unequivocally demonstrated by labeling experiments (*301*) and product substitution patterns. The labeling experiment also confirmed retention of the integrity of the carbene carbon.

Three different mechanisms have been considered for the reaction requiring breakage of bond a (*293, 296*): (a) initial closure to tricyclopentane (**225**) followed by reorganization, (b) insertion into a vinyl hydrogen to give bicyclopentene (**226**), and (c) concerted rearrangement. Both

rearrangement of vinylcyclopropylidenes without vinyl hydrogens and the known thermal stability of bicyclo[2.1.0]pentenes (*302*) are compelling arguments against any mechanism including a bicyclopentene intermediate. The latter is also an argument against mechanism 1 if **225** must pass through **226**. Other arguments against the tricyclopentane as an intermediate in the rearrangement include (a) the absence of the tricyclic hydrocarbons, (b) substitution patterns in the products, and (c) the rearrangement of cisoid vinylcyclopropylidenes such as **227** (*297*). In such cases, cis addition of the carbene to the double bond would give impossibly strained tricyclopentanes represented by **228**. Furthermore, although

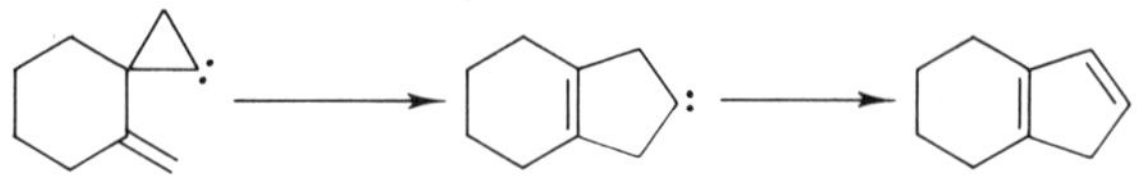

it is sterically possible for tricyclopentanes, such as **225**, to be formed from carbenes that can assume a transoid conformation, concerted open-

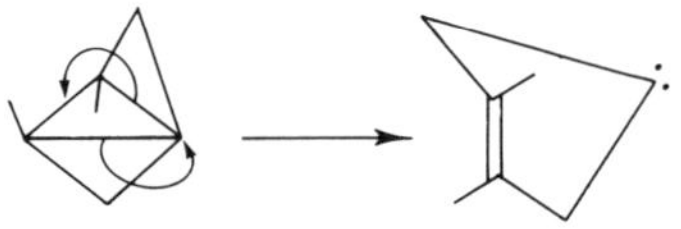

ing to the carbene (retro π^2 to ω^2) would lead to a *trans*-cyclopentenylidene. For these reasons, it is highly unlikely that tricyclopentanes are intermediates in any of these reactions.

Concerted rearrangement (mechanism 3) of vinylcyclopropylidene to cyclopentenylidene is a [1,3] sigmatropic rearrangement, and therefore for this to be an allowed pericyclic reaction at least one inversion is required. To explain the profound accelerating effect of the carbene, it has been suggested (2) that this may be viewed as a ($\pi^2_s + \sigma^2_a$) cycloaddition assisted by the vacant *p* orbital of the singlet carbene in much the same way the vacant orbital of the vinyl cation is believed to assist its ($\pi^2_s + \pi^2_a$) cycloaddition to olefins (303). At a much more sophisticated level,

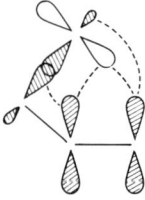

the MINDO/3 approximation developed by Dewar (304) has been applied to this rearrangement (305). In this treatment, the $^1(\sigma)^2$ was assumed to be the reacting state, and points along the electronic hypersurface were computed. From these computations, the following reaction path was predicted. (a) The reaction is initiated by π-complex formation between

3. REARRANGEMENTS OF CARBENES AND NITRENES

the vacant p orbital of the carbene and the double bond; to improve overlap the CH_2 rotates in the direction indicated in **229**. (b) The π complex then collapses to the nonclassical carbene **230**. The details of how this collapse takes place were not reported, but an activation energy

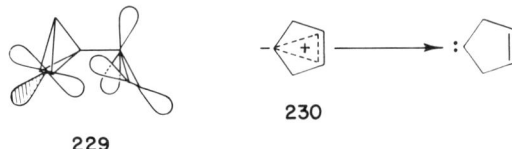

229   **230**

of 13.8 kcal/mol (upper limit) was computed. Computed charge densities in **230** are consistent with a transfer of electron density into the vacant orbital of the carbene. (c) The nonclassical carbene opens to the classical carbene. The hypersurface for this opening is very flat, with the two intermediates at nearly identical energies. However, the authors point out that this type of calculation tends to underestimate ring strain and that more sophisticated *ab initio* calculations may predict **230** to be unstable relative to the classical carbene. Finally, the interesting and reasonable predictions are made that electron-donating groups on the double bond should accelerate the rearrangement, whereas electron-donating atoms in the ring (O and N) should retard it. Neither of these predictions has been tested.

REFERENCES

1. W. Kirmse, "Carbene Chemistry," 2nd ed. Academic Press, New York, 1971.
2. W. M. Jones and U. H. Brinker, *Org. Chem.* **35**, 109 (1977).
3. P. W. Dillon and G. R. Underwood, *J. Am. Chem. Soc.* **99**, 2435 (1977).
4. K. Kleveland and L. Skattebøl, *Acta Chem. Scand., Ser. B* **29**, 191 (1975).
5. M. Christl and M. Lechner, *Angew. Chem., Int. Ed. Engl.* **14**, 765 (1975).
6. M. S. Baird and A. C. Kaura, *Chem. Commun.* p. 356 (1976).
7. M. S. Baird and C. B. Reese, *Tetrahedron* **32**, 2153 (1976).
8. H. J. J. Loozen, W. A. Castenmiller, E. J. M. Buter, and H. M. Buck, *J. Org. Chem.* **41**, 2965 (1976).
9. Y. Okuda, T. Hiyama, and H. Nozaki, *Tetrahedron Lett.* p. 3829 (1977).
10. R. W. Hoffmann and R. Schuttler, *Chem. Ber.* **108**, 844 (1975).
11. H. W. Chang, A. Lautzenheiser, and A. P. Wolf, *Tetrahedron Lett.* p. 6295 (1966); R. F. Peterson, Jr., R. T. K. Baker, and R. L. Wolfgang, *ibid.* p. 4749 (1969); P. B. Shevlin and A. P. Wolf, *J. Am. Chem. Soc.* **92**, 406 (1970).
12. R. Kalish and W. H. Pirkle, *J. Am. Chem. Soc.* **89**, 2781 (1967).
13. J. E. Baldwin and J. A. Walker, *Chem. Commun.* p. 354 (1972).
14. A. L. Wilds, R. L. von Trebra, and N. F. Woolsey, *J. Org. Chem.* **34**, 2401 (1969).
15. A. B. Smith III, *Chem. Commun.* p. 695 (1974).
16. J. P. Lokensgard, J. O'Dea, and E. A. Hill, *J. Org. Chem.* **39**, 3355 (1974).
17. H. E. Zimmerman and R. D. Little, *J. Am. Chem. Soc.* **96**, 4623 (1974).

18. A. B. Smith III, B. H. Toder, and S. J. Branca, *J. Am. Chem. Soc.* **98,** 7456 (1976).
19. S. J. Branca, R. L. Lock, and A. B. Smith III, *J. Org. Chem.* **42,** 3165 (1977).
20. M. Jones, Jr., *Acc. Chem. Res.* **7,** 415 (1974).
21. R. Y. Levina, L. G. Zaitseva, I. B. Avezov, and I. G. Bolesov, *Zh. Org. Khim.* **8,** 1105 (1972).
22. A. Hartmann, W. Welter, and M. Regitz, *Tetrahedron Lett.* p. 1825 (1974).
23. H. Meier and K. P. Zeller, *Angew. Chem., Int. Ed. Engl.* **14,** 32 (1975).
24. D. C. Richardson, M. E. Hendricks, and M. Jones, Jr., *J. Am. Chem. Soc.* **93,** 3790 (1971).
25. S. L. Kammula, H. L. Tracer, P. B. Shevlin, and M. Jones, Jr., *J. Org. Chem.* **42,** 2931 (1977).
26. J. J. Havel and K. H. Chan. *J. Org. Chem.* **42,** 569 (1977), and references cited therein.
27. H. D. Roth and M. L. Manion, *J. Am. Chem. Soc.* **98,** 3392 (1976).
28. H. D. Roth, *Acc. Chem. Res.* **10,** 85 (1977).
29. W. Ando, A. Sekiguchi, J. Ogiwara, and T. Migita, *Chem. Commun.* p. 145 (1975).
30. R. F. C. Brown, F. W. Eastwood, and G. L. McMullen, *Chem. Commun.* p. 328 (1975).
31. R. L. Kreeger and H. Shechter, *Tetrahedron Lett.* p. 2061 (1975).
32. W. Ando, A. Sekiguchi, and T. Migita, *Chem. Lett.* p. 779 (1976).
33. R. F. C. Brown, F. W. Eastwood, S. T. Lim, and G. L. McMullen, *Aust. J. Chem.* **29,** 1705 (1976).
34. M. Regitz, *Angew. Chem., Int. Ed. Engl.* **14,** 222 (1975).
35. H. Eckes and M. Regitz, *Tetrahedron Lett.* p. 447 (1975).
36. R. A. Moss and M. A. Joyce, *J. Am. Chem. Soc.* **99,** 1262 (1977).
37. D. J. Martella, M. Jones, Jr., and P. von R. Schleyer, *J. Am. Chem. Soc.* **100,** 2896 (1978).
38. I. Moritani, Y. Yamamoto, and S. I. Murahashi, *Tetrahedron Lett.* p. 5755 (1968).
39. W. I. Bevan and R. N. Haszeldine, *J. Chem. Soc., Dalton Trans.* p. 2509 (1974).
40. W. I. Bevan, R. N. Haszeldine, J. Middleton, and A. E. Tipping, *J. Chem. Soc., Dalton Trans.* p. 620 (1975).
41. B. W. Philip and J. Keating, *Tetrahedron Lett.* p. 523 (1961).
42. L. Friedman and H. Shechter, *J. Am. Chem. Soc.* **83,** 3159 (1961).
43. J. J. Havel, *J. Org. Chem.* **41,** 1464 (1976).
44. H. E. Zimmerman and J. H. Munch, *J. Am. Chem. Soc.* **90,** 187 (1968).
45. P. B. Sargeant and H. Shechter, *Tetrahedron Lett.* p. 3957 (1964).
46. H. E. Zimmerman, *Org. Chem.* **35,** 53 (1977).
47. W. Kirmse and M. Buschoff, *Chem. Ber.* **100,** 1491 (1967).
48. W. Kirmse and M. Buschoff, *Angew. Chem., Int. Ed. Engl.* **4,** 692 (1965).
49. J. H. Robson and H. Shechter, *J. Am. Chem. Soc.* **89,** 7112 (1967).
50. A. G. Hortmann and A. Bhattacharjya, *J. Am. Chem. Soc.* **98,** 7081 (1976).
51. W. A. Pryor, "Free Radicals," p. 266. McGraw-Hill, New York, 1966.
52. P. de Mayo, "Molecular Rearrangements," p. 416. Wiley (Interscience), New York, 1963.
53. M. Jones, Jr. and W. Ando, *J. Am. Chem. Soc.* **90,** 2200 (1968).
54. J. Moritani, Y. Yamamoto, and S. I. Murahashi, *Tetrahedron Lett.* p. 5697 (1968).
55. M. B. Sohn and M. Jones, Jr., *J. Am. Chem. Soc.* **94,** 8280 (1972).
56. Y. Yamamoto, S. I. Murahashi, and I. Moritani, *Tetrahedron* **31,** 2663 (1975).
57. R. M. Moriarty, J. M. Kliegman, and C. Shovlin, *J. Am. Chem. Soc.* **89,** 5958 (1967).
58. W. Lwowski, *in* "Nitrenes" (W. Lwowski, ed.), p. 217. Wiley (Interscience), New York, 1970.
59. G. R. Felt and W. Lwowski, *J. Org. Chem.* **41,** 96 (1976).

3. REARRANGEMENTS OF CARBENES AND NITRENES

60. H. P. Benecke, *Tetrahedron Lett.* p. 997 (1977).
61. A. Rauk and P. F. Alewood, *Can. J. Chem.* **55**, 1498 (1977).
62. P. A. S. Smith, in "Molecular Rearrangements" (P. de Mayo, ed.), Vol. 1, p. 528ff. Wiley, New York, 1973.
63. F. D. Lewis and W. H. Saunders, Jr., in "Nitrenes" (W. Lwowski, ed.), Chapter 3. Wiley (Interscience), New York, 1970.
64. R. A. Abramovitch and D. P. Vanderpool, *Chem. Commun.* p. 18 (1977).
65. G. Schroeter, *Chem. Ber.* **44**, 1201 (1911).
66. J. H. Boyer and D. S. Straw, *J. Am. Chem. Soc.* **75**, 1642 (1953).
67. C. L. Arcus, R. E. Marks, and M. M. Coombs, *J. Chem. Soc.* p. 4064 (1957).
68. D. L. Knunyants, E. G. Bykhovskaya, and V. N. Frosin, *Dokl. Akad. Nauk SSSR* **132**, 513 (1960).
69. W. H. Saunders, Jr. and E. A. Caress, *J. Am. Chem. Soc.* **86**, 861 (1964).
70. W. Pritzkow and D. Timm, *J. Prakt. Chem.* **32**, 178 (1966).
71. F. D. Lewis and W. H. Saunders, Jr., *J. Am. Chem. Soc.* **90**, 7031 (1968).
72. W. H. Saunders, Jr. and J. C. Ware, *J. Am. Chem. Soc.* **80**, 3328 (1958).
73. F. D. Lewis and W. H. Saunders, Jr., *J. Am. Chem. Soc.* **89**, 645 (1967).
74. G. Smolinsky, *J. Am. Chem. Soc.* **83**, 4483 (1961).
75. G. Smolinsky, *J. Org. Chem.* **27**, 3557 (1962).
76. G. R. Harvey and K. W. Ratts, *J. Org. Chem.* **31**, 3907 (1966).
77. K. Friedrich, *Angew. Chem., Int. Ed. Engl.* **6**, 959 (1967).
78. H. Reimlinger, F. P. Woerner, and D. R. Arnold, *Angew. Chem., Int. Ed. Engl.* **7**, 130 (1968).
79. M. Jones, Jr., S. D. Reich, and L. T. Scott, *J. Am. Chem. Soc.* **92**, 3118 (1970).
80. Y. Yamamoto and I. Moritani, *Tetrahedron* **26**, 1235 (1970).
81. O. S. Tee and K. Yates, *J. Am. Chem. Soc.* **94**, 3074 (1972).
82. N. Bodor and M. J. S. Dewar, *J. Am. Chem. Soc.* **94**, 9103 (1972).
83. E. P. Kyba, *J. Am. Chem. Soc.* **99**, 8330 (1977).
84. J. A. Altmann, I. G. Csizmadia, and K. Yates, *J. Am. Chem. Soc.* **96**, 4196 (1974).
85. J. A. Altmann, I. G. Csizmadia, and K. Yates, *J. Am. Chem. Soc.* **97**, 5217 (1975).
86. C. Trindle and J. K. George, *J. Quant. Chem.* **10**, 21 (1976).
87. J. A. Altmann, O. S. Tee, and K. Yates, *J. Am. Chem. Soc.* **98**, 7132 (1976).
88. D. T. T. Su and E. R. Thornton, *J. Am. Chem. Soc.* **100**, 1872 (1978).
89. E. P. Kyba and A. M. John, *J. Am. Chem. Soc.* **99**, 8329 (1977).
90. J. W. Powell and M. C. Whiting, *Tetrahedron* **12**, 168 (1961).
91. L. Seghers and H. Shechter, *Tetrahedron Lett.* p. 1943 (1976).
92. A. Nickon, F. Huang, R. Weglein, K. Matsuo, and H. Yagi, *J. Am. Chem. Soc.* **96**, 5264 (1974).
93. E. P. Kyba and C. W. Hudson, *J. Am. Chem. Soc.* **98**, 5696 (1976).
94. E. P. Kyba and C. W. Hudson, *J. Org. Chem.* **42**, 1935 (1977).
95. S. Inagaki, F. Fujimoto, and K. Fukui, *J. Am. Chem. Soc.* **98**, 4054 (1976), and references cited therein.
96. P. K. Freeman, T. A. Hardy, J. R. Balyeat, and L. D. Wescott, Jr., *J. Org. Chem.* **42**, 3356 (1977).
97. R. Gleiter and R. Hoffmann, *J. Am. Chem. Soc.* **90**, 5457 (1968).
98. R. A. Moss and U. H. Dolling, *Tetrahedron Lett.* p. 5117 (1972).
99. R. A. Moss and C. T. Ho, *Tetrahedron Lett.* p. 1651 (1976).
100. H. E. Zimmerman and L. R. Sousa, *J. Am. Chem. Soc.* **94**, 834 (1972).
101. M. Schneider and B. Csacsko, *Angew. Chem., Int. Ed. Engl.* **16**, 867 (1977).
102. T. J. Katz, E. J. Wang, and N. Acton, *J. Am. Chem. Soc.* **93**, 3782 (1971).

103. U. Burger and F. Mazenod, *Tetrahedron Lett.* p. 2881 (1976).
104. G. L. Closs and R. B. Larrabee, *Tetrahedron Lett.* p. 287 (1965).
105. D. M. Lemal and K. S. Shim, *Tetrahedron Lett.* p. 3231 (1964).
106. R. A. Moss and J. R. Whittle, *Chem. Commun.* p. 341 (1969).
107. G. N. Fickes and C. B. Rose, *J. Org. Chem.* **37,** 2898 (1972).
108. R. A. Moss, U. H. Dolling, and J. R. Whittle, *Tetrahedron Lett.* p. 931 (1971).
109. R. A. Moss and C. T. Ho, *Tetrahedron Lett.* p. 3397 (1976).
110. M. H. Fisch and H. D. Pierce, *Chem. Commun.* p. 503 (1970).
111. W. T. Brown and W. M. Jones, unpublished results.
112. K. Okumura and S. I. Murahashi, *Tetrahedron Lett.* p. 3281 (1977).
113. M. S. Baird and C. B. Reese, *Tetrahedron Lett.* p. 2895 (1976).
114. T. Antkowiak, D. C. Sanders, G. B. Trimitsis, J. B. Press, and H. Shechter, *J. Am. Chem. Soc.* **94,** 5366 (1972).
115. T. V. Rajan Babu and H. Shechter, *J. Am. Chem. Soc.* **98,** 8261 (1976).
116. J. B. Carlton, R. H. Levin, and J. Clardy, *J. Am. Chem. Soc.* **98,** 6068 (1976).
117. J. B. Press and H. Shechter, *J. Org. Chem.* **40,** 2446 (1975).
118. H. Babad, W. Flemon, and J. B. Wood III, *J. Org. Chem.* **32,** 2871 (1967).
119. C. J. Cardenas, B. A. Shoulders, and P. D. Gardner, *J. Org. Chem.* **32,** 1220 (1967).
120. S. I. Murahashi, K. Okumura, Y. Maeda, A. Sonado, and I. Moritani, *Bull. Chem. Soc. Jpn.* **47,** 2420 (1974).
121. S. I. Murahashi, K. Okumura, T. Kubota, and I. Moritani, *Tetrahedron Lett.* p. 4197 (1973).
122. P. K. Freeman, R. S. Raghavan, and D. G. Kuper, *J. Am. Chem. Soc.* **93,** 5288 (1971).
123. P. K. Freeman, T. A. Hardy, R. S. Raghavan, and D. G. Kuper, *J. Org. Chem.* **42,** 3882 (1977).
124. H. D. Hartzler, *Carbenes* **2,** 43 (1975).
125. C. E. Dykstra and H. F. Schaefer III, *J. Am. Chem. Soc.* **100,** 1378 (1978).
126. O. P. Strausz, R. J. Norstrom, A. C. Hopkinson, M. Schoenborn, and I. G. Csizmadia, *Theor. Chim. Acta* **29,** 183 (1973).
127. G. Kobrich, *Angew. Chem., Int. Ed. Engl.* **6,** 41 (1967).
128. C. Wentrup and W. Reichen, *Helv. Chim. Acta* **59,** 2615 (1976).
129. R. F. Brown, F. W. Eastwood, and G. L. McMullen, *Aust. J. Chem.* **30,** 179 (1977).
130. G. J. Baxter, R. F. C. Brown, F. W. Eastwood, and K. J. Harrington, *Aust. J. Chem.* **30,** 459 (1977).
131. J. Wolinsky, G. W. Clark, and P. C. Thorstenson, *J. Org. Chem.* **41,** 745 (1976).
132. P. J. Stang and M. G. Mangum, *J. Am. Chem. Soc.* **97,** 1459 (1975).
133. P. J. Stang and M. G. Mangum, *J. Am. Chem. Soc.* **97,** 6478 (1973).
134. P. J. Stang and M. G. Mangum, *J. Am. Chem. Soc.* **99,** 2597 (1977).
135. P. J. Stang and D. P. Fox, *J. Org. Chem.* **42,** 1667 (1977).
136. P. J. Stang, J. R. Madsen, M. G. Mangum, and D. P. Fox, *J. Org. Chem.* **42,** 1802 (1977).
137. P. J. Stang and M. G. Mangum, *J. Am. Chem. Soc.* **97,** 3854 (1975).
138. J. Wolinsky, *J. Org. Chem.* **26,** 704 (1961).
139. K. L. Erickson and J. Wolinsky, *J. Am. Chem. Soc.* **87,** 1142 (1965).
140. G. J. Baxter and R. F. C. Brown, *Aust. J. Chem.* **31,** 327 (1978).
141. R. J. Norstrom, H. E. Gunning, and O. P. Strausz, *J. Am. Chem. Soc.* **98,** 1454 (1976).
142. P. Yates and L. Kilmurry, *Tetrahedron Lett.* p. 1739 (1964).
143. D. R. Morton and N. J. Turro, *Adv. Photochem.* **9,** 197 (1974).
144. W. D. Stohrer, P. Jacobs, K. H. Kaiser, G. Wiech, and G. Quinkert, *Fortschr. Chem. Forsch.* **46,** 181 (1974).
145. P. Yates and R. O. Loutfy, *Acc. Chem. Res.* **8,** 209 (1975).

3. REARRANGEMENTS OF CARBENES AND NITRENES 157

146. P. Yates and J. C. L. Tam, *Chem. Commun.* p. 737 (1975).
147. W. C. Agosta and S. Wolff, *J. Am. Chem. Soc.* **98**, 4182 (1976).
148. P. M. Collins, N. N. Opararche, and B. R. Whitton, *Chem. Commun.* p. 292 (1974).
149. A. G. Brook, *Acc. Chem. Res.* **7**, 77 (1974).
150. J. M. Duff and A. G. Brook, *Can. J. Chem.* **51**, 2869 (1973).
151. A. G. Brook and J. M. Duff, *J. Am. Chem. Soc.* **89**, 1454 (1967).
152. A. G. Brook, H. W. Kucera, and R. Pearce, *Can. J. Chem.* **49**, 1618 (1971).
153. A. G. Brook, R. Pearce, and J. B. Pierce, *Can. J. Chem.* **49**, 1622 (1971).
154. W. D. Stohrer, G. Wiech, and G. Quinkert, *Angew. Chem., Int. Ed. Engl.* **13**, 199 (1974).
155. W. D. Stohrer, G. Wiech, and G. Quinkert, *Angew Chem., Int. Ed. Engl.* **13**, 200 (1974).
156. J. A. Altmann, I. G. Csizmadia, K. Yates, and P. Yates, *J. Chem. Phys.* **66**, 298 (1977).
157. J. C. Altmann, I. G. Csizmadia, M. A. Robb, K. Yates, and P. Yates, *J. Am. Chem. Soc.* **100**, 1653 (1978).
158. R. J. Cvetanović and L. C. Doyle, *J. Chem. Phys.* **37**, 543 (1962).
159. E. Tschuikov-Roux, J. R. McNesby, W. M. Jackson, and J. L. Faris, *J. Phys. Chem.* **71**, 1351 (1967).
160. S. J. Cristol and C. S. Illenda, *J. Am. Chem. Soc.* **97**, 5862 (1975).
161. P. J. Kropp, E. J. Reardon, Jr., Z. L. F. Gaibel, K. F. Williard, and J. H. Hattaway, Jr., *J. Am. Chem. Soc.* **95**, 7058 (1973).
162. T. R. Fields and P. J. Kropp, *J. Am. Chem. Soc.* **96**, 7559 (1974).
163. H. G. Fravel, Jr. and P. J. Kropp, *J. Org. Chem.* **40**, 2434 (1975).
164. P. J. Kropp, H. G. Fravel, Jr., and T. R. Fields, *J. Am. Chem. Soc.* **98**, 840 (1976).
165. F. H. Watson, Jr., A. T. Armstrong, and S. P. McGlynn, *Theor. Chim. Acta* **16**, 75 (1970); F. H. Watson, Jr. and S. P. McGlynn, *ibid.* **21**, 309 (1971).
166. Y. Inone, S. Takamuku, and H. Sakurai, *Chem. Commun.* p. 577 (1975).
167. S. S. Hixson, *J. Am. Chem. Soc.* **97**, 1981 (1975).
168. S. S. Hixson, J. C. Tausta, and J. Borovsky, *J. Am. Chem. Soc.* **97**, 3230 (1975).
169. T. H. Chan and D. Massuda, *J. Am. Chem. Soc.* **99**, 936 (1977).
170. R. F. C. Brown, F. W. Eastwood, K. J. Harrington, and G. L. McMullen, *Aust. J. Chem.* **27**, 2393 (1974).
171. R. F. C. Brown, F. W. Eastwood, and G. P. Jackman, *Aust. J. Chem.* **30**, 1757 (1977).
172. C. Wentrup, *Top. Curr. Chem.* **62**, 175 (1976).
173. W. M. Jones, *Acc. Chem. Res.* **10**, 353 (1977).
174. C. Wentrup, *in* "Reactive Intermediates" (R. A. Abramovitch, ed.), Plenum, New York, 1978.
175. H. Hauptman, *Tetrahedron* **32**, 1293 (1976).
176. A. Krantz and J. Laureni, *J. Am. Chem. Soc.* **99**, 4842 (1977).
177. T. T. Coburn and W. M. Jones, *Tetrahedron Lett.* p. 3903 (1973).
178. R. V. Hoffman and H. Shechter, *J. Am. Chem. Soc.* **93**, 5940 (1971).
179. T. L. Gilchrist and D. P. J. Pearson, *J. Chem. Soc., Perkin Trans. 1* p. 1257 (1976).
180. W. N. Washburn and R. Zahler, *J. Am. Chem. Soc.* **98**, 7827 (1976).
181. B. H. Freeman and D. Lloyd, *Tetrahedron* **30**, 2557 (1974).
182. S. F. Dyer, S. Kammula, and P. B. Shevlin, *J. Am. Chem. Soc.* **99**, 8104 (1977).
183. R. A. Olofson, K. D. Lotts, and G. N. Barber, *Tetrahedron Lett.* p. 3381 (1976).
184. A. G. Anastassiou, H. E. Simmons, and F. D. Marsh, *in* "Nitrenes" (W. Lwowski, ed.), p. 305, Chapter 12. Wiley (Interscience), New York, 1970.
185. O. P. Strausz, R. K. Gosavi, A. S. Denest, and I. G. Csizmadia, *J. Am. Chem. Soc.* **98**, 4784 (1976).
186. M. J. S. Dewar and C. A. Ramsden, *Chem. Commun.* p. 688 (1973).

187. R. Hoffmann, G. D. Zeiss, and G. W. van Dine, *J. Am. Chem. Soc.* **90**, 1485 (1968).
188. M. J. S. Dewar and D. Landman, *J. Am. Chem. Soc.* **99**, 6179 (1977).
189. C. Mayor and C. Wentrup, *J. Am. Chem. Soc.* **97**, 7467 (1975).
190. N. M. Lan and C. Wentrup, *Helv. Chim. Acta* **59**, 2068 (1976).
191. J. H. Davis, W. A. Goddard III, and R. G. Bergman, *J. Am. Chem. Soc.* **98**, 4015 (1976).
192. J. H. Davis, W. A. Goddard III, and R. G. Bergman, *J. Am. Chem. Soc.* **99**, 2427 (1977).
193. J. A. Pincock and R. J. Boyd, *Can. J. Chem.* **55**, 2482 (1977).
194. R. L. Tyner, W. M. Jones, N. Y. Ohrn, and J. R. Sabin, *J. Am. Chem. Soc.* **95**, 3765 (1974).
195. M. J. S. Dewar, private communication.
196. B. A. de Graff, D. W. Gillespie, and R. J. Sundberg, *J. Am. Chem. Soc.* **96**, 7491 (1974), footnote No. 27.
197. B. A. de Graff, D. W. Gillespie, and R. J. Sundberg, *J. Am. Chem. Soc.* **96**, 7491 (1974).
198. R. Gleiter, W. Rettig, and C. Wentrup, *Helv. Chim. Acta* **57**, 2111 (1974).
199. A. Krantz, *Chem. Commun.* p. 694 (1972).
200. A. Krantz, *Chem. Commun.* p. 670 (1973).
201. H. Meier and K. P. Zeller, *Angew. Chem., Int. Ed. Engl.* **14**, 32 (1975), ref. 232.
202. K. P. Zeller, *Tetrahedron Lett.* p. 707 (1977).
203. S. A. Matlin and P. G. Sammes, *Chem. Commun.* p. 11 (1972).
204. R. A. Cormier, K. M. Freeman, and D. M. Schnar, *Tetrahedron Lett.* p. 2231 (1977).
205. E. G. Lewars and G. Morrison, *Can. J. Chem.* **55**, 966 (1977).
206. E. G. Lewars and G. Morrison, *Tetrahedron Lett.* p. 501 (1977).
207. J. Fenwick, G. Frater, K. Ogi, and O. P. Strausz, *J. Am. Chem. Soc.* **95**, 124 (1973).
208. I. G. Csizmadia, J. Font, and O. P. Strausz, *J. Am. Chem. Soc.* **90**, 7360 (1968).
209. Z. Majerski and C. S. Redvanly, *Chem. Commun.* p. 694 (1972).
210. K. P. Zeller, *Chem. Commun.* p. 317 (1975).
211. U. Timm, K. P. Zeller, and H. Meier, *Tetrahedron* **33**, 453 (1977).
212. K. P. Zeller, *Angew. Chem., Int. Ed. Engl.* **16**, 781 (1977).
213. D. J. Anderson, T. L. Gilchrist, G. E. Gymer, and C. W. Rees, *J. Chem. Soc., Perkin Trans. 1* p. 550 (1973).
214. T. L. Gilchrist, G. E. Gymer, and C. W. Rees, *J. Chem. Soc., Perkin Trans. 1* p. 555 (1973).
215. T. L. Gilchrist, G. E. Gymer, and C. W. Rees, *Chem. Commun.* p. 835 (1973).
216. T. L. Gilchrist, G. E. Gymer, and C. W. Rees, *J. Chem. Soc., Perkin Trans. 1* p. 1 (1975).
217. T. L. Gilchrist, C. W. Rees, and C. Thomas, *J. Chem. Soc., Perkin Trans. 1* p. 8 (1975).
218. C. Thetaz and C. Wentrup, *J. Am. Chem. Soc.* **98**, 1258 (1976).
219. T. Wooldridge and T. D. Roberts, *Tetrahedron Lett.* p. 2643 (1977).
220. J. I. G. Cadogan, J. T. Sharp, and M. H. Trattles, *Chem. Commun.* p. 900 (1974).
221. A. Krantz and J. Laureni, *J. Am. Chem. Soc.* **96**, 6768 (1974).
222. C. Mayor and W. M. Jones, *J. Org. Chem.* **43**, 4498 (1978).
223. P. Ashkenasi, S. Lupan, A. Schwartz, and M. Cais, *Tetrahedron Lett.* p. 817 (1969).
224. T. H. Ledford and W. M. Jones, unpublished results.
225. W. M. Jones and C. L. Ennis, *J. Am. Chem. Soc.* **91**, 6391 (1969).
226. L. W. Christensen, E. E. Waali, and W. M. Jones, *J. Am. Chem. Soc.* **94**, 2118 (1972).
227. B. L. Duell and W. M. Jones, unpublished results.
228. C. Mayor and W. M. Jones, *Tetrahedron Lett.* p. 3855 (1977).

3. REARRANGEMENTS OF CARBENES AND NITRENES

229. E. E. Waali, J. M. Lewis, D. E. Lee, E. W. Allen, III, and A. K. Chappell, *J. Org. Chem.* **42,** 3460 (1977).
230. K. E. Krajca and W. M. Jones, unpublished results.
231. K. E. Krajca and W. M. Jones, *Tetrahedron Lett.* p. 3807 (1975).
232. M. L. Manion and H. D. Roth, *J. Am. Chem. Soc.* **97,** 6919 (1975).
233. T. T. Coburn and W. M. Jones, *J. Am. Chem. Soc.* **96,** 5218 (1974).
234. J. P. Mykytka and W. M. Jones, *J. Am. Chem. Soc.* **97,** 5933 (1975).
235. W. E. Billups, L. P. Lin, and W. J. Chow, *J. Am. Chem. Soc.* **96,** 4026 (1974).
236. W. E. Billups and L. E. Reed, *Tetrahedron Lett.* p. 2239 (1977).
237. P. H. Gebert, R. W. King, R. A. LaBar, and W. M. Jones, *J. Am. Chem. Soc.* **95,** 2357 (1973).
238. U. H. Brinker and W. M. Jones, *Tetrahedron Lett.* p. 577 (1976).
239. W. M. Jones, R. A. LaBar, U. H. Brinker, and P. H. Gebert, *J. Am. Chem. Soc.* **99,** 6379 (1977).
240. U. H. Brinker, R. W. King, and W. M. Jones, *J. Am. Chem. Soc.* **99,** 3175 (1977).
241. L. Wolff, *Justus Liebigs Ann. Chem.* **394,** 59 (1912).
242. P. A. S. Smith, *in* "Nitrenes" (W. Lwowski, ed.), p. 99. Wiley (Interscience), New York, 1970.
243. R. A. Abramovitch and E. P. Kyba, *in* "The Chemistry of the Azido Group" (S. Patai, ed.), Chapter 5. Wiley (Interscience), New York, 1971.
244. R. Huisgen, D. Vossins, and M. Appl, *Chem Ber.* **91,** 1 (1958).
245. S. E. Carroll, B. Nay, E. F. V. Scriven, H. Suschitzky, and D. R. Thomas, *Tetrahedron Lett.* p. 3175 (1977).
246. R. A. Odum and G. Wolf, *Chem. Commun.* p. 360 (1973).
247. M. Masaki, K. Fukui, and J. Kita, *Bull. Chem. Soc. Jpn.* **50,** 2013 (1977).
248. F. P. Tsui, Y. H. Chang, T. M. Vogel, and G. Zon, *J. Org. Chem.* **41,** 3381 (1976).
249. A. Michaelis and A. Schäfer, *Justus Liebigs Ann. Chem.* **397,** 119 (1913).
250. R. J. Sundberg and R. W. Heintzelman, *J. Org. Chem.* **39,** 2546 (1974).
251. R. A. Abramovitch and B. A. Davis, *Chem. Rev.* **64,** 149 (1964).
252. J. I. G. Cadogan and M. J. Todd, *J. Chem. Soc. C* p. 2808 (1969).
253. R. J. Sundberg, M. Brenner, S. R. Suter, and B. P. Das, *Tetrahedron Lett.* p. 2715 (1970).
254. O. L. Chapman and J. P. LeRoux, *J. Am. Chem. Soc.* **100,** 282 (1978).
255. R. N. Carde and G. Jones, *J. Chem. Soc., Perkin Trans. 1* p. 519 (1975).
256. R. Huisgen and M. Appl, *Chem. Ber.* **91,** 12 (1958).
257. E. B. Norsoph, B. Coleman, and M. Jones, Jr., *J. Am. Chem. Soc.* **100,** 994 (1978).
258. A. Sekiguchi and W. Ando, *Bull. Chem. Soc. Jpn.* **50,** 3067 (1977).
259. W. Ando, A. Sekiguchi, A. J. Rothschild, R. R. Gallucci, M. Jones, Jr., T. J. Barton, and J. A. Kilgour, *J. Am. Chem. Soc.* **99,** 6995 (1977).
260. B. E. Sarver, M. Jones, Jr., and A. M. van Leusen, *J. Am. Chem. Soc.* **97,** 4771 (1975).
261. T. J. Barton, J. A. Kilgour, R. R. Gallucci, A. J. Rothschild, J. Slutsky, A. D. Wolf, and M. Jones, Jr., *J. Am. Chem. Soc.* **97,** 657 (1975).
262. W. Ando, A. Sekiguchi, T. Hagiwara, and T. Migita, *Chem. Commun.* p. 372 (1974).
263. R. C. Joines, A. B. Turner, and W. M. Jones, *J. Am. Chem. Soc.* **91,** 7754 (1969).
264. P. Schissel, M. E. Kent, D. J. McAdoo, and E. Hedaya, *J. Am. Chem. Soc.* **92,** 2147 (1970).
265. C. Wentrup and K. Wilczek, *Helv. Chim. Acta* **53,** 1459 (1970).
266. W. M. Jones, R. C. Joines, J. A. Myers, T. Mitsuhashi, K. E. Krajca, E. E. Waali, T. L. Davis, and A. B. Turner, *J. Am. Chem. Soc.* **95,** 826 (1973).
267. W. J. Baron, M. Jones, Jr., and P. P. Gaspar, *J. Am. Chem. Soc.* **92,** 4739 (1970).

268. E. Hedaya and M. E. Kent, *J. Am. Chem. Soc.* **93**, 3283 (1971).
269. W. D. Crow and M. N. Paddon-Row, *J. Am. Chem. Soc.* **94**, 4746 (1972).
270. W. D. Crow and M. N. Paddon-Row, *Aust. J. Chem.* **25**, 1704 (1973).
271. C. Wentrup, E. Wentrup-Byrne, and P. Müller, *Chem. Commun.* p. 210 (1977).
272. D. L. Muck and W. M. Jones, *J. Am. Chem. Soc.* **88**, 3798 (1966), and references cited therein.
273. P. B. Shevlin and A. P. Wolf, *Tetrahedron Lett.* p. 3987 (1970).
274. R. W. Hoffmann, private communication; this result is also mentioned in Brinker and Jones (*238*).
275. R. W. Hoffman and W. Lilienblum, *Chem. Ber.* **110**, 3405 (1977).
276. R. W. Hoffmann, R. Schuttler, and I. H. Loef, *Chem. Ber.* **110**, 3410 (1977).
277. C. Wentrup and P. Müller, *Tetrahedron Lett.* p. 2915 (1973).
278. U. E. Wiersum and T. Nieuwenhuis, *Tetrahedron Lett.* p. 2581 (1973).
279. R. G. Bergman, *Acc. Chem. Res.* **6**, 25 (1973).
280. W. D. Crow, A. N. Khan, and M. N. Paddon-Row, *Aust. J. Chem.* **28**, 1741 (1975).
281. W. D. Crow and C. Wentrup, *Tetrahedron Lett.* p. 6149 (1968).
282. W. D. Crow and C. Wentrup, *Tetrahedron Lett.* p. 4379 (1967).
283. G. Smolinsky and B. I. Feuer, *J. Org. Chem.* **31**, 3382 (1966).
284. C. Wentrup, *Chem. Commun.* p. 1386 (1969).
285. W. D. Crow, A. N. Khan, M. N. Paddon-Row, and D. S. Sutherland, *Aust. J. Chem.* **28**, 1763 (1975).
286. W. D. Crow and M. N. Paddon-Row, *Aust. J. Chem.* **28**, 1755 (1975).
287. W. D. Crow and C. Wentrup, *Tetrahedron* **26**, 3965 (1970).
288. E. Hedaya, M. E. Kent, D. W. McNeil, F. P. Lossing, and T. McAllister, *Tetrahedron Lett.* p. 3415 (1968).
289. W. D. Crow, A. R. Lea, and M. N. Paddon-Row, *Tetrahedron Lett.* p. 2235 (1972).
290. R. Harder and C. Wentrup, *J. Am. Chem. Soc.* **98**, 1259 (1976).
291. M. S. Baird and C. B. Reese, *Chem. Commun.* p. 523 (1972).
292. W. T. Brown and W. M. Jones, unpublished results.
293. L. Skattebøl, *Chem. Ind. (London)* p. 2146 (1962).
294. L. Skattebøl, *Tetrahedron* **31**, 2175 (1975).
295. M. Z. Nazer, *J. Org. Chem.* **30**, 1737 (1965).
296. L. Skattebøl, *Tetrahedron* **23**, 1107 (1967).
297. R. B. Reinarz and G. J. Fonken, *Tetrahedron Lett.* p. 4591 (1973).
298. R. B. Reinarz and G. J. Fonken, *Tetrahedron Lett.* p. 441 (1974).
299. P. H. Gebert and W. M. Jones, unpublished results.
300. K. H. Holm and L. Skattebøl, *J. Am. Chem. Soc.* **99**, 5480 (1977).
301. K. H. Holm and L. Skattebøl, *Tetrahedron Lett.* p. 2347 (1977).
302. J. I. Brauman, L. E. Ellis, and E. E. van Tamelen, *J. Am. Chem. Soc.* **88**, 846 (1966).
303. R. B. Woodward and R. Hoffmann, "The Conservation of Orbital Symmetry." Verlag Chemie, Weinheim, 1970.
304. R. C. Bingham, M. J. S. Dewar, and D. H. Lo, *J. Am. Chem. Soc.* **97**, 1285 (1975), and references cited therein.
305. W. W. Schoeller and U. H. Brinker, private communication.

ESSAY 4 | **FREE-RADICAL REARRANGEMENTS**

A. L. J. BECKWITH
and K. U. INGOLD

I.	Introduction	162
II.	Measurement of Rates of Unimolecular Radical Reactions	165
III.	Rearrangement by Transfer of a Carbon-Centered Group	168
	A. Group Mobilities	168
	B. Aryl Migration	170
	C. Vinyl Migration	178
	D. Migration of Other Unsaturated Groups	181
IV.	Ring-Closure Reactions	182
	A. Comparisons with Analogous Intermolecular Processes	182
	B. The 5-Hexenyl System	185
	C. 3-Butenyl, 4-Pentenyl, 6-Heptenyl, and 7-Octenyl Radicals	198
	D. Alkynyl Radicals	202
	E. Alkenoxy, Alkenaminyl, and Other Heteroatom-Centered Radicals	203
	F. Alkenylaryl Radicals	209
	G. Intramolecular Aromatic Homolytic Substitution	211
	H. Other Cyclizations	216
V.	Ring-Opening Reactions	220
	A. Cycloalkyl Radicals and Related Species	220
	B. Cycloalkylcarbinyl Radicals	227
	C. Cycloalkyloxy and Similar Heteroradicals	238
VI.	Rearrangement by Transfer of a Heteroatom-Centered Group	241
	A. Group Mobilities	241
	B. Carbon to Carbon Migration	242
	C. Migration to or from Heteroatoms	247
VII.	Isomerization by Transfer of a Halogen Atom	248
	A. 1,2-Migration of Halogen	248
	B. Transfer of Halogen to a More Remote Radical Center	251
VIII.	Isomerization by Hydrogen Atom Transfer	251
	A. Atom Mobilities	251
	B. Carbon to Carbon Migration	253
	C. Carbon to Oxygen Migration	258
	D. Carbon to Nitrogen Migration	264
	E. Oxygen to Carbon Migration	266
	F. Oxygen to Oxygen Migration	268

IX.	ROTATION	270
	A. Allylic Radicals	270
	B. Bridged Radicals	273
X.	INVERSION	276
	A. Carbon-Centered Radicals	276
	B. Heteroatom-Centered Radicals	281
	REFERENCES	283

I. INTRODUCTION

This essay is concerned mainly with organic monoradicals—that is, organic species that contain a single unpaired electron—and their unimolecular reactions to form isomeric monoradicals. Most of the reactions that will be discussed have been studied in solution. However, identical, or analogous, processes can generally be expected (and are occasionally known) to occur in the gas phase and even in the solid state.

In his foreword to "Molecular Rearrangements," de Mayo (*1*) resolved the problem of what constituted a rearrangement by accepting as such the occurrence of any "change in atomic disposition in the molecule (with concomitant bond cleavage, σ or π, and reformation)." The title of the present essay has been chosen in order to continue the fine tradition established by Walling (*2*), Freidlina (*3*), and Wilt (*4*) in their comprehensive and imaginative reviews of the same topic. However, de Mayo's solution does not comfortably accommodate all unimolecular radical "rearrangements." As an extreme example consider the inversion of a pyrimidal, trivalent, silicon-centered radical that has been derived from an optically active molecular precursor:

No bond has been cleaved, and yet the radical has been "rearranged" to form its enantiomer. In this essay, the rather vague term "rearrangement" will be extended to include such processes. However, we intend to distinguish between rearrangements that involve changes in the basic skeleton of the radical and those that do not. The latter will hereafter be referred to as "isomerizations."

Rearrangements can, for convenience, be subdivided into group transfers, ring closures, and ring openings.

Group transfers (Sections III and VI) can be represented by the generalized one-step, or concerted, reaction,

4. FREE-RADICAL REARRANGEMENTS

$$A-B-C_n-\dot{D} \xrightarrow{\quad} \dot{B}-C_n-D-A$$

The most commonly encountered group transfers involve a 1,2 shift ($n = 0$) of certain unsaturated groups such as phenyl. The classic example of such a process, and the first radical rearrangement to be identified, is the neophyl rearrangement, which was discovered in 1944 by Urry and Kharasch (5). More remote group transfers are also known, including 1,4 shifts ($n = 2$) and 1,5 shifts ($n = 3$).

Ring-closure reactions (Section IV) normally involve an intramolecular addition to an unsaturated function. The most commonly encountered ring closure can be represented by the generalized reaction.

$$A{=}B-C_n-\dot{D} \xrightarrow{\quad} \overset{\frown}{\dot{A}-B-C_n-D}$$

Such cyclizations normally yield five- or six-membered rings (6–6b). The $n = 0$ case involves the *formal* migration of a double bond in an allylic-type radical and will not be dealt with in this review. Less common ring closures involve additions to the outer terminus of the unsaturated function,

$$A{=}B-C_n-\dot{D} \xrightarrow{\quad} \overset{\frown}{A-\dot{B}-C_n-D}$$

and additions to coordinatively unsaturated atoms,

$$A-B-C_n-\dot{D} \xrightarrow{\quad} \overset{\frown}{A-\dot{B}-C_n-D}$$

Ring-opening reactions (Section V) are essentially the reverse of the ring-closing processes.

Isomerizations can be conveniently subdivided into atom transfers, rotations, and inversions.

Heteroatom transfers (Section VII) normally involve the migration of a halogen atom, with a 1,2 shift being common. This contrasts with hydrogen transfers for which the true 1,2 shift probably does not occur. More remote migration of halogen, e.g., 1,5 shifts, are also known, so these reactions have features in common with both hydrogen atom and group transfers. This class of isomerizations also encompasses those processes in which the atom bearing the unpaired electron is itself transferred. Although such reactions do not appear to be known, the only requirement for their occurrence would seem to be that the radical in question contain a suitably located atom [e.g., P(III)] or group that is capable of accepting a divalent atom.

$$A-B-C_n-\dot{D} \xrightarrow{\quad} A-B({=}D)-\dot{C}_n$$

$$A{=}B-C_n-\dot{D} \xrightarrow{\quad} A-B-\dot{C}_n \atop \underset{D}{\vee\!\!\vee}$$

Hydrogen atom transfers (Section VIII) can be considered to be intramolecular hydrogen atom abstractions.

$$H-B-C_n-\dot{D} \longrightarrow \dot{B}-C_n-D-H$$

By far the most facile and commonly encountered isomerizations of this type involve a 1,5-hydrogen transfer, which occurs via a six-center cyclic transition state (see also abstractions by excited species in Essay 20).

Rotations (Section IX) involve a change in the conformation* of the whole radical that is brought about by bond rotation. The cis–trans isomerization of allylic radicals provides the most important example of this type of process.

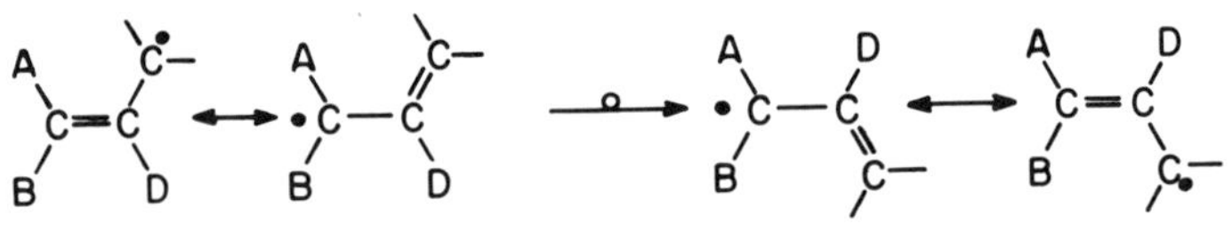

Another example of rotational isomerization that is of some structural and mechanistic interest is provided by certain "bridged" radicals, notably β-bromo-substituted alkyls, which lose their stereochemical identity rather slowly (8, 9). Thus, for example, when $BrCH_2\dot{C}(CH_3)C_2H_5$ is generated from an optically active precursor, the degree of racemization of the products implies that this radical can hold its initial (bromine-bridged) conformation for ca. 10^{-8} sec at ambient temperatures (8, 10).

Inversions (Section X) involve a change in the configuration of the groups attached to the radical center without any concomitant bond breaking or bond making. This type of isomerization is, in general, most easily observed in those Group IV-centered radicals in which the unpaired electron occupies an orbital that has appreciable s character. Such radicals include nonplanar trisubstituted radicals, notably the triorganosilyl and triorganogermyl radicals (11). They also include certain carbon-centered radicals that have strongly electron-withdrawing groups (particularly fluorine atoms and alkoxy groups) attached to the radical center (7, 12–14) or in which the radical center is incorporated in a small ring (7). The majority of vinyl radicals also have "bent" structures (15–16) and therefore can, in principle, undergo isomerization between their Z and E

---

* Following Kochi's suggestion (7), conformation is used herein to describe the geometry of the whole radical, and configuration is used to describe the geometry at the radical center.

4. FREE-RADICAL REARRANGEMENTS 165

isomers. Such isomerizations must be inversions since they occur too rapidly to involve a rotational process in which the π bond is broken.

In this essay the topics discussed most thoroughly are those dictated by our preference for quantitative work and absolute rate measurements. This field of free-radical chemistry is now sufficiently well developed (2-4) that there can be little excuse in current and future work for qualitative statements such as "radical A rearranges faster than radical B." The proper understanding of rearrangements and future work on this topic require that "faster" be quantified.

Radical rearrangements and isomerizations have a growing importance in more general studies of organic reaction mechanisms. This is because the presence, or otherwise, of radical intermediates in a reaction of uncertain mechanism can frequently be probed by using as the potential intermediate a species that, if it is a free radical, will undergo an irreversible and unequivocal rearrangement or isomerization. In one of its simplest manifestations, the reactant would be optically active and a loss of activity in the product could implicate a nonchiral free-radical intermediate (e.g., a planar or rapidly inverting alkyl radical). Once radicals have been implicated in a reaction, the next step is to use as a probe a species that will undergo rearrangement or isomerization at a known rate and, furthermore, at a rate that is competitive with the rate of the elementary step in which the radical is involved, since this allows the rate of the elementary step to be determined. Numerous examples will be described.

II. MEASUREMENT OF RATES OF UNIMOLECULAR RADICAL REACTIONS

Under most conditions, only exothermic unimolecular radical reactions are sufficiently rapid to compete with bimolecular reactions between the radical and its environment. However, this does not imply that the unimolecular process will give the thermodynamically most favored product. Just as in any other chemical system, the rearrangement or isomerization pathway that is followed is that which is kinetically favored. The speed of any type of unimolecular radical reaction can, of course, be varied enormously by appropriate modifications of the structure of the reactant.

There are a variety of methods available for measuring the rates of unimolecular radical reactions. Certain processes can be studied directly, provided that the radical can be observed by ESR spectroscopy. First are those isomerizations that proceed by rotation or inversion at a rate suitable for study by ESR line broadening (7, 13, 17, 18). The ESR time scale for such processes is much faster than the NMR time scale, rate constants having to be about* 10^7 sec^{-1}. In general, free radicals can be detected by ESR spectroscopy in solution at temperatures anywhere between ca. $-150°$ and ca. $100°C$. Since we can assume that most radical isomerizations have a preexponential factor of ca. 10^{13} sec^{-1}, this technique will, in principle, yield rate constants for isomerization that, when extrapolated to room temperature, lie between ca. 10^{10} and ca. 10^5 sec^{-1}. Many radical isomerizations have been investigated in this way,† but only the vinylic inversion process (15, 15a) has actually been employed in any mechanistic or kinetic studies (29).

A procedure that would, at first sight, appear to provide chemically "useful" information in a simple and direct manner would be to follow the rearrangement or isomerization of the radical in question by the technique of kinetic ESR spectroscopy (30, 31). Unfortunately, this procedure is unsuitable for the vast majority of radicals that might, or do, undergo unimolecular reactions. This is because the radicals must generally be produced photochemically at concentrations of ca. 10^{-6} M so that they can be detected and their decay can be monitored when the light is cut off. At such concentrations most radicals decay with *second-order* kinetics by a diffusion-controlled *bimolecular* self-reaction (30). Even when decay does occur with *first-order* kinetics, this does *not* normally imply a *unimolecular* process. Instead, it generally indicates either an attack on solvent (32), starting materials (33), or impurities (34) or, alternatively, the occurrence of a fast radical–dimer equilibrium and a slow decay of the dimer or a slow, irreversible radical–radical reaction (30, 35). However, true unimolecular decay processes can be studied for radicals in which other decay paths have been blocked, generally by steric hindrance (36). This procedure has been employed to study the isomerization by H atom transfer of sterically hindered phenyls (37) (see Section VI). It has also been used to study a few scission reactions, e.g.,

$$(CH_3O)_3\dot{P}OC(CH_3)_3 \longrightarrow (CH_3O)_3P=O + (CH_3)_3C^\bullet \quad (38)$$

$$[(CH_3)_3C]_2\dot{C}=N^\bullet \longrightarrow (CH_3)_3CC\equiv N + (CH_3)_3C^\bullet \quad (39)$$

---

* The rate constant for the process that produces magnetic equivalence is given by $k = 6.22 \times 10^6 \Delta a$ sec^{-1}, where Δa is the difference (in gauss) in the hyperfine splittings of the magnetically inequivalent atoms. See, e.g., Russell et al. (18a).

† For example, axial–equatorial exchange in cyclic radicals, such as cyclohexyl (15, 19), piperidine nitroxide (20), and 1,3-dioxan-2-yl (21); rotations about formally single bonds (7), as in alkoxyalkyls (22, 23), alkylthiylalkyls (23), and alkanoylalkyls (24–25); ligand exchange in phosphoranyl radicals (26); and rotation in certain substituted allyl radicals (27–27c). The allyl radical itself shows no sign of rotation on the ESR time scale at temperatures as high as 280°C (28). This implies that any process exchanging the CH$_2$ protons requires a free energy of activation of more than 17 kcal/mol (28). Barriers to rotation in some benzyl radicals have been measured by this technique (28a).

4. FREE-RADICAL REARRANGEMENTS

Of possibly greater importance in the long run is the application of the well-developed adamantane matrix technique (40) to kinetic studies. Radicals generated in an adamantane matrix are completely isolated from one another. They exist in cavities within the matrix that are of sufficient size that radicals as large as benzyl can rotate freely. This means that sharp, isotropic, ESR spectra are obtained (40). Sustmann and Lübbe (41) have shown that, because of this isolation, it is possible to monitor rearrangements that are much too slow to be examined by kinetic ESR in solution (see Section V). This method holds great promise, and much research on slow unimolecular radical reactions in matrices can be expected in the future.*

Indirect procedures for measuring the rate of a unimolecular radical reaction involve a competition between this process and some other reaction of known, or readily determined, rate. As normally employed, this involves a combination of product analysis with a "rotating-sector" study of an appropriate reaction (30). For example, the ratio of 1-hexene to methylcyclopentane formed in the radical chain reaction between 5-hexenyl bromide and known concentrations of tri-n-butyltin hydride at 40°C was measured (43):

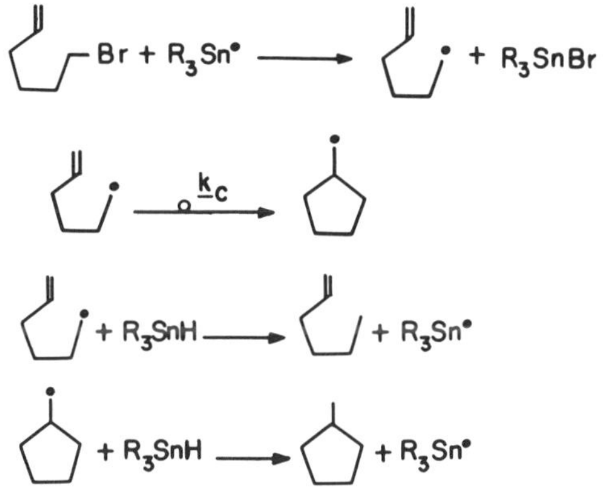

Subsequently, the kinetics of the alkyl halide–tin hydride reaction was studied (44) and the rate constant determined for hydrogen abstraction from the tin hydride by n-hexyl radicals. Since all n-alkyl radicals are expected to be equally reactive, the two sets of data could be combined to obtain $k_c \approx 1 \times 10^5$ sec^{-1} at ambient temperatures for cyclization of the 5-hexenyl radical (see Section IV).

---

* There is, of course, always the danger that the "rearrangement" may actually involve elimination and readdition since the two species cannot diffuse apart in a matrix. It seems likely that this is the mechanism for the otherwise unprecedented rearrangement, $(CF_3)_3CCF_2\cdot \longrightarrow (CF_3)_2\dot{C}CF_2CF_3$, observed in a perfluoroneopentane matrix (42).

Electron spin resonance spectroscopy provides a somewhat simpler technique for the indirect measurement of the rates of unimolecular radical reactions, provided that conditions can be found in which both the unrearranged radical U and the rearranged radical R can be detected simultaneously under steady-state conditions. If the radicals react according to the following scheme,

$$U \xrightarrow{k_r} R$$

$$\left. \begin{array}{l} U + U \xrightarrow{k_t^U} \\ U + R \xrightarrow{k_t^{UR}} \\ R + R \xrightarrow{k_t^R} \end{array} \right\} \text{nonradical products}$$

the usual steady-state treatment (38, 45, 46) yields the equation

$$\frac{1}{[R]} = \frac{2k_t^R[R]}{k_r[U]} + \frac{2k_t^{UR}}{k_r}$$

The relative and absolute concentrations of U and R can be varied by changing the rate of radical production. A plot of 1/[R] against [R]/[U] yields a straight line of slope $2k_t^R/k_r$. The rate constant for the rearrangement is then obtained following direct measurement of $2k_t^R$ by the kinetic ESR method (30) under similar experimental conditions. In certain cases, it can safely be assumed that $k_t^{UR} = k_t^R$, in which case the above equation reduces to

$$k_r/2k_t^R = ([R]^2/[U]) + [R]$$

and only one measurement of the U and R concentrations is necessary. This technique has been quite widely employed to determine rate constants and Arrhenius parameters for radical rearrangements and scissions (25, 38, 45–51), although occasionally only the activation energy has been measured (27b, 52–54). The rate constants that are *actually* measured are generally ca. 10^3 sec^{-1}. However, fast reactions are studied at low temperatures and slow reactions at high temperatures, and the rate constants extrapolated to 25°C have varied from a high of 1.3×10^8 sec^{-1} (51) to a low of 10 sec^{-1} (25).

III. REARRANGEMENT BY TRANSFER OF A CARBON-CENTERED GROUP

A. Group Mobilities

Under "normal" experimental conditions, a radical rearrangement in solution will be only fast enough to compete with the other processes by

4. FREE-RADICAL REARRANGEMENTS

which the radical can be destroyed if its activation energy is less than *ca.* 15 kcal/mol. The favored direction for the rearrangement is determined by the thermodynamic stabilities of the unrearranged and rearranged species. An approximate idea as to which radical rearrangements might occur under appropriate circumstances can be obtained by examining analogous intermolecular processes since the conditions that govern inter- and intramolecular reactions are similar except for the different constraints imposed by acyclic and cyclic transition states.

Intramolecular group transfers are akin to bimolecular homolytic substitutions (S_H2 reactions) (55). Both processes occur readily at carbon, provided that there is a low-lying unfilled orbital available to accept the unpaired electron in the transition state (and/or intermediate). Thus, homolytic substitution at sp^2-hybridized carbon is generally a facile process (56, 57), and so aryl, vinyl, and similar unsaturated groups migrate readily, but homolytic substitution at sp^3-hybridized carbon occurs only under rather special circumstances (55, 58). The lowest-energy pathway by which a radical can "rearrange" a simple alkyl group such as methyl or *t*-butyl involves its elimination and readdition, e.g. (59),

$$(CH_3)_3CCCH_2 \longrightarrow (CH_3)_3\dot{C} + O{=}C{=}CH_2 \xrightarrow{k_{add}} (CH_3)_3CCH_2\dot{C}{=}O$$
(with O double-bonded to central C on left)

In solution, or in the solid phase, it may be difficult to distinguish such a process from a true 1,2 intramolecular migration. Such mechanistic dilemmas can be unambiguously resolved by carrying out the reaction in the gas phase. If this is not possible, it is necessary to determine (by trapping or by "crossover" experiments) whether any of the presumed intermediate radicals or unsaturated molecules "leak out" of their solvent cages. Leakage from the solvent cage during the elimination–readdition reaction will be significant if the analogous intermolecular addition is slower than diffusion-controlled (i.e., if k_{add} is less than *ca.* $10^9 \, M^{-1} \, sec^{-1}$).

The stability of a ring system depends on its internal strain, but its ease of formation also depends on how much time the two reactive centers spend in propinquity (60). This decreases with increasing separation of the two centers since the number of possible conformations that an acyclic system can assume increases with the length of the system. This is one reason that three-membered rings, although subject to considerable strain, are formed more easily than four-membered rings (60).* Group

---

* In some radical reactions three-membered ring formation may be favored by more subtle stereoelectronic factors. For example, the ready formation of epoxides from β-peroxyalkyls (61, 61a) may arise because the half-filled orbital in the reactant and the O—O bond are nearly collinear in the transition state (277, 391a, Section IV, H). It is known that S_Hi reactions [like S_H2 and S_N2 processes (55)] occur by "backside" attack and that a linear transition state is favored (see also Section VI).

transfers by 1,2 shifts (which involve three-membered cyclic transition states) are therefore much more common than 1,3 shifts (four-membered rings) and somewhat more common than 1,4 and 1,5 shifts. This is also true for heteroatom transfers but not for hydrogen atoms since true 1,2 shifts do not occur with hydrogen (see Section VIII).

B. Aryl Migration

1. Neophyl rearrangement and related processes (1,2-aryl shifts)

Urry and Kharasch's (5) pioneering discovery of the neophyl rearrangement,

$$C_6H_5C(CH_3)_2\dot{C}H_2 \longrightarrow C_6H_5CH_2\dot{C}(CH_3)_2$$
$$\quad\quad\quad 1 \quad\quad\quad\quad\quad\quad\quad 2$$

was based on a careful study of the products formed during the cobaltous chloride-catalyzed reaction of neophyl chloride with phenylmagnesium bromide. The occurrence of this Ar_1-3 rearrangement can nowadays be very simply verified by UV photolysis of a di-t-butyl peroxide solution of t-butylbenzene in the cavity of an ESR spectrometer. At room temperature and below, the neophyl radical 1 is observed,

$$(CH_3)_3CO\cdot + C_6H_5C(CH_3)_3 \longrightarrow (CH_3)_3COH + C_6H_5C(CH_3)_2\dot{C}H_2$$
$$\quad\quad\quad\quad\quad\quad\quad\quad\quad\quad\quad\quad\quad\quad\quad\quad\quad\quad\quad\quad\quad 1$$

whereas at higher temperatures the spectrum due to the 2-benzylprop-2-yl radical 2 makes its appearance (50, 52). The rate constants at 25°C and the activation parameters for the neophyl rearrangement and some related rearrangements (25, 44, 50, 62–66) are recorded in Table 1.

There is abundant evidence that the neophyl rearrangement is intramolecular (4). The reaction must therefore proceed through a spiro-[2.5]octadienyl radical (3) as an intermediate or transition state.

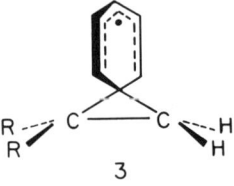

No cyclohexadienyl radical of this structure has yet been observed by ESR. In fact, the reaction of t-butoxy radicals with 4 at $-166°C$ in propane in an ESR spectrometer gave only the 2-phenylethyl radical 5 (64).

4. FREE-RADICAL REARRANGEMENTS

TABLE 1

Kinetic Data for Some 1,2-Aryl Shifts

Unrearranged radical	k at 25°C, sec^{-1}	log(A/sec^{-1})	E, kcal/mol	Ref.
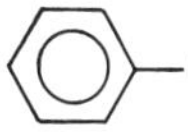 Ph—CMe$_2$ĊH$_2$ [a]	59	11.75[b]	13.6	50[c]
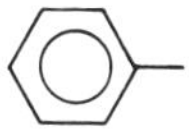 Ph—CMe$_2$ĊH$_2$	1400	11.8	11.8	50[c]
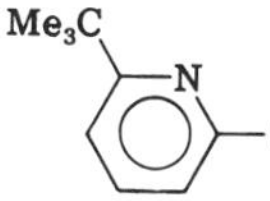 (Me$_3$C-pyridyl)—CMe$_2$ĊH$_2$	800	11.7	12.0	50[c]
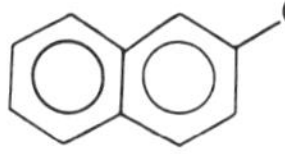 (2-Naphthyl)—CMe$_2$ĊH$_2$	2900	11.75[b]	11.3	50[c]
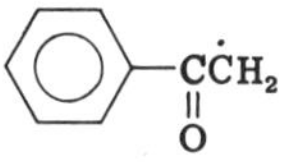 Ph—C(=O)ĊH$_2$	10	11.8[b]	14.7	25[c]
(Ph)$_3$—CĊH$_2$	3.6 × 10^5	11.8[b]	8.5	62[d]
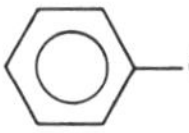 Ph—CH$_2$ĊD$_2$	<0.01	11.8[b]	>19	64
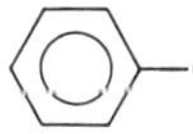 Ph—CMe[OC(O)Me]ĊH$_2$	120	11.8[b]	13.2	65[e]
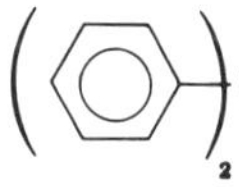 (Ph)$_2$—CMeO·	≥10^6			66[f]

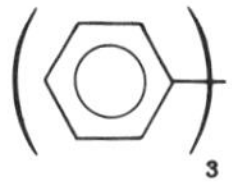

[a] The radical from 1,3-di-t-butylbenzene has the same kinetic parameters.
[b] Assumed (50).
[c] By the steady-state ESR method.
[d] A comprehensive product study of the competition,

$$(C_6H_5)_3CCH_2\cdot \xrightarrow{k_1} (C_6H_5)_2\dot{C}CH_2C_6H_5$$
$$(C_6H_5)_3CCH_2\cdot \xrightarrow{k_2, \; n\text{-Bu}_3\text{SnH}} (C_6H_5)_3CCH_3$$

$$\text{4} \xrightarrow{\text{Me}_3\text{CO}^\bullet} \text{5 (C}_6\text{H}_5\text{-CH}_2\dot{\text{C}}\text{H}_2\text{)}$$

4 → **5**

However, nanosecond flash photolysis of **4** in di-*t*-butyl peroxide at 25°C gave an absorption spectrum ($\lambda_{max} \approx$ 565 nm) similar to that of the cyclohexadienyl radical produced from 1,3- or 1,4-cyclohexadiene by the same technique (*67*). The rate of hydrogen abstraction from **4** was only about twice as large as the rates of abstraction from the two cyclohexadienes. This suggests that abstraction and 2-phenylethyl formation are not concerted (*67*).

Since **3** (R = H) is so short-lived it seems unlikely that **3** (R = CH$_3$) will be observed in the neophyl rearrangement (even if it is an intermediate and not merely a representation of the transition state; see 1,2-acyloxy migration, Section VI). In fact, a recent attempt to detect **3** (R = CH$_3$) by the CIDNP method during a neophyl rearrangement was unsuccessful (*68*).

Neophyl-like rearrangements are facilitated by the following:

1. Stabilization of the intermediate spiro[2.5]octadienyl radical **3** by greater delocalization of the unpaired electron [compare the migratory aptitudes of phenyl and β-naphthyl (Table 1)].

2. Stabilization of **3** by the *gem*-dialkyl (Thorpe–Ingold) effect. Both heterolytic (*69*) and homolytic (*61*) (see also Section IV) ring-closure reactions yielding three-membered rings proceed much more rapidly when methyl groups are introduced into the reactant. This *gem*-dialkyl effect is usually ascribed to compression of the internal angle of the ring and expansion of the external angle as steric strain between the alkyl groups is relieved. The importance of this effect in the neophyl rearrangement does not appear to have been investigated, possibly because it is not easily separated from effects due to the thermodynamic driving forces for the overall reaction (*vide infra*). It would be interesting to compare the rates of the thermoneutral rearrangements of the same Ar group in a set of three isotopically labeled ArCH$_2$ĊH$_2$, ArCHMeĊHMe, and ArCMe$_2$ĊMe$_2$ radicals.

3. Stabilization of the rearranged radical relative to the unrearranged radical by relieving steric strain and by forming a radical with a greater stabilization energy (*36*). That is, the overall driving force for neophyl-like

---

yields $k_1/k_2 = 0.36 M$ at 25°C by extrapolation (*62*). At this temperature, $k_2 = 1 \times 10^6 M^{-1} \text{sec}^{-1}$ (*44*). The assumed value for log(A_1/sec^{-1}) yields log(A_2/M^{-1} sec^{-1}) = 8.5, as would be expected for an H atom abstraction (*63*), and E_2 = 3.4 kcal/mol.

e Based on the products of reaction of this radical with tri-*n*-butyltin hydride at 70°C, taking the same A_2 and E_2 values as in footnote *d*.

f See text.

4. FREE-RADICAL REARRANGEMENTS

rearrangements comes from the relief of strain as the congested primary alkyl is converted to a less crowded tertiary alkyl radical, and the rate is strongly enhanced when the tertiary alkyl can also be further stabilized by resonance [see the rates of rearrangement of $C_6H_5CMe_2\dot{C}H_2$ and $(C_6H_5)_3C\dot{C}H_2$ (Table 1)]. Rearrangements that do not have such thermodynamic driving forces and are not assisted by the *gem*-dialkyl effect are much slower, e.g., rearrangement of isotopically labeled 2-phenylethyl radicals (64, 70) (Table 1) and of the $(C_6H_5)_2CH^{14}\dot{C}HC_6H_5$ radical (71). Strain also manifests itself in the fact that **6** ($n = 2$) rearranges more rapidly than **6** ($n = 3$) (72).

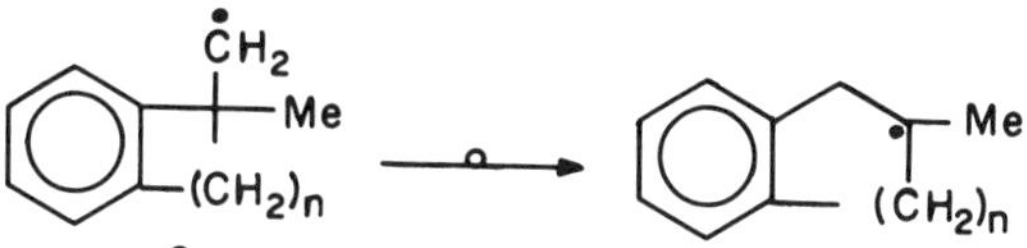

4. Electron-withdrawing groups (particularly *p*-CN and *p*-NO$_2$). This was initially observed for groups attached to the migrating aromatic nucleus (73–75) and was attributed to the transition state having some polar character and being stabilized by charge separation as indicated below, the extent of the polar contribution (i.e., **7b** and **7c**) increasing as X became more electron withdrawing:

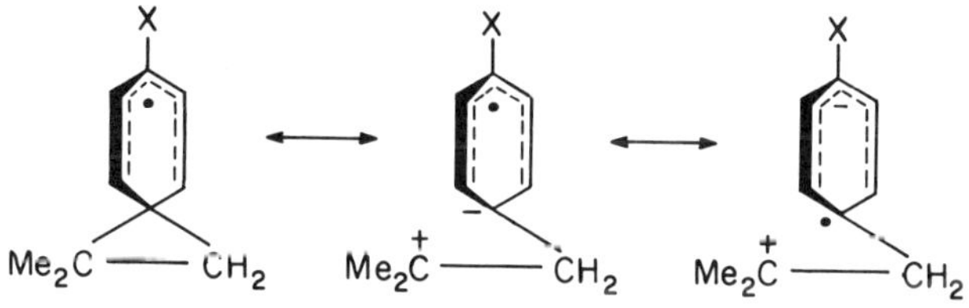

However, substituent effects at the migratory origin in the neophyl-like rearrangement of (9-*p*-X-phenyl-9-fluorenyl)carbinyl radicals (**8**) have shown that this reaction is *also* accelerated by electron-withdrawing X substituents (76).

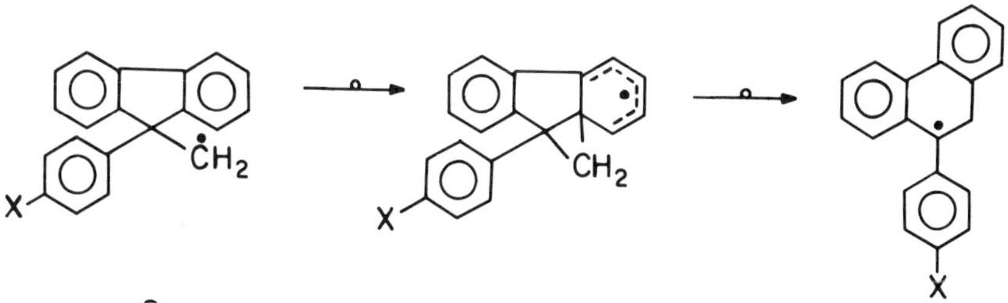

This result militates against a charge-separated transition state similar to one proposed (75) that has partial positive change at the migratory origin. Reversal of the charges in the **8**⎯⎯**9** transition state would be inconsistent with the polar effect observed when X is attached to the migratory aryl (73–75) [e.g., in p-$NO_2C_6H_4C(C_6H_5)_2\dot{C}H_2$ the p-nitrophenyl/phenyl migration ratio is at least 8 (74)]. The "polar" effect of substituents is therefore more apparent than real. It can be attributed to delocalization of the unpaired electron into the substituent, which may stabilize either the spiro[2.5]octadienyl intermediate (when XC_6H_4 migrates) or the rearranged product radical (when XC_6H_4 does not migrate). Any true polar effect on the neophyl rearrangement must be of negligible importance (76).

1,2-Aryl shifts are not restricted to hydrocarbon radicals (77) or even to carbon-centered radicals. Thus, triarylmethoxy radicals rearrange extremely rapid (73, 78, 78a), and aryl migratory aptitudes follow a pattern similar to that observed in carbon–carbon shifts (73, 78a). The 1,1-diphenylethoxy radical **10** rearranges much more rapidly at 30°C than it undergoes β scission (66):

$$(C_6H_5)_2CMeO^\cdot \xrightarrow{\beta\ scission}\!\!\!\!\!\!\!\!\!\!\!\!/\!\!\!\!\!\!\rightarrow (C_6H_5)_2CO + Me^\cdot$$
$$\mathbf{10}\quad\quad\longrightarrow C_6H_5\dot{C}MeOC_6H_5$$

Since the β scission of 1,1-diphenylethoxy should be at least as fast as that for the cumyloxy radical $C_6H_5Me_2CO^\cdot$ ($k_\beta^{30°C} \approx 3 \times 10^4$ sec^{-1}), the rate constant for the diphenylethoxy rearrangement must be greater than 10^6 sec^{-1} at this temperature (30).* For thermodynamic reasons† 1,2-aryl shifts from carbon to oxygen, and from carbon to nitrogen (which appear to be unknown), are expected to be faster than carbon–carbon shifts in structurally analogous radicals. As a corollary, unsuccessful attempts to observe O to C (80)‡ and N to C (82) 1,2-aryl shifts can be attributed, in part at least, to adverse thermodynamics. These difficulties might be overcome by the use of an aryl group with a high migratory ability, e.g., naphthyl. The same would apply to the unsuccessful attempt to observe a 1,2-phenyl shift between sp^2-hybridized carbon atoms (83).

---

* Reliable absolute rate constants for hydrogen abstraction from alkanes by alkoxy radicals are at last available (79). These data could be combined with product studies to obtain rate constants for alkoxy rearrangements and β scissions.

† Oxygen–carbon bonds are stronger than analogous C—C bonds (63). Moreover, oxygen stabilizes an adjacent carbon-centered free radical by ca. 3–5 kcal/mol (63) by virtue of conjugative delocalization of the unpaired electron and the p-type lone pair on oxygen.

‡ At elevated temperatures phenoxymethyl would appear to rearrange to benzyloxy (81). This reaction is endothermic by ca. 8 kcal/mol.

4. FREE-RADICAL REARRANGEMENTS

$$(C_6H_5)_2C=\dot{C}R \; -\!\!\!/\!\!\!\to\!\!\!/\!\!- \; C_6H_5\dot{C}=CRC_6H_5$$

$$R = H, \; p\text{-}CH_3C_6H_4$$

1,2-Aryl shifts from silicon to carbon are also unknown (84). Since analogous reactions involving 1,2-phenyl shifts from silicon to oxygen have been reported (85), e.g.,

$$(C_6H_5)_3SiOO\cdot \longrightarrow (C_6H_5)_2\dot{S}iOC_6H_5$$

the failure to detect an aryl shift from Si to C is rather surprising (84).

Other 1,2-aryl shifts between two heteroatoms appear to be unknown. However, there would seem to be no intrinsic reason why the following four types of rearrangement should not be readily observed using appropriately structured radicals in a suitable environment (including, if necessary, a matrix):

$$R\dot{N}OAr \rightleftharpoons Ar RNO\cdot$$

$$R\dot{N}NR'Ar \rightleftharpoons ArRN\dot{N}R'$$

The chosen structure must, of course, be one that does not allow the radical to decay by fast processes, such as β scission, e.g.,

$$R\dot{N}NR'Ar \longrightarrow RN=NAr + R'\cdot$$

An attempt to observe the rearrangement of $C_6H_5{}^{15}N=\dot{N}$ was unsuccessful (86), although it might, of course, have succeeded with a better migrating group.

It should be apparent not only that there are many neophyl-like rearrangements for which quantitative rate data could be obtained quite readily, but that there are many new classes of neophyl-like rearrangements awaiting discovery.

2. 1,3-, 1,4-, and 1,5-Aryl shifts

These rearrangements have been thoroughly reviewed (4,6–6b). There do not appear to be any authentic 1,3-aryl shifts, but 1,4 and 1,5 shifts occur with considerable facility. In these rearrangements, which are designated Ar_1-5 and Ar_1-6 reactions, respectively, the aryl may migrate between two carbon atoms, between carbon and a heteroatom, or between two heteroatoms. 1,4-Aryl shifts are often accompanied by an Ar_2-6 reaction (see Section IV), which usually yields cyclized products having two (or more) fused rings, i.e., a Tetralin type of structure.

A variety of radicals having an aryl group in the 4 position, i.e., **11**, have been found to undergo 1,4-aryl rearrangements, e.g., $C_6H_5CMe_2(CH_2)_2CH_2\cdot$ (87), $C_6H_5CMe_2(CH_2)_2\dot{C}=O$ (88), $\alpha\text{-}C_{10}H_7(CH_2)_3CD_2\cdot$ (6b, 89), $(C_6H_5)_3CCH_2CO_2\cdot$ (4), $C_6H_5SiMe_2(CH_2)_2CH_2\cdot$ (4, 90, 90a), and $p\text{-}MeC_6H_4SO_2\overline{NCHCH_2}\cdot$ (91). An interesting subclass of such rearrangements involves migration of an aryl group across a cyclohexane ring, e.g. (92),

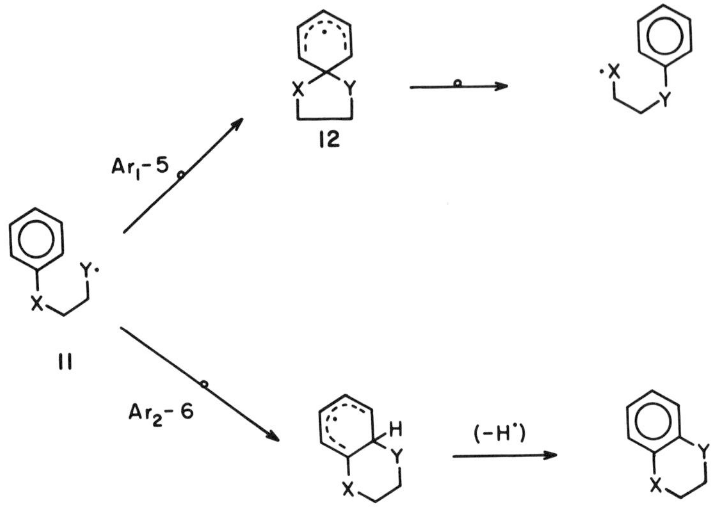

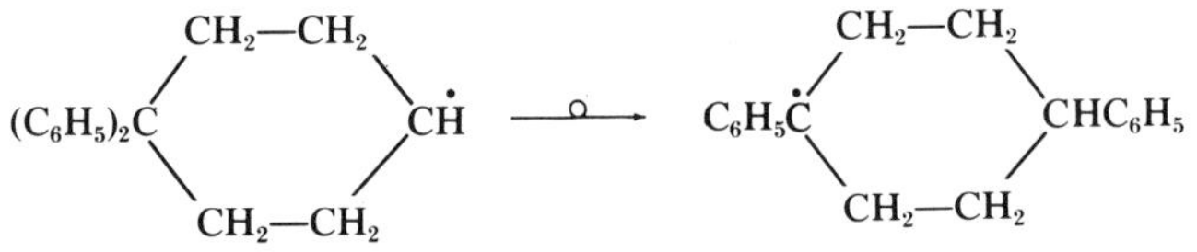

Although there is little quantitative information regarding the rates of 1,4-aryl migrations, these reactions would appear to be reasonably fast when they are strongly exothermic as, for example, in the conversion of a primary alkyl radical to a tertiary (87) (compare the 1,2-aryl shifts). For the transannular rearrangement **13** ⟶ **14**, we estimate from the reported data (92) that the rate constant is ca. $1\text{--}5 \times 10^4$ sec^{-1} at 150°C. Even thermoneutral 1,4 shifts can be fairly rapid with a "good" migrating group. Thus, the rate constant for the migration of the α-$C_{10}H_7(CH_2)_3CD_2^{\cdot}$ has been estimated (6b, 89) to be ca. 10^4 sec^{-1} at 80°C, and the activation energy for the Ar_1-5 cyclization has been estimated (6b, 89) to be about 0.6 kcal/mol less than that for the Ar_2-6 process (see Section IV). By utilizing data concerning the cyclization (both Ar_1-5 and Ar_2-6) of 4-phenylbutyl and its oxidation by cupric ion (93) and accepting a value of $\sim 10^8$ M^{-1} sec^{-1} as the rate constant for the oxidation (94), the rate constant for the cyclization can be calculated to be ca. 5×10^4 sec^{-1} at

4. FREE-RADICAL REARRANGEMENTS

50°C (6). This value must be too high since phenyl would not migrate nearly as rapidly as α-naphthyl. Moreover, the radical resulting from cyclization of 4-phenylbutyl could not be detected by ESR spectroscopy at $-75°$ to $-90°C$ (95, 96) or at 25°C (6), which implies that the rate constant for cyclization is less than 10^3 sec^{-1} at each of these temperatures. This conclusion is consistent with the absence (<2%) of the products of rearrangement of the $C_6H_5(CH_2)_3CD_2\cdot$ radical produced by photolysis of the parent iodide (97). For the 1,4 migration of phenyl from Si to C in $C_6H_5SiMe_2(CH_2)_2CH_2\cdot$, we estimate from the literature (90a) a rate constant of $\sim 3 \times 10^3$ sec^{-1} at 140°C. The reverse of this reaction is three times as rapid.

Rearrangements of the Ar$_1$-5 type must proceed by way of a spirodecadienyl radical intermediate (12) or transition state. Since radicals 12 are less strained than analogous spirooctadienyl radicals (3), they should be longer-lived and hence more easy to detect than 3. Qualitative information about 12 is scarce. The parent radical 15, when generated from the hydrocarbon in cyclohexane at elevated temperatures, yields both n-butylbenzene and Tetralin (98):

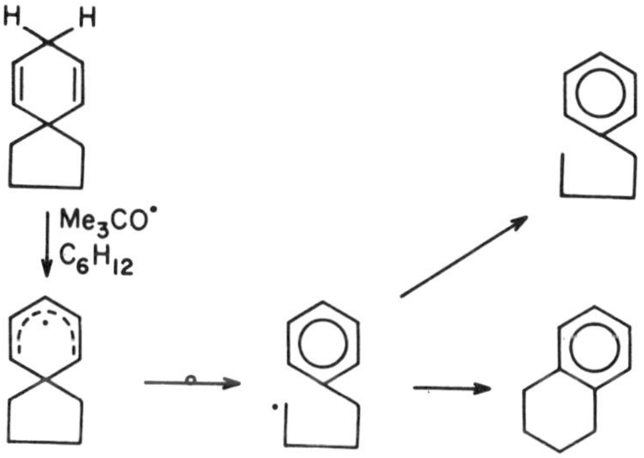

It has been estimated (6b) that 4-phenylbutyl (16) is more stable than 15 by about 2.5 kcal/mol and is less stable than 17 (R = H) by about 6.2 kcal/mol. It is therefore no surprise that the Ar$_2$-6 cyclization of 4-phenylbutyl is not reversible (<1%) up to 200°C (99).

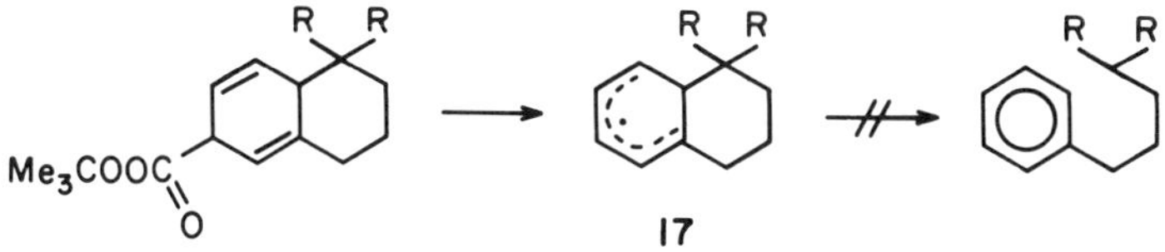

R = H, CH$_3$

When the migratory origin and terminus are both Me_2Si groups, only spiro radicals (**18**) can be observed by ESR spectroscopy (*100*), even at temperatures as high as 50°C. The kinetic stability and persistence of these three spiro radicals should make them useful as models for the intermediates involved in more typical Ar_1-5 ($n = 2$), Ar_1-6 ($n = 3$), and Ar_1-7 ($n = 4$) rearrangements.

n = 2, 3, 4

1,5-Aryl shifts have not received quite as much attention as the 1,4 rearrangements. However, they do appear to follow the same ground rules, and examples are known of aryl migration between two carbon atoms (*92*), between carbon and a heteroatom (*90, 90a, 101*),* and between two heteroatoms (*102*). In cases in which data are available for homologous radicals, the 1,4-aryl migration is sometimes faster [e.g., C to C (*92*), Si to C‡] and sometimes slower [e.g., C to Si (*101*)] than the 1,5 shift. It is clear that a great deal of qualitative and quantitative work remains to be done on 1,4- and 1,5-aryl shifts.

C. Vinyl Migration

The migration of vinyl groups has been less thoroughly explored than the migration of aryl groups. There is a wealth of evidence that 1,2-vinyl shifts in allylcarbinyl radicals (homoallylic rearrangements) occur by way of cyclopropylcarbinyl radicals as discrete intermediates (*103–105*). For example (*103*), decarbonylation of 3-methyl-4-pentenal (**19**) gave 3-methylbutene and 1-pentene (in a ratio that increased with increasing aldehyde concentration) together with trace quantities of the *cis*- and *trans*-1,2-dimethylcyclopropanes. The same products were formed from 2-methyl-4-pentenal (**20**) (see p. 179).

Similarly (*105*), γ,γ-diphenylallylcarbinyl (**21**), and cyclopropyl diphenylcarbinyl (**22**) interconvert rapidly at 125°C relative to the rate at which they react with triethyltin hydride (see p. 179).

Even when products from the intermediate cyclopropylcarbinyl radical cannot be identified, it has been shown that this radical survives long enough for rotation about the exocyclic bond to occur, e.g. (*104a*).

* The 1,5-phenyl shift from carbon to silicon is not mirrored by a 1,4 shift. In the latter case, only the products of cyclization are observed; see also Sakurai *et al.* (*101a*).

‡ See note on data of Wilt *et al.* (*90, 90a*) contained in footnote 146 of Wilt (*4*).

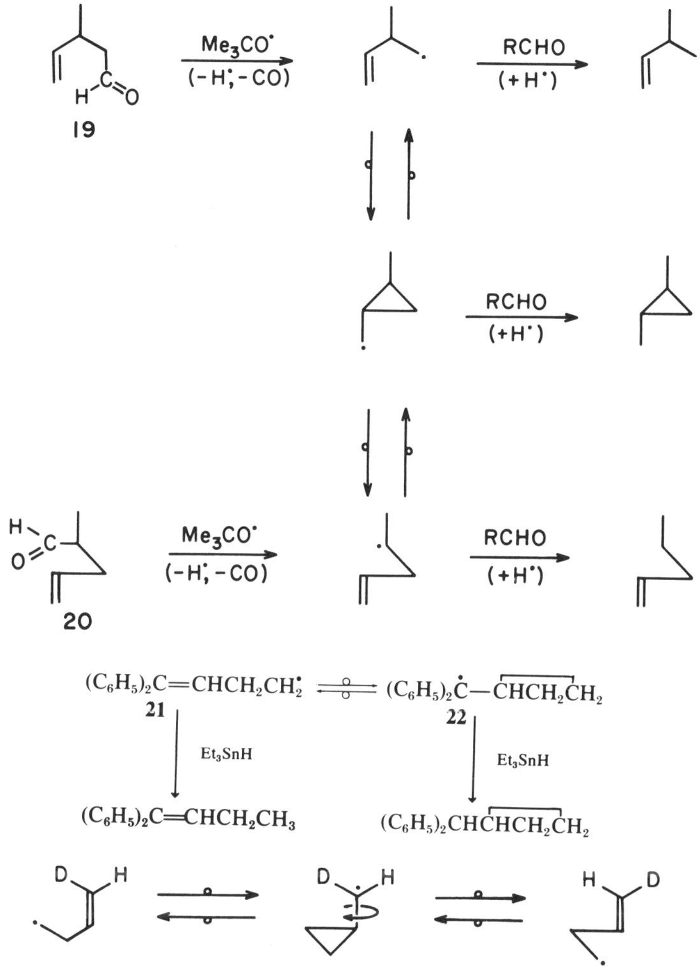

4. FREE-RADICAL REARRANGEMENTS

1,2-Vinyl rearrangements are therefore intimately related to ring-closing (Section IV) and ring-opening (Section V) reactions.

The rate constant $k_{1,3}$, for the ring-closing of CH_2=$CHCH_2CD_2\cdot$ has been measured both by a kinetic ESR method and by the tin hydride technique (243). It can be represented by $\log(k_{1,3}/\text{sec}^{-1}) = (10.3_6 \pm 0.5) - (9.0_9 \pm 0.5)/\theta$, with $k_{1,3} = 4.9 \times 10^3$ sec^{-1} at 25°C. Since the ring-opening of the cyclopropylcarbinyl radical which is formed is a very much faster process (51), the overall rate of the vinyl migration will, if secondary isotope effects are small, be one half (for statistical reasons) of $k_{1,3}$.

There are few quantitative data regarding the rates of 1,2-vinyl rearrangements that go to completion. Thus, it has only recently been shown that the neophyl rearrangement is much slower than the analogous vinyl migration, **23** ⟶ **24** (*106, 107*).

$$H_2C=CHC(CH_3)_2\dot{C}H_2 \longrightarrow H_2C=CHCH_2\dot{C}(CH_3)_2$$
$$\quad\quad\quad\quad\quad\textbf{23} \quad\quad\quad\quad\quad\quad\quad\quad\quad \textbf{24}$$

In the vinyl migration that converts 2-cyclopentenecarbinyl (**25**) to 3-hexenyl (**27**) (*108*), the bicyclic radical **26** can be intercepted with certain spin traps (*109*). The bicyclic radical must be formed reversibly

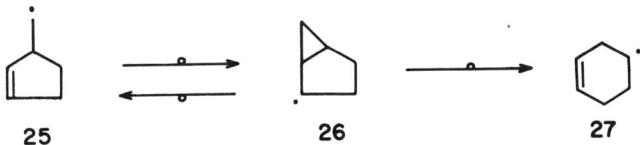

25　　　　　　　**26**　　　　　　　**27**

since the ring opening of **26** (*110*) and related radicals (*110–113*) yields the primary alkyl preferentially (see also Section V). Thus (*110*), reaction of the parent halides of **25, 26,** and **27** with *n*-Bu₃SnH in a 1:1 mole ratio yielded the following products:

$$\textbf{25} \xrightarrow{65°C} 0.87\ \textbf{25H} + 0.13\ \textbf{27H}$$
$$\textbf{26} \xrightarrow{3°C} 0.92\ \textbf{25H} + 0.05\ \textbf{26H} + 0.03\ \textbf{27H}$$
$$\textbf{27} \xrightarrow{65°C} 1.0\ \textbf{27H}$$

Therefore, $k(\textbf{26}\rightarrow\textbf{25})/k(\textbf{26}\rightarrow\textbf{27})$ is *ca.* 30. Moreover, $k(\textbf{26}\rightarrow\textbf{25})$ can be estimated* to be *ca.* $0.5–3.0 \times 10^6$ sec^{-1} at 3°C.

In an analogous vinyl rearrangement, the neat parent bromide of **28** was reacted with equimolar *n*-Bu₃SnH or (C₆H₅)₃SnH (*111*):

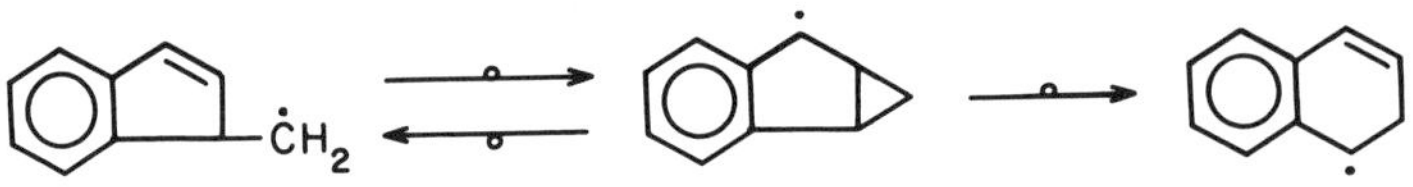

28　　　　　　　**29**　　　　　　　**30**

The products from the two hydrides were similar, which suggests that their composition was determined by the equilibrium composition of the three radicals (*111*), and so the rate constant for the vinyl shift cannot be estimated. However, separate experiments with the parent bromide of **29** suggest that $k(\textbf{29}\rightarrow\textbf{28})/k(\textbf{29}\rightarrow\textbf{30})$ is *ca.* 14 and that $k(\textbf{29}\rightarrow\textbf{28})$ is *ca.* 1–5 $\times 10^5$ sec^{-1}.* The vinyl migration (*110*),

---

* Based on the assumption that the rate constant for H abstraction from *n*-Bu₃SnH by **26** is *ca.* $1–5 \times 10^5$ sec^{-1} [and that for abstraction from (C₆H₅)₃SnH is four times as large] at 25°C; see Carlsson and Ingold (*44*).

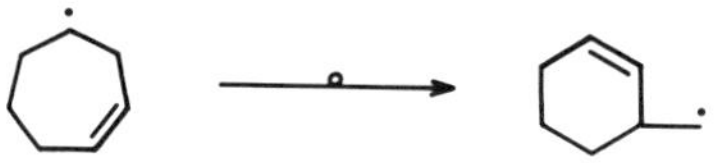

can be estimated to have an overall rate constant of *ca.* $1-5 \times 10^4 \text{ sec}^{-1}$. This reaction is not reversible *(110)*.

A 1,2-vinyl migration from carbon to oxygen with concomitant ring expansion has been observed during the decomposition of 1-alkyl-3,5-di-*t*-butyl-4-oxo-2,5-cyclohexadiene peroxides *(113a)*.

There appear to be few studies specifically devoted to the detection of 1,4- or 1,5-vinyl shifts, probably because there has been more interest in identifying the intermediate cyclic radical (see Section IV).

D. Migration of Other Unsaturated Groups

There is no intrinsic reason why unsaturated groups such as C=O, C≡N, C=N, and C≡C should not also rearrange under appropriate conditions, but few such reactions have been unequivocally identified. Thus, the products formed by the thermolysis of some *t*-butyl perlevulinate esters have been interpreted in terms of an intramolecular 1,2-acyl shift *(114)*, i.e.,

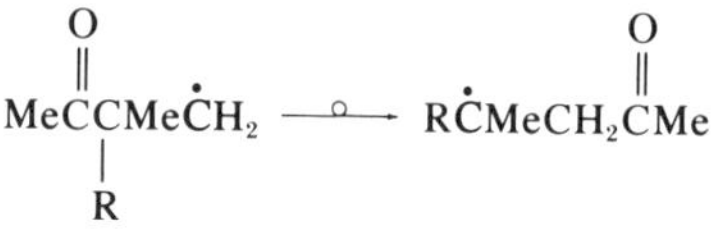

$$\underset{\underset{R}{|}}{\text{MeCCMeĊH}_2}\overset{\text{O}}{\overset{\|}{}} \longrightarrow \text{RĊMeCH}_2\overset{\text{O}}{\overset{\|}{\text{CMe}}}$$

R = CH₃, C₆H₅

However, an elimination–readdition seems somewhat more likely since it has been shown *(115)* that acetyl radicals can be trapped in the following apparent rearrangement:

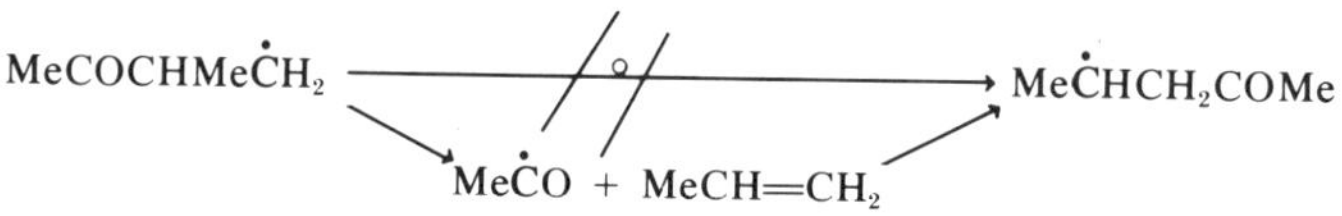

$$\text{MeCOCHMeĊH}_2 \longrightarrow \text{MeĊHCH}_2\text{COMe}$$
$$\searrow \text{MeĊO} + \text{MeCH=CH}_2 \nearrow$$

The rate constant for the purported intramolecular acetyl migration between the O atoms in **31** can be represented by $\log(k/\text{sec}^{-1}) = 12.9 - 11.4/\theta$ *(116)*.

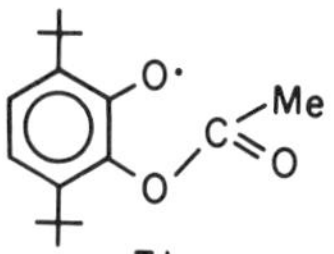

31

The involvement, or otherwise, of "free" acyl radicals in acyl "rearrangements" could be readily determined if phenylacetyl were used in place of the acetyl group. Phenylacetyl undergoes an extremely rapid α scission (25, 117):

$$C_6H_5CH_2\dot{C}O \xrightarrow{k_\alpha} C_6H_5CH_2\cdot + CO$$

$$k_\alpha \approx 5 \times 10^7 \text{ sec}^{-1} \text{ at } 25°C \quad (25)$$

The rate constant for the neophyl-like homopropargyl rearrangement,

$$Me_3CC\equiv CCMe_2\dot{C}H_2 \longrightarrow Me_3CC\equiv CCH_2\dot{C}Me_2$$

has been measured by kinetic ESR spectroscopy (50b) and can be represented by $\log(k/\text{sec}^{-1}) = (12.4 \pm 1.5) - (14.6 \pm 2.4)/\theta$, with $k = 49 \text{ sec}^{-1}$ at 25°C. The rate and Arrhenius parameters for this reaction are similar to those of the neophyl rearrangement (see Section III). The corresponding vinyl migration is a much faster process (106).

Rearrangements involving a 1,4-cyano shift in a steroidal radical (118), and in simple analogues (118a),

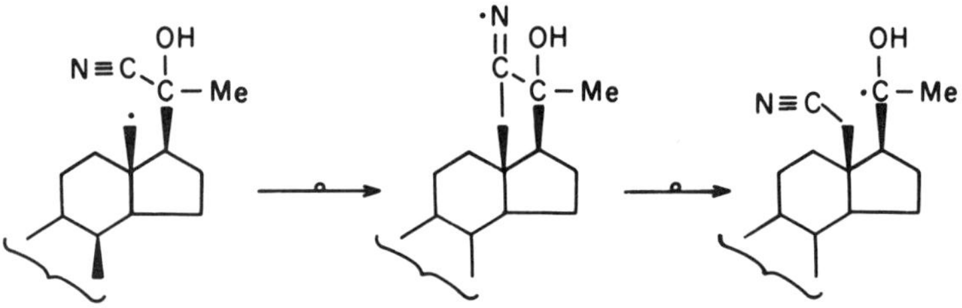

and an expansion of the "unsaturated" cyclopropane ring (119) are worth noting.

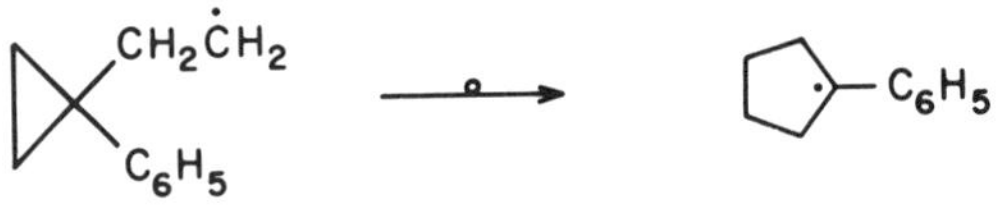

To conclude this section, we note that information on the rearrangement of carbon-centered groups is highly fragmented and is mainly of a qualitative nature.

IV. RING-CLOSURE REACTIONS

A. Comparisons with Analogous Intermolecular Processes

Ring closure occurs when a suitably constituted radical undergoes intramolecular addition, either to an unsaturated function or to a coordinatively unsaturated atom.

4. FREE-RADICAL REARRANGEMENTS

Intermolecular radical addition to multiple bonds has been studied extensively (57, 120, 121) not only because a number of synthetically useful transformations include such a step in their mechanisms, but also because of its importance in vinyl polymerization. Well-known examples include those in which the radical bears its free spin on carbon (R·, Ar·), sulfur (RS·), oxygen (RO·, ROO·), nitrogen ($R_2NH^+_·$), silicon ($R_3Si·$), phosphorus ($R_2P·$), and tin ($R_3Sn·$). The unsaturated reactant may contain a localized multiple bond, e.g., C=C, C≡C, C=O, C≡N, N=N, N=O, or an extended π system such as those in aromatic nuclei (56, 122, 123) or conjugated polyolefins. Most reported cyclizations involve intramolecular reactions of the same groups and radicals (4, 6–6b, 124).

There has been less extensive investigation of radical additions to coordinatively unsaturated atoms, a type of reaction that often constitutes the first step in S_H2 processes (55, 125). An adduct radical is likely to be formed only if the acceptor atom has accessible a relatively low-lying orbital, e.g., S in R_2S, B in R_3B, P in R_3P, and if the newly formed bond is of comparatively high energy.

The preferred kinetic pathway for many radical reactions is that which has the most favorable thermochemistry.* Radical addition to an unsaturated substrate generally, although by no means always, occurs in the direction that affords the most stable possible product; a tertiary alkyl radical is formed in preference to a secondary alkyl radical, and homolytic aromatic substitution proceeeds preferentially at positions of lowest atom-localization energy (56, 122, 123, 127, 128).

The addition process is usually completed by a chain-transfer step. In most cases the orientation of the final product reflects that of the preferred radical adduct, but this is not necessarily so when *fast, reversible* addition is followed by a *slow* transfer step (129). Nor, under these kinetic circumstances, do the relative rates of formation of final products from closely related substrates necessarily reveal the relative rates of the primary addition steps (130).

However, the major reason why predictions of the directions and relative rates of addition reactions based on the energies of initial and final states break down, even in acyclic systems, is because they fail to allow for steric and polar effects (120, 131, 132). Why these are often important is revealed by a consideration of the intimate structure of the transition state. A recent model (32) of the transition complex for addition of methyl radical to ethylene (133) incorporates the three participating carbon atoms at the vertices of a slightly obtuse triangle lying in a plane orthogonal to

---

* The thermochemistry of many radical addition processes can be reliably estimated from thermodynamic data, and methods exist for the similar calculation of approximate kinetic parameters (63, 126).

that of the olefin's σ framework. The complex is dipolar; the incoming radical is usually computed to assume a fractional positive charge (i.e., to behave as a nucleophile), whereas the ethylene moiety becomes slightly negative. The formation of the transition state involves primarily interaction of the semioccupied 2p orbital with one lobe of the unoccupied π^* orbital. This interaction defines both the shape and the polarity of the transition complex. Substituents affect the energy of the complex through their interactions with both the delocalized free spin and the fractional charges and through steric effects developed by the mutual approach and change in configurations of the original radical center and seat of reaction.

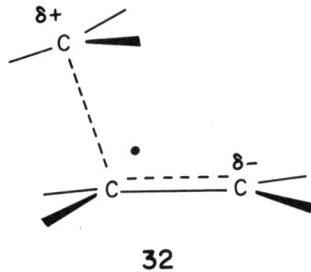

32

Ring-closure reactions respond in the same way as analogous intermolecular processes to thermochemical factors and to steric and polar effects, but they are also subject to additional constraints connected with the formation of cyclic systems (*134, 135*). For example, ring strain makes a direct contribution to ΔH for an intramolecular radical reaction, and this is reflected in the energy of the transition complex. On this basis alone, the relative rates of formation of carbocyclic systems should be in the order six- > seven- > four- > three-membered. However, entropy changes also make a major contribution to the free energy of activation. In general, intramolecular additions occur more readily than their intermolecular analogues, for the latter involve a substantial unfavorable loss of translational entropy, whereas the former involve only the loss of internal rotational degrees of freedom. Furthermore, the entropy change due to loss of rotational freedom becomes increasingly unfavorable with increasing chain length. When both enthalpy and entropy changes are taken into account, it is seen that intramolecular addition occurs most rapidly at ordinary temperatures when the ring formed contains three, five, or six members.

Another effect of some importance in ring closures arises from the presence of alkyl substituents at positions other than those *directly* involved in the reaction. Comparison of substituted acyclic compounds with analogous cyclic systems discloses that there are fewer extra gauche

4. FREE-RADICAL REARRANGEMENTS

interactions due to alkyl substituents in the cycles than there are in the open chains (*136, 137*). Also, the change in hybridization accompanying the formation of small rings relieves steric compression between gem substituents (*136, 137*). Finally, substitution has a favorable effect on the entropy of ring closure (*134*). Consequently, substituted radicals should undergo cyclization more readily than the parent species.

There is another subtle, yet often profound effect that distinguishes intramolecular from intermolecular addition reactions. When two reactive centers reside within the same molecule, the intimate structure of the transition complex must be compatible with the overall structure of the reactant. Cyclization occurs only when attainment of the mandatory disposition of reactive centers does not unduly increase the strain energy of the whole system. If two modes of cyclization are available, the reaction proceeds through that transition complex which is the more readily accommodated. For intramolecular addition in most alkenyl radicals, the required disposition of centers (**32**) is more readily accommodated in the transition complex (**33**) for exo cyclization than that (**34**) for endo cyclization (*6, 138–140*):

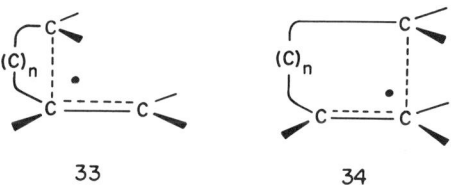

Models and statistical calculations (*141*) show that this constraint is very severe for small rings, but less so for more flexible large rings. Butenyl and pentenyl radicals should therefore undergo specific exo cyclization, but cyclization of longer chains may also afford endo products.

B. The 5-Hexenyl System

1. The 5-hexenyl radical

It has been firmly established that the 5-hexenyl radical, contrary to earlier indications (*142*), undergoes cyclization by intramolecular addition in a highly regiospecific fashion to afford mainly the cyclopentylcarbinyl radical (*43, 47, 95, 96, 143–158*).

In the early work, the 5-hexenyl radical was generated by thermolysis of di-6-heptenoyl peroxide (*143*), by interaction of 6-mercapto-1-hexene with triethyl phosphite (*144*), and by Kolbe electrolysis of 6-heptenoic acid (*145*). In each case cyclic products were formed solely or predominantly (>95%) by 1,5 ring closure. More recently, a wide variety of methods has been employed to generate the 5-hexenyl radical and to study its cyclization (*43, 47, 95, 96, 146–158*). In many experiments, products derived from the cyclohexyl radical were not detected, but careful analysis of the reaction mixture from treatment of 1-bromohex-5-ene with tributylstannane confirmed the earlier observations that 1,6 ring closure is a real, although very minor, reaction pathway at ordinary temperatures (*155*).

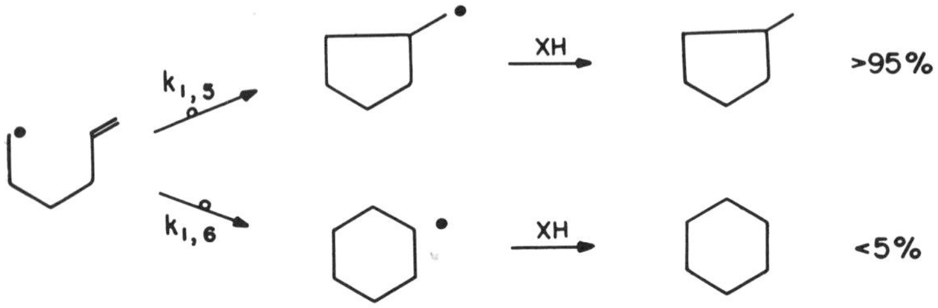

Experiments (*143, 145, 147, 148*) in which cyclopentylcarbinyl and cyclohexyl radicals were separately generated demonstrated the *irreversibility* of each mode of cyclization and the fact that they are discrete species (*95, 96*) and not a π complex (*143, 144*). Since the relevant hydrogen atom transfer steps are fast and essentially quantitative, the relative yields of methylcyclopentane and cyclohexane from the reaction of 1-halohex-5-ene with tributylstannane (*43, 152, 155, 156*) reflect the relative rates of the two ring-closure processes. Clearly, $k_{1,5} \gg k_{1,6}$. Both thermodynamic (*63, 126, 143*) and kinetic (*142, 159*) data indicate that cyclohexyl is more stabilized than cyclopentylcarbinyl. Cyclization of 5-hexenyl radical thus follows, in a highly selective fashion, the thermodynamically less favorable pathway. Why should this intramolecular radical addition behave so differently from many related intermolecular reactions (*57*)?

We adumbrated above (Section IV,A) one possible explanation (*6, 138–140*), namely, that the stereoelectronic requirements of the transition complex can be more readily attained on the 1,5 cyclization pathway than on the 1,6 pathway. That is, transition state **35** has lower strain energy than **36**.

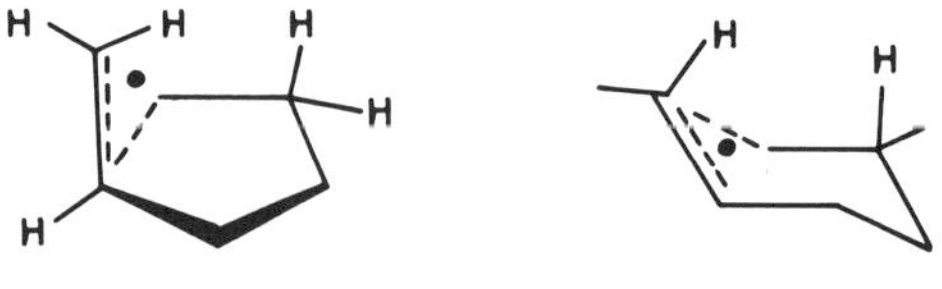

Another possible explanation (*156, 160*) arises from the suggestion (*161*) that six-membered ring formation is disfavored by nonbonded steric interaction between the pseudoaxial proton at the 2 position and the syn proton at C-6 in **36**. This view is supported by the fact that the disubstituted 5-hexenyl radical **37** affords both five- and six-membered cyclic products, whereas its geometric isomer **38**, in which the 2,6 interaction should be more severe, undergoes solely 1,5 cyclization (*156*). On the other hand, the alkenylaryl radical **39**, in which this type of steric interaction is inoperative because it contains no proton at C-2, also undergoes rapid and specific 1,5 ring closure (*139, 162*). Although further work with substituted radicals is required to define more precisely the relative importance of stereoelectronic and steric effects in controlling the direction of cyclization, it seems probable that for the parent 5-hexenyl radical the former is the factor primarily responsible for the observed regioselectivity.

37   **38**   **39**

The more favorable entropy of activation associated with the formation of the smaller possible ring (Section IV,A) has also been advanced (*163, 164*) as an explanation for the preferential 1,5 cyclization of 5-hexenyl radical. Careful scrutiny of the temperature dependence of each mode of ring closure has allowed this hypothesis to be tested (*155*). Determination of the relative yields of cyclohexane and methylcyclopentane from the reaction of 1-bromohex-5-ene with tributylstannane shows that the value of $k_{1,5}/k_{1,6}$ is 0.02 at 65°C but increases with increasing temperature. The usual Arrhenius treatment of the data indicates (*155*), as expected, that the entropy term favors 1,5 cyclization, but its magnitude ($\Delta S^{\ddagger}_{1,5} - \Delta S^{\ddagger}_{1,6}$ = 2.8 gibbs/mol) is such that it makes a minor contribution to the difference in rates of the two reactions at ordinary temperatures. In accord with hypotheses based on steric and/or stereoelectronic effects, the difference in activation enthalpy ($\Delta H^{\ddagger}_{1,6} - \Delta H^{\ddagger}_{1,5}$ = 1.7 kcal/mol) is mainly responsible for the observed regioselectivity of cyclization.

Absolute values of the rate constant for cyclization of 5-hexenyl radical have also been determined. One method (see Section II) is based on the competition between cyclization of 5-hexenyl radical and its hydrogen transfer reaction with tributylstannane. This gives $k_c = 1 \times 10^5$ sec^{-1} at ambient temperatures (*44*). A basically similar method (*149*) involving the reaction of 1-fluoro-5-hexene with sodium naphthalenide sets an upper limit for k_c of 7×10^5 sec^{-1} at 25°C. The kinetic ESR method (see Section

II) has yielded the temperature dependence of k_c (47),

$$\log(k_c/\text{sec}^{-1}) = (10.7 \pm 1.0) - (7.8 \pm 1.0)/\theta$$

and more recently (158),

$$\log(k_c/\text{sec}^{-1}) = (9.5 \pm 1.1) - (6.1 \pm 1.1)/\theta$$

Both sets of data can be extrapolated to yield $k_c = 1 \times 10^5$ sec^{-1} at 25°C. Comparison of this value with that ($k \approx 1 \times 10^3$ M^{-1} sec^{-1}) for a similar intermolecular reaction, the addition of ethyl radicals to 1-heptene at 40°C (165), reveals why cyclization of 5-hexenyl competes so effectively with possible bimolecular processes. The "effective" double-bond concentration for cyclization is ~40 M!

Recently, careful analysis of reaction products (155, 166) and the use of an integrated rate equation (167) have allowed the determination of more accurate values of k_c/k_H, where k_H is the rate constant for hydrogen atom transfer from tributylstannane to the 5-hexenyl radical. Combining the data for the temperature dependence of k_c/k_H and of k_c obtained by the ESR method (47) allows the determination of the kinetic parameters for the hydrogen atom transfer reaction. At 25°C, $k_H = 8 \times 10^5$ M^{-1} sec^{-1}, in fair agreement with earlier estimates (44), while the values of log A (8.3 ± 1.0) and E_a (3.1 ± 0.6 kcal/mol)* are reasonable for a simple metathesis reaction (63). The availability of these kinetic parameters will facilitate the determination of cyclization rate constants for a variety of alkenyl radicals by the stannane method, for it is known that the magnitude of the rate constant for hydrogen transfer is relatively insensitive to the nature of the radical (44).

The proclivity of 5-hexenyl radical for undergoing 1,5 cyclization makes the hexenyl system a useful tool for identifying the nature of reactive intermediates, for the hexenyl cation forms only the six-membered ring (168), whereas the anion undergoes 1,5 ring closure but only very slowly (169). Also, the availability of rate constant values at various temperatures allows the cyclization of 5-hexenyl radical to be used as a kinetic yardstick against which the rates of competing processes can be measured.

A nice example is provided by Kochi's work (151, 170) on the oxidative–elimination reaction of alkyl radicals with cupric ion, a key step in the synthetically useful copper-catalyzed decarboxylation of acids by lead tetraacetate (171). Decomposition of 6-heptenoyl peroxide in the presence of cupric acetate afforded 1,5-hexadiene and methylenecyclo-

---

* See also footnote d, Table 1.

pentane in good combined yield. These results are compatible with the following scheme:

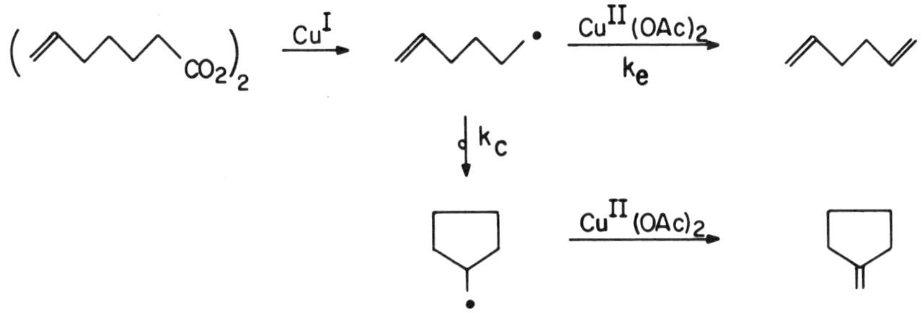

$$(\text{\textasciitilde\textasciitilde} CO_2)_2 \xrightarrow{Cu^I} \text{\textasciitilde\textasciitilde} \cdot \xrightarrow{Cu^{II}(OAc)_2}_{k_e} \text{\textasciitilde\textasciitilde}$$

$$\downarrow k_c$$

$$\bigcirc\!\!\!\!\!\cdot \xrightarrow{Cu^{II}(OAc)_2} \bigcirc\!\!\!\!\!\parallel$$

The absence of cyclohexene from the product mixture indicates that the oxidative–elimination step does not involve the intermediacy of the carbonium ion. Since oxidation of 5-hexenyl radical competes with its cyclization, the ratio of yields of acyclic and cyclic products are linearly related to cupric acetate concentration and to k_e/k_c. Since k_c was known, it was possible to determine k_e [$1.2 \times 10^6 \, M^{-1} \, \text{sec}^{-1}$ at ambient temperatures (*151*)].

Other examples of the utility of 5-hexenyl radical as a mechanistic probe and kinetic standard include the determination of rate constants for spin trapping of primary alkyl radicals (*172*) and for aromatic homolytic substitution (*173*); the elucidation of the mechanisms of the reductive demercuration of alkylmercuric halides (*153*), of the Wittig rearrangement (*174*), of the reaction of alkali benzophenone ketyls with alkyl iodides (*175*), and of the reduction of alkyl halides with sodium naphthalenide and similar reagents (*176–179*); the detection of radical intermediates in reactions of enones (*157, 180*), in the reductive elimination reactions of dialkylmercury compounds (*181*), in the reduction of alkyl halides with zinc and acid (*182*) or chromium (II) salts (*171, 183*), and in the formation and oxidation of Grignard reagents (*184*); and the determination of 1,4 diradical lifetimes (*185*).

2. Substituted 5-hexenyl radicals

The reduction of alkenyl halides with tributylstannane can be employed to determine quantitatively the rates and direction of cyclization of alkyl-substituted 5-hexenyl radicals. Although agreement among data from different laboratories (*152, 154, 156*) is not good and further work is clearly required, some interesting trends can be discerned.

The results given in Table 2 for methyl-substituted species exemplify the behavior of a much wider range of substituted radicals. Surprisingly, alkyl substitution at either the 1 or 6 position has very little effect.

TABLE 2

Relative Rate Constants for Cyclization of Substituted 5-Hexenyl Radicals at 65°–70°C

Radical	$k_{1,5}$ (rel)	$k_{1,6}$ (rel)	$k_{1,6}/k_{1,5}$	Ref.
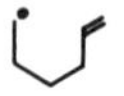	1.0	0.02	0.02	154
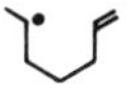	1.4	0.02	0.014	154
	1.4	0.02	0.014	154
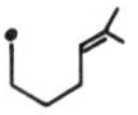	2.4	<0.01	<0.005	154
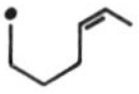	1.0	<0.01	<0.01	156
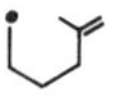	0.022 0.06	0.04 0.16	1.8 2.5	154 156
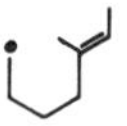	0.11	0.20	1.7	156
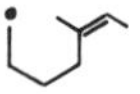	0.16	<0.002	<0.01	156
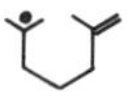	<0.0002	0.02	ca. 100	154

Five-membered ring formation is highly preferred, and values of $k_{1,5}$ are very close to that for 5-hexenyl radical. These results conflict with the simplistic view that the relative rates of closely related radical reactions reflect the relative thermodynamic stabilities of reactants and products. On this basis the rate of 1,5 ring closure should be diminished by substituents at C-1 and enhanced by substituents at C-6.

The absolute rate constant for cyclization of the secondary 6-hepten-2-yl radical can be represented by (50a) $\log(k_c/\text{sec}^{-1}) = (9.8 \pm 0.3) - (6.4 \pm 0.3)/\theta$, which gives $k_c = 1.3 \times 10^5 \text{ sec}^{-1}$ at 25°C.

One possible explanation for the observed kinetic effect of substituents at C-1 or C-5 is that at the transition state their radical-stabilizing effects are counterbalanced by polar factors. Both theoretical considerations (133) and experimental observation (132, 186) suggest that, in alkyl radical addition to an olefin, the attacking reagent behaves as a nucle-

4. FREE-RADICAL REARRANGEMENTS

ophile (see Section IV,A). Consequently, although substituents at C-1 stabilize the acyclic radical by hyperconjugative interaction with the adjacent free spin, they also stabilize the transition state by interaction with the fractional positive charge. Conversely, substituents at the 6 position effect the stability of the transition state both by their favorable interaction with the developing radical and by their unfavorable interaction with the newly formed fractional negative charge. An alternative argument (154) based on the hypothesis that the stabilizing effect of alkyl substituents on radicals arises partly from nonbonded interactions (187) suggests that at the transition state for cyclization there is little change from the original configuration at C-1 and C-5.

Substitution at C-5 has a profound effect on cyclization of the 5-hexenyl radical. Both 5-methyl- and 5-isopropyl-5-hexenyl radicals afford mainly six-membered cyclic products (154, 156). 2,6-Dimethylhept-6-en-2-yl radical, which is fully substituted at both the radical center and the seat of 1,5 ring closure, undergoes exclusive 1,6 cyclization (154). Evaluation of the rate constants (154) shows that this is due not to a major increase in the value of $k_{1,6}$ but to a dramatic decrease in the value of $k_{1,5}$. Intermolecular additions of alkyl radicals to olefins show similar sensitivity to the presence of substituents at the point of attack (188). It appears, therefore, that radical addition reactions are subject to a complex interplay of electronic and steric effects and that attempts to rationalize their regioselectivity in terms of radical stability alone are untenable (189).

5-Hexenyl radicals and related species bearing a substituent at C-1 afford a mixture of cis- and trans-disubstituted compounds by 1,5 ring closure. Preferential formation of the latter would be expected (4) since the transition state for trans cyclization (**40**) is less subject to nonbonded interactions than that for cis cyclization (**41**). Nevertheless, both the hept-6-en-2-yl radical **42a** and the allyloxyprop-2-yl radical **42b** afford mainly the cis products **43**, with $k_{43}/k_{44} = 2.3$ at 65°C in both cases (190). Other substituted radicals behave similarly (191, 192). This outcome has been attributed (190) to the favorable interaction in the transition state for cis cyclization between the hyperconjugatively delocalized orbital of the attacking radical and the π^* orbital of matching symmetry, i.e., **45** (193).

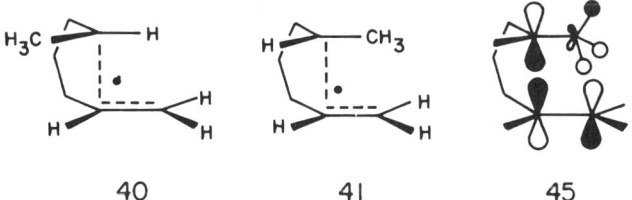

40   41   45

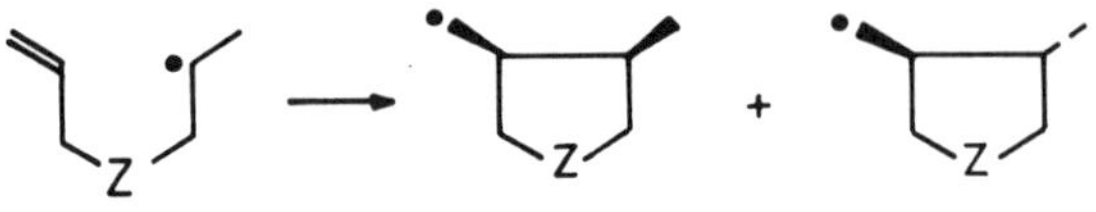

42 → 43 + 44

a Z = CH$_2$
b Z = O

Little quantitative work has been done on 5-hexenyl radicals containing substituents at the 2, 3, or 4 position. However, it has been noted (*194*) that 2,2-dimethylhex-5-enyl radical undergoes 1,5 cyclization about ten times more rapidly than does 5-hexenyl radical and that 3-propylhex-5-enyl radical (*195*) also shows an enhanced rate of ring closure. These reactions provide examples of the *gem*-dialkyl effect on radical cyclization (see Sections III,B,1 and IV,A).

Replacement of the CH$_2$ group in position 3 by an oxygen atom produces a substantial increase in the rate of 1,5 cyclization (*139, 190, 196*). For example, with the secondary alkyls **42a** and **42b**, the rate constant ratio, $(k_{1,5})_b/(k_{1,5})_a$, is 12 at 65°C (*190*) whereas, for the primary alkyls **46a** and **46b**, this ratio can be estimated to be about 30 at 65°C (*196*). The relative yield of the five-membered ring product is also enhanced by the oxygen. Thus, at 65°C, $k_{1,6}/k_{1,5}$ is *ca.* 2 for **46a** (see Table 2) but it is only *ca.* 0.026 for **46b** (*196*).

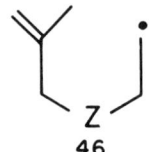

46

a Z = CH$_2$
b Z = O

When either the free spin or the double bond of the 5-hexenyl system resides within a ring, intramolecular addition leads to the formation of bicyclic systems. For example, ring closure of the cyclohexenylbutyl radical **47**, generated by interaction of the appropriate bromide with tributylstannane, affords the 9-decalyl (**48**) and spirodecyl (**49**) radicals (*138*). Formation of **49** is slightly favored ($k_{1,5}/k_{1,6} \sim 1.2$), but both modes of cyclization are considerably slower than that of 5-hexenyl radical (*44*).

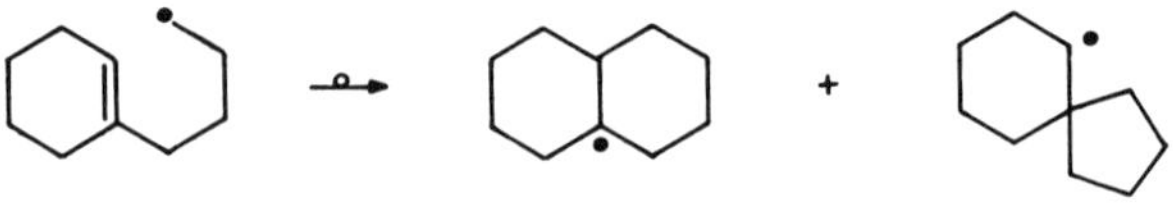

47 → 48 + 49

4. FREE-RADICAL REARRANGEMENTS

In this case the relative yields of *cis*- and *trans*-decalin formed from **48** (cis : trans = 0.17 for reaction with tributylstannane) depend on the nature of the hydrogen atom donor. However, when, as in **50**, the free spin resides in a ring, the stereochemistry of the final products is defined by the cyclization process. Unfortunately, the course of this reaction has not yet

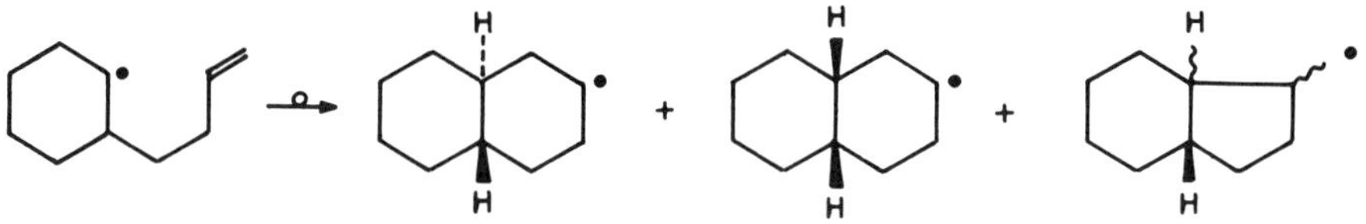

50   **51**

been completely clarified (*167, 197, 198*), but there is general agreement that 1,6 ring closure affords mainly the *trans*-decalyl radical **51**.

Other interesting examples of the formation of bi- or polycyclic products from radicals containing the 5-hexenyl system include ring closure of **52** (*199*) and **53** (*200*). Cyclization of **53** is relatively slow, presumably because of the strain energy associated with formation of the bicycloheptyl ring system.

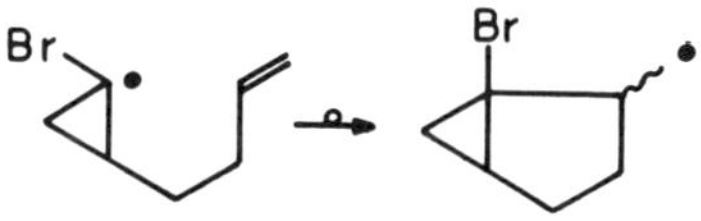

52 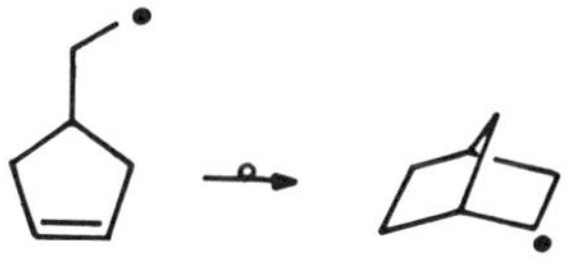   **53**

A convenient and synthetically useful method for generating substituted 5-hexenyl radicals and related species involves radical addition to an appropriate diene.

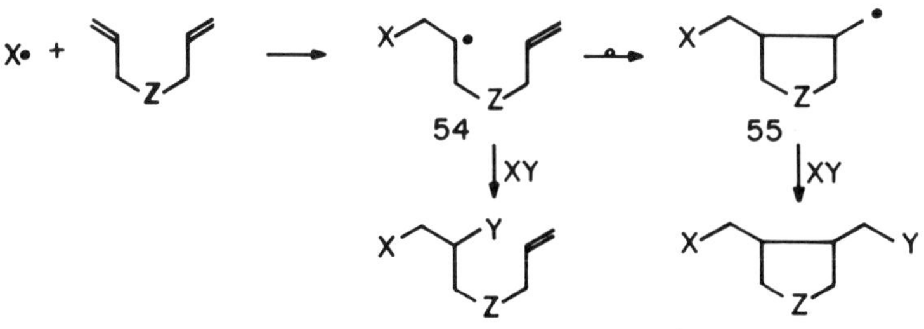

54   **55**
56   **57**

X≡OH (*201, 202*), NH$_2$ (*201, 202*), Ph (*201, 202*), perfluoroalkyl (*203–207*), RS (*192, 208*), CCl$_3$ (*201, 206*), NF$_2$ (*209*), CMe$_2$CO$_2$H (*210*), CMe$_2$CN (*211, 211a*), Cl (*211b*), PPh$_2$ (*211c*).

$Z \equiv CH_2$ (191, 204, 205, 210), $C(CO_2R)_2$ (197, 206, 209, 211b), $CHCO_2R$ (206), NR (201–203, 205, 211, 211a), NCOR (203), O (201, 205).

$(CH_2ZCH_2) \equiv Ph_2PCrPPh_2, Ph_2PMoPPh_2, Ph_2PWPPh_2$ (211c)

The ultimate fates of the radicals **54** and **55** depend on the experimental conditions. When XY is an efficient chain-transfer agent, formation of **56** competes effectively with ring closure. Conversely, in the absence of a good chain-transfer agent, cyclization proceeds readily but **55** then undertakes attack on starting diene to afford telomers and polymers. The fact that monocyclic adducts (**57**) are so frequently obtained in good yield reflects the relatively high rate of ring closure of substituted 5-hexenyl radicals as compared with the rates of intermolecular processes.

Cyclizations of the substituted 5-hexenyl radicals generated by homolytic additions to 1,6-dienes show the same features as those of the simpler systems discussed above: 1,5 ring closure is usually [see, however, Bradney *et al.* (*210*)] preferred (*191–193, 201–211a*), and cis-disubstituted products frequently predominate over trans (*192, 211*). However, substituents at the 2 and 5 positions hinder five-membered ring formation of the primary adduct, thus causing increased yields of acyclic and six-membered cyclic products (*198, 211–212*).

A good example of the effect of substituents is provided by comparison of the reactions of **58** and **59** with ethanethiol (*192*). The former reaction gives good yields of five-membered cyclic products. In the latter case, one initial 5-hexenyl radical (**61**), being substituted at both C-1 and C-5, undergoes solely 1,6 ring closure at a rate that is insufficient to compete effectively with chain transfer. The other primary adduct (**60**), being a 4-pentenyl system, does not cyclize.

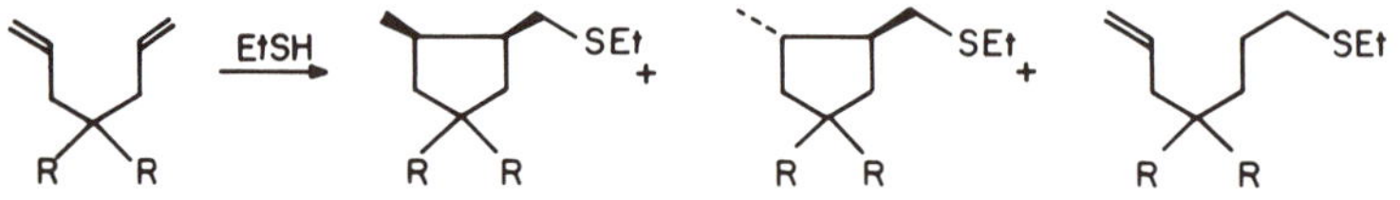

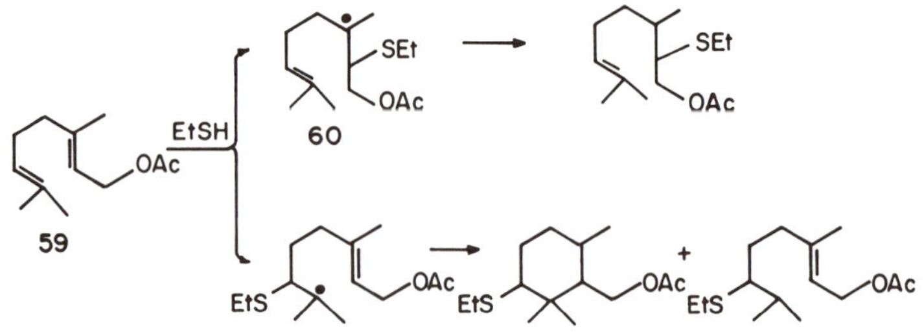

4. FREE-RADICAL REARRANGEMENTS

Other examples of the behavior of substituted dienes include the copper-catalyzed reaction of benzoyl peroxide with geranyl acetate *(213)* and with 6-methylhepta-1,5-diene *(214)* and the addition of a variety of substituted radicals to 3,7-dimethylocta-1,6-diene *(210)*.

When 1,6-dienes are treated with a suitable radical initiator in the absence of chain-transfer agents, cyclopolymers are formed *(215)*. Recent studies suggest that the structures of monocyclic adducts of dienes provide a reliable guide to the structures of their polymers. Contrary to earlier evidence *(216)*, it now appears that most cyclopolymers contain recurring five-membered cyclic units *(217)*; possibly only those formed from 1,6-dienes with substituents at the 2 and 6 positions contain a substantial proportion of six-membered rings *(217)*.

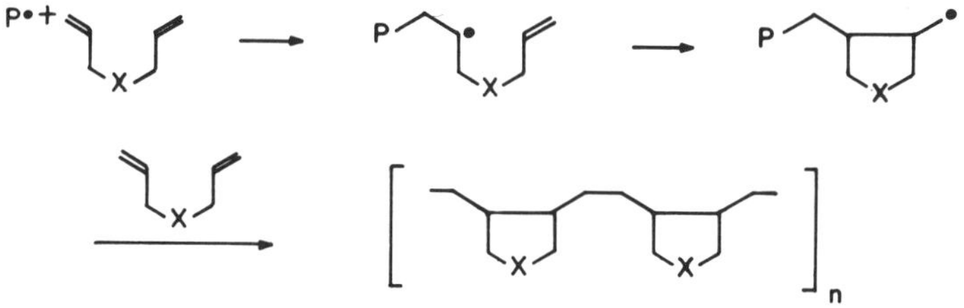

The facility with which some 1,6-dienes undergo cyclopolymerization has prompted the suggestion *(218)* that homoconjugative interaction between the double bonds enhances their reactivity toward radical addition. However, although spectroscopic evidence *(218, 219)* has been adduced in support of this hypothesis, no kinetic manifestations have been uncovered. Thus, when due allowance has been made for statistical factors, 1,6-heptadiene reacts with fluoralkyl radicals at approximately the same rate as 1-heptene *(204)*; similarly, the radicals **62** and **63** undergo 1,5 ring closure at comparable rates *(195)*.

62   63

Radical addition to cyclic 1,5-dienes affords bicyclic products. Thus, treatment of 1,5-cyclooctadiene (**64**) under appropriate conditions with adducts Y—H [Y = CH_2CO_2Ac *(220)*, $SnMe_3$ *(221)*, $COCH_3$ *(222)*, CCl_3 *(222)*, $CONHC(CH_3)_3$ *(222, 223)*, CO_2Me *(223)*, $PO(OEt)_2$ *(223)*, NH_2 *(224)*, and others *(222, 223)*] gives mainly or exclusively *exo-cis*-bicyclo[3.3.0]octane derivatives (**66**). Addition of carbon tetrachloride to **64** affords a product tentatively identified as **67** *(222)*, whereas *cis,trans*-

1,5-cyclodecadiene (**68**) undergoes similar transannular radical additions (*225*), e.g., **68** → **70** + **71**:

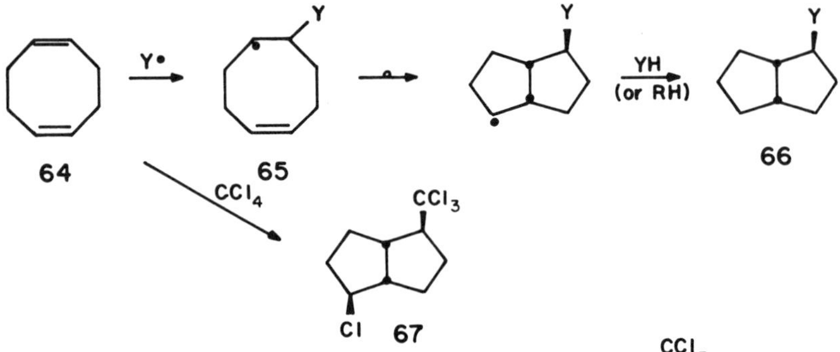

The last reaction is remarkable in that the initial radical addition appears to occur regiospecifically on the *cis*-olefin bond at the 2 position to give **69**, which then undergoes exclusive 1,6 ring closure to afford only cis-fused products **70** and **71**. However, addition of radicals to germacratriene (**72**) affords only derivatives of *trans*-decalin (**73**) in reactions that appear to be completely regio- and stereospecific (*226*), e.g.,

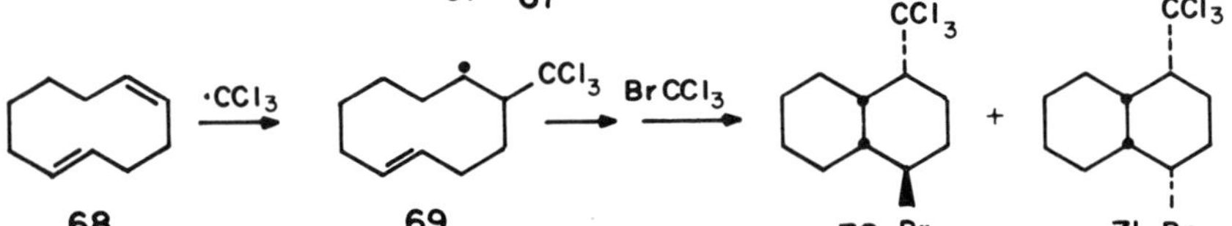

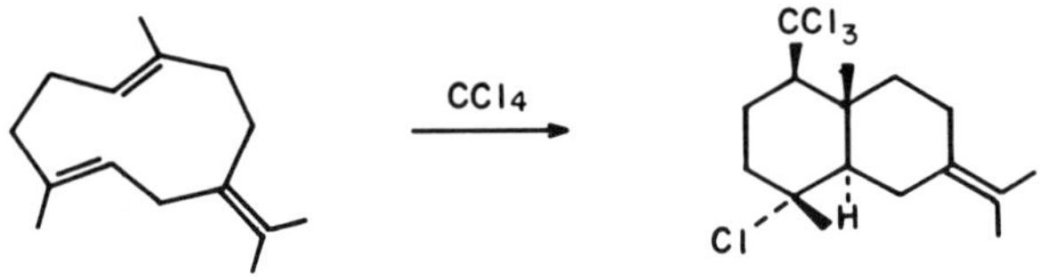

Although concerted mechanisms have been postulated for these and other transannular additions, the fact that some reactions give 1,2 addition products (*224, 227*) or show a dependence of product composition on adduct concentration (*225*) strongly suggests that they proceed in stepwise fashion via substituted hexenyl radicals (e.g., **65**). The regio- and stereoselectivity of transannular addition must therefore be attributed to steric and stereoelectronic effects on the various mechanistic steps.

Polycyclic systems can also be formed by radical additions to suitable dienes (*192, 228*) and polyenes (*229*), e.g., **74** → **75**:

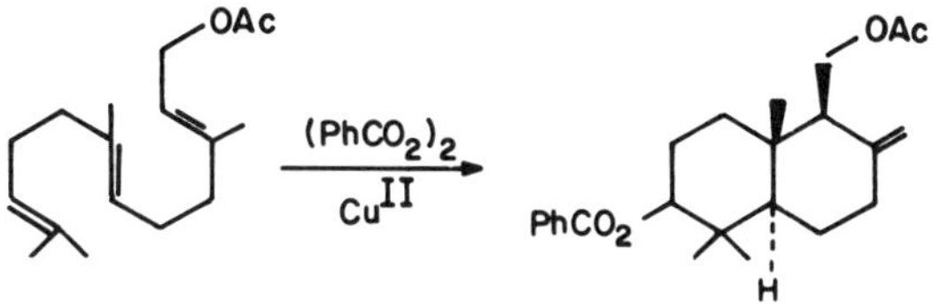

4. FREE-RADICAL REARRANGEMENTS

      74   75

Surprisingly, the above reaction appears to be both structurally and stereochemically specific, for the sole bicyclic product detected was 75.

The examples given above show that the presence of alkyl substituents at the 1 position of 5-hexenyl radical has remarkably little effect on the rate or regioselectivity of ring closure. However, the data summarized in previous reviews (4, 6–6b) and more recent results (230–234), show that this is not the case for radicals containing substituents capable of delocalizing the free spin [e.g., CH_2=$CHCH_2CH_2CH_2\dot{C}R^1R^2$, where either or both R^1 and R^2 are Ar (230, 235–237), CN (232–234, 238), CO_2R (232–234, 238), =O (231)]. Such radicals frequently afford substantial yields of six-membered cyclic products; so, also, do radicals having 2-substituents that can delocalize the free spin [see, e.g., Smith and Butler (196)]. The relative yields of adducts arising from 1,5 and 1,6 ring closure are often dependent on the experimental conditions. In some cases solely six-membered cyclic products are formed, and the reaction is then of preparative value (233). The results obtained (6a, 232–234) and exemplified here for the radical (77) derived from pentenylmalononitrile (76), reflect partial thermodynamic control of the products formed in many of these systems.

The relative yields of cyclic products depend not only on the ratio of rate constants, $k_{1,5}/k_{1,6}$, but also on the competition, $k_f/k_H[HX]$, between ring fragmentation of 79 and its chain transfer with the hydrogen donor HX. Thus, the ratio 81/80 is increased by lowering the reaction temperature or by the addition of a good hydrogen donor. The validity of the mechanistic scheme is supported by the fact that decomposition of the cyclic perester 78 and of its six-membered analogue gave mixtures of products similar to those obtained by cyclization.

Unfortunately, available data do not allow absolute values of $k_{1,5}$, $k_{1,6}$, or $k_{1,5}/k_{1,6}$ to be determined. It appears that $k_{1,5} \gg k_{1,6}$ when the 5-hexenyl radical contains a single stabilizing substituent at the 1 position, but this may not necessarily be the case when there are two such substituents present. Presumably, stereoelectronic effects are less important when the transition complex lies toward the product end of the cyclization reaction coordinate.

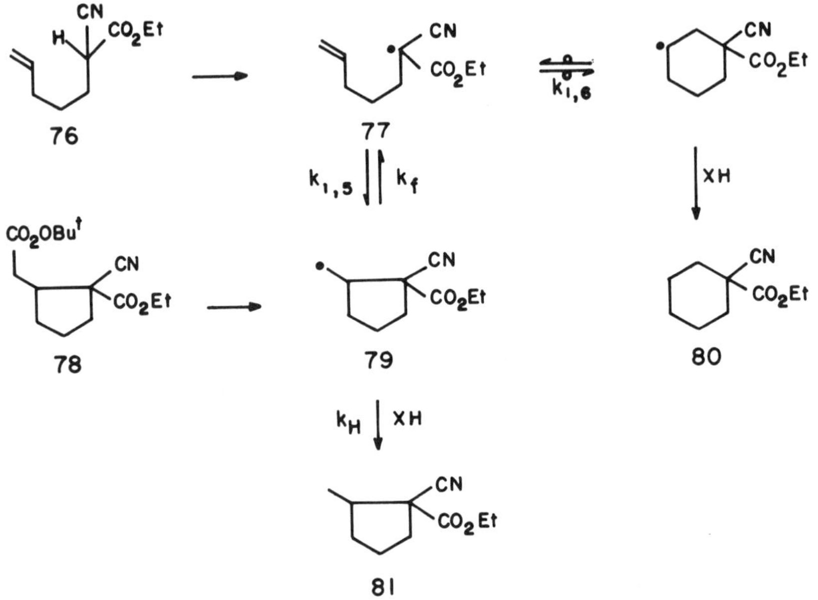

[Scheme showing structures 76, 77, 78, 79, 80, 81 with rate constants $k_{1,5}$, k_f, $k_{1,6}$, k_H, XH]

Ring closure of appropriately constituted radicals containing stabilizing substituents can give rise to bi- and tricyclic products. An example selected from Julia's work (*6a, 124, 239–241*) is:

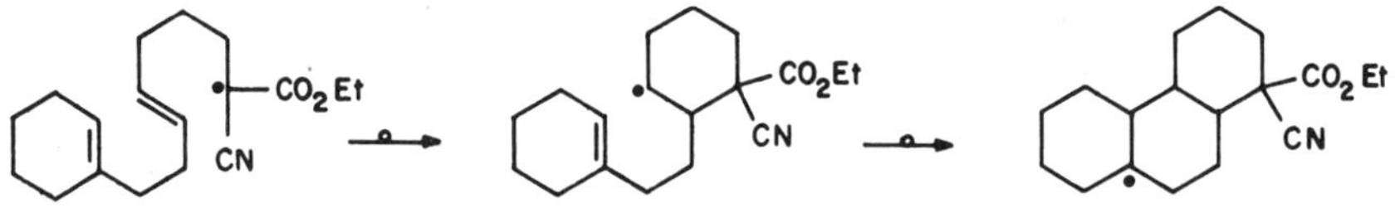

[Scheme showing tricyclic ring closure with CO_2Et and CN substituents]

Although the stereochemistry of ring closure in such cases is not clear, it seems probable that the formation of trans-fused rings is favored. What is surprising is the apparent regioselectivity of cyclization of the intermediate radicals, both of which would have been predicted on the basis of the behavior of similar unsubstituted radicals (e.g., **50**) to undergo preferential 1,5 ring closure.

C. 3-Butenyl, 4-Pentenyl, 6-Heptenyl, and 7-Octenyl Radicals

Thermodynamic criteria (*63, 126, 242, 242a*) suggest that the two possible modes of cyclization of 3-butenyl radical should be approximately equally unfavorable, for each of them is endothermic (because of ring strain) and each involves a decrease in entropy:

4. FREE-RADICAL REARRANGEMENTS

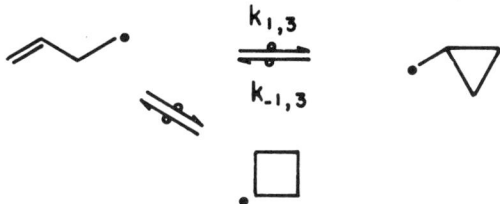

Nevertheless, ring closure occurs in a highly regiospecific manner. No evidence has been found for formation of the cyclobutyl radical from 3-butenyl, whereas the equilibrium between the latter and cyclopropylcarbinyl, although unfavorable, is established relatively rapidly. The difference is attributable mainly to the stereoelectronic demands of the transition state (Section IV,A).

Although numerous examples have been reported of 1,2-vinyl shifts proceeding via cyclopropylcarbinyl radical intermediates (see Section III,C), it has rarely been possible to obtain any cyclic products from acyclic reactants because the equilibrium lies too heavily in favor of open-chain radicals. Recent determinations (51, 243) give log K = log($k_{1,3}/2k_{-1,3}$) = $-2.12 - 3.15/\theta$ where the 2 is present for statistical reasons and $K = 3.8 \times 10^{-5}$ at 25°C. The thermodynamic parameters for this equilibrium are in satisfactory agreement with earlier estimates (242, 242a, 244).

In some compounds with a 3-butenyl system the equilibrium is more favorable and products containing the cyclopropane ring can be isolated in reasonable yield. A good example is provided by the norbornenyl system (245, 245a). Interaction of the halide **82** with tributylstannane affords a mixture of norbornene and nortricyclene in amounts which indicate that the equilibrium **83** ⇌ **84** slightly favors the uncyclized radical **83** (245) and that the rate constant in each direction is ~10^8 sec^{-1} (44). However, when **83** was generated at temperatures above -100°C in an ESR spectrometer, the radical present in the higher concentration was **84** (245a).

Radicals with the bicyclo[3.2.1]octane skeleton (**85**) are more stable than the rearranged tricyclic (**86**) and tetracyclic (**87**) radicals even when the substituent R might be expected to stabilize **87**, e.g., R = C≡N (*246*).

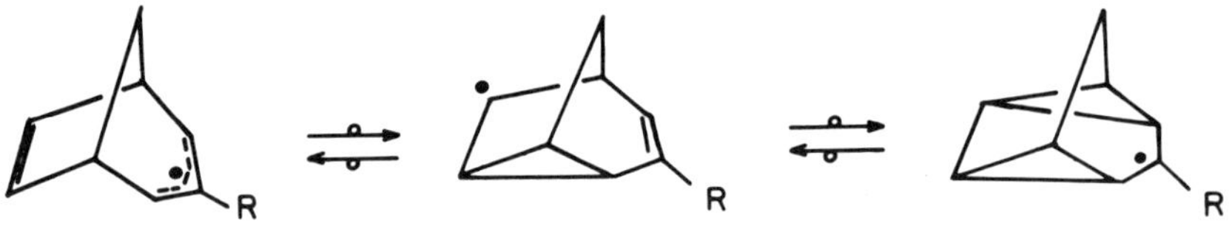

85 ⇌ **86** ⇌ **87**

Although 1,4 ring closure of the 4-pentenyl radical to give cyclobutylcarbinyl meets the stereoelectronic requirements for intramolecular addition, it is disfavored on thermodynamic grounds. Because of the ring

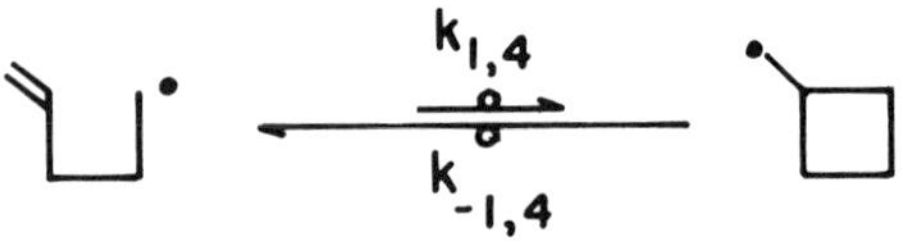

$$\ce{\cdot} \xrightleftharpoons[k_{-1,4}]{k_{1,4}} \ce{\cdot}$$

strain the cyclization is endothermic, and the equilibrium must lie well to the left. At 60°C the ring-opening reaction has $k_{-1,4} \approx 5 \times 10^3$ sec^{-1} (*166*), and consequently it is unlikely that $k_{1,4}$ can be greater than 1.0 sec^{-1}. It is hardly surprising, therefore, that cyclobutane compounds are not obtained from acyclic precursors, the only apparent exception to this rule being the curious formation of a cyclobutane derivative by radical addition of carbon tetrachloride to the diolefin **88** (*247*).

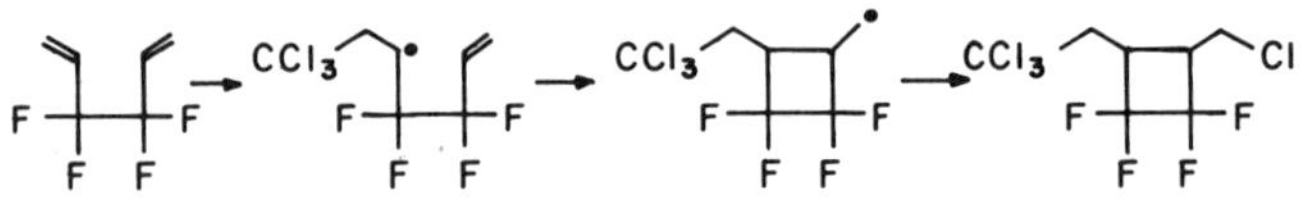

88

The formation of cyclopentyl from 4-pentenyl, although thermodynamically favorable, is inconsistent with the rules for ring closure and has not been observed (*144, 154*). The value of the activation energy has been estimated (*248*) to be 18 ± 3 kcal/mol. However, 1,5 ring closures of some substituted 4-pentenyl radicals have been reported (*210, 237, 249, 250*), e.g.,

The fact that the isolated products are formed by 1,5 cyclization does not necessarily indicate that $k_{1,4} < k_{1,5}$, for there is the strong possibility that these reactions, like those of related hexenyl systems (see Section

4. FREE-RADICAL REARRANGEMENTS

IV,B,2), involve competing equilibria and fall under thermodynamic control.

Stereoelectronic constraints on the regioselectivity of ring closure become less severe with increasing chain length (see Section IV,A). Thus, for cyclization of 6-heptenyl (*155*), $k_{1,6}/k_{1,7} \approx 6$ at 63°C (compare $k_{1,5}/k_{1,6} \approx 50$ for 5-hexenyl). Both $\Delta H^{\ddagger}$ and $\Delta S^{\ddagger}$ are more favorable toward formation of the smaller ring. Although 1,6 ring closure is relatively slow ($k_{1,6} \approx 1 \times 10^4$ sec^{-1} at 65°C), a number of cases of the formation of cyclic products from substituted 6-heptenyl radicals have been reported (*210, 231, 251–256*), e.g. (*252*),

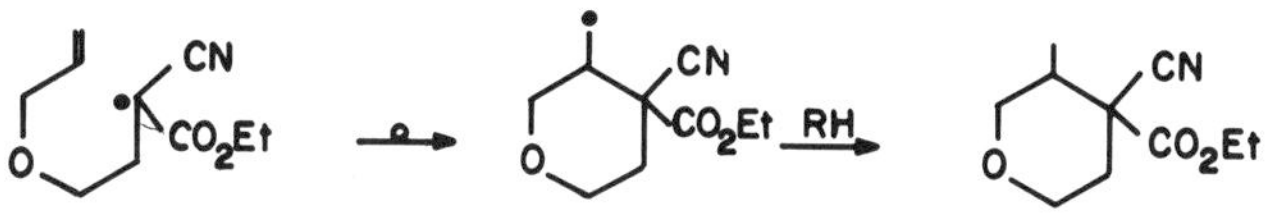

Bi- and polycyclic products can be formed by ring closure of appropriately constituted 6-heptenyl radicals. Thus, **89** generated by cyclization of **62** undergoes further 1,6 and 1,7 ring closure to afford **90** and **91** (*195*). The generation of **93** from **92** involves two successive 1,6 ring closures (*251*), and the formation of **95** by cyclization of **94** constitutes a key step in the synthesis of sativene and copacamphene (*256*).

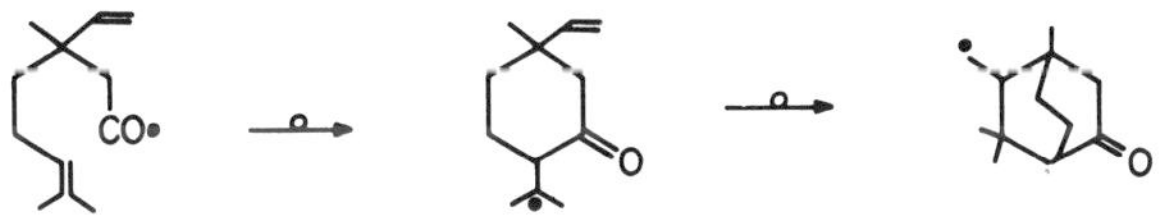

7-Octenyl radicals undergo slow but regiospecific 1,7 ring closure [$k_{1,7} \approx 7 \times 10^2$ sec^{-1} at 65°C (*155*)]. Presumably, the less favorable enthalpy change associated with the formation of the larger ring outweighs the expected less severe stereoelectronic constraints in these flexible systems.

D. Alkynyl Radicals

There have been relatively few studies of intramolecular addition in alkynyl radicals. However, it is clear that the 5-hexynyl radical (*194*) and 6-substituted species [**96** (*257–260*)] generated by reaction of the appropriate halide with tributylstannane (*194, 257, 158*), with lithium biphenyl (*257*), with magnesium (*259*), or by electrolytic reduction (*260*) undergo specific 1,5 ring closure.

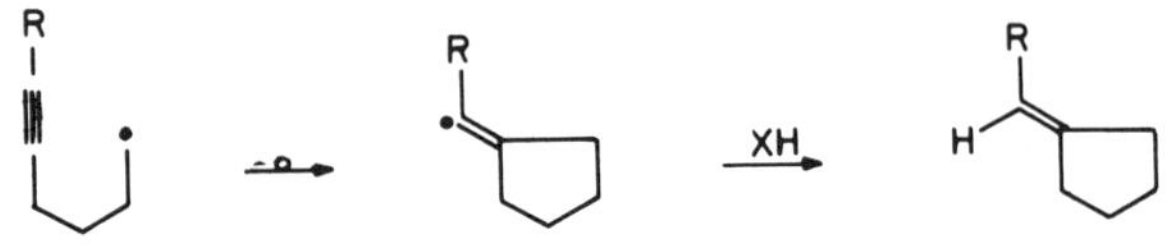

R = H, Ph, n-C_5H_{11}, $Cl(CH_2)_5$

Presumably, the transition state for radical addition to a triple bond is similar in shape to that for olefinic addition, and the regiospecific cyclization of 5-hexynyl radicals is a reflection of stereoelectronic demand. The rate constant [$k_{1,5} \approx 1.0 \times 10^5$ sec^{-1} at 80°C (*194*)] is somewhat lower than that for 5-hexenyl radical [$k_{1,5} \approx 5.2 \times 10^5$ sec^{-1} at 80°C (*47*)], but good yields of cyclic products can be obtained under suitable experimental conditions.

The 7-phenyl-6-heptynyl radical PhC≡C(CH$_2$)$_4$CH$_2$· also undergoes regiospecific cyclization to form the smaller possible ring (*257*), but the rate is clearly less than that of its lower homologue.

Intramolecular addition in the specifically deuteriated 6-ethoxy radical EtOC≡CCD$_2$CH$_2$CH$_2$CH$_2$· affords only the trans product (*258*); clearly, configurational change in the intermediate α-ethoxyvinyl radical **97** occurs too slowly to compete effectively with hydrogen atom transfer from tributylstannane.

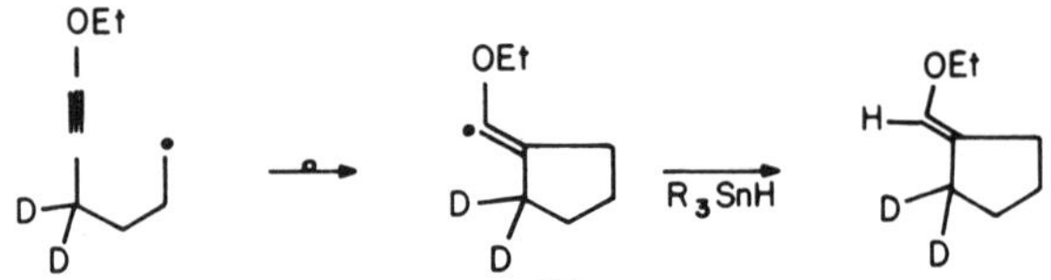

Unlike their olefinic counterparts (Section IV,B,2), alkynyls (**98**) containing stabilizing substituents at the 1 position undergo specific 1,5 ring closure (*124, 261*). However, the homologous heptynyl radical **99** gives both endo and exo cyclization products (*6–6b, 261*), although it is not clear whether this is due to kinetic or thermodynamic factors.

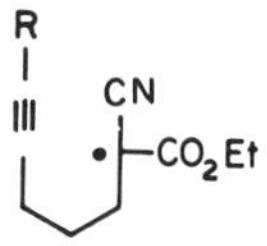

4. FREE-RADICAL REARRANGEMENTS

98

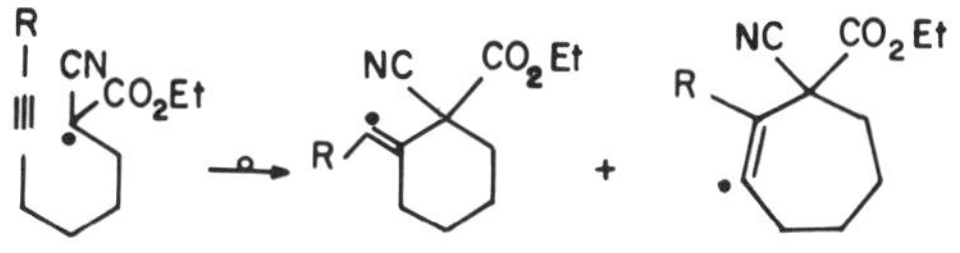

99

The Me$_3$SiC≡C group provides a better terminus than HC≡C for a cyclization which yields a 6-membered ring (*261a*).

Radical addition of iodofluoropropane to 1,6-heptadiyne failed to give cyclic products (*262*). However, with arenethiols, **100** affords, among other products, the cyclopentene derivative **101** (*262*), presumably by the following mechanism:

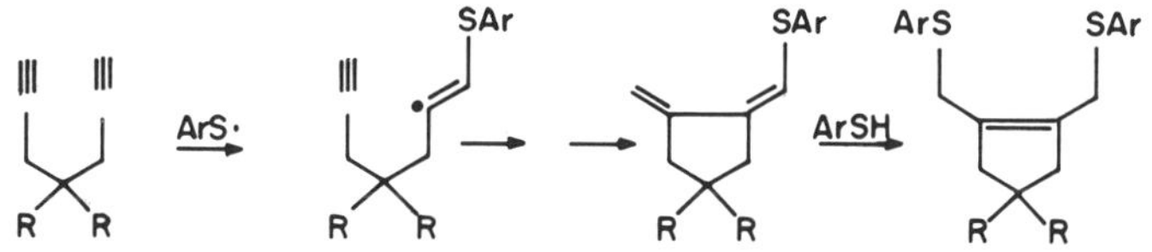

100 → **101**

R = CO$_2$Me

E. Alkenoxy, Alkenaminyl, and Other Heteroatom-Centered Radicals

In their intermolecular reactions with olefins, alkoxy radicals show a marked preference for allylic hydrogen abstraction over addition (*263, 264*). However, the 4-pentenoxy radical **103** and related systems, in which efficient 1,5-hydrogen atom transfer cannot occur, undergo intramolecular addition with high regioselectivity (*163, 264–268*). Thus, photolysis of the nitrite **102** gives good yields (~70%) of the oxime **104** or the nitroso dimer via specific 1,5 ring closure of the intermediate radical **103**. The presence of a methyl substituent at the 1, 4, or 5 position appears to have little effect on the reaction (*265*). For example, in contrast to its alkenyl analogue, which undergoes substantial 1,6 ring closure (*154, 156*), the radical **105** undergoes only 1,5 ring closure. The regioselectivity of these reactions has been attributed to entropy factors (*163*). However, the observed preference for exo cyclization could equally well reflect stereoelectronic factors, although the electronic interactions giving rise to the transition state (*269*) are probably somewhat different from those in carbon radicals. Unfortunately, no accurate rate constants are available, but ESR data (*268*) and product studies (*265*) suggest that cyclization of **103** is very fast ($k > 10^8$ sec^{-1}) and competes effectively with intramolecular hydrogen transfer from an unactivated CH_2 group.

102 103 104

105

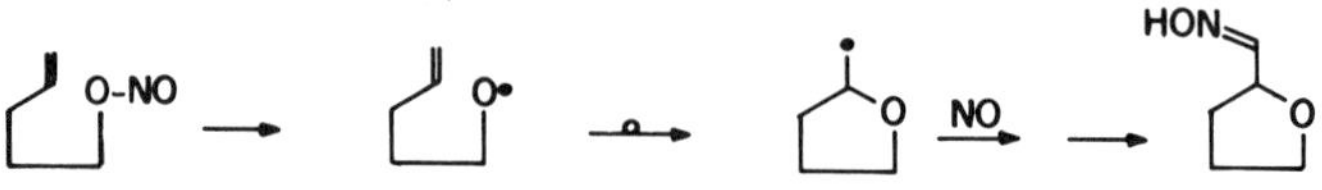

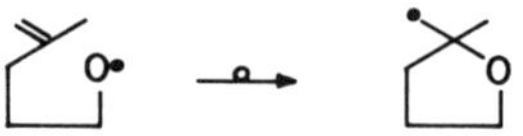

Photolysis of 5-hexenyl nitrite (**106**) affords no cyclic product (*163*, *265*). This result has been ascribed (*265*) to the preferential occurrence of 1,5-hydrogen atom transfer (**107** → **108**). On the other hand, the formation of **110** when the reaction is conducted in the presence of iodine (*163*) appears to indicate the intermediacy of the cyclized radical (**109**). Clearly, further work is required to elucidate the behavior of this system. The alkynoxy radical $HC{\equiv}CCH_2CH_2CH_2O^{\bullet}$ undergoes cyclization, but less efficiently than its olefinic analogue **103** (*270*). Like their carbon analogues, alkenoxy radicals containing cyclic olefinic moieties give rise to bicyclic systems. Thus, photolysis of the nitrite **111** affords, after appropriate work-up, the ketone **112** (*271*).

106 107 109 110 108

111 112

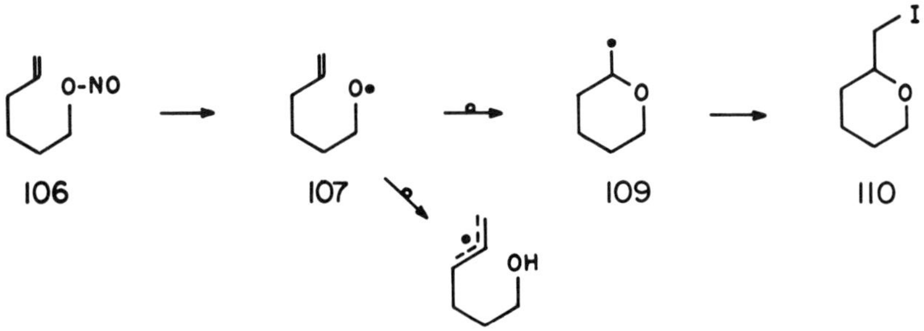

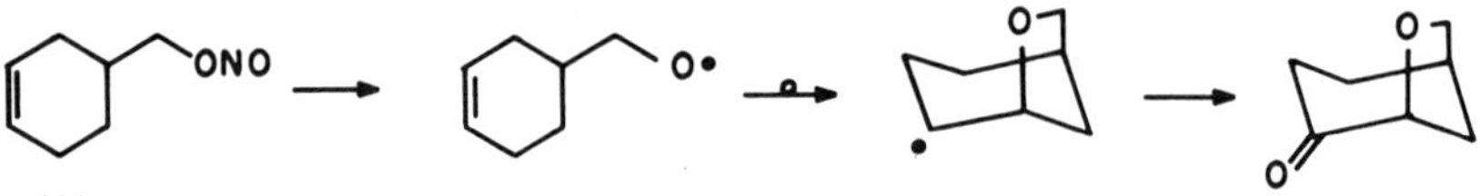

4. FREE-RADICAL REARRANGEMENTS

Cyclization of alkenylperoxy radicals is of particular interest because of the possible occurrence of such a step (**113** → **114**) in the biosynthesis of prostaglandins (*272*) and related endoperoxides of physiological significance (*273*).

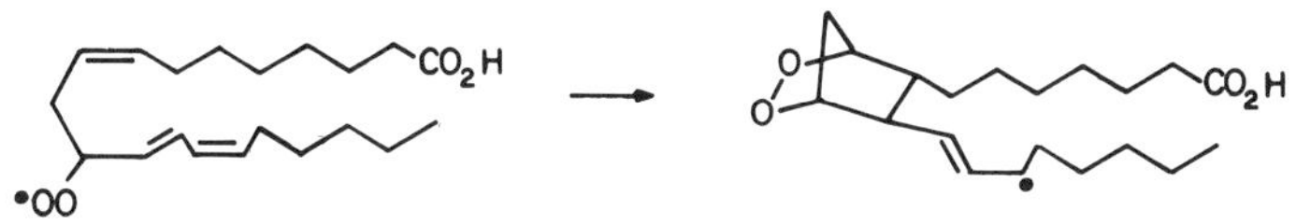

Radicals such as **113** are expected to be intermediates in the autoxidation of suitable olefinic precursors. Indeed, cyclic peroxides have been detected in the autoxidation products of methyl α-linolenate and similar lipids (*274*), and their formation has been invoked to account for the oxidative deterioration of natural rubber (*275*).

Model experiments with simple alkenylperoxy radicals give less complex results (*276, 277*). Thus, treatment of the hydroperoxide **115** with di-*t*-butyl peroxyoxalate or with acetophenone under UV irradiation affords the cyclic peroxide **116**. No six-membered cyclic product was detected. The radical **117** undergoes similarly specific exo cyclization.

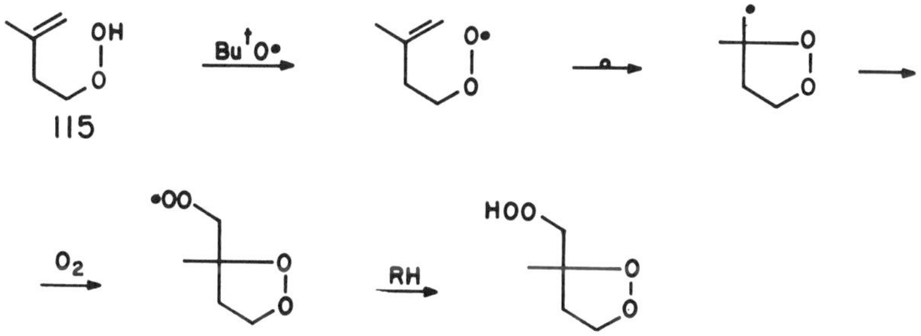

The cyclization reactions of unsaturated aminyl and related radicals generally resemble those of analogous oxygen- and carbon-centered species. Thus, the radical CH_2=$CHCH_2CH_2\dot{N}R$ does not undergo intramolecular addition (*278, 279*), but the two higher homologues (**118,** $n =$ 1, 2) and related species can cyclize and show a preference for exo ring closure (*278–283*). Complexed aminyls and particularly protonated aminyls [aminium radicals, e.g., $CH_2{=}CH(CH_2)_3\overset{+\cdot}{N}HR$] cyclize much more readily (279–287). In suitable systems bicyclic products are formed (283, 286–291a).

118

An attempt (292) to measure the absolute rate constants for cyclization of **118** ($n = 1$; $R = Pr$) by the steady-state ESR technique (see Section II) was unsuccessful because of the low rate of cyclization ($k_c < 10^3$ sec^{-1} at 80°C). Other alkenylaminyls also cyclize much more slowly than their carbon-centered counterparts (292). Aminium radicals, on the other hand, cyclize rapidly. Thus, the Hoffman–Loeffler reaction (see Section VIII,D) does not occur with protonated **118** ($n = 2$; $R = Pr$), despite the presence of an allylic hydrogen atom suitably located for 1,5 migration (281).

Methods for the generation of suitable aminyl and aminium radicals include the thermolysis or photolysis of tetrazenes (279) or N-chloroamines in neutral or acidic media (278, 281–284, 286a, 287, 289, 290), the treatment of chloramines with ferrous, cuprous, titanous, or silver salts (278, 280, 281, 283–285, 288–290), and the photolysis of nitrosamines in acidic solution (286, 287, 291). The nature of the products depends on the method used and on the ability of the solvent to act as an H· donor. The following transformations (280, 290) are illustrative of the scope of the reaction:

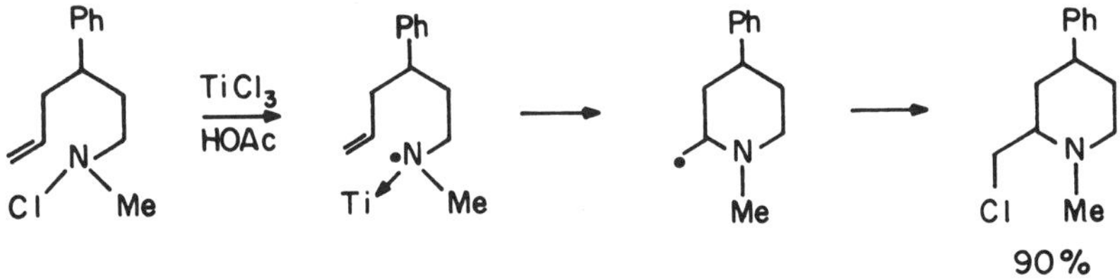

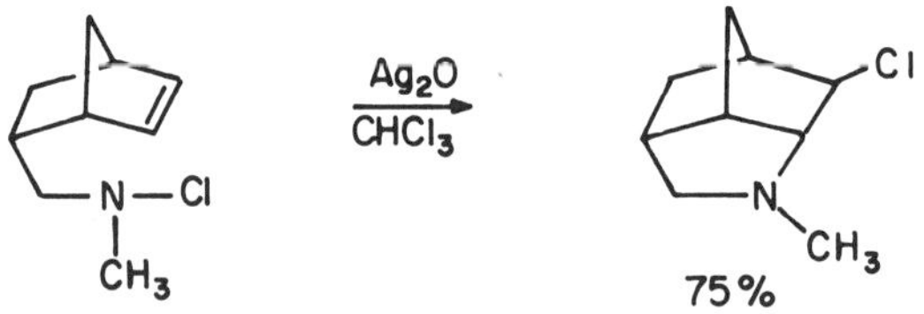

Other N-centered radicals that undergo intramolecular addition include the amidyls (293–294a) e.g., **119** (293) and **120** (294). It has also been

4. FREE-RADICAL REARRANGEMENTS

suggested (*295, 296*) that hydroxyaminyl radicals, e.g., **121** (*296*), cyclize directly. However, for thermodynamic and other reasons, it seems more likely that the observed products arise from the corresponding carbon-centered radical **123**, which could be formed from the nitroxide **122**.*

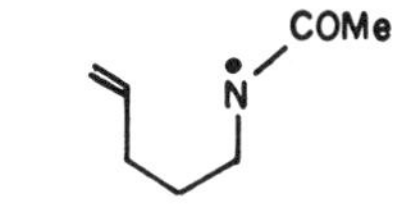

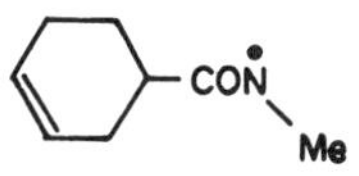

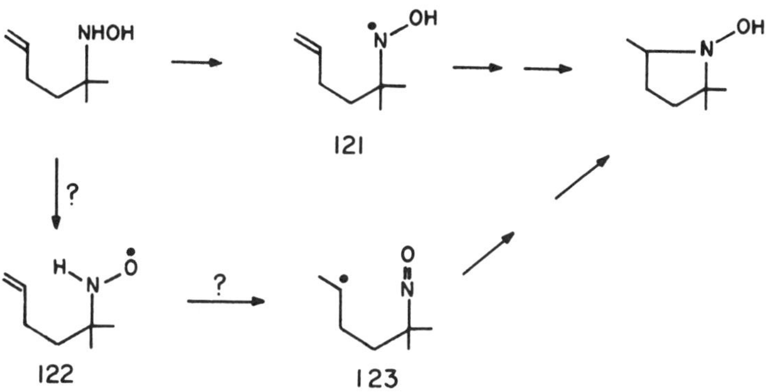

Addition of thiyl radicals to olefins is reversible (*297, 298*), and it is not surprising, therefore, that ring closure of suitable alkenylthiyl radicals (*298–305*), although relatively efficient, lacks regiospecificity. For example, 4-pentene- and 5-hexenethiol (**124**, $n = 2$ and 3, respectively) afford mixtures of exo and endo cyclization products when heated in cyclohexane (*299, 300*). The lower homologue (**124**, $n = 1$) undergoes only endo ring closure (*299*).

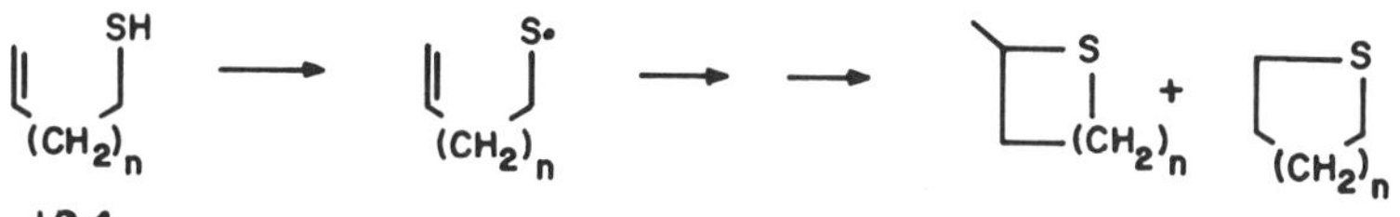

n	Relative yield (%)	
1	0	99
2	24	76
3	85	15

* Analogous intermolecular H transfers by ketyl radicals, $R_2\dot{C}OH$, are known, and so we expect that the **122** → **123** conversion would occur readily [see, however, footnote 11 of House *et al.* (*296*)].

As in the case of the reversible intramolecular addition of carbon radicals (see Section IV,B,2), the outcome of these reactions reflects the interplay of the relative rates of ring closing, ring opening, and chain transfer (by transfer of H· from thiol to cyclic radicals), and it is not surprising, therefore, that the relative yields of products depend on the experimental conditions. There is an increasing preference shown for endo cyclization with increasing reaction temperature or decreasing thiol concentration (*299, 300*). Although various explanations are possible, the most straightforward is that k(exo cyclization) $>$ k(endo cyclization). If so, this implies that ring closure of thiyl radicals, like those of N-, C-, and O-centered radicals, is subject to stereoelectronic control.

Other examples of thiyl radical cyclization include the formation of cyclic compounds containing additional heteroatoms (*298, 300, 301*), the formation of bi- and polycyclic systems (*302–305*), and the cyclization of acetylenic thiols (*306*).

4-Pentenylsilyl radicals can also cyclize (*11, 307*). The methylchloro radical **125a** gives a good yield of cyclized products, with the exo isomer predominating (**127a/126a** = 5.7 at 135°C). However, when the substituents on silicon are bulky (e.g., **125b**) or capable of delocalizing the unpaired electron (e.g., **125c**), the reaction proceeds through the more stable possible radical to afford a preponderance of endo product (**127c/126c** = 0.12 at 125°C).

125   126   127

a R = Me; X = Cl
b R = *i*-Pr; X = Cl
c R = X = Ph

The cyclization of unsaturated phosphorus-centered radicals has been studied by ESR techniques (*308*). The radicals **128** (n = 2) and related phosphoranyl species behave similarly to carbon-centered radicals in that they undergo selective exo cyclization. In the absence of thermodynamic data, it is not clear whether these reactions are under kinetic or thermodynamic control, but it seems likely that stereoelectronic factors are important. Radical **130**, which has a methyl group at the point of attack, does not afford a cyclized radical, nor do phosphoranyl radicals containing longer alkenyl chains (**128**, n = 3 or 4). In these cases the rates of ring closure must be too slow to compete effectively with β scission and other available processes.

4. FREE-RADICAL REARRANGEMENTS

[Structures **128**, **129**, **130**]

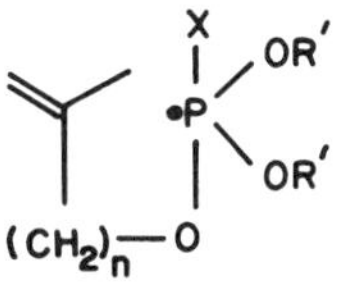

X = OEt, Me; R = $(CH_2)_n CH= CH_2$; R' = $CH_2 CMe= CH_2$

Unlike similar carbon-centered species, the phosphoranyl radicals **128** ($n = 1$) readily undergo 1,4 cyclization to afford the exo product **129** ($n = 1$). Presumably, the ease of this reaction reflects the unusually high stability of four-membered phosphorus-containing rings, a phenomenon nicely exemplified by the formation of the bicyclic radical **132** from **131**.

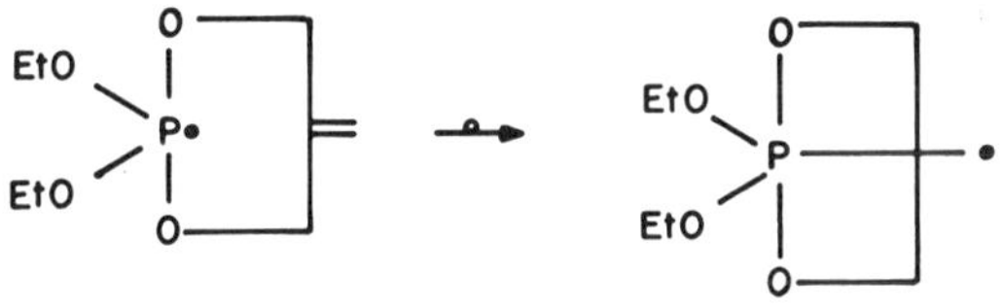

[Structures **131**, **132**]

F. Alkenylaryl Radicals

Suitably constituted alkenylaryl radicals undergo ring closure by intramolecular addition. The only systematic study to have been reported fully (*139, 162*) indicates that radicals **131** undergo stereospecific exo cyclization:

[Structure **133**]

$n = 1$ or 2
$Z = CH_2$, O, NMe

In contrast to analogous hexenyl systems (Section IV,B,2) radical **134** undergoes specific 1,5 ring closure, as does **135**, even though stabilization of 1,6 cyclization (**136**) by interaction of the free spin with adjacent oxygen lone pairs should favor its formation. Radical **137**, like 4-pentenyl, affords no cyclic products. The regiospecificity of these cyclizations has been attributed to stereoelectronic factors, reinforced by the pronounced steric constraints associated with the aromatic ring (*139*).

134

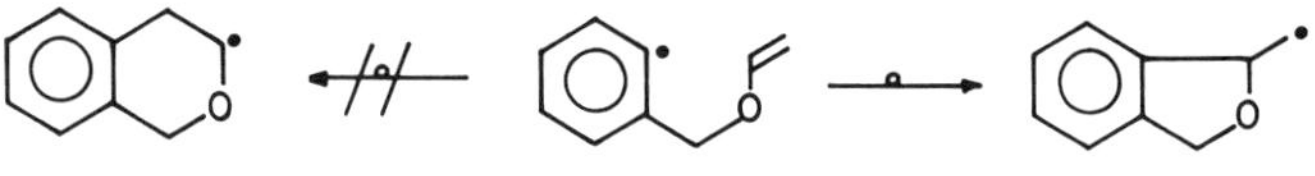

136   **135**

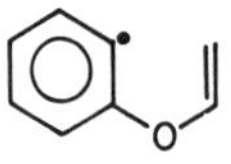

137

Relative rate constants have been determined by the stannane method (see Sections II and IV,B,2). Values of $k_{1,5}/k_H$ for 1,5 cyclization of **133** ($n = 1$) at 130°C lie in the range of 0.5–60 M. The value of k_H is unknown, but it is probably greater than that for 5-hexenyl. A reasonable minimal figure for $k_{1,5}$ for **133** at 130°C is 2.0×10^6 sec^{-1}. At this temperature the calculated value for 5-hexenyl is 1.5×10^6 sec^{-1}. It appears, therefore, that **133** ($n = 1$) cyclize at approximately the same rates as do comparable alkenyl radicals, whereas the higher homologues (**133**, $n = 2$) cyclize more rapidly than 6-heptenyl radicals (*139*).

The facility with which alkenylaryl radicals undergo ring closure suggests that the reaction should be of synthetic utility. This has been demonstrated recently by the synthesis of cephalotaxine (*309*) and various cyclic ketones (*310*), e.g., **139** from the iodoaryl ketone **138**. This reaction, which probably involves the intermediacy of enolate aryl radical ions, has been successfully applied to the formation of six-, seven-, eight-, and ten-membered rings. Somewhat similar reactions have also been reported (*311, 312*).

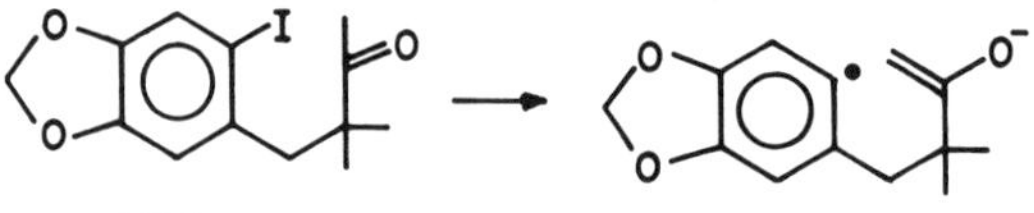

138

G. Intramolecular Aromatic Homolytic Substitution

Ring closure of suitably constituted aryl radicals by intramolecular aromatic homolytic arylation is of considerable synthetic utility and has been extensively studied (*313*). The best-known reaction of this type is the Pschorr phenanthrene synthesis (*314*), e.g., **140 → 143**, involving 1,6 ring closure, but mechanistically similar processes can also give rise to five- (*315–324*), seven- (*325–327*), eight- (*325*), and nine-membered (*328*) rings.

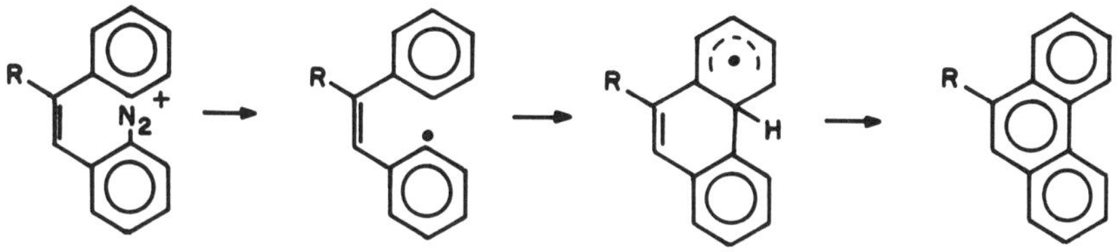

The overall mechanism shows the same features as intermolecular homolytic arylation. The addition step to afford a cyclohexadienyl radical (e.g., **142**) is usually regarded as irreversible (*329*), at least below 200°C (*330*), although this view has been disputed (*331*). The ultimate fate of the primary adduct depends on the nature of the reactant and the experimental conditions. Radicals containing a single atom bridge between the two aromatic nuclei (e.g., **144**) or a sterically rigid two-atom bridge (e.g., **141**) undergo exclusive 1,5 and 1,6 ring closure, respectively. The resultant adducts are then converted to the fully aromatic products by hydrogen atom transfer to a suitable acceptor radical, by addition of $\dot{X}$ to give a cyclohexadiene followed by elimination of HX, or by one-electron oxidation with loss of a proton.

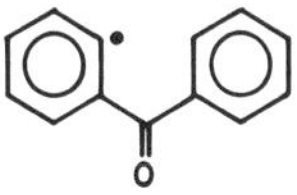

When there is a more flexible two-atom bridge between the two aromatic nuclei, both 1,5 and 1,6 ring closure can occur. The mechanistic complexity of such systems has been revealed by work on benzanilide derivatives. Radical **145** cyclizes to form both **146** and **149** (*332, 333*). Oxidation of the δ-lactam intermediate **149** affords phenanthridones **150**. The fate of **146** depends on the nature of any substituents and the reaction conditions, but major pathways include dimerization, formation, in the presence of oxygen, of spirodienone (**148**), and rearrangement to give **149**. Independent generation of **146** (*334–336*), e.g., by thermolysis of the dimer **147** (*335*), suggests that **146** may lie on the pathway from **145** to **149**.

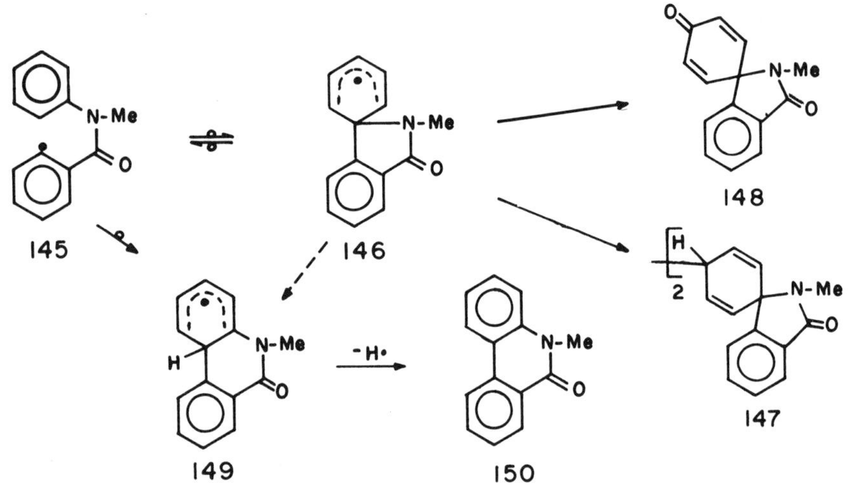

Most of the work in this area has been directed toward the development of improved experimental procedures and new synthetic applications (*337*). Useful methods for generating radicals include the photolysis of iodides (*323, 325, 332, 338, 339*), bromides (*328, 340*), chlorides (*319*), and diazonium salts (*341, 342*); electrolysis of diazonium salts (*324, 343*) or their decomposition induced by copper in acetone (*333*), by cuprous complexes (*317*), or by sodium iodide in various solvents (*334, 344*); and by treatment of halides with methylmagnesium iodide and cobaltous chloride (*318*), with sodamide (*345*), and with sodium naphthalenide (*346*). Recent applications of the method include the synthesis of heterocycles (*316, 323, 347, 348*) and spiro compounds (*321*), particularly in the alkaloid field (*320, 328, 337, 349*).

Little is known about the absolute or relative rates of the various mechanistic steps. However, the success of Pschorr and similar reactions in benzene, in potential H atom donor solvents, and in the presence of iodine indicates that the ring closure step can compete effectively with a number of relatively fast intermolecular reactions. The relative rates of 1,5 and 1,6 ring closure clearly depend on steric and electronic effects, but there is some indication from work on benzanilide derivatives (*333*) and the successful preparation of varous spirodienes that $k_{1,5}$ is often greater than $k_{1,6}$. This has been attributed (*333*) to the necessity for the radical to assume a conformation, e.g., **151**, in which attack can occur along a line almost perpendicular to the aromatic ring. In accord with this view, substituents at the 2 position enhance the ratio of 1,5 to 1,6 ring closure more than those at the 3 and 4 positions (*333*). Finally, the ease of ring opening of the intermediate spirohexadienyl radical should reflect the stability of the product. Thus, a major reaction pathway for **152** involves formation of **153** (*326*), which is stabilized by interaction of the free spin with the adjacent nitrogen lone pair.

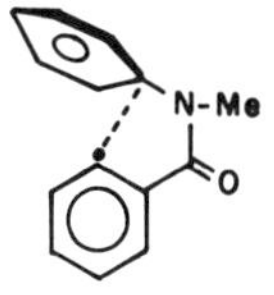

151

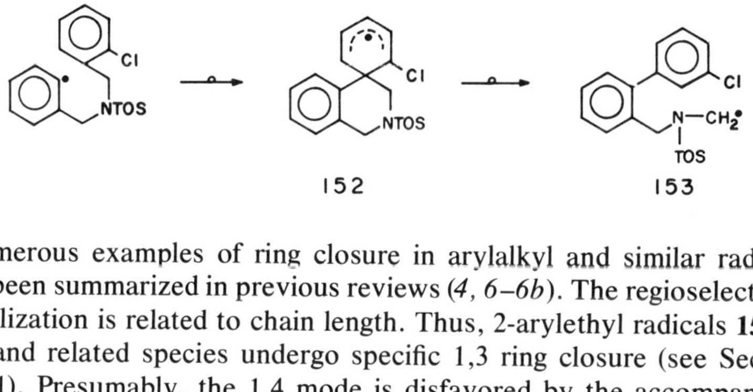

152   **153**

Numerous examples of ring closure in arylalkyl and similar radicals have been summarized in previous reviews (*4, 6–6b*). The regioselectivity of cyclization is related to chain length. Thus, 2-arylethyl radicals **154** ($n = 1$) and related species undergo specific 1,3 ring closure (see Section III,B,1). Presumably, the 1,4 mode is disfavored by the accompanying increase in strain energy.

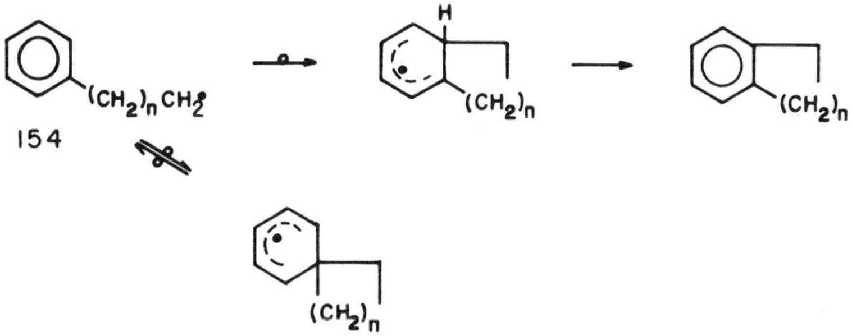

154

In 3-phenylpropyl (**154**, $n = 2$) and similar radicals, 1,4 ring closure involves considerable strain energy and appears not to have been observed. Nor does Ar_2-5 cyclization occur readily in simple alkyl-substituted 3-phenylpropyl radicals, although the formation of indanes from alkylated benzenes at elevated temperatures probably proceeds via radical cyclization (*350*). However, ring closure is facilitated by favorable steric effects, e.g., **155**—o—**156** (*237*). Other examples of Ar_2-5 cyclization include the formation of indanones (*74, 88*), e.g., **157** → **158** → **159** (*88*), and ring closure of $PhOCOCH_2\cdot$ (*351*), 2-arylbenzoyl radicals (*352*), $PhCH=CH\dot{C}H_2$ (*353*), and $Ph\dot{C}(CH_2\dot{C}H_2)CH_2\dot{C}H_2$ (*119*).

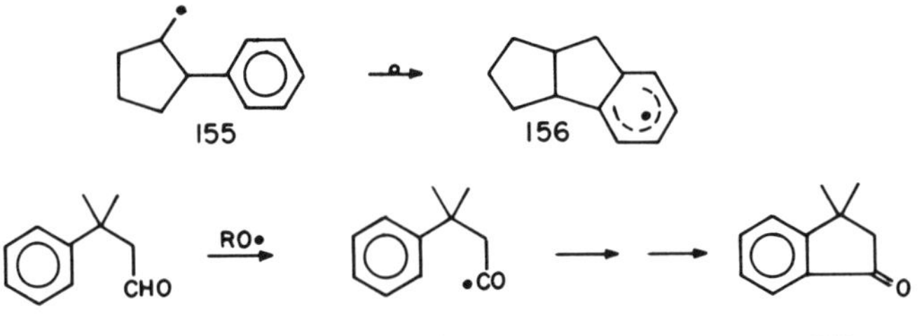

Both Ar_1-5 and Ar_2-6 modes of ring closure are available to 4-phenylbutyl (**154**, $n = 3$) and related radicals (*89, 92, 93, 95–97, 99, 119, 249, 354–358*). The mechanistic features of the system were elucidated by a study of the naphthylbutyl radical **160** (*66, 89, 99*). Determination of the scrambling of deuterium in the products (**163** and **164**) when **160** was labeled with deuterium at the 1 position showed that the formation of the spiro intermediate **161** is relatively fast and reversible. The irreversibility of Ar_2-6 cyclization, at least below 200°C, was demonstrated (*99*) by the generation of the radical **17** ($R = H, CH_3$). No ring-opened product could be detected (see Section III,B,2). Although **161** is a good deal less stable than **162**, the 1,5 ring closure is faster than the 1,6 closure (*89*). Apparently this reaction, like the cyclization of 5-hexenyl, is subject to stereoelectronic effects.

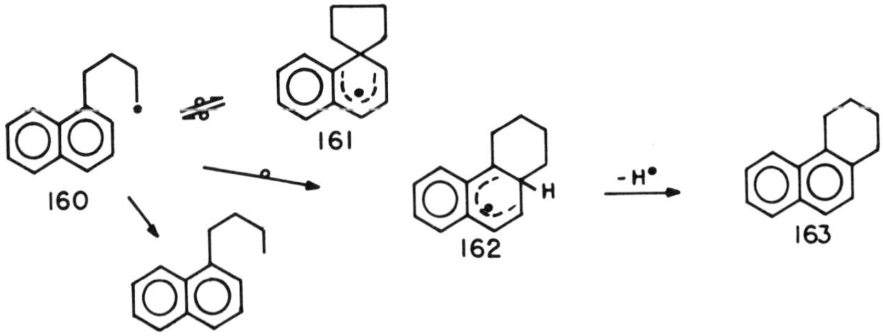

4. FREE-RADICAL REARRANGEMENTS 215

Accurate rate constants for Ar_2-6 and Ar_1-5 ring closure are not yet available. Estimates of the rate constants for cyclization of 4-phenylbutyl (*6a, 93*) and the naphthylbutyl radical **160** (*6a, 89*), give values of about 5 × 10⁴ sec⁻¹ at 50°C and 1 × 10⁴ sec⁻¹ at 80°C, respectively. However, ESR and product studies indicate that the first and possibly the second value are too high (see Section III,B,2).

Normally, the radical formed by Ar_2-6 cyclization (e.g., **162**) is converted to the fully aromatic product (e.g., **163**) by loss of a hydrogen atom or by consecutive one-electron oxidation and loss of a proton. However, a few examples of the formation of dihydroarenes are known, e.g., **165** → **166** (*119*).

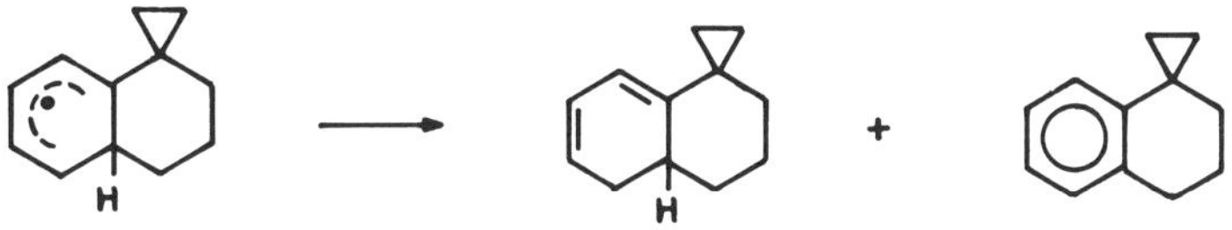

```
     165              166
```

The Ar_2-6 cyclization of suitable radicals constitutes the key step in a number of reactions of potential synthetic utility (*359–361a*), e.g. (*361*),

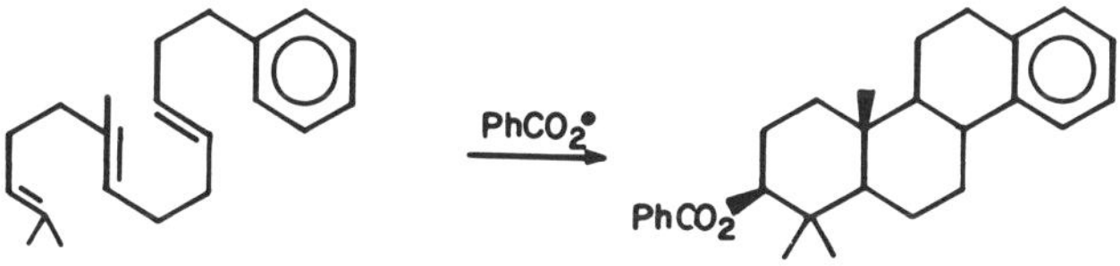

The same is true for certain Ar_2-7 cyclizations (*361b*).

Heteroradicals containing suitably disposed aryl nuclei also undergo intramolecular aromatic substitution. Examples include silyl (*90, 90a, 101, 362*), e.g., $PhCH_2CH_2CH_2\dot{S}iMe_2$, alkoxy (*363, 364*), carboxylate (*93, 365, 366*), acyl (*356*), iminyl, e.g., **167** → **168** (*367*), amidyl (*334, 368–370*), aminium (*371, 371a*), and thiyl (*372*) radicals. Although the evidence is somewhat scanty, it appears that such systems behave similarly to carbon-centered radicals in that 4-aryl-substituted species undergo 1,5 ring closure more rapidly than 1,6 closure. Consequently, 1,5-aryl rearrangements, e.g., **169** ⟶ **170** ⟶ **171** (*373*), compete with Ar_2-6 cyclization. Stable spirocyclohexadienes have occasionally been obtained (*364*).

```
       C=N•                          
       |                            
       R                            R
      167                          168
```

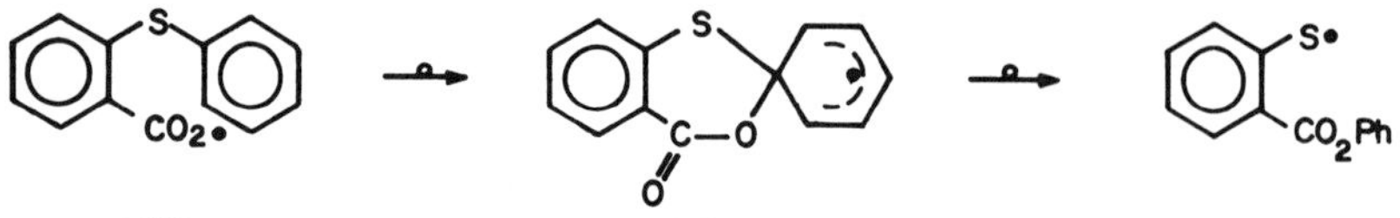

169 → **170** → **171**

H. Other Cyclizations

This section deals with intramolecular addition to multiple bonds containing heteroatoms and to coordinatively unsaturated atoms.

Although individual examples of intramolecular homolytic addition to C=O, C=N, C=S, N=N, and similar unsaturated groups containing a heteroatom have been reported, there have been no comprehensive surveys of such reactions, and no accurate rate measurements have been made.

Ring opening of cycloalkoxy radicals (e.g., **174** ⇌ **173**) has been frequently described (see Section V), but the reverse reaction can also be observed under appropriate conditions. Treatment of the chloroaldehyde **172** with tributylstannane affords cyclohexanol (*374*), presumably via cyclization of **173**. Calculations from the literature data suggest that k_c/k_H ~ 0.5 M, a somewhat *larger* value than that for cyclization of the hexenyl radical. Further work is needed to verify this result and to determine the position of the equilibrium **173** ⇌ **174**.

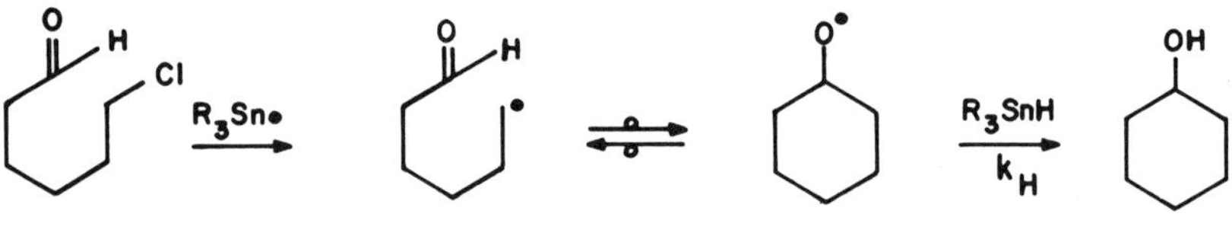

172 → **173** ⇌ **174** → (OH cyclohexanol)

The phenyl-substituted radical **175** also undergoes cyclization, but in the endo mode, to afford **176** (*375*). Here, also, the rate constant appears to be relatively high; the reported data suggest that $k_c/k_H \approx 1.6$. Other possible examples of endo cyclization onto the carbonyl group (*376–379*) include ring closure of **177** (*377*) and the formation of lactones by oxidative decarboxylation of alkanedicarboxylic acids (*379*), e.g., **178** → **179**.

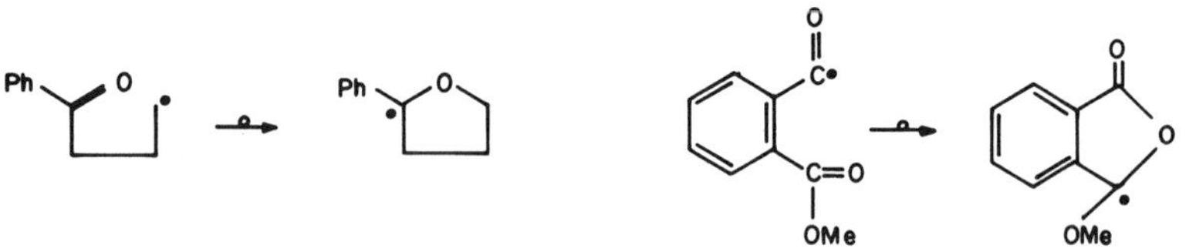

175 → **176**   **177** →

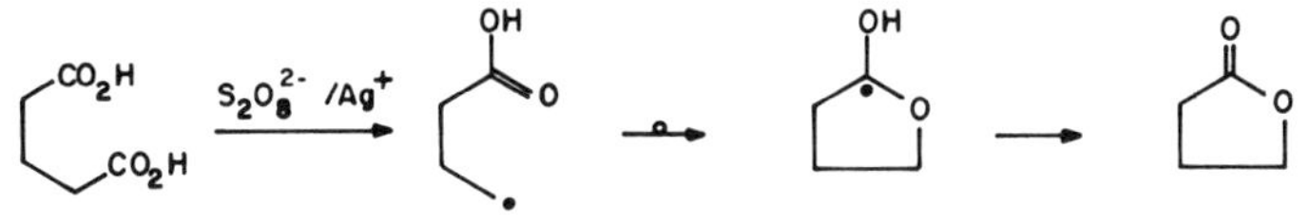

4. FREE-RADICAL REARRANGEMENTS

[Structure 178: cyclopentane with two CO₂H groups] → (S₂O₈²⁻/Ag⁺) → [intermediate with OH] → [radical intermediate 179 with OH] → [lactone]

178 → **179**

Lactone formation via cyclization of the resonance stabilized benzyl radical analogue of **177** has even been observed (*379a*).

Cyclic iminyl radicals formed by intramolecular addition to the cyano group may undergo subsequent ring opening with net migration of C≡N (see Section III,D). However, under suitable conditions the cyclized form is converted by hydrogen atom transfer to the imine, which then undergoes hydrolysis to the ketone, e.g., **180** → **182** (*380*). The cyclization **180** → **181** is slower than the 5-hexenyl cyclization (*380a, 381*). Kinetic ESR spectroscopy gives a rate constant for this rearrangement of 4.0×10^3 sec^{-1} at 25°C and a temperature dependence which can be represented by $\log(k_c^{C\equiv N}/\text{sec}^{-1}) = (9.9 \pm 1.0) - (8.6 \pm 1.0)/\theta$. An interesting series of reactions is involved in the formation of **183** by addition of cyanopropyl radical to diallylamine derivatives (*211a, 382*).

An intriguing variant of the above class of reactions is one in which iminyl abstracts hydrogen *intramolecularly* (i.e., there is a 1,5 H-atom migration), and the resultant carbon-centered radical then undergoes an *endo*-cyclization, adding to the *nitrogen* of the newly formed C=NH group to form a 5-membered azacyclopentane ring system (*382a*).

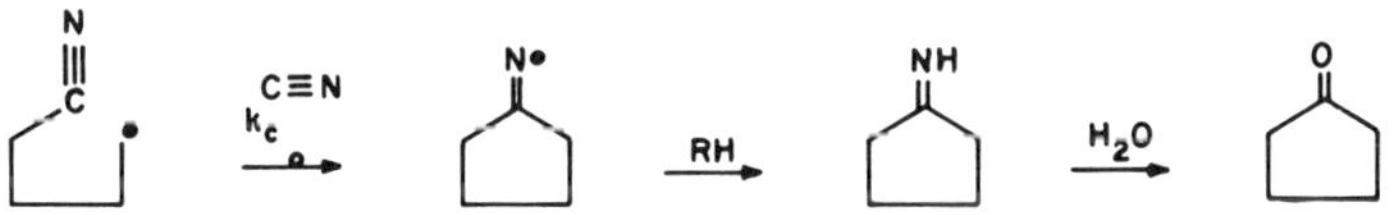

180 [cyclopentyl with C≡N and radical] →k_c **181** [N•] →RH [NH] →H₂O **182** [ketone]

Me—C•(CN)(Me) + [diallylamine with N-Me] → [pyrrolidine with CN] → [bicyclic with N•] →RH→

[bicyclic with NH] →H₂O→ [bicyclic with O]

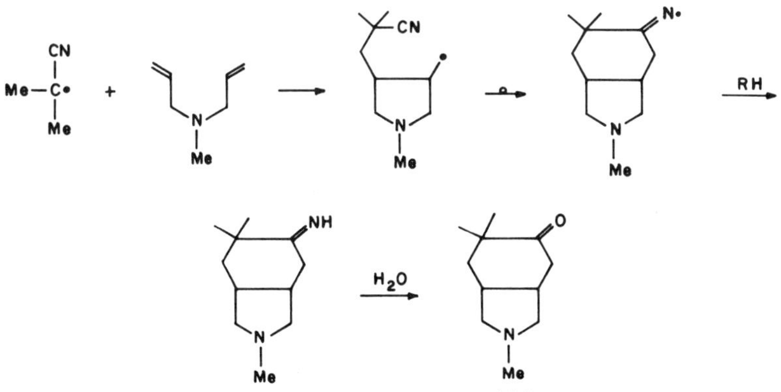

183

The dithio analogues of β-acyloxy radicals, e.g., **184**, do not undergo a reaction equivalent to the acyloxy migration (see Section VI,B,1), but are instead converted to 1,3-dithiolanes (*383*), e.g.,

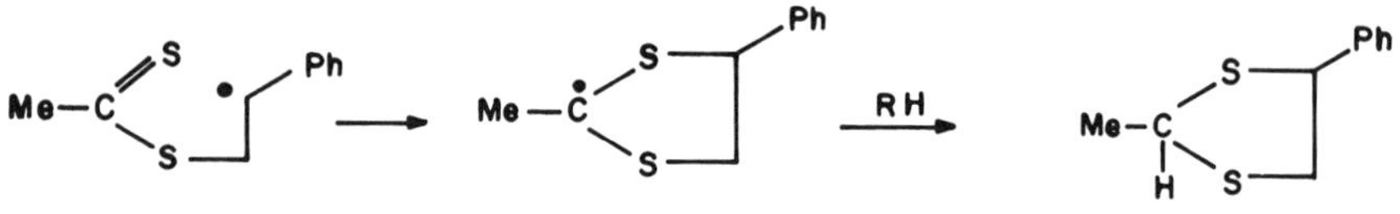

The nitro group is an effective spin trap, and it is not surprising, therefore, that suitably constituted radicals cyclize readily (*384, 385*), e.g., **185**⟶**186** (*384*). The azo group is much less reactive toward homolytic attack. Nevertheless, the aryl **187** does undergo exo cyclization (*386*).

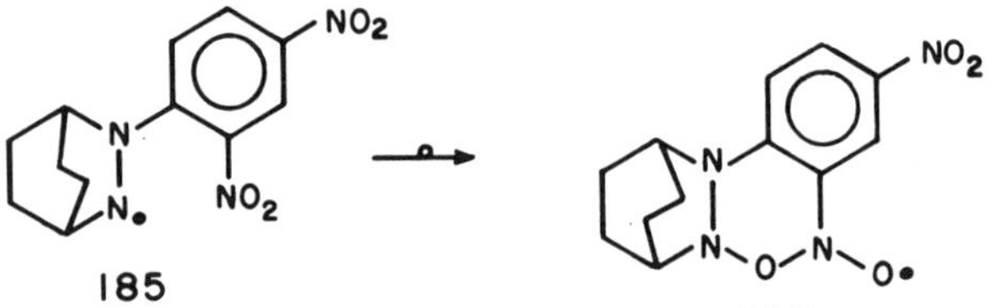

Although intramolecular S_H2 reactions at coordinatively unsaturated atoms might reasonably be expected to proceed through cyclic intermediates, the latter have not been directly observed, although their presence is sometimes implied, e.g., **189** by CIDNP during the thermolysis of **188** (*387*).

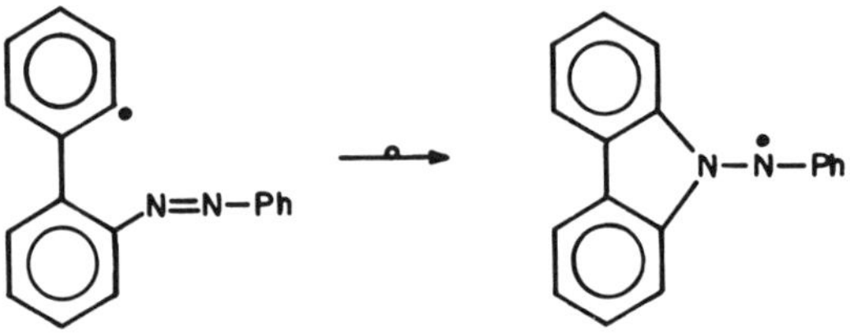

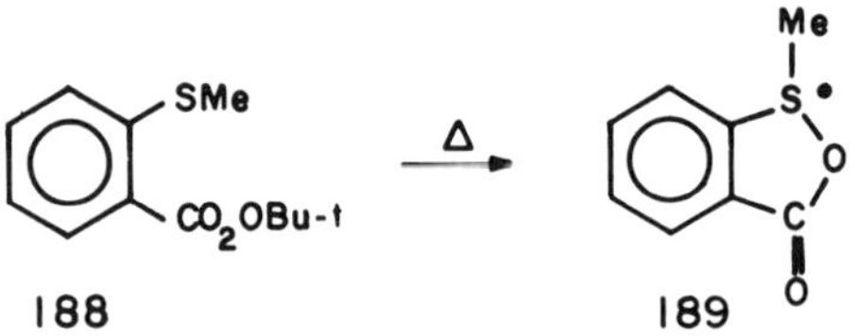

In other instances of the migration of heteroatom-centered groups (see Section VI), it is not clear whether the cyclized forms are true intermediates or transition complexes. Two cases in which the formation of intermediates (**190** and **191**) appears likely are (*388, 389*) the following:

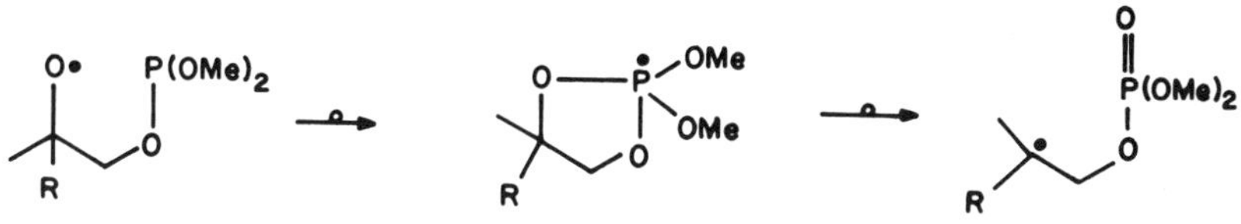

4. FREE-RADICAL REARRANGEMENTS

190

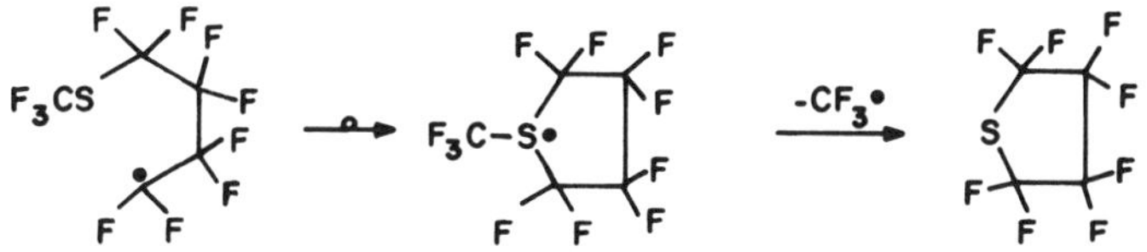

191

An intermediate cyclic sulfur-centered radical may also be involved in the formation of 2,2-dimethyltetrahydrothiophen from $Me_2\dot{C}(CH_2)_3SS(CH_2)_3\text{-}CMe_2CO_2H$ (*389a*). However, when intramolecular attack occurs at carbon (*390, 391*), e.g., **192**⟶**193** (*390*), or oxygen, e.g., **194**⟶**195** (*61*), the reaction probably proceeds through a cyclic transition state:

192   **193**

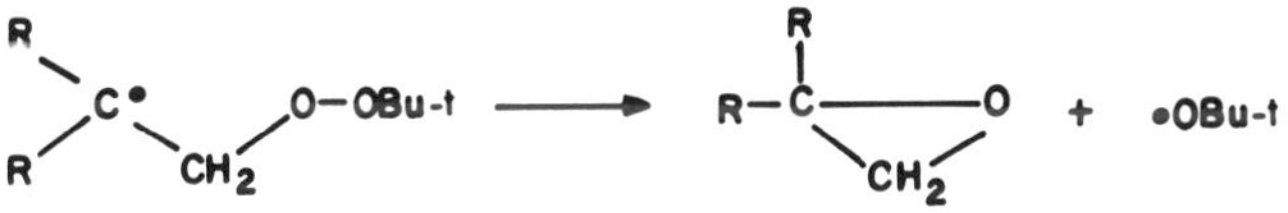

194   **195**

The stereochemistry of the latter class of reactions has been studied by utilizing alkyl radicals (**196**) derived from cyclic peroxides (*277*).

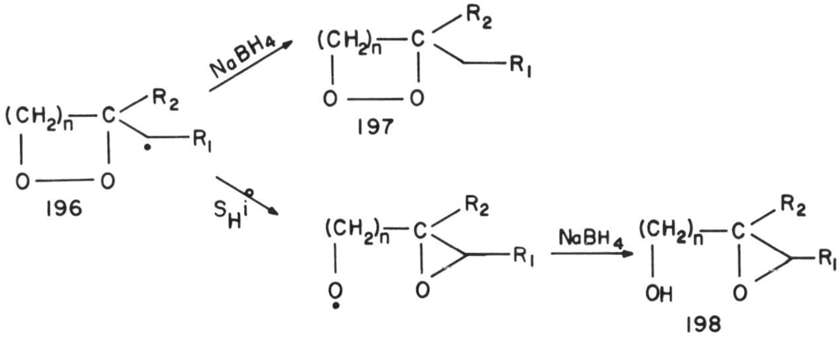

196   **197**   **198**

The product ratios, namely,

	196		197 (%)	198 (%)
$n = 2$;	$R_1 = Et$;	$R_2 = H$	75	25
$n = 2$;	$R_1 = H$;	$R_2 = Me$	90	10
$n = 3$;	$R_1 = Me$;	$R_2 = H$	10	90

indicate that the $S_H i$ reaction is most favored when the carbon radical center and the O—O bond are collinear, i.e., a backside attack on oxygen is preferred. The rate constants for these three $S_H i$ reactions are 7.5×10^4, 1.0×10^4, and 8.7×10^5 sec^{-1}, respectively, at 25°C (*391a*).

V. RING-OPENING REACTIONS

A. Cycloalkyl Radicals and Related Species

The β scission of cyclopropyl is highly exothermic, $\Delta H° \approx -23$ kcal/mol (*63, 126, 392*), because it relieves ring strain and affords the resonance-stabilized allyl radical. Formation of 3-butenyl from the cyclobutyl radical is also exothermic, $\Delta H° \approx$ ca. -5 kcal/mol (*392, 393*). However, each of these thermochemically favorable processes has such a small rate constant that reactions involving cyclopropyl or cyclobutyl radicals in solution usually proceed without the formation of acyclic products (*394–394d*).* Measurements on gas-phase systems show that these ring-opening reactions have unexpectedly high activation energies in the range 20–30 kcal/mol (*242, 393, 397, 398*), as do the endothermic β-scission reactions of cyclopentyl and cyclohexyl radicals (*248, 393, 399, 400*). In comparison, the mildly exothermic ring openings of cyclopropylcarbinyl and cyclobutylcarbinyl have relatively high rate constants and low activation energies (see Section V,B). Clearly, there exists a barrier to the opening of cycloalkyl systems containing a radical center within the ring that does not apply when the radical center is exocyclic.

One explanation (*6, 393*) for this behavior, consistent with the stereoelectronic approach to radical cyclization (see Section IV), rests on the hypothesis that β scission of carbon-centered radicals proceeds most readily when the semioccupied orbital can assume an eclipsed conformation with respect to a β,γ bond. The preferred transition state (see Sections IV,A and V,B) cannot be accommodated within small cycloalkyl radicals without the development of considerable strain since, in such

---

* Although some examples of the apparent opening of cyclobutyl radicals in solution have been reported (*395, 395a*), it is probable that ionic intermediates are involved (see *396*).

4. FREE-RADICAL REARRANGEMENTS

species, the orbital containing the free electron is approximately orthogonal to the plane of the ring.

Theoretical treatments of the cyclopropyl–allyl rearrangement (401, 402) give a somewhat similar picture. At the transition state there is a relatively small degree of rotation of the two CH_2 groups. Although the calculated activation energies for disrotatory and conrotatory ring opening are both large, there appears to be a preference for the disrotatory mode (402).

Some substituted cyclopropyl radicals undergo β scission much more rapidly than the parent radical. Thus, treatment of **199** with tributylstannane affords both the cyclopropane **200** and the ring-opened olefin **201** (92). The data indicate that $k_R/k_H \approx 0.03\ M$ at 78°C. The value of k_H for this system is not known but since it is probably greater than $10^6\ M^{-1}\ \text{sec}^{-1}$, k_R is greater than $10^5\ \text{sec}^{-1}$.

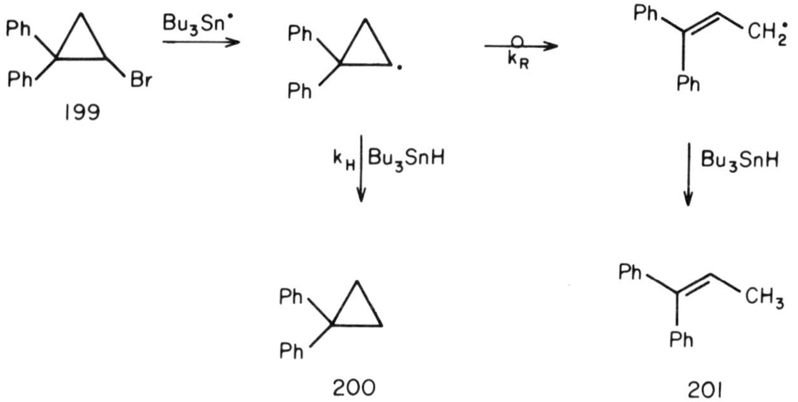

Although it has been suggested (92) that this reaction and closely related processes (403) proceed readily because of the stabilizing effect of the phenyl substituents on the ring-opened products, it is noteworthy that the trialkyl-substituted radical **202** also undergoes ring fission (199).

202

It is difficult to determine the preferred mode of ring opening of substituted cyclopropyls because the initially formed allyl radicals interconvert rapidly (see Section IX). Thus, the mixtures of dienes and other products obtained from the stereoisomeric diphenylcyclopropyl radicals **203** and **204** are complex, and their compositions do not specify unambiguously the mode of ring opening (*404*).

203 204

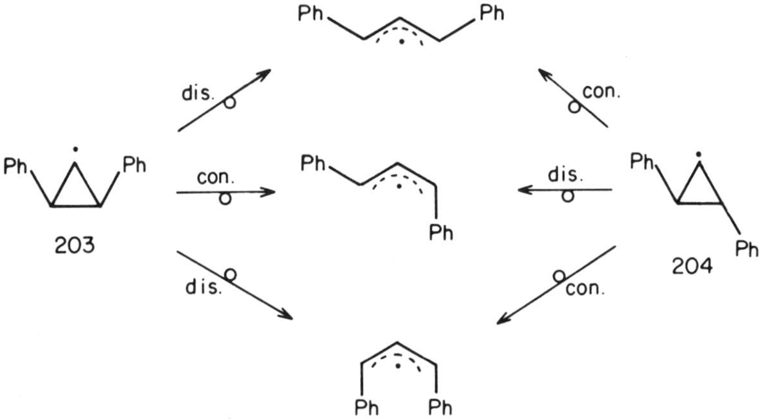

The formation of the cyclohexenyl (**206**) and cyclohexadienyl (**208**) radicals from the bicyclic radicals **205** and **207**, respectively, clearly indicates that ring opening can occur in systems in which geometric constraints preclude the conrotatory mode (*405, 406*). Other examples (*405–409*) include ring opening of **209** (*405*), **210** (*407*), and **211** (*408*). Although no experiment has yet been devised that unambiguously distinguishes between the two modes of ring opening in sterically unconstrained systems, the available results are consistent with the view that there is a preference for the disrotatory mode.

205 206 207 208

209 210 211

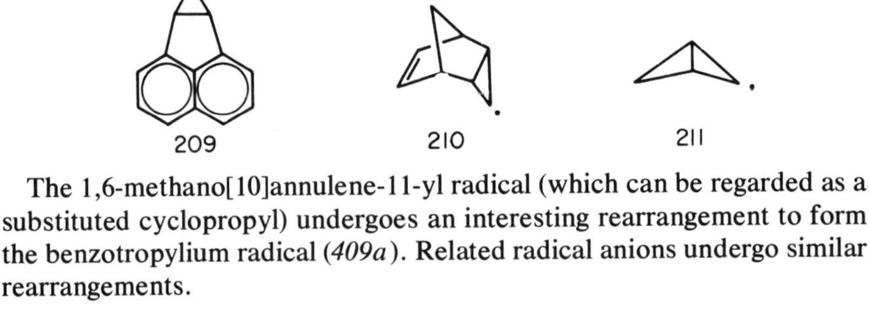

The 1,6-methano[10]annulene-11-yl radical (which can be regarded as a substituted cyclopropyl) undergoes an interesting rearrangement to form the benzotropylium radical (*409a*). Related radical anions undergo similar rearrangements.

4. FREE-RADICAL REARRANGEMENTS

Ring opening of three-membered cyclic radicals containing heteroatoms has also been detected. For example, thermolysis of the percarboxylate **212** gives products derived from the radical **214**, which is, presumably, formed by disrotatory ring opening of *cis*-2,3-diphenylaziridinyl (**213**) (*410*):

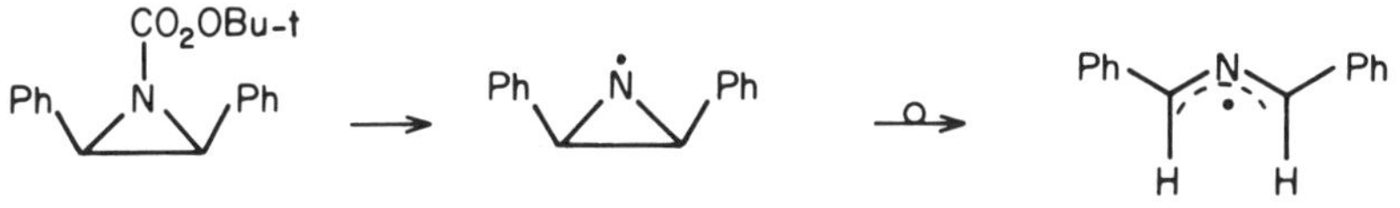

On the other hand, *trans*-2,3-dimethylaziridinyl (**215**) does not rearrange when generated in an ESR spectrometer at temperatures below 40°C (*410*), i.e., $k_R < 10^3$ sec^{-1} at 40°C. The unsubstituted aziridinyl radical **216** is persistent in matrices (methanol, toluene, etc.) at -196°C (*411, 412*). However, it is converted to both CH$_2$=N—CH$_2\cdot$ and HN=CH—CH$_2\cdot$ upon irradiation with visible light (*411, 412*):

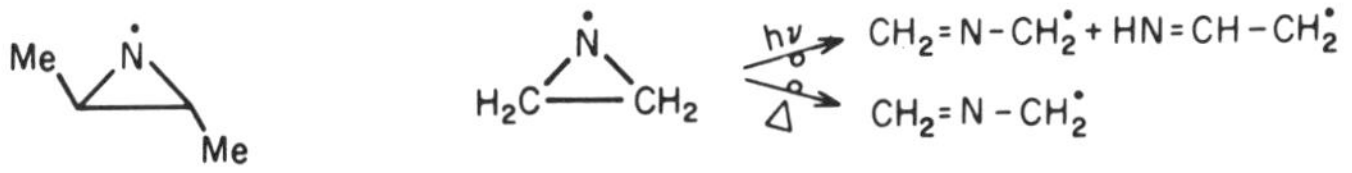

Ring opening of Ph$_2$CON$\cdot$ occurs slowly and affords Ph$_2$C=N—O$\cdot$ (*413*).

The β scission of oxiranyl radicals has been more extensively studied (*59, 414–419*). Electron spin resonance measurements (*59, 418, 419*) indicate that the rates of formation of α-ketoalkyl radicals depend on the number and type of substituents. The parent radical **217a** undergoes ring opening relatively slowly, $k_R < 10^3$ sec^{-1} at 70°C (*59*). The presence of one methyl substituent, i.e., **217b**, causes a substantial rate increase, $k_R = 10^3$ sec^{-1} at *ca.* 0°C, whereas radicals bearing three methyl groups (i.e., **217c**, $k_R = 10^3$ sec^{-1} at *ca.* -45°C) or a *t*-butyl group (i.e., **217d**, $k_R = 10^3$ sec^{-1} at *ca.* -30°C) rearrange even more rapidly (*59*). The temperature dependence for the rearrangement of **217b** gives log(A/sec^{-1}) = 15 ± 1 and E = 15 ± 2 kcal/mol (*59*).

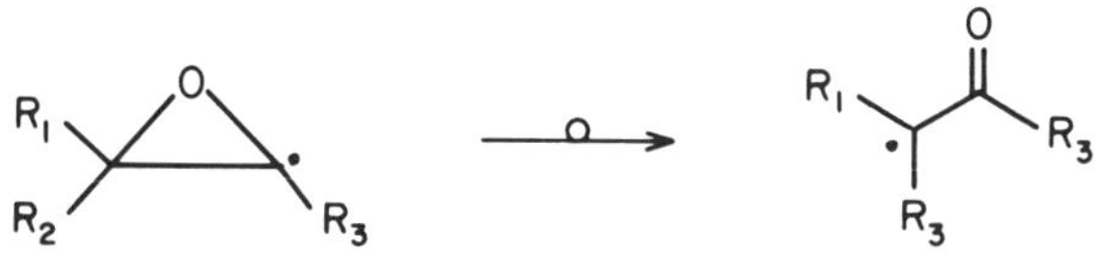

a, R$_1$ = R$_2$ = R$_3$ = H

b, R$_1$ = R$_2$ = H; R$_3$ = Me

c, R$_1$ = R$_2$ = R$_3$ = Me

d, R$_1$ = R$_2$ = H; R$_3$ = *t*-Bu

These results provide support for the view (59) that the intimate mechanisms for ring opening of oxiranyl and cyclopropyl radicals are the same. The relatively high rate constants for rearrangement of oxiranyl radicals can reasonably be attributed to their very favorable thermochemistry [$\Delta H°$ for **217b** → **218b** is −39 kcal/mol (59)], while the observed effects of substituents are consistent with the steric facilitation of disrotatory ring opening. On the other hand, the considerable interaction that occurs in oxiranyl between the unpaired electron and the oxygen lone pairs (14, 420, 421) may provide alternative mechanistic pathways that are not available in cyclopropyl radicals.

Cyclic α-alkoxyalkyl radicals (**219**) derived from oxetane and its higher homologues also undergo β scission to afford ring-opened species (4). For example, treatment of oxepane with *t*-butyl peroxide affords cyclohexanol (374), the formation of which is thought to involve recyclization of **220** ($n = 4$) generated by ring opening of **219** ($n = 4$) (see Section IV, H). Such reactions, however, are relatively slow and generally do not compete effectively with available intermolecular processes. Thus, ESR observations (420–422) of 2-oxetanyl (**219**, $n = 1$) or its higher homologues reveal no signals from the ring-opened forms, and such reactions as acyloxylation by cupric carboxylates (423) or addition to olefins (424, 425) usually afford products containing the intact ether ring. When **219** ($n = 2$) is generated in the presence of butynedioic acid, addition to afford **221** is followed by 1,5-hydrogen transfer and then by ring opening (426).

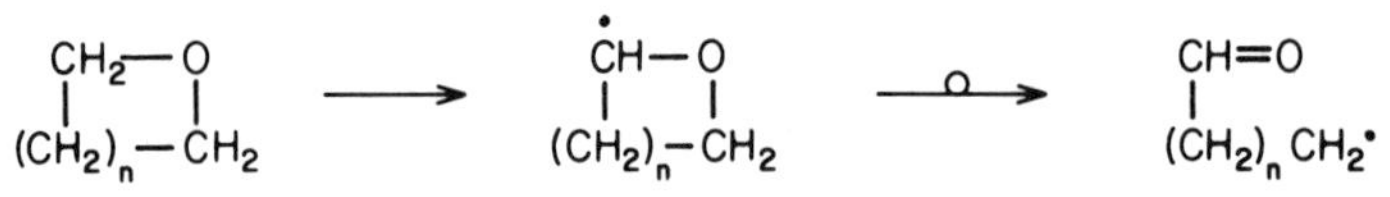

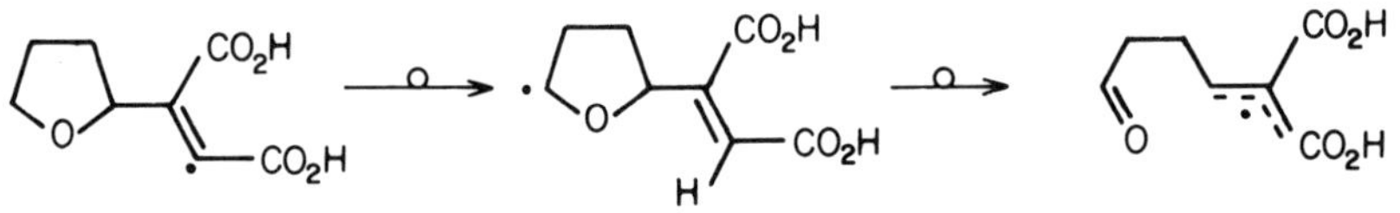

Radicals generated by hydrogen abstraction from cyclic acetals (e.g., **222–224**) appear to be somewhat more susceptible to β scission than their monooxa analogues. They can be readily observed by ESR spectroscopy at room temperature and below (14, 21, 49, 420, 427) but, as the temperature is increased, signals from the rearranged radicals (e.g., **225**) become apparent (49). As expected, the unsymmetric **222** fragments regiospe-

4. FREE-RADICAL REARRANGEMENTS

cifically to afford the more stabilized product radical, MeOCOCH$_2$ĊMe$_2$. The rate constants for ring opening of **222**, **223**, and **224** at 72°C can be estimated to be ca. 7×10^2, 6×10^4, and 6×10^3 sec^{-1}, respectively. Although there is some doubt about the absolute values for these rate constants (49), their relative order must be correct. Thus, 1,3-dioxanyl radicals undergo β scission more rapidly than their five-membered counterparts, but both cyclic species are less reactive than analogous acyclic radicals, such as (Me$_3$CO)$_2$ĊH. These results have been rationalized in stereoelectronic terms (49). The preferred coplanar arrangement of the semioccupied orbital and the C—O bond undergoing scission can be readily achieved in (Me$_3$CO)$_2$ĊH but not in **224**. The six-membered radical **223** represents an intermediate situation.

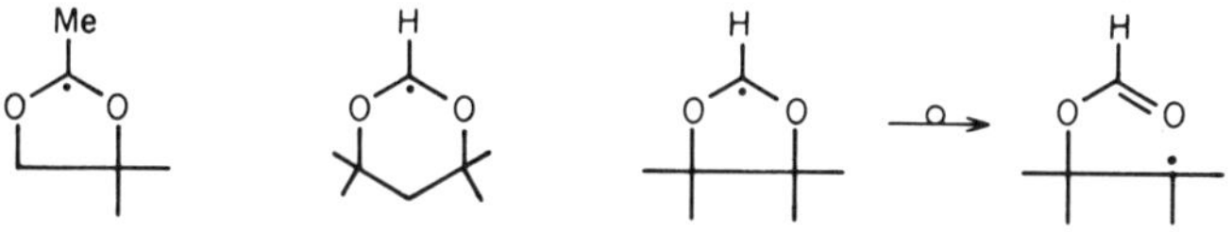

222   223   224   225

The products of radical reactions of 1,3-dioxacyclanes depend on the relative rates of β scission and intermolecular processes. For example, the perester reaction (423), which involves a fast ligand transfer from cupric carboxylate, converts 1,3-dioxane to a product containing the intact ether ring (428):

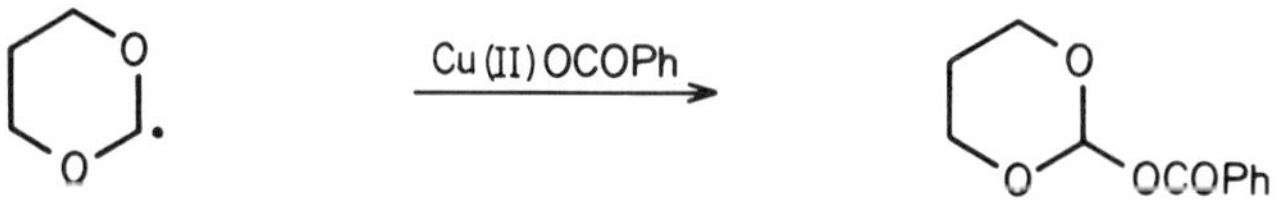

Similarly, **226** adds to 1-octene at 30°C (429), but at higher temperatures (430) ring opening of **226** competes with addition, and the product mixture contains compounds derived from both **227** and C$_6$H$_{13}$ĊH(CH$_2$)$_3$OCHO. Other examples (431–433) of reactions in which ring opening competes with intermolecular processes include the formation of dodecyl formate in 38–47% yield during the free-radical addition of 1,3-dioxepan to 1-octene (431).

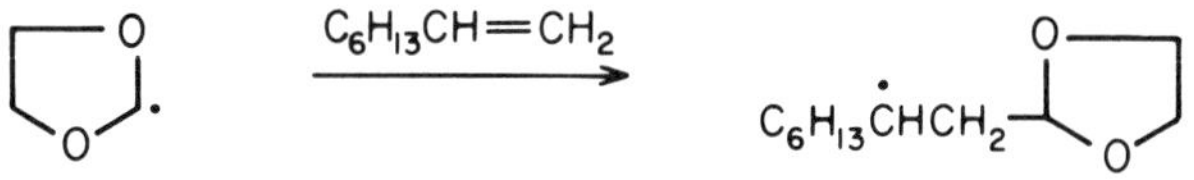

226   227

Radicals formed by hydrogen abstraction from the 4 position in cyclic acetals and ketals can undergo ring opening followed by β scission (432).

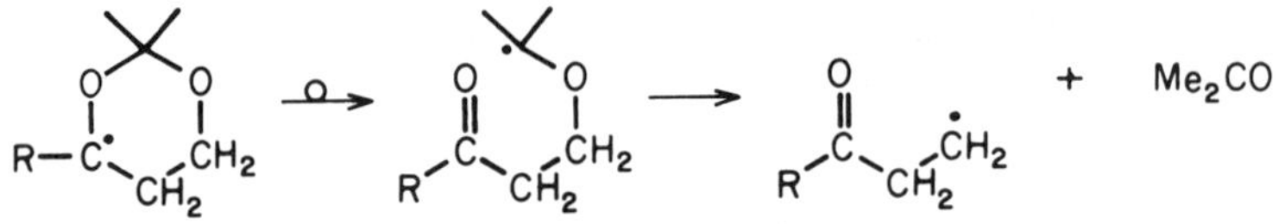

For such radicals as **228** the formation of lactone by alkyl loss competes with ring opening (*379a, 434*). The former process is usually preferred, presumably because the exocyclic O—C bond achieves overlap with the semioccupied orbital more readily than does the endocyclic O—C bond.

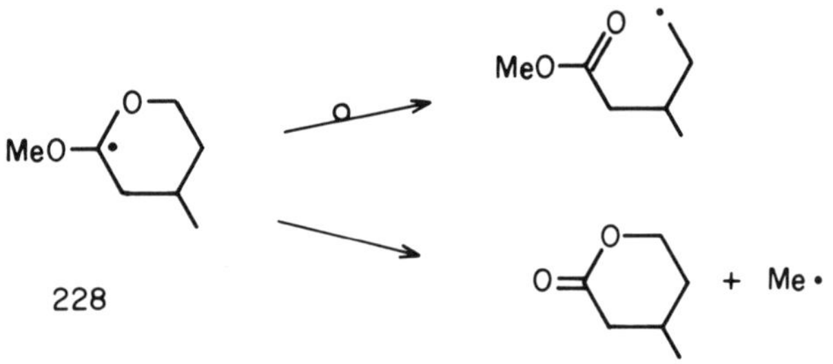

There have been few reports of the opening of cyclic radicals containing sulfur. Treatment of propylene sulfide with di-*t*-butyl peroxide affords, among other products, thioacetone (*415*). The proposed mechanism involves formation and ring opening of **229**. However, the report (*435*) that thietane (**230**) is attacked by *t*-butoxy radical at sulfur to afford **231** suggests that the reactions of propylene sulfide and similar compounds should be reexamined.

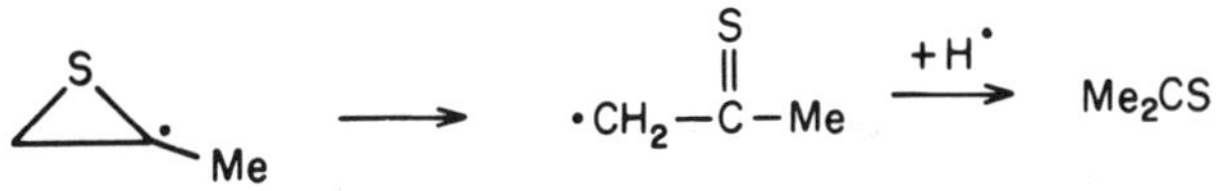

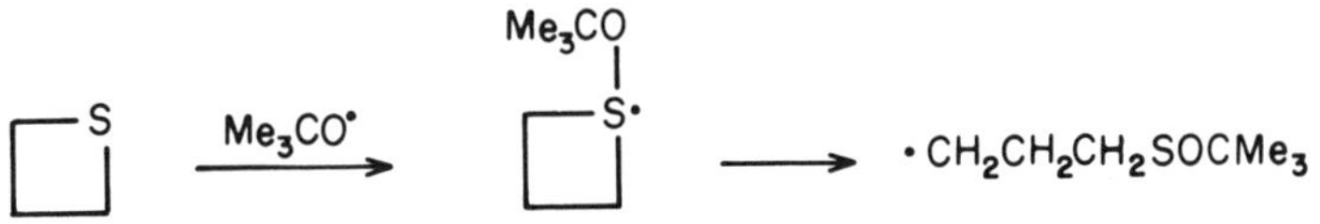

A heavily substituted azetidin-2-yl radical has been found to undergo ring-opening (*435a*).

Cyclic radicals that are resistant to thermal ring opening may undergo photolytic cleavage. Thus, UV irradiation of 2-pyridyl (**232**) in an argon matrix affords **233** (*436*). 3-Pyridyl and 4-pyridyl behave similarly.

4. FREE-RADICAL REARRANGEMENTS

$$232 \xrightarrow{h\nu} 233$$

B. Cycloalkylcarbinyl Radicals

The β scission of cyclopentylcarbinyl and higher cycloalkylcarbinyl radicals, to give ring-opened products, is endothermic (*63, 126*) and usually occurs too slowly to compete effectively with intermolecular reactions except when suitable substituents stabilize the product molecule. However, the ring openings of both cyclopropylcarbinyl and cyclobutylcarbinyl are exothermic [$\Delta H°$ = −5.1 and *ca.* −4.0 kcal/mol, respectively (*63, 126, 399, 437*)] because of the relief of ring strain. These reactions are relatively rapid (*4*).

Extensive studies of the cyclopropylcarbinyl–allylcarbinyl rearrangement **234** ⇌ **235** have revealed a number of mechanistically interesting features. It is clear that **234** and **235** are discrete chemical entities and do not have a nonclassical structure. Cyclopropylcarbinyl generated in an ESR cavity at temperatures of −140°C and below shows the expected spectrum (*51, 438*). Between −140° and −100°C spectra for both **234** and **235** are observed, whereas above −100°C only **235** can be detected. Some substituted cyclopropylcarbinyl radicals and related species behave similarly (*7, 22, 439, 440*), but in other cases only the ring-opened forms can be detected even below −140°C (*7, 22, 439–446*). The ESR spectrum of **234** indicates that the bisected conformation **236** is preferred to the perpendicular conformation **237** (*7, 22, 138, 139, 446*). This conformational preference may indicate the existence of a favorable interaction between the unpaired electron and the ring orbitals (*7, 22, 438, 439, 446–457*), but any such stabilization is probably small [1.4 kcal/mol (*437*)].

$$\triangleright\!-CH_2\cdot \xrightarrow{k_R} \diagup\!\!\diagdown\!\!-CH_2\cdot$$

234   235   236   237

Kinetic ESR data for the β scission of cyclopropylcarbinyl yield the Arrhenius parameters $\log(A/\text{sec}^{-1})$ = 12.48 ± 0.85 and E = 5.95 ± 0.57 kcal/mol, giving k_R = 1.3 × 10^8 sec^{-1} at 25°C (*51*). However, the preexponential factor should probably be 10$^{13.0}$ sec^{-1} (*51*), in which case k_R at 25°C would be 2 × 10^8 sec^{-1}. Earlier estimates of k_R by indirect methods for cyclopropylcarbinyl (*458*) and related radicals (*44, 459–461*) were in the approximate range 0.3 × 10^7 to 1.0 × 10^8 sec^{-1}.

An important feature of the cyclopropylcarbinyl–allylcarbinyl rearrangement is its sensitivity to stereoelectronic effects. Thus, the steroid radical **238** and its isomer **240** each undergo specific fission to afford **239** and **241**, respectively (*459*). Other examples (*110–112, 462–467*) of radicals that, like **240**, preferentially afford the less stabilized of the possible product radicals include **26** (*110*) and **29** (*111*) (see Section III,C). Examination of models reveals that in each case the β,γ bond that cleaves preferentially is that which is most nearly in the eclipsed conformation with respect to the C_α semioccupied orbital.

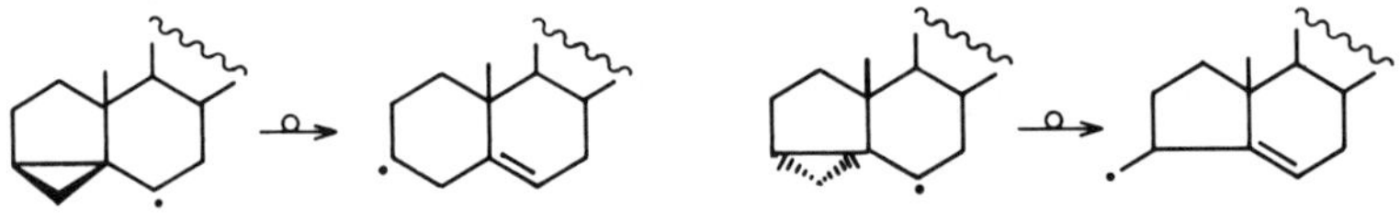

238   239   240   241

One explanation (*6, 459, 467*) for these observations is that the transition state (**242**) for the β scission of an alkyl radical should allow maximal interaction between the semioccupied orbital and the σ* orbital of the bond undergoing change. This hypothesis is consistent with the experimental evidence, accords with the stereoelectronic approach to radical cyclization (see Section IV,A), and explains the reluctance of cyclopropyl and cyclobutyl radicals to undergo ring opening (see Section V,A).

242

This hypothesis is also consistent with the fact that many conformationally mobile derivatives of cyclopropylcarbinyl preferentially afford the more stabilized product radical [e.g., **243** (*468*) and **244** (*440*)]. On the other hand, a high degree of overlap between the semioccupied orbital and the β,γ bond cannot be mandatory since radicals that are fixed in a bisected conformation also undergo ring opening, e.g., **245** ($n = 1, 2,$ or 3) (*454*).

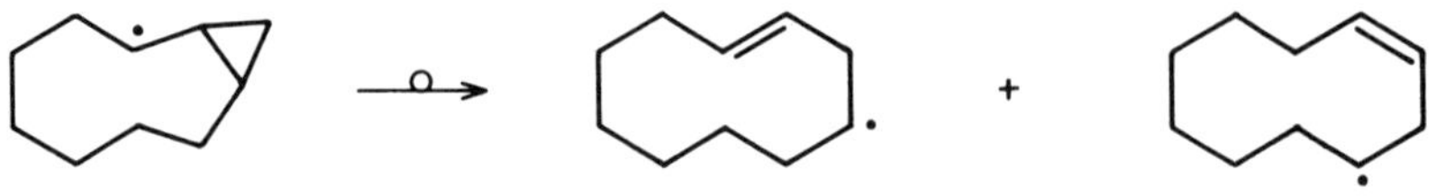

243

4. FREE-RADICAL REARRANGEMENTS

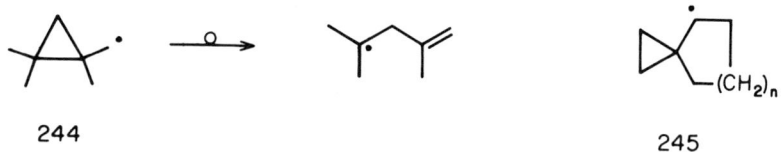

244   245

Two cyclopropane rings must cleave in the interesting rearrangement of **246** to **247** (*469*). The 7-norbornadienyl radical is not an intermediate in this reaction.

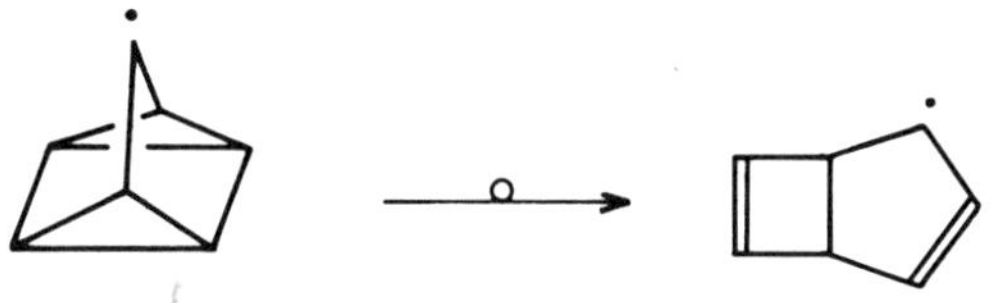

246 → 247

cis-2-Alkyl-substituted cyclopropylcarbinyls yield the thermodynamically favored secondary alkyl, e.g., **248**—○—**249**, but, under conditions favoring kinetic control, their trans isomers anomalously give the primary alkyl, e.g., **250**—○—**251** (*442–445, 470–473*). When conditions of thermodynamic control are employed, the trans isomers give products derived from the secondary alkyl **249** (*442, 443, 472–475*).

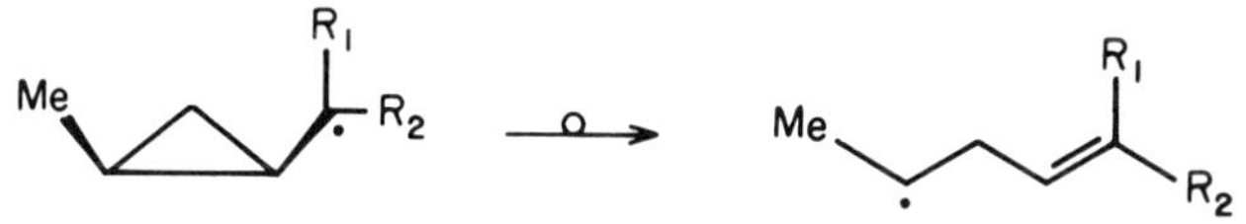

248 → 249

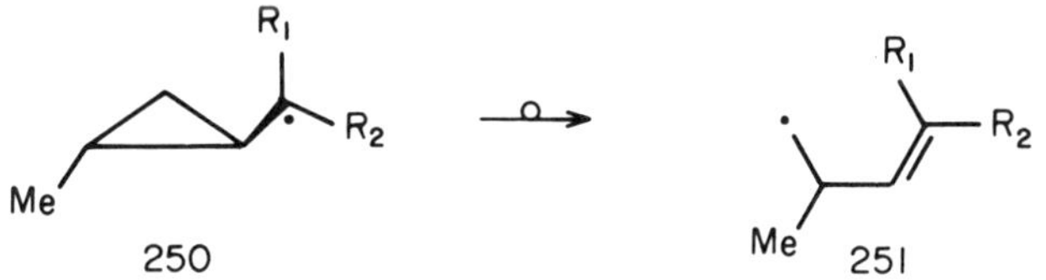

250 → 251

$R_1 = R_2 = H$
$R_1 = OH; R_2 = Me$
$R_1 = OSnBu_3; R_2 = Me$

No completely satisfactory explanation for the curious behavior of *trans*-2-alkylcyclopropylcarbinyl radicals has yet been advanced. Possibly polar effects are important. It has been suggested (*441, 442, 470, 471*) that the oxygen-substituted radicals **250b** and **250c** afford the primary radicals **251b** and **251c** because of the contribution of the polar structure

252 to the transition state. At first sight, this explanation is inapplicable to species such as **250a** which contain no polar substituent. If, however, the transition state involves interaction of the unpaired electron with a σ* orbital, as illustrated in **242**, it should have dipolar character even in the absence of polar substituents. Cis radicals **248** give the thermodynamically favored secondary alkyl because steric factors cause them to adopt conformation **253** in preference to **254**.

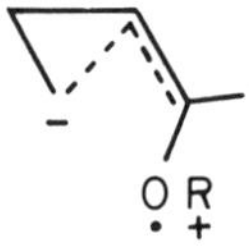

 252 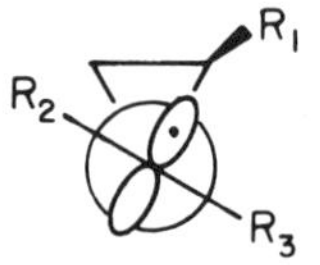   253 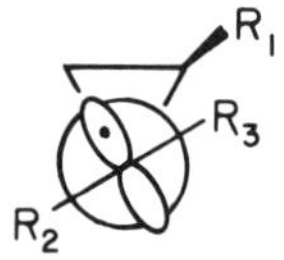   254

Fragmentations of cyclopropylcarbinyl radicals bearing a single substituent or two dissimilar substituents at the radical center are subject to the further stereochemical complexity of geometric isomerism about the double bonds of the products. For example, reduction of (1-chloroethyl)-cyclopropane with tributylstannane affords a mixture of cis- and trans-pent-2-ene via the rearrangement of **255** to **256** and **257** (*166*). The predominance of the trans product (trans/cis = 2.2) may be attributed to the lower energy of the conformation **258** as compared with conformation **259**, in which nonbonded interactions are more severe.

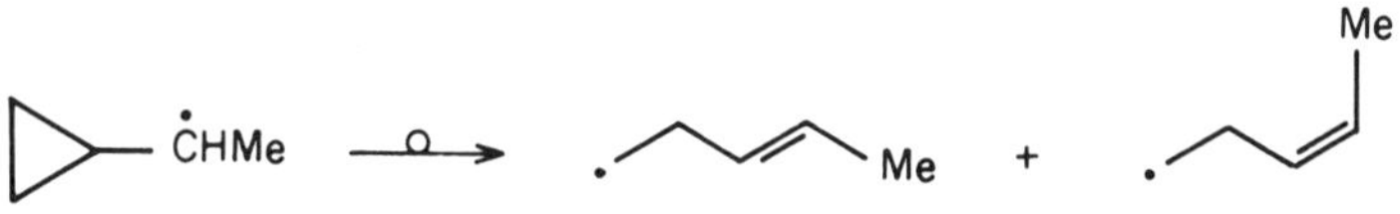

 255   256   257

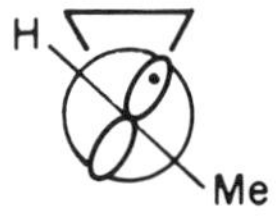

 258 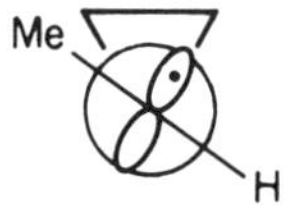   259

The formation of pairs of geometrically isomeric substituted allylcarbinyl radicals from suitable precursors has been observed by ESR spectroscopy (*439, 444, 445*). For example (*445*), at low temperatures each of the two isomeric radicals (**261** and **262**) formed by rearrangement of **260** can be detected. At higher temperatures **263** is also observed. The temperature dependence of the concentrations of the various radicals shows that **263** is formed only from **261**.

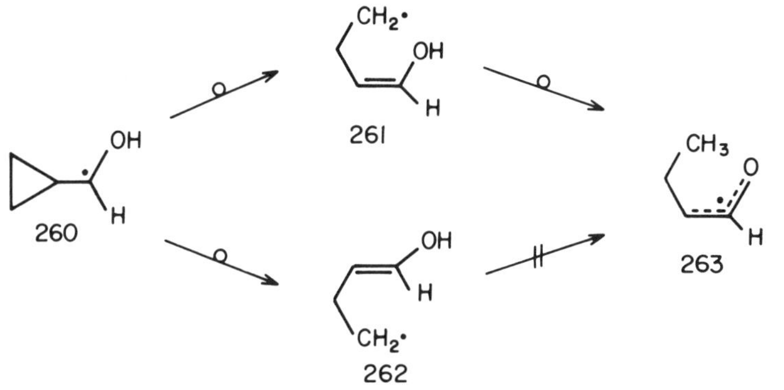

4. FREE-RADICAL REARRANGEMENTS

[Scheme showing 260 → 261, 262 → 263]

Because of their high rate constants, the rearrangements of cyclopropylcarbinyls often compete effectively with intermolecular reactions. Products containing the intact cyclopropyl ring are obtained in good yield only when the radical is generated in the presence of a high concentration of an efficient chain-transfer agent. Thus, in the radical addition of thiols (*451, 476, 477*), halomethanes (*451*), methyl hypochlorite (*478*), iodobenzene dichloride (*478*), and trialkylstannanes (*479*) to 2-cyclopropylpropene (**264**) the distribution of products varies according to the chain-transfer ability of the addend XY and its concentration.

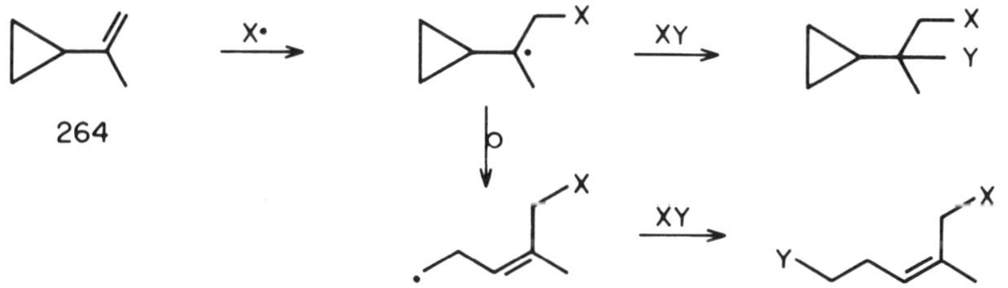

[Scheme showing 264 reactions with X• and XY]

Radicals containing 1-cyclopropylallyl and related systems are less prone to undergo rearrangement than their saturated analogues. In the case of **265**, this unreactivity was originally attributed to conformational effects (*119*). However, conformationally mobile radicals such as **266** (*157*) also show markedly reduced rates of rearrangement (*157, 480, 481*), and it seems likely, therefore, that the low reactivity of such radicals is due mainly to resonance stabilization.

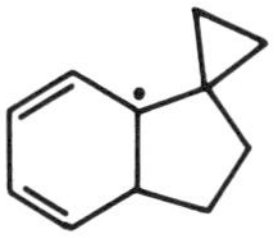

[Structures 265 and 266]

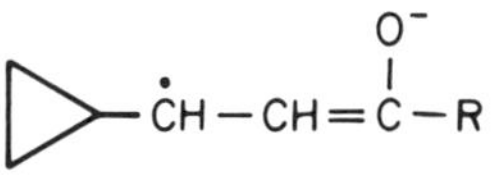

Some bicyclic cyclopropylallyls undergo rearrangements that can be formally regarded as pericyclic reactions. When generated in solution at 130°C, **267** affords benzene via an electrocyclic ring opening (*463*). Electron spin resonance studies of **267** in an adamantane matrix (*41, 482*) show that the reaction **267**⟶**268** has $\Delta G^{\ddagger} = 14.5$ kcal/mol at $-50°C$. Generation of a deuterium-labeled radical revealed a more rapid sigmatropic shift (**269**⟶**270**). It seems probable that this rearrangement, and the related isomerization of **271** to its exo isomer (*465*), are not truly pericyclic but involve ring-opened intermediates such as **272** (*482a*). A more complex example or apparent valence isomerization is provided by the rearrangement of **273** (*483*).

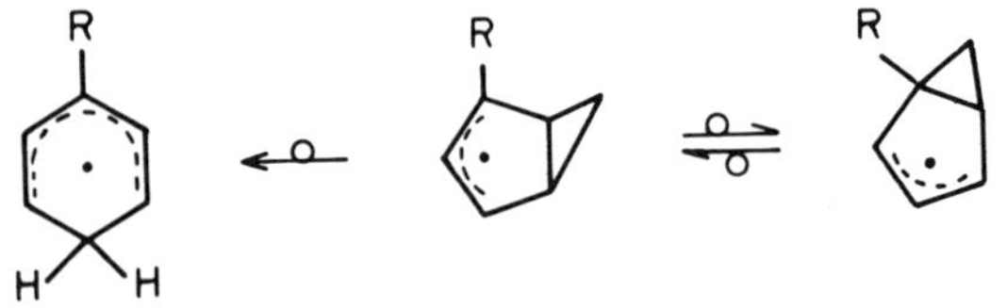

268 R = H

267 R = H
269 R = D

270 R = D

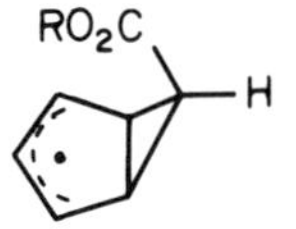

271

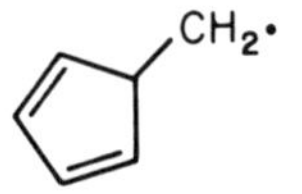

272

273

Some semidione radical anions containing the cyclopropane ring also appear to undergo valence isomerization (*484*). Thus, the rearrangement of **274** to **276** is believed to proceed via electrocyclic reactions of the radical dianion **275** (*485*). It is not clear, however, whether the formation of **279** by electrolytic reduction of the dione **277** (*484*) involves valence isomerization of the radical anion **278** or the analogous dianion.

We noted previously (Section IV,C) that the equilibrium **234**⇌**235** lies heavily in favor of the open-chain form and that this also appears to be the case for a wide range of substituted cyclopropylcarbinyls. However, in some polycyclic systems, favorable steric factors shift the position of equilibrium toward the cyclized form. In the best known of such cases,

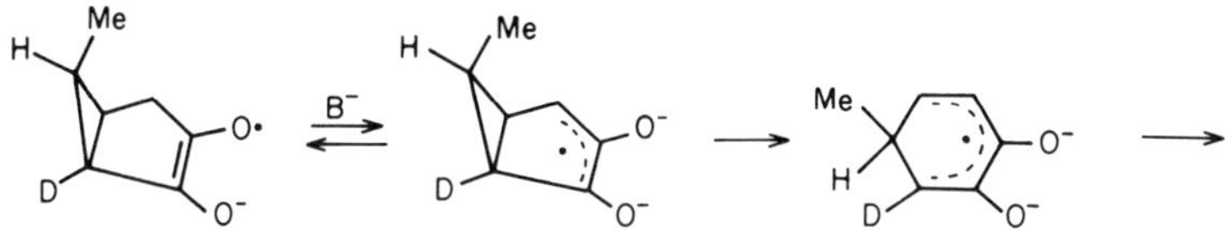

4. FREE-RADICAL REARRANGEMENTS

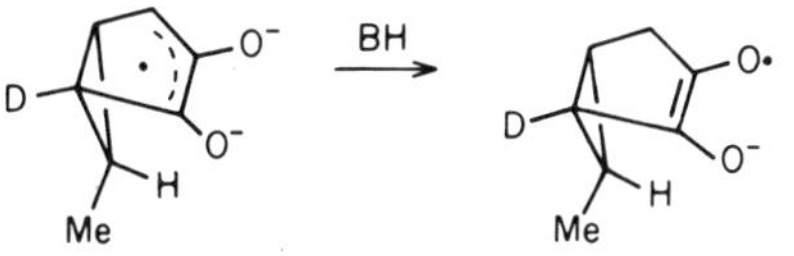

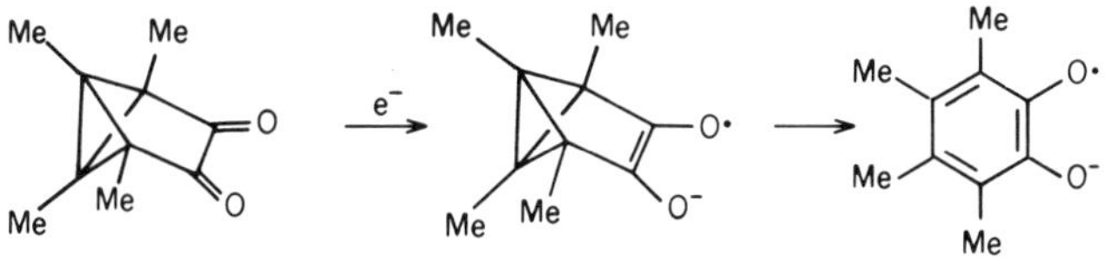

the norbornenyl–tricyclyl radical system **83** ⇌ **84**, the rate constant for ring opening is similar to that for cyclopropylcarbinyl, but the rate constant for cyclization is abnormally high (44). Rearrangements of norbornenyl and tricyclyl radicals and similar species have been extensively investigated (4, 486); although sometimes complex, recent examples (246, 407, 461, 487, 488), e.g., **280** ⟶ **281** (488), show no unexpected mechanistic features.

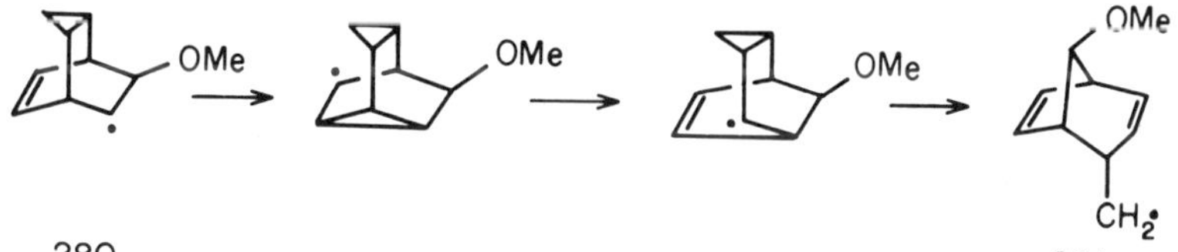

The ring-opening reactions of some substituted bicyclobutylcarbinyl radicals also show no unusual features (488a).

A number of radicals related to cyclopropylcarbinyl undergo similar ring-opening reactions. They include the vinyl radicals **282** (489) and **283** (490) and various heterocyclic species (418, 444, 446, 491–493):

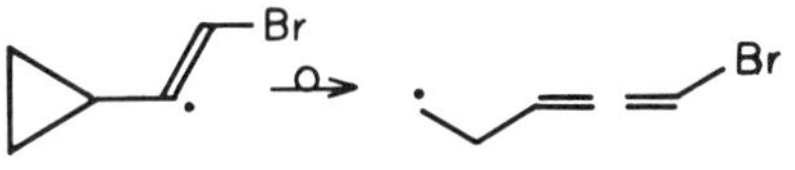

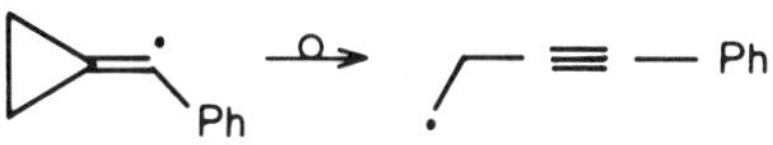

With radicals (**284**) derived from glycidols and their *O*-trialkyltin derivatives, β cleavage of the C—O bond in the ring is followed by a 1,5 transfer of hydrogen or of the trialkyltin group, respectively, from enoxyl oxygen to alloxyl oxygen (*444, 493*):

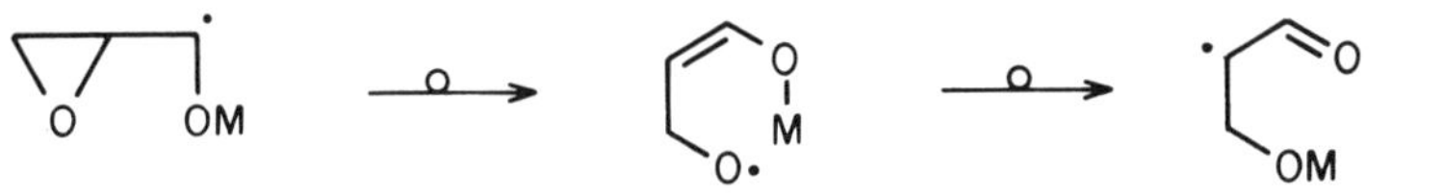

284

M = H, Bu₃Sn

The overall process is similar to the rearrangement (**260** ⟶ **261** ⟶ **263**) observed with cyclopropylhydroxymethyl radicals. In all cases the experimental data indicate a high rate of rearrangement. For example, when **285** was generated in the ESR cavity at −116°C, the only radical detected was **286** arising from consecutive ring opening and hydrogen transfer (*444*). A signal for **287** was obtained at temperatures lower than −136°C, but at higher temperatures the spectrum of the rearranged radical $\dot{C}H_2CH_2N{=}CH_2$ appeared (*446*).

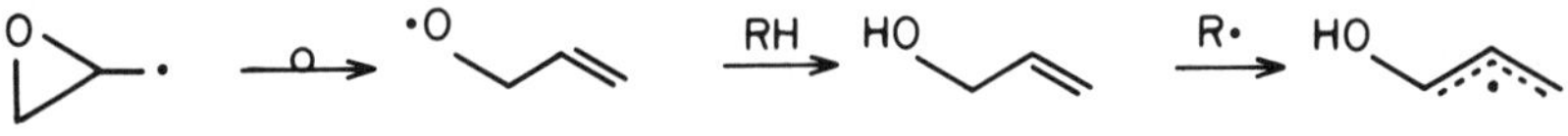

285   **286**

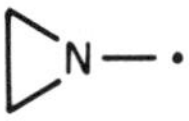

287

The cyclopropylcarbinyl–allylcarbinyl rearrangement is a useful mechanistic probe for the detection of radical intermediates and the estimation of the rates of competing processes. Thus, the formation of cyclopropylmethyl bromide in high yield from the reaction of copper bromide with cyclopropylacetyl peroxide (*492*) indicates the absence of cationoid intermediates and allows the rate constant for ligand-transfer oxidation of cyclopropylcarbinyl to be determined (*170, 492*). Other examples include the detection of radical mechanisms in reactions of lithium dimethyl cuprate with enones (*157, 180, 494*) and of metalate anions with alkyl halides (*495*), in lead tetraacetate oxidations (*395a*), in electrolysis (*496*), and in various photochemical reactions (*467, 497, 498*) and the determination of the lifetimes of photochemically generated diradicals (*498*).

Despite the many satisfactory applications of this rearrangement as a mechanistic probe, there are reactions for which there is good evidence

4. FREE-RADICAL REARRANGEMENTS

for alkyl radical participation *except* when cyclopropylcarbinyl is involved. Thus, numerous tests indicate that the reduction of alkyl halides by (η^5-cyclopentadienyl)tricarbonyl hydridovanadate, ($\eta^5 - C_5H_5$)-$V(CO)_3H^-$, proceeds by a radical route *(499)*. However, cyclopropylcarbinyl bromide is anomalous, the methylcyclopropane/1-butene ratio being almost independent of the hydride concentration *(499)*. This bromide is probably reduced by a two-electron mechanism *(499)*. Another possible example of anomalous behavior occurs in the oxidation of alcohols in aqueous solution by such metal ions as Cr(IV) *(500, 500a)* and Ce(IV) *(501)*. Various lines of evidence [e.g., deuterium kinetic isotope effects *(500, 500a, 502)*, the low reactivity of *t*-butanol *(500, 500a)*, the effect of oxygen *(503)*, and the initiation of polymerization *(500, 500a, 504)*] suggest that these reactions proceed by one-electron oxidations with cleavage of an α C—H bond, i.e.,

$$M^{(n+1)+} + R_1R_2CHOH \longrightarrow M^{n+} + H^+ + R_1R_2\dot{C}OH$$
$$M^{(n+1)+} + R_1R_2\dot{C}OH \longrightarrow M^{n+} + H^+ + R_1R_2CO$$

However, cyclopropylcarbinol gives such a high yield of cyclopropylaldehyde when oxidized by chromic acid *(500, 500a)* or ceric ammonium nitrate *(501)* that other mechanistic pathways should be considered for this alcohol, if for no other.

The intimate mechanism of the reduction of cyclopropyl ketones by dissolving metals *(505)* has been extensively investigated *(474, 505–507)*. Most experiments have been carried out on rigid or semirigid systems in which the direction of ring cleavage is determined primarily by stereoelectronic factors. However, in some conformationally mobile systems, e.g., **288** *(474, 506)*, the major product (**289**) is that formed by scission of the less substituted β,γ bond, a result that has been interpreted *(474, 506)* as demonstrating the intermediacy of **290** or **291** but not of **292**. The recent observations *(442–445, 470–473)* that ring opening of some monosubstituted cyclopropylcarbinyl radicals shows a similar preference (*vide supra*) indicate that this conclusion may not be valid and that reduction of cyclopropyl ketones does not necessarily involve anionic ring cleavage.

Rearrangements of cyclobutylcarbinyl radicals have been less extensively studied than those of their cyclopropylcarbinyl analogues (4). The best-known examples of such processes involve ring opening of the intermediates (**293** and **294**) formed by radical addition to α- and β-pinene (4, 508, 509). The salient features of the reaction are that the more highly substituted β,γ bond specifically undergoes scission and that it proceeds sufficiently rapidly to prevent the formation of vicinal adducts except when highly efficient chain-transfer agents, e.g., thiols (508), are employed. A more complex example of a cyclobutylcarbinyl ring opening is involved in the formation of **296** from **295** (510).

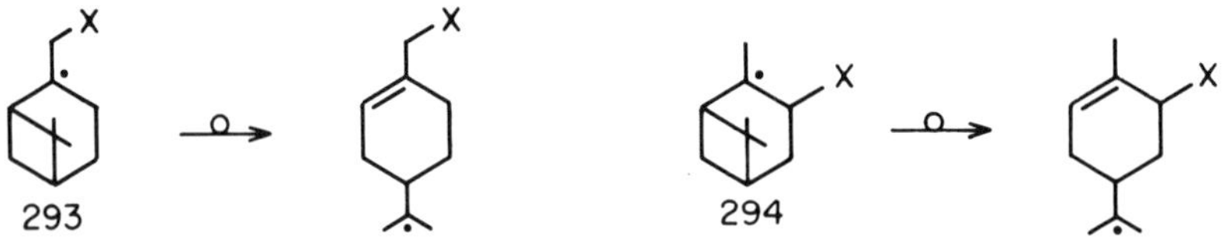

Monocyclic radicals containing the cyclobutylcarbinyl system undergo ring opening much more slowly than do related cyclopropylcarbinyl species. Thus, **297a** can be observed by ESR spectroscopy (511) under conditions that give only the spectra of ring-opened forms of cyclopropylcarbinyl radicals. Similarly, reactions of cyclobutylcarbinyl halides with tributylstannane give good yields of cyclobutane derivatives (166, 441, 512–516) under conditions in which the cyclopropane analogues afford only butenyl compounds. The stannane reaction has been employed to obtain kinetic data for ring opening of **297b** (166, 512). The Arrhenius parameters are $\log(A/\text{sec}^{-1}) \approx 11.7$ and $E \approx 12.2$ kcal/mol, which gives $k \approx 5.6 \times 10^2 \text{ sec}^{-1}$ at 25°C. The difference in activation energies for the rearrangement of cyclobutylcarbinyl and cyclopropylcarbinyl is too large to be attributed to the *ca.* 1.1 kcal/mol (*vide supra*) difference in reaction enthalpies. Presumably, steric or electronic factors, as yet undefined, preferentially stabilize the transition state for the cyclopropyl system. Nevertheless, the two transition states must be of basically similar structure since the ring openings of cyclobutylcarbinyls show the expected stereoelectronic effect. Thus, **298** undergoes specific scission of the bond that is more nearly eclipsed by the semioccupied orbital and thus affords the thermodynamically less stable product (166, 512, 515).

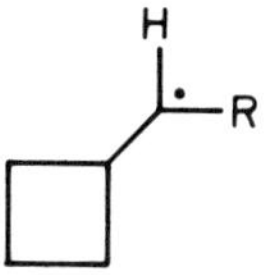

4. FREE-RADICAL REARRANGEMENTS

297a R = OH
297b R = H

298

The relative rates of ring opening of a number of simple derivatives of cyclobutylcarbinyl radical are recorded in Table 3. The main features revealed by these data are (a) that the more highly substituted β,γ bond undergoes scission, (b) that cis-2-methylcyclobutylcarbinyl radical rearranges more rapidly than its trans isomer, and (c) that *trans*-olefins are formed in preference to *cis*-olefins.

TABLE 3
Relative Rate Constants for β Scission of Cyclobutylcarbinyl Radicals (166,512)

Radical	k (1,2 fission)	k (1,4 fission)
Cyclobutylcarbinyl (CBC)	0.5	0.5
α-Methyl-(CBC)	0.45	0.45 (trans/cis = 3)
α,α-Dimethyl-(CBC)	0.4	0.4
1-Methyl-(CBC)	0.25	0.25
3,3-Dimethyl-(CBC)	0.3	0.3
trans-2-Methyl-(CBC)	3.3	0.4
cis-2-Methyl-(CBC)	20	0.4
2,2-Dimethyl-(CBC)	150	<0.8
1,2,2-Trimethyl-(CBC)	190	<0.3
2,2,3,3-Tetramethyl-(CBC)	66	<0.3

Finally, the allylic cyclobutenylcarbinyl radical **247** when generated in an adamantane matrix, rearranges to **299** ($\Delta G^{\ddagger}$ = 21.5 kcal/mol) by what is formally an electrocyclic disrotatory ring opening (*517*).

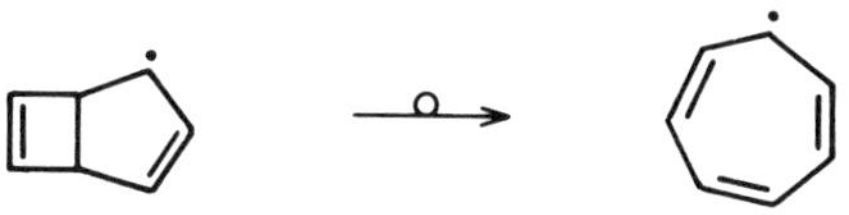

247

299

C. Cycloalkyloxy and Similar Heteroradicals

Rearrangements of cyclopropyloxy (**300**) and cyclobutyloxy (**301**), being highly exothermic, proceed with facility. Thus, cyclopropyl nitrite (*518*) and cyclobutyl nitrite (*519*) each undergo clean decomposition to afford ring-opened nitroso compounds or products derived therefrom. Similarly, thermolysis of cyclobutyl hypochlorite gives 4-chlorobutanal, which then undergoes further reactions (*520*). The ring opening of alkyl-substituted cyclopropyl nitrites, e.g., **302**, involves specific scission of the most highly substituted bond, and the resultant radical (**303**) can be trapped with CCl_3Br (*518*).

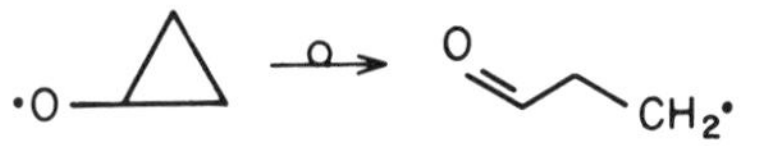

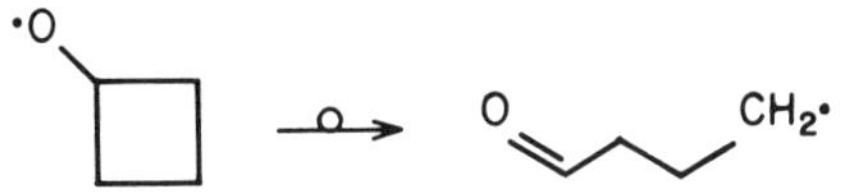

Many oxidation reactions of cyclopropanols and cyclobutanols also afford ring-opened products by free-radical mechanisms (*500, 521–534*). Cyclopropanone hemiacetals, e.g., **304**, are particularly susceptible to oxidation by oxygen (*530, 531*), peroxides (*532*), and metal ions (*533*). The homolytic nature of such reactions is supported by ESR (*534, 535*) and CIDNP measurements (*532*). The ring opening of *cis*- and *trans*-2-ethyl-cyclopropyloxy radicals (**305** and **306**, respectively) does not appear to show the unusual regioselectivity that is observed with the corresponding substituted cyclopropylcarbinyl radicals referred to above (*535*). The main product from both **305** and **306** (R = ethyl) is the secondary alkyl radical (*535*).

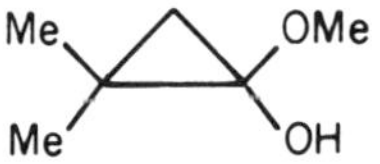

The main question concerning the above reactions is whether **300** and **301** actually participate as discrete intermediates. No product containing an intact cycloalkyl ring that could reasonably be formed directly from **300** or **301** has been detected, and, indeed, the formation of cyclic prod-

4. FREE-RADICAL REARRANGEMENTS

ucts in some reactions of cyclobutanols is regarded as evidence of two-electron oxidation (*524, 536*). Furthermore, the rates of reactions expected to generate cyclopropyloxy or cyclobutyloxy radicals are enhanced by substituents capable of stabilizing ring-opened radicals (*518, 524*). It has therefore been concluded that **300, 301,** and similar radicals are not true intermediates in such reactions, which are believed to proceed in a concerted fashion directly to ring-opened product radicals (*518, 524, 531*).

Reactions of cyclopentyloxy and higher cycloalkyloxy radicals (*537*) present no such mechanistic ambiguities. Here, hydrogen atom transfer, either intermolecular or intramolecular (see Section VIII), often competes effectively with β scission, and both cyclic and acyclic products are obtained. For example, decomposition of cyclopentyl hypochlorite in cyclohexane (*538*) affords mixtures of cyclopentanol and 5-chloropentanal in relative yields, which vary with cyclohexane concentration as expected on the basis of the following mechanism:

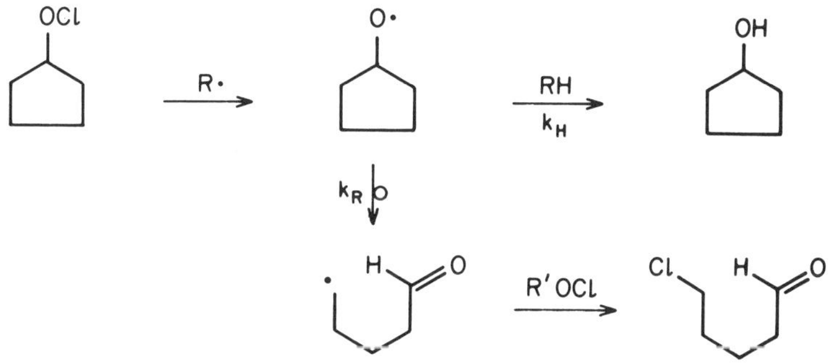

This type of experiment allows the ratio k_R/k_H to be determined. As expected, ring scission in 2-substituted cycloalkyloxy radicals usually occurs preferentially at that bond which affords the most stabilized product radical. An unusual exception to this rule, as yet unexplained, is the formation of **310** from **307** (*539*), presumably via scission of **308** to give **309**:

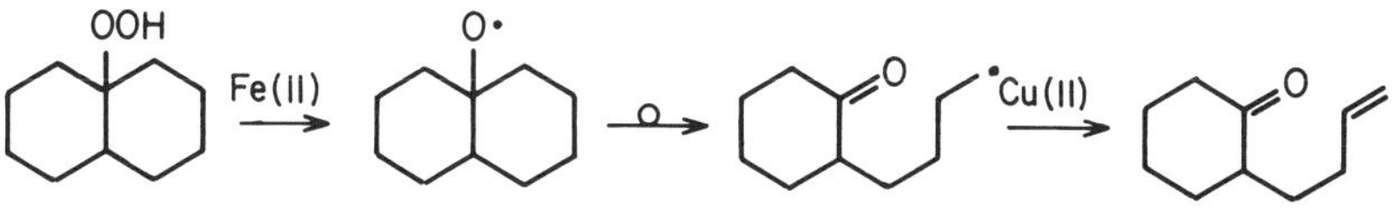

307   308   309   310

The relative rates of fragmentation of aryl-substituted acyclic alkoxy radicals vary according to the electron-donating power of the substituent (*538*). It appears, therefore, that the reaction involves a dipolar transition state ($C^{\delta+}$---$\dot{C}$=$O^{\delta-}$) and that the observed regioselectivity of ring opening of alkyl-substituted cycloalkyloxy radicals reflects not only the stabilizing effect of the substituents on the new radical center, but also their electronic effects (538).

Although absolute rate constants for rearrangements of cycloalkyloxy radicals are not yet available (79), it is clear that they must be relatively large. Calculations from literature data (374) indicate that k for cyclization of $\dot{C}H_2(CH_2)_4CH$=O to cyclohexyloxy is about 5×10^5 sec^{-1} at ordinary temperatures. Since the equilibrium between the two radicals probably favors the open-chain form (63, 126), k for ring scission is probably greater than 10^6 sec^{-1}. The rate constants for rearrangement of cyclopentyloxy and cycloheptyloxy radicals are expected to be larger because of the more favorable enthalpy changes (63, 126).

Cycloalkyloxy radicals can be readily generated by a wide variety of methods, and their β scission to afford ring-opened products is of considerable synthetic utility (371a, 540–543). The following are examples:

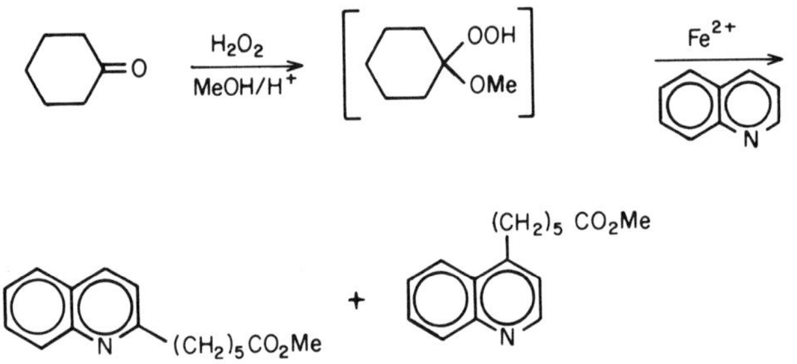

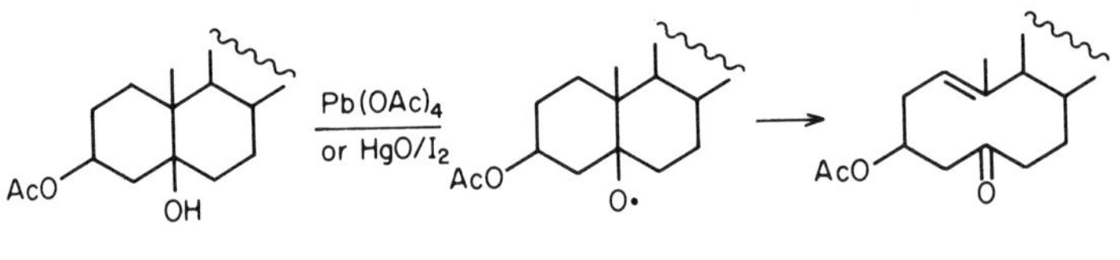

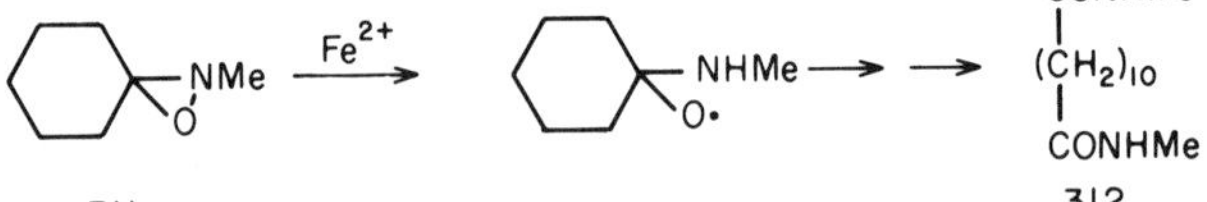

Oxy radicals bearing heterocyclic rings also undergo β scission (544, 545). One recent example, e.g., 313 → 314 (545) [which provides an extremely useful method for the hydroalkylation of olefins (545)], appears to involve concerted O—H bond breaking and loss of nitrogen.

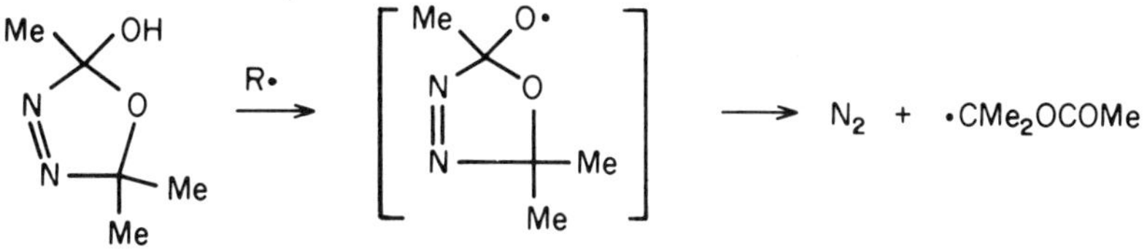

4. FREE-RADICAL REARRANGEMENTS

[Structure 313] → [R•] → [intermediate] → N_2 + •CMe_2OCOMe [314]

The conversion **311** → **312** shown above may equally well involve β scission of the cycloalkylamino radical **315** (*546*). Instances of similar rearrangements are known. For example, photolysis of a mixture of cyclopropylamine and di-*t*-butyl peroxide in an ESR cavity does not give the spectrum of **316** even at −120°C but gives instead the spectrum of $\dot{C}H_2CH_2CH=NH$ (*446, 547*). Under similar conditions, *N*-alkyl cyclopropylamines also give spectra of ring-opened primary alkyl radicals (*292*). In contrast, the *N*-cyclobutyl-*N-n*-propylamino radical can be detected at −100°C (*292*). However, at higher temperatures the cyclobutyl ring opens and an *n*-alkyl radical is observed (*292*). The rate constant for this last rearrangement can be represented by $\log(k/\text{sec}^{-1}) = (12.8 \pm 1.5) - (10.5 \pm 1.5)/\theta$, with $k = 1.2 \times 10^5 \text{ sec}^{-1}$ at 25°C.

Some cycloalkylimino radicals, e.g., **317** (*548*), also undergo ring opening (see Sections III,D and IV,H). The rate constant for the ring-opening of the cyclobutylimino radical is greater than 10^3 sec^{-1} at −73°C (*381*).

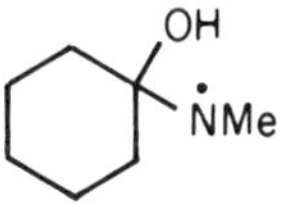

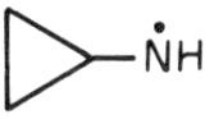

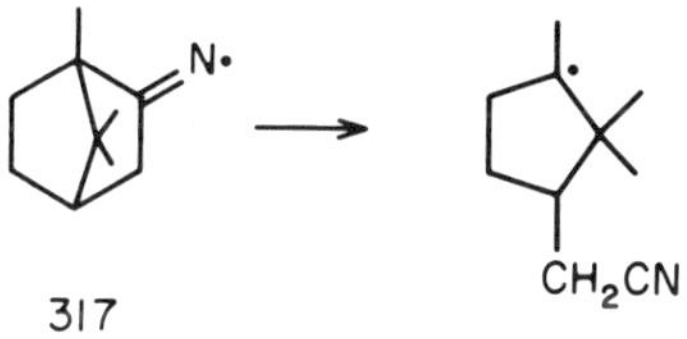

[Structures 315, 316, 317]

VI. REARRANGEMENT BY TRANSFER OF A HETEROATOM-CENTERED GROUP

A. Group Mobilities

The ease of migration of a heteroatom-centered group (and of a heteroatom; see Section VII) should parallel the reactivity of the heteroatom in S_H2 processes. Such reactions occur most readily at multivalent atoms that have a low-lying, unfilled orbital available to accept the unpaired electron (*55, 125*). Thus, with trialkylboranes in noncoordinating media the boron $2p_z$ orbital is unoccupied, and S_H2 reactions occur with great facility unless suppressed by complexing (*55, 125*). In contrast, trialkylamines, which have no vacant orbital on nitrogen, do not undergo S_H2 reactions readily. However, trialkylphosphines have vacant, low-lying 3d orbitals, and so they undergo a wide variety of S_H2 reactions. In general, it is found that S_H2 reactions at heteroatoms take place with increasing facility on moving away from Group IV along any row of the periodic table, e.g., N < O; Si(IV) < P(III) < S(II) (<Cl), and on moving down any particular group, e.g., O << S < Se; Si < Ge < Sn < Pb. The same ordering is expected to apply to radical rearrangements. However, many apparent intramolecular rearrangements actually involve an elimination–readdition pathway (*vide infra*).

B. Carbon to Carbon Migration

1. 1,2-Acyloxy migration

Suitably constituted β-acyloxyalkyl radicals rearrange by 1,2 migration of the acyloxy group (*65, 421, 549–551*). The mechanism of this intramolecular transfer is of particular interest because it does not appear to have a direct intermolecular analogue. This rearrangement was originally observed during a study of radical additions to the acetate of 2-methylbut-3-en-2-ol (*549*). The same type of rearrangement was observed

$$Me_2C(OAc)CH=CH_2 \xrightarrow{R\cdot} Me_2C(OAc)\dot{C}HCH_2R \xrightarrow{RH} Me_2C(OAc)CH_2CH_2R$$

$$Me_2\dot{C}CH(OAc)CH_2R \xrightarrow{RH} Me_2CHCH(OAc)CH_2R$$

in the decarbonylation of 3-acetoxy-3-methylbutanal (*550*). The relative yield of the rearranged product increased as the aldehyde concentration decreased. These data have been used to estimate a rearrangement rate constant k_r of ca. 3×10^3 sec^{-1} at 75°C (*421*).

$$Me_2C(OAc)CH_2CHO \xrightarrow[(-CO)]{(-H)} Me_2C(OAc)\dot{C}H_2 \xrightarrow{RCHO} Me_2C(OAc)CH_3$$

$$\downarrow k_r$$

$$Me_2\dot{C}CH_2OAc \xrightarrow{RCHO} Me_2CHCH_2OAc$$

There are four distinct mechanistic pathways by which this class of rearrangements might proceed. Route a was considered unlikely in view of the rapid decarboxylation of acyloxy radicals (*550*). Route d could be ruled out because generation of $\dot{C}H_2CMe_2OAc$ in an ESR spectrometer under conditions in which the rearranged radical, $Me_2\dot{C}CH_2OAc$, could also be detected showed no trace of the cyclized intermediate radical, $OCH_2CMe_2O\dot{C}Me$, while this radical, generated independently under similar conditions, showed no sign of ring opening (*421*).* Experiments

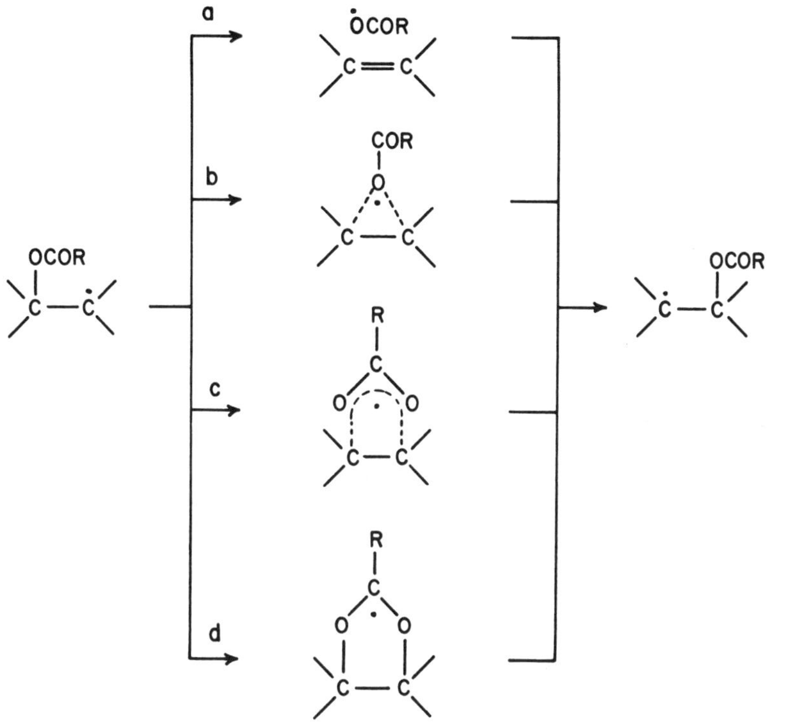

4. FREE-RADICAL REARRANGEMENTS

[Scheme showing routes a, b, c, d for rearrangement of β-acyloxyalkyl radical]

using ^{18}O-labeled material have shown that rearrangement inverts the labeling pattern (65), i.e.,

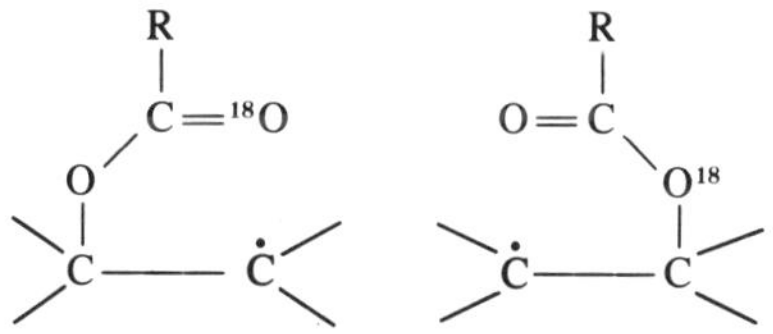

[Structures showing ^{18}O labeling inversion]

This rules out routes a and b. The rearrangement of β-acyloxyalkyl radicals must therefore proceed by the concerted five-membered cyclic transition state, route c (65). This conclusion received added support from kinetic and product studies in which the radicals were generated by reaction of β-bromoalkyl esters with tri-n-butyltin hydride (65). These data also yield rate constants for some of these 1,2-acyloxy migrations (see Table 4).

* Radicals of this class do undergo ring opening at higher temperatures (49).

TABLE 4

Rate Constants[a] for the Rearrangement MeCR^1(OCOR^2)ĊH$_2$ ⟶ MeĊR^{1}CH$_2$OCOR^2

Temp. (°C)	R^1	R^2	k_r (sec^{-1})
75	Me	Me	2.7×10^3
75	Me	Me(CH$_2$)$_5$	1.6×10^3
70	Me	Ph	1.8×10^3
70	Ph	Me	1.9×10^{4b}

[a] Rate constants for the competitive H atom abstraction from n-Bu$_3$SnH by the primary alkyl were calculated taking log(A/M^{-1} sec^{-1}) = 8.5 and E = 3.4 kcal/mol (see footnote d, Table 1). This yielded k_r values that are lower than originally (65) estimated.

[b] This radical also undergoes a 1,2-phenyl shift (see Table 1).

The concerted mechanism of path c raises two questions. First, why do the 2-substituted 1,3-dioxolan-2-yl radicals (of path d) not lie on the reaction coordinate? Second, what is the difference between such species and the suggested cyclic transition state? Although there is no clear answer to the first question, it has been suggested that the answer to the second lies in the geometries of the two intermediates (65). The cyclic transition state would be expected to have the five annular atoms and the 2-substituent lying in one plane, whereas ESR studies have shown that the radical centers in 1,3-dioxolan-2-yl radicals are pyramidal (14, 420, 421).

Several reactions analogous to the 1,2-acyloxy migration are also known (552).

2. Organothiyl migrations

There have been numerous reports that organothiyl groups undergo 1,2 migrations (553–556). This subject has also been reviewed (3). "Proof" that such processes are intramolecular rearrangements has usually been supplied by the composition of the products. However, it seems more probable that the majority of these 1,2 migrations actually involve an elimination–readdition sequence since olefins are readily isomerized by organothiyl radicals. This conclusion (557) is supported by the observation that β-organothiylalkyl radicals lose the RS· moiety extremely readily (57, 557–559) and by the fact that organic disulfides always appear to be formed concurrently (553–556):

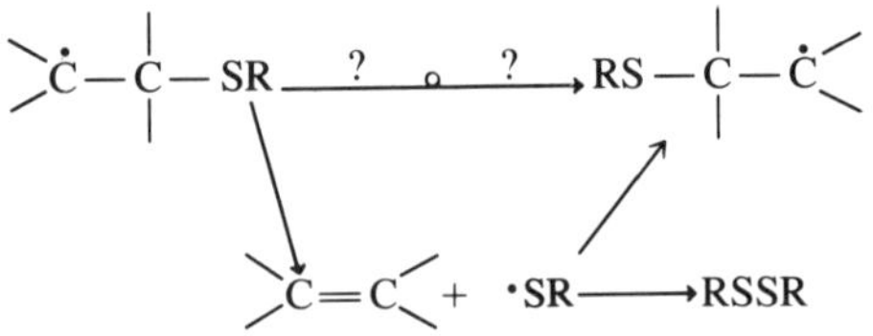

$$\text{>Ċ–C–SR} \xrightarrow{?} \xrightarrow{?} \text{RS–C–Ċ<}$$
$$\searrow \qquad \nearrow$$
$$\text{>C=C<} + \cdot\text{SR} \longrightarrow \text{RSSR}$$

4. FREE-RADICAL REARRANGEMENTS

The interesting rearrangement

$$(C_6H_5)_2P\dot{C}HCH_2SR \rightleftharpoons (C_6H_5)_2\dot{P}(SR)CH=CH_2$$

probably also involves an elimination–readdition of the RS· radical *(560)*.

Although intramolecular migrations of organothiyl groups to remote atoms do not appear to have been identified, there is no doubt that such processes would occur with suitably constituted radicals. For example, radical **318** partitions between substitution ($S_H i$) at sulfur and reduction by hydrogen atom donors *(561, 562)*. The same is true of radical **319** *(561, 562)*.

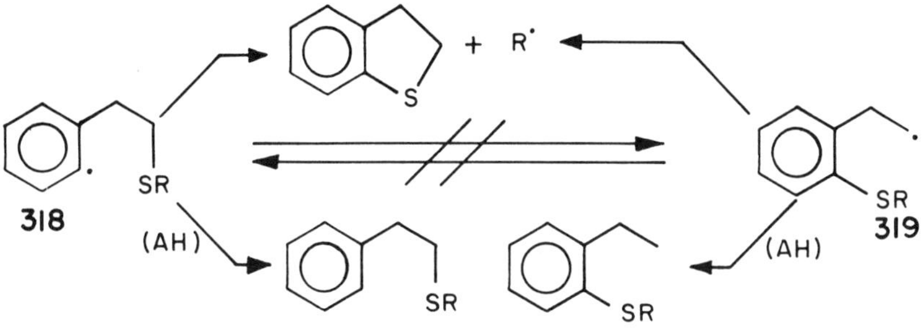

318 R = CH_3, CF_3, C_2H_5, $(CH_3)_3C$, $C_6H_5CH_2$, C_6H_5
319 R = CH_3, $C_6H_5CH_2$, C_6H_5

These results imply that there is a preferred linear geometry for substitution at bivalent sulfur, with the exocyclic "inversion" pathway being preferred to the endocyclic "retention" path. This is consistent with the formation of a trigonal bipyramidal sulfuranyl radical (**320**) in which the sulfur has expanded its coordination number and its electronic octet and in which there is a preference for apical attack and apical leaving with a rate of pseudorotation that is slow with respect to decomposition. A variety of sulfuranyl radicals have been identified by ESR spectroscopy *(563)*, and labeling experiments have indeed confirmed the apical attack of radicals at bivalent sulfur *(564)*. Sulfuranyl radicals are also involved in

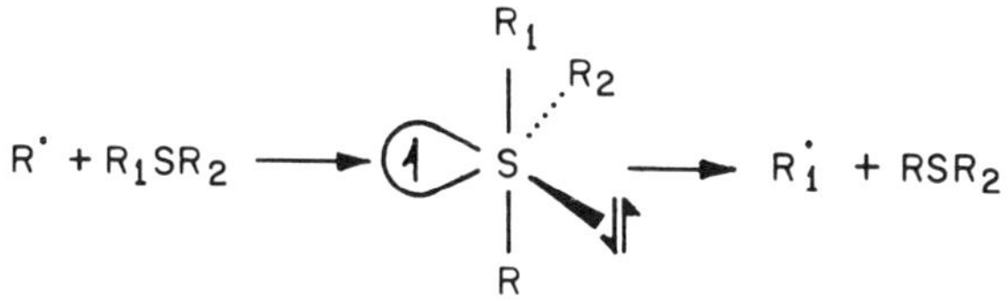

$$R· + R_1SR_2 \longrightarrow \text{(320)} \longrightarrow R_1· + RSR_2$$

320

the homolytic decomposition and rearrangements of certain sulfur-containing peresters *(387, 565)*. Finally, it is worth noting that the sulfoxide of **318** (R = CH$_3$), but not the sulfone, undergoes an intramolecular substitution similar to that of **318**. In this case, an intermediate sulfuranyloxyl radical with a structure similar to **320** (but with an oxygen atom in place of the lone pair) must be involved. There is ESR evidence that sulfuranyloxyl radicals do indeed have a trigonal bipyramidal structure (566).

The requirement for a linear transition state in S$_H$i and in S$_H$2 reactions at sulfide and sulfoxide sulfur helps to explain why the 1,2 migration of organothiyl groups occurs preferentially by an elimination–readdition route.

3. Other migrations

Intramolecular migrations of other heavy-atom-centered groups (as well as 1,2 migrations occurring by elimination–readdition) have received almost no attention. Such processes should be relatively facile for suitably constituted radicals containing Se(II), P(III), Sn(IV), etc., since these atoms can expand their valence shell to accommodate the extra electron. In contrast, intramolecular 1,2 migrations of OH, OR, or NH$_2$ groups should be extremely slow because such rearrangements would require one-electron occupancy of an antibonding molecular orbital. The elimination–readdition path would also have a high activation energy because of the strength of C—O and C—N bonds. It is therefore somewhat surprising to discover that such rearrangements may be occurring in aqueous media. When vicinal diols and related compounds (567) such as amino alcohols (568) react with HO• in water, the initial radicals are readily converted to α-carbonyl radicals. At low pH, an acid-catalyzed elimination of water has been proposed (567), e.g.,

$$\text{HOCH}_2\text{CH}_2\text{OH} \xrightarrow{\cdot\text{OH}} \text{HOCH}_2\dot{\text{C}}\text{HOH} \xrightleftharpoons{\text{H}^+} \text{H}_2\overset{+}{\text{O}}\text{CH}_2\dot{\text{C}}\text{HOH} \xrightarrow{-\text{H}_2\text{O}} \dot{\text{C}}\text{H}_2\text{CH}=\overset{+}{\text{O}}\text{H} \xrightarrow{-\text{H}^+} \dot{\text{C}}\text{H}_2\text{CH}=\text{O}$$

$$\dot{\text{C}}\text{H}_2\text{CH(OH)}_2 \updownarrow$$

A molecular orbital theoretical study of some analogous enzyme-mediated rearrangements has led to the proposal that these reactions may involve the intramolecular 1,2 migration of the protonated group, because protonation is calculated to facilitate migration (569), e.g.,

$$\underset{\underset{\dot{\text{H}}_2\text{C}-\text{CH(OH)}}{|}}{\overset{\text{H}}{\text{H}-\text{O}^+}} \longrightarrow \underset{\dot{\text{H}}_2\text{C}\cdots\cdots\text{CH(OH)}}{\overset{\text{H}\diagdown\diagup\text{H}}{\text{O}}} \longrightarrow \underset{\underset{\text{H}_2\dot{\text{C}}-\text{CH(OH)}}{|}}{\overset{\text{H}}{{}^+\text{O}-\text{H}}} \longrightarrow \text{H}_2\dot{\text{C}}\text{CH}=\text{O}$$

4. FREE-RADICAL REARRANGEMENTS

C. Migration to or from Heteroatoms

Kinetic studies have shown that the phenoxyl **321** decays with second-order kinetics (*570*). The purported (*571*) 1,3 shift of the Me₃Si group does not, therefore, occur:

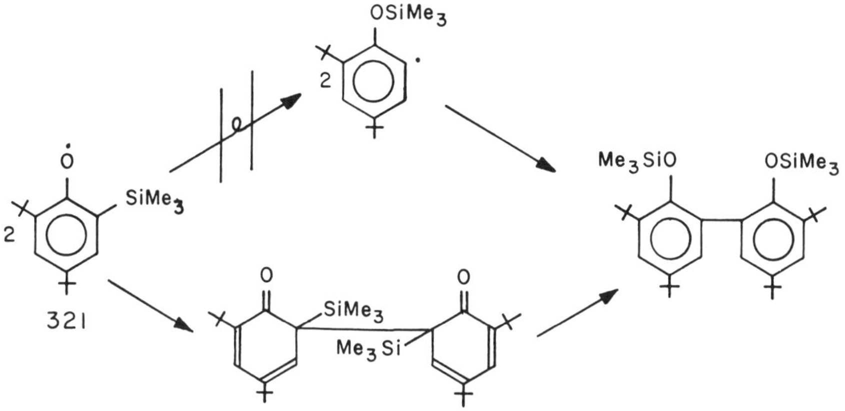

The thermal rearrangement of alkyl(silylmethyl) ethers (**322** ⟶ **323**) involves free radicals. However, it is migration of the Me₃Si group that initiates homolysis to form the ketyl and alkyl, which then combine (*572*):

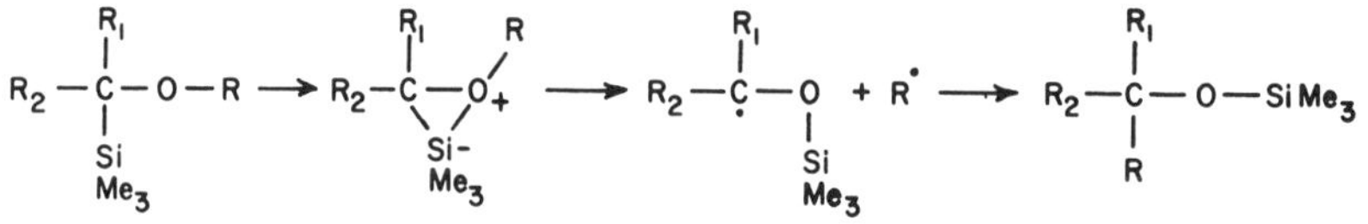

$$R_2-\underset{\underset{Me_3}{Si}}{\overset{R_1}{C}}-O-R \longrightarrow R_2-\underset{\underset{Me_3}{Si-}}{\overset{R_1}{C}}\underset{O^+}{\diagup}{}^R \longrightarrow R_2-\underset{\underset{Me_3}{Si}}{\overset{R_1}{C}}-\overset{\cdot}{O} + R^\cdot \longrightarrow R_2-\underset{R}{\overset{R_1}{C}}-O-SiMe_3$$

322 **323**

The 1,2 migration of silicon-centered groups between heteroatoms is, however, well documented. For example, there is facile migration from silicon to sulfur, with Me₃SiMe₂Si migrating more rapidly than Me₃Si (*573*).

$$Me_3SiMe_2SiMe_2Si\overset{\cdot}{S} \longrightarrow Me_2\overset{\cdot}{Si}SSiMe_2SiMe_3$$

$$(Me_3Si)_2MeSi\overset{\cdot}{S} \longrightarrow Me_3SiMe\overset{\cdot}{Si}SSiMe_3$$

Arrhenius parameters for the thermoneutral 1,4 migrations of triorganosilyl and triorganogermyl groups between the oxygen atoms of benzil (*574*) and 3,6-di-butyl-1,2-benzoquinone (*575*) have been measured by ESR line width effects, e.g. (*574*),

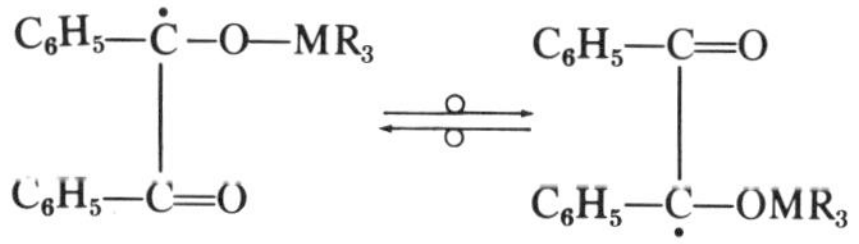

$$\begin{array}{c} C_6H_5-\overset{\cdot}{C}-O-MR_3 \\ | \\ C_6H_5-C=O \end{array} \rightleftharpoons \begin{array}{c} C_6H_5-C=O \\ | \\ C_6H_5-\overset{\cdot}{C}-OMR_3 \end{array}$$

R$_3$M	log(A/sec^{-1})	E (kcal/mol)
Et$_3$Si	12.3	9.4
Ph$_3$Si	11.6	5.9
Et$_3$Ge	12.2	7.0
Ph$_3$Ge	11.1	3.7

Similar reactions occur between acenaphthoquinone and triorganogermyl or triorganotin radicals (*575a*).

An exothermic 1,5 migration of the tributyltin group between the oxygen atoms in the ring-opened radical derived from **284** (M = Bu$_3$Sn) has also been reported (*493*) (see Section V,B).

Also known are 1,2 migrations of R$_3$Si and R$_3$Ge groups from oxygen to nitrogen, e.g. (*576*),

$$Me_3SiNOSiMe_3 \longrightarrow (Me_3Si)_2NO^{\bullet}$$

VII. ISOMERIZATION BY TRANSFER OF A HALOGEN ATOM

A. 1,2 Migration of Halogen

1. Chlorine

The 1,2 shift of chlorine in β-chloroalkyl radicals was first recognized in 1951 (*577, 578*). This isomerization (the Nesmeyanov rearrangement) has been identified in numerous reaction systems and has been thoroughly reviewed (*3, 4, 8, 579*). In this section, attention will be focused on the isomerization of a few simple β-chloroalkyls.

The reaction of *t*-butyl chloride with *t*-butyl hypobromite yields bromides derived from both the unrearranged radical **324** and the rearranged radical **325** (*8, 580, 581*):

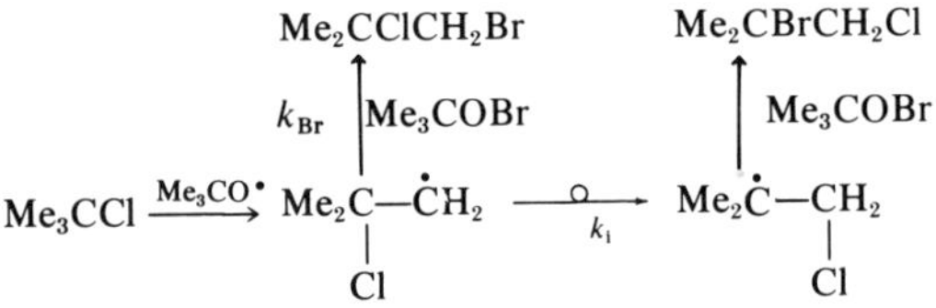

$$Me_3CCl \xrightarrow{Me_3CO^{\bullet}} \underset{\underset{324}{Cl}}{Me_2\overset{\bullet}{C}-CH_2} \xrightarrow{k_i} \underset{\underset{325}{Cl}}{Me_2\overset{\bullet}{C}-CH_2}$$

with $k_{Br} \uparrow Me_3COBr$ giving Me_2CClCH_2Br and $Me_3COBr \uparrow$ giving Me_2CBrCH_2Cl

Studies on the trapping of optically active β-bromoalkyls by the hypobromite (see Section IX) indicate that k_{Br} is nearly diffusion-controlled (*581*). On this basis, the product ratio yields a value for k_i of *ca*. 10^{10} sec^{-1} at 50°C and an activation energy E_i of 3–5 kcal/mol (*8, 580, 581*). Forma-

4. FREE-RADICAL REARRANGEMENTS

tion of **324** in an ESR spectrometer yields only the rearranged radical **325**, even at temperatures as low as $-130°C$ (*582–584*). This result implies that $k_i \geq ca.\ 10^4\ \text{sec}^{-1}$ at this temperature, which is quite consistent with the hypobromite studies (i.e., even assuming that A_i is as large as $10^{13}\ \text{sec}^{-1}$, the hypobromite data yield $E_i \sim 4.5\ \text{kcal/mol}$ and hence $k_i \sim 10^6\ \text{sec}^{-1}$ at $-130°C$). Reaction of the less reactive radical trapping agent, *t*-butyl hypochlorite, with $Me_3C^{36}Cl$ under conditions similar to the hypobromite reaction yielded 90% of the rearranged product, $Me_2C(Cl)CH_2{}^{36}Cl$ (*8, 580, 581*). The foregoing experiments all point to the fact that the Nesmeyanov rearrangement of $Me_2C(Cl)\dot{C}H_2$ is in intramolecular isomerization.

The 1,2 migration of chlorine is also extremely rapid in $Me_2C(Cl)\dot{C}HMe$ (*583*) and in $Cl_3C\dot{C}H_2$ (*583, 584*), only the isomerized radicals being observed by ESR spectroscopy at temperatures of *ca*. $-120°$ to $-140°C$. However, $Me_2C(Cl)\dot{C}HCl$ showed no tendency to rearrange (*583*). Radical additions to 3,3,3-trichloropropene appear always to yield rearranged products (*3, 4, 579*).

$$R\cdot + CH_2\!\!=\!\!CHCCl_3 \longrightarrow RCH_2\dot{C}HCCl_3 \stackrel{\circ}{\longrightarrow} RCH_2CHCl\dot{C}Cl_2$$

The ESR spectra of β-chloroalkyls show that these radicals prefer a conformation in which the chlorine is eclipsed by the $C_\alpha\ 2p_z$ orbital (*7*). The β-carbon in $ClCH_2\dot{C}H_2$ (**326**) is distorted in a direction that moves H_β closer to the nodal plane and the chlorine closer to the semioccupied orbital; that is, this radical can be considered to be unsymmetrically "bridged":

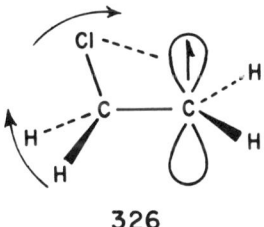

326

This geometry is obviously ideal for a 1,2 migration of chlorine. However, it must be added that a suitable conformation does not ensure an intramolecular migration. For example, β-alkylthioalkyl radicals also adopt an eclipsed conformation (*7*), but they appear to rearrange by an elimination–readdition sequence (see Section VI,B,2).

Product studies (*3, 4, 8, 579*) and the spin-trapping technique (*585*) have shown that there is wide variety in the rates at which different β-chloroalkyls isomerize. There can be no doubt that many of these reactions will be suitable for quantitative investigation by ESR spectroscopy.

Finally, although the isomerization **327** —⚬→ **328** occurs by a free-radical chain, it does not involve a 1,2 migration of chlorine from carbon to silicon in the intermediate Me$_2$ṠiCH$_2$Cl radical, as originally proposed (*586*). Instead, this reaction involves two consecutive chain processes, all steps of which are bimolecular (*587*):

$$Me_2SiHCH_2Cl \xrightarrow{\;\;\circ\;\;} Me_3SiCl$$

$$327 \qquad\qquad\qquad 328$$

Chain 1:

$$Me_2\dot{S}iCH_2Cl + Me_2SiHCH_2Cl \longrightarrow Me_2SiClCH_2Cl + Me_2SiHCH_2\cdot$$

$$Me_2SiHCH_2\cdot + Me_2SiHCH_2Cl \longrightarrow Me_3SiH + Me_2\dot{S}iCH_2Cl$$

Chain 2:

$$Me_3Si\cdot + Me_2SiClCH_2Cl \longrightarrow Me_3SiCl + Me_2SiClCH_2\cdot$$

$$Me_2SiClCH_2\cdot + Me_3SiH \longrightarrow Me_3SiCl + Me_3Si\cdot$$

The Me$_2$SiClCH$_2\cdot$ radical is not chlorine-bridged (*587a*).

2. Bromine

The 1,2 migration of bromine occurs even more rapidly than do analogous chlorine shifts (*3, 4, 8, 579*). For example (*8*), the chlorination of *t*-butyl bromide with *t*-butyl hypochlorite or chlorine yields the rearranged 1-chloro-2-bromo-2-methylpropane exclusively. In certain cases, 1,2 shifts of bromine may occur by elimination and readdition (*8*). Certainly the loss of bromine is a facile process since even the BrCH$_2$CH$_2\cdot$ radical cannot be detected by ESR spectroscopy at $-120°$C in solution (*7*), which implies that elimination occurs with a rate constant $\geqslant 10^4$ sec^{-1} at this temperature. However, there is strong evidence that the isomerization *can* be intramolecular. Thus, in the chlorination of optically active 2-bromobutane with *t*-butyl hypochlorite, a minor product is 1-bromo-2-chlorobutane (**331**) *with a high optical purity and retained configuration* (*8, 580*). This result requires not only that the bromine shift be intramolecular (**329** —⚬→ **330**), but also that the rearranged 1-bromobut-2-yl radical **330** be bromine-"bridged" (see Section IX,B) for a sufficient length of time for it to be trapped by the hypochlorite:

ACTIVE **329** —⚬→ ACTIVE **330** —Me$_3$COCl→ ACTIVE **331**

4. FREE-RADICAL REARRANGEMENTS

The relative importance of inter- and intramolecular isomerization must be dependent on the nature of the β-bromoalkyl radical and on the experimental conditions.

B. Transfer of Halogen to a More Remote Radical Center

Isomerizations in this class have received very little attention (*588*) compared with the analogous reaction involving hydrogen transfer (see Section VIII). Most experimental searches for such isomerizations have yielded only negative results. For example (*589*), the potential 1,3 shift of iodine in $I(CH_2)_2CD_2\cdot$ is not fast enough to compete with the trapping of this radical by CCl_4 at 95°C, and the activation energy for isomerization was estimated to be >11 kcal/mol greater than that for the intermolecular identity reaction $n\text{-}Pr\cdot + n\text{-}PrI \rightarrow n\text{-}PrI + n\text{-}Pr\cdot$. Similarly, when the three radicals $Cl_3C(CH_2)_nCH_2\cdot$ ($n = 1, 2,$ and 3) were generated in an ESR spectrometer at 20°C, in no case was there any sign of the isomerized $Cl_2\dot{C}(CH_2)_nCH_2Cl$ radical (*584*), implying that k_i is less than 10^3 sec^{-1} at this temperature. There would seem to be no intrinsic reason why halogen atoms should not undergo 1,5 and similar migrations, provided that there is sufficient driving force. Indeed, such isomerizations might provide a convenient route for the quantitative study of halogen abstraction by a number of heteroatom-centered radicals for which there are few or no absolute rate data in solution.*

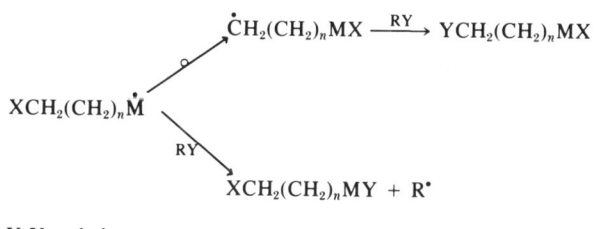

X,Y = halogen

VIII. ISOMERIZATION BY HYDROGEN ATOM TRANSFER

A. Atom Mobilities

. Intramolecular hydrogen atom transfers have been thoroughly discussed in several excellent and extensive reviews (*2–4, 6, 592–597*). As with their intermolecular counterparts, the transition state that appears to

---

* For example, there do not appear to be any absolute rate data for halogen abstraction by trialkylsilyl radicals in solution (*11, 590*), although there are some data on chlorine abstraction by $Me_3Si\cdot$ in the gas phase (*591*).

be preferred has an approximately collinear arrangement of the bond being broken and the bond being formed. Thus, those hydrogen atom isomerizations in which the transition state would have to deviate farthest from linearity are either unknown (1,2 migrations) or are extremely uncommon (1,3 and, to a lesser extent, 1,4 migrations). A reasonable approach to a linear transition state is possible in 1,5 migrations. These occur via a distorted six-membered cyclic transition state and, as Wilt (*4*) noted, "1,5 shifts dominate this area of organic radical rearrangements."

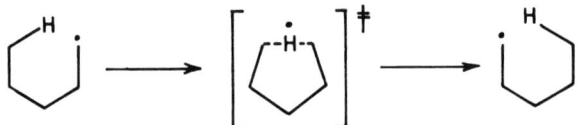

Although an acyclic radical can achieve a nearly linear transition state in 1,6-hydrogen migrations and in migrations to certain more remote atoms, such "long-range" isomerizations are much less common than the 1,5 migrations. This is due to adverse entropic effects since conformers that would allow isomerization to occur are formed with decreasing probability as the radical center and the bond to be broken become farther apart.

The rule (that 1,5 migration predominates) can break down when a rigid geometry holds a "remote" atom in close juxtaposition to the radical center. This phenomenon is most often observed in cyclic and multicyclic radicals. However, it is significant that 1,10 and 1,11 migrations have been observed starting with long, straight-chain, primary alkyl radicals (*598*). Furthermore, the 1,6 migration is not quite as unimportant as is frequently supposed. In two careful studies, Lefort and co-workers (*598, 599*) showed that the ratio of 1,5- to 1,6-hydrogen migration, $r = k_{1,5}/k_{1,6}$, is *ca.* 3.3 for $CH_3(CH_2)_nCH_2^\cdot$ (*598*) and *ca.* 10 for $CH_3(CH_2)_nO^\cdot$ (*599*) ($n \geq 5$). A value for r of *ca.* 15 had peviously been reported (*600*) for $CH_3(CH_2)_{n-1}C(CH_3)_2O^\cdot$ radicals. The fact that the value of r for the tertiary alkoxys is larger than that for the primary alkoxys can probably be attributed to the Thorpe–Ingold effect of the *gem*-methyl groups (see Section III,B,1). Lefort's values of r for both the C to C and C to O isomerizations were independent of the temperature, except in the case of the *n*-heptyl radical (*598*). This means that the activation energies for 1,5- and 1,6-hydrogen migrations are virtually equal. The preponderance of 1,5 over 1,6 migration is due to a more favorable entropy of activation [more favorable by 1.6 gibbs/mol for the alkyl (*598*) and by 4.4 gibbs/mol for the alkoxy isomerization (*599*)]. Differences in C—C and C—O bond lengths suggest that the seven-atom cyclic transition state that is required for the 1,6 migration is slightly less easily achieved by the alkoxy than by the alkyl radical (*599*).

4. FREE-RADICAL REARRANGEMENTS

Although the absolute rate constants for most 1,5-hydrogen migrations in solution are unknown, it would appear that the activation energies for these isomerizations are generally comparable to those found for analogous intermolecular abstractions, i.e., $E_{1,5} \approx E_H$:

$$R_1 \cdot \xrightarrow{k_{1,5}} R_2 \cdot$$

$$R_1 \cdot + R_2H \xrightarrow{k_H} R_1H + R_2 \cdot$$

Because the transition state for the isomerization is conformationally very restrictive, the preexponential factor is "low" (63, 601, 602) in comparison with the "normal" value of ca. 10^{13} sec^{-1} found for many other types of unimolecular reaction. One way in which $k_{1,5}$ and k_H can be compared is to calculate an "effective" concentration for the hydrogen atoms that are available for migration from the rate constant ratio $k_{1,5}/k_H$. This ratio is generally in the range 1–100 M. Under competitive conditions, therefore, $R_1 \cdot$ will react intramolecularly except at very high R_2H concentrations.

B. Carbon to Carbon Migration

1. 1,2- and 1,3-Migrations

There are no authentic 1,2-hydrogen atom migrations in monoradicals in solution (2–4, 6, 592), although evidence in favor of such a process in the gas phase at 500°–600°C has been presented (603). Such reactions *may* occur in diradicals (604). Authentic 1,3 migrations are exceedingly rare. Reutov and co-workers (605–608) have adduced evidence favoring the isomerization of 1-propyl radicals isotopically labeled at the 1 position. The radicals were generated from labeled butyryl peroxide in halogenated solvents at elevated temperatures, and the position of the label in the propyl halide product was determined. However, the extent of the rearrangement was always very small, and it shows a rather disturbing dependence on the nature of the isotopic label. For example (606), in CCl$_4$ at 80°C the yields of the rearranged chlorides were 4.0, 1.3, and 0.9% when the radicals initially generated were $CH_3CH_2^{14}CH_2 \cdot$, $CH_3CH_2CD_2 \cdot$, and $CH_3CH_2CT_2 \cdot$, respectively. Moreover, the $(CH_3)_2CHCH_2CH_2 \cdot$ radical did not rearrange under similar conditions (608). Taken at face value, these results imply that the symmetry of the transition state for a 1,3-hydrogen migration is very important, more so in fact than the thermodynamic driving force.

The one unequivocal 1,3-hydrogen migration from carbon to carbon (609, 610) is actually exceedingly exothermic (by ca. 20–24 kcal/mol) and must have a very asymmetric transition state. This is the rearrangement of o-tolyl to benzyl (332 ⟶ 333), which was first detected in a CIDNP study of the thermolysis of o-toluoyl peroxide in hexachloroacetone (609):

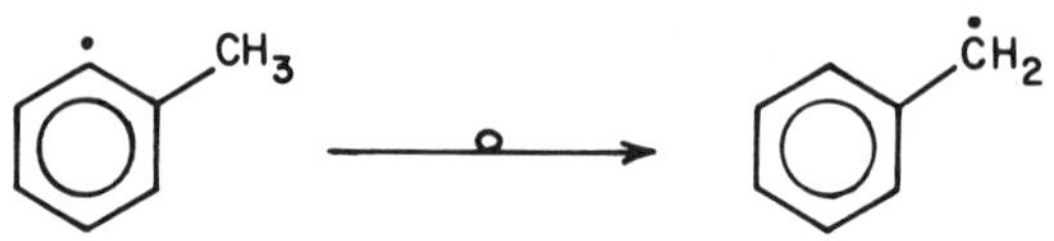

332 → **333**

Small amounts of polarized benzyl chloride were produced and were attributed to rearrangement of **332** within the solvent cage (*609*). This rearrangement has also been studied in the gas phase, with **332** being generated by the attack of sodium or barium atoms on *o*-chlorotoluene (*610*).

Some *formal* 1,3-hydrogen migrations to a vinylic radical center in aqueous media have been reported (*426*). These reactions are highly exothermic, but their detailed mechanism remains to be clarified. Thus, the isomerization **334**→**335**, which does not shift the formal site of the unpaired electron, is believed to involve a base-catalyzed tautomerism, whereas the isomerization **336**→**337** may actually have occurred indirectly via an intermolecular H atom abstraction (*426*).

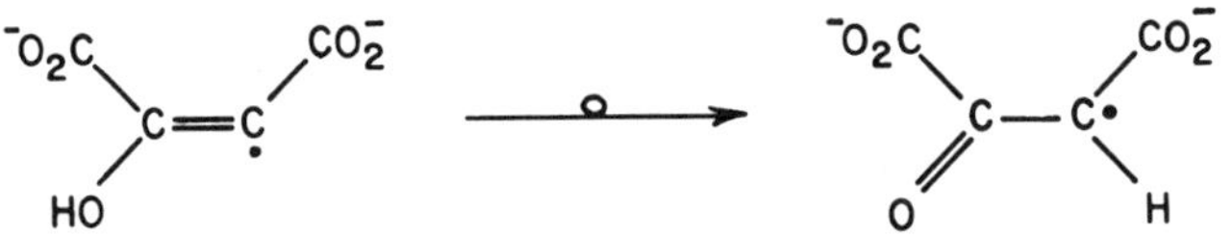

334 → **335**

336 → **337**

It has been claimed that 1,3-hydrogen migration contributes substantially to the telomerization of propylene with methyl chloroacetate and methyl propionate (*594*), e.g.,

$$CH_3\dot{C}HCH_2CHClCO_2CH_3 \longrightarrow CH_3CH_2CH_2\dot{C}ClCO_2CH_3$$

Similar isomerizations have been claimed in related systems (*611, 612*). In our opinion, the evidence that these isomerizations are actually 1,3-intramolecular migrations, rather than intermolecular hydrogen transfers, is not unequivocal.

4. FREE-RADICAL REARRANGEMENTS

2. 1,4-Migrations

These reactions are very uncommon in solution (4). They do not occur unless the isomerization is strongly exothermic. For example (613), a search for the approximately thermoneutral isomerization of the 2,3,5,6-tetramethylbenzyl radical under a variety of conditions was unsuccessful.

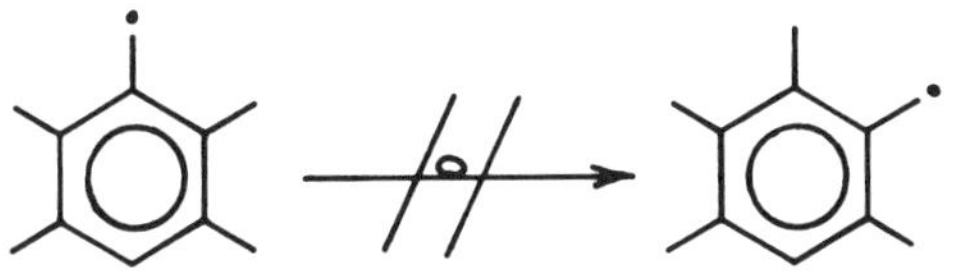

The most interesting 1,4-hydrogen atom migration yet to be observed in solution occurs during isomerization of the persistent 2,4,6-tri-*t*-butyl-phenyl **338** to the 3,5-di-*t*-butylneophyl radical **339** (37):

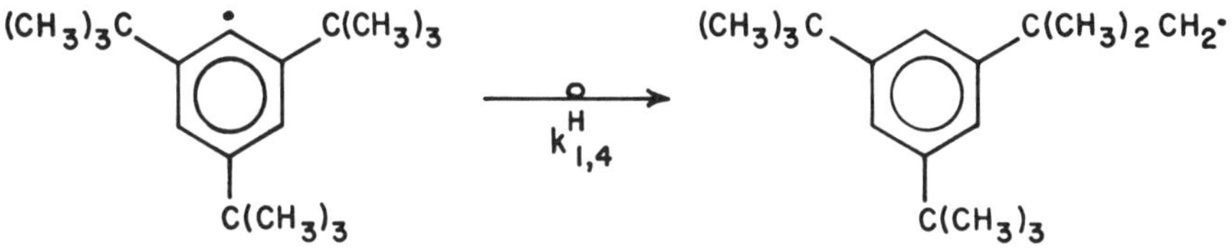

338 → 339

This isomerization provided the first authentic example of quantum mechanical tunneling in an intramolecular hydrogen atom transfer; that is, the hydrogen is transferred "through" the potential barrier rather than "over" it. The occurrence of tunneling was indicated by four kinetic phenomena. First, the kinetic deuterium isotope effect, $k^H_{1,4}/k^D_{1,4}$, was much greater at all temperatures than the classical "maximal" value (which is calculated by assuming that the maximal isotope effect will occur when all zero-point energy is lost in the transition state). Thus, the "maximal" isotope effects at $-30°$ and $-150°C$ were calculated to be 17 and 260 respectively, whereas the measured values were 80 and 13,000. Second, the Arrhenius plots were strongly curved, with the temperature dependence of the isomerization becoming less as the temperature was lowered. The third and fourth pieces of evidence for tunneling were the values of the Arrhenius preexponential factors and the activation energies. That is, although the normal Arrhenius equation was inapplicable, it was possible, using only a limited temperature range, to draw straight lines through the data in the usual way. The A factors and activation energies obtained in this way were very different from those that would be expected if the reaction followed the laws of classical mechanics (37).

Analogous isomerizations involving 1,4-hydrogen migration by quantum mechanical tunneling occur with other di-ortho-alkylated aryls, *provided* that the radicals are too sterically hindered to react with the surrounding solvent *and provided* that they have no accessible hydrogens that would allow decay to occur by a 1,5-migration (*37, 614*). Examples of such aryl radicals are 2,4,6-tri(1′-adamantyl)phenyl (*37*) and octamethyloctahydroanthracen-9-yl (*614*). In the case of 2,4,6-tri-*t*-butylphenyl the unimolecular isomerization rate is the same in solution as it is in matrices, and the reaction has been studied at very low temperatures (*614*).

The isomerization of *o*-methoxyphenyl to phenoxymethyl, which has been observed in an argon matrix at liquid helium temperatures, also involves a 1,4-hydrogen migration (*615*).

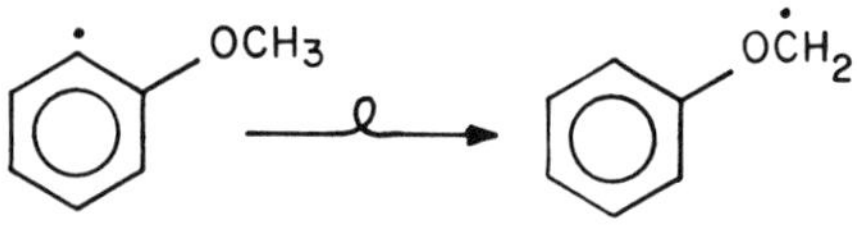

The biradical, $PhĊ(OH)CH_2CH_2OĊHCH_3$, decays largely by an unexpected 1,4-hydrogen migration (*615a*).

3. 1,5- and 1,6-Migrations

Both isomerizations are well known (*3, 4, 6, 592, 594*) and occur with considerable facility when the reaction is exothermic. In connection with the previous section, it is worth noting that many 1,5-migrations of aliphatic hydrogen to an aryl carbon have been reported (*37, 162, 616–619*), and they all appear to be extremely fast. Thus, 2,4,6-trineopentylphenyl isomerizes more rapidly than it abstracts hydrogen from tetramethylgermane (*37*).

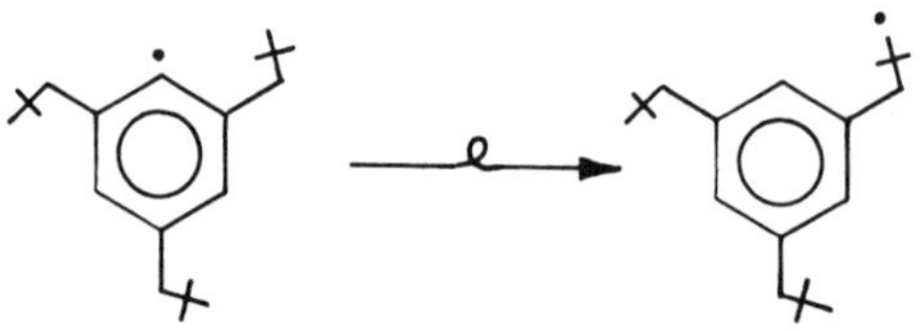

It was shown by deuterium labeling that the *o*-aryl radical derived from *N,N*-dimethylbenzamide (**342**) isomerizes more rapidly than it rotates about its carbonyl C—N bond (*616*). It was subsequently shown that conformer **340** of this radical reacts predominantly by internal hydrogen atom transfer, whereas conformer **341** decays, predominantly or exclusively, by intermolecular hydrogen abstraction (*617*). Interconversion of **340** and **341** must be relatively slow:

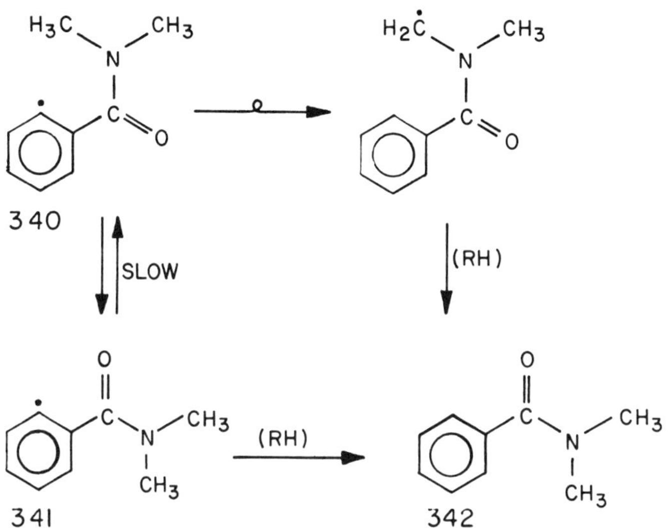

4. FREE-RADICAL REARRANGEMENTS

[Structures 340, 341, 342 with SLOW and (RH) arrows]

Conformational control of these fast reactions is also indicated by the fact that the corresponding radical (**343**) from di-n-propylbenzamide shows a four-fold preference for 1,6- over 1,5-hydrogen migration, whereas **344** isomerizes by 1,5-, 1,6-, and also a highly unusual 1,7-hydrogen migration (*619*).

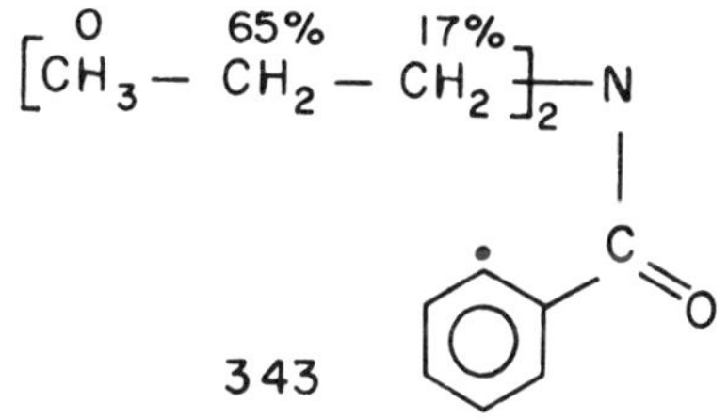

[Structures 343 and 344 with percentages: 65%, 17% for 343 and 7%, 38%, 20% for 344]

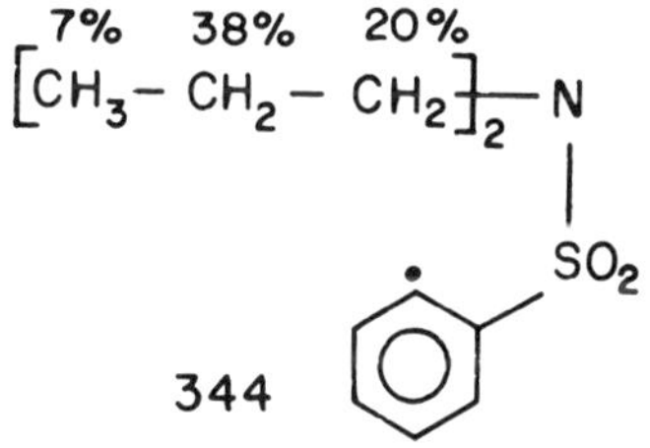

Several other facile ($k > 10^2$ sec^{-1}) isomerizations involving a 1,5 aliphatic hydrogen migration to an aryl radical center have been identified by ESR spectroscopy and by product analysis (*162, 618*). Analogous highly exothermic migrations to a vinyl radical center have also been reported, e.g., **345** ⟶ **346** (*620*) and **347** ⟶ **348** (*426*).

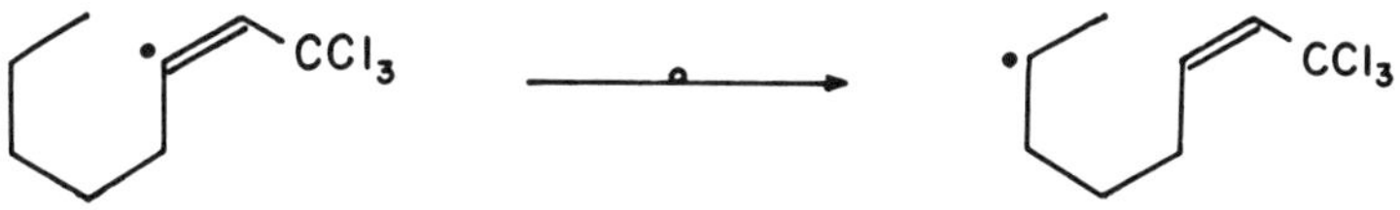

[Structures 345 and 346]

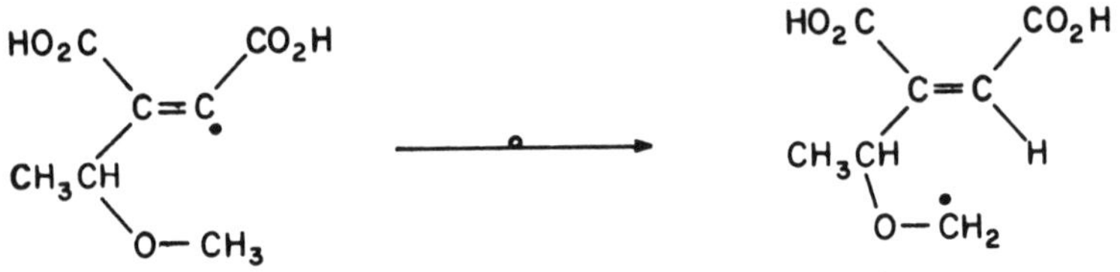

[Structures 347 → 348]

Both 1,5- and 1,6-hydrogen migrations are quite commonly encountered in free-radical polymerizations and telomerizations (*594, 621, 622*) and in the isomerization of primary *n*-alkyl radicals (*598*) (see Section VIII,A) and 1,1-dichloro-1-alkyl radicals (*623*). More interesting, perhaps, are the intramolecular hydrogen transfers that occur across cyclic radicals e.g., 349——350 (*624*), 351——352 (*625, 626*), and 353——354 (*431*).

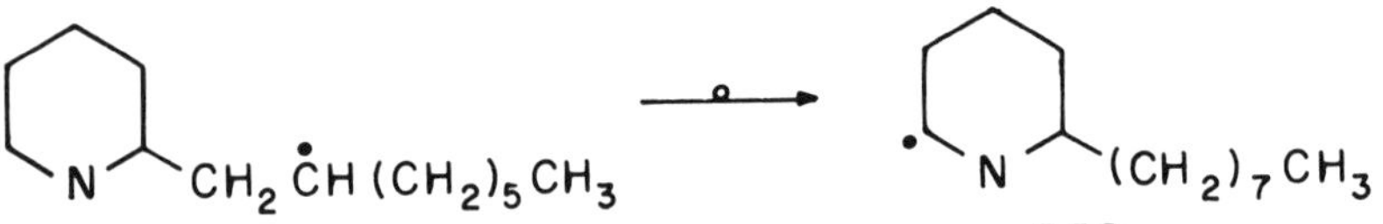

[Structures 349 → 350, 351 → 352, 353 → 354]

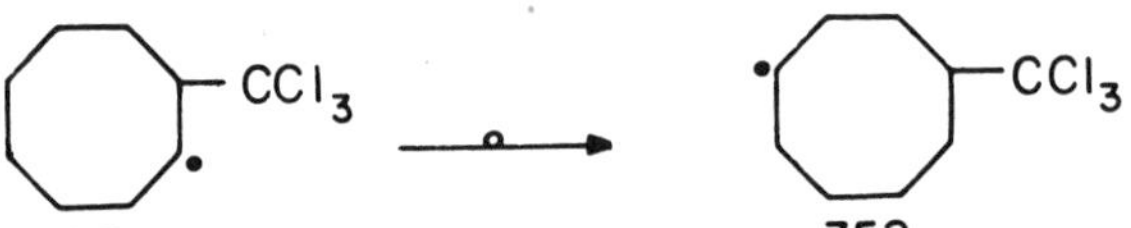

C. Carbon to Oxygen Migration

1. 1,2-Migrations

The study of organic radicals in nonaqueous solvents is generally straightforward. However, results obtained in aqueous media are often difficult to interpret unambiguously. An excellent example of this phenomenon is provided by the 1,2-hydrogen migration. This does not occur from C to C, nor is there any sound evidence for C to O migration in nonaqueous media. However, numerous *formal* C to O 1,2-migrations have been reported for alkoxy radicals generated in water as the solvent (*268, 419, 627–629*). The alkoxy radicals cannot themselves be detected by ESR spectroscopy in any solvent (*630*). In nonaqueous media the

4. FREE-RADICAL REARRANGEMENTS

detected radicals are formed either by intermolecular processes (e.g., hydrogen atom abstraction from a solvent molecule) or by β scission of the alkoxy. In contrast, when primary or secondary alkoxy radicals are generated in water, the corresponding hydroxyalkyl radical is detected (268, 419, 627–629), e.g., **355** ⎯⎯→ **356** (268) and **357** ⎯⎯→ **358** (629):

$$CH_3CH_2CH_2O^{\bullet} \xrightarrow{H_2O} CH_3CH_2\dot{C}HOH$$

$$\text{355} \qquad\qquad\qquad \text{356}$$

$$\underset{\text{357}}{\bigcirc\!-\!\dot{O}} \xrightarrow{H_2O} \underset{\text{358}}{\bigcirc\!\overset{\bullet}{-}\!OH}$$

The 1,2-migration is inherently less efficient in secondary than in primary alkoxys (629). With primary alkoxys having a sufficient chain length, 1,5 migration may compete with the 1,2-migration. For example, $CH_3(CH_2)_3O^{\bullet}$ gave $\dot{C}H_2(CH_2)_3OH$ and $CH_3(CH_2)_2\dot{C}HOH$ in approximately equal concentrations, whereas $CH_3(CH_2)_5O^{\bullet}$ gave only $CH_3CH_2\dot{C}H(CH_2)_3OH$ (268). The alkoxy can also be trapped by intramolecular addition to a double bond without any competing 1,2-migration (268) (see Section IV,E).

Primary alkoxy radicals can also be trapped intermolecularly both by addition to the aci anion of nitromethane, $CH_2:NO_2^-$, and by hydrogen abstraction from added CH_3OH. The isomerization is not acid-catalyzed, but it does require water. Two mechanisms for isomerization, which differ essentially only in the timing of events, have been suggested (268):

$$RCH_2O^{\bullet} + H_2O \xrightleftharpoons{-OH^-} RCH_2\overset{\bullet}{\overset{+}{O}}H \xrightarrow{-H^+} R\dot{C}HOH$$

$$RCH\overset{\dot{O}:\searrow H}{\underset{H}{\diagdown\!\!\!\diagup}}_{OH} \longrightarrow R\overset{-}{\overset{\bullet+}{C}}H\overset{\bullet+}{O}H \longleftrightarrow R\dot{C}HOH$$

Competition between the isomerization of 1-propoxy,

$$CH_3CH_2CH_2O^{\bullet} + H_2O \xrightarrow{k_{1,2}} CH_3CH_2\dot{C}HOH + H_2O$$

and its abstraction from methanol ($k_H = 2.6 \times 10^5\ M^{-1}\ sec^{-1}$),

$$CH_3CH_2CH_2O^{\bullet} + CH_3OH \xrightarrow{k_H} CH_3CH_2CH_2OH + \dot{C}H_2OH$$

gave a bimolecular rate constant for the 1,2 migration, $k_{1,2} = 1.4 \times 10^5\ M^{-1}\ sec^{-1}$ (268). The pseudo-first-order rate constant in pure water would be $8 \times 10^6\ sec^{-1}$.

2. 1,5 and Other Migrations

Since "an extraordinary number of such hydrogen transfers are known" (4), only a few representative reactions will be discussed in this section.

Alkoxy radicals of suitable structure readily undergo 1,5-hydrogen migrations (2–4, 6, 592, 593, 595) both in aqueous (268, 419, 627–629) and in nonaqueous (599, 600) media (*vide supra*). These isomerizations are exothermic ($D[RO—H] \sim 104$ kcal/mol) and rapid. The rate constants for the reactions $RCH_2(CH_2)_3O^\cdot \longrightarrow R\dot{C}H(CH_2)_3OH$ have been estimated to be *ca.* 8×10^6 sec^{-1} for R = H and $>10^8$ sec^{-1} for R = ethyl (268). There is a strong preference for 1,5- over 1,6-migration (599, 600) (see Section VIII,A) even when the latter position has been "activated" by a phenyl group. For example, in the reaction,

$$R(CH_2)_4C(CH_3)_2O^\cdot \longrightarrow RCH_2\dot{C}H(CH_2)_2C(CH_3)_2OH + R\dot{C}H(CH_2)_3C(CH_3)_2OH$$

in CCl$_4$ at 0°C, $r = k_{1,5}/k_{1,6} = 14.6$ for R = CH$_3$ and $r = 9.5$ for R = C$_6$H$_5$ (600).

The Barton reaction (537, 631–633) is the best known and by far the most important of these isomerizations (593). In this reaction, alkoxy radicals are generated by the photolysis of nitrite esters (which are readily prepared from alcohols). This reaction is of great synthetic utility, particu-

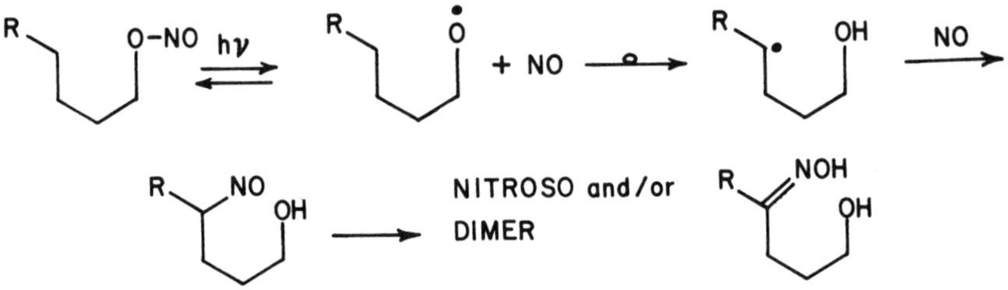

larly in steroid synthesis, because it provides an efficient way to functionalize, by an intramolecular process, an unactivated site even including an unactivated methyl. Whether hydrogen migration will occur is determined by stereochemical factors. This can be nicely illustrated by comparison of the alkoxy radicals derived (although not by a Barton reaction) from *trans*- and *cis*-3,3,5-trimethylcyclohexanol (634). The *trans*-alkoxy **359** reacts mainly by a 1,5-hydrogen migration from the axial 3-methyl. This is because the alcohol (and presumably the alkoxy) favor a conformation appropriate for this isomerization (634):

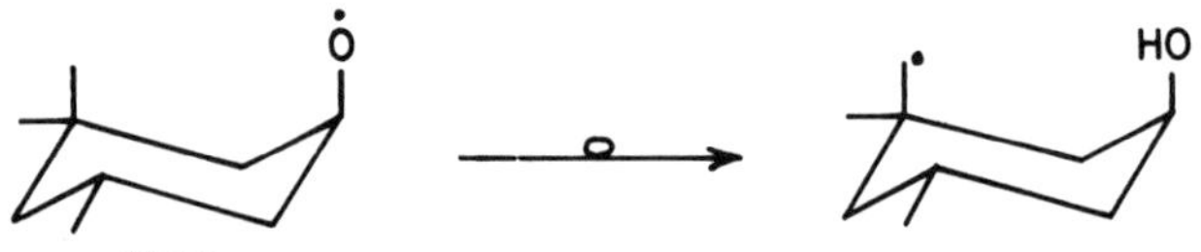

4. FREE-RADICAL REARRANGEMENTS

However, the *cis*-alkoxy **360** favors a conformation that is inappropriate for a 1,5-hydrogen migration from any methyl group, and this reaction is not observed (*634*):

360

The regioselectivity of many 1,5-hydrogen migrations from oxygen to carbon is remarkable even for two competing 1,5-migrations (*537, 635, 636*). The extraordinary structural sensitivity of the transition state in a Barton reaction was demonstrated in a study of intramolecular deuterium isotope effects during the 1,5-hydrogen migration in apollan-11-oxy (**361**) and the related radicals **362, 363,** and **364** (*637*):

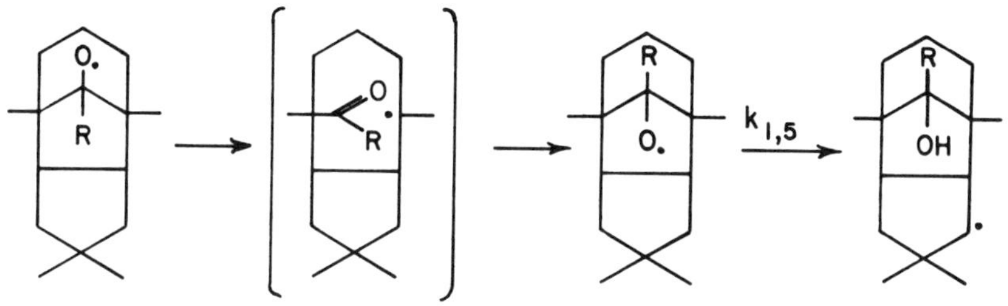

361 R=H
363 R=CH$_3$

362 R=H
364 R=CH$_3$

The epimeric alkoxy radical pairs **361** and **362**, and **363** and **364**, gave identical products and identical deuterium isotope effects, which indicates that **361** and **363** epimerize (presumably by the ring-opened radical) before they react. However, the secondary alkoxy radicals **361** and **362** gave a primary kinetic isotope effect, $k^H_{1,5}/k^D_{1,5} = 4.3$, whereas the tertiary alkoxy radicals **363** and **364** gave $k^H_{1,5}/k^D_{1,5} = 1.1$. This result is all the more remarkable because the locked skeleton in these tricyclic radicals ensures that the secondary and tertiary alkoxy centers cannot differ much in their spatial locations. It is also noteworthy that $k^H_{1,5}/k^D_{1,5}$ is about 4.7 for the analogous 1,5-hydrogen migration in *acyclic* secondary alkoxy radicals (*638*). Unfortunately, the isotope effect for this reaction in acyclic tertiary alkoxys does not appear to have been measured.

The other major class of 1,5-hydrogen migrations from carbon to oxygen involves peroxy radicals. Unless the C—H bond is "activated," this isomerization will be endothermic ($D[ROO-H] \sim 88$ kcal/mol) and hence will proceed more rapidly in the reverse direction. However, even endothermic isomerizations can be readily observed during the autoxidation of suitable substrates because the carbon-centered radical **365** will be rapidly trapped by the molecular oxygen present in the system. The critical oxygen pressure required to trap all **365** that are formed will, presumably, increase as the peroxy radical isomerization becomes more endothermic. The overall process can be represented as follows:

$$R_1R_2C-X-CR_3R_4 \underset{}{\overset{k_{1,5}}{\rightleftarrows}} R_1R_2\overset{}{C}-X-\overset{\cdot}{C}R_3R_4 \xrightarrow{O_2} R_1R_2C-X-CR_3R_4$$

(with O–O· and H substituents on left; O–O–H in middle; HOO and OO· on right — labeled **365** for middle species)

$$k_H \downarrow RH \qquad\qquad k'_H \downarrow RH$$

$$R_1R_2C-X-CR_3R_4 \qquad\qquad R_1R_2C-X-CR_3R_4$$
(OOH and H substituents) (HOO and OOH substituents)

X = CH$_2$, O, etc.

The ratios of the rate constants for peroxy isomerization and intermolecular hydrogen abstraction are highly dependent on the structure of the substrate being oxidized. This can be illustrated by contrasting $k_{1,5}/k_H$ ratios for tertiary and secondary hydrogens in autoxidizing ethers and alkanes. In the autoxidation of diisopropyl ether and dibenzyl ether at 30°C, similar values were found for $k_{1,5}/k_H$, namely, 2.1 and 2.9 M, respectively (639). It was shown that the oxygen pressure was sufficient to trap all **365**. In these reactions, the C—H bond that is cleaved may not be "activated" by the neighboring ethereal oxygen because, in the transition state, it must lie approximately in the nodal plane of the oxygen's p-type lone pair (14). The thermochemistry of these isomerizations is therefore unknown. In the autoxidation of alkanes yielding tertiary peroxy radicals which can isomerize by migration of tertiary hydrogen, e.g., 2,4-dimethylpentane (640, 641) and 2,4,6-trimethylheptane (642), there is a high yield of bifunctional [and even trifunctional (642)] products. Thus, for neat 2,4-dimethylpentane at 100°C and 1 atm O$_2$, $k_{1,5}/k'_H = 77$ M (641), it being assumed that all **365** were trapped. This isomerization would be slightly endothermic ($D[C_{tert}-H] \approx 91$ kcal/mol). In contrast, alkanes yielding secondary peroxy radicals which must isomerize by a strongly endothermic migration of a secondary hydrogen ($D[C_{sec}-H] \approx 95$ kcal/mol), e.g., n-octane (643), n-hexadecane (644), and n-octadecane (645), give relatively low yields of bifunctional products. Thus, $k_{1,5}/k_H = 1.1$ M for neat n-hexadecane at 160°C and 1 atm O$_2$, again assuming that all **365** were trapped (644). A more detailed study found $(k_{1,5}/H)/(k_H/H)$ equal to 4.8 M at 120°C (644a).

4. FREE-RADICAL REARRANGEMENTS

The reasonable suggestion has been made (646) that difunctional products are disfavored for the straight-chain alkanes because the isomerization of their peroxys will be more readily reversible than will be the case with the branched alkanes. However, this does not explain the similarities in the $k_{1,5}/k_H$ ratios for n-alkanes and the primary and secondary alkyl ethers. Moreover, others (641) have calculated that all **365** should be trapped by oxygen under the experimental conditions normally used for alkane autoxidations. Unfortunately, there is little experimental evidence to confirm or refute their conclusions. Although Thorpe–Ingold effects (see Section III,B) should favor isomerization of those peroxys in which a tertiary hydrogen migrates, it is not obvious why this effect should operate in alkanes but not in ethers. One complicating factor in the alkane autoxidations is that the product balances tend to be very much poorer for the n-alkanes than for the branched alkanes. It has been suggested (647) that with n-alkanes some peroxy isomerization may remain undetected even when all **365** are trapped by oxygen because of a second isomerization involving migration of the hydrogen attached to the carbon bearing the hydroperoxy group. This will certainly be the most labile hydrogen in the peroxy derived from **365**, i.e.,

$$\text{RCHCH}_2\text{CHR} \longrightarrow \text{R}\overset{\cdot}{\text{C}}\text{CH}_2\text{CR} \longrightarrow \text{products}$$
$$\phantom{\text{RCHCH}_2}\text{O} \phantom{\text{CHR}} \text{O} \phantom{\longrightarrow \text{R}\overset{\cdot}{\text{C}}}\text{O} \phantom{\text{CH}_2}\text{O}$$
$$\phantom{\text{RCHCH}_2}\text{OH} \phantom{\text{CHR}} \cdot\text{O} \phantom{\longrightarrow \text{R}\overset{\cdot}{\text{C}}}\text{OH} \phantom{\text{CH}_2}\text{OH}$$

In fact, the ratio of the rate constants (per H) for this isomerization and for intermolecular abstraction (k_H) in autoxidizing n-hexadecane at 120°C has been estimated to be 270 M (644a).

A proper understanding of the effect of reagent structure, oxygen pressure, and temperature on the intra-/intermolecular autoxidation of organic substrates will require a great deal more work.

This section is concluded with an interesting aroyloxy isomerization (609),

and a reported acyloxy isomerization (119),

which is rather surprising in view of the normally rapid decarboxylation of such radicals.

D. Carbon to Nitrogen Migration

These isomerizations are nearly as common as the carbon to oxygen migrations (4). Although the protonated aminyl radical (aminium radical) $R_2\overset{\cdot}{N}H^+_{}$ has received most attention, hydrogen migration from carbon to nitrogen has been identified for nitrogen-centered radicals derived from amines, $R_2\overset{\cdot}{N}$; amides, $RCO\overset{\cdot}{N}R$; imides, $(RCO)_2\overset{\cdot}{N}$; and sulfonamides, $RSO_2\overset{\cdot}{N}R$ (4).

In aqueous media, alkylaminyl radicals undergo a *formal* 1,2-hydrogen migration (*648*). Since this does not occur in nonaqueous solvents (*649, 650*), water is probably directly involved in the reaction (*268*), which means that it is not a true intramolecular isomerization (see Section VIII, C,1).

Bicyclic amines are formed in good yield by reaction of *N*-chloroazacyclooctane and *N*-chloroazacyclononane with metallic silver (*651*). A short-chain reaction involves *neutral* aminyl radicals isomerizing by a transannular 1,5-hydrogen migration, e.g.,

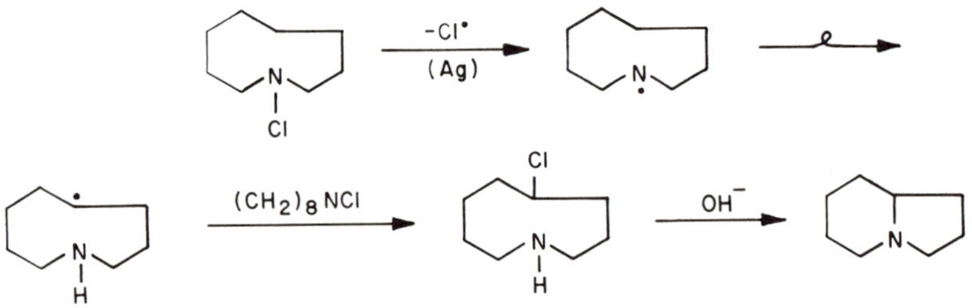

These reactions may have synthetic utility when the strongly acidic conditions of the Hoffmann–Löffler reaction (*vide infra*) cannot be used. Analogous isomerizations leading to cyclic products occur with amidyl radicals derived from *N*-haloamides, e.g. (*652*),

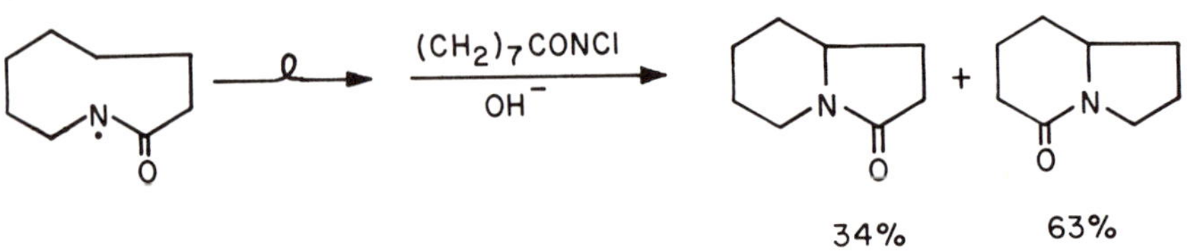

34% 63%

Although 1,5-hydrogen migration does not appear to have been observed in acyclic aminyl radicals, it is a very common reaction of amidyl (*596, 653–657*); imidyl (*658*), and sulfonamidyl (*596, 659–661*) radicals (*4*). A rare 1,4-migration to an amidyl nitrogen has even been reported (*662*):

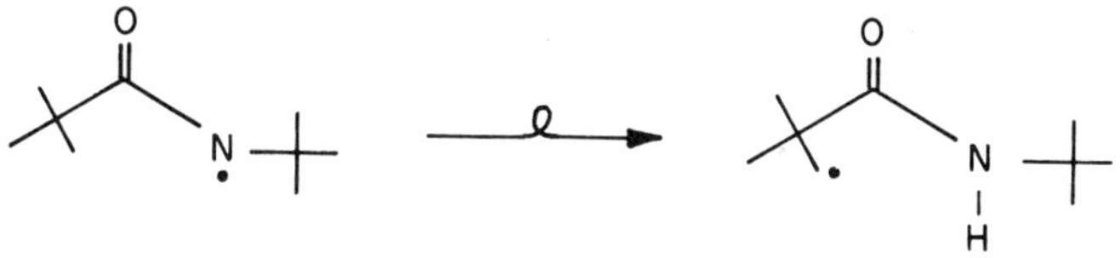

4. FREE-RADICAL REARRANGEMENTS

The reactivity of nitrogen-centered radicals is obviously enhanced by the presence of neighboring electron-withdrawing groups. The ultimate in electron deficiency is achieved by protonation of the nitrogen. The resultant aminium radicals are extremely reactive (286, 286a, 596)* in intermolecular H abstractions (286a, 371a, 596, 665–669) and intermolecular additions (286, 286a, 371a, 596, 668–671). However, from a synthetic viewpoint, probably their most important reaction is the intramolecular 1,5-hydrogen migration—the well-known Hoffmann–Löffler reaction, **366** → **367** (4, 286a, 596, 597).

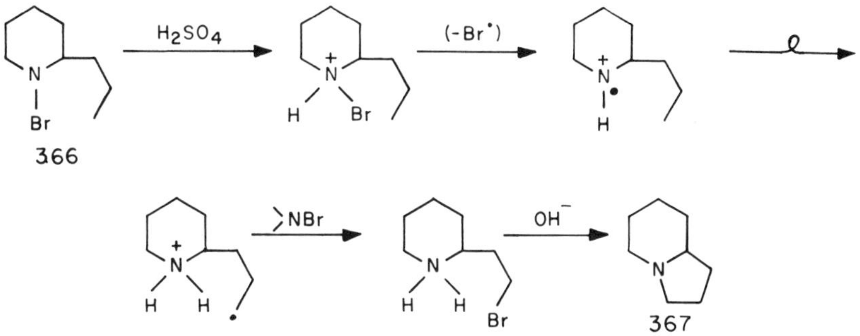

The Hoffmann–Löffler reaction is generally carried out by heating or photolyzing an N-chloramine in a strong acid such as 4 M H_2SO_4/ CH_3COOH or neat CF_3COOH. Yields of cyclized product are generally over 80% and, although 1,5-migration is favored, 1,6-migration can occur (just as is the case with alkoxy radicals), e.g.,

$$R(CH_2)_5 \overset{+\cdot}{N}HR' \longrightarrow R\dot{C}H_2CH(CH_2)_3 \overset{+}{N}H_2R' + R\dot{C}H(CH_2)_4 \overset{+}{N}H_2R'$$

R	R'	$k_{1,5}/k_{1,6}$	Reference
C_6H_5	CH_3	1.65	672
CH_3	CH_3	4.0	673
CH_3	$CH_3(CH_2)_5$	21.7	673

The primary kinetic isotope effect, $k^H_{1,5}/k^D_{1,5}$, for the Hoffmann–Löffler reaction has been determined using N-chloro-5-deuterio-2-

---

* $D[H_3\overset{+}{N}-H] = 123$, $D[CH_3CH_2\overset{+}{N}H_2-H] = 103$, and $D[(CH_3)_2\overset{+}{N}H-H] = 94.5$ kcal/mol (663); $D[H_2N-H] = 110$, $D[CH_3NH-H] = 103$, and $D[(CH_3)_2N-H] = 95$ kcal/mol (664). There appear to be some inconsistencies in these bond strengths.

hexylamine (*638*). The isotope effect in the derived aminium radical **368** is appreciably smaller than for the Barton reaction of the corresponding alkoxy **369**. It is, in fact, more nearly equal to the isotope effect found when the corresponding alcohol loses H_2O following ionization in a mass spectrometer—a reaction that appears to involve a protonated alkoxy, **370** (*674*). However, the stereoselectivity for these three 1,5-hydrogen migrations, $k^a_{1,5}/k^b_{1,5}$ is highest for the aminium radical (*638*).

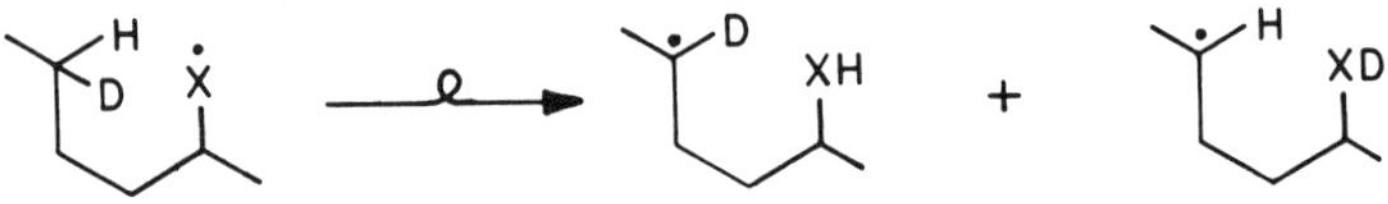

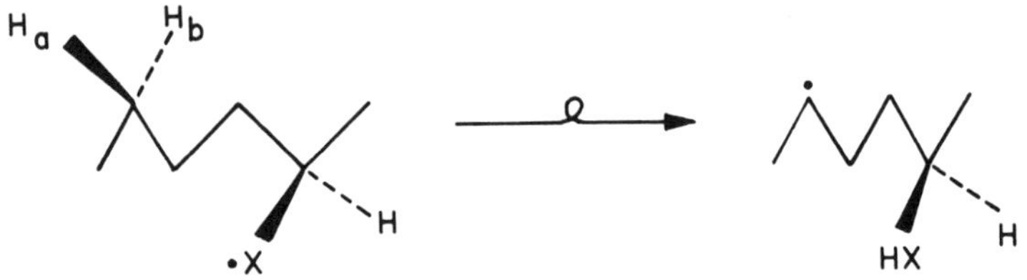

	X·	$k^H_{1,5}/k^D_{1,5}$	$k^a_{1,5}/k^b_{1,5}$
368	$NH_2^{+\cdot}$	1.2	1.5
369	O·	4.7	1.2
370	$OH^{+\cdot}$	1.1	1.1

In aminium radicals of suitable structure the Hoffmann–Löffler reaction is favored over intermolecular H atom abstractions (*666, 670a, 675*). An interesting intramolecular competition has shown that aminium radicals cyclize by addition to a suitably located double bond very much more readily than they isomerize by abstraction of a primary hydrogen (*281*):

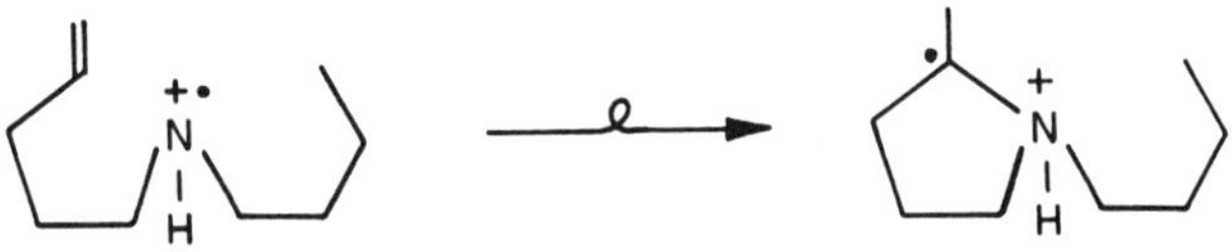

Amidyl radicals also exhibit a strong preference for intramolecular addition rather than intramolecular H migration (*294*) (see Section IV,E).

E. Oxygen to Carbon Migration

Since O—H bonds are stronger than structurally related C—H bonds (*63*), this type of isomerization is rather uncommon (see, however, peroxy radicals in Section VIII,C,2). The one example of a 1,3-migration, **371** —∘— **372** (*615*), involves the breaking of a weak phenolic O—H bond $(D[C_6H_5O—H] \approx 88$ kcal/mol).

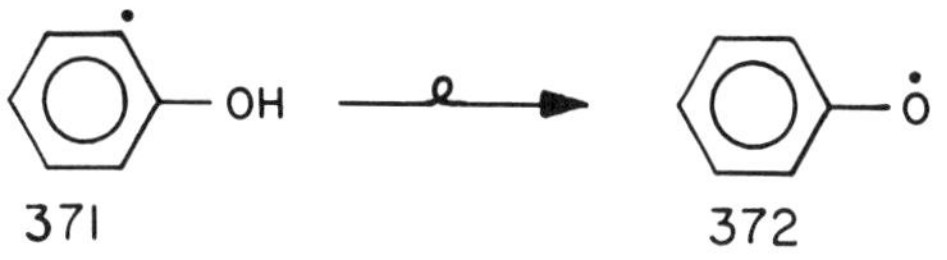

4. FREE-RADICAL REARRANGEMENTS

$$\text{371} \xrightarrow{\rho} \text{372}$$

Several 1,5-migrations in which a weak enolic O—H bond—$D[\text{CH}_2=\text{CHO—H}] \approx 81$ kcal/mol (*444*)—is broken have been reported (*426, 444, 445*) (see also Section V,B). The ring opening of 1-cyclopropyl-1-hydroxyalkyl radicals (**373**) provides a general route to the class of radicals (**374**) that show this type of isomerization (*444, 445*). The formation of **373** may proceed either by H abstraction from the secondary alcohol (*444, 445*) or by H addition to the ketone, photoexcited to its triplet state (*444*):

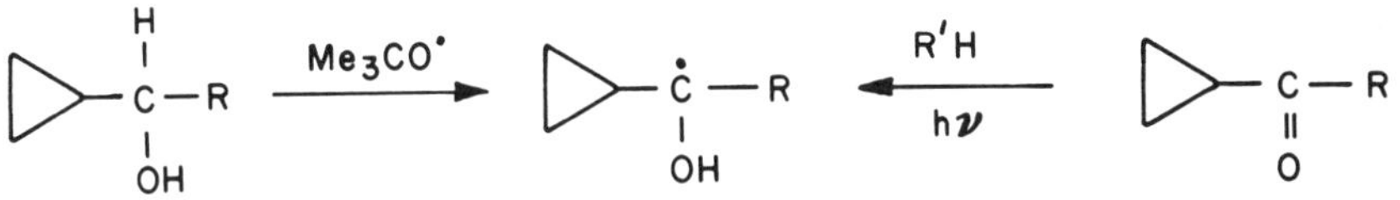

373

The ring openings of **373** occur at low temperatures (except for the resonance-stabilized **373** having R = C_6H_5), and the homoallylic radical **374** is observed by ESR spectroscopy. Near room temperature, a 1,5 migration of the enolic hydrogen yields the alkanoylalkyl radical **375** (*444, 445*). For R = H, both *cis*- and *trans*-**374** are observed, but *only* the cis radical isomerizes (*445*).

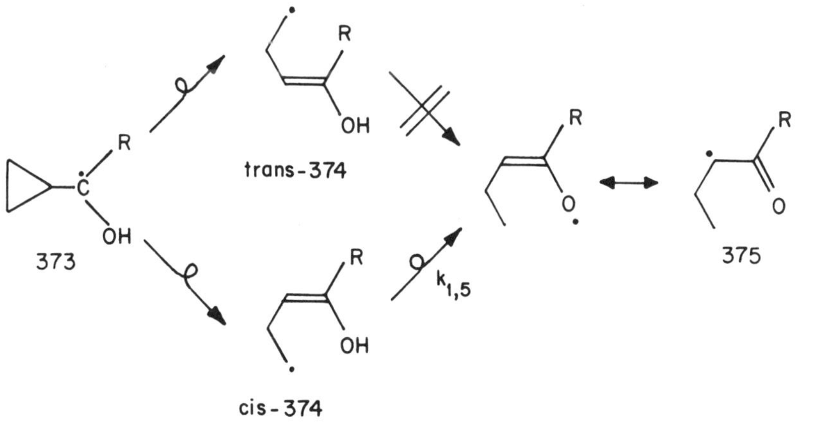

For R = CH_3, only *cis*-**374** could be detected at low temperatures and, at temperatures from *ca.* $-60°$ to $+30°C$, the rate constant for its isomerization could be represented by $\log(k_{1,5}/\text{sec}^{-1}) = 8 - 4.8/\theta$ (*445*). The behavior of **373** having R = cyclopropyl was similar (445), and so was that of the radical derived from *trans*-(2-methylcyclopropyl)ethanol (444). However, the aziridinylcarbinol **376** gave only the rearranged, ring-opened radical **377** and not the isomerized radical, possibly because the conformation of **377** prevented the 1,5-hydrogen migration (444).

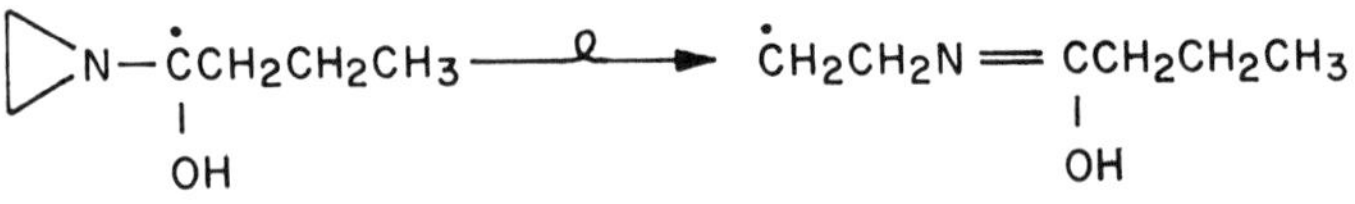

$$\underset{376}{\triangleright N-\overset{\cdot}{C}CH_2CH_2CH_3} \xrightarrow{\varrho} \underset{377}{\overset{\cdot}{C}H_2CH_2N=\underset{OH}{C}CH_2CH_2CH_3}$$

Radicals that also undergo 1,5-hydrogen migration from oxygen to carbon have been generated by addition of the hydroxyl radical to propynoic acid, e.g., (426),

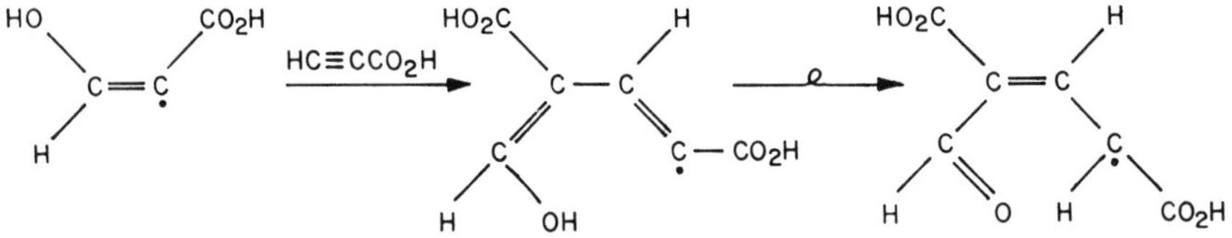

$$\underset{H}{\overset{HO}{\diagdown}}C=\underset{\cdot}{\overset{CO_2H}{\diagup}} \xleftarrow{HC\equiv CCO_2H} \cdots \xrightarrow{\varrho} \cdots$$

F. Oxygen to Oxygen Migration

In connection with the foregoing, radical **284** (M = H) undergoes a 1,5-hydrogen migration from enoxyl oxygen to alloxyl oxygen (444, 493) (see Section IV,B).

An intramolecular 1,5-hydrogen migration that would be endothermic were it not concerted with C—O bond homolysis is believed to be responsible for the first-order decay of the peroxy radical derived from isopropanol (676):

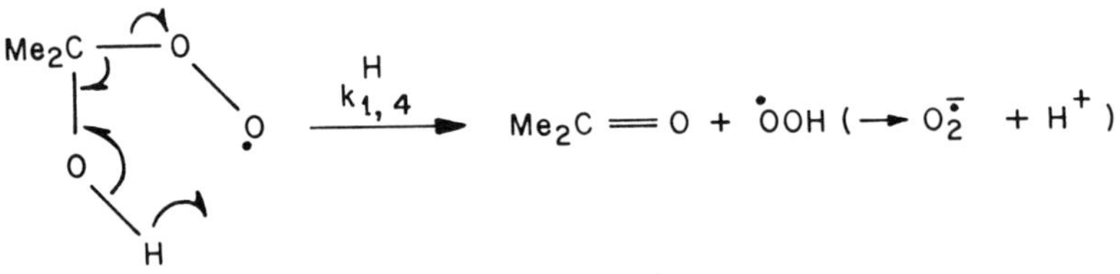

$$\cdots \xrightarrow{k_{1,4}^H} Me_2C=O + \overset{\cdot}{O}OH \; (\rightarrow O_2^{\overline{\cdot}} + H^+)$$

This reaction has an activation energy of 13.5 kcal/mol and, at 22°C, $k_{1,4}^H = 665 \text{ sec}^{-1}$ and $k_{1,4}^H/k_{1,4}^D = 3.5$.

Intermolecular hydrogen atom transfers between identical oxygen-centered radicals (i.e., RO· + HOR → ROH + ·OR) and other nearly thermoneutral intermolecular hydrogen transfers between oxygen-

4. FREE-RADICAL REARRANGEMENTS

centered radicals generally occur rapidly and have very low activation energies *(677)*. The same appears to be true for intramolecular reactions involving hydrogen migration from oxygen to oxygen. For example, the identity reaction,

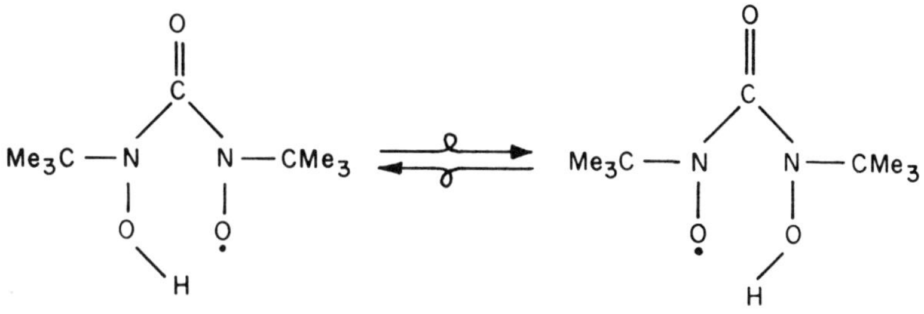

which involves a 1,5-hydrogen migration, has been shown by ESR spectroscopy to have an activation energy of 2.7 kcal/mol in toluene *(678)*. The interesting 1,4-hydrogen migration,

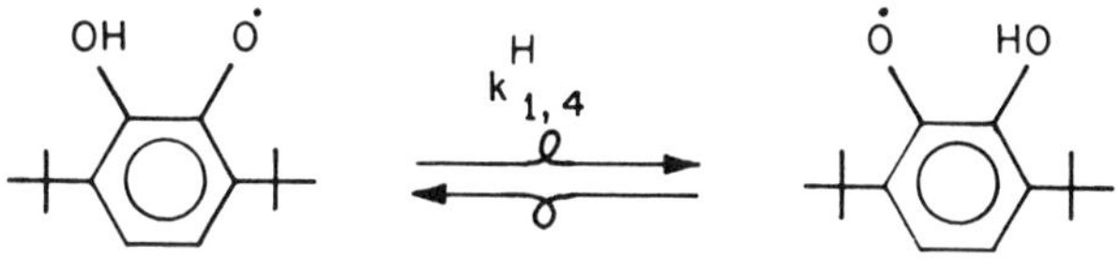

has been extensively investigated by ESR *(679–681)*. It has an activation energy that varies from 2.9 kcal/mol in alkanes and in toluene to 5.3 kcal/mol in the strongly hydrogen-bonding solvents methanol and acetone *(679, 681)*. However, the rate constant for this isomerization, $k^H_{1,4}$, varies less than might be expected because there is a compensating change in the Arrhenius preexponential factor *(681)*. Thus, in *n*-heptane *(679)*,

$$\log(k^H_{1,4}/\text{sec}^{-1}) = 11.52 - 2.9/\theta$$

and, at 20°C, $k^H_{1,4} = 2.3 \times 10^9$ sec^{-1}, whereas in methanol *(681)*,

$$\log(k^H_{1,4}/\text{sec}^{-1}) = 12.85 - 5.3/\theta$$

and, at 20°C, $k^H_{1,4} = 0.78 \times 10^9$ sec^{-1}. The reaction also exhibits an appreciable deuterium kinetic isotope effect, e.g., in *n*-heptane *(679)*,

$$\log(k^D_{1,4}/\text{sec}^{-1}) = 11.78 - 4.5/\theta$$

and, at 20°C, $k^D_{1,4} = 0.26 \times 10^9$ sec^{-1}.

Hydrogen atom migrations between pairs of heteroatoms other than oxygen have not, apparently, been investigated.

IX. ROTATION

A. Allylic Radicals

1. Allyl and substituted allyls

The rotational cis–trans isomerization of allylic radicals is of considerable interest both for intrinsic reasons and because it does, in principle, provide a simple way to measure the rates at which these resonance-stabilized radicals react with various molecules. With "slow" radical traps, isomeric allylic radicals equilibrate and yield an identical mixture of products, but this does not occur with "fast" traps (2). Thus, in the allylic chlorination of *cis*- and *trans*-2-butenes with *t*-butyl hypochlorite at 40°C, the 1-chloro-2-butenes are formed with complete retention of the cis–trans stereochemistry (682). This implies that simple cis and trans allylic radicals are configurationally stable in this reaction. On the other hand, in the reduction of *cis*- and *trans*-1-chloro-2-butenes with triphenyltin hydride at 80°C, the 2-butenes produced indicate that interconversion of the isomeric allylic radicals is competitive with hydrogen abstraction from the tin hydride (683, 684). Other reactions are known in which allyl radicals suffer a partial loss of their cis–trans stereochemistry (685–687).

The allyl resonance stabilization energy, E_s (allyl), which is defined as the difference in the primary C—H bond dissociation energy between propane and propylene (63), amounts to 12–14 kcal/mol (688). It has been suggested (689) that the barrier to rotation of the allyl radical should amount to the sum of E_s (which is required to "undo" the stabilization) plus ca. 4 kcal/mol for rotation of the vinyl group [estimated to be the same as for an ethyl group (689)] in the radical. For the butenyl radical **378**, for which $E_s \leq E_s$ (allyl), this procedure yields a rate constant for rotation that can be represented by $\log(k/\text{sec}^{-1}) = 13.5 - 16.7/\theta$ (689). The data (683)

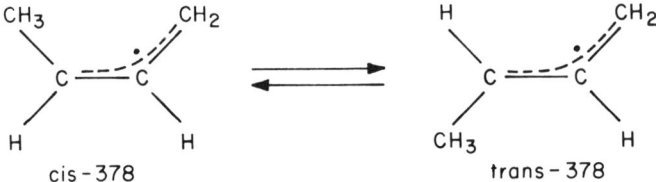

cis-378   trans-378

on the tin hydride reduction of 1-chloro-2-butenes were shown to be consistent with this estimate (689).*

There can be no doubt that the activation energy for the isomerization

---

* The rate constant for H abstraction from $(C_6H_5)_3SnH$ can be represented by $\log(k/M^{-1} \text{sec}^{-1}) = 8.2 - 7.8/\theta$ (689). Thus, the activation energy for H abstraction from $(C_6H_5)_3SnH$ by the butenyl radical is estimated to be ca. 4 kcal/mol greater than that for H abstraction from $n\text{-Bu}_3SnH$ by a primary alkyl radical (see Table 1, footnote *d*).

of allylic radicals is several kilocalories per mole greater than their stabilization energy. Thus, the free energy of activation $\Delta G^{\ddagger}$ for the isomerization of butenyl radicals at 126°C in the gas phase has been reported to be 21 ± 3 kcal/mol (*686*). Of greater significance is an ESR study of the allyl radical at elevated temperatures (*28*). Since there was no indication of any line width effects, even at 280°C, any process exchanging the syn and anti terminal protons in allyl must have a rate constant of less than 2×10^6 sec^{-1} at this temperature. This corresponds to $\Delta G^{\ddagger} > 17$ kcal/mol (*28*) (see Table 5). Similarly, $\Delta G^{\ddagger}$ for 1,1-dimethylallyl was shown to be greater than 14 kcal/mol (*28*).

Radical **379** was the first allyl to have its rotational barrier measured by ESR line width effects (*27*). Barriers for a number of other substituted allyls have now been determined by this technique (*27–27c*), by kinetic ESR sepctroscopy under steady-state conditions (*27b, 27c, 54*) (see Section II), and by product studies (*690*). The data are summarized in Table 5. It is apparent that allyl rotational barriers may be lowered in a variety of ways. Thus, in **379** the two cyano groups are presumed to stabilize the localized radical [a cyano group stabilizes a neighboring carbon radical center by 8 ± 1 kcal/mol (*691*)] generated in the orthogonal transition state to a greater extent than they stabilize the delocalized ground state (*27*). There also appears to be a small steric acceleration since $\Delta G^{\ddagger}$ for **379** is 1.7 kcal/mol smaller than that for $(NC)_2CCMeCH_2\cdot$ (see Table 5). The important role that steric destabilization of the ground state plays in accelerating allyl isomerizations was recognized by Walling and Thaler (*682*). They observed that the reaction of *t*-butyl hypochlorite with *cis*-4,4-dimethyl-2-pentene gave both *cis*- and *trans*-1-chloro-4,4-dimethyl-2-pentene at 40°C, although the pure cis product could be obtained at 78°C. The low barrier in 1,1,3,3-tetrafluoroallyl has been attributed primarily to ground state destabilization arising from electron–electron repulsion between the fluorine pπ lone pairs and the doubly occupied allyl π molecular orbital (*27a*). The further lowering of the barrier upon halogen substitution at the 2 position was interpreted (*27a*) in terms of stabilization of the transition state, the structure of which is analogous to the conformation adopted by β-chloroalkyl radicals (see Section VII,A,1).

379

TABLE 5
Activation Parameters for the Isomerization of Some Allylic Radicals as Measured by ESR Spectroscopy[a]

Radical	$\Delta G^{\ddagger}$ (kcal/mol)	E (kcal/mol)	Ref.
H_2CCHCH_2	>17		28
$H_2CC(CMe_3)CH_2$	>12		27[c]
$Me_3CCHCHCH_2$[b]		10.3[c]	54
Me_2CCHCH_2	>14		28
$NCCHCHCH_2$[b,d]		9.9[c]	27[b]
$(NC)_2CCMeCH_2$	10.7		27[b]
$(NC)_2CC(CMe_3)CH_2$	9.0	9.6[e]	27[b]
$EtOOCCHC(COOEt)CH_2$[d]		5.5[c]	27[c]
F_2CCHCF_2	7.2		27[a]
F_2CCFCF_2	6.1		27[a]
$F_2CCClCF_2$	4.5		27[a]
$C_6H_5CHCHCHC_6H_5$[f]	>3.7 (<9)		690

[a] Lineshape analysis unless otherwise specified.
[b] Syn → anti.
[c] Steady-state kinetics.
[d] Anti → syn.
[e] $\log(A/\text{sec}^{-1}) = 13.2$.
[f] $Z,E \to E,E$ by product analysis.

The (Z,E)-1,3-diphenylallyl radical **380** also has a very low barrier to rotation since the isomerization **380** ⟶ **381** competes with the diffusion-controlled dimerization of **380** even at low temperatures (*690*). As with **379**, the small barrier for this isomerization ($3.7 < \Delta G^{\ddagger}_{67°C} \leq 9$ kcal/mol) (*690*) can be attributed to stabilization of the localized radical generated in the orthogonal transition state.

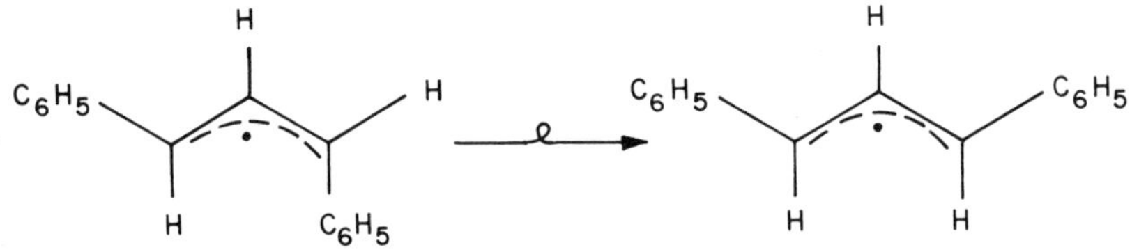

2. Alkanoylalkyls

Alkanoylalkyl radicals (**382**) are related to allyl radicals, but the contribution of the canonical structure having the unpaired electron on oxygen (**383**) is only *ca.* 15% (*24a, 692*). Activation energies for rotation about the $R_1C(O)-CR_2R_3$ bond have been determined by ESR lineshape analy-

4. FREE-RADICAL REARRANGEMENTS

[structures 382 and 383 shown]

sis for several radicals. These barriers to rotation are smaller than in most allyls and are generally in the range 8–10 kcal/mol, e.g., 9.4 for CH_3COCH_2 (**24**), 8.6 for $(CH_3)_3CCOCH_2$ (**24a**), 8.3 for $C_6H_5COCH_2$ (**25**), and 9.5 kcal/mol for $C_6H_5COC(CH_3)_2$ (**25**).

B. Bridged Radicals

The subject of bridged free radicals was comprehensively reviewed in 1972 (9) and again in 1973 (8). There are enormous complexities involved in obtaining unequivocal evidence for the participation of bridged radicals **384*** in reacting systems (9). From the viewpoint of the present essay the most interesting aspect of "bridging" relates to the surprisingly high barrier to rotation about the C_α—C_β bond in acyclic β-haloalkyl radicals **385**, and it is in this context that the subject of bridging will be discussed.

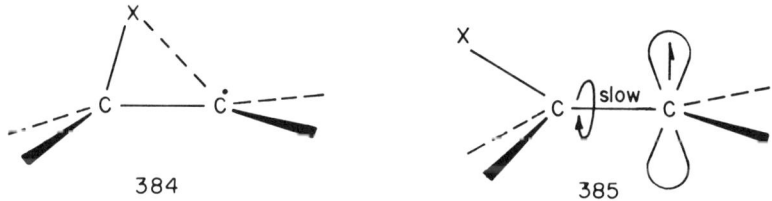

There is now abundant evidence that a bromine which is β and antiperiplanar to a C—H bond weakens that bond by *ca.* 2–3 kcal/mol; i.e., the bromine provides anchimeric assistance to the homolytic rupture of the C—H bond (*8, 10, 580, 581, 694–699*). Furthermore, the β-bromoalkyl radicals that are produced are slow to lose their stereochemical identity (*8, 10, 580, 581, 694, 696*). For example, photobromination of optically active 1-bromo-2-methylbutane (**386**) at Br_2 concentrations equal to or greater than 0.05 *M* gave 1,2-dibromo-2-methylbutane having a high optical purity and retained configuration (*8, 581*). If the system contains a large

---

* Some distortion of C_β from a tetrahedral geometry, which causes X to move closer to C_α, is commonly assumed and is generally supported by the ESR spectra of "bridged" radicals (*7, 693*) (see also Section VII,A,1).

amount of deuterium bromide, a second product is 1-bromo-1-deuterio-2-methylbutane with an optical purity nearly identical to that of the starting material (8, 694). This control of product stereochemistry can be attributed to the formation of a discrete, bridged, free-radical intermediate that has a lifetime long enough for it to be trapped with high stereoselectivity by molecular bromine or DBr without loss of optical purity (8). The trapping agent attacks the bridged intermediate from the backside (8).

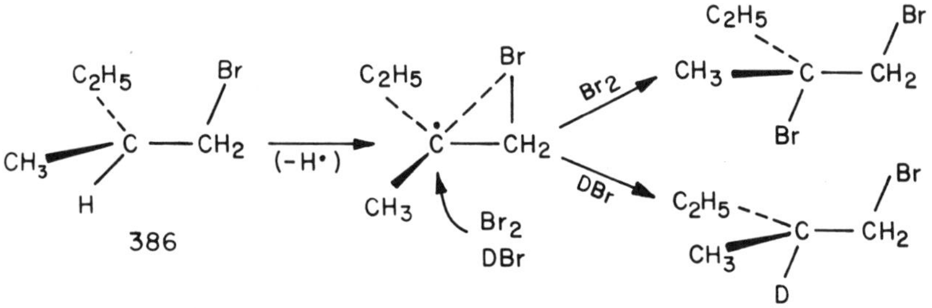

At Br_2 concentrations of less than ca. 0.05 M the optical activity of the dibromide starts to decrease (581). On the reasonable assumption that the reaction of the radical with Br_2 is diffusion-controlled, the experimental data yield a half-life of ca. 10^{-8} sec for racemization of this bromine bridged radical (8, 581). An identical half-life has been obtained by analysis of the optical activity of 1-bromo-2-methylbutan-2-ol (**387**) formed by autoxidation of active **386** and reduction of the intial hydroperoxide (10).

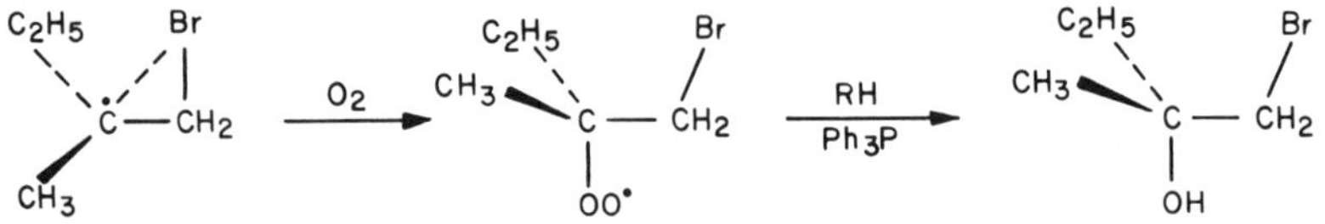

In the brominations of **386**, the optical purity of the dibromide increased with decreasing temperature (8). An activation energy for rotation about the $C_2H_5(CH_3)\overset{.}{C}$—$CH_2Br$ bond of ca. 4 kcal/mol is obtained if it is assumed that the activation energy for radical trapping is equal to that for diffusion (i.e., ca. 2 kcal/mol).

Bridged 2-methyl-1-bromobut-2-yl can also be trapped before racemization by Me_3COBr, but Me_3COCl and Cl_2 must be less reactive toward alkyl radicals since the products are racemic (8). Experiments with other optically active 1-halo-2-methylbutanes indicate that the chloro and iodo radicals are also bridged with racemization half-lives of ca. 10^{-10} and 10^{-5} sec, respectively (8). There is no evidence for bridging in the corresponding fluoroalkyl.

4. FREE-RADICAL REARRANGEMENTS

Bromination studies on 2-bromobutane tend to support the conclusions reached from the 2-methyl-1-bromobutane work (*580, 696*). The most recent kinetic data (*696*) indicate that the intermediate $CH_3CHBr\dot{C}HCH_3$ radical is not a symmetric species (i.e., it is not **388**), nor does the bromine move very rapidly between the two central carbons. These conclusions are consistent with those reached via a CIDNP study of the $BrCH_2CH_2\cdot$ radical (*700*).

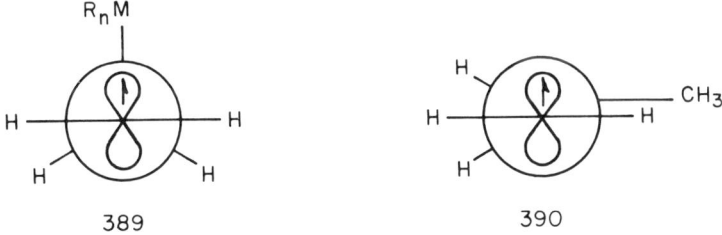

388

There can be no doubt that β-haloalkyl radicals derived from optically active precursors will continue to provide a useful method for studying very fast radical–molecule reactions. It is likely that some other β substituents can perform a similar function, albeit somewhat less efficiently than bromine (*559*), since it has been shown by ESR spectroscopy that β-substituted ethyls, $R_nMCH_2\dot{C}H_2$, adopt an eclipsed conformation (**389**) when M is from rows 2, 3, or 4 of the periodic table (*7, 701*).* In some of these radicals the barrier to rotation about the C_α–C_β bond is several kilocalories per mole greater than that for the 1-propyl radical, which itself prefers to adopt a staggered conformation, **390** (*7*). Evidently, rather subtle factors may control conformation; e.g., $CF_3OCH_2CH_2\cdot$ is staggered but $CF_3OCH_2\dot{C}HCH_3$ is eclipsed (*703*).

389   **390**

R_nM = Me_3Si, Me_3Ge, Me_3Sn, MeS, Cl, etc.

---

* The preferred conformation of $BrCH_2CH_2\cdot$ is unknown because the ESR spectrum of this radical in solution has not been observed. Electron spin resonance spectra attributed to a variety of β-bromoalkyls have been observed in matrices, but the interpretation of these data has been a matter of controversy (*702*). The present authors believe that both the chemical behavior of β-bromoalkyls and the ESR data on other $R_nMCH_2CH_2\cdot$ radicals in solution imply that $BrCH_2CH_2\cdot$ will have a fairly strong preference for the eclipsed conformation.

X. INVERSION

A. Carbon-Centered Radicals

Methyl is planar, and other acyclic radicals derived from saturated hydrocarbons are either planar or, perhaps, very slightly bent with an extremely small barrier to inversion (7, 704–706). The degree of bending at C_α and the inversion barrier can be increased by ring strain (7) and by attaching electronegative heteroatoms, particularly fluorine (707) or oxygen (14), directly to the α-carbon.

1. Cyclopropyl and substituted cyclopropyls

INDO calculations on cyclopropyl suggest an out-of-plane angle of ca. 30°–35° for the CH_α bond, with the ESR hyperfine coupling constants for the syn and anti β protons nearly equal (7). This implies that the observed absence of lineshape effects in the ESR spectrum of cyclopropyl at temperatures down to $-120°C$ does not automatically require a rapid inversion ($k_{inv} = 10^8$–10^{10} sec^{-1}) at the radical center as originally assumed (15).

A large number of chemical trapping experiments have been carried out with a wide variety of substituted cyclopropyls to determine the extent to which the radicals retain their stereochemical integrity (394, 394b, 394c, 708–728). The tin hydride–cyclopropyl halide reaction has been employed most commonly (708–717), but other methods of generating and trapping cyclopropyl radicals have also been used (394–394c, 718–728). The results with tin hydrides have varied through the entire spectrum of possibilities, from complete inversion (708) through partial inversion (709, 710), complete configurational equilibration (711, 712), and partial retention (713–716), all the way to complete retention (710, 714, 716, 717). A similar spectrum of results has been obtained in other systems (394–394b, 718–728). Fortunately, the detailed picture is not quite as confusing as the above summary might suggest. Inversion appears always to be due to steric constraints, which prevent pyramidal cyclopropyls from reacting until they have inverted. Equilibration is observed for cyclopropyls α-substituted with groups, such as cyano or methoxycarbonyl, that can delocalize the unpaired electron and, by so doing, partially or completely flatten the radical center (712). Equilibration of pyramidal cyclopropyls is also favored at elevated temperatures and when the trap is insufficiently reactive or is present in low concentration. Under similar experimental conditions the degree of retention is very dependent on the nature of the α substituent. Retention is least, although it is observable under appropriate conditions (719–722, 725), for cyclopropyl radicals having hydrogen or an alkyl group attached to the α-carbon, retention being apparently somewhat greater with an α-alkyl than with an α-hydrogen [see (*721*) and (*728*) and also the oxiranyl inversion data in Table 6]. Retention occurs most readily with an α-fluorine substituent (*394b, 714, 716, 717, 727*) and somewhat less readily with an α-chlorine (*715, 727*) or α-methoxy (*394b*) substituent. However, inversion or equilibration can occur even with an α-chlorine (*394b, 708, 723*) or an α-methoxy (*726*).

The high configurational stability of α-fluorocyclopropyls that is deduced from the trapping experiments has been confirmed by ESR spectroscopy (*729–731*). The spectrum of each inversion isomer of 1-fluoro-2,3-*cis*-dimethylcyclopropyl (**391** and **392**) has been observed separately at −108°C (*729*), which means that these radicals are not interconverting on the ESR time scale (*ca.* 10^{-6} sec). The rate of inversion of 1-fluoro-2,3-*trans*-dimethylcyclopropyl is also slow ($k < 7 \times 10^6$ sec^{-1} at −108°C) (*730*), but 2,3-*cis*-, 2,3-*trans*-, and 2,2-dimethylcyclopropyl all invert rapidly ($k > 8 \times 10^7$ sec^{-1} at −99°C). Rather surprisingly, the more stable form of 2,3-*cis*-dimethylcyclopropyl has the α-H on the same side of the ring as the two methyl groups (*730*). The sterically more crowded 2,2-di-*t*-butyl-3,3-difluorocyclopropyl radical is planar (*731*).

Several attempts have been made to estimate the rates of inversion of substituted cyclopropyls from the extent of their trapping in some particular reaction, particularly their reduction by tin hydrides. Since there are

TABLE 6

Absolute Temperatures at which the Rate of Inversion of the Oxiranyl Radical is *ca.* 10^7 sec^{-1} (*59*)

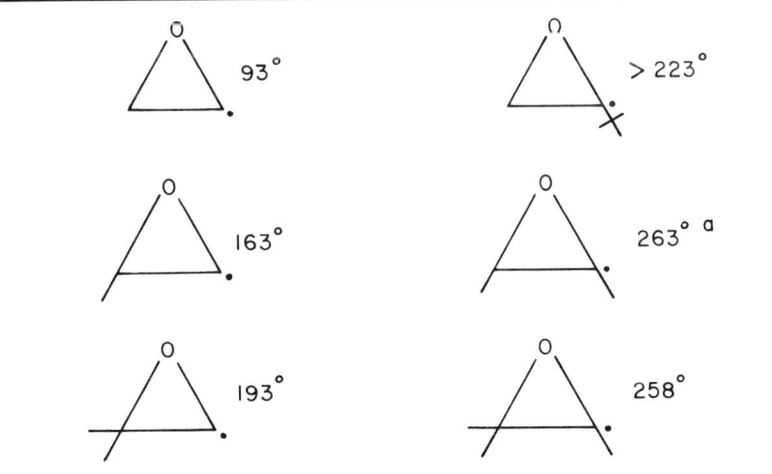

[a] Fischer (*733*).

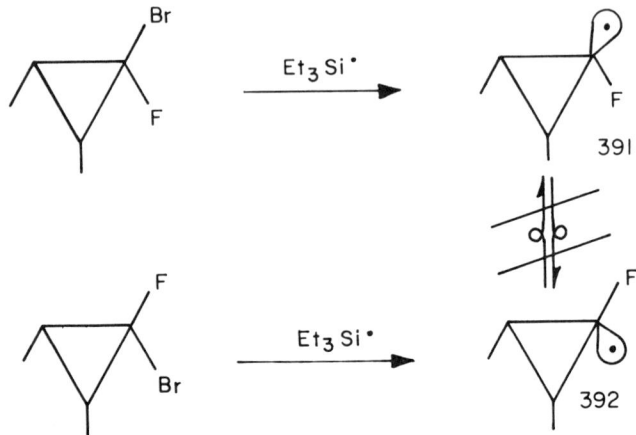

no rate constants for reaction of any cyclopropyl with any tin hydride, these calculations have little value. However, there is no intrinsic reason why these rate constants could not be measured using the technique employed for following the reaction of simple alkyls with tin hydrides (44). If the rates of inversion of α-halocyclopropyls were known, they would provide a valuable probe for additional studies of these interesting and reactive radicals.

2. Oxiranyl and substituted oxiranyls

The pyramidal nature of the oxiranyl radical has been firmly established by ESR spectroscopy (7, 59, 418, 420, 421). Chemical trapping of cis- and trans-2,3-dimethyloxiranyl (732) by both triphenyltin hydride (4 M at 30°C) and t-butyl hypochlorite (ca. 4 M at 0°C) suggests that these radicals are better able to retain their configuration than the analogous cyclopropyls, e.g.,

A complete ESR study of several alkyl-substituted oxiranyl radicals has provided an interesting picture of their behavior (59). The rates of inversion at C_α decreased with increasing alkyl substitution (see Table 6), whereas the rates of ring opening to produce α-ketoalkyls, 217 ⟶ 218, increased (see Section V,A). The effect of alkyl substitution at C_α on the inversion rate was ascribed to changes in the bending zero-point vibrational energies, with there being little or no change in the barrier to inversion. However, alkyl substitution at C_β does increase the barrier since the rate constants for inversion of oxiranyl (59) can be represented by

$$\log(k_{inv}/\text{sec}^{-1}) = (11.9 \pm 2) - (2 \pm 1)/\theta$$

and that for inversion of cis-2,3-dimethyloxiranyl (59, 733) by

$$\log(k_{inv}/\text{sec}^{-1}) = (11.8 \pm 0.8) - (5.8 \pm 1.4)/\theta$$

3. Other cyclic radicals

Reaction of 2-fluoro- and 2-chloroaziridin-2-yl radicals with tri-n-butyltin hydride at room temperature is stereospecific and occurs with complete retention of configuration (734), e.g.,

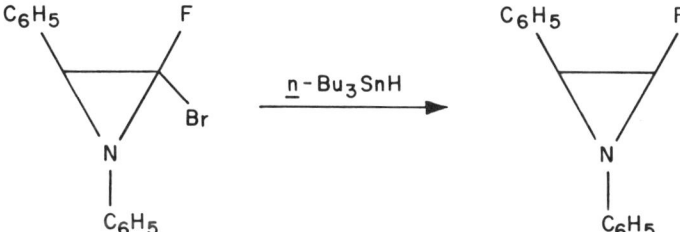

Substituted α-fluorocyclobutyl radicals are either planar or invert more rapidly than they can be trapped by triphenyltin hydride (715).

The 2-methyl-2,3-dioxolan-2-yl radical 393 is severely bent at C_α (421, 427, 708, 735, 736). Its ESR spectrum provides evidence that rotation about the C_α—CH_3 bond is restricted (735, 736) [as is also the case with certain other bent ethyl radicals (737)] and that inversion at C_α is fairly slow (427, 735, 736), with $\Delta H^\ddagger_{inv} = 5.6 \pm 0.2$ kcal/mol and $\Delta S^\ddagger_{inv} = 1.0$

± 1.0 gibbs/mol (*427, 736*). This *cis*- and *trans*-2,4-dimethyl-1,3-dioxolan-2-yl radicals have been trapped by methyl acrylate at −78°C with partial retention of their configuration (*738*).

Freshly generated 9-decalyl radicals undergo fast conformational changes (*739–742*). Thus, a cis source of 9-decalyl gives a short-lived (10^{-8}–10^{-9} sec) species, which changes into the same radical as is obtained from a trans source but which can yield cis products on reaction with oxygen (*739*) or *cis*-9-decalylcarbinyl hypochlorite (*740*), although not with tri-*n*-butyltin hydride (*741*). The thermodynamically more stable radical obtained from a trans source yields both cis and trans products on reaction with all these reagents.

4. Vinyl radicals

It has been shown by ESR spectroscopy that vinyl radicals can be classified as either "bent" σ radicals (**394**) or as "linear" π radicals (**395**). Structure **394**, in which the unpaired electron is in an orbital with substantial s character, is adopted by vinyl and by 1-methylvinyl. The *Z—E* isomerization of these radicals is rapid; for vinyl the rate constant for inversion lies between 3×10^7 and 3×10^9 sec^{-1} at −180°C, whereas 1-methylvinyl inverts somewhat more slowly (*7, 15*). The degree of bending and the inversion barrier are expected to become greater with more electronegative R_1 substituents, such as fluorine and methoxy. Structure **395** has been identified for a sterically crowded vinyl (*16*) and for vinyls having R_1 groups capable of delocalizing the unpaired electron, e.g., C_6H_5 (*743*) and COOH (*744*).

$$\underset{394}{\overset{R_3}{\underset{R_2}{>}}C=C\overset{}{\underset{R_1}{<}}} \rightleftharpoons \underset{}{\overset{R_3}{\underset{R_2}{>}}C=C\overset{R_1}{<}} \qquad \underset{395}{\overset{R_3}{\underset{R_2}{>}}C=C-R_1}$$

Chemical trapping studies complement the ESR evidence. Thus, scavenging of *cis*- and *trans*-1-methoxy-2-methylvinyl radicals by cumene at 110°C occurs with complete retention of their stereochemistry (*745, 746*). Under similar conditions, the scavenging of the cis and trans isomers of 1-chloro- and 1-bromo-2-phenylvinyls is partly stereospecific (*747*), but 1-methyl-2-phenylvinyl, 2-methyl-1-phenylvinyl, and 1,2-diphenylvinyl radicals generated from cis and trans precursors yield products that have completely lost their stereochemical memories (*748–750*). However, isomeric 1,2-dialkylvinyls can be trapped with some stereochemical retention by the use of very efficient scavengers, such as the naphthalene radical anion

4. FREE-RADICAL REARRANGEMENTS

(751), tri-*n*-butyltin hydride *(752, 753)*, and oxygen *(754)*. They can also be trapped electrochemically *(755)* and during photolysis of the parent iodides *(756)*. The 2-phenylvinyl radical generated from a trans source at 76°C isomerizes to yield the equilibrium mixture of vinyl bromides in CCl_3Br *(757)*.

In general, the configurational stability of free radicals increases with the s character of the semioccupied orbital *(758)*. Vinyl radicals, being sp^2-hybridized, are configurationally more stable than analogously substituted (α-Cl and α-MeO) cyclopropyls, which are *ca.* $sp^{2.5}$-hybridized *(759)*. The influence of substituents on the geometry of these radicals has become a subject of some theoretical interest *(394b, 711, 716, 760, 761)*.

B. Heteroatom-Centered Radicals

1. Silyl, germyl, and stannyl radicals

In contrast to $CH_3\cdot$ and $Me_3C\cdot$, both of which are approximately planar, the $MH_3\cdot$ and $Me_3M\cdot$ radicals of the remaining elements, M, in Group IV,B are approximately tetrahedral *(7, 11, 762–764)*. Racemic products are *generally* formed when a radical is generated at an asymmetric carbon atom from an optically active precursor. However, this is not the case with radicals generated from an optically active silane or germane, both of which *generally* yield products that are optically active and have retained the configuration of the starting compound *(11, 765)*. Thus, when chiral silyl or germyl radicals are generated from 1-naphthylphenylmethylsilane *(765, 766)*, 1-napthylphenylacetylsilane *(767, 768)*, and 1-naphthylphenylmethylgermane *(769)*, in CCl_4 at temperatures from 25° to 80°C, the chlorosilanes and chlorogermanes produced are optically active and have retained a large part of the stereochemistry of the parent compounds. The degree of retention decreases as the CCl_4 is diluted with benzene or cyclohexane *(765)*.

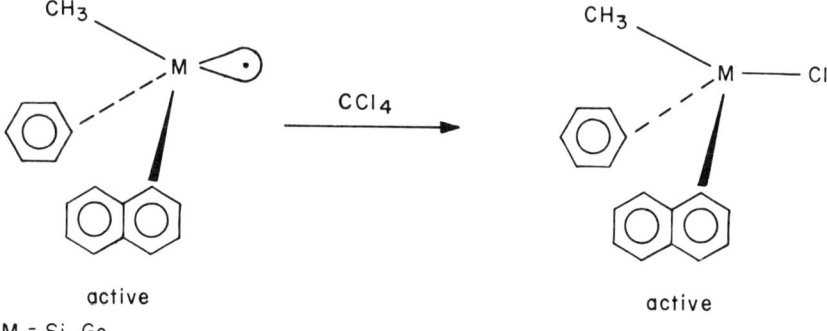

M = Si, Ge

A number of other optically active silanes give optically active chlorides under similar conditions (765). However, Ph₃SiSi*(Ph)(Me)H gave a chloride that was optically inactive (765) or nearly so (770). This result nicely complements ESR studies which show that the silyl radicals Me₃Ṡi, Me₃SiṠiMe₂, (Me₃Si)₂ṠiMe, and (Me₃Si)₃Ṡi become more planar, and presumably invert more rapidly, along this series (771).

Optically active 1-naphthylphenylmethylsilyl radicals have been trapped with phenylazide (772). They have also been generated photochemically from the active disilylmercury compound and found to yield disilane with retained configuration (773).

The stereochemically distinguishable pair of silyl radicals produced from cis- and trans-1-methyl-4-t-butyl-1-silacyclohexane react stereospecifically with CCl₄ to yield chlorosilanes with almost complete retention of configuration (11, 774). In similar reactions with other chlorinated solvents the configurational purity of the chlorosilane was found to decrease along the series CCl₄ < CHCl₂CCl₃ < CH₂ClCCl₃ < CHCl₃ (11, 774).

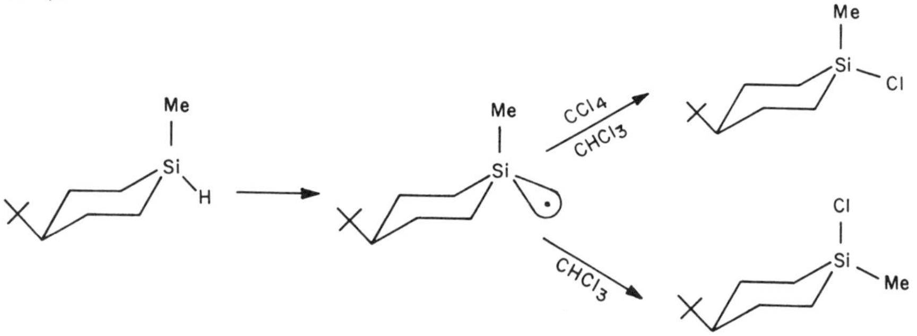

The free-radical chlorination of some silacyclobutanes (775) and of a silaacenaphthene (776) in CCl₄ also occurs with retention of configuration.

Precursors of some potential chiral tin-centered radicals have been prepared (777), and it has been shown that optically active methylneophylphenyltin hydride can be converted to the active (although unstable) chloride by a radical chain reaction in carbon tetrachloride (778).

The rates of inversion of none of the radicals mentioned above have been determined, nor are the rates of their reaction with CCl₄ or CHCl₃ known, although they must be rapid (779). The potential utility of these species in quantitative investigations of silyl, germyl, and stannyl radical chemistry has yet to be thoroughly exploited.

2. Iminoxy radicals

The rates of Z–E isomerization of several iminoxy radicals have been measured (780–783). Rates of inversion are highly dependent on the

4. FREE-RADICAL REARRANGEMENTS

$$R_1R_2C=N-O^\bullet \rightleftharpoons R_1R_2C=N-O^\bullet$$

substituents. Thus, at ambient temperatures, rate constants for inversion vary from low values implied for the configurationally stable acetyltrifluoracetyliminoxy (780), through a value of ca. 4×10^{-4} sec^{-1} for $R_1 = C_6H_5$, $R_2 = p\text{-}CH_3C_6H_4$ (781) and >30 sec^{-1} for $R_1 = C_6H_5$, $R_2 = C_6H_5CO$ (782), to a high of ca. 5×10^5 sec^{-1} for $R_1 = R_2 = (CH_3)_3C$ (783). These results do not provide any coherent pattern of reactivity. Further work on this isomerization is obviously required, particularly since the long-held view that the radical having $R_1 = C_6H_5$, $R_2 = CH_3$ exists entirely in the Z form has been shown by Mackor (784) to be incorrect. This radical actually exists in both the Z and E forms, with the E isomer favored over the Z, especially at low temperatures (784).

3. Phosphoranyl radicals

Finally, although it is not an inversion in the usual sense, we note for the sake of completeness that many phosphoranyl radicals undergo intramolecular ligand exchange, e.g., **396** ⟶ **397**. These free-radical isomerizations have been extensively investigated by product studies (785) and by ESR spectroscopy (786, 787).

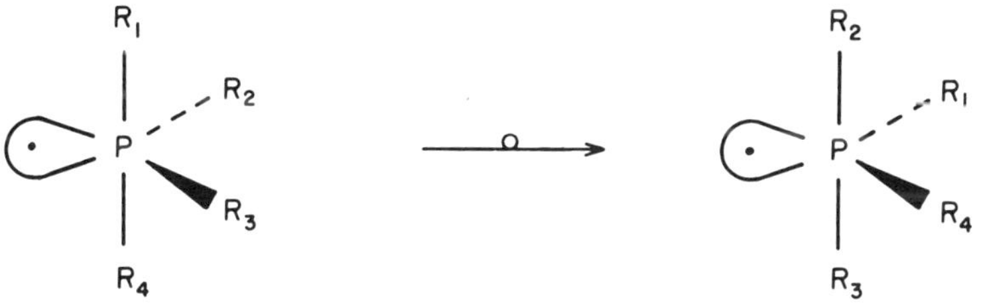

396 ⟶ **397**

REFERENCES

1. P. de Mayo, in "Molecular Rearrangements" (P. de Mayo, ed.), Part I, p. vii. Wiley (Interscience), New York, 1963.
2. C. Walling, in "Molecular Rearrangements" (P. de Mayo, ed.), Part I, Chapter 7. Wiley (Interscience), New York, 1963.
3. R. Kh. Freidlina, Adv. Free-Radical Chem. **1**, 211 (1965).
4. J. W. Wilt, in "Free Radicals" (J. K. Kochi, ed.), Vol. 1, Chapter 8. Wiley, New York, 1973.

5. W. H. Urry and M. S. Kharasch, *J. Am. Chem. Soc.* **66,** 1438 (1944).
6. For recent reviews of radical ring-closure reactions, see A. L. J. Beckwith, *in* "Essays on Free Radical Chemistry," Chem. Soc., Spec. Publ. No. 24, p. 239. Chem. Soc., London, 1970.
6a. M. Julia, *Acc. Chem. Res.* **4,** 386 (1971).
6b. M. Julia, *Pure Appl. Chem.* **40,** 553 (1974).
7. J. K. Kochi, *Adv. Free-Radical Chem.* **5,** 189 (1975).
8. P. S. Skell and K. J. Shea, *in* "Free Radicals" (J. K. Kochi, ed.), Vol. 2, Chapter 26. Wiley, New York, 1973.
9. L. Kaplan, "Bridged Free Radicals." Dekker, New York, 1972.
10. J. A. Howard, J. H. B. Chénier, and D. A. Holden, *Can. J. Chem.* **55,** 1463 (1977).
11. H. Sakurai, *in* "Free Radicals" (J. K. Kochi, ed.), Vol. 2, Chapter 25. Wiley, New York, 1973.
12. L. Kaplan, *in* "Free Radicals" (J. K. Kochi, ed.), Vol. 2, Chapter 18. Wiley, New York, 1973.
13. H. Fischer, *in* "Free Radicals" (J. K. Kochi, ed.), Vol. 2, Chapter 19. Wiley, New York, 1973.
14. G. Brunton, K. U. Ingold, B. P. Roberts, A. L. J. Beckwith, and P. J. Krusic, *J. Am. Chem. Soc.* **99,** 3177 (1977).
15. R. W. Fessenden and R. H. Schuler, *J. Chem. Phys.* **39,** 2147 (1963).
15a. E. L. Cochran, F. J. Adrian, and V. A. Bowers, *J. Chem. Phys.* **40,** 213 (1964); F. J. Adrian and M. Karplus, *ibid.* **41,** 56 (1964); R. W. Fessenden, *J. Phys. Chem.* **71,** 74 (1967); P. H. Kasai and E. B. Whipple, *J. Am. Chem. Soc.* **89,** 1033 (1967); S. Nagai, S. Ohnishi, and I. Nitta, *Chem. Phys. Lett.* **13,** 379 (1972).
16. D. Griller, J. W. Cooper, and K. U. Ingold, *J. Am. Chem. Soc.* **97,** 4269 (1975).
17. J. E. Werz and J. R. Bolton, "Electron Spin Resonance". McGraw-Hill, New York, 1972.
18. J. A. Ladd and W. H. Wardale, *in* "Internal Rotation in Molecules" (W. J. Orville-Thomas, ed.), Chapter 5. Wiley, New York, 1974.
18a. G. A. Russell, G. R. Underwood, and D. C. Lini, *J. Am. Chem. Soc.* **89,** 6636 (1967).
19. S. Ogawa and R. W. Fessenden, *J. Phys. Chem.* **41,** 994 (1964).
20. R. E. Rolfe, K. D. Sales, and J. H. P. Utley, *Chem. Commun.* p. 540 (1970).
21. C. Gaze and B. C. Gilbert, *J. Chem. Soc., Perkin Trans.* 2 p. 754 (1977).
22. P. J. Krusic, P. Meakin, and J. P. Jesson, *J. Phys. Chem.* **75,** 3438 (1971).
23. I. Biddles, A. Hudson, and J. T. Wiffen, *Tetrahedron* **28,** 867 (1972); D. Forrest, K. U. Ingold, and D. H. R. Barton, *J. Phys. Chem.* **80,** 915 (1977).
24. G. Golde, K. Möbius, and W. Kaminski, *Z. Naturforsch., Teil A* **24,** 1214 (1969).
24a. D. M. Camaioni, H. F. Walter, J. E. Jordan, and D. W. Pratt, *J. Am. Chem. Soc.* **95,** 7978 (1973).
25. G. Brunton, H. C. McBay, and K. U. Ingold, *J. Am. Chem. Soc.* **99,** 4447 (1977).
26. P. J. Krusic and P. Meakin, *Chem. Phys. Lett.* **18,** 347 (1973); R. W. Dennis and B. P. Roberts, *J. Organomet. Chem.* **47,** C8 (1973); I. H. Elson, M. J. Parrott, and B. P. Roberts, *Chem. Commun.* p. 586 (1975); R. W. Dennis and B. P. Roberts, *J. Chem. Soc., Perkin Trans.* 2 p. 140 (1975); J. W. Cooper and B. P. Roberts, *ibid.* p. 808 (1976); J. W. Cooper, M. J. Parrott, and B. P. Roberts, *ibid.* p. 730 (1977); R. W. Dennis, I. H. Elson, B. P. Roberts, and R. C. Dobbie, *ibid.* p. 889.
27. R. Sustmann and H. Trill, *J. Am. Chem. Soc.* **96,** 4343 (1974).
27a. B. E. Smart, P. J. Krusic, P. Meakin, and R. C. Bingham, *J. Am. Chem. Soc.* **96,** 7382 (1974).
27b. R. Sustmann, H. Trill, F. Vahrenholt, and D. Brandes, *Chem. Ber.* **110,** 255 (1977).

4. FREE-RADICAL REARRANGEMENTS

27c. R. Sustmann, H. Trill, and D. Brandes, *Chem. Ber.* **110**, 245 (1977).
28. P. J. Krusic, P. Meakin, and B. E. Smart, *J. Am. Chem. Soc.* **96**, 6211 (1974).
28a. M. S. Conradi, H. Zeldes, and R. Livingston, *J. Phys. Chem.* **83**, 2160 (1979).
29. See, e.g., P. S. Skell and R. G. Allen, *J. Am. Chem. Soc.*, **86**, 1559 (1964); A. A. Oswald, K. Griesbaum, B. E. Hudson, Jr., and J. M. Bregman, *ibid.* p. 2877; J. A. Kampmeier and G. Chen, *ibid.* **87**, 2608 (1965); J. A. Kampmeier and R. M. Fantazier, *ibid.* **88**, 1959 (1966); G. M. Whitesides and C. P. Casey, *ibid.* p. 4541; L. A. Singer and N. P. Kong, *ibid.* p. 5213; R. M. Fantazier and J. A. Kampmeier, *ibid.* p. 5219; O. Simamura, K. Tokumaru, and H. Yui, *Tetrahedron Lett.* p. 5141 (1966); E. I. Heiba and R. M. Dessau, *J. Am. Chem. Soc.* **89**, 2238 (1967); G. D. Sargent and M. W. Browne, *ibid.* p. 2788; E. I. Heiba and R. M. Dessau, *ibid.* p. 3772; L. A. Singer and N. P. Kong, *ibid.* p. 5251; R. C. Neuman, Jr. and G. D. Holmes, *J. Org. Chem.* **33**, 4317 (1968); G. M. Whitesides, C. P. Casey, and J. K. Krieger, *J. Am. Chem. Soc.* **93**, 1379 (1971); P. G. Webb and J. A. Kampmeier, *ibid.* p. 3730; M. S. Liu, S. Soloway, D. K. Wedegaertner, and J. A. Kampmeier, *ibid.* p. 3809; E. J. Panek, L. R. Kaiser, and G. M. Whitesides, *ibid.* **99**, 3708 (1977).
30. K. U. Ingold, *in* "Free Radicals" (J. K. Kochi, ed.), Vol. 1, Chapter 2. Wiley, New York, 1973.
31. K. Adamic, D. F. Bowman, T. Gillan, and K. U. Ingold, *J. Am. Chem. Soc.* **93**, 902 (1971).
32. S. Icli, V. J. Nowlan, P. M. Rahimi, C. Thankachan, and T. T. Tidwell, *Can. J. Chem.* **55**, 3349 (1977).
33. A. R. Bassindale, A. J. Bowles, M. A. Cook, C. Eaborn, A. Hudson, R. A. Jackson, and A. E. Jukes, *Chem. Commun.* p. 559 (1970); G. D. Mendenhall and K. U. Ingold, *J. Am. Chem. Soc.* **95**, 3422 (1973); G. D. Mendenhall, D. Griller, D. Lindsay, T. T. Tidwell, and K. U. Ingold, *ibid.* **96**, 2441 (1974).
34. S. Brownstein, J. Dunogues, D. Lindsay, and K. U. Ingold, *J. Am. Chem. Soc.* **99**, 2073 (1977).
35. S. A. Weiner and L. R. Mahoney, *J. Am. Chem. Soc.* **94**, 5029 (1972); J. L. Brokenshire, J. R. Roberts, and K. U. Ingold, *ibid.* p. 7040; V. Malatesta and K. U. Ingold, *ibid.* **96**, 3949 (1974); R. A. Kaba and K. U. Ingold, *ibid.* **98**, 7375 (1976).
36. D. Griller and K. U. Ingold, *Acc. Chem. Res.* **9**, 13 (1976).
37. G. Brunton, D. Griller, L. R. C. Barclay, and K. U. Ingold, *J. Am. Chem. Soc.* **98**, 6803 (1976).
38. G. B. Watts, D. Griller, and K. U. Ingold, *J. Am. Chem. Soc.* **94**, 8784 (1972).
39. D. Griller, G. D. Mendenhall, W. Van Hoof, and K. U. Ingold, *J. Am. Chem. Soc.* **96**, 6068 (1974).
40. D. E. Wood and R. V. Lloyd, *J. Chem. Phys.* **52**, 3840 (1970); **53**, 3932 (1970); D. E. Wood, R. V. Lloyd, and D. W. Pratt, *J. Am. Chem. Soc.* **92**, 4115 (1970); D. W. Pratt, J. J. Dillon, R. V. Lloyd, and D. E. Wood, *J. Phys. Chem.* **75**, 3486 (1971); R. V. Lloyd and D. E. Wood, *Mol. Phys.* **20**, 735 (1971); D. E. Wood, R. V. Lloyd, and W. A. Lathan, *J. Am. Chem. Soc.* **93**, 4145 (1971); D. E. Wood, L. F. Williams, R. F. Sprecher, and W. A. Lathan, *ibid.* **94**, 6241 (1972); D. E. Wood, C. A. Wood, and W. A. Lathan, *ibid.* p. 9280; L. F. Williams, M. B. Yim, and D. E. Wood, *ibid.* **95**, 6475 (1973); S. DiGregorio, M. B. Yim, and D. E. Wood, *ibid.* p. 8455; R. V. Lloyd and D. E. Wood, *ibid.* **96**, 659 (1974); **97**, 5986 (1975); J. E. Jordan, D. W. Pratt, and D. E. Wood, *ibid.* **96**, 5588 (1974); M. B. Yim and D. E. Wood, *ibid.* **97**, 1004 (1975); **98**, 2053 (1976).
41. R. Sustmann and F. Lübbe, *J. Am. Chem. Soc.* **98**, 6037 (1976).
42. M. B. Yim and D. E. Wood, *J. Am. Chem. Soc.* **98**, 3457 (1976).

43. C. Walling, J. H. Cooley, A. A. Ponaras, and E. J. Racah, *J. Am. Chem. Soc.* **88,** 5361 (1966).
44. D. J. Carlsson and K. U. Ingold, *J. Am. Chem. Soc.* **90,** 7047 (1968).
45. D. Griller and B. P. Roberts, *J. Chem. Soc., Perkin Trans. 2* p. 747 (1972).
46. A. G. Davies, D. Griller, and B. P. Roberts, *J. Chem. Soc., Perkin Trans. 2* p. 993 (1972).
47. D. Lal, D. Griller, S. Husband, and K. U. Ingold, *J. Am. Chem. Soc.* **96,** 6355 (1974).
48. H. Schuh, E. J. Hamilton, Jr., H. Paul, and H. Fischer, *Helv. Chim. Acta* **57,** 2011 (1974).
49. M. J. Perkins and B. P. Roberts, *J. Chem. Soc., Perkin Trans. 2* p. 77 (1975).
50. B. Maillard and K. U. Ingold, *J. Am. Chem. Soc.* **98,** 1224 and 4692 (1976).
50a. Y. Maeda and K. U. Ingold, *J. Am. Chem. Soc.* **101,** 4975 (1979).
50b. K. U. Ingold and J. Warkentin, *Can. J. Chem.* **58,** 348 (1980).
51. B. Maillard, D. Forrest, and K. U. Ingold, *J. Am. Chem. Soc.* **98,** 7024 (1976).
52. E. J. Hamilton, Jr. and H. Fischer, *Helv. Chim. Acta* **56,** 795 (1973).
53. S. Steenken, H. -P. Schuchmann, and C. von Sonntag, *J. Phys. Chem.* **79,** 763 (1975).
54. R. Sustmann and D. Brandes, *Chem. Ber.* **109,** 354 (1976).
55. K. U. Ingold and B. P. Roberts "Free-Radical Substitution Reactions." Wiley, New York, 1971.
56. G. H. Williams, "Homolytic Aromatic Substitution." Pergamon, Oxford, 1960; D. H. Hey, *Adv. Free-Radical Chem.* **2,** 47 (1967); M. J. Perkins, *in* "Free Radicals" (J. K. Kochi, ed.), Vol. 2, Chapter 16. Wiley, New York, 1973; R. Bolton and G. H. Williams, *Adv. Free-Radical Chem.* **5,** 1 (1975); F. Minisci, *Top. Curr. Chem.* **62,** 1 (1976).
57. P. I. Abell, *in* "Free Radicals" (J. K. Kochi, ed.), Vol. 2, Chapter 13. Wiley, New York, 1973.
58. R. A. Jackson and M. Townson, *Tetrahedron Lett.* p. 193 (1973); R. S. Iyer and F. S. Rowland, *Chem. Phys. Lett.* **21,** 346 (1973); G. G. Maynes and D. E. Applequist, *J. Am. Chem. Soc.* **95,** 856 (1973); R. F. Drury and L. Kaplan, *ibid.* p. 2217; J. H. Espenson and T. D. Sellers, Jr., *ibid.* **96,** 94 (1974); J. B. Levy and R. C. Kennedy, *ibid.* p. 4791; J. Z. Chrzastowski, C. J. Cooksey, M. D. Johnson, B. L. Lockman, and P. N. Steggles, *ibid.* **97,** 932 (1975); C. J. Upton and J. H. Incremona, *J. Org. Chem.* **41,** 523 (1976).
59. H. Itzel and H. Fischer, *Helv. Chim. Acta* **59,** 880 (1976).
60. F. D. Gunstone, "Guidebook to Sterochemistry." Longmans, Green, New York, 1975.
61. A. J. Bloodworth, A. G. Davies, I. M. Griffin, B. Muggleton, and B. P. Roberts, *J. Am. Chem. Soc.* **96,** 7599 (1974).
61a. J. R. Nixon, M. A. Cudd, and N. A. Porter, *J. Org. Chem.* **43,** 4048 (1978).
62. M. L. Poutsma and P. A. Ibarbia, *Tetrahedron Lett.* p. 3309 (1972).
63. S. W. Benson, "Thermochemical Kinetics," 2nd ed. Wiley, New York, 1976.
64. D. Griller, unpublished results.
65. A. L. J. Beckwith and C. B. Thomas, *J. Chem. Soc., Perkin Trans. 2* p. 861 (1973).
66. J. A. Howard and K. U. Ingold, *Can. J. Chem.* **47,** 3797 (1969).
67. J. C. Scaiano, unpublished results.
68. P. B. Shevlin and H. J. Hansen, *J. Org. Chem.* **42,** 3011 (1977).
69. E. L. Eliel, "Stereochemistry of Carbon Compounds," p. 197. McGraw-Hill, New York, 1962.
70. L. H. Slaugh, *J. Am. Chem. Soc.* **81,** 2262 (1959).
71. W. A. Bonner and F. D. Mango, *J. Org. Chem.* **29,** 29 (1964).

4. FREE-RADICAL REARRANGEMENTS 287

72. J. W. Wilt and W. W. Pawlikowski, Jr., *J. Org. Chem.* **40**, 3641 (1975).
73. P. D. Bartlett and J. D. Cotman, Jr., *J. Am. Chem. Soc.* **72**, 3095 (1950).
74. D. Y. Curtin and J. C. Kauer, *J. Org. Chem.* **25**, 880 (1960).
75. C. Rüchardt, *Chem. Ber.* **94**, 2599, 2609 (1961); C. Rüchardt and S. Eichler, *ibid.* **95**, 1921 (1962); C. Rüchardt and H. Trautwein, *ibid.* **96**, 160 (1963); **98**, 2478 (1965); C. Rüchardt and R. Hecht, *ibid.* pp. 2460 and 2471.
76. R. N. Goerner, Jr., P. N. Cote, and B. M. Vittimberga, *J. Org. Chem.* **42**, 19 (1977).
77. See, e.g., R. G. Gasanov and R. Kh. Freidlina, *Izv. Akad. Nauk SSSR, Ser. Khim.* p. 484 (1977).
78. H. Wieland, *Ber. Dtsch. Chem. Ges.* **44**, 2553 (1911); M. S. Kharasch, A. Fono, and W. Nudenberg, *J. Org. Chem.* **15**, 763 (1950).
78a. M. S. Kharasch, A. C. Poshkus, A. Fono, and W. Nudenberg, *J. Org. Chem.* **16**, 1458 (1951).
79. R. D. Small, Jr. and J. C. Scaiano, *J. Am. Chem. Soc.* **100**, 296 (1978); H. Paul, R. D. Small, Jr., and J. C. Scaiano, *ibid.* p. 4520. S. K. Wong, *ibid.* **101**, 1235 (1979).
80. J. W. Wilt and M. P. Stumpf, *J. Org. Chem.* **30**, 1256 (1965).
81. M. F. R. Mulcahy, B. G. Tucker, D. J. Williams, and J. R. Wilmshurst, *Chem. Commun.* p. 609 (1965); A. H. K. Yousufzai, *Proc. Pak. Acad. Sci.* **13**, 13 (1976); C. J. Collins, W. H. Roark, V. F. Raaen, and B. M. Benjamin, *J. Am. Chem. Soc.* **101**, 1877 (1979); C. J. Collins, V. F. Raaen, B. M. Benjamin, P. H. Maupin, and W. H. Roark, *ibid.* p. 5009.
82. B. M. Vittimberga and M. L. Herz, *J. Org. Chem.* **35**, 3694 (1970).
83. J. K. Haynes, Jr. and J. A. Kampmeier, *J. Org. Chem.* **37**, 4167 (1972).
84. J. W. Wilt, O. Kolewe, and J. F. Kraemer, *J. Am. Chem. Soc.* **91**, 2624 (1969).
85. R. L. Dannley and G. Jalics, *J. Org. Chem.* **30**, 3848 (1965); A. K. Shubber and R. L. Dannley, *ibid.* **36**, 3784 (1971).
86. N. A. Porter and J. G. Green, *Tetrahedron Lett.* p. 2667 (1975).
87. S. Winstein, R. Heck, S. Lapporte, and R. Baird, *Experientia* **12**, 138 (1956).
88. W. H. Urry, D. J. Trecker, and H. D. Hartzler, *J. Org. Chem.* **29**, 1663 (1964). See also, F. J. L. Aparicio, J. F. Sanchez, and M. T. P. L. Espinosa, *An. Quim.* **74**, 294 (1978).
89. J. C. Chottard and M. Julia, *Tetrahedron* **28**, 5615 (1972).
90. J. W. Wilt and C. F. Dockus, *J. Am. Chem. Soc.* **92**, 5813 (1970).
90a. For recent work leading to some revision of the earlier conclusions, see J. W. Wilt, W. K. Chwang, C. F. Dockus, and N. M. Tomiuk, *J. Am. Chem. Soc.* **100**, 5534 (1978).
91. R. Loven and W. N. Speckamp, *Tetrahedron Lett.* p. 1567 (1972).
92. J. W. Wilt, R. A. Dabek, and K. C. Welzel, *J. Org. Chem.* **37**, 425 (1972).
93. J. K. Kochi and R. D. Gilliom, *J. Am. Chem. Soc.* **86**, 5251 (1964).
94. J. K. Kochi and R. V. Subramanian, *J. Am. Chem. Soc.* **87**, 4855 (1965).
95. J. K. Kochi and P. J. Krusic, *J. Am. Chem. Soc.* **91**, 3940 (1969).
96. D. J. Edge and J. K. Kochi, *J. Am. Chem. Soc.* **94**, 7695 (1972).
97. J. L. Charlton and G. J. Williams, *Tetrahedron Lett.* p. 1473 (1977).
98. M. Julia and B. Malassiné, *Tetrahedron Lett.* p. 987 (1971).
99. M. Julia and B. Malassiné, *Tetrahedron* **30**, 695 (1974).
100. H. Sakurai, I. Nozue, and A. Hosomi, *J. Am. Chem. Soc.* **98**, 8279 (1976).
101. H. Sakurai and A. Hosomi, *J. Am. Chem. Soc.* **92**, 7507 (1970).
101a. H. Sakurai, A. Hosomi, and M. Kumada, *Tetrahedron Lett.* p. 1757 (1969).
102. D. F. DeTar and A. Hlynsky, *J. Am. Chem. Soc.* **77**, 4411 (1955).
103. L. K. Montgomery, J. W. Matt, and J. R. Webster, *J. Am. Chem. Soc.* **89**, 923 (1967).
104. L. K. Montgomery and J. W. Matt, *J. Am. Chem. Soc.* **89**, 934 (1967).

104a. L. K. Montgomery and J. W. Matt, *J. Am. Chem. Soc.* **89**, 3050 and 6556 (1967).
105. T. A. Halgren, M. E. H. Howden, M. E. Medof, and J. D. Roberts, *J. Am. Chem. Soc.* **89**, 3051 (1967).
106. J. Warkentin, unpublished results.
107. A. L. J. Beckwith, unpublished results.
108. L. H. Slaugh, *J. Am. Chem. Soc.* **87**, 1522 (1965).
109. P. Schmid and Y. Maeda, unpublished results.
110. E. C. Friedrich and R. L. Holmstead, *J. Org. Chem.* **37**, 2550 (1972).
111. E. C. Friedrich and R. L. Holmstead, *J. Org. Chem.* **36**, 971 (1971).
112. E. C. Friedrich and R. L. Holmstead, *J. Org. Chem.* **37**, 2546 (1972).
113. R. S. Boikess, M. Mackay, and D. Blithe, *Tetrahedron Lett.* p. 401 (1971).
113a. A. Nishinaga, K. Nakamura, and T. Matsuura, *Tetrahedron Lett.* p. 3557 (1978).
114. C. L. Karl, E. J. Maas, and W. Reusch, *J. Org. Chem.* **37**, 2834 (1972).
115. F. Bertini, T. Caronna, L. Grossi, and F. Minisci, *Gazz. Chim. Ital.* **104**, 471 (1974).
116. A. I. Prokof'ev, N. A. Malysheva, N. N. Bubnov, S. P. Solodovnikov, I. S. Belostotskaya, V. V. Ershov, and M. I. Kabachnik, *Dokl. Akad. Nauk SSSR* **229**, 1128 (1976).
117. W. K. Robbins and R. H. Eastman, *J. Am. Chem. Soc.* **92**, 6077 (1970); M. J. Perkins and B. P. Roberts, *J. Chem. Soc., Perkin Trans. 2* p. 297 (1974).
118. J. Kalvoda, C. Meystre, and G. Anner, *Helv. Chim. Acta* **49**, 424 (1966); J. Kalvoda, *ibid.* **51**, 267 (1968).
118a. D. S. Watt, *J. Am. Chem. Soc.* **98**, 271 (1976); G. I. Nikishin, E. I. Troyanskii, D. S. Velibekova, and Yu. N. Ogibin, *Izv. Akad. Nauk SSSR, Ser. Khim.* p. 2430 (1978).
119. M. P. Doyle, P. W. Raynolds, R. A. Barents, T. R. Bade, W. C. Danen, and C. T. West, *J. Am. Chem. Soc.* **95**, 5988 (1973).
120. J. M. Tedder and J. C. Walton, *Acc. Chem. Res.* **9**, 183 (1976); *Adv. Phys. Org. Chem.* **16**, 86 (1978).
121. F. Minisci, *Acc. Chem. Res.* **8**, 165 (1975).
122. G. H. Williams, *in* "Essays on Free Radical Chemistry," Chem. Soc., Spec. Publ. No. 24, p. 25. Chem. Soc., London, 1970.
123. M. Tiecco, *Org. Chem., Ser. Two* **10**, 25 (1975).
124. M. Julia, *Rec. Chem. Prog.* **25**, 1 (1964).
125. A. G. Davies and B. P. Roberts, *in* "Free Radicals" (J. K. Kochi, ed.), Vol. 1, Chapter 10. Wiley, New York, 1973.
126. H. E. O'Neal and S. W. Benson, *in* "Free Radicals" (J. K. Kochi, ed.), Vol. 2, Chapter 17. Wiley, New York, 1973.
127. M. J. S. Dewar, "The Molecular Orbital Theory of Organic Chemistry," p. 299. McGraw-Hill, New York, 1969.
128. M. Szwarc and J. H. Binks, *Theor. Org. Chem., Pap. Kekule Symp., 1958*, p. 262 (1959); A. L. J. Beckwith and M. J. Thompson, *J. Chem. Soc.* p. 73 (1961).
129. F. R. Mayo, *J. Am. Chem. Soc.* **84**, 3964 (1962).
130. L. H. Gale, *J. Org. Chem.* **34**, 81 (1969).
131. D. S. Ashton, D. J. Shand, J. M. Tedder, and J. C. Walton, *J. Chem. Soc., Perkin Trans. 2* p. 320 (1975); L. T. Vertommen, J. M. Tedder, and J. C. Walton, *J. Chem. Res. (S)* p. 18 (1977); M. H. Treder, H. Kratzin, H. Lübbecke, C. -Y. Yang, and P. Boldt, *ibid.* p. 165 (1977).
132. T. Caronna, A. Citterio, M. Ghirardini, and F. Minisci, *Tetrahedron* **33**, 793 (1977).
133. H. Fujimoto, S. Yamabe, T. Minato, and K. Fukui, *J. Am. Chem. Soc.* **94**, 9205 (1972); see also S. Nagase, K. Takatsuka, and T. Fueno, *ibid.* **98**, 3838 (1976); J. R. Hoyland, *Theor. Chim. Acta* **22**, 229 (1971). For more sophisticated calculations leading to

somewhat different conclusions, see M. J. S. Dewar and S. Olivella, *J. Am. Chem. Soc.* **100**, 5290 (1978).
134. E. L. Eliel "Stereochemistry of Carbon Compounds," p. 188. McGraw-Hill, New York, 1962.
135. E. L. Eliel, N. L. Allinger, S. J. Angyal, and G. A. Morrison, "Conformational Analysis," p. 191. Wiley, New York, 1965.
136. N. L. Allinger and V. Zalkow, *J. Org. Chem.* **25**, 701 (1960).
137. P. von R. Schleyer, *J. Am. Chem. Soc.* **83**, 1368 (1961).
138. A. L. J. Beckwith, G. E. Gream, and D. L. Struble, *Aust. J. Chem.* **25**, 1081 (1972).
139. A. L. J. Beckwith and W. B. Gara, *J. Chem. Soc., Perkin Trans. 2* p. 795 (1975).
140. J. E. Baldwin, *Chem. Commun.* p. 734 (1976).
141. S. D. Hamann, A. Pompe, D. H. Solomon, and T. H. Spurling, *Aust. J. Chem.* **29**, 1975 (1976).
142. G. R. Butler and R. J. Angelo, *J. Am. Chem. Soc.* **79**, 3128 (1957); C. S. Marvel and R. D. Vest, *ibid.* p. 5771; S. Arai, S. Sato, and S. Shida, *J. Chem. Phys.* **33**, 1277 (1960); A. S. Gordon and S. R. Smith, *J. Phys. Chem.* **66**, 521 (1962).
143. R. C. Lamb, P. W. Ayers, and M. K. Toney, *J. Am. Chem. Soc.* **85**, 3483 (1963).
144. C. Walling and M. S. Pearson, *J. Am. Chem. Soc.* **86**, 2262 (1964).
145. R. F. Garwood, C. J. Scott, and B. C. L. Weedon, *Chem. Commun.* p. 14 (1965).
146. J. F. Garst, P. W. Ayers, and R. C. Lamb, *J. Am. Chem. Soc.* **88**, 4260 (1966).
147. R. C. Lamb, P. W. Ayers, M. K. Toney, and J. F. Garst, *J. Am. Chem. Soc.* **88**, 4261 (1966).
148. J. K. Kochi and T. W. Bethea, III, *J. Org. Chem.* **33**, 75 (1968).
149. J. F. Garst and F. E. Barton, II, *Tetrahedron Lett.* p. 587 (1969).
150. R. A. Sheldon and J. K. Kochi, *J. Am. Chem. Soc.* **92**, 4395 (1970).
151. C. L. Jenkins and J. K. Kochi, *J. Am. Chem. Soc.* **94**, 843 (1972).
152. C. Walling and A. Cioffari, *J. Am. Chem. Soc.* **94**, 6059 (1972).
153. R. P. Quirk and R. E. Lea, *Tetrahedron Lett.* p. 1925 (1974); *J. Am. Chem. Soc.* **98**, 5973 (1976).
154. A. L. J. Beckwith, I. A. Blair, and G. Phillipou, *Tetrahedron Lett.* p. 2251 (1974).
155. A. L. J. Beckwith and G. Moad, *Chem. Commun.* p. 472 (1974).
156. M. Julia, C. Descoins, M. Baillarge, B. Jacquet, D. Uguen, and F. A. Groeger, *Tetrahedron* **31**, 1737 (1975).
157. H. O. House and P. D. Weeks, *J. Am. Chem. Soc.* **97**, 2778 (1975).
158. P. Schmid, D. Griller, and K. U. Ingold, *Int. J. Chem. Kinet.* **11**, 333 (1979).
159. H. Hart and D. P. Wyman, *J. Am. Chem. Soc.* **81**, 4891 (1959).
160. M. Julia and M. Maumy, *Bull. Soc. Chim. Fr.* p. 1603 (1968).
161. N. le Bel, quoted in Julia *et al.* (*6b, 156*).
162. A. L. J. Beckwith and W. B. Gara, *J. Chem. Soc., Perkin Trans. 2* p. 593 (1975).
163. R. D. Rieke and N. A. Moore, *Tetrahedron Lett.* p. 2035 (1969); *J. Org. Chem.* **37**, 413 (1972).
164. D. Capon and C. W. Rees, *J. Chem. Soc., Annu. Rep.* **61**, 221 (1964).
165. D. G. L. James and T. Ogawa, *Can. J. Chem.* **43**, 640 (1965).
166. G. Moad, Ph.D. Thesis, University of Adelaide (1976).
167. A. L. J. Beckwith and G. Phillipou, *Chem. Commun.* p. 280 (1973).
168. P. D. Bartlett, W. D. Clossen, and T. J. Cogdell, *J. Am. Chem. Soc.* **87**, 1308 (1965).
169. W. C. Kossa, Jr., T. C. Rees, and H. G. Richey, Jr., *Tetrahedron Lett.* p. 3455 (1971).
170. J. K. Kochi, *in* "Free Radicals,' (J. K. Kochi, ed.), Vol. 1, Chapter 11. Wiley, New York, 1973.

171. R. A. Sheldon and J. K. Kochi, *Org. React.* **19**, 279 (1972).
172. P. Schmid and K. U. Ingold, *J. Am. Chem. Soc.* **99**, 6434 (1977); **100**, 2493 (1978).
173. A. Citterio, F. Minisci, O. Porta, and G. Sesana, *J. Am. Chem. Soc.* **99**, 7960 (1977); A. Citterio, *Tetrahedron Lett.* p. 2701 (1978); A. Citterio, A. Arnoldi, and F. Minisci *J. Org. Chem.* **44**, 2674 (1979).
174. J. F. Garst and C. D. Smith, *J. Am. Chem. Soc.* **98**, 1526 (1976).
175. J. F. Garst and C. D. Smith, *J. Am. Chem. Soc.* **98**, 1520 (1976).
176. J. F. Garst, *in* "Free Radicals" (J. K. Kochi, ed.), Vol. 1, Chapter 9. Wiley, New York, 1973.
177. J. F. Garst and F. E. Barton, II, *J. Am. Chem. Soc.* **96**, 523 (1974); J. F. Garst and J. T. Barbas, *ibid.* p. 3239.
178. J. F. Garst, R. D. Roberts, and J. A. Pacifici, *J. Am. Chem. Soc.* **99**, 3528 (1977); J. F. Garst, J. A. Pacifici, C. C. Felix, and A. Nigam, *J. Am. Chem. Soc.* **100**, 5974 (1978).
178a. D. E. Bergbreiter and J. M. Killough *J. Am. Chem. Soc.* **100**, 2126 (1978).
179. J. W. Sease and R. C. Reed, *Tetrahedron Lett.* p. 393 (1975).
180. H. O. House, *Acc. Chem. Res.* **9**, 59 (1976).
181. W. A. Nugent and J. K. Kochi, *J. Am. Chem. Soc.* **98**, 5405 (1976); *J. Organomet. Chem.* **124**, 327 (1977).
182. N. O. Brace and J. E. Van Elswyck, *J. Org. Chem.* **41**, 766 (1976).
183. J. K. Kochi and J. W. Powers, *J. Am. Chem. Soc.* **92**, 137 (1970).
184. C. Walling and A. Cioffari, *J. Am. Chem. Soc.* **92**, 6609 (1970); H. W. H. J. Bodewitz, C. Blomberg, and F. Bickelhaupt, *Tetrahedron* **31**, 1053 (1975); A. I. Meyers, R. Gabel, and E. D. Mihelich, *J. Org. Chem.* **43**, 1372 (1978); F. A. Davis, P. A. Mancinelli, K. Balasubramanian, and U. K. Nadir, *J. Am. Chem. Soc.* **101**, 1044 (1979).
185. P. J. Wagner and K. -C. Liu, *J. Am. Chem. Soc.* **96**, 5952 (1974).
186. B. Giese and J. Meister, *Angew. Chem., Int. Ed. Engl.* **16**, 178 (1977).
187. C. Rüchardt, *Angew. Chem., Int. Ed. Engl.* **9**, 830 (1970).
188. B. Giese and J. Meixner, *Tetrahedron Lett.* p. 2779 (1977).
189. V. Bonacic-Koutecky, J. Koutecky, and L. Salem, *J. Am. Chem. Soc.* **99**, 842 (1977).
190. A. L. J. Beckwith, I. Blair, and G. Phillipou, *J. Am. Chem. Soc.* **96**, 1613 (1974).
191. N. O. Brace, *J. Org. Chem.* **32**, 2711 (1967).
192. M. E. Kuehne and R. E. Damon, *J. Org. Chem.* **42**, 1825 (1977).
193. R. Hoffmann, C. C. Levin, and R. A. Moss, *J. Am. Chem. Soc.* **95**, 629 (1973).
194. A. L. J. Beckwith and T. Lawrence, unpublished results.
195. A. L. J. Beckwith and G. Moad, *J. Chem. Soc., Perkin Trans. 2* p. 1726 (1975).
196. T. W. Smith and G. B. Butler, *J. Org. Chem.* **43**, 6 (1978).
197. A. L. J. Beckwith and G. Phillipou, unpublished results.
198. B. Jacquet, unpublished results, quoted in Julia (*6a*).
199. C. Descoins, M. Julia, and H. van Sang, *Bull. Soc. Chim. Fr.* p. 4087 (1971).
200. J. W. Wilt, S. N. Massie, and R. B. Dabek, *J. Org. Chem.* **35**, 2803 (1970).
201. A. L. J. Beckwith, A. K. Ong, and D. H. Solomon, *J. Macromol. Sci., Chem.* **9**, 115 (1975).
202. A. L. J. Beckwith, A. K. Ong, and D. H. Solomon, *J. Macromol. Sci., Chem.* **9**, 125 (1975).
203. N. O. Brace, *J. Org. Chem.* **36**, 3187 (1971).
204. N. O. Brace, *J. Org. Chem.* **38**, 3167 (1973); **44**, 212 (1979).
205. N. O. Brace, *J. Org. Chem.* **31**, 2879 (1966).
206. N. O. Brace, *J. Org. Chem.* **34**, 2441 (1969).
207. N. O. Brace, *J. Polym. Sci., Part A-1* **8**, 2091 (1970).

4. FREE-RADICAL REARRANGEMENTS

208. J. I. G. Cadogan, M. Grunbaum, D. H. Hey, A. S. H. Ong, and J. T. Sharp, *Chem. Ind., (London)* p. 422 (1968).
209. S. F. Reed, Jr., *J. Org. Chem.* **33**, 1861 (1968).
210. M. A. M. Bradney, A. D. Forbes, and J. Wood, *J. Chem. Soc., Perkin Trans. 2* p. 1655 (1973).
211. D. G. Hawthorne, S. R. Johns, D. H. Solomon, and R. I. Willing, *Chem. Commun.* p. 982 (1975).
211a. D. G. Hawthorne, S. R. Johns, and R. I. Willing, *Aust. J. Chem.* **29**, 315 (1976).
211b. W. R. Dolbier, Jr., and O. T. Garza, *J. Org. Chem.* **43**, 3848 (1978).
211c. R. L. Keiter, Y. Y. Sun, J. W. Brodack, and L. W. Cary, *J. Am. Chem. Soc.* **101**, 2638 (1979).
212. A. L. J. Beckwith, D. G. Hawthorne, and D. H. Solomon, *Aust. J. Chem.* **29**, 995 (1976).
213. R. Breslow, J. T. Groves, and S. S. Olin, *Tetrahedron Lett.* p. 4717 (1966).
214. M. Julia and D. Mansuy, *C. R. Hebd. Seances Acad. Sci., Ser. C* **274**, 408 (1972).
215. For reviews of cyclopolymerization, see D. H. Solomon and D. G. Hawthorne, *J. Macromol. Sci., Rev. Macromol. Chem.* **15**, 143 (1976); C. L. McCormick and G. B. Butler, *ibid.* **8**, 201 (1972); G. C. Corfield, *Chem. Soc. Rev.* **1**, 523 (1972); W. E. Gibbs and J. M. Barton, *in* "Kinetics and Mechanism of Polymerization" (G. E. Ham, ed.), Vol. 1, Part 1, Chapter 2. Dekker, New York, 1967.
216. G. B. Butler, A. Crawshaw, and W. L. Miller, *J. Am. Chem. Soc.* **80**, 3615 (1958); A. Crawshaw and G. B. Butler, *ibid.* p. 5464; G. B. Butler, *J. Polym. Sci.* **48**, 279 (1960).
217. S. R. Johns, R. I. Willing, S. Middleton, and A. K. Ong, *J. Macromol. Sci., Chem.* **10**, 875 (1976), and references cited; D. G. Hawthorne and D. H. Solomon, *ibid.* p. 923.
218. G. B. Butler and B. Iachia, *J. Macromol. Sci., Chem.* **3**, 803 (1969); G. B. Butler and W. L. Miller, *ibid.* p. 1493; G. B. Butler and S. Kimura, *ibid.* **5**, 181 (1971).
219. G. B. Butler and T. W. Brooks, *J. Org. Chem.* **28**, 2699 (1963); G. B. Butler and M. A. Raymond, *ibid.* **30**, 2410 (1965).
220. G. Pregaglia and G. Gregario, *Chim. Ind. (Milan)* **45**, 1065 (1963).
221. R. H. Fish, H. G. Kuivila, and I. J. Tyminski, *J. Am. Chem. Soc.* **89**, 5861 (1967).
222. R. Dowbenko, *Tetrahedron* **20**, 1843 (1964).
223. L. Friedman, *J. Am. Chem. Soc.* **86**, 1885 (1964).
224. R. P. A. Sneeden, *Synthesis* p. 259 (1971).
225. J. G. Traynham and H. H. Hsieh, *J. Org. Chem.* **38**, 868 (1973).
226. E. D. Brown, T. W. Sam, J. K. Sutherland, and A. Torre, *J. Chem. Soc., Perkin Trans. 1* p. 2326 (1975).
227. M. C. Lasne and A. Thuillier, *C. R. Hebd. Seances Acad. Sci., Ser. C* **273**, 1258 (1971).
228. A. G. Yurchenko, L. A. Zosim, and N. L. Dovgan, *Zh. Org. Khim.* **10**, 1996 (1974).
229. R. Breslow, S. S. Olin, and J. T. Groves, *Tetrahedron Lett.* p. 1837 (1968).
230. M. Barreau and M. Julia, *Tetrahedron Lett.* p. 1537 (1973); M. Julia, B. Mansour, and D. Mansuy, *ibid.* p. 3443 (1976).
231. Z. Cekovic, *Tetrahedron Lett.* p. 749 (1972).
232. M. Julia and M. Barreau, *C. R. Hebd. Seances Acad. Sci., Ser. C* **280**, 957 (1975).
233. M. Julia and M. Maumy, *Org. Synth.* **55**, 57 (1976).
234. M. Julia, J. M. Salard, and J. C. Chottard, *Bull. Soc. Chim. Fr.* p. 2478 (1973).
235. R. Perrey, Thesis, University of Paris (1969), cited in Julia (*6a*).
236. C. Walling and A. Cioffari, *J. Am. Chem. Soc.* **94**, 6064 (1972).
237. H. Pines, N. C. Sih, and D. B. Rosenfield, *J. Org. Chem.* **31**, 2255 (1966).
238. M. Julia and M. Maumy, *Bull. Soc. Chim. Fr.* pp. 2415 and 2427 (1969).

239. M. Julia, J. -M. Surzur, L. Katz, and F. Le Goffic, *Bull. Soc. Chim. Fr.* p. 1116 (1964).
240. M. Julia, F. Le Goffic and L. Katz, *Bull. Soc. Chim. Fr.* p. 1122 (1964).
241. M. Julia and F. Le Goffic, *Bull. Soc. Chim. Fr.* p. 1129 (1964).
242. A. S. Gordon, S. R. Smith, and C. M. Drew, *J. Chem. Phys.* **36**, 824 (1962).
242a. D. F. McMillan, D. M. Golden, and S. W. Benson, *Int. J. Chem. Kinet.* **3**, 359 (1971).
243. A. Effio, D. Griller, K. U. Ingold, A. L. J. Beckwith, and A. K. Serelis, *J. Am. Chem. Soc.* **102**, 1734 (1980).
244. W. P. L. Carter and D. C. Tardy, *J. Phys. Chem.* **78**, 1245 (1974).
245. C. R. Warner, R. J. Strunk, and H. G. Kuivila, *J. Org. Chem.* **31**, 3381 (1966).
245a. Y. Sugiyama, T. Kawamura, and T. Yonezawa, *Chem. Lett.* p. 639 (1978).
246. R. Sustmann and R. W. Gellert, *Chem. Ber.* **111**, 388 (1978).
247. P. Piccardi, P. Massardo, M. Modena, and E. Santoro, *J. Chem. Soc., Perkin Trans. 1* p. 982 (1973).
248. K. W. Watkins and D. K. Olsen, *J. Phys. Chem.* **76**, 1089 (1972).
249. J. W. Wilt, L. L. Maravetz, and J. F. Zawadzki, *J. Org. Chem.* **31**, 3018 (1966).
250. M. Julia and F. Le Goffic, *Bull. Soc. Chim. Fr.* pp. 1550 and 1555 (1965).
251. M. Chatzopoulos and J. -P. Montheard, *C. R. Hebd. Seances Acad. Sci., Ser. C* **280**, 29 (1975).
252. J. M. Surzur and G. Torri, *Bull. Soc. Chim. Fr.* p. 3070 (1970).
253. E. Van Bruggen, *Recl. Trav. Chim. Pays-Bas* **87**, 1134 (1968).
254. A. S. Atavin, B. A. Trofimov, G. M. Gavrilova, and I. M. Korotaeva, *Zh. Obshch. Khim.* **41**, 804 (1971).
255. P. D. Gokhale, A. P. Joshi, R. Sahni, V. G. Naik, N. P. Damodaran, U. R. Nayak, and S. Dev, *Tetrahedron* **32**, 1391 (1976).
256. P. Bakuzis, O. O. S. Campos, and M. L. F. Bakuzis, *J. Org. Chem.* **41**, 3261 (1976).
257. J. K. Crandall and D. J. Keyton, *Tetrahedron Lett.* p. 1653 (1969); S. A. Dodson and R. D. Stipanovic, *J. Chem. Soc., Perkin Trans. 1* p. 410 (1975).
258. T. Ohnuki, M. Yoshida, and O. Simamura, *Chem. Lett.* p. 797 (1972); T. Ohnuki, M. Yoshida, O. Simamura, and M. Fukuyama, *ibid.* p. 999.
259. H. R. Ward, *J. Am. Chem. Soc.* **89**, 5517 (1967).
260. W. M. Moore and D. G. Peters, *Tetrahedron Lett.* p. 453 (1972); W. M. Moore, A. Salajegheh, and D. G. Peters, *J. Am. Chem. Soc.* **97**, 4954 (1975); B. C. Willet, W. M. Moore, A. Salajegheh, and D. G. Peters, *J. Am. Chem. Soc.* **101**, 1162 (1979).
261. M. Julia and C. James, *C. R. Hebd. Seances Acad. Sci.* **255**, 959 (1962).
261a. G. Büchi and H. Wüest, *J. Org. Chem.* **44**, 546 (1979).
262. M. E. Kuehne and W. H. Parsons, *J. Org. Chem.* **42**, 3408 (1977).
263. J. K. Kochi, in "Free Radicals" (J. K. Kochi, ed.), Vol. 2, Chapter 23. Wiley, New York, 1973.
264. For a survey of the cyclization reactions of alkenoxy radicals, see J. -M. Surzur and M. P. Michele, *Bull. Soc. Chim. Fr.* p. 1861 (1973).
265. J. -M. Surzur, M. -P. Bertrand, and R. Nouguier, *Tetrahedron Lett.* p. 4197 (1969).
266. P. Tordo, M. -P. Bertrand, and J. -M. Surzur, *Tetrahedron Lett.* p. 3399 (1970).
267. J. -M. Surzur, P. Cozzone, and M. -P. Bertrand, *C. R. Hebd. Seances Acad. Sci., Ser. C* **267**, 908 (1968).
268. B. C. Gilbert, R. G. G. Holmes, H. A. H. Laue, and R. O. C. Norman, *J. Chem. Soc., Perkin Trans. 2* p. 1047 (1976).
269. M. -P. Bertrand and J. -M. Surzur, *Tetrahedron Lett.* p. 3451 (1976).
270. J. -M. Surzur, C. Dupuy, M. P. Bertrand, and R. Nouguier, *J. Org. Chem.* **37**, 2782 (1972).
271. R. Nouguier and J. -M. Surzur, *Bull. Soc. Chim. Fr.* p. 2399 (1973).

272. For recent reviews of prostaglandin biosynthesis, see B. Samuelsson, *Fed. Proc., Fed. Am. Soc. Exp. Biol.* **31**, 1442 (1972); J. W. Hinman, *Annu. Rev. Biochem.* **41**, 161 (1972); M. P. L. Caton, *Prog. Med. Chem.* **8**, 317 (1971); T. O. Oesterling, W. Morozowich, and T. J. Roseman, *J. Pharm. Sci.* **61**, 1861 (1972); K. C. Nicolaou, G. P. Gasic, and W. E. Barnette, *Angew. Chem., Int. Ed. Eng.* **17**, 293 (1978).
273. M. Hamberg and B. Samuelsson, *Proc. Natl. Acad. Sci. U.S.A.* **71**, 3400 (1974); M. Hamberg, P. Hedqvist, K. Strandberg, J. Svensson, and B. Samuelsson, *Life Sci.* **16**, 451 (1975).
274. N. A. Porter and M. O. Funk, *J. Org. Chem.* **40**, 3614 (1975); W. A. Pryor and J. P. Stanley, *ibid.* p. 3615.
275. R. L. Huang, S. H. Goh, and S. H. Ong, in "The Chemistry of Free Radicals," p. 134. Arnold, London, 1974.
276. N. A. Porter, M. O. Funk, D. Gilmore, R. Isaac, and J. Nixon, *J. Am. Chem. Soc.* **98**, 6000 (1976).
277. N. A. Porter, J. R. Nixon, and D. W. Gilmore, in "Organic Free Radicals" (W. A. Pryor, ed.), ACS Symp. Ser. 69, Chapter 6. Am. Chem. Soc., Washington, D.C., 1978. See also, A. L. J. Beckwith and R. D. Wagner *J. Am. Chem. Soc.* **101**, 7099 (1979).
278. J. -M. Surzur, L. Stella, and P. Tordo, *Bull. Soc. Chim. Fr.* p. 1425 (1975).
279. C. J. Michejda, D. H. Campbell, D. H. Sieh, and S. R. Koepke, in "Organic Free Radicals" (W. A. Pryor, ed.), ACS Symp. Ser. 69, Chapter 18. Am. Chem. Soc., Washington, D.C., 1978.
280. L. Stella, B. Raynier, and J. -M. Surzur, *Tetrahedron Lett.* p. 2721 (1977).
281. J. -M. Surzur, L. Stella, and P. Tordo, *Bull. Soc. Chim. Fr.* p. 1429 (1975).
282. J. -M. Surzur, L. Stella, and P. Tordo, *Tetrahedron Lett.* p. 3107 (1970).
283. J. -M. Surzur and P. Tordo, *C. R. Hebd. Seances Acad. Sci., Ser. C* **263**, 446 (1966).
284. J. -M. Surzur, L. Stella, and R. Nouguier, *Tetrahedron Lett.* p. 903 (1971).
285. J. -M. Surzur, L. Stella, and P. Tordo, *Bull. Soc. Chim. Fr.* p. 115 (1970).
286. Y. L. Chow, *Acc. Chem. Res.* **6**, 354 (1973).
286a. Y. L. Chow, W. C. Danen, S. F. Nelsen, and D. H. Rosenblatt, *Chem. Rev.* **78**, 243 (1978).
287. R. A. Perry, S. C. Chen, B. C. Menon, K. Hanaya, and Y. L. Chow, *Can. J. Chem.* **54**, 2385 (1976).
288. J. -M. Surzur and L. Stella, *Tetrahedron Lett.* p. 2191 (1974).
289. J. W. Bastable, J. D. Hobson, and W. D. Riddell, *J. Chem. Soc., Perkin Trans. 1* p. 2205 (1972).
290. R. Tadayoni, J. Lacrampe, A. Heumann, R. Furstoss, and B. Waegell, *Tetrahedron Lett.* p. 735 (1975); R. Furstoss, R. Tadayoni, and B. Waegell, *Nouv. J. Chim.* **1**, 167 (1977); R. Furstoss, R. Tadayoni, and B. Waegell, *J. Org. Chem.* **42**, 2844 (1977).
291. Y. L. Chow, R. A. Perry, B. C. Menon, and S. C. Chen, *Tetrahedron Lett.* p. 1545 (1971); Y. L. Chow, R. A. Perry, and B. C. Menon, *ibid.* p. 1549.
291a. G. Grethe, H. L. Lee, T. Mitt, and M. R. Uskoković, *J. Am. Chem. Soc.* **100**, 581 (1978).
292. Y. Maeda and K. U. Ingold, *J. Am. Chem. Soc.* **102**, 328 (1980).
293. M. E. Keuhne and D. A. Horne, *J. Org. Chem.* **40**, 1287 (1975).
294. Y. L. Chow and R. A. Perry, *Tetrahedron Lett.* p. 531 (1972); see also E. Flesia, A. Croatto, P. Tordo, and J. -M. Surzur, *ibid.* p. 535.
294a. P. Mackiewicz and R. Furstoss, *Tetrahedron* **34**, 3241 (1978); P. Mackiewicz, R. Furstoss, and B. Waegell, *J. Org. Chem.* **43**, 3746 (1978); J. Lessard, R. Cote, P. Mackiewicz, R. Furstoss, and B. Waegell, *ibid.* p. 3750.
295. W. B. Motherwell and J. S. Roberts, *Chem. Commun.* p. 328 (1972).

296. H. O. House, D. T. Manning, D. G. Melillo, L. F. Lee, O. R. Haynes, and B. E. Wilkes, *J. Org. Chem.* **41,** 855 (1976).
297. For leading references, see J. L. Kice, in "Free Radicals" (J. K. Kochi, ed.), Vol. 2, Chapter 24. Wiley, New York, 1973.
298. M. -P. Crozet, J. -M. Surzur, and C. Dupuy, *Tetrahedron Lett.* p. 2031 (1971).
299. J. -M. Surzur, M. -P. Crozet, and C. Dupuy, *C. R. Hebd. Seances Acad. Sci., Ser. C.* **264,** 610 (1967).
300. J. -M. Surzur, M. -P. Crozet, and C. Dupuy, *Tetrahedron Lett.* p. 2025 (1971).
301. J. -M. Surzur and M. -P. Crozet, *C. R. Hebd. Seances Acad. Sci., Ser. C* **268,** 2109 (1969)
302. J. -M. Surzur, R. Nouguier, M. -P. Crozet, and C. Dupuy, *Tetrahedron Lett.* p. 2035 (1971).
303. R. Nouguier and J. -M. Surzur, *Tetrahedron* **32,** 2001 (1976).
304. Y. Makisumi and A. Murabayashi, *Tetrahedron Lett.* p. 2453 (1969).
305. D. N. Jones, D. A. Lewton, J. D. Msonthi, and R. J. K. Taylor, *J. Chem. Soc., Perkin Trans. 1* p. 2637 (1974).
306. J. -M. Surzur, C. Dupuy, M. -P. Crozet, and N. Aimar, *C. R. Hebd. Seances Acad. Sci., Ser. C* **269,** 849 (1969).
307. H. Sakurai, T. Hirose, and A. Hosomi, *Abstr., 26th Annu. Meet. Chem. Soc. Jpn.* Vol. III, p. 1001 (1972).
308. A. G. Davies, M. J. Parrott, and B. P. Roberts, *Chem. Commun.* p. 27 (1974); *J. Chem. Soc., Perkin Trans. 2* p. 1066 (1976).
309. M. F. Semmelhack, B. P. Chong, R. D. Stauffer, T. D. Rogerson, A. Chong, and L. D. Jones, *J. Am. Chem. Soc.* **97,** 2507 (1975).
310. M. F. Semmelhack and T. M. Barger, *J. Org. Chem.* **42,** 1481 (1977).
311. I. Tse and V. Snieckus, *Chem. Commun.* p. 505 (1976).
312. H. Iida, S. Aoyogi, and C. Kabayashi, *J. Chem. Soc., Perkin Trans. 1* p. 120 (1977); H. Iida, Y. Yuasa, and C. Kibayashi, *Tetrahedron Lett.* p. 3817 (1978); *J. Org. Chem.* **44,** 1236 (1979).
313. For a review of intramolecular homolytic arylation, see R. A. Abramovitch, *Adv. Free-Radical Chem.* **2,** 87 (1967).
314. A. J. Floyd, S. F. Dyke, and S. E. Ward, *Chem. Rev.* **76,** 509 (1976).
315. A. Fozard and C. K. Bradsher, *J. Org. Chem.* **32,** 2966 (1967).
316. L. Benati, P. C. Montevecchi, A. Tundo, and G. Zanardi, *J. Chem. Soc., Perkin Trans. 1* p. 1276 (1974); L. Benati and P. C. Montevecchi, *J. Org. Chem.* **42,** 2025 (1977).
317. A. H. Lewin and R. J. Michl, *J. Org. Chem.* **38,** 1126 (1973); **39,** 2261 (1974); A. H. Lewin, N. C. Petersen, and R. J. Michl, *ibid.* p. 2747.
318. M. Tiecco, *Chem. Commun.* p. 555 (1965).
319. W. A. Henderson, Jr., J. R. Lopresti, and A. Zweig, *J. Am. Chem. Soc.* **91,** 6049 (1969); W. A. Henderson, Jr. and A. Zweig, *Tetrahedron Lett.* p. 625 (1969).
320. T. Kametani, S. Shibuya, R. Charubala, M. S. Premila, and B. R. Pai, *Heterocycles* **3,** 439 (1975).
321. S. Beveridge and J. L. Huppatz, *Aust. J. Chem.* **23,** 781 (1970); **25,** 1341 (1972).
322. A. P. Gray, V. M. Dipinto, and I. J. Solomon, *J. Org. Chem.* **41,** 2428 (1976).
323. R. D. Youssefyeh and M. Weisz, *Tetrahedron Lett.* p. 4317 (1973).
324. F. F. Gadallah, A. A. Cantu, and R. M. Elofson, *J. Org. Chem.* **38,** 2386 (1973).
325. P. W. Jeffs, J. F. Hansen, and G. A. Brine, *J. Org. Chem.* **40,** 2883 (1975).
326. J. L. Huppatz, *Aust. J. Chem.* **26,** 1307 (1973).
327. O. Yonemitsu, T. Tokyama, M. Chaykovsky, and B. Witkop, *J. Am. Chem. Soc.* **90,** 776 (1968).

4. FREE-RADICAL REARRANGEMENTS

328. K. Ito and H. Tanaka, *Chem. Pharm. Bull.* **22,** 2198 (1974).
329. E. L. Eliel, M. Eberhardt, O. Simamura, and S. Meyerson, *Tetrahedron Lett.* p. 749 (1962); J. Saltiel and H. C. Curtis, *J. Am. Chem. Soc.* **93,** 2056 (1971); R. A. Jackson, *Chem. Commun.* p. 573 (1974); S. Vidal, J. Court, and J. M. Bonnier, *Tetrahedron Lett.* p. 2023 (1976).
330. D. J. Atkinson, M. J. Perkins, and P. Ward, *J. Chem. Soc. C* p. 3240 (1971).
331. M. Kobayashi, H. Minato, and N. Kobari, *Bull. Chem. Soc. Jpn.* **42,** 2738 (1969); R. Henriquez, A. R. Morgan, P. Mulholland, D. C. Nonhebel, and G. G. Smith, *Chem. Commun.* p. 987 (1974); R. Henriquez and D. C. Nonhebel, *Tetrahedron Lett.* pp. 3855 and 3857 (1975).
332. D. H. Hey, G. H. Jones, and M. J. Perkins, *J. Chem. Soc. C* p. 116 (1971); *J. Chem. Soc., Perkin Trans. 1* p. 1150 (1972).
333. D. H. Hey, G. H. Jones, and M. J. Perkins, *J. Chem. Soc., Perkin Trans. 1* p. 1155 (1972).
334. D. H. Hey, G. H. Jones, and M. J. Perkins, *J. Chem. Soc., Perkin Trans. 1* p. 118 (1972).
335. D. H. Hey, G. H. Jones, and M. J. Perkins, *J. Chem. Soc., Perkin Trans. 1* p. 105 (1972).
336. D. H. Hey, G. H. Jones, and M. J. Perkins, *J. Chem. Soc., Perkin Trans. 1* p. 113 (1972).
337. For a recent review of photochemical methods of ring closure in the field of alkaloid synthesis, see T. Kametani and K. Fukumoto, *Acc. Chem. Res.* **5,** 212 (1972).
338. S. M. Kupchan and H. C. Wormser, *J. Org. Chem.* **30,** 3792 (1965).
339. G. De Luca, G. Martelli, P. Spagnolo, and M. Tiecco, *J. Chem. Soc. C* p. 2504 (1970).
340. M. S. Premila and B. R. Pai, *Indian J. Chem.* **13,** 13 (1975); M. S. Premila, B. R. Pai, and P. C. Parthasarathy, *ibid.* p. 945.
341. T. Kametani, M. Koizumi, and K. Fukumoto, *J. Chem. Soc. C* p. 1792 (1971).
342. T. Kametani, T. Sugahara, and K. Fukumoto, *Tetrahedron* **27,** 5367 (1971).
343. R. M. Elsofson and F. F. Gadallah, *J. Org. Chem.* **36,** 1769 (1971).
344. B. Chauncey and E. Gellert, *Aust. J. Chem.* **22,** 993 (1969); **23,** 2503 (1970).
345. S. V. Kessar, S. Narula, S. S. Gandhi, and U. K. Nadir, *Tetrahedron Lett.* p. 2905 (1974).
346. P. R. Singh and R. Kumar, *Indian J. Chem.* **11,** 692 (1973).
347. D. Berney and K. Schuh, *Helv. Chim. Acta* **59,** 2059 (1976).
348. L. Benati, P. C. Montevecchi, A. Tundo, and G. Zanardi, *J. Org. Chem.* **41,** 1331 (1976).
349. For a review, see T. Kametani and K. Fukumoto, *J. Heterocycl. Chem.* **8,** 341 (1971).
350. H. Pines and J. T. Arrigo, *J. Am. Chem. Soc.* **79,** 4958 (1957).
351. A. I. Feinstein and E. K. Fields, *J. Org. Chem.* **37,** 118 (1972).
352. D. B. Denney and P. P. Klemchuk, *J. Am. Chem. Soc.* **80,** 3289 (1958).
353. W. S. Trahanovsky and C. C. Ong, *J. Am. Chem. Soc.* **92,** 7174 (1970).
354. C. Descoins, M. Julia, and H. van Sang, *Bull. Soc. Chim. Fr.* p. 2037 (1972).
355. K. N. V. Duong, A. Gaudemer, M. D. Johnson, R. Quillivic, and J. Zylber, *Tetrahedron Lett.* p. 2997 (1975); N. Mariaggi, J. Cadet, and R. Teoule, *Tetrahedron* **32,** 2385 (1976).
356. J. I. Okogun and K. S. Okwute, *Chem. Commun.* p. 8 (1975).
357. P. S. Dewar, A. R. Forrester, and R. H. Thomson, *J. Chem. Soc. C* p. 3950 (1971).
358. J. J. Köhler and W. N. Speckamp, *Tetrahedron Lett.* pp. 631 and 635 (1977).
359. B. D. Baigrie, J. I. G. Cadogan, J. Cook, and J. T. Sharp, *Chem. Commun.* p. 1318 (1972).

360. E. I. Heiba and R. M. Dessau, *J. Am. Chem. Soc.* **94**, 2888 (1972).
361. J. Y. Lallemand, M. Julia, and D. Mansuy, *Tetrahedron Lett.* p. 4461 (1973); see also M. Julia and D. Mansuy, *Bull. Soc. Chim. Fr.* p. 2684 (1972); D. Mansuy and M. Julia, *ibid.* p. 2689.
361a. A. R. Forrester, M. Gill, R. J. Napier, and R. H. Thomson, *J. Chem. Soc., Perkin Trans. 1* p. 632 (1979).
361b. H. Iida, T. Takarai, and C. Kibayashi, *J. Org. Chem.* **43**, 975 (1978).
362. H. Sakurai, A. Hosomi, and M. Kumada, *Tetrahedron Lett.* p. 1757 (1969).
363. M. P. Doyle, L. H. Zuidema, and T. R. Bade, *J. Org. Chem.* **40**, 1454 (1975).
364. A. Goosen and C. W. McCleland, *Chem. Commun.* p. 655 (1975).
365. P. M. Brown, J. Russell, R. H. Thomson, and A. G. Wylie, *J. Chem. Soc. C* p. 842 (1968).
366. D. I. Davies and C. Waring, *J. Chem. Soc. C* p. 1639 (1967).
367. A. R. Forrester, M. Gill, J. S. Sadd, and R. H. Thomson, *Chem. Commun.* p. 291 (1975); A. R. Forrester, M. Gill, and R. H. Thomson, *ibid.* p. 677 (1976); A. R. Forrester, M. Gill, J. S. Sadd, and R. H. Thomson, *J. Chem. Soc., Perkin Trans. 1* p. 612 (1979).
368. A. R. Forrester, A. S. Ingram, and R. H. Thomson, *J. Chem. Soc., Perkin Trans. 1* p. 2847 (1972).
369. S. A. Glover, A. Goosen, and H. A. H. Laue, *J. Chem. Soc., Perkin Trans. 1* p. 1647 (1973); S. A. Glover and A. Goosen, *ibid.* p. 2353 (1974); p. 1348 (1977).
370. A. R. Forrester, M. Gill, E. M. Johansson, C. J. Meyer, and R. H. Thomson, *Tetrahedron Lett.* p. 3601 (1977).
371. F. Minisci and R. Galli, *Tetrahedron Lett.* p. 2531 (1966).
371a. F. Minisci, *Synthesis* p. 1 (1973).
372. W. Ando, T. Oikawa, K. Kishi, T. Saiki, and T. Migita, *Chem. Commun.* p. 704 (1975).
373. P. M. Brown, P. S. Dewar, A. R. Forrester, and R. H. Thomson, *J. Chem. Soc., Perkin Trans. 1* p. 2842 (1972).
374. F. Flies, R. Lalande, and B. Maillard, *Tetrahedron Lett.* p. 439 (1976).
375. L. W. Menapace and H. G. Kuivila, *J. Am. Chem. Soc.* **86**, 3047 (1964).
376. T. A. Pudova, F. K. Velichko, L. V. Vinogradova, and R. Kh. Freidlina, *Izv. Akad. Nauk SSSR, Ser. Khim.* p. 116 (1975).
377. K. Praefcke, *Tetrahedron Lett.* p. 973 (1973).
378. E. K. Starostin, S. I. Moryasheva, and G. I. Nikishin, *Izv. Acad. Nauk SSSR, Ser. Khim.* p. 2048 (1974).
379. Yu. N. Ogibin, E. I. Troyanskii, and G. I. Nikishin, *Izv. Akad. Nauk SSSR, Ser. Khim.* pp. 2239 and 2522 (1974); Yu. N. Ogibin, E. I. Troyanskii, G. I. Nikishin, O. S. Chizhov, and V. I. Kadentsev, *ibid.* p. 967 (1975); Yu. N. Ogibin, E. I. Troyanskii, G. I. Nikishin, V. I. Kadentsev, E. D. Lubuzh, and O. S. Chizhov, *ibid.* p. 2518 (1976).
379a. C. M. Rynard, C. Thankachan, and T. T. Tidwell, *J. Am. Chem. Soc.* **101**, 1196 (1979).
380. Yu. N. Ogibin, E. I. Troyanski, and G. I. Nikishin, *Izv. Akad. Nauk SSSR, Ser. Khim.* p. 1461 (1975); p. 843 (1977).
380a. D. Griller, P. Schmid, and K. U. Ingold, *Can. J. Chem.* **57**, 831 (1979).
381. B. P. Roberts and J. N. Winter, *J. Chem. Soc., Perkin Trans. 2* p. 1353 (1979).
382. D. G. Hawthorne and D. H. Solomon, *J. Macromol. Sci., Chem.* **9**, 149 (1975).
382a. D. D. Tanner and P. M. Rahimi, *J. Org. Chem.* **44**, 1674 (1979); A. R. Forrester, M. Gill, and R. H. Thomson, *J. Chem. Soc., Perkin Trans. 1* p. 621 (1979).
383. G. Levesque, G. Tabak, F. Outurquin, and J. -C. Gressier, *Bull. Soc. Chim. Fr.* p. 1156 (1976).

4. FREE-RADICAL REARRANGEMENTS 297

384. M. L. Heymann and J. P. Snyder, *J. Am. Chem. Soc.* **97**, 4416 (1975).
385. A. L. J. Beckwith and M. D. Lawton, *J. Chem. Soc., Perkin Trans 2* p. 2134 (1973).
386. L. Benati, G. Placucci, P. Spagnolo, A. Tundo, and G. Zanardi, *J. Chem. Soc., Perkin Trans. 1* p. 1684 (1977).
387. W. Nakanishi, S. Koike, M. Inoue, Y. Ikeda, H. Iwamura, Y. Imahashi, H. Kihara, and M. Iwai, *Tetrahedron Lett.* p. 81 (1977).
388. C. E. Griffin, W. G. Bentrude, and G. M. Johnson, *Tetrahedron Lett.* p. 969 (1969).
389. G. Haran and D. W. A. Sharp, *J. Chem. Soc., Perkin Trans. 1* p. 34 (1972).
389a. J. E. Baldwin and T. S. Wan, *Chem. Commun.* p. 249 (1979).
390. P. K. Freeman, T. D. Ziebarth, and R. S. Raghavan, *J. Am. Chem. Soc.* **97**, 1875 (1975).
391. S. G. Bayliss, R. L. Failes, J. S. Shapiro, and E. S. Swinbourne, *J. Chem. Soc., Faraday Trans. 1* **74**, 776 (1978).
391a. N. A. Porter and J. R. Nixon, *J. Am. Chem. Soc.* **100**, 7116 (1978). See also references 61 and 61a.
392. D. I. Schuster and J. D. Roberts, *J. Org. Chem.* **27**, 51 (1962).
393. S. E. Stein and B. S. Rabinovitch, *J. Phys. Chem.* **79**, 191 (1975).
394. J. T. Groves and K. W. Ma, *J. Am. Chem. Soc.* **96**, 6527 (1974).
394a. K. Herwig, P. Lorenz, and C. Rüchardt, *Chem. Ber.* **108**, 1421 (1975).
394b. H. M. Walborsky and P. C. Collins, *J. Org. Chem.* **41**, 940 (1976).
394c. K. Kobayashi and J. B. Lambert, *J. Org. Chem.* **42**, 1254 (1977).
394d. M. R. Detty and L. A. Paquette, *J. Org. Chem.* **43**, 1118 (1978).
395. J. K. Kochi and H. E. Mains, *J. Org. Chem.* **30**, 1862 (1965); T. N. Shatkina and O. A. Reutov, *Zh. Org. Khim.* **10**, 873 (1974).
395a. M. Lj. Mihailović, J. Bošnjak, and Ž. Čeković, *Helv. Chim. Acta* **59**, 475 (1976).
396. D. C. Nonhebel and J. C. Walton, *Org. React. Mech.* p. 69 (1974).
397. G. Greig and J. C. J. Thynne, *Trans. Faraday Soc.* **62**, 3338 (1966); **63**, 1369 (1967); J. A. Kerr, A. Smith, and A. F. Trotman-Dickenson, *J. Chem. Soc. A* p. 1400 (1969).
398. R. Walsh, *Int. J. Chem. Kinet.* **2**, 71 (1970); A. S. Gordon, *ibid.* p. 75.
399. W. P. L. Carter and D. C. Tardy, *J. Phys. Chem.* **78**, 1573 (1974).
400. H. E. Gunning and R. L. Stock, *Can J. Chem.* **42**, 357 (1964); A. S. Gordon, *ibid.* **43**, 570 (1965).
401. R. B. Woodward and R. Hoffmann, *J. Am. Chem. Soc.* **87**, 395 (1965); H. C. Longuet-Higgins and E. W. Abrahamson, *ibid.* p. 2045.
402. E. Haselbach, *Helv. Chim. Acta* **54**, 2257 (1971); M. J. S. Dewar and S. Kirschner, *J. Am. Chem. Soc.* **93**, 4290 and 4291 (1971); M. J. S. Dewar, *Fortschr. Chem. Forsch.* **23**, 1 (1971); G. Szeimies and G. Boche, *Angew. Chem., Int. Ed. Engl.* **10**, 912 (1971); D. T. Clark and D. B. Adams, *Nature (London), Phys. Sci.* **233**, 121 (1971); L. Farnell and W. G. Richards, *Chem. Commun.* p. 334 (1973); P. Merlet, S. D. Peyerimhoff, R. J. Buenker, and S. Shih, *J. Am. Chem. Soc.* **96**, 959 (1974); M. J. S. Dewar and S. Kirschner, *ibid.* p. 5244; S. Beran and R. Zahradnik, *Collect. Czech. Chem. Commun.* **41**, 2303 (1976).
403. H. M. Walborsky and J. -C. Chen, *J. Am. Chem. Soc.* **92**, 7573 (1970); J. -C. Chen, *Tetrahedron Lett.* p. 3669 (1971).
404. S. Sustmann, C. Rüchardt, A. Bieberbach, and G. Boche, *Tetrahedron Lett.* p. 4759 (1972); S. Sustmann and C. Rüchardt, *ibid.* p. 4765; *Chem. Ber.* **108**, 3043 (1975).
405. A. Barmetler, C. Rüchardt, R. Sustmann, S. Sustmann, and R. Verhülsdonk, *Tetrahedron Lett.* p. 4389 (1974). See also, R. P. Corbally, M. J. Perkins, and A. P. Elnitski, *J. Chem. Soc., Perkin Trans. 1* p. 793 (1979).
406. R. Sustmann and F. Lübbe, *Chem. Ber.* **109**, 444 (1976).

407. R. Sustmann and R. W. Gellert, *Chem. Ber.* **109**, 345 (1976).
408. P. J. Krusic, J. P. Jesson, and J. K. Kochi, *J. Am. Chem. Soc.* **91**, 4566 (1969).
409. P. Weyerstahl, R. Mathias, and G. Blume, *Tetrahedron Lett.* p. 611 (1973).
409a. F. Gerson, W. Huber, and K. Müllen, *Helv. Chim. Acta* **62**, 2109 (1979).
410. S. Sustmann, R. Sustmann, and C. Rüchardt, *Chem. Ber.* **108**, 1527 (1975).
411. B. G. Dzantiev, A. T. Koritskii, R. G. Kostyanovskii, Yu. M. Rumyantsev, and A. V. Shishkov, *Khim. Vys. Energ.* **9**, 180 (1975).
412. D. Cordischi and R. Di Blasi, *Can. J. Chem.* **47**, 2601 (1969).
413. R. F. Hudson, A. J. Lawson, and K. A. F. Record, *Chem. Commun.* p. 322 (1975).
414. T. J. Wallance and R. J. Gritter, *Tetrahedron* **19**, 657 (1963).
415. E. C. Sabatino and R. J. Gritter, *J. Org. Chem.* **28**, 3437 (1963).
416. W. Reusch, C. K. Johnson, and J. A. Manner, *J. Am. Chem. Soc.* **88**, 2803 (1966); A. Padwa and N. C. Das, *J. Org. Chem.* **34**, 816 (1969).
417. D. R. Paulson, A. S. Murray, D. Bennett, E. Mills, Jr., V. O. Terry, and S. D. Lopez, *J. Org. Chem.* **42**, 1252 (1977).
418. G. Behrens and D. Schulte-Frohlinde, *Angew. Chem., Int. Ed. Engl.* **12**, 932 (1973).
419. A. J. Dobbs, B. C. Gilbert, H. A. H. Laue, and R. O. C. Norman, *J. Chem. Soc., Perkin Trans. 2* p. 1044 (1976).
420. A. J. Dobbs, B. C. Gilbert, and R. O. C. Norman, *J. Chem. Soc. A* p. 124 (1971); *J. Chem. Soc., Perkin Trans. 2* p. 786 (1972).
421. A. L. J. Beckwith and P. K. Tindal, *Aust. J. Chem.* **24**, 2099 (1971).
422. B. C. Gilbert and M. Trenwith, *J. Chem. Soc., Perkin Trans. 2* p. 1083 (1975), and references cited.
423. D. J. Rawlinson and G. Sosnovsky, *Synthesis* p. 1 (1972).
424. M. I. Shuikin and B. L. Lebedew, *Z. Chem.* **6**, 459 (1966); T. N. Abroskina, A. D. Sorokin, R. V. Kudryavtsev, and Yu. A. Cheburkov, *Izv. Akad. Nauk SSSR, Ser. Khim.* p. 1823 (1974).
425. For a review of the free-radical chemistry of cyclic ethers, see R. J. Gritter, *in* "The Chemistry of the Ether Linkage" (S. Patai, ed.), Chapter 9. Wiley, New York, 1967.
426. W. T. Dixon, J. Foxall, G. H. Williams, D. J. Edge, B. C. Gilbert, H. Kazarians-Moghaddam, and R. O. C. Norman, *J. Chem. Soc., Perkin Trans. 2* p. 827 (1977).
427. S. O. Kobayashi and O. Simamura, *Chem. Lett.* p. 699 (1973).
428. C. Berglund and S. -O. Lawesson, *Ark. Kemi* **20**, 225 (1963).
429. I. Rosenthal and D. Elad, *J. Org. Chem.* **33**, 805 (1968).
430. R. Lalande, B. Maillard, and M. Cazaux, *Tetrahedron Lett.* p. 745 (1969); B. Maillard, M. Cazaux, and R. Lalande, *Bull. Soc. Chim. Fr.* p. 467 (1971).
431. B. Maillard, M. Cazaux, and R. Lalande, *C. R. Hebd. Seances Acad. Sci., Ser. C* **279**, 701 (1974).
432. B. Maillard, M. Cazaux, and R. Lalande, *Bull. Soc. Chim. Fr.* p. 183 (1975).
433. For further examples, see B. Maillard, M. Cazaux, and R. Lalande, *Bull. Soc. Chim. Fr.* p. 1368 (1973); D. L. Rakhmankulov, S. S. Zlotskii, V. N. Uzikova, and Ya. M. Paushkin, *Dokl. Akad. Nauk SSSR* **218**, 156 (1974); S. S. Zlotskii, E. Kh. Kravets, V. S. Martem'yanov, and D. L. Rakhmankulov, *Zh. Org. Khim.* **11**, 1982 (1975); D. L. Rakhmankulov, V. N. Uzikova, S. S. Zlotskii, and G. Ya. Estrina, *ibid.* p. 2223; V. V. Zorin, S. S. Zlotskii, V. P. Nayanov, and D. L. Rakhmankulov, *Zh. Prikl. Khim. (Leningrad)* **50**, 1131 (1977).
434. T. Yamagishi, T. Yoshimoto, and K. Minami, *Tetrahedron Lett.* p. 2795 (1971); K. Hayday and R. D. McKelvey, *J. Org. Chem.* **41**, 2222 (1976); C. Bernasconi, L. Cottier, and G. Descotes, *Bull. Soc. Chim. Fr.* p. 101 (1977).
435. J. S. Chapman, J. W. Cooper, and B. P. Roberts, *Chem. Commun.* p. 407 (1976).

4. FREE-RADICAL REARRANGEMENTS

435a. C. A. Whitesitt and D. K. Herron, *Tetrahedron Lett.* p. 1737 (1978).
436. P. H. Kasai and D. McLeod, Jr., *J. Am. Chem. Soc.* **94**, 720 (1972).
437. W. J. Hehre, *J. Am. Chem. Soc.* **95**, 2643 (1973). P. Bischof, *Tetrahedron Lett.* p. 1291 (1979); M. J. S. Dewar and S. Olivella, *J. Am. Chem. Soc.* **101**, 4958 (1979).
438. J. K. Kochi, P. J. Krusic, and D. R. Eaton, *J. Am. Chem. Soc.* **91**, 1877 (1969).
439. J. K. Kochi, P. J. Krusic, and D. R. Eaton, *J. Am. Chem. Soc.* **91**, 1879 (1969).
440. K. S. Chen, D. J. Edge, and J. K. Kochi, *J. Am. Chem. Soc.* **95**, 7036 (1973).
441. P. M. Blum, A. G. Davies, M. Pereyre, and M. Ratier, *Chem. Commun.* p. 814 (1976); M. Castaing, M. Pereyre, M. Ratier, P. M. Blum, and A. G. Davies, *J. Chem. Soc., Perkin Trans. 2* p. 287 (1979).
442. A. G. Davies, J. Y. Godet, B. Muggleton, and M. Pereyre, *Chem. Commun.* p. 813 (1976); A. G. Davies, B. Muggleton, J. Y. Godet, M. Pereyre, and J. C. Pommier, *J. Chem. Soc., Perkin Trans. 2* p. 1719 (1976).
443. M. Castaing, M. Pereyre, M. Ratier, P. M. Blum, and A. G. Davies, *J. Chem. Soc., Perkin Trans. 2* p. 589 (1979).
444. A. G. Davies and B. Muggleton, *J. Chem. Soc., Perkin Trans. 2* p. 502 (1976).
445. H. Itzel and H. Fischer, *Tetrahedron Lett.* p. 563 (1975).
446. W. C. Danen and C. T. West, *J. Am. Chem. Soc.* **96**, 2447 (1974).
447. W. C. Danen, *J. Am. Chem. Soc.* **94**, 4835 (1972).
448. J. C. Martin and J. W. Timberlake, *J. Am. Chem. Soc.* **92**, 978 (1970), and references cited.
449. T. Shono and I. Nishiguchi, *Tetrahedron* **30**, 2173 (1974).
450. E. C. Friedrich, *J. Org. Chem.* **34**, 1851 (1969).
451. E. S. Huyser and J. D. Taliaferro, *J. Org. Chem.* **28**, 3442 (1963).
452. C. Walling and P. S. Fredricks, *J. Am. Chem. Soc.* **84**, 3326 (1962).
453. D. E. Applequist and J. A. Landgrebe, *J. Am. Chem. Soc.* **86**, 1543 (1964).
454. M. Suzuki, S. -I. Murahashi, A. Sonoda, and I. Moritani, *Chem. Lett.* p. 267 (1974).
455. D. F. McMillen, D. M. Golden, and S. W. Benson, *Int. J. Chem. Kinet.* **3**, 359 (1971).
456. D. C. Neckers and A. P. Schaap, *J. Org. Chem.* **32**, 22 (1967).
452. W. J. Hehre, *J. Am. Chem. Soc.* **94**, 6592 (1972).
458. R. Kaptcin, *J. Am. Chem. Soc.* **94**, 6262 (1972).
459. A. L. J. Beckwith and G. Phillipou, *Aust. J. Chem.* **29**, 123 (1976).
460. S. J. Cristol and R. V. Barbour, *J. Am. Chem. Soc.* **90**, 2832 (1968).
461. V. M. A. Chambers, W. R. Jackson, and G. W. Young, *Chem. Commun.* p. 1275 (1970); *J. Chem. Soc. C* p. 2075 (1971).
462. P. K. Freeman, F. A. Raymond, J. C. Sutton, and W. R. Kindley, *J. Org. Chem.* **33**, 1448 (1968); R. S. Boikess, M. Mackay, and D. Blithe, *Tetrahedron Lett.* p. 401 (1971).
463. P. K. Freeman, M. F. Grostic, and F. A. Raymond, *J. Org. Chem.* **36**, 905 (1971).
464. E. Müller, *Tetrahedron Lett.* p. 1835 (1974).
465. R. Sustmann and F. Lübbe, *Tetrahedron Lett.* p. 2831 (1974).
466. W. G. Dauben, L. Schutte, and E. J. Deviny, *J. Org. Chem.* **37**, 2047 (1972); G. W. Shaffer, *ibid.* **38**, 2842 (1973).
467. W. G. Dauben, L. Schutte, R. E. Wolf, and E. J. Deviny, *J. Org. Chem.* **34**, 2512 (1969).
468. R. W. Thies and D. D. McRitchie, *J. Org. Chem.* **38**, 112 (1973).
469. Y. Sugiyama, T. Kawamura, and T. Yonezawa, *J. Am. Chem. Soc.* **100**, 6525 (1978). See also, *idem. Chem. Lett.* p. 639 (1978).
470. J. -Y. Godet and M. Pereyre, *C. R. Hebd. Seances Acad. Sci., Ser. C* **273**, 1183 (1971); **277**, 211 (1973).
471. J. -Y. Godet and M. Pereyre, *J. Organomet. Chem.* **40**, C23 (1972).

472. J. -Y. Godet, M. Pereyre, J. -C. Pommier, and D. Chevolleau, *J. Organomet. Chem.* **55,** C15 (1973).
473. J. -Y. Godet and M. Pereyre, *Bull. Soc. Chim. Fr.* p. 1105 (1976).
474. A. J. Bellamy, E. A. Campbell, and I. R. Hall, *J. Chem. Soc., Perkin Trans. 2* p. 1347 (1974).
475. W. G. Dauben and R. E. Wolf, *J. Org. Chem.* **35,** 2361 (1970).
476. I. S. Lishanskii, N. D. Vinogradova, A. G. Zak, A. B. Zvyagina, A. M. Guliev, O. S. Fomina, and A. S. Khachaturov, *Zh. Org. Khim.* **10,** 493 (1974).
477. I. S. Lishanskii, N. D. Vinogradova, A. M. Guliev, A. G. Zak, A. B. Zvyagina, O. S. Fomina, and A. S. Khachaturov, *Dokl. Akad. Nauk SSSR* **179,** 882 (1968).
478. D. F. Shellhamer, D. B. McKee, and C. T. Leach, *J. Org. Chem.* **41,** 1972 (1976).
479. M. Ratier and M. Pereyre, *Tetrahedron Lett.* p. 2273 (1976).
480. T. A. B. M. Bolsman and T. J. de Boer, *Tetrahedron,* **29,** 3579 (1973); **31,** 1019 (1975).
481. J. C. Martin, J. E. Schultz, and J. W. Timberlake, *Tetrahedron Lett.* p. 4629 (1967).
482. R. Sustmann and F. Lübbe, *Abstr., Colloq. Int. CNRS, Radicaux Libres Org., 1977* p. 158 (1977); *Chem. Ber.* **112,** 42 (1979).
482a. F. Lübbe and R. Sustmann, *Chem. Ber.* **112,** 57 (1979).
483. H. -P. Löffler, *Chem. Ber.* **104,** 1981 (1971).
484. G. A. Russell, K. Schmidt, C. Tanger, E. Goettert, M. Yamashita, Y. Kosugi, J. Siddens, and G. Senatore, *in* "Organic Free Radicals" (W. A. Pryor, ed.), ACS Symp. Ser. 69, Chapter 23, and references cited. Am. Chem. Soc., Washington, D.C., 1978.
485. G. A. Russell, J. T. McDonnell, P. R. Whittle, R. S. Givens, and R. G. Keske, *J. Am. Chem. Soc.* **93,** 1452 (1971).
486. V. A. Azovskaya and E. N. Prilezhaeva, *Russ. Chem. Rev. (Engl. Transl.)* **41,** 516 (1972); D. I. Davies, *in* "Essays on Free Radical Chemistry," Chem. Soc., Spec. Publ. No. 24, p. 201. Chem. Soc., London, 1970.
487. T. A. Halgren, J. L. Firkins, T. A. Fujimoto, H. H. Suzukawa, and J. D. Roberts, *Proc. Natl. Acad. Sci. U.S.A.* **68,** 3216 (1971); R. G. Pews and T. E. Evans, *Chem. Commun.* p. 1397 (1971); T. G. Burrowes and W. R. Jackson, *Aust. J. Chem.* **28,** 639 (1975); R. K. Bansal, A. W. McCulloch, P. W. Rasmussen, and A. G. McInnes, *Can. J. Chem.* **53,** 138 (1975); A. W. McCulloch, A. G. McInnes, D. G. Smith, and J. A. Walter, *ibid.* **54,** 2013 (1976); B. Giese and K. Jay, *Chem. Ber.* **110,** 1364 (1977); R. Sustmann and R. W. Gellert, *ibid.* **111,** 42 (1978).
488. E. Müller, *Chem. Ber.* **109,** 3804 (1976).
488a. J. Elzinga and H. Hogeveen, *J. Org. Chem.* **44,** 2381 (1979).
489. D. F. Shellhamer and M. L. Oakes, *J. Org. Chem.* **43,** 1316 (1978).
490. J. L. Derocque and F. B. Sundermann, *J. Org. Chem.* **39,** 1411 (1974).
491. S. K. Pradhan and V. M. Girijavallabhan, *Tetrahedron Lett.* p. 3103 (1968); *Chem. Commun.* p. 591 (1975).
492. C. L. Jenkins and J. K. Kochi, *J. Am. Chem. Soc.* **94,** 856 (1972).
493. A. G. Davies and M. -W. Tse, *J. Organomet. Chem.* **155,** 25 (1978).
494. H. O. House and K. A. J. Snoble, *J. Org. Chem.* **41,** 3076 (1976); H. O. House, W. C. McDaniel, R. F. Sieloff, and D. Vanderveer, *ibid.* **43,** 4316 (1978).
495. P. J. Krusic, P. J. Fagan, and J. S. Filippo, Jr., *J. Am. Chem. Soc.* **99,** 250 (1977); J. S. Filippo, Jr., J. Silbermann, and P. J. Fagan, *ibid.* **100,** 4834 (1978).
496. S. Torii, T. Okamoto, G. Tanida, H. Hino, and Y. Kitsuya, *J. Org. Chem.* **41,** 166 (1976).
497. See, for example, J. J. McCullough and P. W. W. Rasmussen, *Chem. Commun.* p. 387 (1969); H. E. Zimmerman and T. W. Fletchner, *J. Am. Chem. Soc.* **92,** 6931 (1970); J. K. Crandall, J. P. Arrington, and C. F. Mayer, *J. Org. Chem.* **36,** 1428 (1971); S. Moon

and H. Bohm, *ibid.* p. 1434; N. Shimizu, M. Ishikawa, K. Ishikura, and S. Nishida, *J. Am. Chem. Soc.* **96**, 6456 (1974); A. B. Smith, L. Brodsky, S. Wolff, and W. C. Agosta, *Chem. Commun.* p. 509 (1975); I. M. Takakis and W. C. Agosta, *J. Org. Chem.* **44**, 1294 (1979); *J. Am. Chem. Soc.* **101**, 2383 (1979).
498. P. J. Wagner, *Abstr., Colloq. Int. CNRS, Radicaux Libres Org.*, *1977* p. 174 (1977).
499. R. J. Kinney, W. D. Jones, and R. G. Bergman, *J. Am. Chem. Soc.* **100**, 7902 (1978).
500. J. Roček and A. E. Radkowsky, *J. Am. Chem. Soc.* **90**, 2986 (1968).
500a. M. Rahman and J. Roček, *J. Am. Chem. Soc.* **93**, 5455 (1971); J. Roček, private communication.
501. L. B. Young and W. S. Trahanovsky, *J. Org. Chem.* **32**, 2349 (1967).
502. J. S. Littler, *J. Chem. Soc.* p. 4135 (1959).
503. P. M. Nave and W. S. Trahanovsky, *J. Am. Chem. Soc.* **92**, 1120 (1970).
504. P. M. Nave and W. S. Trahanovsky, *J. Am. Chem. Soc.* **90**, 4755 (1968).
505. S. W. Staley, *Sel. Org. Transform.* **2**, 309 (1972).
506. R. Fraisse-Jullien and C. Frejaville, *Bull. Soc. Chim. Fr.* p. 4449 (1968); W. G. Dauben and R. E. Wolf, *J. Org. Chem.* **35**, 374 (1970).
507. T. Norin, *Acta Chem. Scand.* **19**, 1289 (1965); W. G. Dauben and E. J. Deviny, *J. Org. Chem.* **31**, 3794 (1966); S. A. Monti, D. J. Bucheck, and J. C. Shepard, *ibid.* **34**, 3080 (1969); Y. Bessière-Chrétien and M. M. El Gaied, *Bull. Soc. Chim. Fr.* p. 2189 (1971); E. Piers and P. M. Worster, *J. Am. Chem. Soc.* **94**, 2895 (1972); M. M. El Gaied and Y. Bessière-Chrétien, *Bull. Soc. Chim. Fr.* p. 1351 (1973); L. N. Mander, R. H. Prager, and J. V. Turner, *Aust. J. Chem.* **27**, 2645 (1974); S. W. Staley and A. S. Heyn, *Tetrahedron* **30**, 3671 (1974). See also, L. Mandell, J. C. Johnston, and R. A. Day, Jr., *J. Org. Chem.* **43**, 1616 (1978).
508. J. A. Claisse, D. I. Davies, and L. T. Parfitt, *J. Chem. Soc. C* p. 258 (1970).
509. For recent examples, see M. Cazaux and R. Lalande, *Bull. Soc. Chim. Fr.* pp. 1887 and 1894 (1972); M. Julia, D. Mansuy, and J. Y. Lallemand, *ibid.* p. 2695; H. H. Quon, T. Tezuka, and Y. L. Chow, *Chem. Commun.* p. 428 (1974); R. Lalande, M. -J. Bourgeois, and B. Maillard, *J. Heterocycl. Chem.* **12**, 509 (1975).
510. P. K. Freeman and T. D. Ziebarth, *J. Org. Chem.* **41**, 949 (1976).
511. P. M. Blum, A. G. Davies, and R. A. Henderson, *Chem. Commun.* p. 569 (1978).
512. A. L. J. Beckwith, *Abstr., Colloq. Int. CNRS, Radicaux Libres Org.*, *1977* p. 7 (1977).
513. E. A. Hill, A. T. Chen, and A. Doughty, *J. Am. Chem. Soc.* **98**, 167 (1976).
514. J. Y. Godet and M. Pereyre, *J. Organomet. Chem.* **77**, C1 (1974).
515. E. A. Hill, R. J. Theissen, C. E. Cannon, R. Miller, R. B. Guthrie, and A. T. Chen, *J. Org. Chem.* **41**, 1191 (1976).
516. L. Kaplan, *J. Org. Chem.* **33**, 2531 (1968).
517. R. Sustmann and D. Brandes, *Tetrahedron Lett.* p. 1791 (1976).
518. C. H. DePuy, H. L. Jones, and D. H. Gibson, *J. Am. Chem. Soc.* **90**, 5306 (1968); **94**, 3924 (1972).
519. P. Kabasakalian and E. R. Townley, *J. Org. Chem.* **27**, 2918 (1962).
520. D. B. Denney and J. W. Hanifin, Jr., *J. Org. Chem.* **29**, 732 (1964).
521. For reviews of cyclopropanol chemistry, see C. H. DePuy, *Acc. Chem. Res.* **1**, 33 (1968); D. H. Gibson and C. H. DePuy, *Chem. Rev.* **74**, 605 (1974).
522. C. H. DePuy, H. L. Jones, and W. M. Moore, *J. Am. Chem. Soc.* **95**, 477 (1973).
523. C. H. DePuy and R. J. Van Lanen, *J. Org. Chem.* **39**, 3360 (1974); see also H. L. Jones, quoted in DePuy (*521*).
524. J. Roček and D. E. Aylward, *J. Am. Chem. Soc.* **97**, 5452 (1975).
525. M. Lj. Mihailović, Ž. Čeković, V. Andrejević, R. Matić, and D. Jeremić, *Tetrahedron* **24**, 4947 (1966).

526. J. Roček and A. E. Radkowsky, *J. Am. Chem. Soc.* **95,** 7123 (1973); F. Hasan and J. Roček, *ibid.* **96,** 534 (1974); K. B. Wiberg and S. K. Mukherjee, *ibid.* p. 6647.
527. J. Roček and A. E. Radkowsky, *J. Org. Chem.* **38,** 389 (1973).
528. K. Meyer and J. Roček, *J. Am. Chem. Soc.* **94,** 1209 (1972).
529. Y. Ito, S. Fujii, and T. Saegusa, *J. Org. Chem.* **41,** 2073 (1976).
530. D. B. Priddy and W. Reusch, *Tetrahedron Lett.* p. 2637 (1970); J. F. Pazos, J. G. Pacifici, G. O. Pierson, D. B. Sclove, and F. D. Greene, *J. Org. Chem.* **39,** 1990 (1974).
531. B. H. Bakker, H. Steinberg, and T. J. de Boer, *Recl. Trav. Chim., Pays-Bas* **95,** 274 (1976).
532. B. H. Bakker, G. J. A. Schilder, T. R. Bok, H. Steinberg, and T. J. de Boer, *Tetrahedron* **29,** 93 (1973); B. H. Bakker, T. R. Bok, H. Steinberg, and T. J. de Boer, *Recl. Trav. Chim. Pays-Bas* **96,** 31 (1977).
533. S. E. Schaafsma, E. J. F. Molenaar, H. Steinberg, and T. J. de Boer, *Recl. Trav. Chim. Pays-Bas* **87,** 1301 (1968).
534. S. E. Schaafsma, H. Steinberg, and T. J. de Boer, *Recl. Trav. Chim. Pays-Bas* **85,** 70 and 73 (1966).
535. P. M. Blum, Ph.D. Thesis, University of London (1977); P. M. Blum and A. G. Davies, unpublished results.
536. G. M. Whitesides, J. S. Sadowski, and J. Lilburn, *J. Am. Chem. Soc.* **96,** 2829 (1974).
537. For reviews, see M. Akhtar, *Adv. Photochem.* **2,** 263 (1964); R. H. Hesse, *Adv. Free-Radical Chem.* **3,** 83 (1969); also Wilt (*4*).
538. C. Walling and R. T. Clark, *J. Am. Chem. Soc.* **96,** 4530 (1974).
539. A. L. J. Beckwith and W. B. Gara, unpublished results; see also C. Rüchardt and H. -J. Quadbeck-Seeger, *Chem. Ber.* **102,** 3525 (1969).
540. R. M. Black and G. B. Gill, *Chem. Commun.* p. 972 (1970).
541. F. Minisci, R. Galli, V. Malatesta, and T. Caronna, *Tetrahedron* **26,** 4083 (1970).
542. M. Akhtar and S. Marsh, *J. Chem. Soc. C* p. 937 (1966); M. Lj. Mihailović, Lj. Lorenc, M. Gašić, M. Rogić, A. Melara, and M. Stefanović, *Tetrahedron* **22,** 2345 (1966).
543. E. Schmitz and D. Murawski, *Chem. Ber.* **98,** 2525 (1965); see also D. St. C. Black and K. G. Watson, *Aust. J. Chem.* **26,** 2515 (1973).
544. B. C. Gilbert, R. O. C. Norman, and R. C. Sealy, *J. Chem. Soc., Perkin Trans. 2* p. 308 (1975).
545. P. Knittel and J. Warkentin, *Can. J. Chem.* **53,** 2275 (1975); **54,** 1341 (1976); see also D. W. K. Yeung and J. Warkentin, *ibid.* pp. 1345 and 1349.
546. E. Schmitz, H. Striegler, H. -U. Heyne, K. -P. Hilgetag, H. Dilcher, and R. Lorenz, *J. Prakt. Chem.* **319,** 274 (1977).
547. See also E. E. J. Dekker, J. B. F. N. Engberts, and T. J. de Boer, *Tetrahedron Lett.* p. 2651 (1969).
548. L. J. Winters, J. F. Fischer, and E. R. Ryan, *Tetrahedron Lett.* p. 129 (1971); W. D. Crow and A. N. Khan, *Aust. J. Chem.* **29,** 2289 (1976).
549. J. -M. Surzur and P. Teissier, *C. R. Hebd. Seances Acad. Sci., Ser. C* **264,** 1981 (1967); *Bull. Soc. Chim. Fr.* p. 3060 (1970).
550. D. D. Tanner and F. C. P. Law, *J. Am. Chem. Soc.* **91,** 7535 (1969).
551. S. Julia and R. Lorne, *C. R. Hebd. Seances Acad. Sci., Ser. C* **273,** 174 (1971).
552. See, e.g., T. T. Tidwell *in* "Organic Free Radicals" (W. A. Pryor, ed.), ACS Symp. Ser. 69, p. 102. Am. Chem. Soc., Washington, D.C., 1978.
553. R. Kh. Freidlina, A. B. Terent'ev, and V. G. Petrova, *Dokl. Akad. Nauk SSSR* **149,** 860 (1963); **151,** 866 (1963); **152,** 637 (1963); A. B. Terent'ev and R. G. Petrova, *Izv. Akad. Nauk SSSR, Ser. Khim.* p. 2153 (1963); A. B. Terent'ev and R. Kh. Freidlina, *Dokl. Akad. Nauk SSSR* **158,** 679 (1964).
554. A. B. Terent'ev, *Izv. Akad. Nauk SSSR, Ser. Khim.* p. 1258 (1965).

555. R. G. Petrova, T. D. Churkina, Sh. A. Karapet'yan, and R. Kh. Freidlina, *Izv. Akad. Nauk SSSR, Ser. Khim.* p. 2517 (1974).
556. R. G. Petrova and R. Kh. Freidlina, *Izv. Akad. Nauk SSSR, Ser. Khim.* p. 1805 (1975).
557. S. N. Lewis, J. J. Miller, and S. Winstein, *J. Org. Chem.* **37**, 1478 (1972).
558. C. Walling and W. Helmreich, *J. Am. Chem. Soc.* **81**, 1144 (1959).
559. T. E. Boothe, J. L. Greene, Jr., and P. B. Shevlin, *J. Am. Chem. Soc.* **98**, 951 (1976).
560. D. H. Brown, R. J. Cross, and D. Millington, *Inorg. Nucl. Chem. Lett.* **11**, 783 (1975).
561. J. A. Kampmeier, R. B. Jordan, M. S. Liu, H. Yamanaka, and D. J. Bishop, *in* "Organic Free Radicals" (W. A. Pryor, ed.), ACS Symp. Series 69, p. 275. Am. Chem. Soc., Washington, D.C., 1978.
562. R. B. Jordan, *Diss. Abstr. Int. B* **34**, 1041 (1973).
563. J. S. Chapman, J. W. Cooper, and B. P. Roberts, *Chem. Commun.* p. 835 (1976); B. C. Gilbert, C. M. Kirk, R. O. C. Norman, and H. A. H. Laue, *J. Chem. Soc., Perkin Trans. 2* p. 497 (1977); W. B. Gara and B. P. Roberts, *J. Organomet. Chem.* **135**, C20 (1977); W. B. Gara, B. P. Roberts, B. C. Gilbert, C. M. Kirk, and R. O. C. Norman, *J. Chem. Res.*, 152, (1977).
564. B. P. Roberts, *Abstr., Colloq. Int. CNRS, Radicaux Libres Org., 1977* p. 135 (1977).
565. P. Livant and J. C. Martin, *J. Am. Chem. Soc.* **98**, 7851 (1976).
566. J. R. Morton and K. F. Preston, *J. Chem. Phys.* **58**, 2657 (1973); B. C. Gilbert, C. M. Kirk, and R. O. C. Norman, *J. Chem. Res.* 173 (1977); W. B. Gara and B. P. Roberts, *J. Magn. Reson.* **27**, 509 (1977).
567. B. C. Gilbert, J. P. Larkin, and R. O. C. Norman, *J. Chem. Soc., Perkin Trans. 2* p. 794 (1972), and references cited; see also C. Walling and R. A. Johnson, *J. Am. Chem. Soc.* **97**, 2405 (1975).
568. T. Foster and P. R. West, *Can. J. Chem.* **51**, 4009 (1973); **52**, 3589 (1974).
569. B. T. Golding and L. Radom, *J. Am. Chem. Soc.* **98**, 6331 (1976).
570. G. A. Razuvaev, N. S. Vasileiskaja, and D. V. Muslin, *J. Organomet. Chem.* **7**, 531 (1967).
571. I. L. Khrzhanovskaya and N. S. Vasileiskaya, *Izv. Akad. Nauk SSSR, Ser. Khim.* p. 572 (1972).
572. M. T. Reetz, M. Kliment, and N. Greif, *Chem. Ber.* **111**, 1083 and 1095 (1978).
573. C. G. Pitt and M. S. Fowler, *J. Am. Chem. Soc.* **90**, 1928 (1968).
574. A. Alberti and A. Hudson, *Chem. Phys. Lett.* **48**, 331 (1977).
575. A. I. Prokof'ev, T. I. Prokof'eva, N. N. Bubnov, S. P. Solodovnikov, I. S. Belostotskaya, V. V. Ershov, and M. I. Kabachnik, *Dokl. Akad. Nauk SSSR* **239**, 1367 (1978); Tetrahedron, **35**, 2471 (1979). For a general review of related processes see: N. N. Bubnov, S. P. Solodovnikov, A. I. Prokof'ev, and M. I. Kabachnik, *Russ. Chem. Revs.* **47**, 549 (1978).
575a. A. Alberti and A. Hudson, *J. Chem. Soc., Perkin Trans. 2* p. 1098 (1978).
576. R. West and P. Boudjouk, *J. Am. Chem. Soc.* **95**, 3983 (1973).
577. A. N. Nesmeyanov, R. Kh. Freidlina, and V. I. Firstov, *Izv. Akad. Nauk SSSR, Otd. Khim. Nauk* p. 505 (1951); A. N. Nesmeyanov, R. Kh. Freidlina, and L. I. Zakharkin, *Dokl. Akad. Nauk SSSR* **81**, 199 (1951).
578. W. H. Urry and J. R. Eiszner, *J. Am. Chem. Soc.* **73**, 2977 (1951); **74**, 5822 (1952).
579. R. Kh. Freidlina, F. K. Velichko, E. Ts. Chukovskaya, M. Ya. Khorlina, B. A. Krentsel, D. E. Il'ina, N. V. Kruglova, L. S. Mayants, and R. G. Gasanov, "Metody Elementoorganicheskoi Khimii. Khlor. Alifaticheskie Soedineniya." Nauka, Moscow, 1973.
580. P. S. Skell, R. R. Pavlis, D. C. Lewis, and K. J. Shea, *J. Am. Chem. Soc.* **95**, 6735 (1973).
581. P. S. Skell, *Chem. Soc., Spec. Publ.* **19**, 131 (1964).

582. K. S. Chen, I. H. Elson, and J. K. Kochi, *J. Am. Chem. Soc.* **95**, 5341 (1973).
583. J. Cooper, A. Hudson, and R. A. Jackson, *Tetrahedron Lett.* p. 831 (1973).
584. K. S. Chen, D. Y. H. Tang, L. K. Montgomery, and J. K. Kochi, *J. Am. Chem. Soc.* **96**, 2201 (1974).
585. R. G. Gasanov, I. I. Kandror, and R. Kh. Freidlina, *Tetrahedron Lett.* p. 1485 (1975); R. G. Gasanov, I. I. Kandror, M. Ya. Khorlina, and R. Kh. Freidlina, *Izv. Akad. Nauk SSSR, Ser. Khim.* p. 1758 (1976); M. Ya. Khorlina, *ibid.* p. 2122; R. G. Gasanov and R. Kh. Freidlina, *ibid.* p. 2242 (1977); R. G. Gasanov, T. T. Vasil'eva, and R. Kh. Freidlina, *Izv. Akad. Nauk SSSR, Ser. Khim.* p. 817 (1978); R. Kh. Freidlina, I. I. Kandror, and R. G. Gasanov, *Usp. Khim.* **47**, 508 (1978).
586. D. Atton, S. A. Bone, and I. M. T. Davidson, *J. Organomet. Chem.* **39**, C47 (1972).
587. I. N. Jung and W. P. Weber, *J. Org. Chem.* **41**, 946 (1976).
587a. R. V. Lloyd, *J. Phys. Chem.* **83**, 276 (1978).
588. W. C. Danen, *Methods Free-Radical Chem.* **5**, 1 (1974).
589. R. F. Drury and L. Kaplan, *J. Am. Chem. Soc.* **94**, 3982 (1972).
590. See also, R. Aloni, A. Horowitz, and L. A. Rajbenbach, *Int. J. Chem. Kinet.* **8**, 673 (1976), and references cited.
591. P. Cadman, G. M. Tisley, and A. F. Trotman-Dickenson, *J. Chem. Soc., Faraday Trans. 1* **69**, 914 (1973).
592. A. L. J. Beckwith, *Org. Chem., Ser. One* **10**, 1 (1973).
593. K. Heusler and J. Kalvoda, *Angew. Chem., Int. Ed. Engl.* **3**, 525 (1964).
594. R. Kh. Freidlina and A. B. Terent'ev, *Acc. Chem. Res.* **10**, 9 (1977).
595. R. H. Hesse, *Adv. Free-Radical Chem.* **3**, 83 (1969).
596. R. S. Neale, *Synthesis* p. 1 (1971).
597. M. E. Wolff, *Chem. Rev.* **63**, 55 (1963).
598. J. Y. Nedelec and D. Lefort, *Tetrahedron* **31**, 411 (1975). See also, A. G. Shostenko, V. E. Myshkin, and V. Kim, *React. Kinet. Catal. Lett.* **10**, 311 (1979); R. G. Gasanov, *Izv. Akad. Nauk SSSR, Ser. Khim.* p. 891 (1979).
599. J. Y. Nedelec, M. Gruselle, A. Triki, and D. Lefort, *Tetrahedron* **33**, 39 (1977).
600. C. Walling and A. Padwa, *J. Am. Chem. Soc.* **85**, 1597 (1963).
601. K. J. Mintz and D. J. LeRoy, *Can. J. Chem.* **51**, 3534 (1973), and references cited.
602. K. W. Watkins, *J. Phys. Chem.* **77**, 2938 (1973), and references cited.
603. A. S. Gordon, D. C. Tardy, and R. Ireton, *J. Phys. Chem.* **80**, 1400 (1976).
604. S. S. Hixson and J. C. Tausta, *J. Org. Chem.* **42**, 2191 (1977).
605. O. A. Reutov and T. N. Shatkina, *Tetrahedron* **18**, 305 (1962).
606. O. A. Reutov, *Pure Appl. Chem.* **7**, 203 (1963).
607. O. A. Reutov, G. M. Ostapchuk, K. Uteniyazov, and E. V. Binshtok, *Zh. Org. Khim.* **4**, 1869 (1968).
608. Yu. G. Bundel, I. Yu. Levina, A. M. Krzhizhevskii, and O. A. Reutov, *Izv. Akad. Nauk SSSR, Ser. Khim.* p. 2583 (1968).
609. B. C. Childress, A. C. Rice, and P. B. Shevlin, *J. Org. Chem.* **39**, 3056 (1974).
610. D. M. Brenner, G. P. Smith, and R. N. Zare, *J. Am. Chem. Soc.* **98**, 6707 (1976).
611. M. A. Churilova, D. I. Povolotskii, and A. B. Terent'ev, *Izv. Akad. Nauk SSSR, Ser. Khim.* p. 2497 (1975).
612. R. G. Gasanov, A. B. Terent'ev, and R. Kh. Freidlina, *Izv. Akad. Nauk SSSR, Ser. Khim.* pp. 545 and 2628 (1977).
613. R. Criegee and R. Huber, *Chem. Ber.* **105**, 1972 (1972).
614. G. Brunton, J. A. Gray, D. Griller, L. R. C. Barclay, and K. U. Ingold, *J. Am. Chem. Soc.* **100**, 4197 (1978).
615. P. H. Kasai and D. McLeod, Jr., *J. Am. Chem. Soc.* **96**, 2338 (1974).

4. FREE-RADICAL REARRANGEMENTS

615a. P. J. Wagner and C. Chiu, *J. Am. Chem. Soc.* **101**, 7134 (1979).
616. T. Cohen, C. H. McMullen, and K. Smith, *J. Am. Chem. Soc.* **90**, 6866 (1968).
617. T. Cohen, K. W. Smith, and M. D. Swerdloff, *J. Am. Chem. Soc.* **93**, 4303 (1971).
618. A. L. J. Beckwith and W. B. Gara, *J. Am. Chem. Soc.* **91**, 5689 and 5691 (1969).
619. S. H. Pines, R. M. Purick, R. A. Reamer, and G. Gal, *J. Org. Chem.* **43**, 1337 (1978).
620. E. I. Heiba and R. M. Dessau, *J. Am. Chem. Soc.* **89**, 3772 (1967).
621. N. A. Kuz'mina, E. C. Chukovskaya, and R. Kh. Freidlina, *Chem. Commun.* p. 315 (1976).
622. V. E. Myshkin, A. G. Shostenko, P. A. Zagorets, and A. I. Pchelkin, *Zh. Org. Khim.* **13**, 696 (1977).
623. T. T. Vasil'eva and R. Kh. Freidlina, *Izv. Akad. Nauk SSSR, Ser. Khim.* p. 2373 (1976). T. T. Vasil'eva, N. V. Kruglova, V. I. Dostovalova, and R. Kh. Freidlina, *ibid.* p. 1856 (1977).
624. M. J. Bourgeois, R. Lalande, and B. Maillard, *C. R. Hebd. Seances Acad. Sci., Ser. C.* **285**, 393 (1977).
625. J. G. Traynham and T. M. Couvillon, *J. Am. Chem. Soc.* **87**, 5806 (1965); **89**, 3205 (1967); J. G. Traynham, T. M. Couvillon, and N. S. Bhacca, *J. Org. Chem.* **32**, 529 (1967).
626. H. Matsumoto, T. Nakano, K. Takasu, and Y. Nagai, *J. Org. Chem.* **43**, 1734 (1978).
627. V. M. Berdnikov, N. M. Bazhin, V. K. Fedorov, and O. V. Polyakov, *Kinet. Katal.* **13**, 1093 (1972).
628. B. C. Gilbert, H. A. H. Laue, R. O. C. Norman, and R. C. Sealy, *J. Chem. Soc., Perkin Trans. 2* p. 1040 (1976).
629. B. C. Gilbert, R. G. G. Holmes, and R. O. C. Norman, *J. Chem. Res. (M)* p. 0101 (1977).
630. M. C. R. Symons, *J. Am. Chem. Soc.* **91**, 5924 (1969).
631. D. H. R. Barton, J. M. Beaton, L. E. Geller, and M. M. Pechet, *J. Am. Chem. Soc.* **82**, 2640 (1960); **83**, 4076 (1961).
632. D. H. R. Barton and J. M. Beaton, *J. Am. Chem. Soc.* **82**, 2641 (1960).
633. D. H. R. Barton, *Pure Appl Chem.* **16**, 1 (1968); D. H. R. Barton, R. H. Hesse, M. M. Pechet, and L. C. Smith, *J. Chem. Soc., Perkin Trans. 1* p. 1159 (1979).
634. J. Bošnjak, V. Andrejević, Ž. Čeković, and M. Lj. Mihailović, *Tetrahedron* **28**, 6031 (1972).
635. M. Lj. Mihailović, S. Gojković, and S. Konstantinović, *Tetrahedron* **29**, 3675 (1973).
636. M. P. Kullberg and B. Green, *Chem. Commun.* p. 637 (1972).
637. A. Nickon, R. Ferguson, A. Bosch, and T. Iwadare, *J. Am. Chem. Soc.* **99**, 4518 (1977).
638. M. M. Green, J. M. Moldowan, M. W. Armstrong, T. L. Thompson, K. J. Sprague, A. J. Hass, and J. J. Artus, *J. Am. Chem. Soc.* **98**, 849 (1976).
639. J. A. Howard and K. U. Ingold, *Can. J. Chem.* **48**, 873 (1970).
640. F. F. Rust, *J. Am. Chem. Soc.* **79**, 4000 (1957).
641. T. Mill and G. Montorsi, *Int. J. Chem. Kinet.* **5**, 119 (1973).
642. D. E. Van Sickle, *J. Org. Chem.* **37**, 755 (1972).
643. D. E. Van Sickle, T. Mill, F. R. Mayo, H. Richardson, and C. W. Gould, *J. Org. Chem.* **38**, 4435 (1973).
644. L. R. Mahoney, S. Korcek, R. K. Jensen, and M. Zinbo, *Prepr., Div. Petrol. Chem., Am. Chem. Soc.* **21**, 852 (1976).
644a. R. K. Jensen, S. Korcek, L. R. Mahoney, and M. Zinbo, *J. Am. Chem. Soc.* **101**, 7574 (1979).
645. N. K. Voskresenskaya and B. G. Freidin, *Zh. Prikl. Khim. (Leningrad)* **50**, 1315 (1977).
646. S. W. Benson, *J. Am. Chem. Soc.* **87**, 972 (1965).

647. L. R. Mahoney, private communication.
648. N. H. Anderson and R. O. C. Norman, *J. Chem. Soc. B* p. 993 (1971).
649. W. C. Danen and T. T. Kensler, *J. Am. Chem. Soc.* **92,** 5235 (1970).
650. J. R. Roberts and K. U. Ingold, *J. Am. Chem. Soc.* **93,** 6686 (1971); **95,** 3228 (1973).
651. O. E. Edwards, D. Vocelle, and J. W. ApSimon, *Can. J. Chem.* **50,** 1167 (1972).
652. O. E. Edwards, J. M. Paton, H. H. Benn, R. E. Mitchell, C. Watanatada, and K. N. Vohra, *Can. J. Chem.* **49,** 1648 (1971).
653. D. H. R. Barton and A. L. J. Beckwith, *Proc. Chem. Soc., London* p. 335 (1963); D. H. R. Barton, A. L. J. Beckwith, and A. Goosen, *J. Chem. Soc.* p. 181 (1965).
654. A. L. J. Beckwith and J. E. Goodrich, *Aust. J. Chem.* **18,** 747 (1965).
655. Y. L. Chow and T. C. Joseph, *Chem. Commun.* p. 490 (1969).
656. R. S. Neale, N. L. Marcus, and R. G. Schepers, *J. Am. Chem. Soc.* **88,** 3051 (1966).
657. M. Benn and K. N. Vohra, *Can. J. Chem.* **54,** 136 (1976).
658. R. C. Petterson and A. Wambsgans, *J. Am. Chem. Soc.* **86,** 1648 (1964).
659. S. M. Verma and R. C. Srivastava, *Bull. Chem. Soc. Jpn.* **40,** 2184 (1967).
660. M. Okahara, T. Ohashi, and S. Komori, *Tetrahedron Lett.* p. 1629 (1967); *J. Org. Chem.* **33,** 3066 (1968).
661. R. S. Neale and N. L. Marcus, *J. Org. Chem.* **34,** 1808 (1969).
662. R. A. Johnson and F. D. Greene, *J. Org. Chem.* **40,** 2186 (1975),
663. R. H. Staley, M. Taagepera, W. G. Henderson, I. Koppel, J. L. Beauchamp, and R. W. Taft, *J. Am. Chem. Soc.* **99,** 326 (1977).
664. D. M. Golden, R. K. Solly, N. A. Gac, and S. W. Benson, *J. Am. Chem. Soc.* **94,** 363 (1972).
665. B. C. Gilbert and P. R. Marriott, *J. Chem. Soc., Perkin Trans. 2* p. 987 (1977).
666. J. Spanswick and K. U. Ingold, *Can. J. Chem.* **48,** 546 and 554 (1970).
666a. V. Malatesta and K. U. Ingold, *J. Am. Chem. Soc.* **95,** 6400 (1973).
667. N. C. Deno, *Methods Free-Radical Chem.* **3,** 135 (1972); N. C. Deno, E. J. Jedziniak, L. A. Messer, and E. S. Tomezsko, *Prepr., Div. Petrol. Chem., Am. Chem. Soc.* **21,** 426 (1976); N. C. Deno and E. J. Jedziniak, *Tetrahedron Lett.* p. 1259 (1976).
668. F. Minisci, G. P. Gardini, and F. Bertini, *Can. J. Chem.* **48,** 544 (1970).
669. A. J. Cessna, S. E. Sugamori, R. W. Yip, M. P. Lau, R. S. Snyder, and Y. L. Chow, *J. Am. Chem. Soc.* **99,** 4044 (1977).
670. F. Minisci, *Top. Curr. Chem.* **62,** 3 (1976); F. Minisci, R. Galli, and M. Cecere, *Tetradhedron Lett.* p. 4663 (1965); p. 3163 (1966).
670a. F. Minisci, R. Galli, and R. Bernardi, *Tetrahedron Lett.* p. 699 (1966).
671. R. S. Neale and R. L. Hinman, *J. Am. Chem. Soc.* **85,** 2666 (1963); R. S. Neale, *ibid.* **86,** 5340 (1964); *Tetrahedron Lett.* p. 483 (1966); *J. Org. Chem.* **32,** 3263 (1967); R. S. Neale and N. L. Marcus, *ibid.* p. 3273; **33,** 3457 (1968).
672. R. S. Neale, M. R. Walsh, and N. L. Marcus, *J. Org. Chem.* **30,** 3683 (1965).
673. S. Wawzonek and T. P. Culbertson, *J. Am. Chem. Soc.* **82,** 441 (1960).
674. M. M. Green, J. G. McGrew, II, and J. M. Moldowan, *J. Am. Chem. Soc.* **93,** 6700 (1971).
675. F. Minisci, R. Galli, R. Bernardi, and M. Perchinumno, *Chim. Ind. (Milan)* **50,** 453 (1968).
676. E. Bothe, G. Behrens, and D. Schulte-Frohlinde, *Z. Naturforsch., Teil B* **32,** 886 (1977).
677. See, e.g., R. W. Kreilick and S. I. Weissman, *J. Am. Chem. Soc.* **84,** 306 (1962); **88,** 2645 (1966); M. R. Arick and S. I. Weissman, *ibid.* **90,** 1654 (1968); L. R. Mahoney and M. A. DaRooge, *ibid.* **92,** 890 and 4063 (1970); **94,** 7002 (1972); J. A. Howard and E.

4. FREE-RADICAL REARRANGEMENTS

Furimsky, *Can. J. Chem.* **51**, 3738 (1973); D. Griller and K. U. Ingold, *J. Am. Chem. Soc.* **96**, 630 (1974).
678. H. G. Aurich and K. Stork, *Tetrahedron Lett.* p. 555 (1972).
679. A. I. Prokof'ev, N. N. Bubnov, S. P. Solodovnikov, and M. I. Kabachnik, *Tetrahedron Lett.* p. 2479 (1973).
680. A. I. Prokof'ev, N. N. Bubnov, S. P. Solodovnikov, I. S. Belostotskaya, and V. V. Ershov, *Dokl. Akad. Nauk SSSR* **210**, 361 (1973).
681. A. I. Prokof'ev, A. S. Masalimov, N. N. Bubnov, S. P. Solodovnikov, and M. I. Kabachnik, *Izv. Akad. Nauk SSSR, Ser. Khim.* p. 310 (1976).
682. C. Walling and W. Thaler, *J. Am. Chem. Soc.* **83**, 3877 (1961).
683. D. B. Denney, R. M. Hoyte, and P. T. MacGregor, *Chem. Commun.* p. 1241 (1967).
684. R. M. Hoyte and D. B. Denney, *J. Org. Chem.* **39**, 2607 (1974).
685. W. A. Thaler, A. A. Oswald, and B. E. Hudson, Jr., *J. Am. Chem. Soc.* **87**, 311 (1965).
686. R. J. Crawford, J. Hamelin, and B. Strehlke, *J. Am. Chem. Soc.* **93**, 3810 (1971).
687. D. C. Montague, *Int. J. Chem. Kinet.* **5**, 513 (1973).
688. See, e.g., W. von E. Doering and G. H. Beasley, *Tetrahedron* **29**, 2231 (1973); W. R. Roth, G. Ruf, and P. W. Ford, *Chem. Ber.* **107**, 48 (1974); see also, F. A. Houle and J. L. Beauchamp, *J. Am. Chem. Soc.* **100**, 3290 (1978).
689. D. M. Golden, *Int. J. Chem. Kinet.* **1**, 127 (1969).
690. G. Boche and D. R. Schneider, *Angew. Chem., Int. Ed. Engl.* **16**, 869 (1977).
691. See, e.g., W. von E. Doering, G. Horowitz, and K. Sachdev, *Tetrahedron* **33**, 273 (1977).
692. D. M. Camaioni, H. F. Walter, and D. W. Pratt, *J. Am. Chem. Soc.* **95**, 4057 (1973).
693. See, however, L. M. Stock and M. R. Wasielewski, *J. Am. Chem. Soc.* **97**, 5620 (1975).
694. K. J. Shea and P. S. Skell, *J. Am. Chem. Soc.* **95**, 283 (1973).
695. D. D. Tanner, Y. Kosugi, R. Arhart, N. Woda, T. Pace, and T. Ruo, *J. Am. Chem. Soc.* **98**, 6275 (1976); D. D. Tanner, T. Pace, Y. Kosugi, E. V. Blackburn, and T. Ruo, *Tetrahedron Lett.* p. 2413 (1976).
696. D. D. Tanner, E. V. Blackburn, Y. Kosugi, and T. C. S. Ruo, *J. Am. Chem. Soc.* **99**, 2714 (1977).
697. E. S. Lewis and C. C. Shen, *J. Am. Chem. Soc.* **99**, 3055 (1977).
698. See, however, K. Ody, A. Nechvatal, and J. M. Tedder, *J. Chem. Soc., Perkin Trans. 2* p. 521 (1976).
699. See also W. C. Danen and K. A. Rose, *J. Org. Chem.* **40**, 619 (1975).
700. J. H. Hargis and P. B. Shevlin, *Chem. Commun.* p. 179 (1973).
701. D. Griller and K. U. Ingold, *J. Am. Chem. Soc.* **96**, 6715 (1974).
702. See, e.g., R. V. Lloyd, D. E. Wood, and M. T. Rogers, *J. Am. Chem. Soc.* **96**, 7130 (1974); R. V. Lloyd and D. E. Wood, *ibid.* **97**, 5986 (1975); A. R. Lyons, G. W. Neilson, S. P. Mishra, and M. C. R. Symons, *J. Chem. Soc., Faraday Trans. 2* **71**, 363 (1975); D. Nelson and M. C. R. Symons, *Tetrahedron Lett.* p. 2953 (1975); D. E. Wood and R. V. Lloyd, *ibid.* p. 345 (1976); A. R. Rossi and D. E. Wood, *J. Am. Chem. Soc.* **98**, 3452 (1976); M. C. R. Symons and I. G. Smith, *J. Chem. Soc., Perkin Trans. 2* p. 1362 (1979).
703. K. S. Chen and J. K. Kochi, *J. Am. Chem. Soc.* **96**, 1383 (1974).
704. For a "bent" view, see J. B. Lisle, L. F. Williams, and D. E. Wood, *J. Am. Chem. Soc.* **98**, 227 (1976); P. J. Krusic and P. Meakin, *ibid.* p. 228. For a "flat" view, see T. A. Claxton, E. Platt, and M. C. R. Symons, *Mol. Phys.* **32**, 1321 (1976); L. Bonazzola, N. Leray, and J. Roncin, *J. Am. Chem. Soc.* **99**, 8348 (1977).
705. D. Griller, K. U. Ingold, P. J. Krusic, and H. Fischer, *J. Am. Chem. Soc.* **100**, 6750 (1978).

706. D. Griller and K. F. Preston, *J. Am. Chem. Soc.* **101**, 1975 (1979); D. Griller, P. R. Marriott, and K. F. Preston, *J. Chem. Phys.* **71**, 3703 (1979).
707. R. W. Fessenden and R. H. Schuler, *J. Chem. Phys.* **43**, 2704 (1965).
708. J. Hatem and B. Waegell, *Tetrahedron Lett.* p. 2019 (1973).
709. L. J. Altman and B. W. Nelson, *J. Am. Chem. Soc.* **91**, 5163 (1969).
710. P. Warner and S. -L. Lu, *Tetrahedron Lett.* p. 4665 (1976).
711. L. J. Altman and J. C. Vederas, *Chem. Commun.* p. 895 (1969).
712. T. Ando, K. Wakabayashi, H. Yamanaka, and W. Funasaka, *Bull. Chem. Soc. Jpn.* **45**, 1576 (1972).
713. L. J. Altman and T. R. Erdman, *Tetrahedron Lett.* p. 4891 (1970).
714. T. Ando, H. Yamanaka, F. Namigata, and W. Funasaka, *J. Org. Chem.* **35**, 33 (1970).
715. L. J. Altman and R. C. Baldwin, *Tetrahedron Lett.* p. 2531 (1971).
716. T. Ishihara, E. Ohtani, and T. Ando, *Chem. Commun.* p. 367 (1975).
717. T. Ando, F. Namigata, H. Yamanaka, and W. Funasaka, *J. Am. Chem. Soc.* **89**, 5719 (1967).
718. D. E. Applequist and A. M. Peterson, *J. Am. Chem. Soc.* **82**, 2372 (1960).
719. H. M. Walborsky and J. -C. Chen, *J. Am. Chem. Soc.* **89**, 5499 (1967).
720. H. M. Walborsky, F. P. Johnson, and J. B. Pierce, *J. Am. Chem. Soc.* **90**, 5222 (1968).
721. J. Jacobus and D. Pensak, *Chem. Commun.* p. 400 (1969).
722. M. J. S. Dewar and J. M. Harris, *J. Am. Chem. Soc.* **91**, 3652 (1969).
723. L. A. Singer and J. Chen, *Tetrahedron Lett.* p. 939 (1971).
724. M. Schlosser, G. Heinz, and L. V. Chau, *Chem. Ber.* **104**, 1921 (1971).
725. H. M. Walborsky and J. -C. Chen, *J. Am. Chem. Soc.* **93**, 671 (1971); see also Walborsky and Chen references (*403*).
726. T. Ando, A. Yamashita, M. Matsumoto, T. Ishihara, and H. Yamanaka, *Chem. Lett.* p. 1133 (1973).
727. T. Ishihara, K. Hayashi, T. Ando, and H. Yamanaka, *J. Org. Chem.* **40**, 3264 (1975).
728. G. Boche and D. R. Schneider, *Tetrahedron Lett.* p. 2327 (1978).
729. T. Kawamura, M. Tsumura, and T. Yonezawa, *Chem. Commun.* p. 373 (1977).
730. T. Kawamura, M. Tsumura, Y. Yokomichi, and T. Yonezawa, *J. Am. Chem. Soc.* **99**, 8251 (1977).
731. V. Malatesta, D. Forrest, and K. U. Ingold, *J. Am. Chem. Soc.* **100**, 7073 (1978).
732. L. J. Altman and R. C. Baldwin, *Tetrahedron Lett.* p. 981 (1972).
733. H. Fischer, private communication.
734. H. Yamanaka, J. Kikui, K. Teramura, and T. Ando, *J. Org. Chem.* **41**, 3794 (1976).
735. C. Gaze and B. C. Gilbert, *J. Chem. Soc., Perkin Trans. 2* p. 1161 (1977).
736. C. Gaze, B. C. Gilbert, and M. C. R. Symons, *J. Chem. Soc., Perkin Trans. 2* p. 235 (1978).
737. P. Meakin and P. J. Krusic, *J. Am. Chem. Soc.* **95**, 8185 (1973); K. S. Chen and J. K. Kochi, *ibid.* **96**, 794 (1974); *Can. J. Chem.* **52**, 3529 (1974); K. S. Chen, P. J. Krusic, P. Meakin, and J. K. Kochi, *J. Phys. Chem.* **78**, 2014 (1974).
738. S. O. Kobayashi and O. Simamura, *Chem. Lett.* p. 695 (1973).
739. P. D. Bartlett, R. F. Pincock, J. H. Rolston, W. G. Schindel, and L. A. Singer, *J. Am. Chem. Soc.* **87**, 2590 (1965); **96**, 6818 (1974).
740. F. D. Greene and N. N. Lowry, *J. Org. Chem.* **32**, 875 (1967).
741. F. D. Greene and N. N. Lowry, *J. Org. Chem.* **32**, 882 (1967).
742. N. -T. Giac and C. Rüchardt, *Chem. Ber.* **110**, 1095 (1977).
743. J. E. Bennett and J. A. Howard, *Chem. Phys. Lett.* **9**, 460 (1971).
744. G. W. Neilson and M. C. R. Symons, *J. Chem. Soc., Perkin Trans. 2* p. 1405 (1973).

4. FREE-RADICAL REARRANGEMENTS

745. M. S. Liu, S. Soloway, D. K. Wedegaertner, and J. A. Kampmeier, *J. Am. Chem. Soc.* **93**, 3809 (1971).
746. See also T. Ohnuki, M. Yoshida, and O. Simamura, *Chem. Lett.* p. 797 (1972).
747. L. A. Singer and N. P. Kong, *Tetrahedron Lett.* p. 643 (1967); *J. Am. Chem. Soc.* **89**, 5251 (1967).
748. J. A. Kampmeier and R. M. Fantazier, *J. Am. Chem. Soc.* **88**, 1959 and 5219 (1966).
749. L. A. Singer and N. P. Kong, *Tetrahedron Lett.* p. 2089 (1966); *J. Am. Chem. Soc.* **88**, 5213 (1966).
750. L. A. Singer and J. Chen, *Tetrahedron Lett.* p. 4849 (1969).
751. G. D. Sargent and M. W. Browne, *J. Am. Chem. Soc.* **89**, 2788 (1967).
752. H. G. Kuivila, *Acc. Chem. Res.* **1**, 299 (1968).
753. See, however, G. M. Whitesides, C. P. Casey, and J. K. Krieger, *J. Am. Chem. Soc.* **93**, 1379 (1971).
754. E. J. Panek, L. R. Kaiser, and G. M. Whitesides, *J. Am. Chem. Soc.* **99**, 3708 (1977).
755. A. J. Fry and M. A. Mitnick, *J. Am. Chem. Soc.* **91**, 6207 (1969).
756. R. C. Neuman, Jr. and G. D. Holmes, *J. Org. Chem.* **33**, 4317 (1968).
757. L. A. Singer and S. S. Kim, *Tetrahedron Lett.* p. 1705 (1973).
758. A. D. Walsh, *Discuss. Faraday Soc.* **2**, 18 (1947); L. Pauling, *J. Chem. Phys.* **51**, 2767 (1969).
759. K. Mislow "Introduction to Stereochemistry," p. 19. Benjamin, New York, 1965; K. B. Wiberg, *Tetrahedron* **24**, 1083 (1968).
760. R. C. Bingham and M. J. S. Dewar, *J. Am. Chem. Soc.* **95**, 7180 and 7182 (1973).
761. C. U. Pittman, Jr., L. D. Kispert, and T. B. Patterson, Jr., *J. Phys. Chem.* **77**, 494 (1973).
762. S. W. Bennett, C. Eaborn, A. Hudson, R. A. Jackson, and K. D. J. Root, *J. Chem. Soc. A* p. 348 (1970).
763. J. Reffy, *J. Organomet. Chem.* **97**, 151 (1975).
764. H. Sakurai, K. Mochida, and M. Kira, *J. Am. Chem. Soc.* **97**, 929 (1975); *J. Organomet. Chem.* **124**, 235 (1977).
765. L. H. Sommer and L. A. Ulland, *J. Org. Chem.* **37**, 3878 (1972).
766. H. Sakurai, M. Murakami, and M. Kumada, *J. Am. Chem. Soc.* **91**, 519 (1969).
767. A. G. Brook and J. M. Duff, *J. Am. Chem. Soc.* **91**, 2118 (1969).
768. See, however, N. A. Porter and P. M. Iloff, Jr., *J. Am. Chem. Soc.* **96**, 6200 (1974).
769. H. Sakurai and K. Mochida, *Chem. Commun.* p. 1581 (1971).
770. H. Sakurai and M. Murakami, *Chem. Lett.* p. 7 (1972).
771. J. Cooper, A. Hudson, and R. A. Jackson, *Mol. Phys.* **23**, 209 (1972).
772. F. A. Carey and C. W. Hsu, *Tetrahedron Lett.* p. 3885 (1970).
773. C. Eaborn, R. A. Jackson, D. J. Tune, and D. R. M. Walton, *J. Organomet. Chem.* **63**, 85 (1973); see also V. M. Vodolazskaya, B. V. Fedot'ev, Yu. I. Baukov, O. A. Kruglaya, and N. S. Vyazankin, *ibid.* p. C5.
774. H. Sakurai and M. Murakami, *Bull. Chem. Soc. Jpn.* **49**, 3185 (1976); **50**, 3384 (1977).
775. B. G. McKinnie, N. S. Bhacca, F. K. Cartledge, and J. Fayssoux, *J. Am. Chem. Soc.* **96**, 2637 (1974); *J. Org. Chem.* **41**, 1534 (1976).
776. D. N. Roark and L. H. Sommer, *J. Am. Chem. Soc.* **95**, 969 (1973).
777. See, e.g., M. Gielen, C. Hoogzand, S. Simon, Y. Tondeur, T. Van den Eynde, and M. Van de Steen, *Adv. Chem. Ser.* **157**, 249 (1976); M. Gielen and Y. Tondeur, *J. Organomet. Chem.* **127**, C75 (1977); **128**, C25 (1977); *Chem. Commun.* p. 81 (1978).
778. M. Gielen and Y. Tondeur, *Nouv. J. Chim.* **2**, 117 (1978).
779. A. Hudson and H. A. Hussain, *Mol. Phys.* **16**, 199 (1969).

780. W. H. Wolodarsky and J. K. S. Wan, *Spectrosc. Lett.* **6,** 429 (1973); see also B. L. Booth, D. J. Edge, R. N. Haszeldine, and R. G. G. Holmes [*J. Chem. Soc., Perkin Trans. 2* p. 7 (1977)] for a related observation.
781. T. S. Dobashi, D. R. Parker, and E. J. Grubbs, *J. Am. Chem. Soc.* **99,** 5382 (1977); see also J. A. Villarreal, T. S. Dobashi, and E. J. Grubbs, *J. Org. Chem.* **43,** 1890 (1978); J. A. Villarreal and E. J. Grubbs, *ibid.* p. 1896.
782. R. O. C. Norman and B. C. Gilbert, quoted in Dobashi *et al.* references *(781).*
783. K. U. Ingold and S. Brownstein, *J. Am. Chem. Soc.* **97,** 1817 (1975).
784. A. Mackor, *J. Org. Chem.* **43,** 3241 (1978).
785. For leading references, see W. G. Bentrude, in "Free Radicals" (J. K. Kochi, ed.), Vol. 2, Chapter 22. Wiley, New York, 1973; *in* "Organic Free Radicals" (W. A. Pryor, ed.), ACS Symp. Ser. 69, Chapter 20. Am. Chem. Soc., Washington, D.C., 1978.
786. For leading references, see P. J. Krusic and P. Meakin, *Chem. Phys. Lett.* **18,** 347 (1973); R. W. Dennis and B. P. Roberts, *J. Organomet. Chem.* **47,** C8 (1973); *J. Chem. Soc., Perkin Trans. 2* p. 140 (1975); J. W. Cooper and B. P. Roberts, *ibid.* p. 808 (1976); J. W. Cooper, M. J. Parrott, and B. P. Roberts, *ibid.* p. 730 (1977); R. W. Dennis, I. H. Elson, B. P. Roberts, and R. C. Dobbie, *ibid.* p. 889; B. P. Roberts, *Adv. Free-Radical Chem.* (in press); A. Nakanishi and W. G. Bentrude, *J. Am. Chem. Soc.* **100,** 6271 (1978); A. Nakanishi, K. Nishikida, and W. G. Bentrude, *ibid.* pp. 6398 and 6403; R. S. Hay and B. P. Roberts, *J. Chem. Soc., Perkin Trans. 2* p. 770 (1978).
787. One of us (KUI) would like to thank the staff of the Chemistry Department of the University of St. Andrews, Scotland, for their hospitality and for many helpful discussions on free-radical rearrangements during a visit in 1977. We would both like to thank Mrs. W. Mazur for her careful typing and proofreading of this essay.

Added in Proof:

The following papers describe important advances in the chemistry of free-radical rearrangements: (1) A. L. J. Beckwith and T. Lawrence, *J. Chem. Soc., Perkin Trans. 2* p. 1535 (1979). Effect of nonbonded interactions (methyl substitution at positions 2 and 5) on the regioselectivity of 5-hexenyl cyclization is described. See Section IV,B; pp. 186–187; 189–192; (2) J.-L. Stein, L. Stella, and J. M. Surzur, *Tetrahedron Lett.* p. 287 (1980). A preferred *endo*-cyclization by a nitrogen-centered radical is reported. See Section IV,E; pp. 205–206; (3) N. A. Porter, M. A. Cudd, R. W. Miller, and A. T. McPhail, *J. Am. Chem. Soc.* **102,** 414 (1980). The stereochemical course of an S_H2 reaction at a peroxide bond is described. See Section IV,H; pp. 219; (4) D. Brandes, F. Lange, and R. Sustmann, *Tetrahedron Lett.* p. 261, p. 275 (1980). Further studies on radical rearrangements in matrices are reported. See Section V,B; pp. 232 and 229; (5) C. Berti and M. J. Perkins, *J. Chem. Soc., Chem. Comm.* p. 1167 (1979). Intramolecular hydrogen abstraction by an acyl nitroxide is faster for a 1,6 or 1,7 than for a 1,5 hydrogen migration.

ESSAY 5 | HYPOTHETICAL BIRADICAL PATHWAYS IN THERMAL UNIMOLECULAR REARRANGEMENTS

JEROME A. BERSON

I.	Introduction	311
	A. Biradical Hypothesis	311
	B. Scope of This Essay	312
II.	1,1 Biradicals (Carbenes)	313
	A. Thermal Rearrangements of Alkynes	313
	B. Thermal Rearrangements of Cyclopropenes . . .	315
	C. Ketocarbene–Oxirene System	322
III.	1,3 Biradicals	324
	A. Trimethylenes in the Pyrolysis of Cyclopropanes	324
	B. Trimethylenemethanes in the Pyrolysis of Methylenecyclopropanes	334
	C. Spiropentane and Methylenespiropentane Rearrangements	348
IV.	1,4 Biradicals	353
	A. From Olefin Dimerizations, Cyclobutane Reversions, and Diazene Decompositions	353
	B. Hypothetical 1,4 Biradical Intermediates in the Cope Rearrangement	358
V.	Biradicals in [1,3] and [1,5] Sigmatropic Rearrangements: Some Comments on the Principle of Least Motion	372
	References	383

I. INTRODUCTION

A. Biradical Hypothesis

It is not difficult to imagine why the notion of a biradical intermediate has been so attractive to chemists who have tried to conceptualize the events in thermal unimolecular isomerizations. Since bonds must be broken to achieve the overall structural or stereochemical transformation, the simplest kind of bond-breaking reaction, gas-phase homolysis of a diatomic molecule, becomes a natural model. This reaction seems to reduce chemistry to its essentials. There is only one vibrational degree of freedom, and, superficially at least, there are very few mechanistic questions to ask. Unimolecular dissociation consists of vibrational excitation into the energy region where separation of the atoms to infinity no longer is opposed by a restoring force.*

Straightforward analogy would suggest that the initiating step of a thermal unimolecular isomerization is a homolysis to a biradical in which the odd-electron centers are at "infinity." This is equivalent to the assumption that the centers act independently. Although most chemists would agree that there must be some interaction between the centers, it is not easy to predict its magnitude or its experimental consequences. Attempts to fit the biradical hypothesis to experimental observations on thermal rearrangements have necessitated the attribution of a whole set of properties to these elusive species, but there are few instances in which independent confirmation of their existence is available.

B. Scope of This Essay

For the present purpose, the term "rearrangement" is taken to include the alteration of either stereochemistry or connectivity. This essay concerns those cases of thermal rearrangement in which biradical intermediates either have been postulated or are worthy of discussion as plausible alternatives. Since the number of reactions universally accepted as proceeding by nonbiradical (concerted) mechanisms is small, an attempt to cover all of the biradical cases would require a treatment of virtually every known thermal rearrangement. In fact, the coverage is selective and illustrative rather than comprehensive. However, extensive references to the primary and review literature are supplied and will enable the reader to become well acquainted with the field.

This essay is organized according to the number of bonds separating the odd-electron centers in the putative biradical intermediates. The system of nomenclature of the biradicals themselves should be self-evident. Somewhat arbitrarily, olefin cis–trans isomerizations via orthogonal ethylenes (1,2 biradicals) have been excluded (see Essay 3), but some rearrangements via carbenes (1,1 biradicals) have been included.

The focus of attention is on biradicals that can be generated by cleavage of a single carbon–carbon bond. In some instances, the "same" or at least closely related species can result from a decomposition reaction in which a stable fragment is ejected, as, for example, in the thermolysis of cyclic diazenes to molecular nitrogen and a biradical.

As a guide in evaluating the credibility of individual biradical inter-

---

* Of course, it is true that, at a deeper level, many details remain to be explored, even in the "simple" act, because the mechanistic questions actually have not disappeared but merely have been extended to include the energy-transfer process itself.

5. HYPOTHETICAL BIRADICAL PATHWAYS

mediates, there will be frequent reference to thermochemical arguments. In principle, if the biradical lies energetically below the transition state, it is a permissible (but not required) intermediate, but, if it lies substantially above, it is largely excluded from the mechanism. Biradical mechanisms, then, although difficult to prove, appear to be easy to disprove. Unfortunately, even this is too optimistic a view, as we shall see, because the values of the biradical energies needed for application of the exclusionary form of the guidelines often cannot be reliably estimated.

Other criteria for the detection of biradical intermediates include chemical capture, stereochemical behavior, nuclear polarization effects, and some special kinetic techniques. The reliability of these tests varies from case to case, and, as a rule, it is not possible to apply all of them simultaneously.

II. 1,1 BIRADICALS (CARBENES)

A. Thermal Rearrangements of Alkynes

The insight that a vinylidene (*1*) might be a thermally accessible intermediate in the gas-phase pyrolysis of acetylenes originated in observations by Brown and Harrington (*2*) of the behavior of the ^{13}C-labeled benzylidene derivative of Meldrum's acid (**1**). The major product is phenylacetylene, in which part of the label has migrated from its original site at the phenyl-bearing carbon. A plausible working hypothesis involves phenylvinylidene (**2**), which can give either of the labeled products (**3** or **4**). The labeled phenylacetylene suffers further randomization of the

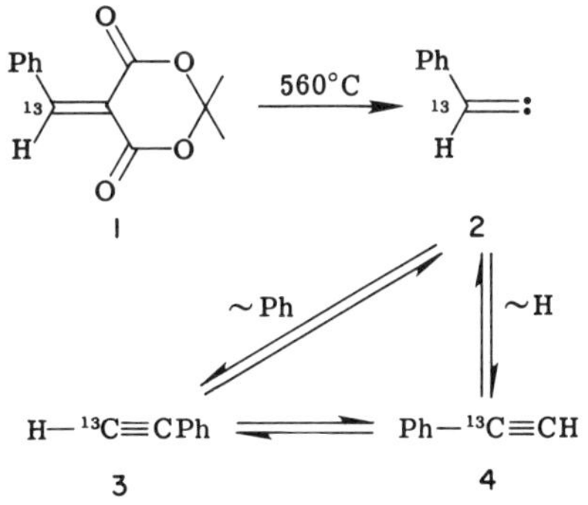

isotope between the acetylenic positions upon pyrolysis, eventually reaching a 50:50 distribution, presumably by reversible generation of the carbene **2**.

Further evidence for carbene intermediates in acetylene pyrolyses comes from intramolecular trapping experiments (*2–3*). Thus, 2-ethynylbiphenyl (**5**) gives benzazulene (**6**) (*2*), and σ-ethynyltoluene (**7**) gives indene (**8**) (*3*). The authors (*3–3a*) scrupulously point out that the **3** ⇌ **4** rearrangement formally does not require a carbene intermediate and could involve a dyatropic mechanism (*4*) (simultaneous rearrangement of Ph and H). However, this seems a less economical hypothesis than the carbene pathway, which accounts for both the rearrangement and the trapping results.

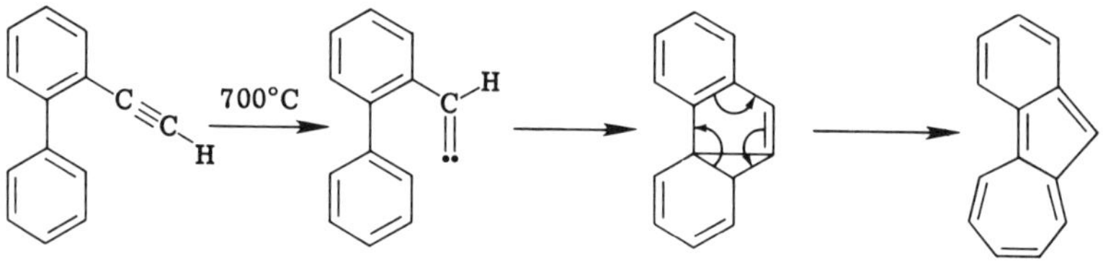

The superficially startling rearrangement of 1-methyl-1-ethynylcyclohexane to toluene and benzene has been studied by deuterium labeling and seems to be reasonably interpreted as an alkyne → carbene rearrangement followed by well-precedented further steps (*3a*).

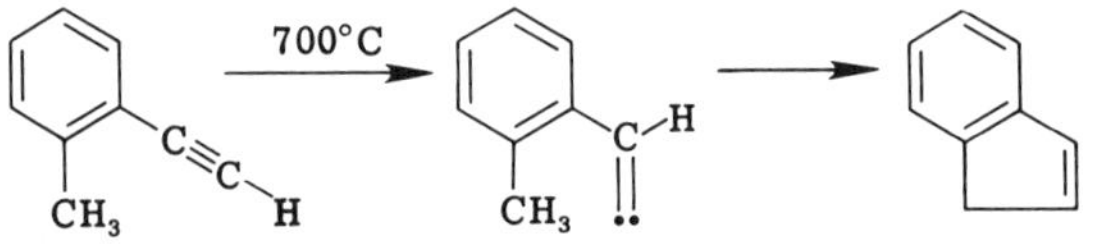

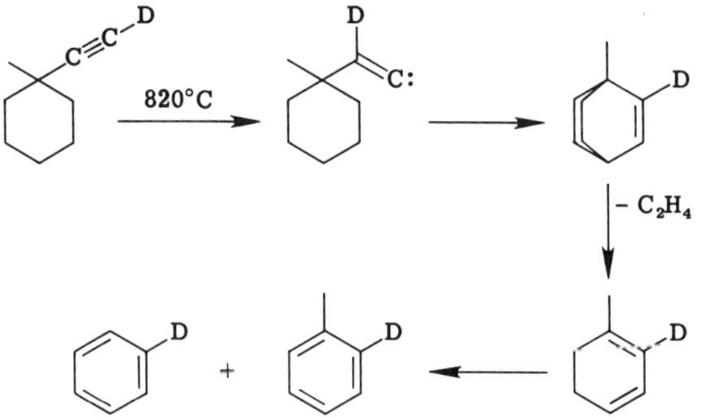

Although the heat of formation (ΔH_f°) of phenylacetylene can be calculated with some confidence from group additivities (*5*), that of phenylvinylidene (**2**) cannot. The usual procedure (*5–8*) for calculating the ΔH_f° value for a biradical (e.g., trimethylene) derived by hypothetical C—C scission of a carbocyclic ring (e.g., cyclopropane) involves adding twice the energy of the terminal C—H bonds of the corresponding alkane (e.g., propane) to the ΔH_f° of the alkane and subtracting from the sum the bond dissociation energy of H_2. The basic assumption (5–8) is that the odd-electron centers are far enough apart to act independently. This method leads to anomalies even when the free-electron centers are separated, as in trimethylene (**9**) (9, 10) and cyclopenta-1,3-diyl (**10**) (11), as we shall

$$H_2\overset{\cdot}{C}\diagup\overset{CH_2}{\diagdown}\overset{\cdot}{C}H_2 \qquad \overset{\cdot}{\diagup}\overset{\frown}{\diagdown}\overset{\cdot}{\diagdown}$$

          9   10

see later (Sections III,A and III,B). There is no reason to think that its accuracy should improve in the case of a carbene, to which the "infinite" separation model cannot be applied. In fact, the ΔH_f° of **2** is calculated as 147 kcal/mol by breaking the two terminal C—H bonds of styrene ($\Delta H_f^\circ = 35.3$ kcal/mol) on the assumption that both are worth 108 kcal/mol, which is the C—H bond energy of ethylene (5). Since the ΔH_f° for phenylacetylene is 78 kcal/mol, the endothermicity of the acetylene → vinylidene reaction on this basis is 69 kcal/mol. An activation energy $\Delta H^\ddagger \sim 46$ kcal/mol can be roughly estimated from the approximate rate data reported (2) for the isotope-position rearrangement **3** ⇌ **4**. The ΔH_f° of the transition state for the rearrangement is 78 + 46 = 124 kcal/mol, or 23 kcal/mol below the calculated ΔH_f° of the carbene **2**. Taken at face value, this would seem to exclude the carbene **2** as an intermediate, but, in the present writer's opinion, the result merely demonstrates that the ΔH_f° calculation should not be trusted in such cases. Note that, if it is assumed that the carbene mechanism is correct, the energetic discrepancy is opposite in direction to those observed with **9** and **10** (9–11).

A different (and perhaps more realistic) picture of the singlet vinylidene–acetylene energy surface, at least for the parent C_2H_2 system, is given by a quantum mechanical calculation using the method of self-consistent electron pairs (12). This predicts that the singlet ground state of vinylidene is only 40 kcal/mol more energetic than acetylene and that the transition state for the hydrogen shift is 8.6 kcal/mol above vinylidene. It is gratifying that the activation energy for the acetylene to vinylidene reaction thus predicted is 48.6 kcal/mol, in good agreement with the above estimate of about 46 kcal/mol, based upon the experimental data in the phenylacetylene case.

B. Thermal Rearrangements of Cyclopropenes

The gas-phase pyrolysis of the unsubstituted parent cyclopropene (**11**) gives methylacetylene as the major product (12, 12a) along with small amounts of allene (13a). Observations suggestive of carbene intermediates in cyclopropene pyrolyses include the rearrangement of 1,2,3-triphenyl-3-vinylcyclopropene (**12**) to 1,2,3-triphenylcyclopentadiene (**13**) (*13b*) and of tetraphenylcyclopropene (**14**) to the indene **16**, presumably via

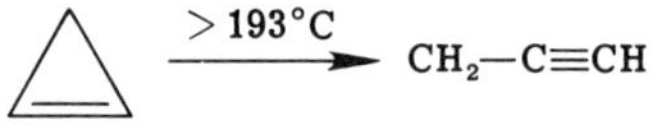

$$\triangle \xrightarrow{>193°C} CH_2-C\equiv CH$$

11

the proximate intermediate **15** (*14*). An apparent intermolecular capture of a carbene intermediate occurs in the pyrolysis of 3,3-dimethylcyclopropene in the presence of an olefin, which yields a cyclopropane by olefin–carbene addition (*15*).

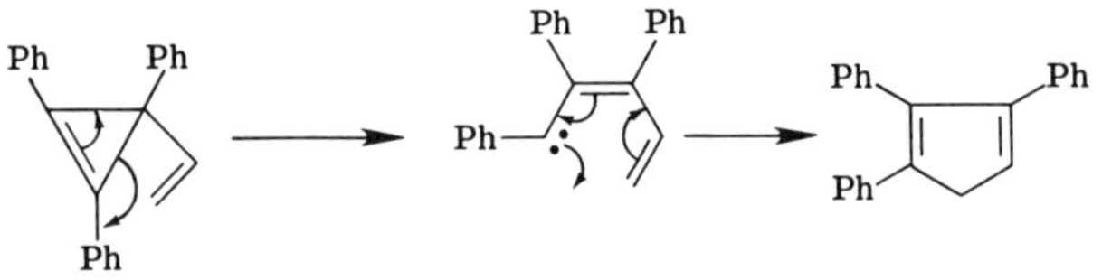

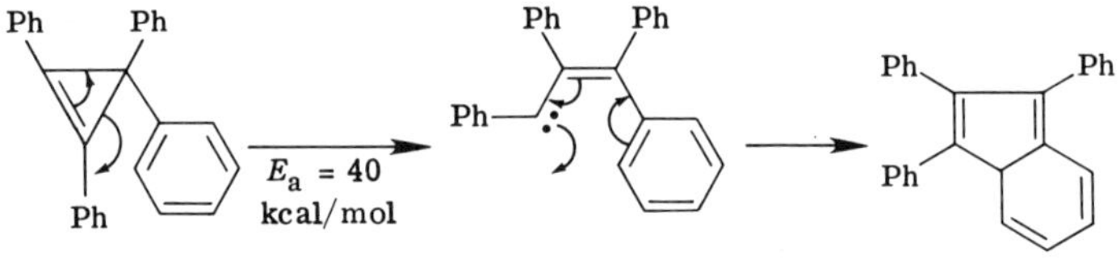

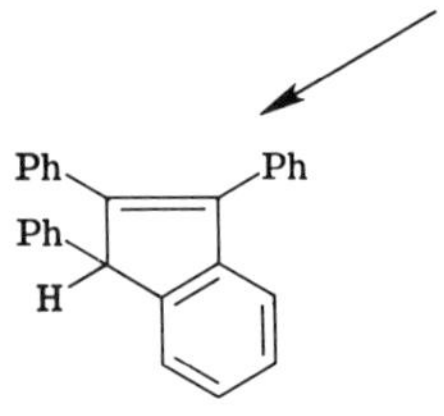

The reverse reaction, ring closure of vinylcarbenes to cyclopropenes (**17**), is known also and occurs with high efficiency when the carbene (or carbenoid) is generated from an alkenyldiazomethane or an allylic chloride (*17, 18*).

Bergman and co-workers (*16*) perceived the implication of these results—that a reversible cyclopropene ⇌ vinylcarbene rearrangement might be detectable in an appropriately constituted molecule—and prepared optically active 1,3-diethylcyclopropene (**18**) as a test system. Two

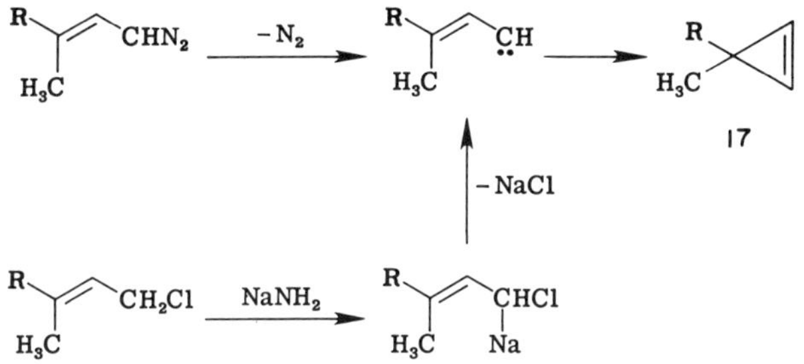

5. HYPOTHETICAL BIRADICAL PATHWAYS

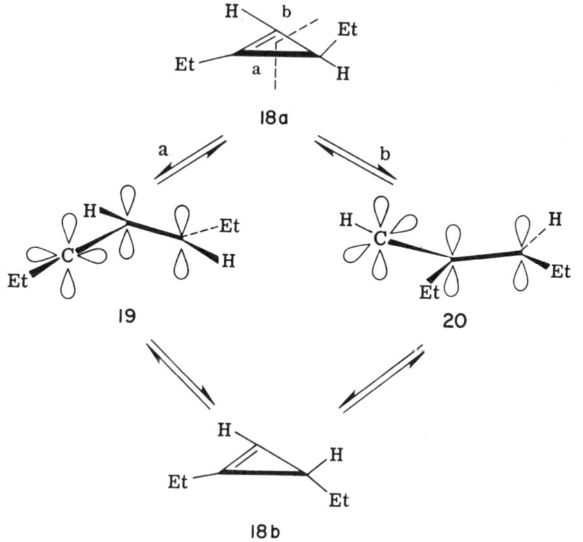

intermediate carbenes (**19** and **20**) might result from cleavage of a ring single bond of **18**. In either case, the carbene should become achiral, because the terminal carbon must rotate to take advantage of allylic resonance. (In the drawings, the —CHEt group of the carbene is arbitrarily assigned the configuration with Et "outside," but the argument would not be altered if H and Et were interchanged.) Pyrolysis of **18** at 161°–190° results not only in conversion to rearrangement products (2,4-heptadienes and 3-heptyne, resulting from hydrogen shift and cleavage of bond a), but also in racemization. The rate constant for the enantiomerization **18a** → **18b** is about nine times the rate constant for overall conversion to acyclic products. Thus, if a common carbene intermediate is involved, its reclosure to give the cyclopropene is much faster than its intramolecular hydrogen shift to give dienes and alkyne. The authors (*16*) assume that racemization is not caused by a 1,3-sigmatropic hydrogen shift, since this reaction does not occur in the thermolysis of 1- or 3-methylcyclopropene (*16a*).

These experiments seem to offer at least presumptive evidence for a carbene intermediate in cyclopropene pyrolyses. As in the case of the alkyne rearrangements, however, there are difficulties in developing a convincing argument for thermochemical compatibility. In one approach to this question (*16*) an attempt is made to calculate the ΔH_f° of the hypothetical carbene **19** from the ΔH_f° of *trans*-3-heptene (**21**), −17.4 kcal/mol. The addition of 105 kcal/mol [ethylene C—H bond energy less 3 kcal/mol alkyl substituent effect (*5–8*)] breaks the C-3—H bond to give the vinyl radical **22**, $\Delta H_f^\circ = 87.6$ kcal/mol. To reach the carbene **19** and H_2, one adds another 84 kcal/mol to break the allylic C—H bond and subtracts the H—H bond energy, 104 kcal/mol. This gives H_f° for **19** as 87.6 + 84 − 104 = 68 kcal/mol, which is 6 kcal/mol less than the sum of ΔH_f° of the reactant **18** (42 kcal/mol) and the activation energy (32 kcal/mol) and hence permits **19** to be an intermediate with a 6 kcal/mol barrier to ring closure (Fig. 1).

It should be noted that the calculation assumes that the energy needed

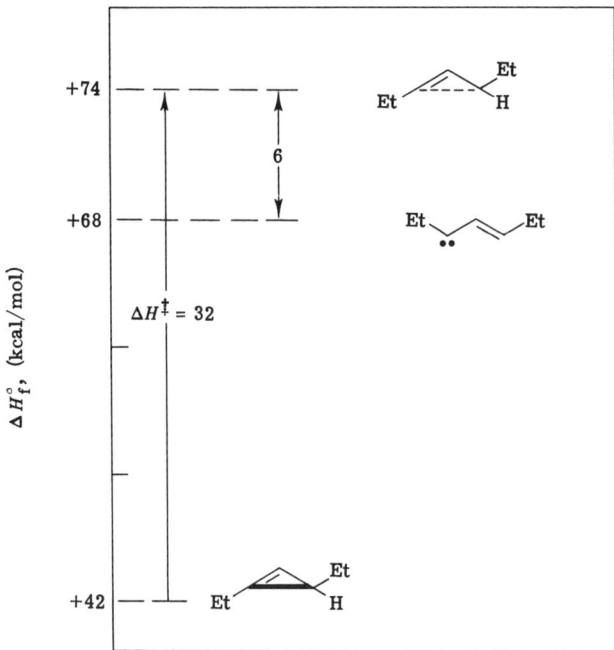

Fig. 1 Hypothetical thermochemical relationships in the thermolysis of 1,3-diethylcyclopropane (*16*).

5. HYPOTHETICAL BIRADICAL PATHWAYS

$$\underset{\substack{\Delta H_f^\circ \ -17.4 \\ (\text{kcal/mol}) \\ \mathbf{21}}}{\text{Et}\diagdown\!\!\!\diagup\!\!\!\overset{H}{\underset{H}{\overset{|}{C}}}\!\!\!\diagdown\!\!\!\text{Et}} \xrightarrow{105 \text{ kcal/mol}} \underset{\substack{+87.6 \\ \mathbf{22}}}{\text{Et}\diagdown\!\!\!\diagup\!\!\!\overset{H\ \ \text{Et}}{\underset{H}{\overset{\bullet}{C}}}} \xrightarrow[-104 \text{ kcal/mol}]{84 \text{ kcal/mol}} \underset{\substack{+68 \\ \mathbf{19}}}{\text{Et}\diagdown\!\!\!\diagup\!\!\!\overset{\bullet\bullet}{\text{CHEt}}} + H_2$$

to break the C—H bond of the vinyl radical **22** is just the standard secondary allylic C—H bond energy (5–8) and that it is not affected by the simultaneous presence in **22** of an odd-electron center 1,3-related to the allylic C—H bond. Another perspective on this can be shown by an alternative calculation of ΔH_f° for **19**, which, because thermochemical cycles are path independent, should give the same value if the same electronic state of the biradical is produced. One starts by removing the allylic hydrogen of **21** to give **23**, using the unperturbed secondary allylic C—H energy. To achieve **19**, one must remove a second hydrogen, which now is bonded to an sp² carbon. If one assumes the energy of this bond to

$$\underset{\substack{\Delta H_f^\circ \ -17.4 \\ (\text{kcal/mol}) \\ \mathbf{21}}}{\text{Et}\diagdown\!\!\!\diagup\!\!\!\overset{H}{\underset{H}{\overset{|}{C}}}\!\!\!\diagdown\!\!\!\text{Et}} \xrightarrow{84 \text{ kcal/mol}} \underset{\substack{+67 \\ \mathbf{23}}}{\text{Et}\diagdown\!\!\!\diagup\!\!\!\overset{\bullet}{\underset{H}{\overset{|}{}}}\!\!\!\diagdown\!\!\!\text{Et}} \xrightarrow[-104 \text{ kcal/mol}]{105 \text{ kcal/mol (?)}} \underset{\substack{+68\ (?) \\ \mathbf{19}}}{\text{Et}\diagdown\!\!\!\diagup\!\!\!\overset{\bullet\bullet}{\underset{H}{}}\!\!\!\diagdown\!\!\!\text{Et}} + H_2$$

be 105 kcal/mol, as in the case of **21** → **22**, one would calculate $\Delta H_f^\circ = 68$ kcal/mol, the same value as before. However, this procedure neglects to take into account the fact that, in the second calculation, the first and second C—H bonds are being broken at the *same* (or an equivalent) carbon. If one may use the energies of sequential breaking of C—H bonds in methane as a model, one might expect a substantial perturbation from this circumstance. For example, the C—H bond energy for the step $CH_4 \to CH_3$ is 104 kcal/mol (5), but that for $CH_3 \to CH_2$ is 118 kcal/mol (19–24). There is no obvious way to estimate how much this "same atom" perturbation should amount to in the reaction **23** → **19**, but adoption of the full 14 kcal/mol effect seen in CH_2 would place ΔH_f° of **19** at 68 + 14 = 82 kcal/mol, which is above the transition state energy (see Fig. 1). Alternatively, one might argue that the "same-atom" effect should be less than 14 kcal/mol. Removal of the second hydrogen (in the step $CH_3 \to CH_2$) should be more difficult than removal of the first (in $CH_4 \to CH_3$), simply because of the change in hybridization. The sp² C—H bond in CH_3 should be about 10 kcal/mol stronger than the sp³ C—H bond in CH_4 [compare ethylene and ethane (5)], which would reduce the "same-atom" effect to only 4 kcal/mol. This would place ΔH_f° of **19** at 72 kcal/mol, or 2 kcal/mol below the transition state (Fig. 1). These calculations are certainly loose enough for **19** to be a plausible intermediate, but, by the same token, they define the height of the ring-closure barrier only roughly.

Extensive theoretical studies (25) by the generalized valence bond method explore a number of subtle problems concerning the spin states, electron configurations, and geometries of various vinylcarbene species. In the context of the thermolyses of cyclopropenes, the major conclusion is that the singlet intermediate **24** is the state most important in the isomerization. The possibility that **24** could decay to the triplet **25**, which

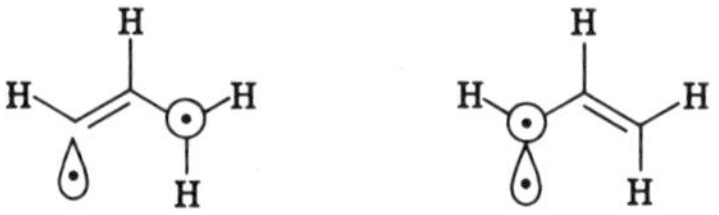

is calculated to be 12 kcal/mol more stable, also is considered, but no opinion is offered as to the probability that this event may occur during the thermolysis (25). One wonders whether the racemization of **18** may not involve a *direct* conversion to a triplet intermediate by a spin-forbidden pathway. Recent theoretical speculation (26–28) suggests that such reactions may be feasible when the ground state of the intermediate formed is of different spin multiplicity than the reactant (e.g., $^1CH_2N_2 \rightarrow {}^1N_2 + {}^3CH_2$). Were this the case for the racemization of **18**, we should expect a low value for A, the Arrhenius preexponential term (negative $\Delta S^{\ddagger}$), because of the less than unit probability of intersystem crossing from singlet to triplet energy surfaces (29). Although the rate measurements should be treated with reserve, because the reaction may have a heterogeneous component (30), it is nevertheless intriguing that the A values for the racemization and structural isomerization of **18** (16) are in fact only 10^{-3}–10^{-4} of those for typical cyclopropane pyrolyses (8).

The interesting rearrangements of bicyclopropenyls to benzenes [e.g., **26** → **27** (31)] are exothermic by about 120 kcal/mol (32). An initial suggestion (33) that vinylcarbenes might be involved in the mechanism was incorporated into a later model (34) for the aromatization of 3,3′-tetramethylenebicyclopropenyl (**28**) (Scheme 1), which gives tetralin (**29**)

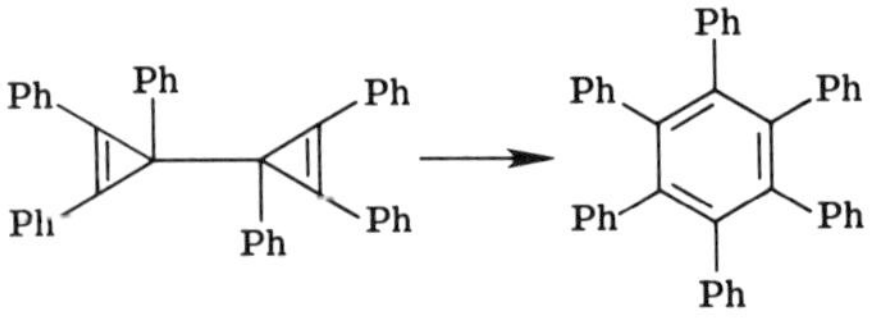

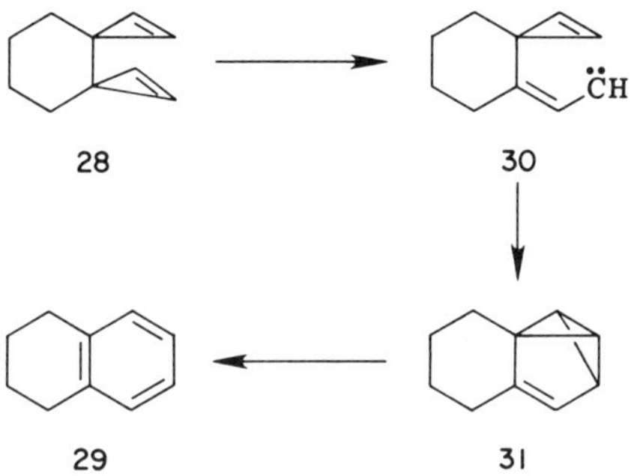

5. HYPOTHETICAL BIRADICAL PATHWAYS

Scheme 1

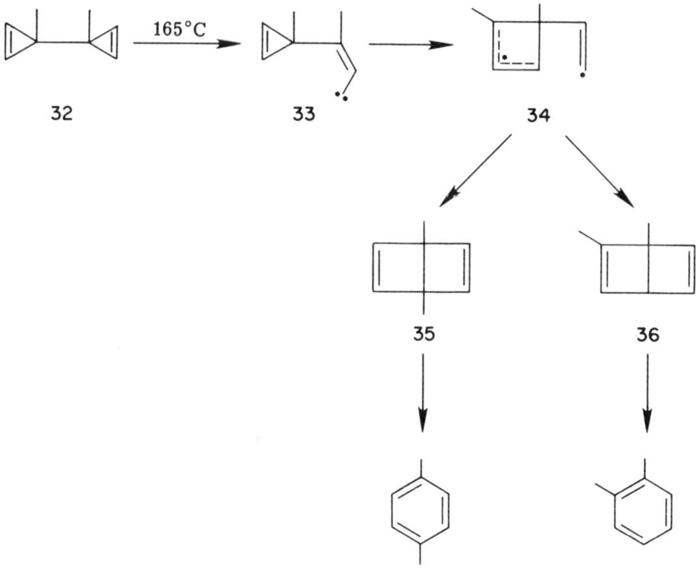

upon pyrolysis, hypothetically by way of the vinylcarbene **30** and the benzvalene **31**. Once the vinylcarbene is formed, however, other paths for the later stages of the aromatization can be postulated. An example is the gas-phase aromatization of 3,3'-dimethylbicyclopropenyl (**32**) to *o*- and *p*-xylene. Ring expansion of the initial vinylcarbene **33** to the cyclobutenyl biradical **34** is thought to be followed by ring closure to Dewar benzenes **35** and **36**, which then aromatize to xylenes (*32*). Additional, indirect evidence for the Dewar benzene mechanism in this case is provided by the observation of chemiluminescence when the reaction is conducted in the liquid phase, a phenomenon that is associated with the aromatization of Dewar benzenes, but not of prismanes or benzvalenes (*35*). Some spectacular further examples of vinylcarbene–cyclopropene rearrangements are described in Essay 3.

C. Ketocarbene–Oxirene System

The ketocarbene–oxirene pair shows a close structural resemblance to the vinylcarbene–cyclopropene case. The photochemical Wolff rearrangement of α-diazo ketones **37** to ketenes **38** competes with another rearrangement, which can be detected by isotopic labeling (*36–39*). For example, azibenzil-carbonyl-^{13}C (**37a**) undergoes photolysis in cyclopen-

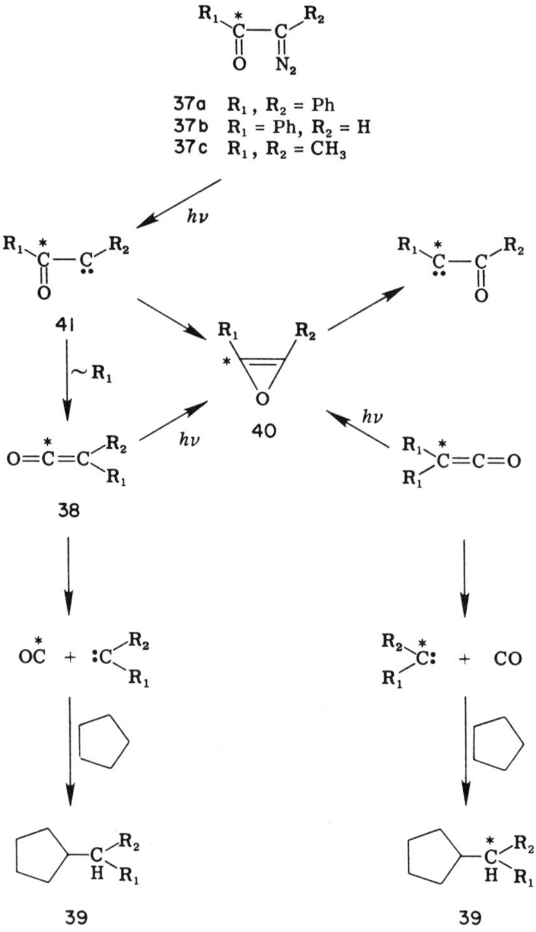
37a R_1, R_2 = Ph
37b R_1 = Ph, R_2 = H
37c R_1, R_2 = CH_3
hν
41
~R_1
40
hν
hν
38
39
39

tane solution to give CO and diphenylcyclopentylmethane (**39**), which are derived from the Wolff rearrangement product, diphenylketene (**38a**), by secondary photolysis (*37*). The isotopic compositions of the CO and **39**

5. HYPOTHETICAL BIRADICAL PATHWAYS

reveal that position exchange of the label occurs during the photolyses of both the diazo compound **37a** (~60% scrambling) and the ketene **38a** (~30% scrambling). The data are expressed as percentages of the product formed with complete mixing of the label. No scrambling is observed in the *thermolysis* of **37a** or **38a** (*36*). Similar results are obtained with **37b** (*37*) and other systems (*38, 39*).

Quasi-symmetric species other than the oxirene **40** (for example, the diazene **42**) might be envisaged as accounting for the scrambling in the products of photolysis of the diazo ketone. It would be useful to know in

$$R_1 - *C \underset{O}{\overset{N=N}{\underset{\diagdown \diagup}{-}}} C - R_2$$

42

this connection whether diazo ketone starting material recovered from partial photolysis has suffered any scrambling. However, it is more difficult to propose an alternative in the case of the ketene. Certainly, those who resist the blunt assertion (*37*) that oxirenes are implicated in these reactions "beyond doubt" must provide some reason for rejecting mechanistic parsimony.

According to the current view of the mechanism (*37, 40*), the photolysis of the diazo ketone **37** gives rise to an excited ketocarbene (**41**), which rearranges in part to the ketene **38** and in part to the oxirene **40**. The relative stabilities of the oxirene and the ketocarbene and the magnitude of the thermal barrier to their interconversion are matters of some controversy. Extended Hückel MO calculations (*41*) place oxirene 30 kcal/mol higher in energy than formylcarbene and show no barrier to ring opening, whereas *ab initio* SCF—MO (*42*), MINDO/3 (*43*), and NDDO (*43*) methods assign oxirene an energy that is *lower* than that of the ketocarbene by 0.4, 18.2, and 20.6 kcal/mol, respectively. More recently (*40*), an SCF—MO calculation using a contracted double-zeta basis set showed the oxirene to be 11.8 kcal/mol higher in energy but protected from ring opening by a barrier of 7.3 kcal/mol. All of the later calculations suggest that oxirene or a derivative might be observed spectroscopically at low temperature or perhaps even be captured chemically. In fact, the report (*44*) that photolysis of the diazoamide **43** in methanol gives the oxirane **44**, presumably by capture of the oxirene **45**, lends credence to this view.

If the ΔH_f° of oxirene really is 11 kcal/mol higher than that of the corresponding ketocarbene, there is a sharp contrast with the structurally similar case of cyclopropene, which has ΔH_f° about 26 kcal/mol *lower* than that of its corresponding form, vinylcarbene (Fig. 1). One wonders how much of this 37 kcal/mol discrepancy should be attributed to specific destabilization of oxirene by its antiaromatic (4n) π-electron system.

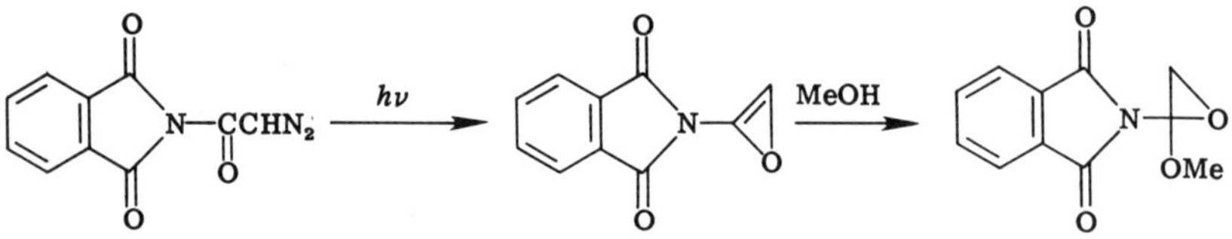

43 → **45** → **44**

III. 1,3 BIRADICALS

A. Trimethylenes in the Pyrolysis of Cyclopropanes

1. Noninteracting biradical model

The structural rearrangements and stereomutations of cyclopropanes (**46**) have been frequently reviewed (*5–10, 45–50*), as befits the most

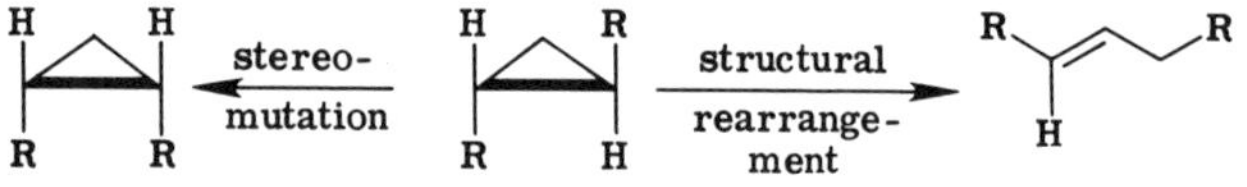

46

intensely studied of all unimolecular reactions. Trimethylene biradicals (**47**) are postulated (*5–8*) as the common intermediates for both processes. The evidence supporting this postulate is circumstantial and consists

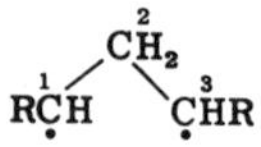

47

largely of the demonstration that the estimated ΔH_f° of the biradical is lower than that of the rate-determining transition state. It will be recognized that this corresponds to a permissive rather than an obligatory basis for a mechanistic assignment. Moreover, there are now several reasons for doubting the validity of the "infinite separation" assumption on which the thermochemical estimates rest. These are given elsewhere in some detail (*9, 45*), but the essential reasons are briefly restated here.

The ΔH_f°, 67 kcal/mol, of unsubstituted trimethylene (**47**, R = H) can be calculated (*5–8*) by adding twice the C-1—H bond dissociation energy of propane (2 × 98 kcal/mol) to the ΔH_f° of propane (−25 kcal/mol) and subtracting from the sum the bond dissociation energy of H_2 (104 kcal/mol) (Fig. 2). The ΔH_f° values of the transition states for the thermal stereomutation (76.3 kcal/mol) and structural isomerization (77.8 kcal/mol) of cyclopropane-1,2-d_2 (**46**, R = D) are obtained by adding the observed (*51–52*) activation enthalpies to the ΔH_f° of cyclopropane (13 kcal/mol). The hypothetical biradical (**47**, R = H) thus resides in a rather

5. HYPOTHETICAL BIRADICAL PATHWAYS

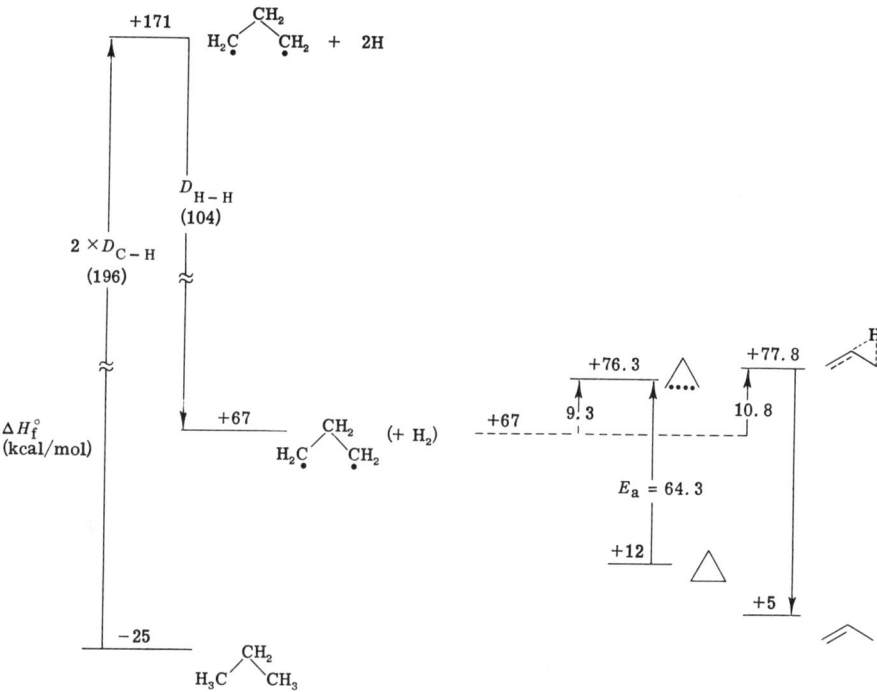

Fig. 2 Measured and thermochemically calculated energies of species in the pyrolysis of cyclopropene.

deep energy well, protected from cyclization to cyclopropane or from hydrogen shift to propylene by barriers of about 9 and 11 kcal/mol, respectively.

The assumption that the odd-electron centers at C-1 and C-3 of **47** do not interact is implicit in this method of calculating ΔH_f°, which uses identical values for the bond dissociation energies of C-1—H in propane and C-3—H in propyl radical. The model gives no reason, therefore, to assume that the barrier heights opposing internal rotation about the C-1—C-2 and C-2—C-3 bonds of **47** (R = D) should be different from the values of a few tenths of a kilocalorie per mole found (53, 54) in ordinary primary monoradicals (RCH$_2$·). Once formed by fission of a ring bond, biradical **47** should become stereochemically randomized before it recyclizes.

In the case of a cyclopropane bearing real rather than isotopic substituents, the internal rotation rate should slow down because of the increased moment of inertia (6). If it is correct to assume (5–8) that the rate of cyclization of **47** is relatively insensitive to substituents, then the stereochemical behavior in heavily substituted examples should approach that of a single rotation of C-1 (or C-3). [Single rotation also could be a consequence of a different pathway involving a transition state with an "expanded ring" (*51a, 55*).]

2. Quantum mechanical model of cyclopropane thermolysis

The observations (*56, 57*) of a stereochemical "crossover" phenomenon in the pyrolysis of 3,5-disubstituted pyrazolines stimulated a consideration of the mechanism from a different viewpoint. The loss of molecular nitrogen in the pyrolysis of a cis-3,5-disubstituted pyrazoline, **48**, for example, might be expected to give a trimethylene intermediate that could cyclize to a cis-disubstituted cyclopropane, if internal rotations were slow, or to a mixture of *cis*- and *trans*-cyclopropanes, if internal rotations were fast. In the latter case, the *cis*- and *trans*-pyrazolines **48** and **49** should give identical mixtures of cyclopropanes. The experimental facts, however, are inconsistent with either of these models, since the stereochemistry of the cyclopropane product in each case is predominantly (~3:1) *opposite* to that of the pyrazoline (Fig. 3) (*56*). Obviously, a stereorandomized trimethylene cannot be the sole intermediate. In fact, the stereochemistry of deazetation of pyrazolines, even after further, extensive investigation (*58*), still is not completely understood. Moreover, it is clear (*58*) that at least a portion of the products from the pyrazoline are formed by pathways that are not the same as those involved in the cyclopropane thermolysis. In particular, these processes cannot be described completely by the simple assumption of *any* common trimethylene intermediate, stereorandomized or not. Nevertheless, the rationalization of the "crossover" phenomenon (*56*) as the result of the conrotatory closure of a "π-cyclopropane" (**50**), a trimethylene in which

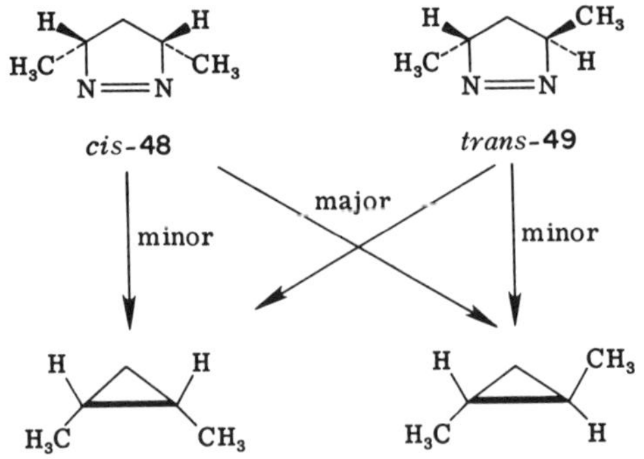

Fig. 3 Stereochemical crossover in the pyrolysis of 3,5-disubstituted pyrazolines (*56*).

5. HYPOTHETICAL BIRADICAL PATHWAYS

the terminal groups lie in the carbon plane, stimulated a theoretical examination (59) of the properties of various trimethylene species.

Consider three idealized trimethylene rotamers, 0,0, 0,90, and 90,90, where the numbers refer to the degrees of arc by which each of the terminal methylenes is twisted out of the carbon plane. An extended Hückel calculation (59) shows that, in the 0,0 form ("π-cyclopropane"), there is hyperconjugative interaction between one of the filled bonding orbitals of the central CH_2 group at C-2 and the p orbitals at C-1 and C-3.

This interaction has two main consequences. First, it splits the degeneracy of the in-phase and out-of-phase π combinations by selective destabilization of the in-phase level. The out-of-phase π combination is of the wrong symmetry to interact with either combination of the C-2 CH_2 orbitals and, to a first-order approximation, remains unperturbed. When

In phase   Out of phase

the C-1—C-2—C-3 bond angle of the 0,0 species exceeds 100°, the out-of-phase combination therefore is the highest occupied molecular orbital. By orbital symmetry (60), overlap of the C-1 and C-3 lobes of like sign would lead to conrotatory ring closure. Moreover, the calculations (59) show that whereas the rotation of one CH_2 group out of the plane (process 0,0 → 0,90) requires a substantial activation energy, which should preserve the stereochemical relationships between the labels in a substituted case, *no barrier* should oppose the synchronous conrotatory closure of the 0,0 species to cyclopropane. These theoretically deduced properties of the 0,0 species are just those needed to provide at least a qualitative rationalization of that part of the pyrazoline thermolysis which is responsible for the "crossover" effect.

The second consequence of the hyperconjugation of the C-2 CH_2 group and the p orbitals is the overall stabilization of the 0,0 form relative to the 0,90 and 90,90 rotamers when the C-1—C-2—C-3 angle is ~125°. This suggests the possibility that the 0,0 species may be on the pathway for the cyclopropane pyrolysis itself (59).

Several quantum mechanical calculations much more elaborate than Hoffmann's original extended Hückel study (59) also indicate that the π-cyclopropane should lie on the energetically favored stereomutation pathway (61–63). Microscopic reversibility requires that, if the 0,0 intermediate closes to cyclopropane by a synchronous double rotation, it also must be formed by such a process. This predicts that, if reaction occurs preferentially at the C-1—C-2 bond of an optically active trans-1,2-disubstituted cyclopropane (51), as would be expected if substituents weaken the bond (5–8), the rate constant for enantiomerization (two-center epimerization) should exceed the rate constant for diastereomerization (one-center epimerization). (Note that this conclusion is independent of whether the coupled rotations are con- or disrotatory.)

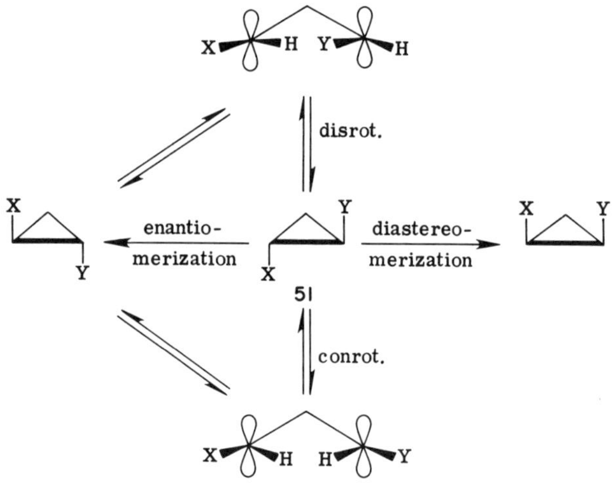

The predictions of the quantum mechanical model thus differ from those of the thermochemical–kinetic one in two important respects. First, the motions of the terminal CH_2 groups of trimethylene should be coupled, not independent. Second, the trimethylene should correspond to a rather flat region of the energy surface near the transition state, not to a deep local minimum.

In a formal sense, the electron pair in the filled C—H bonding level of the developing π-cyclopropane is analogous to a lone pair in the cyclopropyl anion 52 or its isoelectronic analogues, aziridine (53) and oxirane (54). The latter three systems should undergo (60) thermal ring cleavage to the allyl anion **55**, azomethine **56**, or carbonyl ylid species **57**, respectively, with the same orbital symmetry-controlled synchronous conrotatory stereospecificity predicted for cyclopropane itself. In fact, there is strong experimental evidence favoring this pathway in substituted examples of both aziridine *(64)* and oxirane *(65)*. However, the conjugative

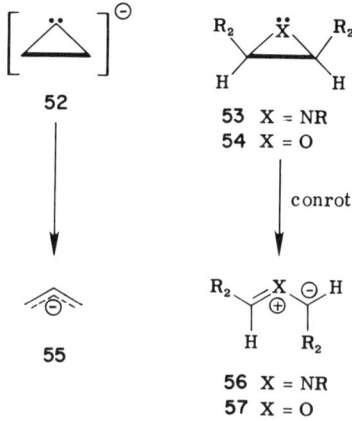

interaction between the nonbonding heteroatom orbitals and the terminal p orbitals in the developing species **56** and **57** is stronger than the hyperconjugative interaction involving the C—H bonding orbital in π-cyclopropane *(59, 66)*. The prediction of coupled motions in cyclopropane ring opening, therefore, depends on a rather delicate quantum mechanical effect, one that barely survives the most sophisticated calculation *(63)*, in which the double-rotation transition state is predicted to lie only 0.6 kcal/mol below that of the single-rotation pathway. Since the calculations pertain to cyclopropane itself, it would not be surprising if substituents could perturb the energetic ordering of the transition states for the competing pathways, and some calculations at the CNDO level are in agreement with this conjecture *(67)*. Moreover, the behavior of even the parent compound may not be controlled simply by the transition state energy differences, but may be influenced by dynamic effects, which depend on the vibrational–rotational energy distribution of the energized molecules and its interaction with the energy hypersurface. In one treatment of such effects by a semiclassical model it is concluded that they should reinforce the preference for double rotation in the cyclopropane case *(68, 69)*.

3. Experimental studies of cyclopropane stereomutations

Examples of each of the extreme mechanisms as well as of mixed behavior have been described in the literature. A stereorandom intermediate (**61**) provides the most convincing explanation of the thermolysis of *anti*-vinylcyclopropane-2,3-*cis*-d_2 (**58**) *70*). In the gas phase at 325°C the other stereoisomers, syn–cis (**60**) and both enantiomers of trans (**59a** and **59b**), are formed in the ratio of 1 part **60** to 2 parts **59a** + **59b**. This ratio is independent of the extent of reaction between 0.6 and 10 half-lives for equilibration of the label, which is consistent with the intervention of

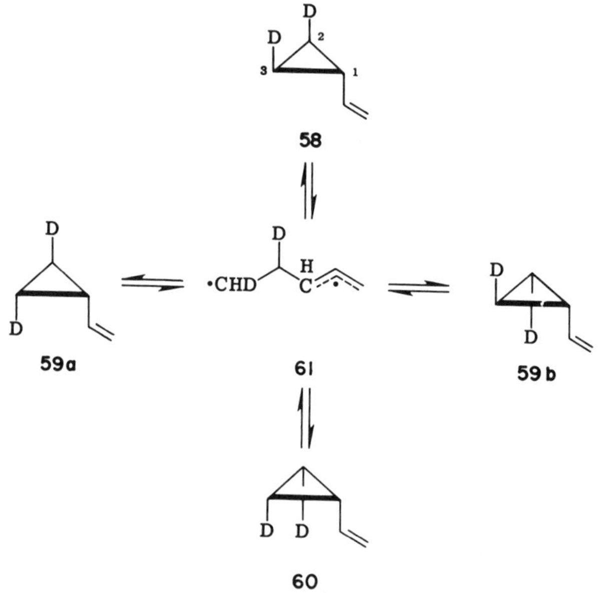

an intermediate having the symmetry of **61**. An interpretation based on synchronous double rotations, although not rigorously excluded, requires fortuitous, and unlikely, equality of the C-2—C-3 rotation with the C-1—C-2 and C-1—C-3 rotations. Although the hypothetical intermediate (**61**) is shown here as having been derived by cleavage of the C-1—C-3 (≡ C-1–C-2) bond, which should be the weakest because of allylic stabilization of one radical site, this is not a necessary assumption. The same kinetic behavior would result if the reaction involved cleavage at C-2–C-3, provided only that internal rotations in the biradical were fast relative to cyclization.

In the most common experimental approach to the stereomutation of cyclopropanes, the rate constants for enantiomerization are compared with those for diastereomerization of an optically active 1,2-disubstituted cyclopropane (**62**). The following scheme enumerates the centers that must suffer epimerization to achieve each overall transformation:

5. HYPOTHETICAL BIRADICAL PATHWAYS

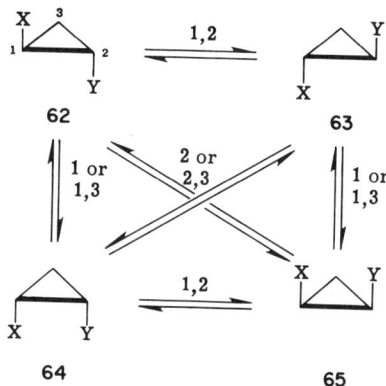

In general, there is an ambiguity in the kinetic analysis of such systems, because there are ten mechanistic rate constants but only six experimentally observable ones. To circumvent this difficulty, it often has been assumed that only the disubstituted bond reacts, since the resulting biradical intermediate should be more effectively stabilized than the one that would result from cleavage of a monosubstituted bond. Although the argument tacitly adopts a particular mechanism and therefore is circular, it probably is safe, at least when X and/or Y are strongly conjugating substituents, because explicit searches in the thermolyses of especially designed cyclopropanes have shown "wrong-bond" cleavage to be absent (71–73).

Table 1 (70–73j) summarizes the known relative values of the rate constants for one-center (k_1 and k_2) and two-center (k_{12}) epimerizations. In some cases, the introduction of symmetry elements or of extra substitution permits a kinetic analysis which is independent of the assumption that reaction occurs only at the most substituted bond. Obviously, a wide range of behavior is permitted, and one of the outstanding challenges in the theory of unimolecular reactions now becomes the interpretation of the variations evident in Table 1. It is perhaps a source of some gratification to theoreticians that the experimental results in the case of cyclopropane itself, the system they have studied most carefully, correspond to the predicted double epimerization. However, we have only the previously mentioned CNDO calculations (67) to explain why, of all the substituted cases, only phenylcyclopropane conforms to this mechanism. The difference in behavior between phenylcyclopropane (two-center epimerization) and vinylcyclopropane (random intermediate), compounds that structurally seem so similar, is especially puzzling.

Aside from vinylcyclopropane, Table 1 does not contain a clear exam-

TABLE 1
Relative Values[a] of Rate Constants for Single and Double Epimerizations in Cyclopropane Stereomutations

Cyclopropane	Relative rate constant			Predominant mechanism	Ref.
	k_1	k_2	k_{12}		
1,2-Dideuterio[b]	1	1	>7[c]	Double epimerization	73a, 73b
1-Phenyl-2-deuterio[b]	1	<2	>12[c]	Double epimerization	71, 73b, 73c
1,1,2,2-Tetramethyl-d_6	1	1	0.15	Single epimerization	73d
1-Vinyl-2,3-d_2[b]		Statistical		Random intermediate	70
1-Ethyl-2-methyl					
Trans	1	1	1.1	Mixed	73e, 73f
Cis	1.1	1	1	Mixed	
1-Cyano-2-isopropenyl					
Trans	3.0	1	3.3	Mixed	73g
Cis	2.1	1	1.2	Mixed	
1,2-Diphenyl	1	1	3.3	Random intermediate (?)	73h
1-Cyano-2-phenyl-1,3-Dideuterio[b,d]	2.3	1	1.8	Mixed	72
1,2-Diphenyl-1-carbomethoxy	13	1	0.4	Single epimerization	73i
1,2-Dimethoxy-3-methyl	1	1	0.5	Mixed	73j
1-Cyano-2-phenyl					
Trans	2.2	1	0.7	Mixed	73
Cis	2.2	1	0.35	Mixed	

[a] Normalized to the slower of the single-epimerization values.
[b] Values do not depend on the assumption that only the C-1–C-2 bond reacts.
[c] The value depends on the magnitude assumed for the secondary isotope effect. For discussion, see (72, 73b).
[d] The experiment establishes that k_3, k_{13}, and k_{23} are essentially zero.

ple of reaction by way of a stereorandomized intermediate. According to the Benson–O'Neal thermochemical–kinetic analysis (*5–8*), the ratio of rates of internal rotation and cyclization of the trimethylene biradical should increase as the terminal substituents become smaller. In the case of cyclopropane itself, the ratio should be large enough to ensure stereorandomization. A direct test of this prediction comes from the study of optically active *trans*-1,2-dideuteriocyclopropane (*73a, 73b*), in which the double-rotation, random intermediate, and single-rotation mechanisms, respectively, predict ratios of k_i/k_α, the rate constants for trans–cis interconversion and optical deactivation, of 1.05, 1.53, and 2.00 (corrected for a secondary isotope effect). The observed (*73a, 73b*) value of 1.07 ± 0.04 corresponds to essentially pure double rotation and leaves little room for a large contribution by the hypothetical random intermediate.

Nevertheless, the intriguing question remains as to why the stereomutation of cyclopropane so carefully avoids the stereorandom intermediate in favor of a transition state 9 kcal/mol higher in energy. In a general way there can be two answers to this question. The first is that, although the enthalpy of formation of stereorandom trimethylene permits its participation, there is for some reason a very unfavorable entropy factor. This is the explanation given by Goddard *et al.* (*62*), whose generalized valence bond quantum mechanical calculations find the species in a very inaccessible region of phase space. Note that, if this interpretation is correct, the Arrhenius plots of the stereorandom and double-rotation mechanisms should cross, and, at some temperature below the region (422°C) where the latter predominates, stereorandom behavior should be observed. Because the absolute rate at temperatures below the crossing point would be too slow to observe during a single investigator's life span, this test might be suitable for inclusion in a time capsule.*

The second explanation is that there is a systematic error in the thermochemical–kinetic scheme that tends to overemphasize the stability of biradicals. Contributions to such an effect could arise, for example, from the neglect of spin correlation (*74*) or from a bond dissociation energy for the C-3—H bond of propyl radical that is about 9 kcal/mol greater than that of the C-1—H bond of propane (*9*). If the heat of formation of trimethylene biradicals were actually 9 kcal/mol greater than that calculated (*5–8*) by bond additivity, internal rotation and ring closure would be predicted to become competitive, and the apparent anomaly deduced from the absence of the stereorandom mechanism would vanish.

---

* The author is assuming that the temperature in the capsule will not be disturbed by nearby nuclear perturbations (Ed.).

The interpretation of the difficulty as an enthalpy rather than an entropy problem seems more plausible, in view of a similar anomaly in the case of cyclopenta-1,3-diyl (**10**) (*11*). The additivity assumption (*5–8*) combined with the activation energy (*75*) for stereomutation of a labeled bicyclo-[2.1.0]pentane places the biradical **10** in a potential well of 9 kcal/mol (*7*). Yet the experimental barrier for ring closure of the diyl, observed by ESR spectroscopy of the triplet at low temperatures (5.5–20 K) is 2 kcal/mol (*11*). Apparently, here again the additivity assumption overestimates the stability of the biradical.

10

Despite intense experimental and theoretical effort directed toward understanding cyclopropanes, these simple molecules obdurately continue to hide the secrets of their behavior. Among the unanswered questions that remain, perhaps the following two are the most obvious: (a) Does the reaction path for the structural isomerization of cyclopropane to propene intersect that for the stereomutation? (b) Is the two-center epimerization conrotatory or disrotatory? It seems probable that the responses to these challenges will require new experimental approaches far more powerful than those available heretofore. If mechanistic research activity declines, the causes will not include an exhaustion of the supply of significant questions.

B. Trimethylenemethanes in the Pyrolysis of Methylenecyclopropanes *(75)*

1. Scope of the reaction

The thermal structural isomerizations of methylenecyclopropanes were discovered by E. F. Ullman, who observed the rearrangement of the trans and cis forms of "Feist's ester," **66** and **67** (X = CO_2Me) (*76–77*). Many other examples of this methylenecyclopropane reorganization were quickly found, including the rearrangement of the parent compound (see Table 2) (*76–89*).

2. Mechanism: role of trimethylenemethanes

The simplest formulation of the rearrangement involves cleavage at C-2—C-3 and rebonding of C-2 or C-3 to the exocyclic methylene carbon. It is convenient to identify the unique carbon atom (C-3), which is a ring member in both the reactant and product as the "pivot carbon" (*78*). The possibility that a trimethylenemethane biradical **70** (or a zwitterionic

5. HYPOTHETICAL BIRADICAL PATHWAYS

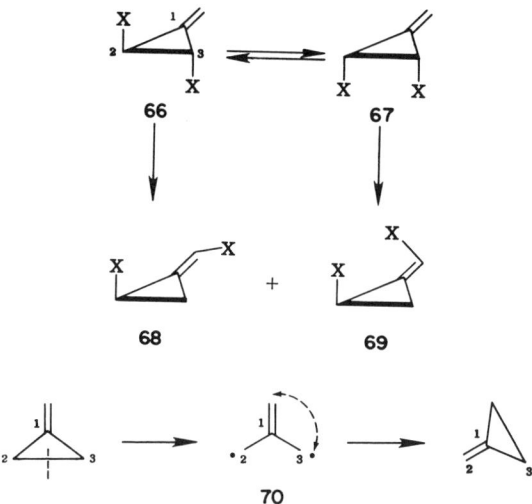

variant) might be an intermediate has been a matter of concern since the earliest studies of the rearrangement, and a series of stereochemical and kinetic experiments have been designed as mechanistic tests for it.

Ullman (*76a*) found that optically active trans ester **66** gave not only the rearrangement products **68** and **69**, but also a partially racemized trans ester, by two-center epimerization. The racemization might result from two successive one-center epimerizations via the cis diester **67**, but, in fact, this indirect pathway cannot account for more than a small fraction of the reaction. The cis diester **67** epimerizes to trans diester **66**, but it gives rearranged products of **68** and **69** 7.5 times as fast. Since the rate of racemization of the trans diester is greater than 1/7.5 times the rate of rearrangement, there must be a mechanism for the racemization of the trans diester that is independent of mere reversible formation of the cis (*76a*).

Similar results are observed in the pyrolyses of optically active 2,3-dimethylmethylenecyclopropane (**71**) (*80, 81*), where partial racemization of the reactants again competes with rearrangement. A plausible mechanism for the racemization of **66** would invoke the achiral planar trimethylenemethane **72**, the p orbitals of which are explicitly shown here to emphasize the geometry.

TABLE 2
Some Methylenecyclopropane Rearrangements

Reactant	Temp (°C)	Products	Ref.
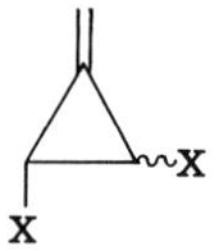 X = CO$_2$Me	210	Cis ⇌ trans 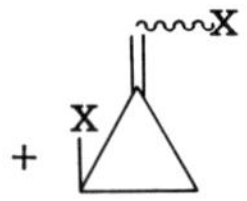 +	76–78
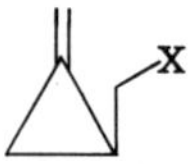 X = CO$_2$Me	250	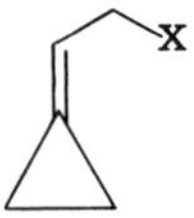	77
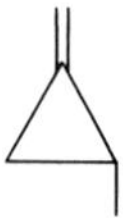	197–234		79
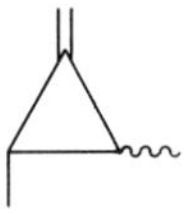	170–225	Cis ⇌ trans 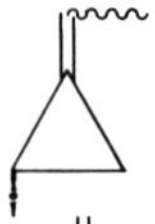 +	80, 81
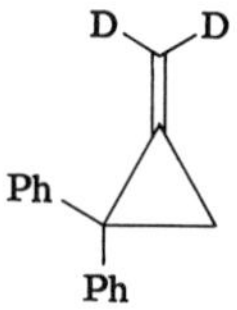	80	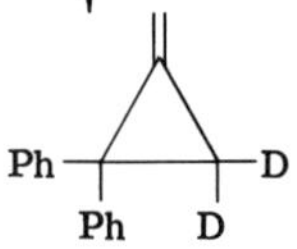	82
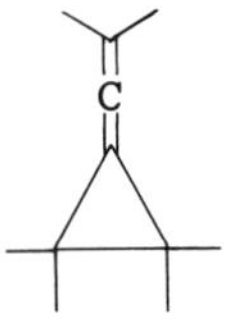	360	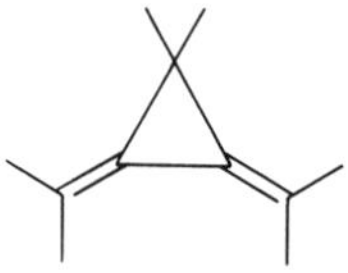	83
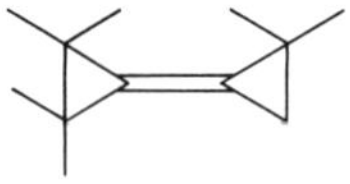	400 (flow)	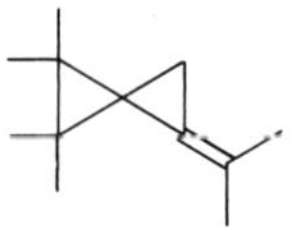	83
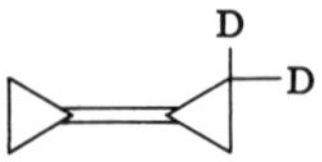	210	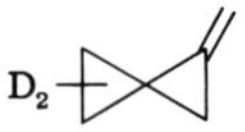	84

5. HYPOTHETICAL BIRADICAL PATHWAYS

TABLE 2 (Continued)

Reactant	Temp (°C)	Products	Ref.
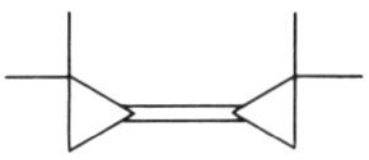	170–200	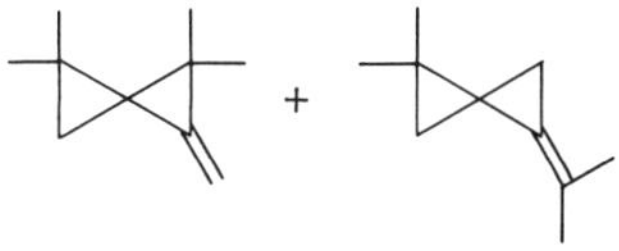	85
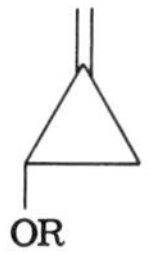	60–90	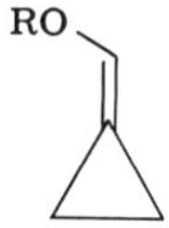	86
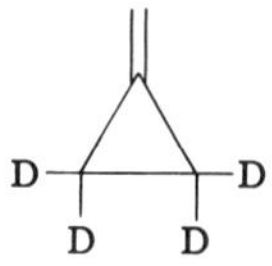	235	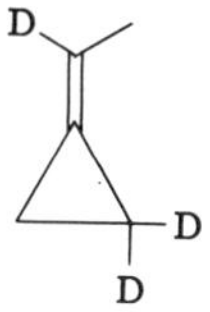	87
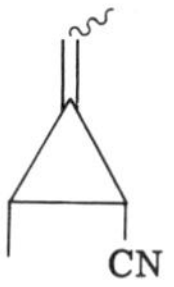	110	Cis ⇌ trans Syn ⇌ anti 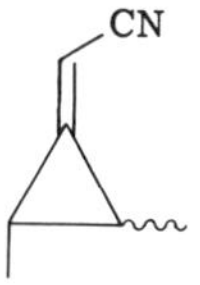	88
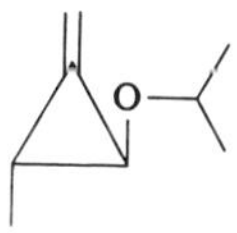	<100	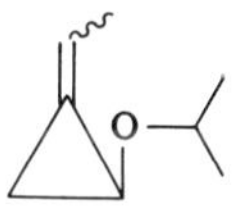	89

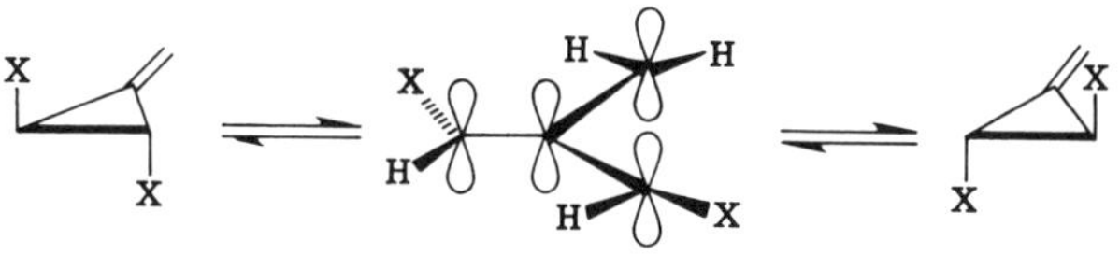

Two formal points should be made about the planar trimethylenemethane type of intermediate. First, there must be pathways to the rearranged products **68** and **69** from **66** (and to the rearranged products from **71** shown in Table 2) that do not involve such species, since a substantial portion of the starting enantiomeric purity survives in the products (*76a, 78, 81*).

Second, if the pyrolysis of the parent compound methylenecyclopropane (**65**, X = H) also occurs, in part, by way of an axially symmetric planar trimethylenemethane (**72**, X = H) (D_{3h} point group), the latter species may be invoked as an intermediate but not as a transition state.*

The predominant pathways in the rearrangements of **71** (*80, 81*) and **66** (*78*) result in inversion of the configuration of the "pivot carbon." However, a complete analysis of the stereochemistry requires the elucidation of the sense of rotation at two other sites: the carbons corresponding to the exocyclic methylene groups of the reactant and product; this information is not yet available.

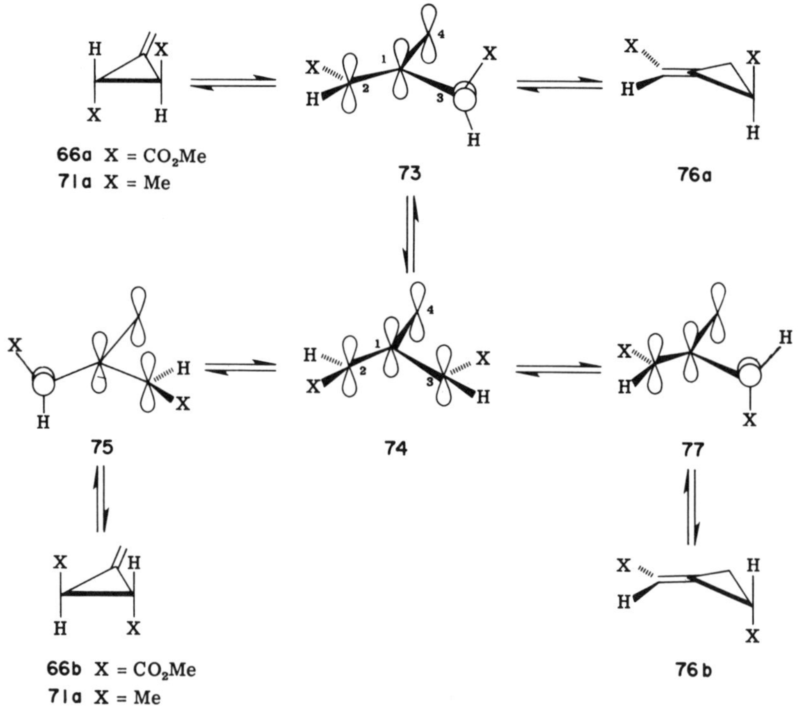

66a X = CO$_2$Me
71a X = Me

73

76a

75

74

77

66b X = CO$_2$Me
71a X = Me

76b

If we accept the most popular mechanism for the methylenecyclopropane rearrangement, which postulates various rotational isomers of a trimethylenemethane singlet biradical as the key intermediates, the racemization of the reactants that is observed in the pyrolyses of **66a** and

* This exclusion derives from the symmetry properties of D_{3h} **72** (X = H), which do not permit it to fulfill the Murrell–Laidler rule (*90*) that the transition state force constant matrix must have one and only one negative eigenvector (*91*). Also, the singlet state of D_{3h} **72** (X = H) has a partially filled pair of degenerate nonbonding molecular orbitals and therefore is subject to a Jahn–Teller distortion (*92*).

5. HYPOTHETICAL BIRADICAL PATHWAYS

71a can be formulated as shown here. The planar, achiral species **74** may be formed by an internal rotation about the unique C-1—C-3 bond of the bisected species **73**. The reverse of this process at the equivalent site, C-1—C-2, in **74** leads to another bisected species **75**, which can cyclize to the enantiomeric methylenecyclopropane series (**66b**, **71b**). This is a mechanical and undoubtedly oversimplified model, since, for example, there is no experimental evidence that species **73–75** represent true energy minima. Nevertheless, in the framework of the trimethylenemethane mechanism, the racemization suggests that internal rotation, converting the bisected to the planar species, is competitive with ring closure. This process also could lead to partial racemization in the formation of the product **76** by way of the planar → bisected conversion **74** → **77**.

Two closely related but independent studies similarly suggest that the energy difference between the planar and bisected singlet forms of trimethylenemethane derivatives is not large. At 164°C, the optically active 3-*endo*-methyl-6-ethylidenebicyclo[3.1.0]hexane (*endo*-**78a**) suffers apparent isomerization of the ring methyl group, presumably by cleavage of the bridge bond, to give the planar trimethylenemethane **79a**. This can recyclize in the inverted sense at both bridge sites and lead to the doubly epimerized *exo*-**78a** product (*93*). Were this the only reaction, *exo*-**78a** would be formed and *endo*-**78a** would be recovered with the same enantiomeric purity as the starting *endo*-**78a**, since the intermediate **79a** is chiral. In fact, however, extensive racemization occurs, forming both *endo*-**78b** and *exo*-**78b**, plausibly by torsion about the exocyclic double bond in an achiral bisected species (**80**). Interpreted on this basis, the kinetic data show that the transition state energy for the enantiomerization (*endo*-**78a** ⇌ *endo*-**78b**) is about 2 kcal/mol lower than that for the diastereomerization (*endo*-**78a** ⇌ *exo*-**78a**). The authors (*93*) propose that the bisected species **80** lies in a small potential well, whereas the planar trimethylenemethane **79** is the transition state for diastereomerization, which implies that the actual energy separation between the planar and bisected singlet forms of this trimethylenemethane is slightly more than the 2 kcal/mol transition state energy separation. The bisected species actually lies *below* the planar one. This conclusion is in accord with all of the recent theoretical calculations on the parent, unsubstituted trimethylenemethane, which predict a small energy difference in favor of the bisected species (*94–96*).

A small energy separation between the planar and bisected forms of trimethylenemethane seems, at first glance, to violate the conventional assumption that a π-conjugated system will be destabilized if it is twisted out of planarity. We can estimate crudely the energy cost of such twisting from simple Hückel theory. The Hückel π-electron energy levels of planar

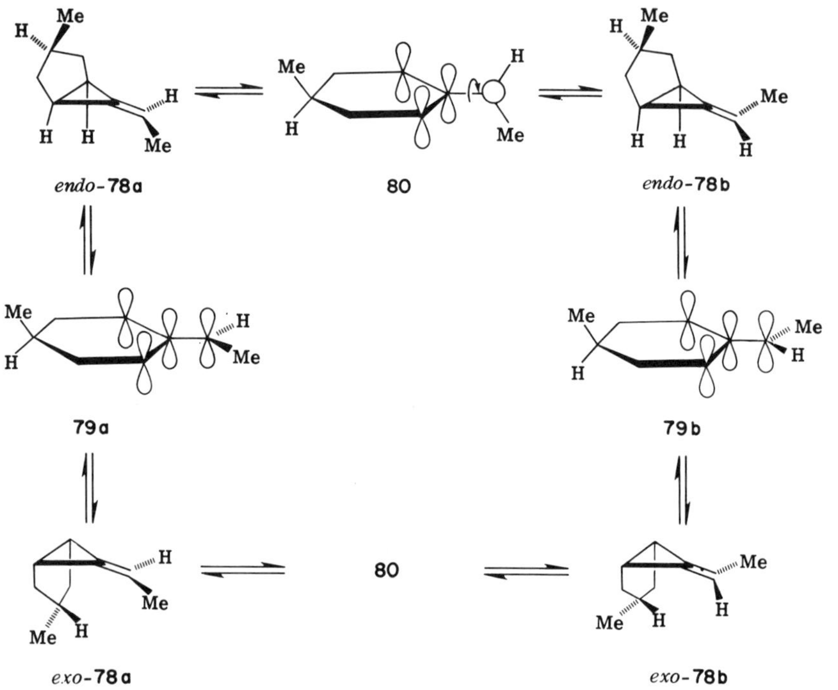

endo-78a 80 *endo*-78b

79a 79b

exo-78a 80 *exo*-78b

trimethylenemethane occur at α (degenerate nonbonding pair), $\alpha + \sqrt{3}\beta$ (bonding), and $\alpha - \sqrt{3}\beta$ (antibonding) (97–99). The bisected form is π-electronically represented by an allyl system (energies: α, $\alpha \pm \sqrt{2}\beta$) and an isolated p orbital (energy: α). The occupation of the two sets of levels by the four π electrons available leads to energies of $4\alpha + 2\sqrt{3}\beta$ and $4\alpha + 2\sqrt{2}\beta$, respectively. Thus, the planar form should be more stable than the bisected one by about $0.64\,\beta$ unit. Any reasonable choice of a value for the resonance integral β leads to a substantial calculated energy gap. For these crude purposes, one might justify a β value of 15 kcal/mol [from the experimental resonance energy of allyl radical (100), ~12 kcal/mol, and the Hückel value, $0.83\,\beta$]. Deplanarization of trimethylenemethane to the bisected form, therefore, might cost ~10 kcal/mol of Hückel delocalization energy.

Overbalancing the π-conjugation energy, however, is an electron repulsion effect (95, 95a, 101), which simple Hückel theory ignores and which operates here because trimethylenemethane, like many other π-biradical systems, is "nondisjoint"; that is, the two degenerate nonbonding orbitals have nonzero coefficients at common sites (95, 95a).

Similar results are observed in the pyrolyses of the stereoisomeric 2-methyl-6-ethylidenebicyclo[3.1.0]hexanes **81–82**: these interconvert at 187°C, all four isomers being formed from any one (81). The interconver-

5. HYPOTHETICAL BIRADICAL PATHWAYS

sions **81a** ⇌ **82b** and **81b** ⇌ **82a** are bridgehead double inversions,

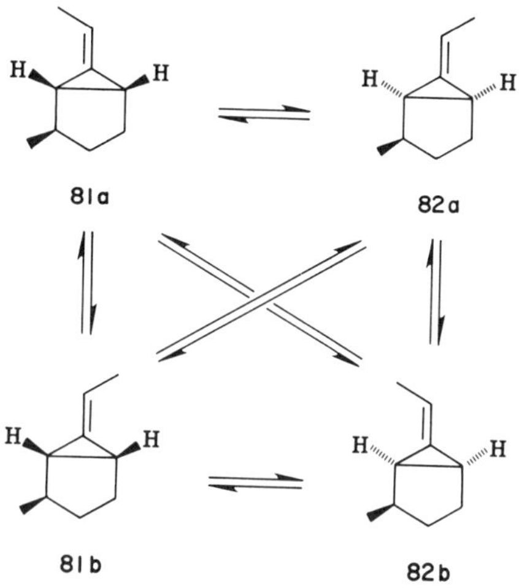

whereas **81a** ⇌ **81b** and **82a** ⇌ **82b** are exocyclic torsions. Reactions **81a** ⇌ **82a** and **81b** ⇌ **82b** require both processes to occur. A detailed kinetic analysis (*81*) shows that all 12 first-order rate constants are of similar magnitude, the ratio of the largest and smallest values being less than 2. The authors (*81*) prefer to interpret these and related data, and in particular the small secondary kinetic isotope effect of deuterium (k_H/k_D = 1.08 for **83** versus **84**), in terms of rate-determining ring opening to a planar trimethylenemethane, followed by a competition between rotation about the exocyclic bond and reclosure, the rotation being the faster process by a factor of 6. This picture differs slightly from the suggestion of two competing direct pathways offered for the **78** series (*93*).

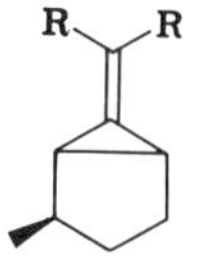

83 R = H
84 R = D

3. The Question of Concert

In attempting to decide whether the methylenecyclopropane rearrangements are concerted reactions, we are faced with a problem that occurs frequently in sigmatropic rearrangements. The activation energies are not obviously lower than one would expect for a nonconcerted, biradical mechanism (*75, 79*), and the assignment depends on the interpretation of the stereochemical results. For example, the thermal rearrangement of racemic *E-trans*-1-ethylidene-2-cyano-3-methylcyclopropane (shown here as the *S,S* enantiomer **85c**) occurs by direct primary kinetic routes to each of the diastereomers **85a, 85b,** and **85d** (*88*). In principle, this could be a consequence of a competition between a concerted orbital symmetry-allowed reaction (*60*), leading to **85b**, and two concerted orbital symmetry-forbidden ones (*102, 103*), leading to **85a** and **85d**. Alternatively, the results could be explained by stepwise reactions via trimethylenemethane biradical intermediates, as the authors prefer on grounds of

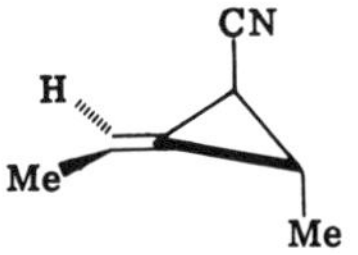

85c 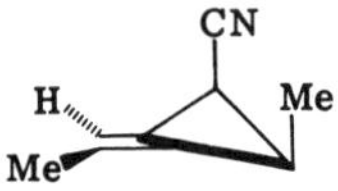   **85d** 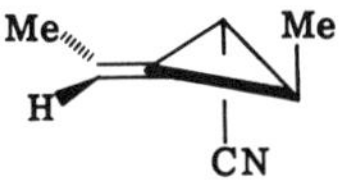   **85a** 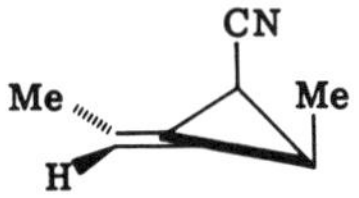   **85b**

supposed mechanistic economy (*88*). However, the suggested biradical mechanism consists of competing paths from **85c** to two separate bisected trimethylenemethane species. At least one of these paths must bifurcate again to account for the three products. The biradical mechanism thus may well be correct, but its use does not result in any conceptual simplification. Both the concerted and stepwise schemes represent competitions among three identifiable pathways and, therefore, are of equivalent complexity.

4. Further Comments on the Structures and Energies of Hypothetical Intermediates in the Methylenecyclopropane Rearrangements

If biradicals are involved in these processes, it is difficult to devise structures for them other than trimethylenemethanes. A superficially attractive alternative might be a carbenacyclobutane (**86**). This species presumably is involved in the thermal decomposition of the sodium salt of cyclobutanone *p*-toluenesulfonylhydrazone (**87**), which gives methylenecyclopropane as the major product (*104*). The first step of the

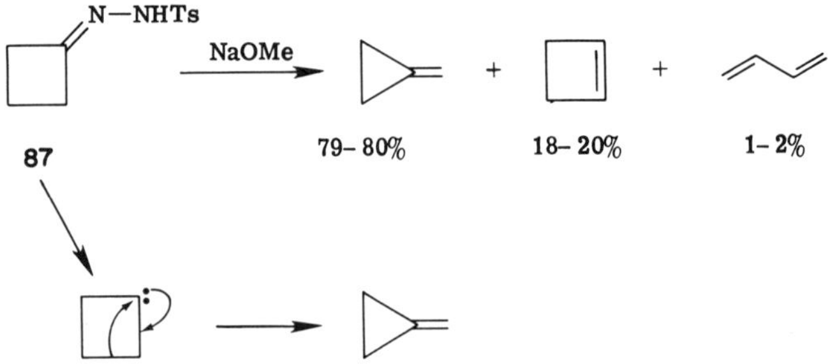

87   79–80%   18–20%   1–2%

86

5. HYPOTHETICAL BIRADICAL PATHWAYS

methylenecyclopropane rearrangement then might be formulated as the reverse of that reaction:

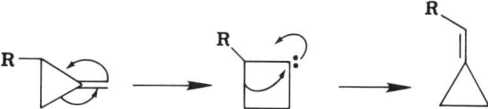

Because of the uncertainty of estimating the ΔH_f° of carbenes, which we already have mentioned, it would be difficult to rule out such a mechanism on thermochemical grounds. However, two other factors make it implausible. If the cyclobutene and butadiene formed from **87** are true carbene products, one might expect to see small amounts of them in methylenecyclopropane pyrolysis mixtures also, but no such examples are described in the literature. Moreover, although one can account for trans ⇌ cis interconversion in the pyrolysis of optically active *trans*-2,3-dimethylmethylenecyclopropane (**71**) by postulating inversion of configuration in the formation of the carbenacyclobutane (**71** → **88** → **89**), some independent process is needed to explain the racemization, i.e., (+)-trans ⇌ (−)-trans, observed (*81*) as a primary kinetic process.

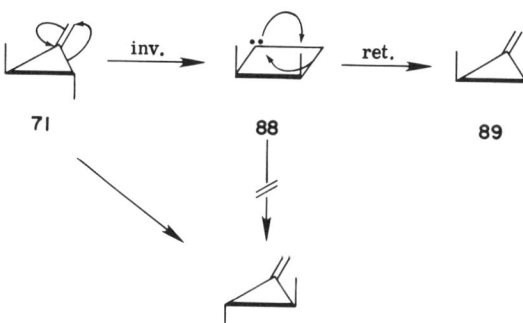

Dowd and Chow (*105*) point out that the straightforward application of the bond additivity scheme (*5–8*) to the calculation of ΔH_f° for the unsubstituted trimethylenemethane leads to a thermodynamic anomaly. Figure 4 shows this procedure. The ΔH_f° value for trimethylenemethane is obtained by adding to the known ΔH_f° of isobutylene twice the dissociation energy of the allylic C—H bond and subtracting from the result the bond energy of H_2. This places the hypothetical trimethylenemethane at 63 kcal/mol on the ΔH_f° scale. It is true that the two C—H bond energies should be slightly different, since one bond is weakened by the allylic resonance energy, and the other is weakened by the difference between the allylic and trimethylenemethanic resonance energies. The discrepancy from the assumption of equal energies would be only about 3–5 kcal/mol and would not affect the argument materially. One also can calculate the

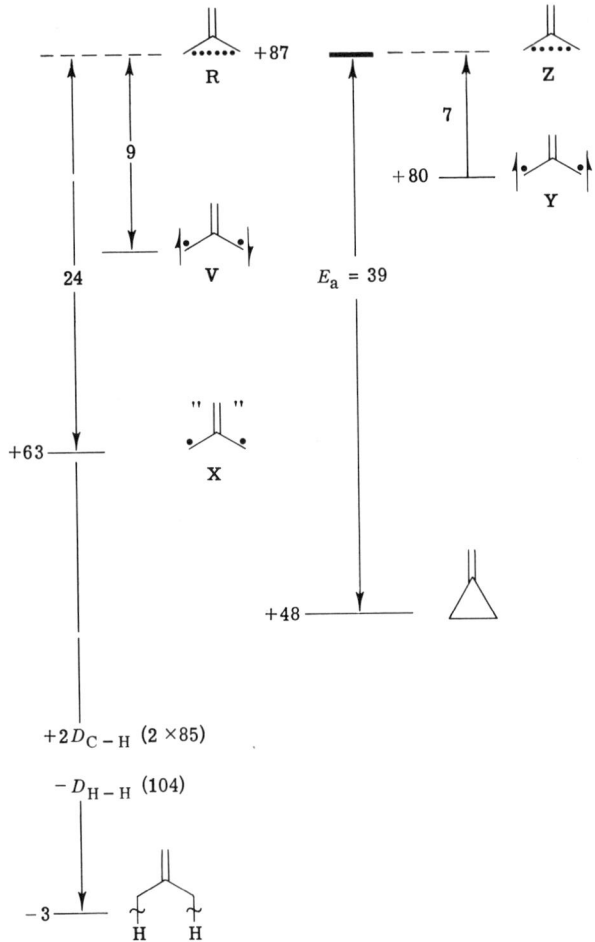

Fig. 4 Thermochemical relationships among points on the trimethylenemethane energy surface. For reasons given in the text, the absolute values of the energies of species V, X, Y, and Z are uncertain. The energies are in kilocalories per mole.

ΔH_f° of the transition state for ring closure of singlet trimethylenemethane as 87 kcal/mol by adding to the known ΔH_f° of methylenecyclopropane the activation energy, 39 kcal/mol (87), for the rearrangement of 2,2,3,3-tetradeuteriomethylenecyclopropane. This would leave the "trimethylenemethane biradical" (X, Fig. 4) in a potential well 24 kcal/mol deep, which should permit its isolation at room temperature!

The anomaly has not yet been fully resolved, but the outlines of the problem can at least be stated. Like all biradicals so generated, the "trimethylenemethane biradical" X is of unspecified configuration and spin state. The thermal rearrangement of methylenecyclopropane presumably is a singlet reaction, so that one needs a way to estimate ΔH_f° of singlet trimethylenemethane. That the parent trimethylenemethane (106), as well as a number of derivatives (107), have triplet ground states is established by ESR spectroscopy of matrix-isolated specimens. Adopting the suggestion (74) that the energy of the "biradical" generated by bond additivity will be lower than that of the real singlet biradical by three-fourths of the singlet–triplet separation, and applying the theoretically calculated (94–96d, 101) singlet–triplet gap of about 14–20 kcal/mol, one can adjust the energy of X upward by 9–15 kcal/mol, which would place singlet trimethylenemethane (species V) about 9–15 kcal/mol below the transition state for cyclization.

A direct measurement (105) of the activation energy for cyclization of *triplet* trimethylenemethane gives the value 7 kcal/mol. In Fig. 4, we have arbitrarily subtracted this from the ΔH_f° of the thermal rearrangement transition state R and obtained the dubious value of 80 kcal/mol for ΔH_f° of triplet trimethylenemethane, Y. The uncertainty in this procedure arises because it is not clear that the ring closure of triplet and singlet trimethylenemethane pass over a common point of the energy surface. In fact, although the straightforward mechanism, triplet trimethylenemethane → singlet trimethylenemethane → methylenecyclopropane, meets the requirement that spin inversion occur somewhere along the reaction coordinate, it creates conflict between the theoretical calculations (94–96d, 101), which suggest that the first step should be rate determining with an activation energy ≥14–20 kcal/mol, and the experimental value, $E_a \cong 7$ kcal/mol. One conceivable resolution of the problem might be developed along the lines indicated in Fig. 5. Because the total spins of the reactant and product ground and excited states exchange values during the reaction, there is an "intended" surface crossing. Depending on the probability of transitions from the triplet to the singlet surface in the vicinity of the crossing, it may be possible to cyclize triplet trimethylenemethane without overcoming the full 20 kcal/mol singlet–triplet energy separation between the open trimethylenemethane species. So far, however, preliminary quantum chemical calculations have failed to substantiate this hypothesis (96e), and the discrepancy remains unexplained.

Much more work is needed to complete the exploration of the methylenecyclopropane energy surface. For example, although there is convincing evidence that vibrationally excited methylenecyclopropane is a product of the gas-phase methylenation of allene (108–110), little is known about the effects of vibrational excitation on the stereochemistry or mechanism of the methylenecyclopropane rearrangement.

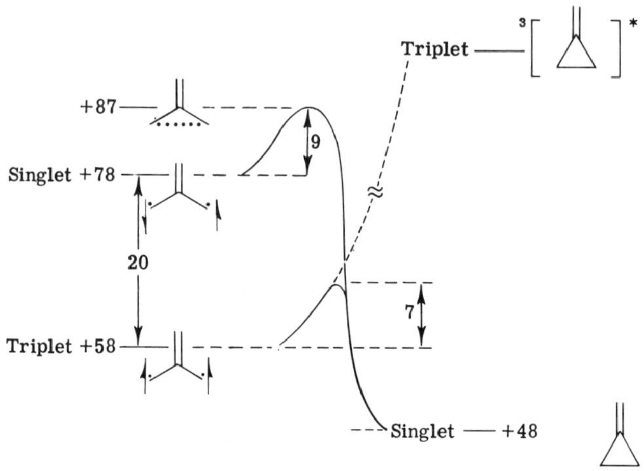

Fig. 5 A conceivable energy surface for trimethylenemethane. The values are in kilocalories per mole. Closure of the triplet trimethylenemethane may occur not by prior population of singlet trimethylenemethane but instead by a surface crossing.

The prototypical 1,3 biradical trimethylene and its substituted derivatives have not been observed to dimerize to cyclohexanes. In contrast, there are sporadic reports of the formation of 1,4-dimethylenecyclohexanes by dimerization of trimethylenemethanes (*111*). This seems to be a characteristic reaction of the triplet form of the biradical [**90** → **91** (*112*), **92** → **93** (*113–115*)], and the absence of such dimers from the products of pyrolysis of methylenecyclopropanes such as **78**, **81**, and the examples listed in Table 2 provides evidence that the rearrangements are singlet-state reactions. Similarly, the very efficient and spin-selective capture of the triplet **92** by molecular oxygen (*116, 117*), combined with the absence of any

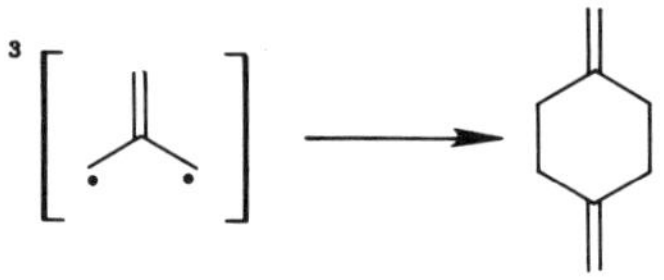

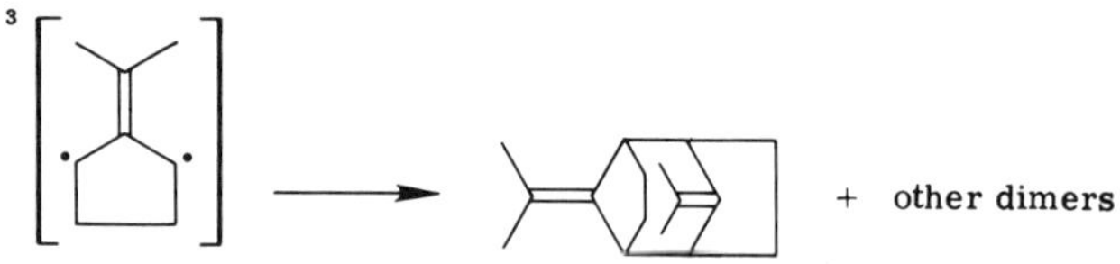

5. HYPOTHETICAL BIRADICAL PATHWAYS

effect of oxygen on the course of the 2,3-dicyanomethylenecyclopropane rearrangement (*118*), suggests that the thermal rearrangement does not enter the triplet manifold.

This apparent restriction to "pure singlet" behavior of the putative trimethylenemethane intermediates generated in methylene–cyclopropane rearrangements has caused some perplexity, because it is not obvious why the singlet biradical should not eventually cross over to the more stable triplet. Recent work (*119–119b*) suggests a resolution of the anomaly.

Methylenecyclopropane rearrangement **93a** → **93b** (*119*) and stereomutation **93c** → **93d** → **93e** (*119b*) are observed at temperatures near −40°C in the highly strained bicyclo[3.1.0]hex-1-ene and 5-alkylidenebicyclo-[2.1.0]pentane series, presumably by way of singlet trimethylenemethanes of the 2-alkylidenecyclopenta-1,3-diyl type, **93h**.

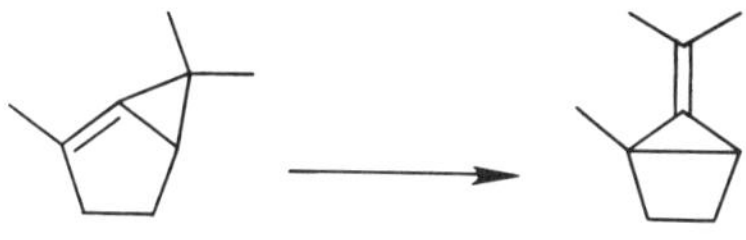

93a → 93b

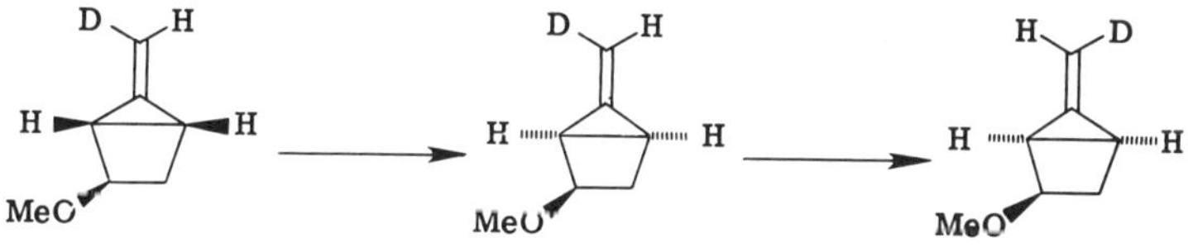

93c → 93d → 93e

93f : R_1 = Me; R_2 = H
93g : R_1 = H; R_2 = OMe

93h

However, in contrast to the pyrolytic stereomutations of the 6-alkylidenebicyclo[3.1.0]hexanes **78** and **81**, which occur in the temperature range above 160°C and are not accompanied by appreciable amounts of 1,4-dimethylenecyclohexane dimers, compounds **93f** and **93g** both form dimers in quantitative yield at temperatures of −40 to +5°C, slightly above their stereomutation reaction temperature. There is evidence (*119a, 119b*) that the Arrhenius plots for stereomutation (higher E_a and A) and dimerization (lower E_a and A) are not parallel. The much lower temperature at which reaction occurs in the 5-alkylidenebicyclo[2.1.0]-pentane series (**93a** and **93c–93d**) then would make the two rates approach each other and would permit dimerization via intersystem crossing to compete more effectively.

C. Spiropentane and Methylenespiropentane Rearrangements

Spiropentanes, like cyclopropanes, undergo both structural isomerization (*120–123*) (e.g., **94** → **95**) and stereomutation (*122–124b*) (e.g., **96** → **97**). The mechanism proposed for the structural isomerization is of special interest because it involves two successive bond cleavages. First, a

$E_a = 54.5$ kcal/mol$^{(119, 121)}$

peripheral bond (*123, 124a*) breaks to give the 1,3 biradical **98**, which, by fission of a β,γ bond, is converted to the allylically stabilized 1,4 biradical **99**. The latter species also is postulated as an intermediate in the degenerate rearrangement of methylenecyclobutane-5,5-d_2 to the 2,2-d_2 isomer (*125, 126*) (see Section V).

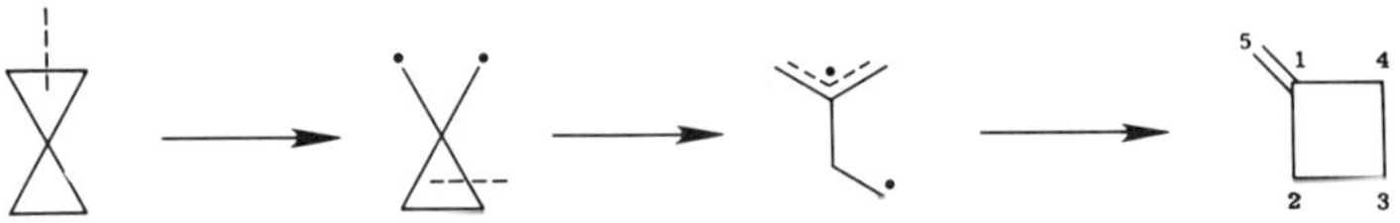

A thermochemical energy surface interrelating spiropentane, methylenecyclobutane, and the fragmentation products ethylene and allene can be constructed (*125*) by combining known heats of formation and activation energies with a heat formation of the allylic biradical **99** calculated from bond additivity (Fig. 6). The highest point on the surface is the

5. HYPOTHETICAL BIRADICAL PATHWAYS

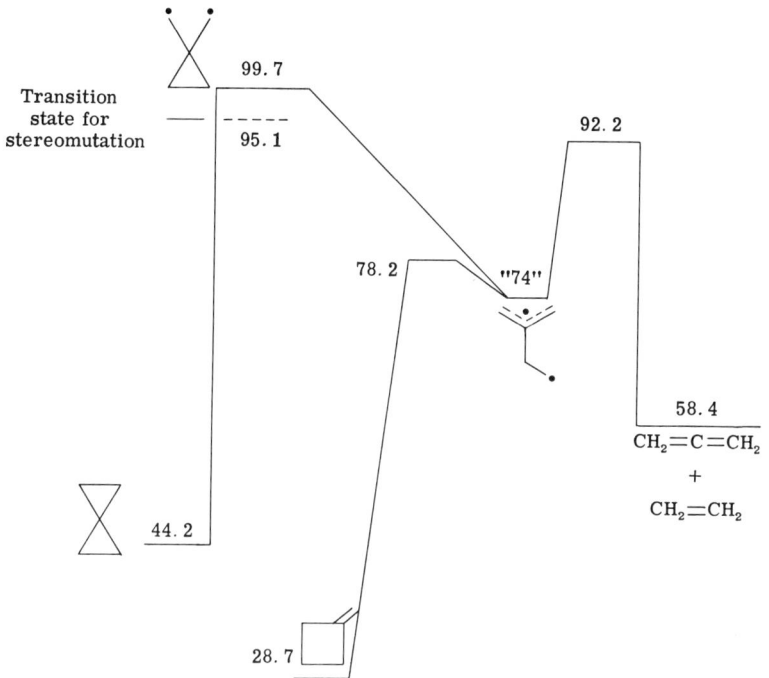

Fig. 6 Thermochemical energy surface for methylenecyclobutane and spiropentane. Energies are ΔH_f° values in kilocalories per mole (*125, 126*).

transition state for the spiropentane rearrangement, somewhat arbitrarily shown here as the cyclopropyl-1,1 biscarbinyl biradical **98** itself, although it should be emphasized that this species could lie below the transition state if there is a finite activation energy for rupture of the second bond.

Another transition state that is presumably related in structure to the "peripheral" biradical **98** may be located at $\Delta H_f^\circ = 95.1$ kcal/mol (dashed line in Fig. 6) by the addition of E_a for the stereomutation *cis-* ⇌ *trans-*spiropentane-1,2-d_2 (50.9 kcal/mol) to ΔH_f° of spiropentane (*124*). That this reaction is a stereomutation and does not involve a masked degenerate structural isomerization seems clear from the observation (*125*) that 1,1,2,2-tetradeuteriospiropentane (**94a**) at elevated temperature does not give the 1,1,4,4 isomer (**94b**) (*125*).

The possibility that the two formally sequential bond cleavage steps may occur in concert has been examined in the rearrangement of 1-carbomethoxy- 2,4- dimethylspiropentane (**100**) to dimethylcarbethoxy-methylidenecyclobutanes **101** (*122*). The stereochemistry of this reaction is consistent with predominant contributions from orbital symmetry-controlled pathways ($_2\sigma_s + {_2\sigma_a}$), although steric factors also could be involved (*122*).

[structures **100** and **101** with CO$_2$Me group]

100   101

A kinetic analysis (*124b*) of the stereomutations of the four isomeric 1,2,4-trimethylspiropentanes **102–105** provides the rate constants shown in Fig. 7. There are two ratios of double to single inversion rates for each isomer, and although when taken at face value these do not show an impressive preference for either pathway, the authors (*124b, 124c*) argue that steric effects tend to retard double inversion in these systems. Correction for such effects by an involved but self-consistent procedure leads the authors to suggest that there is an inherent preference of about threefold for double inversion and that this process is conrotatory from the trans isomers (**102** and **103**) but disrotatory from the cis (**104** and **105**).

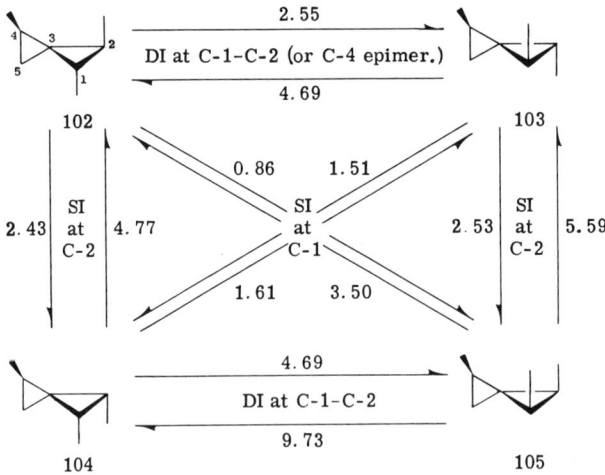

Fig. 7 Interconversions of the 1,2,4-trimethylspiropentanes. The rate constants shown are in units of 10^{-6} sec^{-1} and refer to pyrolyses at 561.7 K and 150 torr. The abbreviations SI and DI denote single and double inversion, respectively (*124b*).

One might hope to generate species similar to the hypothetical spiropentane biradicals **98** and **106** by deazetation of the two pyrazolines **107** and **108**. The fascinating results of such experiments (*127*) suggest that

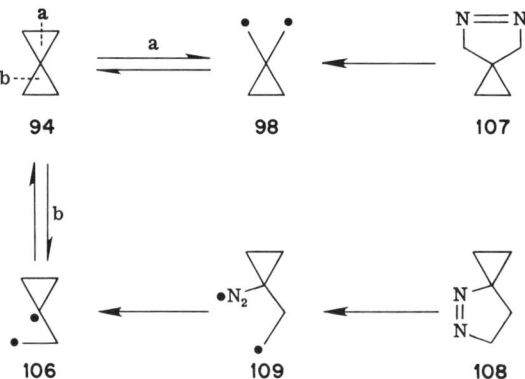

the symmetric azo compound **107** decomposes in the gas phase by simultaneous cleavage of both C—N bonds. In the transition state (*127a*), coupling of the N—N stretching vibration with vibrations in the organic fragment is weak, so that the N_2 molecule is formed vibrationally excited and carries off the excess energy of the decomposition, leaving a thermalized organic fragment. In the case of the unsymmetric compound **108**, the two C—N bonds differ in strength, and the deazetation may occur stepwise over a diazenyl biradical intermediate **109**. This species can lose N_2 by way of a transition state in which the N—N and organic skeletal vibrations are strongly coupled (*127a*). The N_2 molecule departs with little excess vibrational energy, and the organic fragment is born vibrationally excited. The experimental evidence (*127*), especially the pressure dependence of the product distribution, supports the idea that, although spiropentane **94** is formed from either of the pyrazolines, only that from the unsymmetric pyrazoline **108** is vibrationally excited.

At 320°C, methylenespiropentane (**110**) rearranges to both 1,2- and 1,3-bismethylenecyclobutane (**111** and **112**) in the kinetic ratio of about 7:1 (*128*). A plausible mechanism for the formation of **111** would involve the tetramethyleneethane biradical **113**, which has been postulated as the common intermediate in other thermal processes, including the dimerization of allenes and the degenerate thermal rearrangement of 1,2-bismethylenecyclobutane (*126, 126a*). The triplet form of **113** has been observed by ESR spectroscopy in the products of the photolysis of 3,4-bismethylenecyclopentanone at low temperature (*129*).

However, the competitive formation of **112** in the pyrolysis of **110** is quite surprising and has no such simple explanation. It has been suggested that this reaction may involve cleavage of the presumably very strong

110 113 111

112

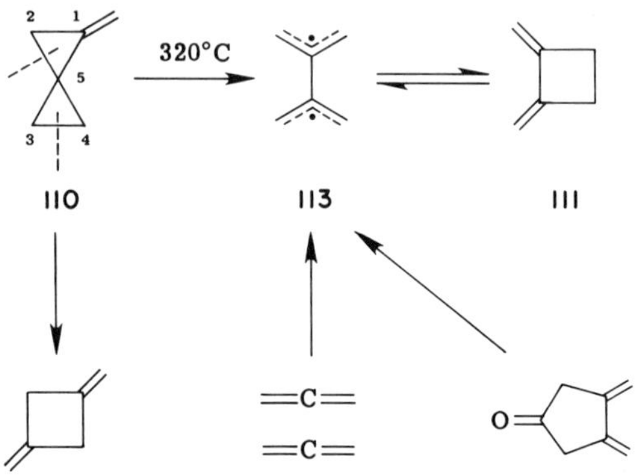

C-1—C-5 bond as well as the C-3—C-4 bond at some stage, giving the allylic–vinylic radical **114** as an intermediate (*128, 128a*).

110 114 112

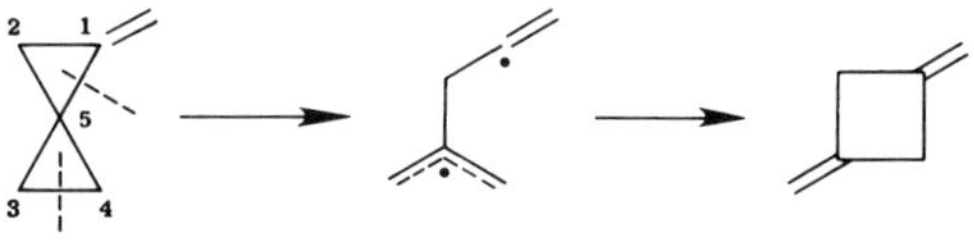

In an analogous system (**115**) (*130, 131*), this type of reaction does not compete with the presumed tetramethyleneethane pathway. Thus, the pyrolysis of **115** gives the product **116** resulting from methylenecyclopropane rearrangement first. This substance undergoes reversible Diels–Alder retrogression to **117** at intermediate temperatures and cleavage to the 1,2-bismethylenecyclobutane **118** and the bismethylenehexadiene **119** at higher temperatures, but no evidence has been reported for the formation of either of the 1,3-bismethylenecyclobutanes **120** or **121**.

115 116 117

118 119 120 121

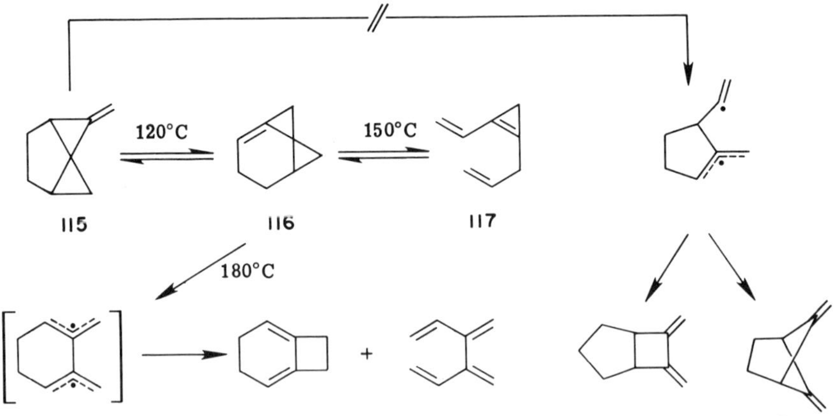

IV. 1,4 BIRADICALS

A. From Olefin Dimerizations, Cyclobutane Reversions, and Diazene Decompositions

Early kinetic studies of the thermal cycloreversion (e.g., **122** → **123** + **124** + **125**) of cyclobutanes and a large body of subsequent observations have been most readily interpreted by a stepwise mechanism via a biradical intermediate. The argument can be formulated on thermochemical–kinetic grounds and is supported by a wealth of experimental detail. Thus, the cycloreversions usually (*134–134b*) [but not always (*135*)] are only feebly stereospecific and are accompanied by stereomutation of the start-

ing cyclobutane (*134, 134a*) [e.g., **122** ⇌ **126** (*134*)]. Moreover, the cycloaddition of two olefins usually occurs with only partial retention of the olefin stereochemistry in the cyclobutane product (*136, 136a*).

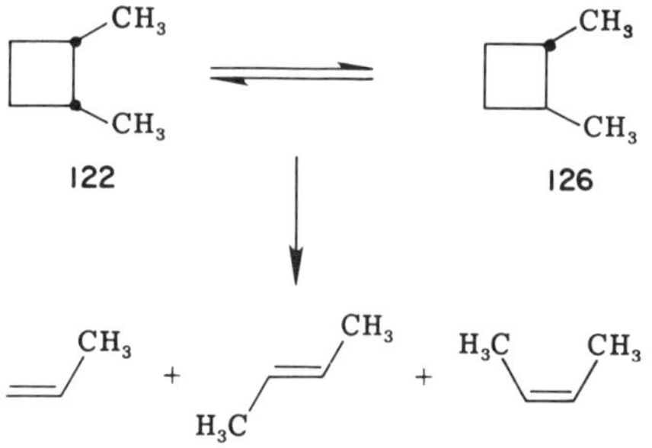

Thermochemical estimates by the usual additivity procedures (*6–8*) place the tetramethylene biradical **127** in a potential well protected from cleavage to two ethylenes and from closure to cyclobutane by barriers of 6 and 4.4 kcal/mol, respectively. In contrast to the case of the trimethylene biradical (see Section III,A,2), where theory fails to find a local minimum, the tetramethylene minimum has some theoretical support. Although quantum mechanical calculations at the extended Hückel level (*137*) show only a broad, flat energy plateau in the region of the tetramethylene structure, a later SCF calculation at the STO-3G level with 15-dimensional configuration interaction does show two true minima, for the gauche and extended conformations (*137a*). The barriers to dissoci-

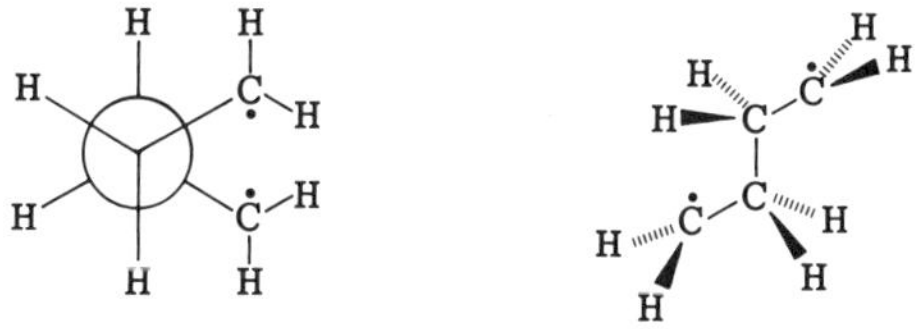

gauche-**127**   *trans*-**127**

ation and cyclization, 3.6 and ⩾2.0 kcal/mol, respectively, are somewhat smaller than those calculated thermochemically, but their difference is the same. Much additional theoretical discussion of tetramethylene is available (*138, 138a*).

Some elegant work by Dervan, Uyehara, and Santilli (*139–141*) provides evidence that thermal decomposition of a 1,2-diazene or a 1,1-diazene occurs in part by a concerted [σ + σ + σ] cycloreversion (*142*) and in part through a tetramethylene biradical intermediate (*141a*). It is especially significant that the properties of this biradical, as measured by the relative rates deduced for internal rotation, fragmentation, and cyclization, are *identical* with those of the biradical implicated in the cyclobutane thermolyses (*139*). Scheme 2 incorporates the mechanistic interrelationships in the case of the 3-methylpenta-1,4-diyl biradical **128** generated from *cis*- and *trans*-1,2-dimethylcyclobutanes **129** and **130** and from *cis*- and *trans*-3,4-dimethyltetrahydropyridazines **131** and **132**. Note that propene, a major product from cyclobutanes **129** and **130**, presumably by way of a different biradical, 2,4-hexadiyl, is not included in the scheme since it is not formed from diazenes **131** and **132** and provides no information relevant to the intermediacy of biradical **128**.

A steady-state analysis (*139*) of the product distributions from pyrolyses of the two cyclobutanes **129** and **130**, and the two tetrahydropyridazines

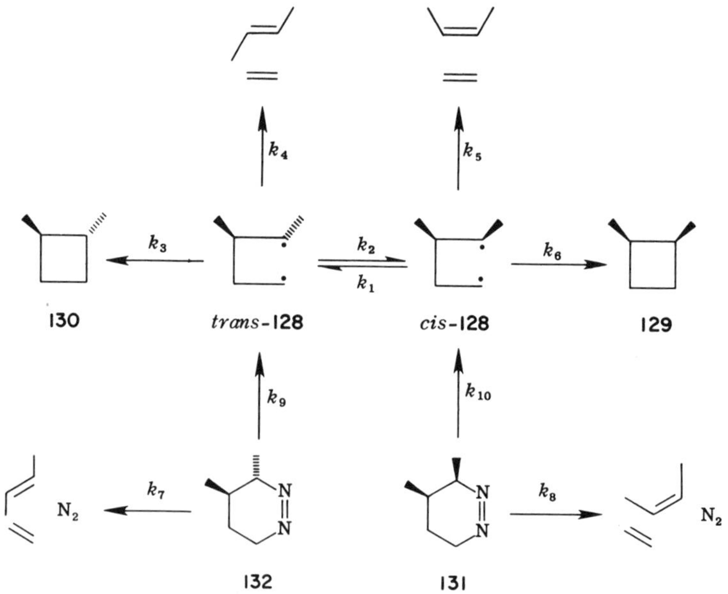

Scheme 2

5. HYPOTHETICAL BIRADICAL PATHWAYS

131 and **132** at 712 K permits the assignment of the following rate constant ratios from *trans*-**128** and *cis*-**128**, respectively: fragmentation versus cyclization, $k_4/k_3 = 1.63$, $k_5/k_6 = 1.79$; cyclization versus internal rotation, $k_3/k_2 = 1.86$, $k_6/k_1 = 0.72$. The ratios of concerted three-bond decomposition versus the biradical pathway from the *trans*- and *cis*-tetrahydropyridazines **132** and **131** are $k_7/k_9 = 0.47$ and $k_8/k_{10} = 0.56$.

From these, one can predict that the kinetically controlled ratio *trans*/*cis*-2-butenes for those portions of the product *derived from the biradical pathway* from **132** and **131**, respectively, should be 88/12 and 28/72. These values are in good agreement with those (88/12 and 36/64) observed by Gerberich and Walters in their study (*134*) of the decomposition of **130** and **129** at a similar temperature (698 K). The agreement is understandable if it is assumed that the cyclobutane pyrolyses occur exclusively by the biradical path and that the hydrocarbon-derived and diazene-derived biradicals have identical properties. Although alternative interpretations might be put forward, for example, a diazenyl biradical intermediate (**133**) that just happens to mimic the behavior of the hydrocarbon diyl, the common intermediate is at least an attractive working hypothesis.

133

The diyls *cis*-**128** and *trans*-**128** of Scheme 2 also seem to be generated in the cycloadditions of ethylene to *cis*- and *trans*-2-butene. The ratios of rate constants for internal rotation, fragmentation, and cyclization deduced (*136a*) from the kinetic data and the assumption that biradical formation from each 2-butene is stereospecific are in good agreement with those obtained (*139*) from Scheme 2. Although this agreement could be taken as evidence in support of Scheme 2, some caution in its interpretation is advisable. As has been pointed out (*139*), if it is assumed that *cis*-**128** and *trans*-**128** of Scheme 2 are formed from the cyclic precursors (cyclobutane or tetrahydropyridazine) in a gauche-like conformation, then either the olefin + olefin reactions also must lead directly to the same conformations or some additional assumptions are necessary. There is no obvious reason why a major portion of a stepwise olefin + olefin reaction initially should not proceed through an extended conformation (**134**). To mimic the behavior of the gauche biradical *cis*-**128**, species **134** must suffer rotation about the C-2—C-3 bond at a rate that is at least competitive with the rate of C-3—C-4 bond rotation. This requirement leads to an apparent contradiction. Thus, a model for the C-2—C-3 bond rotational barrier

134 (extended)   *cis*-**128** (gauche)

would be that of an alkane, about 3–4 kcal/mol, whereas the C-3—C-4 rotational barrier, *if the two radical centers of* **134** *do not interact*, would be comparable to that of a secondary alkyl radical, less than 1.2 kcal/mol (*143*). This suggests that the extended biradical **134** should lose stereospecificity readily and therefore could not account for the observed agreement in stereochemical results. Among the several rationalizations that might be invoked to preserve the idea of a common intermediate, we mention only two. It is possible, perhaps even likely (*138a*), that through-bond coupling of the radical centers in the tetramethylenes is significant. This could change the rotational barriers appreciably from those of simple (or oversimplified) models. Alternatively, it may be that, although the initial approach in most olefin + olefin dimerizations leads to the extended biradical **134**, fragmentation of **134** back to two olefins is much faster than the rate of internal rotation. This would act as a mechanistic filter, since only the olefin dimerizations that give the gauche biradical *cis*-**128** could lead to product. In the absence of experimental or theoretical verification, these speculations do not provide a clear exit through which the common biradical hypothesis of the olefin dimerization can escape jeopardy.

A rather compelling case now can be constructed supporting the idea that substitution slows down the rate of internal rotation of the radical sites of tetramethylenes. This effect was predicted by O'Neal and Benson (*6*), and, in principle, it consists of two contributions, an increased rotational barrier height, which adds to the activation energy, and a decreased entropy of activation. The latter contribution may be understood if internal rotation is taken to be the reaction coordinate, so that this degree of freedom is lost in the transition state. Larger substituents will increase the moment of inertia of the rotating moiety and thereby make the activation entropy more negative (*6*).

Internal rotational rates of biradicals are not directly observable by the types of experiments described here, but, as we have seen, their competition with fragmentation and cyclization can be measured. Tables 3 and 4 (*139, 141, 144*), contain data for fragmentation and cyclization versus internal rotation in a series of putative tetramethylenes. The increased ratio caused by increasing substitution in either instance might be attributable to a decreased rotational rate and/or an increased reference rate (cleavage or cyclization). It is difficult to see, however, why substitution should change the reference rates monotonically in the same direction. A

5. HYPOTHETICAL BIRADICAL PATHWAYS

TABLE 3

Fragmentation/Internal Rotation Rate Constant Ratios (k_f/k_r) from the Biradical-Derived Portion of the Pyrolyses of Cyclobutanes and Diazenes

Reactant[a]	Temp (°C)	k_f/k_r Uncorrected[b]	k_f/k_r Corrected[b,c]	Ref.
(D,D-diazene)	439	0.2	0.2	141
Cis (methyl diazene)	439	1.3	1.3	139
Trans	439	3.0	3.0	139
Cis (cyclobutane)	~400	(~1.3)	(~1.3)	6
Trans	~400	(~4.0)	(~4.0)	6
Cis (substituted cyclobutane)	~400	(~0.67)	(~1.3)	6, 134
Trans	~400	(~1.3)	(~2.6)	6, 134
CD_3, CH_3, CH_3, CD_3 cyclobutane	401	>22	>44	135

[a] All reactions in the gas phase.
[b] Ratios in parentheses calculated using the assumptions of O'Neal and Benson (6).
[c] Normalized by doubling the ratios for the symmetric cases, where the k_r process is inherently twice as probable as elsewhere.

more plausible rationale would ascribe at least the qualitative direction of the effects in both series to a common cause, a decline in the internal rotational rate.

So far, we know virtually nothing about how such rotations occur. With reference to the trimethylene series (see Section III), we may hope for answers soon to questions such as the following: Are the rotations coupled? If so, do they occur by a conrotatory or a disrotatory pathway?

TABLE 4

Retention/Inversion Product Ratios (R/I) and Cyclization/Rotation Rate Constant Ratios (k_c/k_r) in the Formation of Cyclobutane Products from 1,2-Diazenes

Reactant	Temp (°C)	R/I	k_c/k_r Uncorrected	k_c/k_r Corrected[a]	Ref.
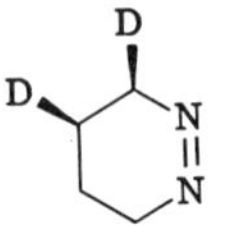 (D,D-dimethyl diazene)[b]	439	1.3	0.08	0.08	141
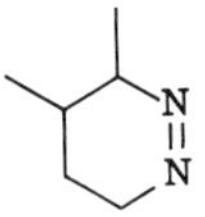 (dimethyl diazene)[b]					
Trans	439	7.8	1.9	1.9	139
Trans	306	8.8	1.7	1.7	139
Cis	439	2.3	0.7	0.7	139
Cis	306	3.7	0.9	0.9	139
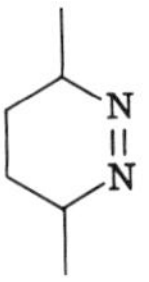 (dimethyl diazene)[b]					
Trans	439	1.7	0.7	1.4	139
Trans	306	2.2	1.2	2.4	139
Cis	439	1.7	0.7	1.4	139
Cis	306	1.9	1.0	2.0	139
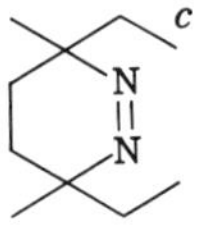 (tetramethyl diazene)[c]					
Meso	148	>49	>48	>96	144
dl	148	>49	>48	>96	144

[a] Normalized by doubling the ratios for the symmetric cases, where the k_r process is inherently twice as probable as elsewhere.
[b] Gas phase.
[c] Solution.

Does fragmentation of a cyclobutane to two olefins occur from the gauche or from the extended tetramethylene?

B. Hypothetical 1,4 Biradical Intermediates in the Cope Rearrangement

The Cope (e.g., **135a** → **136a**) and Claisen (e.g., **135b** → **136b**) rearrangements occupy important positions in the history of thermal reorgani-

5. HYPOTHETICAL BIRADICAL PATHWAYS

zations because they were among the first reactions for which there was convincing evidence of a concerted mechanism (*145*). Although ascribing

135a X = CH$_2$
135b X = O

136a X = CH$_2$
136b X = O

motives to the behavior of chemists is dangerous, one is tempted to argue that the esthetic pleasure afforded by the spectacular molecular acrobatics of some polyallylic systems may contribute to the intense and continuing interest in [3,3] sigmatropic rearrangements. To one unfamiliar with those processes, the stereospecific transformation **137 → 138** (*146*), for example, and the observation (*147*) that bullvalene (**139**) at 180°C shows only one NMR resonance because its 10!/3 = 1,239,600 structurally identical valence tautomers are interconverting rapidly (*148*) must be sources of both delight and bewilderment. Fortunately, the field is blessed with excellent, recent reviews (*145, 149, 150*), so that one may concentrate here on the question of biradical intermediates. These fall into two

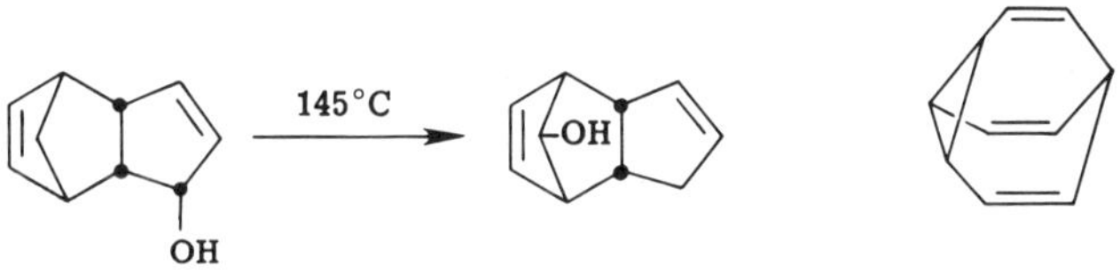

137   138   139

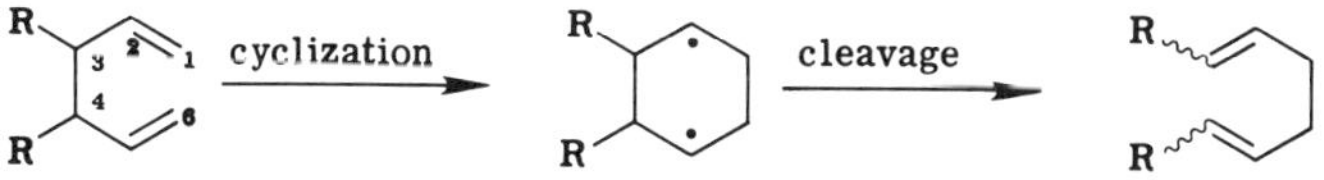

groups: one is generated in a "cleavage–cyclization" mechanism the first step of which is homolysis of the bisallylic 3,4 bond of a cyclic Cope system, the intramolecular analogue of the dissociative radical pair mechanism observed in some acyclic cases (*151, 152*); the other is generated in a "cyclization–cleavage" mechanism the first step of which is formation of the 1,6 bond. The "cleavage–cyclization" mechanism gives a 1,*n* biradical in which the value of *n* is a function of the structure of the starting material, whereas the biradical in the "cyclization–cleavage" mechanism is necessarily a 1,4-cyclohexadiyl. As we shall see, the latter mechanism sometimes is difficult to distinguish from a concerted process.

1. Quasi-dissociative "cleavage–cyclization" mechanism

If C-1 and C-6 of the hexa-1,5-diene are kept apart by the substrate structure, a geometry appropriate to a pericyclic transition state can be reached only at the cost of strain energy. The energy benefit of a concerted reaction may not be enough to compensate for the strain. In these circumstances, a stepwise mechanism via biradical intermediates may become predominant. One may subdivide such substrates into two groups, which for convenience may be termed *epimerically* or *geometrically* unfavorable. These differ in the sense that an epimerically unfavorable substrate, in principle, can be converted by stereomutation or structural rearrangement to an isomer with a configuration epimerically favorable to concerted Cope rearrangement, whereas a geometrically unfavorable substrate cannot.

a. Epimerically Unfavorable Cope Rearrangements. Obviously, it is easier to bring the ends of the bisallylic system of a 1,2-dialkenylcycloalkane together when the alkenyl groups are in a cis (**140**) rather than a trans (**141**) relationship. Nevertheless, formal Cope rearrangements of *trans*-1,2-dialkenylcycloalkanes to *cis,cis*-cycloalkadienes (**142**) do occur, for example, **141b** → **142b** [R = H *(153)*, R = CH_3 *(154)*, or R = n − Bu

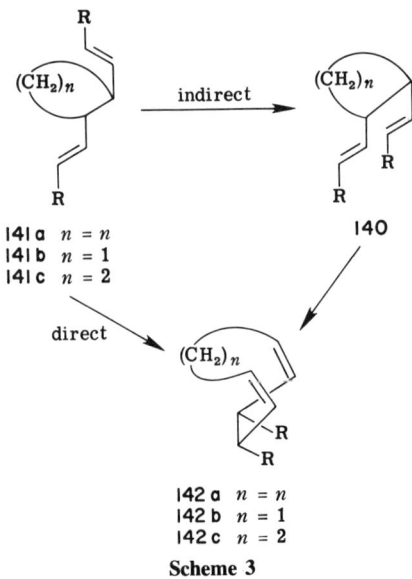

Scheme 3

5. HYPOTHETICAL BIRADICAL PATHWAYS

Scheme 4

(*154a*)], **141c** → **142c** [R = H (*155*, *156*) or R = CH$_3$ (*157–157b*)] (Scheme 3), **143a** → **144a** (*158*), **143b** → **144b** (*159*), and **143c** → **144c** (*160*) (Scheme 4). Such reactions raise the question as to whether their mechanism is *indirect* (**141** → **140** → **142**), requiring prior epimerization to the *cis*-1,2-dialkenylcycloalkene, or *direct*, perhaps by ring closure of an intermediate bisallylic biradical with two cis allylic groups, e.g., **147**.

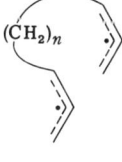

147

Although the indirect mechanism is usually assumed for the monocyclic *trans*-1,2-dialkenylcycloalkanes **141** (Scheme 3), its kinetic demonstration is difficult because at the temperature needed to induce pyrolysis of the trans compound, the intermediate cis isomer **140** often is so reactive that it does not accumulate in detectable amounts. However, when one or both of the alkenyl side chains have cis substituents, (CT or CC in Scheme 5 and Table 5), the [3,3] sigmatropic rearrangement of the cis isomer is retarded. This permits the indirect mechanism to manifest itself. Kinetic analyses (*157a, b*) of the time dependence of the (first-order) decay of the tTT and cTT compounds and of the (non-first-order) behavior of the other

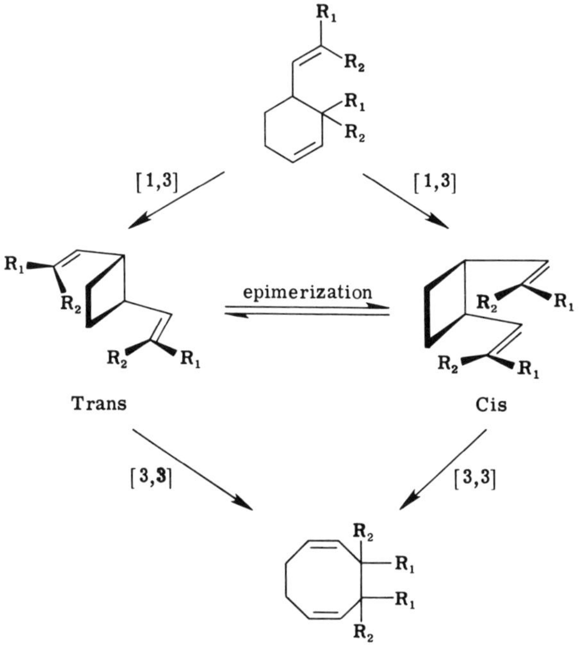

Scheme 5

four isomers show (Table 5) that, in the CT and CC cases, the direct [3,3] sigmatropic rearrangements of the epimerically unfavorable trans isomers have rate constants indistinguishable from zero. Thus, the indirect mechanism dominates here and is a plausible pathway in those cases in Scheme 3 where explicit kinetic verification is not yet available. On the other hand, the origin of the [3,3] sigmatropic rearrangement products from the

TABLE 5

Rate Constants of Thermal Rearrangements of the 1,2-Dipropenylcyclobutanes at 146.5° (157a,b)

Reactant configuration		Rate constant for process ($\times 10^5$ sec)		
Ring	Alkenes	Epimerization	[1,3] sigmatropic	[3,3] sigmatropic
Trans (t)	Trans,trans (TT)	0.436	0.950	~0
Cis (c)	Trans,trans (TT)	~0	~0	544
Trans (t)	Cis,trans (CT)	0.72	1.32	~0
Cis (c)	Cis,trans (CT)	6.70	3.44	6.07
Trans (t)	Cis,cis (CC)	2.0	0.088	~0
Cis (c)	Cis,cis (CC)	11.7	0.17	2.6

5. HYPOTHETICAL BIRADICAL PATHWAYS 363

substrates of Scheme 4 is not entirely clear. These systems are discussed further in Section IV,B,2.

Presumably, the reason that the direct mechanism is so insignificant in the epimerically unfavorable cases is that most of the ring-opening, biradical-producing events occur via a sterically uncrowded anti transition state. The resulting biradical has at least one trans allylic group (e.g., **149**, Scheme 6). Because allylic radicals have a rather high configurational stability (*161*), **149** can only cyclize to the highly strained *trans,trans*-cycloocta-1,5-diene **151** or, more plausibly, to the epimerized *cis*-1,2-dialkenylcyclobutane **150**. To achieve the direct mechanism, the syn,syn biradical **153–154** is needed. This can close to the unstrained *cis,cis*-

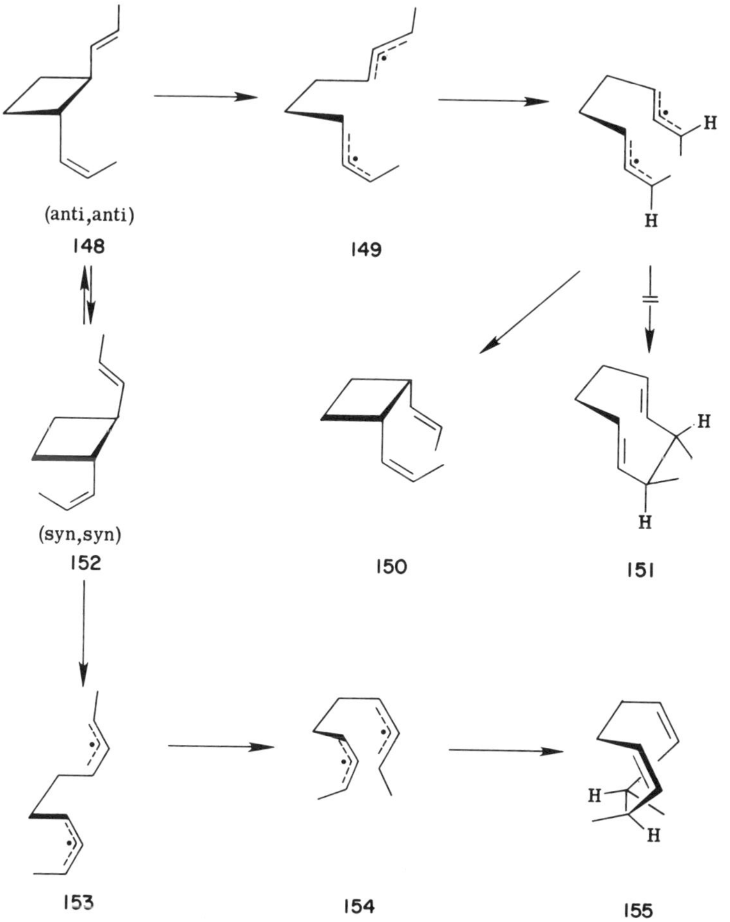

Scheme 6

cycloocta-1,5-diene **155**, but its formation requires that a high-energy, sterically congested transition state from the syn,syn reactant **152** be surmounted.

  b. Geometrically Unfavorable Cope Rearrangements. A substantial list of examples of this process is accumulating (Table 6) (*162–172*). In most cases, the distinction between a quasi-dissociative mechanism and one in which some C-1—C-6 bonding exists in the transition state has not bee made. In principle, secondary deuterium isotope effects could be usefully applied here. Because a cyclic Cope transition state would be partially bonded at C-1 and C-6, the C—H bond orbitals at these atoms should have more p character than those in the reactant, and this rehybridization should cause an inverse isotope effect, $k_H/k_D < 1$ (*173*). One can observe this in geometrically favorable cases where the transition state plausibly may be assumed to be cyclic (*174, 175*). For example, in the rearrangement of **156a**, k_H/k_D is 0.94 at 80°C for two deuteriums (*174*).

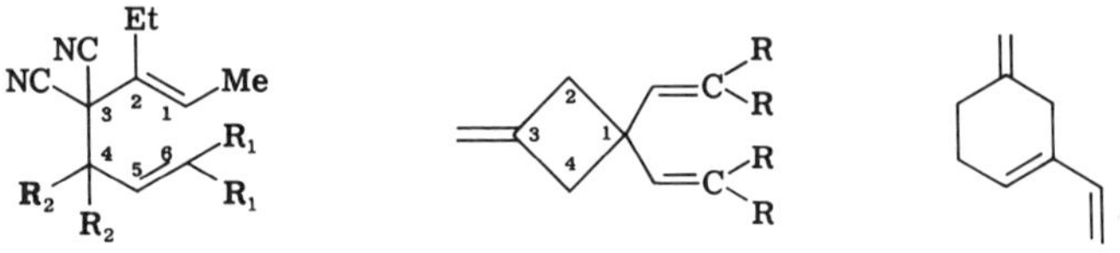

156a $R_1 = D; R_2 = H$
156b $R_1 = H; R_2 = D$

157a $R = H$
157b $R = D$

158

When the positions of the label are shifted to C-4, as in **156b**, the isotope effect becomes "normal," $k_H/k_D = 1.19$ (*174*).

  The major product of the thermal rearrangement of 1,1-divinyl-3-methylenecyclobutane **157a** is 3-vinyl-1-methylenecyclohex-3-ene **158** (Table 6). Unfortunately, the labeling pattern in the deuteriated analogue **157b** does not allow a structural distinction between a Cope rearrangement, joining the ring and side-chain-unsaturated methylene groups, and a [1,3] sigmatropic rearrangement, joining a saturated ring methylene to the side chain (**159**). However, the observed isotope effect for four deuteriums, $k_{157a}/k_{157b} = 1.05$ at 100°C, is normal rather than inverse, which suggests that in

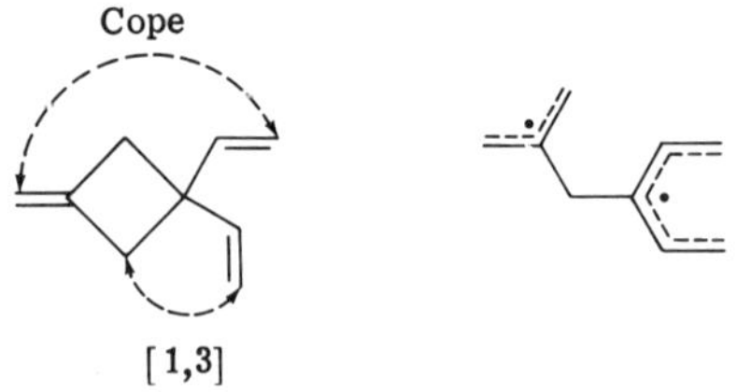

Cope

[1,3]

159

160

5. HYPOTHETICAL BIRADICAL PATHWAYS

TABLE 6

Geometrically Unfavorable Cope Rearrangements

Reactant	Temp (°C)	Products	Ref.
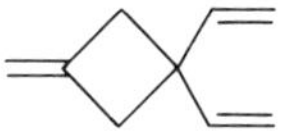	86–121	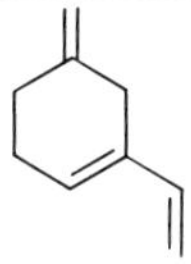	162
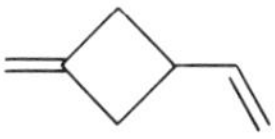	175–211	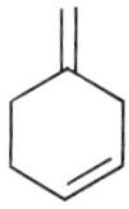	162
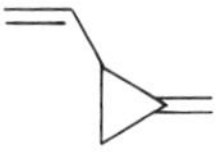	80	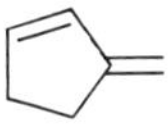	163
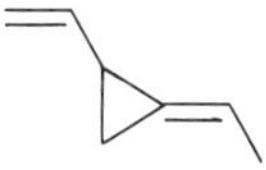	—	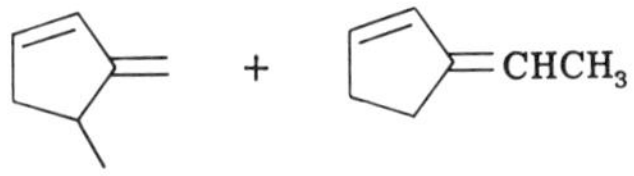 + =CHCH$_3$	164
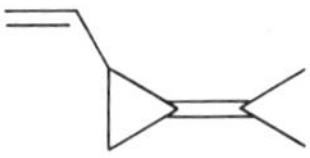	99–118	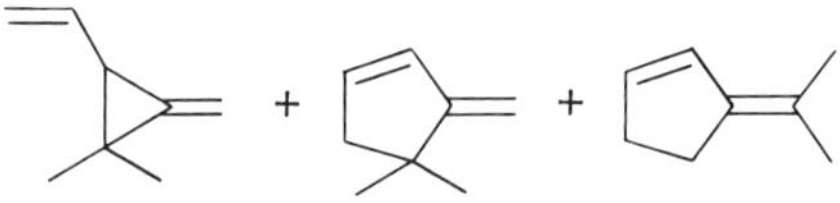 + +	165
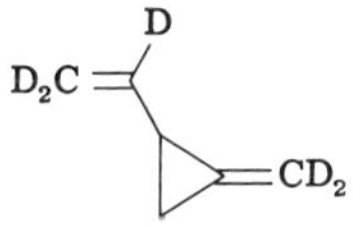	60–90	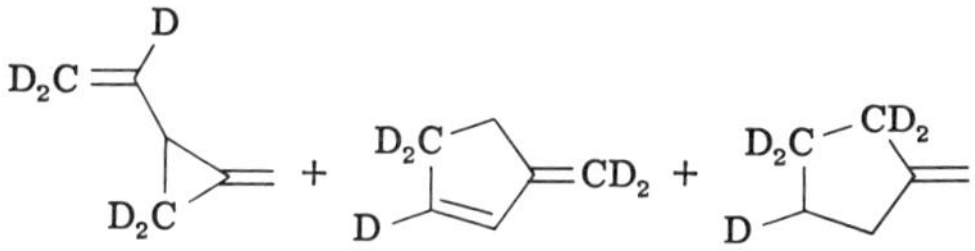 + +	166
	300	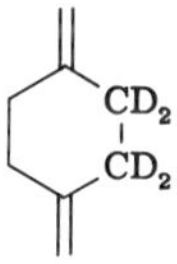 only	167
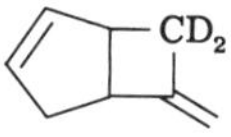	200	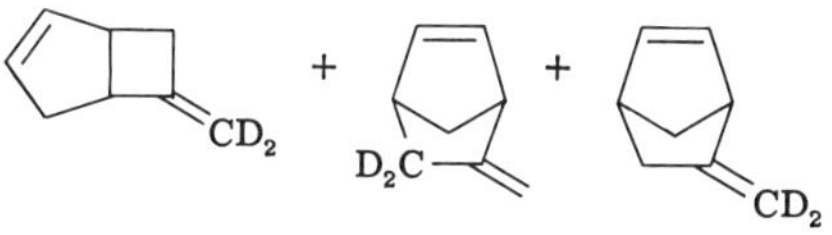 + +	168
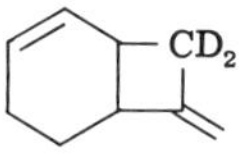	188	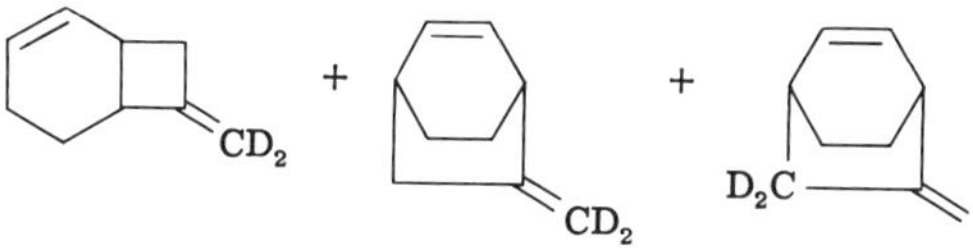 + +	169
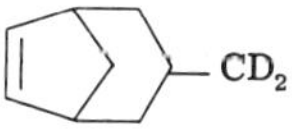	450–500 (flow)	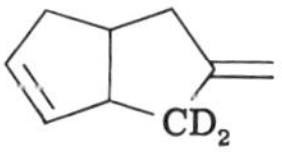 only	170

TABLE 6 (Continued)

Reactant	Temp (°C)	Products	Ref.
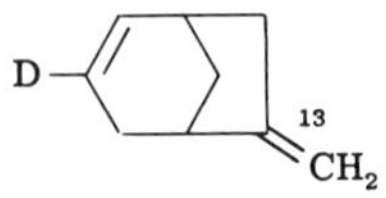 (structure: D-substituted bicyclic with $^{13}CH_2$)	309	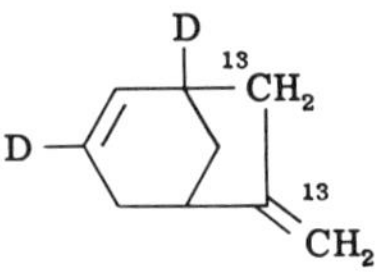 (structure: D-substituted bicyclic with $^{13}CH_2$ groups)	171
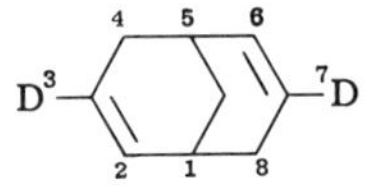 (structure: D^3-labeled bicyclic with positions 1–8, D at 7)	590–600 (flow)	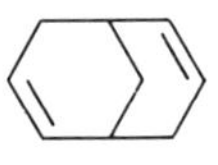 (bicyclic) labeled at 1, 3, 5, 7	172
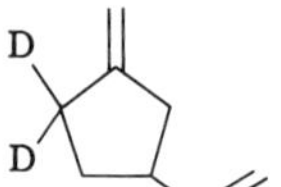 (structure: D,D-substituted methylenecyclopentane with vinyl)	261–304	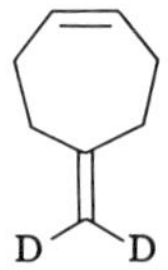 (cycloheptadiene with D,D)	169a

either case little bonding to the vinylic terminus has been achieved in the transition state (*162*). The mechanism is reasonably formulated with a biradical intermediate (**160**) (*162*).

A structural labeling experiment is available in a related case (*166*), the rearrangement of the deuteriated vinylmethylenecyclopropane **161** to methylenecyclopentene, which gives both position-labeled isomers **162** and **163** in *equal amounts*. The complete loss of regiospecificity cannot be blamed on the competing degenerate rearrangement of the starting material, **161** ⇌ **164**; this is not fast enough to account for all of it. Again, a biradical mechanism seems the most economical interpretation, although a competition between [1,3] and [3,3] rearrangements, with fortuitously equal rates, cannot be ruled out. If the biradical mechanism operates, the intermediate must have time-averaged or actual symmetry, as in **165** (*166*), where the allyl groups occupy mutually perpendicular planes.

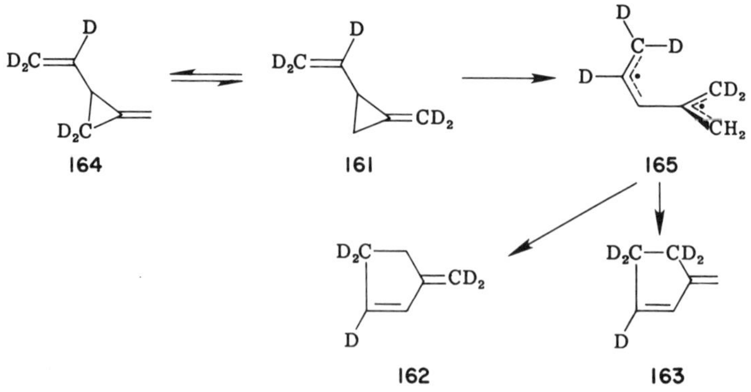

5. HYPOTHETICAL BIRADICAL PATHWAYS

One should not conclude from these results, however, that rotationally free biradicals are invariably the intermediates in geometrically unfavorable Cope rearrangements. Two examples of highly regiospecific behavior occur in the rearrangements of 1,4-bismethylenecyclohexane (**166**) (*167*) and 3-methylenebicyclo[3.2.1]oct-6-ene (**167**) (*170*):

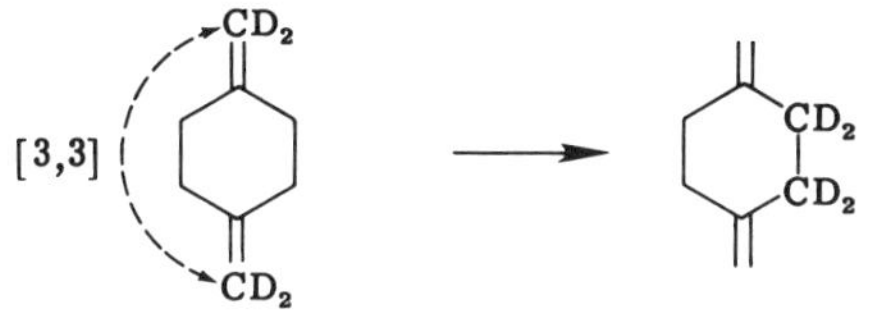

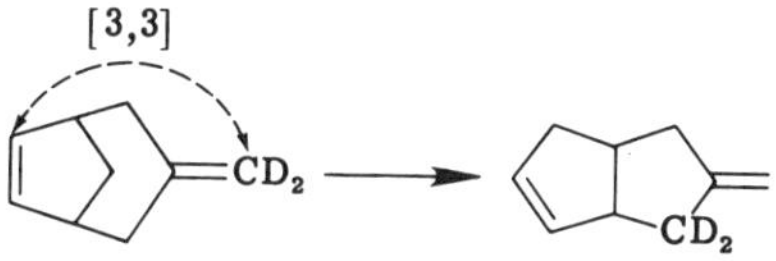

Inspection of scale molecular models does not offer strong support for the idea that the approach of the ends of the hexadiene system, at least in the ground state of either **166** or **167**, is much closer than in that of **159** or **161**. The differences in behavior must have a more subtle origin. Perhaps the molecular distortions needed to achieve the [3,3] transition state are less severe in the larger-ring examples because the strain can be distributed over a larger number of C—C bonds in the region of the molecule between the reacting termini. Information on the secondary kinetic isotope effect in the rearrangement of **166** would be of interest, since it would provide a test of C-1—C-6 bonding in the transition state of a geometrically unfavorable, but highly specific, Cope system.

A borderline situation seems to be exemplified by the automerization of 6-methylenebicyclo[3.2.1]oct-2-ene (**168**) (Scheme 7) (*171*). At 344°C, the optically active, ^{13}C-labeled (■ = $^{13}CH_2$) hydrocarbon A suffers both racemization and position mixing of the label. A kinetic analysis, facilitated by reresolution of the partially equilibrated recovered compound, assigns the rate constants for the two [1,3] sigmatropic reactions as $k_7 \cong k_1 = 1.26 \times 10^{-5}$ sec^{-1} and for the Cope rearrangement as $k_c = 8.26 \times 10^{-5}$ sec^{-1}. The major reaction is still the regiospecific Cope rearrangement, but weak competition from two formal [1,3] sigmatropic processes now has emerged. Because each of the latter processes represents only a small fraction of the total reaction, it is difficult to be sure that k_7 and k_1 are exactly equal. If they are, however, the rearrangement mechanism could involve competition between a regiospecific Cope process and a biradical

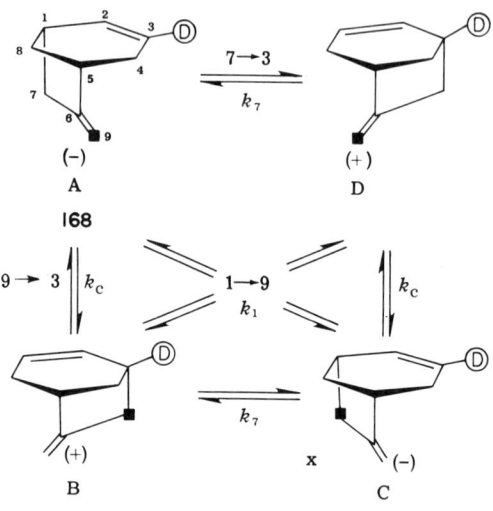

Scheme 7

pathway passing over a symmetric intermediate. Candidates for the latter might include a fully randomized species derived from **168** by cleavage of the C-7—C-8 bond and rapid rotation about the C-5—C-6 bond. This species could cyclize to B, C, and D with equal probability, thus ensuring $k_1 = k_7$. Other entities with the proper bilateral symmetry include **169** and **170**, but not **171**. The bisected geometry of **171** could lead only fortuitously to $k_1 = k_7$.

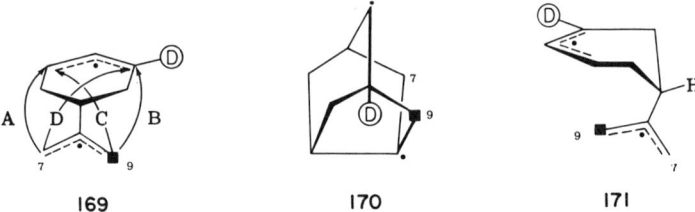

A molecular model of **168** suggests that deformation of the structure to bring together the ends of the Cope hexadiene system (C-3 and C-9) might be more difficult than the corresponding deformation of **167**. This could account for the change from regiospecific reaction in **167** to the **168** case, in which the Cope process barely survives.

2. "Cyclization–cleavage" mechanism

Although the possibility of a cyclohexa-1,4-diyl intermediate was broached in connection with the rearrangement of cyclodeca-1,5-diene at an early stage (*176*), the idea has been subjected to scrutiny only recently.

5. HYPOTHETICAL BIRADICAL PATHWAYS

In the case of 1,1-dideuteriohexa-1,5-diene itself (**172**), thermochemical additivity assumptions lead to a ΔH_f° for cyclohexa-1,4-diyl (**173**) (55–56 kcal/mol) that lies very close to the ΔH_f° of the transition state for the degenerate rearrangement (ΔH_f° of **172** + $\Delta H^\ddagger$ = 20.1 + 33.7 = 53.8 kcal/mol) (*177*). Considerable controversy has developed concerning the details of the reaction coordinate. Does it contain a true local energy minimum (**173**) or does only a single concerted transition state (**174**) intervene?

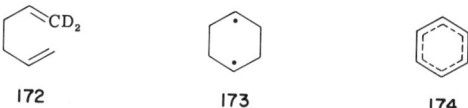

Theoretical support for the diyl **173** comes from qualitative arguments (*178*) as well as MINDO calculations (*179, 180*), but counterarguments are not lacking. It has been suggested (*149, 150*) that the ΔH_f° calculated from additivity tables is too low and that application of a 3–4 kcal/mol correction would raise it to 58–60 kcal/mol, placing **173** well above the Cope transition state and hence excluding it from the main mechanism. Although such corrections may well be appropriate in some cases (see Section III), their general validity has not been established, so that the diyl intermediate conceivably might survive this objection.

Gajewski and Conrad (*175*), however, adopt another line of reasoning that adds a good deal of strength to the challenge. They cast the argument, not in terms of ΔH_f° but of ΔG values on the grounds that $\Delta G^\ddagger$ differences are the operative determinants of the relative rates through the diyl and concerted channels. This leads to the energy diagram shown in Fig. 8, which applies to the experimental temperature 523 K.

The energy of the diyl **173** is estimated from additivity tables, but the accuracy of the estimate is not crucial here. The exposition depends instead on the free energy of the transition state marked "cleavage," which Gajewski and Conrad evaluate from an important study by Goldstein and Benzon (*181*) of the pyrolysis of bicyclo[2.2.0]hexane-2,3,5,6-d_4 (**175**). At 523 K, the stereospecifically labeled bicyclic compound undergoes both stereomutation by double inversion at the bridgeheads and stereospecific ring opening to hexa-1,5-diene of the indicated configuration (Scheme 8, **176**). Goldstein and Benzon prefer to interpret the cleavage as an orbital symmetry-allowed concerted [$_2\sigma_s + _2\sigma_a$] cycloreversion, but an alternative biradical formulation is proposed by Paquette and Schwartz (*182*) and by Roth and Martin (*183*) for similar predominances of the **176** product configuration in pyrolyses of the diester **177** and the hydrocarbon **178** to the dienes **184** and **185**, respectively (Scheme 8). Initial cleavage of the bridge bond of the exo substrate gives the boat biradical

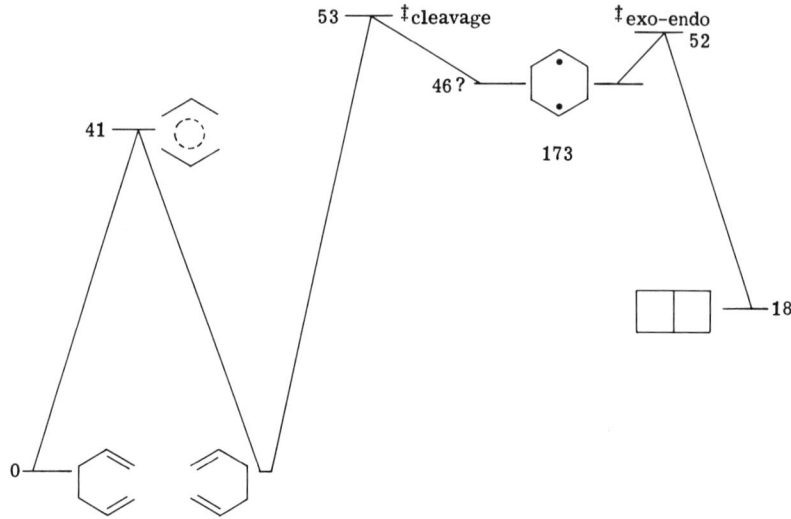

Fig. 8 Relative free-energy relationships among reactants and high-energy species. The values are in kilocalories per mole at 523 K (*175*).

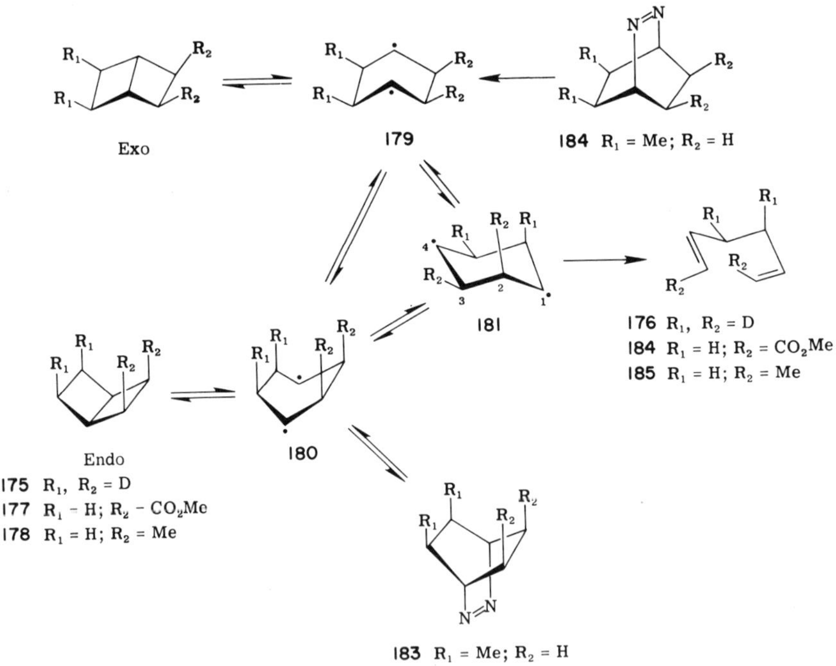

Scheme 8

5. HYPOTHETICAL BIRADICAL PATHWAYS

179, which can undergo conformational isomerization either to another boat biradical, **180**, which is the precursor of the *endo*-bicyclo[2.2.0]hexane, or to the chair biradical **181**, which cleaves to the diene **176**. The conformation of the chair biradical **181** (*cf.* **187**) places one of the two cis R groups equatorial and the other axial, and good overlap with the odd-electron orbitals facilitates cleavage of the C-2—C-3 bond. Diene formation from the boat biradical **180** (*cf.* **186**), which could lead to a different product stereochemistry, is unfavorable because the relevant alignment is poor. Some support for the biradical intermediate is provided

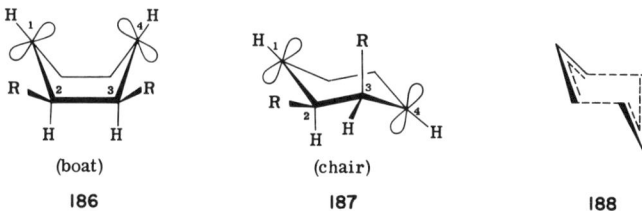

(boat)　　　　　(chair)
186　　　　　**187**　　　　　**188**

by the observations (*183*) that the pyrolyses of the bicyclic azo compounds **182** and **183** or the bicyclic hydrocarbons *endo*- or *exo*-**178** give nearly the same mixtures of products.

If one adopts the biradical formulation of the Goldstein–Benzon results, and if the rate-determining step in the cleavage reaction lies between the chair diyl **181** and the product **176** (Scheme 8), then Fig. 7 shows immediately that the chair diyl (**173** ≡ **181**) cannot be involved in the main mechanistic pathway of the Cope rearrangement, since the transition state energy from **173** is 12 kcal/mol higher than that of the Cope rearrangement (*175*). However, as Alder has pointed out (*175a*), this argument does not hold if the rate-determining step in the bicyclo[2.2.0]hexane cleavage is not a bond cleavage but rather a conformational isomerization of boat (**180**) to chair (**181**) diyl. This would permit the subsidiary transition state between **181** and **176** to be much lower in energy than Fig. 7 shows and would be compatible with a chair diyl intermediate in the Cope rearrangement.

The famous experiment of Doering and Roth (*184*) demonstrated the highly stereospecific conversions of *meso*- and *d,l*-3,4-dimethylhexa-1,5-diene to *cis,trans*- and *trans,trans*-octa-2,6-diene. Although not uniquely definitive of a species with a chair geometry on the upper reaches of the Cope rearrangement energy surface (*75, 181*), the results are conveniently so formulated (**188**). Moreover, an independent study of optically active 3-phenyl-3-methyl-hepta-2,5-diene (*181a*) definitely favors the chair. If we accept the arguments (*149, 150, 175*) that the chair biradical **187** is involved in the bicyclo[2.2.0]hexane cleavage, but not in the Cope rearrangement, there must be two bisallylic entities with the chair conformation, **187** and **188**, the latter being lower in free energy by 12 kcal/mol.

It is a matter of great interest that Dewar and his co-workers, the leading proponents of the cyclization–cleavage mechanism on theoretical grounds (*180, 185*), should, at the same time, find experimental reasons for rejecting it (*186*). Dewar and Wade (*186*) observed that the enhancement of the rearrangement rate by phenyl substitutions at C-2 and at C-4 of the hexadiene system is multiplicative, being roughly equal to the product of the C-2 and C-4 effects. They interpreted this to be consistent only with significant C-3—C-4 bond weakening in the transition state and concluded that the biradical, if formed at all, "must represent a very minor depression in the potential surface and so be essentially irrelevant." Presumably, these authors soon will reconcile the present ambiguity in their position.

Other studies of substituent (*149, 150, 187*) and secondary isotope (*175*) effects on the rate of the Cope rearrangement can be made to conform to a self-consistent interpretation if the reaction is viewed as having a variable transition state structure. Both C-1—C-6 bonding and C-3—C-4 cleavage are well developed in the unsubstituted hexa-1,5-diene transition state. Substitution at C-2 and C-5 shifts the transition state to a structure with strong C-1—C-6 bonding but little C-3—C-4 cleavage, whereas substitution at C-3 and C-4 favors a transition state with weaker bonding at that site (*149, 150, 175*).

An intriguing class of formal Cope rearrangements occurs in the pyrolyses of 1,2-bismethylenecyclobutanes, in which tetramethyleneethane biradicals may be involved. Space limitations preclude an extensive discussion here, but the interested reader may consult the excellent review by Gajewski (*75*).

V. BIRADICALS IN [1,3] AND [1,5] SIGMATROPIC REARRANGEMENTS: SOME COMMENTS ON THE PRINCIPLE OF LEAST MOTION

Sigmatropic rearrangements have provided many experimental evaluations of the ability of orbital symmetry forces to control the stereochemical course of thermal reactions. The subject has been extensively reviewed (*49, 60, 75, 103, 188–195*), and the present discussion concerns primarily the nature and role of biradicals in these processes.

Orbital symmetry-allowed [1,3] sigmatropic rearrangements require the migrant group to invert its configuration if the allylic framework participates suprafacially. Retention of configuration is allowed only if the

5. HYPOTHETICAL BIRADICAL PATHWAYS

rearrangement occurs antarafacially. The antarafacial retention [1,3] reaction has eluded detection despite several deliberate searches for it, including one (Scheme 9, **193**) in which inversion of the migrant carbon was sterically prohibited. The record of such attempts is as follows: (1) **189 → 190** and (2) **191 → 192**, at most a small antarafacial component of the

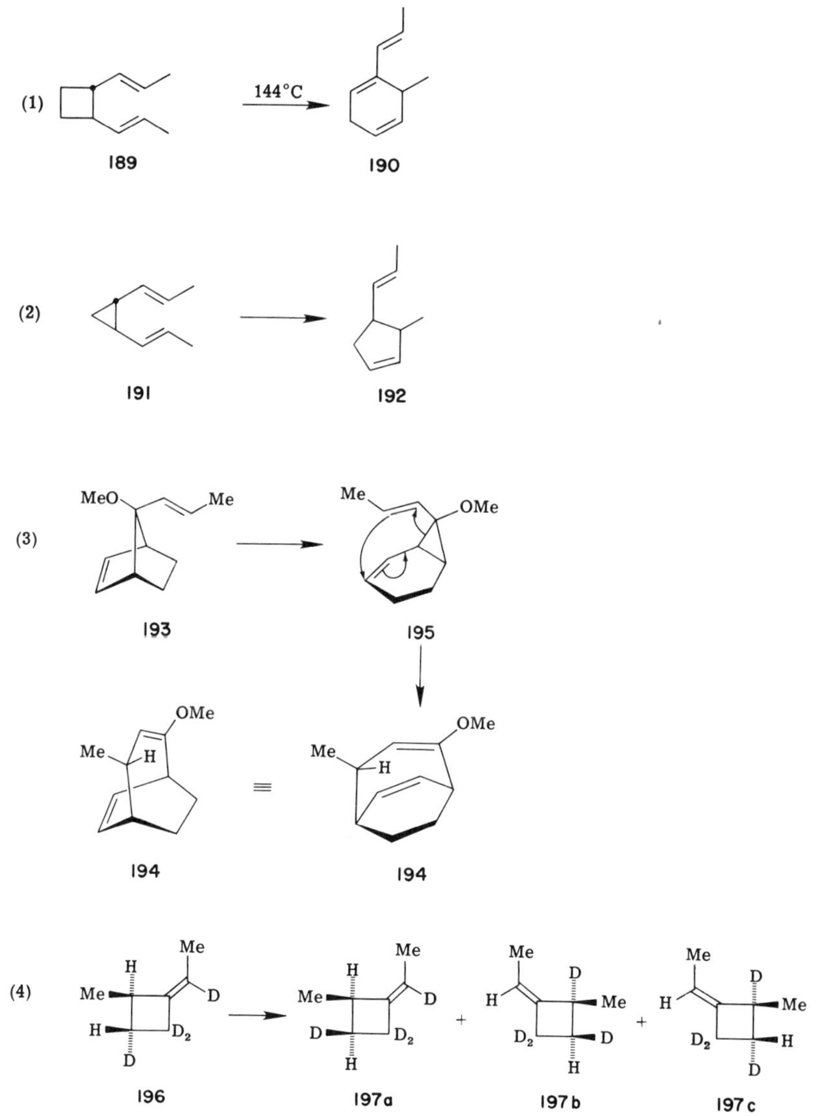

Scheme 9

mechanism (*157b, 196*); (3) **193** → **194**, formal antarafacial migration, probably by another mechanism, via **195** (*159*); (4) **196** → **197**, experimental data interpreted (*197*) in terms of a large antarafacial contribution but later shown (*198*) to be consistent with little.

Despite its classification as an orbital symmetry-"forbidden" reaction (*199*) with an "antiaromatic" transition state (*200, 201*), suprafacial retention [1,3] sigmatropic rearrangement occurs readily in many instances (*157–157b, 202–202b*). Whether such reactions should be considered concerted has been a matter of much discussion (*188–195, 202–207d*), the details of which need not be repeated here. It may be useful to point out, however, that "antiaromaticity" must be defined with reference to a model. The electronic energy of the concerted suprafacial retention transition state **198**, calculated by the most primitive orbital interaction scheme (*202b, 203*), *is* higher than that of its suprafacial inversion counterpart **199**, but the important comparison is with the noninteracting radical pair **200**, which serves as a model for the biradical transition state. Treatment of the primitive interaction scheme **198** by simple Hückel one-electron theory, as well as some more sophisticated molecular orbital calculations on related systems at the INDO (*206, 207*), Mulliken–Wolfsberg–Helmholtz (*207*), and *ab initio* [STO-3G (*207a*), STO-4G (*207*), and 4-31G (*207a*)] levels, suggest that substantial stabilization of the "forbidden antiaromatic" transition state relative to the biradical model can occur by "subjacent orbital control." An additional factor that can reduce the calculated energy of the forbidden pathway is configuration

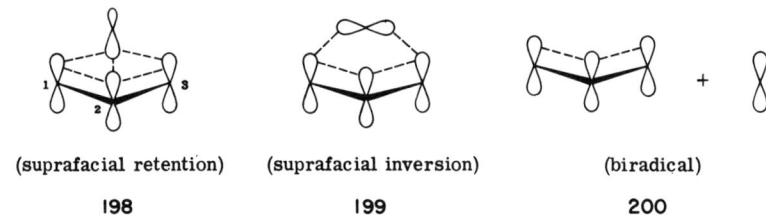

(suprafacial retention) (suprafacial inversion) (biradical)

198 **199** **200**

interaction (*195b,c, 207a*). At the primitive level (*202b, 203*), the "antiaromaticity" of **198** is associated with nodeless ("Hückel") overlap of the orbitals in a four-electron system. The unfavorable nature of this array can be mitigated by transannular overlap between the migrant carbon and C-2 of the allylic framework, since, in effect, this converts a cyclobutadienoid to a bicyclobutadienoid system. In reactant structures that enhance this migrant–C-2 overlap while preserving migrant carbon–C-1–C-3 overlap, the forbidden concerted pathway should become relatively more important, and a body of experimental data now can be interpreted in this way (*202a–202b*).

5. HYPOTHETICAL BIRADICAL PATHWAYS

CNDO calculations (*204*) suggest that the degenerate rearrangement of methylenecyclobutane **201** is nonconcerted. In this system, steric factors permit only feeble interaction of the migrant carbon with the terminal carbons of the allylic system (see **202**). The rearrangement of methylenecyclobutane therefore is not a suitable model for the forbidden concerted pathway postulated in other systems (see *203*).

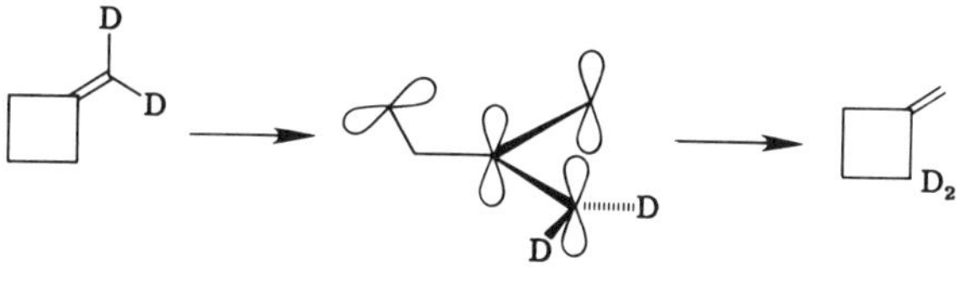

201   **202**

The "allowed" suprafacial inversion pathway dominates in many rearrangements of bicyclo[3.2.0]hept-2-enes (**203**) to norborn-2-enes (**205**) when steric factors do not preclude it. The reaction occurs when the 7-substituent is deuterium (**203a**) (*188, 202c*) or when it occupies an exo configuration in the reactant (**203b,d**) (*202–202b, 205*). If the 7-substituent

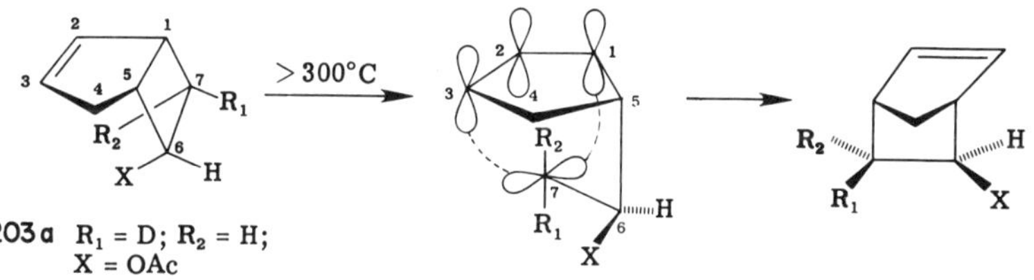

203a $R_1 = D; R_2 = H;$
 $X = OAc$
203b $R_1 = Me; R_2 = H;$
 $X = OAc$
203c $R_1 = H; R_2 = Me;$
 $X = OAc$
203d $R_1 = OMe; R_2 = X = H$

204   **205**

is endo (**203c**) (*202–202b*), the preferred mode becomes suprafacial retention, because the transition state for the suprafacial inversion pathway **204** is destabilized by the severe repulsive interaction generated when R_2 is thrust into the cyclopentane ring.

It may be useful to examine the assertion that the stereochemical course of the suprafacial inversion reaction is "controlled" by orbital symmetry. What is meant, of course, is that not only do the qualitative, approximate, Woodward–Hoffmann rules for comparing the energies of competing transition states predict a preference for this pathway, but also the actual physical source of the observed result is the better bonding in the "allowed" transition state. We do well to preserve a conceptual distinction between an agreement with prediction and an ascription of cause and effect until other interpretations have been scrutinized. In this context, it is important to evaluate the report (*208*) that the suprafacial inversion rearrangement **203a** → **205a** corresponds to a "least-motion" pathway.

At first glance this conclusion seems incredible, because our prejudice is to describe the suprafacial inversion pathway as one that requires two internal rotations, about C-5—C-6 and C-6—C-7, whereas the suprafacial retention reaction requires only one rotation, C-5—C-6. The anomaly is only apparent, however. Its origin lies in the definition (*208*) of "motion," which merely is the change in the atomic coordinates (relative to a fixed center of mass) between the reactant and product. In this definition, the *actual atomic trajectories* are straight lines, not internal rotations, and no attention is paid to whether the "least-motion" pathway entails such severe distortions of bond angles and distances as to render the process energetically infeasible. An inspection of models reveals that straight-line "least-motion" conversion of the atomic coordinates of **203a** to those of **205a** must involve such distortions.

Although the authors (*208*) are careful not to claim that the **203a** → **205a** rearrangement actually follows the "least-motion" pathway, the distinction between "corresponds to" and "follows" seems to have been blurred in commentary elsewhere. Thus, the statement (*209*) that the **203a** → **205a** rearrangement "is consistent with either orbital symmetry *or* least motion control" cannot stand if the word "control" is to retain its conventional meaning.

It is possible to detect [1,5] sigmatropic rearrangements of carbon in the pyrolysis of "tropilidenes," the latter term being a general designation for a cycloheptatriene–norcaradiene valency tautomeric pair. Early examples include the rearrangements of x,7,7-trisubstituted tropilidenes **206** (*189, 210*) and **207** (*211*) by the mechanism shown in Scheme 10. The key reaction is a stepwise circumambulation of C-7 in the norcaradiene form (**208** ⇌ **209** ⇌ **210**), each step of which is a [1,5] sigmatropic rearrangement.

These rearrangements are necessarily suprafacial for steric reasons, and orbital symmetry (*199*) therefore requires that the migrant carbon retain configuration in each step. A proposal (*188, 189, 210*) for an experimental test employs chiral tropilidenes.

A 3,7,7-trisubstituted tropilidene in which the 7-substituents differ (symbolized by one small and one large dot in Figs. 9 and 10) can exist in optically active form. Fast, reversible conversion to a norcaradiene prepares the molecule for the skeletal rearrangement of Scheme 10. Note that, although either of two diastereomeric norcaradienes can be formed, the overall chirality of the system is not affected by the choice.

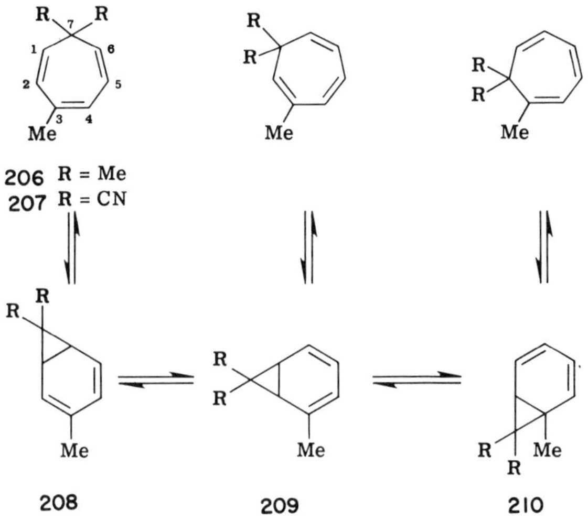

5. HYPOTHETICAL BIRADICAL PATHWAYS

Scheme 10

Any epimerization at C-7 would interchange the large and small dots and racemize the reactant. In detail, one might imagine that this could occur by way of a biradical intermediate formed by cleavage of *either* of the bonds to C-7 in the norcaradiene, or perhaps by way of a species in which *both* bonds to C-7 are stretched, as in a carbene–aromatic complex.

In the absence of such epimerization, one may distinguish between orbital symmetry-"forbidden" and orbital symmetry-"allowed" pathways. If rearrangement occurs by the "forbidden" pathway with inversion at each step, the breaking and forming bonds attach to opposite faces

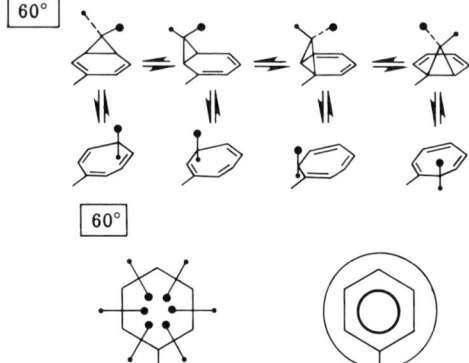

Fig. 9 Circumambulatory [1,5] sigmatropic rearrangement of a chiral tropilidene with inversion of configuration in each step (*189*).

of the migrant carbon. In this process, a substituent (large dot) originally "inside" remains inside, and an "outside" one (small dot) remains outside with respect to the six-membered ring of the norcaradiene. A complete 360° circuit around the ring requires six steps, during which the original large-small "inside–outside" configuration is repeated six times, corresponding to motion with a 60° periodicity, as shown in Fig. 9. Using notation that we introduced (*189*), we show the changes for the molecules themselves and then in abstractions of the motion of C-7 and its attached groups. The abstractions represent, respectively, a series of snapshots of top views of the motion at rest points corresponding to stable intermediates and a blurred time exposure of the motion. The 60° mechanism produces racemization, since passage by *backside bonding* over the site of the lone ring substituent interconverts two enantiomeric norcaradienes. The racemization may be confirmed readily by reference to the abstractions for the 60° mechanism, each of which has a plane of symmetry.

The alternative "allowed" retention pathway would give rise to threefold periodicity (120° mechanism, Fig. 10), since C-7 uses the same face to make the new bond. This requires a twist at each step and causes an *in–out alternation* of the disposition of a given substituent. Passage over the site of the lone ring substituent now does *not* interconvert enantiomers. Thus, in the 120° mechanism, an infinite number of circuits can be achieved with perfect preservation of optical purity. There are two con-

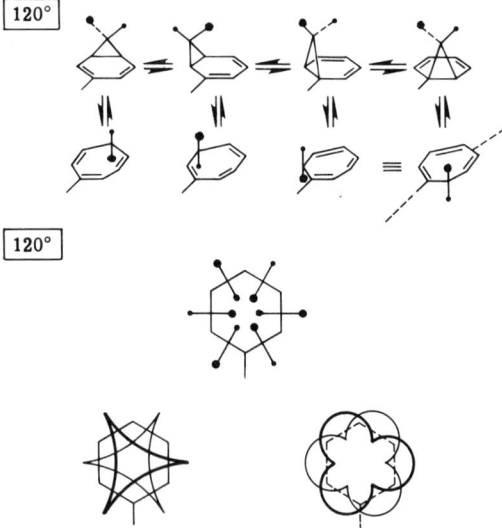

Fig. 10 Circumambulatory [1,5] sigmatropic rearrangement of a chiral tropilidene with retention of configuration in each step (*189*).

5. HYPOTHETICAL BIRADICAL PATHWAYS

ceptually distinct subcategories of the 120° mechanism—one with short swoops and the other with more flamboyant swoops—corresponding to the two ways of connecting like-sized substituents in the time-average exposure. Neither kind of motion has a time-average mirror plane.

The two completed experimental tests of the stereochemistry of the norcaradiene rearrangements give conflicting results. Klärner and his co-workers (*212*) studied the reactions of optically active methyl 2,7-dimethyltropilidene-7-carboxylate **211** in benzene at 180°C. Scheme 11 is

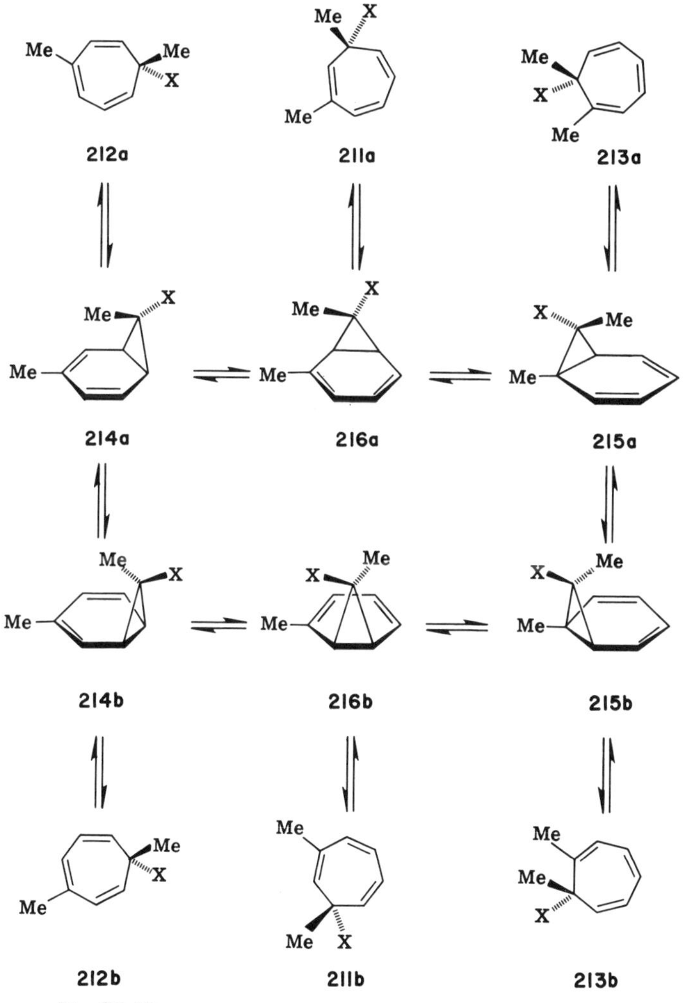

Scheme 11

X = CO_2Me

formulated with exclusive inversion in each step of the [1,5] sigmatropic rearrangement, in accord with the kinetic analysis (*212*) of the rates of interconversion and racemization. In particular, the optical purities of the products and recovered reactants are in agreement with the assignment of substantial rate constants for enantiomerizations **214a** → **214b** and **215a** → **215b**, which can occur by migration with inversion across the symmetry plane containing the ring methyl group. Enantiomerization **216a** → **216b** in this mechanism cannot occur directly and requires three sequential steps, which agrees with the assignment of a zero value to the rate constant for the direct process.

It is important to note that the zero rate constant for the direct path **216a** → **216b** also excludes any significant component in which the reactant **211a** suffers one-center epimerization (and hence racemization) without rearrangement. Thus, **211a** of undiminished optical purity can be recovered from a partial pyrolysis after about 30% conversion to **212** and **213**.

Studies of the norcaradiene rearrangements in these cases are complicated by a homodienyl hydrogen shift reaction giving a 1,4-cyclohexadiene, e.g., **211** → **217**, examples of which also were observed in the pyrolysis of **206** (*210*). This process is reversible, not only to the starting material, but also to rearranged norcaradiene product **218** (≡**212**), thus providing a pathway independent of the [1,5] sigmatropic carbon shift for achieving rearrangement *with inversion*. However, Klärner (*212*) was able

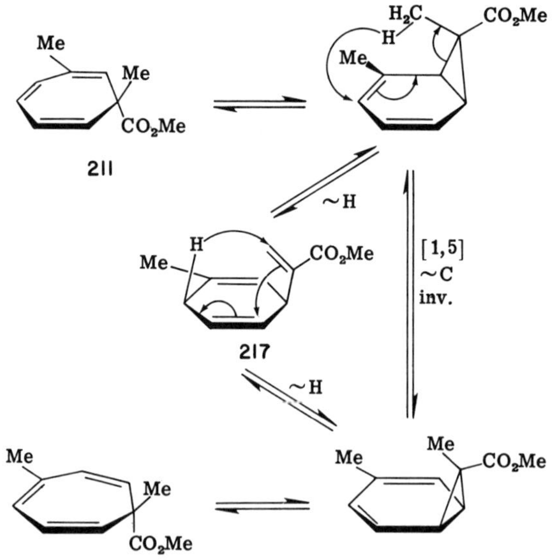

5. HYPOTHETICAL BIRADICAL PATHWAYS

to show that the contribution of the homodienyl hydrogen shift in his system was small.*

Baldwin and Broline (214) determined the rate constants for three processes, one-center epimerization (k_e), rearrangement with inversion (k_i), and rearrangement with retention k_r, by following the racemization and deuterium scrambling in 2-deuterio-3,7-dimethyl-7-methyoxymethyltropilidene (**219**-2-*d*, Scheme 12). The values ($\times 10^6$ sec) at 223.4°C were $k_e = 4.5$, $k_r = 2.9$, and $k_i = 0.5$. Although a direct determination of the contribution to skeletal rearrangement from a homodienyl hydrogen shift was not reported, this process, if present, would appear as a component of k_i, so that the small value observed for this rate constant may be taken as an upper limit for the [1,5] carbon shift with inversion. Retention thus is the dominant pathway for rearrangement ($\geqslant 6:1$ preference) in this system. That one-center epimerization competes so effectively with rearrangement seems surprising. Note that not all of the k_e reaction can involve cleavage of a cyclopropane ring bond, for example, in **220**, to give a stereorandomized biradical **224**. Such a species should have given **222** at the same rate as **221** (Scheme 12), which would have resulted in $k_r \geqslant k_e$.

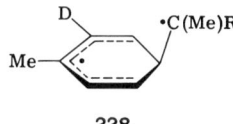

228

The reasons for the discordant results, **212a** inversion (Scheme 11), **219** retention (Scheme 12), are not immediately apparent. However, there seems to be no basis for the implication (214) that Klärner failed to recognize the possibility of one-center epimerization and that the assignment of an inversion mechanism to his example, therefore, is suspect. The experiments already described show that one-center epimerization cannot be a major factor in the Klärner experiment, and we must look elsewhere for the origin of the differences in behavior. One obvious difference between the two systems is the substituent on the migrant carbon, CO_2CH_3 in the case of Scheme 11 (**212**) versus CH_2OCH_3 in the case of Scheme 12 (**214**).

Very recently, Klärner and Brassel (214a) have shown that this structural difference does not affect the stereochemistry of the migration. They

---

* In an incomplete, unpublished study (213), it was observed that optically active 1-deuterio-3-carbomethoxy-7-methyl-7-methoxymethyltropilidene racemized thermally at about the same rate as it suffered degenerate deuterium scrambling. This result is consistent with [1,5] sigmatropic rearrangement with inversion, but the interpretation is uncertain because the contributions of the homodienyl hydrogen shift and the one-center epimerization are not yet known.

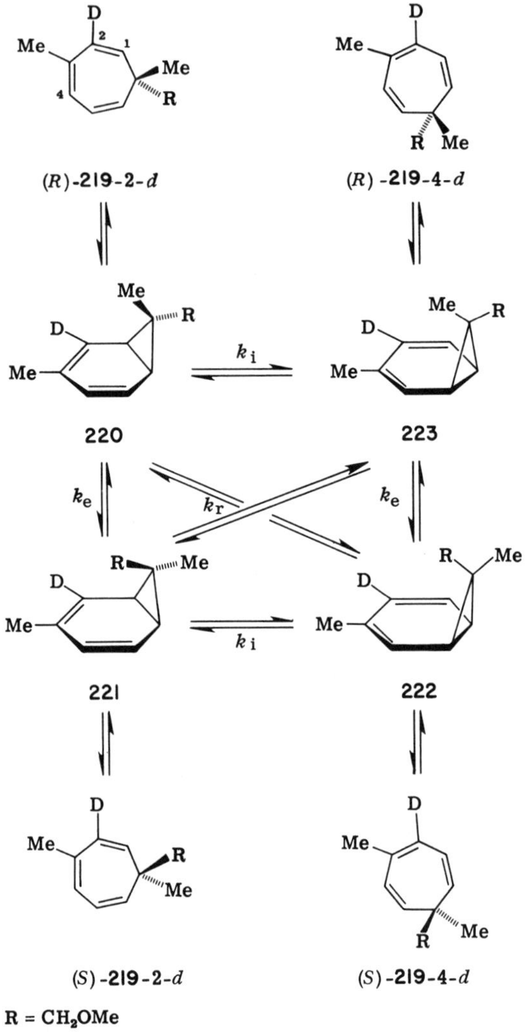

Scheme 12

R = CH₂OMe

replaced the CO_2CH_3 group of **211a** (Scheme 11, X = CO_2CH_3) with CH_2OCH_3. Rearrangement of optically active **211a** (X = CH_2OCH_3) to **212a** and **213a** (X = CH_2OCH_3) occurred with highly specific (>95%) *inversion*, just as in the case of **211a** (X = CO_2CH_3). A most significant observation was the recovery of **211a** (X = CH_2OCH_3) with undiminished optical activity after 30% conversion to rearrangement products. This excludes any important component of direct interconversion **211a → 211b** (by way of **216a** → **216b**) in the CH_2OCH_3-substituted case. Such an interconversion would be required by the one-center epimerization that Baldwin and Broline (*214*) reported to be the dominant reaction in the pyrolysis of the very closely related system of Scheme 12.

The only remaining structural difference between the Klärner and the Baldwin molecules thus is in the location of the ring methyl group, which is at C-2 in Scheme 11 (X = CH_2OCH_3) and at C-3 in Scheme 12. Klärner and Brassel are skeptical that this could cause a complete switch in mechanism. They point out that in the Baldwin–Broline experiment, the analysis for the four components of **219** (Scheme 12) required the separation of the NMR resonance of the proton at position 4 into the two components characteristic of the *R* and *S* enantiomers of **219** by means of a chiral shift reagent. Although Baldwin and Broline (*214*) reported success in this detail, attempts to repeat the procedure so far have been unavailing (*214a, 215*).

REFERENCES

1. For reviews of the chemistry of vinylidenes, see H. D. Hartzler, in "Carbenes," (R. A. Moss and M. Jones, Jr., eds.) Vol. II, Chapter 2, Wiley, New York, 1975; P. J. Stang, *Chem. Rev.* **78**, 383 (1978).
2. R. F. C. Brown and K. J. Harrington, *Chem. Commun.* p. 1175 (1972).
2a. R. F. C. Brown, K. J. Harrington, and G. L. McMullen, *Chem. Commun.* p. 123 (1974).
3. R. F. C. Brown, F. W. Eastwood, K. J. Harrington, and G. L. McMullen, *Aust. J. Chem.* **27**, 2393 (1974).
3a. R. F. C. Brown, F. W. Eastwood, and G. P. Jackman, *Aust. J. Chem.* **30**, 1757 (1977).
4. M. Reetz, *Angew. Chem.* **11**, 129 (1972).
5. S. W. Benson, "Thermochemical Kinetics" 2nd ed. Wiley, New York, 1976.
6. H. E. O'Neal and S. W. Benson, *J. Phys. Chem.* **72**, 1866 (1968).
7. H. E. O'Neal and S. W. Benson, *Int. J. Chem. Kinet.* **2**, 423 (1970).
8. S. W. Benson and H. E. O'Neal, "Kinetic Data on Gas Phase Unimolecular Reactions," NSRDS-NBS 21. U.S. Department of Commerce, Washington, D.C., 1970.
9. J. A. Berson, *Annu. Rev. Phys. Chem.* **28**, 111 (1977).
10. J. A. Berson, L. D. Pedersen, and B. K. Carpenter, *J. Am. Chem. Soc.* **98**, 122 (1976).
11. S. Buchwalter and G. L. Closs, *J. Am. Chem. Soc.* **97**, 3857 (1975); **101**, 4688 (1979).
12. C. E. Dykstra and H. F. Schaefer III, *J. Am. Chem. Soc.* **100**, 1378 (1978); H. F. Schaefer, *Acc. Chem. Res.* **12**, 288 (1979).
13. K. B. Wiberg and W. J. Bartley, *J. Am. Chem. Soc.* **82**, 6375 (1960); R. Srinivasan, *ibid.* **91**, 6250 (1969).
13a. I. M. Bailey and R. Walsh, *J. Chem. Soc., Faraday Trans. 1* p. 1146 (1978).
13b. R. Breslow, in "Molecular Rearrangements" (P. de Mayo, ed.), Vol. 1, p. 233 Wiley (Interscience), New York, 1963.
14. M. A. Battiste, B. Halton, and R. H. Grubbs, *Chem. Commun.* p. 907 (1967).
15. G. L. Closs, private communication, cited in York *et al.* (*16*).
16. E. J. York, W. Dittmar, J. R. Stevenson, and R. G. Bergman, *J. Am. Chem. Soc.* **95**, 5680 (1973).

16a. R. Srinivasan, *J. Am. Chem. Soc.* **91**, 6250 (1969).
17. G. L. Closs and L. E. Closs, *J. Am. Chem. Soc.* **85**, 99 (1963).
18. G. L. Closs, L. E. Closs, and W. A. Böll, *J. Am. Chem. Soc.* **85**, 3796 (1963).
19. Calculated from the heat of formation of triplet CH_2 [+91.9 kcal/mole[20]] and the singlet-triplet energy separation[21-24a] which gives $\Delta H°$ for $^1CH_2$ as ~100 kcal/mole.
20. W. A. Chupka and C. Lifshitz, *J. Chem. Phys.* **48**, 1109 (1968).
21. H. M. Frey and G. J. Kennedy, *J. Chem. Soc., Chem. Commun.* p. 233 (1975).
22. J. W. Simons and R. Curry, *Chem. Phys. Lett.* **38**, 171 (1976).
23. F. Lahmani, *J. Phys. Chem.* **80**, 2623 (1976).
23a. R. K. Lengel and R. N. Zare, *J. Am. Chem. Soc.* **100**, 7495 (1978).
24. See, however, P. F. Zittel, G. B. Ellison, S. Y. O'Neill, E. Herbst, W. C. Lineberger, and W. P. Reinhardt, *J. Am. Chem. Soc.* **98**, 3731 (1976).
24a. R. R. Corderman and W. C. Lineberger, *Annu. Rev. Phys. Chem.* **30**, 347 (1979).
24b. W. T. Borden and E. R. Davidson, *Annu. Rev. Phys. Chem.* **30**, 125 (1979).
25. J. H. Davis, W. A. Goddard, III, and R. G. Bergman, *J. Am. Chem. Soc.* **98**, 4015 (1976); **99**, 2427 (1977).
26. E. Halevi, R. Pauncz, I. Schek, and H. Weinstein, *Jerusalem Symp. Quantum Chem. Biochem.* **6**, 167 (1974).
27. C. D. Duncan and C. Trindle, *Tetrahedron Lett.* p. 2251 (1977).
28. S. N. Datta, C. D. Duncan, H. O. Pamuk, and C. Trindle, *J. Phys. Chem.* **81**, 923 (1977).
29. R. B. Cundall, *Prog. React. Kinet.* **2**, 165 (1964), and references cited therein.
30. Footnote 20 of York *et al.* (*16*).
31. R. Breslow, P. Gal, H. W. Chang, and L. J. Altman, *J. Am. Chem. Soc.* **87**, 5139 (1965).
32. J. H. Davis, K. J. Shea, and R. G. Bergman, *J. Am. Chem. Soc.* **99**, 1499 (1977).
33. R. Weiss and S. Andral, *Angew. Chem., Int. Ed. Engl.* **12**, 150 and 152 (1973).
34. I. J. Landheer, W. H. De Wolf, and F. Bickelhaupt, *Tetrahedron Lett.* p. 349 (1975); p. 2813 (1974).
35. N. J. Turro, G. B. Schuster, R. G. Bergman, K. J. Shea, and J. H. Davis, *J. Am. Chem. Soc.* **97**, 4758 (1975).
36. I. G. Csizmadia, J. Font, and O. P. Strausz, *J. Am. Chem. Soc.* **90**, 7360 (1968); D. E. Thornton, R. K. Gosavi, and O. P. Strausz, *ibid.* **92**, 1768 (1970); G. Frater and O. P. Strausz, *ibid.* p. 6654.
37. J. Fenwick, G. Frater, K. Ogi, and O. P. Strausz, *J. Am. Chem. Soc.* **95**, 124 (1973).
38. R. L. Russell and F. S. Rowland, *J. Am. Chem. Soc.* **92**, 7508 (1970); D. C. Montague and F. S. Rowland, *ibid.* 93, 5381 (1971).
39. S. A. Matlin and P. G. Sammes, *J. Chem. Soc., Chem. Commun.* p. 11 (1972); *J. Chem. Soc., Perkin Trans. 1* p. 2623 (1972).
40. O. P. Strausz, R. K. Gosavi, A. S. Denes, and I. G. Czizmadia, *J. Am. Chem. Soc.* **98**, 4784 (1976).
41. I. G. Csizmadia, H. D. Gunning, R. K. Gosavi, and O. P. Strausz, *J. Am. Chem. Soc.* **95**, 133 (1973).
42. A. C. Hopkinson, *J. Chem. Soc., Perkin Trans. 2* p. 794 (1973).
43. M. J. S. Dewar and C. A. Ramsden, *J. Chem. Soc., Chem. Commun.* p. 688 (1973).
44. P. Heinrich and H. Meier, as cited by H. Meier and K.-P. Zeller, *Angew. Chem. Int. Ed. Engl.* **14**, 32 (1975).
45. R. G. Bergman, in "Free Radicals" (J. Kochi, ed.), Vol 1, p. 191. Wiley, New York, 1973.
46. P. J. Robinson and K. A. Holbrook, "Unimolecular Reactions." Wiley (Interscience), New York, 1972.
47. W. Forst, "Theory of Unimolecular Reactions." Academic Press, New York, 1973.

48. K. J. Laidler and L. F. Loucks, *Comp. Chem. Kinet.* **5,** 1 (1972).
49. M. R. Willcott, R. L. Cargill, and A. B. Sears, *Prog. Phys. Org. Chem.* **9,** 25 (1972).
50. H. M. Frey and R. Walsh, *Chem. Rev.* **69,** 103 (1969).
51. B. S. Rabinovitch, E. W. Schlag, and K. B. Wiberg, *J. Chem. Phys.* **28,** 504 (1958).
51a. E. W. Schlag and B. S. Rabinovitch, *J. Am. Chem. Soc.* **82,** 5996 (1960).
52. T. S. Chambers and G. B. Kistiakowsky, *J. Am. Chem. Soc.* **56,** 399 (1934).
53. P. J. Krusic, P. Meakin, and J. P. Jesson, *J. Phys. Chem.* **75,** 3438 (1971).
54. H. Fischer, *in* "Free Radicals" (J. Kochi, ed.), Vol. 2, p. 435. Wiley, New York, 1973.
55. F. T. Smith, *J. Chem. Phys.* **29,** 235 (1958).
56. R. J. Crawford and A. Mishra, *J. Am. Chem. Soc.* **87,** 3768 (1965); **88,** 3963 (1966).
57. D. E. McGreer, N. W. K. Chiu, M. G. Vinje, and K. C. K. Wong, *Can. J. Chem.* **43,** 1407 (1965).
58. T. C. Clarke, L. A. Wendling, and R. G. Bergman, *J. Am. Chem. Soc.* **97,** 5638 (1975).
59. R. Hoffmann, *J. Am. Chem. Soc.* **90,** 1475 (1968).
60. R. B. Woodward and R. Hoffmann, "The Conservation of Orbital Symmetry." Academic Press, New York, 1970.
61. A. K. Q. Siu, W. M. St. John, III, and E. F. Hayes, *J. Am. Chem. Soc.* **92,** 7249 (1970).
62. P. J. Hay, W. J. Hunt, and W. A. Goddard, III, *J. Am. Chem. Soc.* **94,** 638 (1972).
63. J. A. Horsley, Y. Jean, C. Moser, L. Salem, R. M. Stevens, and J. S. Wright, *J. Am. Chem. Soc.* **94,** 279 (1972); L. Salem, *Acc. Chem. Res.* **4,** 422 (1971); Y. Jean, L. Salem, J. S. Wright, J. A. Horsley, C. Moser, and R. M. Stevens, *Proc. Int. Congr. Pure Appl. Chem., Spec. Lect.* **23,** No. 1, 197 (1971).
64. R. Huisgen, W. Scheer, and H. Huber, *J. Am. Chem. Soc.,* **89,** 1753 (1967); R. Huisgen, *Proc. Int. Cong. Pure Appl. Chem., Spec. Lect.* **23,** No. 1, 175 (1971); A. Dahmen, H. Hamburger, R. Huisgen, and L. Markowski, *Chem. Commun.* p. 1192 (1971).
65. J. C. Pommelet, N. Manisse, and J. Chuche, *Tetrahedron* **28,** 3929 (1972); H. J. H. McDonald and R. J. Crawford, *Can. J. Chem.* **50,** 428 (1972).
66. E. F. Hayes and A. K. Q. Siu, *J. Am. Chem. Soc.* **93,** 2090 (1971).
67. A. Gavezzotti and M. Simonetta, *Tetrahedron Lett.* p. 4155 (1975).
68. Y. Jean and X. Chapuisat, *J. Am. Chem. Soc.* **96,** 6911 (1974).
69. X. Chapuisat and Y. Jean, *J. Am. Chem. Soc.* **97,** 6325 (1975).
70. M. R. Willcott, III and V. H. Cargle, *J. Am. Chem. Soc.* **91,** 4310 (1969).
71. J. T. Wood, J. S. Arney, D. Cortès, and J. A. Berson, *J. Am. Chem. Soc.* **100,** 3855 (1978).
72. J. E. Baldwin and C. Carter, *J. Am. Chem. Soc.* **100,** 3942 (1978); *ibid.,* **101,** 1325 (1979).
73. W. von E. Doering and E. A. Barsa, *Tetrahedron Lett.* p. 2495 (1978).
73a. J. A. Berson and L. D. Pedersen, *J. Am. Chem. Soc.* **97,** 238 (1975).
73b. J. A. Berson, L. D. Pedersen, and B. K. Carpenter, *J. Am. Chem. Soc.* **98,** 122 (1976).
73c. J. A. Berson, L. D. Pedersen, and B. K. Carpenter, *J. Am. Chem. Soc.* **97,** 240 (1975).
73d. J. A. Berson and J. M. Balquist, *J. Am. Chem. Soc.* **90,** 7343 (1968).
73e. W. Carter and R. G. Bergman, *J. Am. Chem. Soc.* **90,** 7344 (1968).
73f. R. G. Bergman and W. Carter, *J. Am. Chem. Soc.* **91,** 7411 (1969).
73g. W. von E. Doering and K. Sachdev, *J. Am. Chem. Soc.* **96,** 1168 (1974).
73h. R. J. Crawford and T. R. Lynch, *Can. J. Chem.* **46,** 1457 (1968).
73i. A. Chmurny and D. J. Cram, *J. Am. Chem. Soc.* **95,** 4237 (1973).
73j. W. Kirmse and M. Zeppenfeld, *J. Chem. Soc., Chem. Commun.* p. 124 (1977).
74. M. J. S. Dewar and S. Kirschner, *J. Am. Chem. Soc.* **96,** 5246 (1974).
75. For a review of the earlier literature, see J. J. Gajewski, *in* "Mechanisms of Molecular Migrations" (B. Thyagarajan, ed.), Vol. 3, p. 11. Wiley, New York, 1971.
76. E. F. Ullman, *J. Am. Chem. Soc.* **81,** 5386 (1959).

76a. E. F. Ullman, *J. Am. Chem. Soc.* **82,** 505 (1960).
77. E. F. Ullman and W. J. Fanshawe, *J. Am. Chem. Soc.* **83,** 2379 (1961).
78. W. von E. Doering and H. D. Roth, *Tetrahedron* **26,** 2825 (1970).
79. J. P. Chesick, *J. Am. Chem. Soc.* **84,** 3250 (1962).
80. J. J. Gajewski, *J. Am. Chem. Soc.* **90,** 7178 (1968); **93,** 4450 (1971).
81. J. J. Gajewski and S. K. Chou, *J. Am. Chem. Soc.* **99,** 5696 (1977).
82. J. C. Gilbert and J. R. Butler, *J. Am. Chem. Soc.* **92,** 2168 (1970).
83. D. R. Paulson, J. K. Crandall, and C. A. Bunnell, *J. Org. Chem.* **35,** 3708 (1970).
84. W. R. Dolbier, Jr. and J. H. Alonso, *J. Am. Chem. Soc.* **95,** 4421 (1973).
85. W. R. Dolbier, Jr., K.-Y. Akiba, M. Bertrand, A. Bezaguet, and M. Santelli, *J. Chem. Soc., Chem. Commun.* p. 717 (1970).
86. W. Kirmse and H.-R. Murawski, *J. Chem. Soc., Chem. Commun.* p. 122 (1977).
87. W. D. Slafer, A. D. English, D. O. Harris, D. F. Shellhamer, M. J. Meshishnek, and D. H. Aue, *J. Am. Chem. Soc.* **97,** 6638 (1975).
88. W. von E. Doering and L. Birladeanu, *Tetrahedron* **29,** 499 (1973).
89. T. C. Shields, B. A. Shoulders, J. F. Krause, C. L. Osborn, and P. D. Gardner, *J. Am. Chem. Soc.* **87,** 3026 (1965).
90. J. N. Murrell and K. J. Laidler, *Trans. Faraday Soc.* **64,** 371 (1968).
91. J. W. McIver and R. E. Stanton, *J. Am. Chem. Soc.* **94,** 8618 (1972).
92. E. R. Davidson and W. T. Borden, *J. Am. Chem. Soc.* **98,** 2053 (1976).
93. W. R. Roth and G. Wegener, *Angew. Chem. Int. Ed. Engl.* **14,** 758 (1975).
94. J. H. Davis and W. A. Goddard, III, *J. Am. Chem. Soc.* **98,** 303 (1976); **99,** 4242 (1977).
95. W. T. Borden and L. Salem, *J. Am. Chem. Soc.* **95,** 932 (1973); W. T. Borden and E. R. Davidson, *ibid.* **98,** 4587 (1976).
95a. W. T. Borden, *J. Am. Chem. Soc.* **97,** 2906 (1975).
96. D. R. Yarkony and H. F. Schaefer, III, *J. Am. Chem. Soc.* **96,** 3754 (1974).
96a. D. A. Dixon, R. Foster, T. A. Halgren, and W. N. Lipscomb, *J. Am. Chem. Soc.* **100,** 1359 (1978).
96b. D. M. Hood, R. M. Pitzer, and H. F. Schaefer III, *J. Am. Chem. Soc.* **100,** 2227 (1978).
96c. D. M. Hood, H. F. Schaefer III, and R. M. Pitzer, *J. Am. Chem. Soc.* **100,** 8009 (1978).
96d. For a review, see Borden and Davidson (*23b*).
96e. E. R. Davidson, K. Tanaka, and W. T. Borden, as cited by Hood, *et al.* (*96c*).
97. W. A. Moffitt, as cited by C. A. Coulson, *J. Chem. Phys.* **45,** 243 (1949); W. Moffitt, *Trans. Faraday Soc.* **45,** 373 (1949); H. C. Longuet-Higgins, *J. Chem. Phys.* **18,** 265 (1950).
98. P. Dowd, *Acc. Chem. Res.* **5,** 242 (1972).
99. J. A. Berson, *Acc. Chem. Res.* **11,** 446 (1978).
100. W. von E. Doering and G. Beasley, *Tetrahedron* **29,** 2231 (1973); K. W. Egger, D. M. Golden, and S. W. Benson, *J. Am. Chem. Soc.* **96,** 6211 (1964).
101. See also M. J. S. Dewar and J. A. Wasson, *J. Am. Chem. Soc.* **93,** 3081 (1971); W. Hehre, L. Salem, and M. R. Willcott, III, *ibid.* **96,** 4328 (1974).
102. J. A. Berson and L. Salem, *J. Am. Chem. Soc.* **94,** 8917 (1972).
103. J. A. Berson, *Acc. Chem. Res.* **5,** 402 (1972).
104. L. Friedman and H. Shechter, *J. Am. Chem. Soc.* **82,** 1002 (1960).
105. P. Dowd and M. Chow, *J. Am. Chem. Soc.* **99,** 6438 (1977).
106. R. J. Baseman, D. W. Pratt, M. Chow, and P. Dowd, *J. Am. Chem. Soc.* **98,** 5726 (1976).
107. M. S. Platz, J. M. McBride, R. D. Little, J. J. Harrison, A. Shaw, S. E. Potter, and J. A. Berson, *J. Am. Chem. Soc.* **98,** 5725 (1976).
108. H. M. Frey, *Trans. Faraday Soc.* **57,** 951 (1961).
109. W. von E. Doering and J. C. Gilbert, *Tetrahedron, Suppl.* **7,** 397 (1966).

110. W. von E. Doering, J. C. Gilbert, and P. A. Leermakers, *Tetrahedron* **24,** 6863 (1968).
111. Dowd (*98*) and Berson (*99*) and references cited therein.
112. J. J. Gajewski, A. Yeshurun, and E. J. Bair, *J. Am. Chem. Soc.* **94,** 2138 (1972).
113. J. A. Berson, R. J. Bushby, J. M. McBride, and M. Tremelling, *J. Am. Chem. Soc.* **93,** 1544 (1971).
114. G. L. Closs, *J. Am. Chem. Soc.* **93,** 1546 (1971).
115. M. S. Platz and J. A. Berson, *J. Am. Chem. Soc.* **98,** 6743 (1976).
116. J. A. Berson, C. D. Duncan, and L. R. Corwin, *J. Am. Chem. Soc.* **96,** 6175 (1974); J. A. Berson, L. R. Corwin, and J. H. Davis, *ibid.* p. 6177.
117. R. M. Wilson and F. Geiser, *J. Am. Chem. Soc.* **100,** 2225 (1978).
118. W. von E. Doering and S. Buchwalter, unpublished observations cited in Borden (*95a*).
119. R. F. Salinaro and J. A. Berson, *J. Am. Chem. Soc.* **101,** 7094 (1979).
119a. M. Rule, M. G. Lazzara, and J. A. Berson, *J. Am. Chem. Soc.* **101,** 7091 (1979).
119b. M. G. Lazzara, J. J. Harrison, M. Rule, and J. A. Berson, *J. Am. Chem. Soc.* **101,** 7092 (1979).
120. M. C. Flowers and H. M. Frey, *J. Chem. Soc.* p. 5550 (1961).
120a. M. C. Flowers and A. R. Gibbons, *J. Chem. Soc.* **B** p. 612 (1971).
121. P. J. Burkhardt, *Diss. Abstr.* **23,** 1524 (1962).
122. J. J. Gajewski, *J. Am. Chem. Soc.* **92,** 3688 (1970).
122a. J. J. Gajewski and L. T. Burka, *J. Am. Chem. Soc.* **94,** 8865 (1972).
123. W. R. Roth and K. Enderer, *Justus Liebigs Ann. Chem.* **730,** 82 (1969).
124. J. C. Gilbert, *Tetrahedron* **25,** 1459 (1969).
124a. J. J. Gajewski and L. T. Burka, *J. Am. Chem. Soc.* **94,** 8857 (1972).
124b. J. J. Gajewski and R. Weber, *J. Am. Chem. Soc.* **99,** 8054 (1977).
124c. J. J. Gajewski, R. J. Weber, and M. J. Chang, *J. Am. Chem. Soc.* **101,** 2100 (1979).
125. W. von E. Doering and J. C. Gilbert, *Tetrahedron, Suppl.* **7,** 397 (1966).
126. For a review of the early work in this field, see J. J. Gajewski (*75*, p. 18ff.).
126a. A recent review of allene dimerizations is given by L. Ghosez and M. J. O'Donnell, in "Pericyclic Reactions" (A. P. Marchand and R. E. Lehr, eds.) Vol. II, p. 109 ff. Academic Press, New York, 1977.
127. K. K. Shen and R. G. Bergman, *J. Am. Chem. Soc.* **99,** 1655 (1977).
127a. S. H. Bauer, *J. Am. Chem. Soc.* **91,** 3688 (1969).
128. W. R. Dolbier, Jr., *Tetrahedron Lett.*, 393 (1968).
128a. W. R. Dolbier, Jr., K. Akiba, J. M. Riemann, C. A. Harmon, M. Bertrand, A. Bezaguet, and M. Santelli, *J. Am. Chem. Soc.*, **92,** 3933 (1970).
129. P. Dowd, *J. Am. Chem. Soc.*, **92,** 1066 (1970).
130. W. R. Roth and G. Erker, *Angew. Chem. Intl. Ed. Engl.* **12,** 505 (1973).
131. W. Grimme and J.-J. Rother, *Angew. Chem. Intl. Ed., Engl.*, **12,** 505 (1973).
132. F. Kern and W. D. Walters, *J. Am. Chem. Soc.*, **75,** 6196 (1953).
133. S. W. Benson and P. S. Nangia, *J. Chem. Phys.* **38,** 18 (1963).
134. H. R. Gerberich and W. D. Walters, *J. Am. Chem. Soc.*, **83,** 3935, 4884 (1961).
134a. A. T. Cocks, H. M. Frey, and I. D. R. Stevens, *Chem. Commun.*, 458 (1969).
134b. J. E. Baldwin and P. W. Ford, *J. Am. Chem. Soc.*, **91,** 7192 (1969).
135. J. A. Berson, D. C. Tompkins, and G. Jones, II, *J. Am. Chem. Soc.*, **92,** 5799 (1970).
136. A. Padwa, W. Koehn, J. Masaracchia, C. L. Osborn, and D. J. Trecker, *J. Am. Chem. Soc.*, **93,** 3633 (1971); P. D. Bartlett, G. M. Cohen, S. P. Elliott, K. Hummel, R. A. Minns, C. M. Sharts, and J. Y. Fukunaga, *ibid.*, **94,** 2899 (1972); P. D. Bartlett, K. Hummel, S. P. Elliott, and R. A. Minns, *ibid.*, **94,** 2898 (1972); P. D. Bartlett and J. B. Mallet, *ibid.*, **98,** 143 (1976); W. von E. Doering and C. A. Guyton, *J. Am. Chem. Soc.*, **100,** 3229 (1978).
136a. G. Scacchi, C. Richard, and M. H. Bach, *Int. J. Chem. Kinet.*, **9,** 525 (1977).

137. R. Hoffmann, S. Swaminathan, B. G. Odell, and R. Gleiter, *J. Am. Chem. Soc.* **92**, 7091 (1970).
137a G. A. Segal, *J. Am. Chem. Soc.* **96**, 7892 (1974).
138. J. S. Wright and L. Salem, *J. Am. Chem Soc.* **94**, 322 (1972); M. J. S. Dewar and S. Kirschner, *ibid.* **96**, 5246 (1974); H. Fujimoto and T. Sugiyama, *ibid.* **99**, 15 (1977); N. D. Epiotis and S. Shaik, *ibid.* **100**, 9 (1978).
138a. L. M. Stephenson, T. A. Gibson, and J. I. Brauman, *J. Am. Chem. Soc.* **95**, 2849 (1973).
139. P. B. Dervan and T. Uyehara, *J. Am. Chem. Soc.* **98**, 1262 (1976); P. B. Dervan, T. Uyehara, and D. S. Santilli, *ibid.*, **101**, 2069 (1979).
140. P. B. Dervan and T. Uyehara, *J. Am. Chem. Soc.* **98**, 2003 (1976); *ibid.*, **101**, 2069 (1979).
141. D. S. Santilli and P. B. Dervan, *J. Am. Chem. Soc.* **101**, 3663 (1979).
141a. See also J. S. Chickas, *J. Org. Chem.* **44**, 780 (1979).
142. Cf. J. A. Berson and S. S. Olin, *J. Am. Chem. Soc.* **91**, 777 (1969); J. A. Berson, E. W. Petrillo, Jr., and P. Bickart, *ibid.* **96**, 636 (1974); J. A. Berson, S. S. Olin, E. W. Petrillo, Jr., and P. Bickart, *Tetrahedron* **30**, 1639 (1974).
143. P. J. Krusic, P. Meakin, and J. P. Jesson, *J. Phys. Chem.* **75**, 3438 (1971); Review: H. Fischer, *in* "Free Radicals" (J. Kochi, ed.), Vol. 2, p. 435. Wiley, New York, 1973.
144. P. D. Bartlett and N. A. Porter, *J. Am. Chem Soc.* **90**, 5317 (1968).
145. S. J. Rhoads and N. R. Raulins, *Org. React.* **22**, 1 (1970).
146. R. B. Woodward and T. J. Katz, *Tetrahedron* **5**, 70 (1959).
147. G. Schröder, *Angew. Chem., Int. Ed. Engl.* **2**, 481 (1963).
148. W. von E. Doering and W. R. Roth, *Tetrahedron* **19**, 715 (1963).
149. R. Wehrli, D. Bellus, H.-J. Hansen, and H. Schmid, *Chimia* **30**, 416 (1976).
150. R. Wehrli, H. Schmid, D. Bellus, and H.-J. Hansen, *Helv. Chim. Acta* **60**, 1325 (1977).
151. D. C. Wigfield and K. Taymaz, *Tetrahedron Lett.* p. 3121 (1975).
152. M. J. Goldstein and M. de Camp, *J. Am. Chem. Soc.* **96**, 7356 (1974).
153. E. Vogel, *Angew. Chem.* **72**, 4 (1960); M. Arai and R. J. Crawford, *Can. J. Chem.* **50**, 2158 (1972).
154. C. Ullenius, P. W. Ford, and J. E. Baldwin, *J. Am. Chem. Soc.* **94**, 5910 (1972).
154a. W. Pickenhagen, F. Näf, G. Ohloff, P. Müller, and J.-C. Perlberger, *Helv. Chim. Acta* **56**, 1868 (1973).
155. G. S. Hammond and C. D. DeBoer, *J. Am. Chem. Soc.* **86**, 899 (1964).
156. D. J. Trecker and J. P. Henry, *J. Am. Chem. Soc.* **86**, 902 (1964).
157. J. A. Berson and P. B. Dervan, *J. Am. Chem. Soc.* **94**, 7597 (1972).
157a. J. A. Berson and P. B. Dervan, *J. Am. Chem. Soc.* **94**, 8949 (1972).
157b. J. A. Berson, P. B. Dervan, R. Malherbe, and J. A. Jenkins, *J. Am. Chem. Soc.* **98**, 5937 (1976).
158. J. A. Berson and M. Jones, Jr., *J. Am. Chem. Soc.* **86**, 5017 and 5019 (1964).
159. J. A. Berson, T. Miyashi, and G. Jones, II, *J. Am. Chem. Soc.* **96**, 3468 (1974).
160. J. A. Berson and E. J. Walsh, Jr., *J. Am. Chem. Soc.* **90**, 4732 (1968).
161. R. J. Crawford, J. Hamelin, and B. Strehlke, *J. Am. Chem. Soc.* **93**, 3810 (1971); P. J. Krusic, P. Meakin, and B. E. Smart, *ibid.* **96**, 6211 (1974); C. Walling and W. Thaler, *ibid.* **83**, 3877 (1961).
162. W. R. Dolbier, Jr. and G. J. Mancini, *Tetrahedron Lett.* p. 2141 (1975).
163. T. C. Shields, W. E. Billups, and A. R. Lepley, *J. Am. Chem. Soc.* **90**, 4749 (1968).
164. T. C. Shields and W. E. Billups, *Chem. Ind. (London)* p. 619 (1969); K. H. Leavell, W. E. Billups, and E. S. Lewis, *Abst. IUPAC Congr. 23rd, 1971,* Pap. No. 108 (1971).
165. A. S. Kende and E. E. Riecke, *J. Am. Chem. Soc.* **94**, 1397 (1972).
166. J. C. Gilbert and D. P. Higley, *Tetrahedron Lett.* p. 2075 (1973).

5. HYPOTHETICAL BIRADICAL PATHWAYS

167. J. J. Gajewski, L. K. Hoffman, and C. N. Shih, *J. Am. Chem. Soc.* **96**, 3705 (1974).
168. D. Hasselmann, *Tetrahedron Lett.* p. 3465 (1972); p. 3739 (1973); see also D. Hasselmann, *Angew. Chem., Int. Ed. Engl.* **14**, 257 (1975).
169. R. W. Holder and R. E. Voorhees, *Abstr. 173rd Meet., Am. Chem. Soc., 1977* ORGN 80 (1977).
169a. S. J. Rhoades, J. M. Watson, and J. G. Kambouris, *J. Am. Chem. Soc.* **100**, 5151 (1978).
170. J. Japenga, M. Kool, and G. W. Klumpp, *Tetrahedron Lett.* p. 1029 (1975).
171. J. A. Berson and J. M. Janusz, *J. Am. Chem. Soc.* **96**, 5939 (1974); **100**, 2237 (1978).
172. R. Bishop, W. Parker, and I. Watt, *Tetrahedron Lett.* p. 4345 (1977).
173. A. Streitwieser, R. H. Jagow, R. C. Fahey, and S. Suzuki, *J. Am. Chem. Soc.* **80**, 2326 (1958).
174. K. Humski, R. Malojcic, S. Borcic, and D. E. Sunko, *J. Am. Chem. Soc.* **92**, 6534 (1970).
175. J. J. Gajewski and N. E. Conrad, *J. Am. Chem. Soc.* **100**, 6268 and 6269 (1978).
175a. R. Alder, personal communication. January 3, 1979.
176. C. A. Grob, H. Link, and P. Schiess, *Helv. Chim. Acta* **46**, 483 (1963).
177. W. von E. Doering, V. G. Toscano, and G. H. Beasley, *Tetrahedron* **27**, 5299 (1971).
178. J. W. McIver, Jr., *Acc. Chem. Res.* **7**, 72 (1974).
179. A. Komornicki and J. W. McIver, Jr., *J. Am. Chem. Soc.* **98**, 4553 (1976).
180. M. J. S. Dewar, G. P. Ford, M. L. McKee, H. R. Zepa, and L. E. Wade, *J. Am. Chem. Soc.* **99**, 5069 (1977).
181. M. J. Goldstein and M. S. Benzon, *J. Am. Chem. Soc.* **94**, 5119 and 7147 (1972).
181a. R. K. Hill and N. W. Gilman, *J. Chem. Soc. Chem. Commun.*, 619 (1967).
182. L. A. Paquette and J. A. Schwartz, *J. Am. Chem. Soc.* **92**, 3215 (1970).
183. W. R. Roth and M. Martin, *Tetrahedron Lett.* p. 3865 (1967); W. R. Roth, unpublished data, cited by R. G. Bergman, in "Free Radicals" (J. K. Kochi, ed.), p. 229. Wiley, New York, 1973.
184. W. von E. Doering and W. R. Roth, *Tetrahedron* **18**, 67 (1962).
185. M. J. S. Dewar, *Ciba Found. Symp.* **53** (New Ser.), 107 (1978).
186. M. J. S. Dewar and L. E. Wade, *J. Am. Chem. Soc.* **99**, 4417 (1977).
187. H. Kessler and W. Ott, *J. Am. Chem. Soc.* **98**, 5014 (1976); A. Busch and H. M. R. Hoffmann, *Tetrahedron Lett.* p. 2379 (1976).
188. J. A. Berson, *Acc. Chem. Res.* **1**, 152 (1968).
189. J. A. Berson and M. R. Willcott, III, *Rec. Chem. Prog.* **27**, 139 (1966).
190. T. L. Gilchrist and R. C. Storr, "Organic Reactions and Orbital Symmetry" Cambridge Univ. Press, London and New York, 1972.
191. N. Trong Anh, "Les Règles de Woodward-Hoffmann." Edisience, Paris, 1970.
192. C. W. Spangler, *Chem. Rev.* **76**, 187 (1976).
193. K. N. Houk, in "Pericyclic Reactions" (A. P. Marchand and R. E. Lehr, eds.), Vol. 2, p. 182. Academic Press, New York, 1976; J. E. Baldwin, *ibid.* p. 273.
194. N. Epiotis, in "Essays in Molecular Rearrangements" (P. de Mayo, ed.). Academic Press, New York, 1980.
195. R. E. Lehr and A. P. Marchand, in "Pericyclic Reactions" (A. P. Marchand and R. E. Lehr, eds.), Vol. 1, p. 1. Academic Press, New York, 1977.
196. J. E. Baldwin and K. E. Gilbert, *J. Am. Chem. Soc.* **98**, 8283 (1976).
197. J. E. Baldwin and R. H. Fleming, *J. Am. Chem. Soc.* **95**, 5249, 5256, and 5261 (1973).
198. J. J. Gajewski, *J. Am. Chem. Soc.* **98**, 5254 (1976).
199. R. B. Woodward and R. Hoffmann, "The Conservation of Orbital Symmetry." Academic Press, New York, 1970.
200. H. E. Zimmerman, *J. Am. Chem. Soc.* **88**, 1564 and 1566 (1966).

201. M. J. S. Dewar, *Tetrahedron, Suppl.* **8**, 75 (1966).
202. J. A. Berson and G. L. Nelson, *J. Am. Chem. Soc.* **92**, 1096 (1970).
202a. J. A. Berson and R. W. Holder, *J. Am. Chem. Soc.* **95**, 2037 (1973).
202b. J. A. Berson, *Acc. Chem. Res.* **5**, 406 (1972).
202c. J. A. Berson and G. L. Nelson, *J. Am. Chem. Soc.* **89**, 5503 (1967).
203. J. A. Berson and L. Salem, *J. Am. Chem. Soc.* **94**, 8917 (1972).
204. W. W. Schoeller, *J. Am. Chem. Soc.* **99**, 5919 (1977).
205. S. R. Wilson and D. T. Mao, *J. Chem. Soc., Chem. Commun.* p. 479 (1978).
206. J. R. de Dobbelaere, J. M. F. van Dijk, J. W. de Haan, and H. M. Buck, *J. Am. Chem. Soc.* **99**, 392 (1977).
207. N. D. Epiotis, R. L. Yates, and F. Bernardi, *J. Am. Chem. Soc.* **97**, 4198 (1975).
207a. W. J. Bouma, M. A. Vincent, and L. Radom, *Int. J. Quant. Chem.* **14**, 767 (1978).
207b. Cf. also A. J. P. Devaquet, and W. J. Hehre, *J. Am. Chem. Soc.* **96**, 3644 (1974).
207c. W. A. M. Castenmiller and H. M. Buck, *Tetahedron* **35**, 397 (1979).
207d. T. Minato, S. Inagaki, and K. Fukui, *Bull. Chem. Soc.* Japan **50**, 1651 (1977).
208. J. A. Altmann, O. S. Tee, and K. Yates, *J. Am. Chem. Soc.* **98**, 7132 (1976).
209. D. W. Jones, *Annu. Rep. Prog. Chem., Sect. B* **73**, 43 (1976).
210. J. A. Berson and M. R. Willcott, III, *J. Am. Chem. Soc.* **87**, 2751 and 2752 (1965); **88**, 2494 (1966).
211. J. A. Berson, P. W. Grubb, R. A. Clark, D. R. Hartter, and M. R. Willcott, III, *J. Am. Chem. Soc.* **89**, 4076 (1966).
212. F.-G. Klärner, *Angew. Chem., Int. Ed. Engl.* **13**, 268 (1974); F.-G. Klärner, S. Yaslak, and M. Wette, *Chem. Ber.* **112**, 1168 (1979).
213. R. T. Hansen and J. A. Berson, unpublished; R. T. Hansen, Ph.D. Thesis, Yale University, New Haven, Connecticut (1976).
214. J. E. Baldwin and B. M. Broline, *J. Am. Chem. Soc.* **100**, 4599 (1978).
214a. F.-G. Klärner and B. Brassel, *J. Am. Chem. Soc.*, in press; B. Brassel, Diplomarbeit, Universität Bochum, 1979.
215. **Acknowledgments.** I am grateful to Professors R. Alder, R. G. Bergman, P. B. Dervan, J. J. Gajewski, and F.-G. Klärner for their helpful comments on various sections of this essay and for their willing provision of information before publication.

ESSAY 6 | REARRANGEMENTS IN CARBANIONS

D. H. HUNTER
J. B. STOTHERS
E. W. WARNHOFF

I.	Migration of Saturated Groups	392
	A. Hydrogen	392
	B. Alkyl Groups	394
	C. Heteroatoms	397
II.	Migration of Unsaturated Carbon	400
	A. Olefins and Acetylenes	401
	B. Aromatic Compounds	403
III.	Migrations in Doubly Bonded Oxygen Compounds	410
	A. Homoenolization	410
	B. Favorskii Rearrangement	437
	C. Ramberg–Bäcklund Reaction	461
	References	465

A discussion of the rearrangements of carbanions requires the definition of two terms: rearrangement and carbanion. Varieties of carbanion rearrangements are possible, do occur, and have been regularly reviewed (1–5). However, this essay is limited to migrations of groups along saturated carbon chains, with emphasis on 1,2 migrations (**1**), but also includes examples of 1,n migrations (**2**). In 1,n migrations in carbanions,

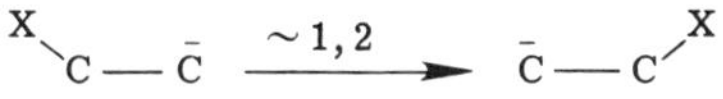

$$\underset{}{\overset{X}{\diagdown}}C-\bar{C} \xrightarrow{\sim 1,2} \bar{C}-C\overset{X}{\diagup}$$

1

$$\underset{\smile}{\overset{X}{\diagdown}}C \quad \bar{C} \xrightarrow{\sim 1,n} \bar{C} \quad \underset{\smile}{C}\overset{X}{\diagup}$$

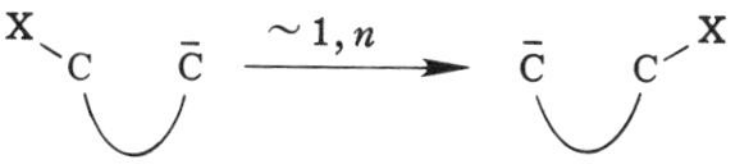

2

X = H, R, Ar, C=C, C=O, B, N, O, Al, Si, P, etc.

hydrogen and saturated carbon are first considered, followed by two other types of migrating groups: heteroatoms and unsaturated carbon. In the former only the main group elements are included; examples of unsaturated carbon include C=C, aryl, and carbonyl functions.

Within these limitations there have been a number of recent comprehensive reviews of migrations in carbanions: 1,n-aryl migrations (6), a survey (7) of organomagnesium migrations (which emphasizes the details of 1,n migrations of C=C functions), and a review of the earlier work on 1,n migrations of organomagnesiums including organoalkalies (8). Except for the case of carbonyl groups, there are very few examples of 1,n migrations involving other atoms or groups, and reviews have not been devoted to individual cases. The present essay (a) reviews in some detail migrations of the carbonyl group, for which there are several examples but no recent summary, (b) includes an overview of aryl and C=C 1,n migrations, and (c) discusses briefly the migrations of other species. For the carbonyl group both homoenolization and the Favorskii rearrangement are discussed, and a brief comparison with the sulfonyl analogue of the Favorskii rearrangement, the Ramberg–Bäcklund reaction, is included.

The definition of carbanion is a more complex problem, for under most circumstances a carbanion is actually an organometallic compound. Carbon–metal bonds have a wide range of polarities, from bonds perhaps best regarded as ionic to bonds that are essentially covalent. Furthermore, for any particular carbon–metal bond, the solvent can greatly affect the nature of the bonding interaction, producing changes from covalent bonds or ion pairs to dissociated ions. In the selection of material for discussion such distinctions have been ignored. Thus, arbitrarily, most examples formally involve alkaline earth and alkali metal cations in the processes here viewed as "carbanion" reactions.

I. MIGRATION OF SATURATED GROUPS

A. Hydrogen

The ease of 1,2-hydrogen migration in carbonium ions compared with its absence in carbanions has been commonly cited as an example of the constraints of orbital symmetry in organic reactions. Indeed, alkyllithiums, such as t-butyllithium, seem to be indefinitely stable to rearrangement, and to our knowledge there are no examples of 1,2 migration of hydrogen along a carbon chain. The 1,3 and more remote migrations along saturated chains do not suffer similar orbital symmetry constraints (Scheme 1) and may be viewed as intramolecular proton-transfer reactions. However, examples of 1,3 and other 1,n migrations of hydrogen along alkyl chains are as rare as 1,2 migrations.

To put these negative results in perspective, it should be recalled that alkyllithiums do not abstract protons from alkanes (t-butyllithium is sold as a solution in pentane, and there is no apparent conversion to isobutane

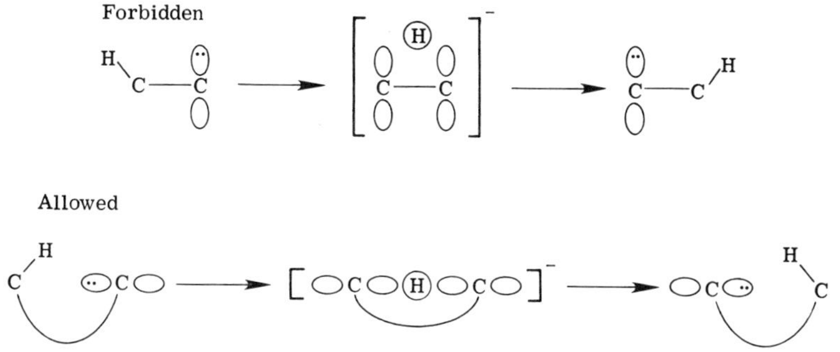

6. REARRANGEMENT IN CARBANIONS

Scheme 1

and a pentyllithium) and do so only very slowly from alkenes or toluenes. Although this is, presumably, a thermodynamically favored process, the activation barriers are too high for intramolecular or intermolecular migrations in simple saturated systems. Thus, alkyllithiums make very poor models for comparisons of carbanions with carbonium ions. Even the much more reactive alkylsodiums or alkylpotassiums do not abstract protons from alkanes.

As the proton-transfer reaction becomes more and more exothermic, the intermolecular process becomes more facile. Thus, triphenylmethane is readily deprotonated by n-butyllithium at $-30°C$ in tetrahydrofuran (9). Also, intramolecular proton transfers should occur when the driving force becomes large enough (Scheme 2). Although clear-cut examples are apparently lacking, such processes have been postulated as part of more complex interconversions (10). At present, there seems no reason to doubt the applicability of orbital symmetry demands to 1,2 migrations of hydrogen, but the experimental evidence on the point is not highly convincing.

Although not specifically within the scope of this chapter, there have been a number of studies concerned with 1,n migrations of the sigmatropic type, and these have been reviewed elsewhere (3–5). In summary, there is no example to date of the allowed [1,4] suprafacial shift of

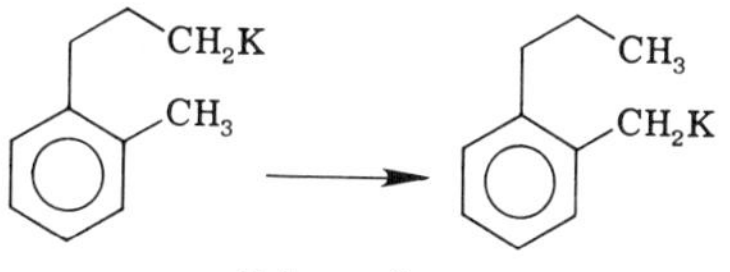

Scheme 2

hydrogen (e.g., **3** ⇸ **4**) (*11*), but there are several examples of [1,6] shifts (e.g., **5** → **6**) (*12*), presumably with antarafacial stereochemistry. There is even an example of a photochemical [1,6] shift in a cyclic system (**7** → **8**) (*13*) where only suprafacial migration is feasible.

B. Alkyl Groups

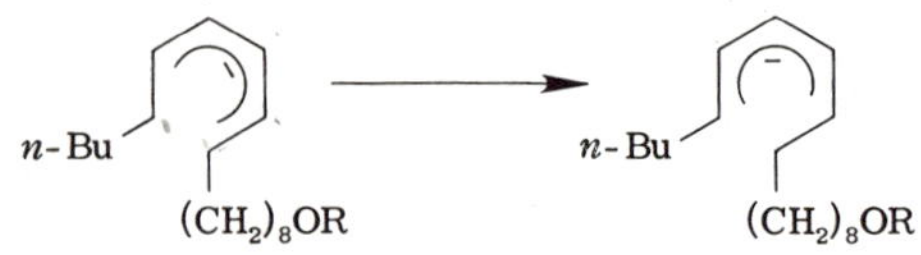

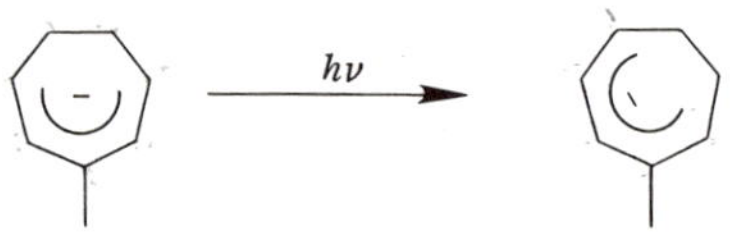

An alkyl group has an alternative pathway for migration that is not available to hydrogen. Suprafacial* migration of R with retention encounters the same symmetry constraints as that of hydrogen, but, in principle, migration with inversion at R is an allowed process (Scheme 3). Although the geometric constraints seem prohibitive, Berson (*14*) showed that similar transition states are attainable in neutral species. Nevertheless, examples of potentially concerted carbanionic 1,2-alkyl migrations are very limited.

As part of some elegant studies of carbanionic 1,2-aryl migrations (*6*), 1,2-benzyl migration in the 2,2,3-triphenylpropyl system was observed (Scheme 4). It was also shown that this is an elimination–readdition process by trapping of the intermediate 1,1-diphenylethylene with a variety of reagents: [α-^{14}C]benzyllithium, isopropyllithium, and cesium and potassium. An apparent 1,2-alkyl shift that converts cyclopropane (**9**) to

---

* Antarafacial 1,2 migration of hydrogen or R with retention is a formally allowed process. However, because of the presence of a nodal plane, at some point during the migration a nonbonding state must occur. In effect, the antarafacial 1,2 migration becomes an elimination–readdition process.

6. REARRANGEMENT IN CARBANIONS

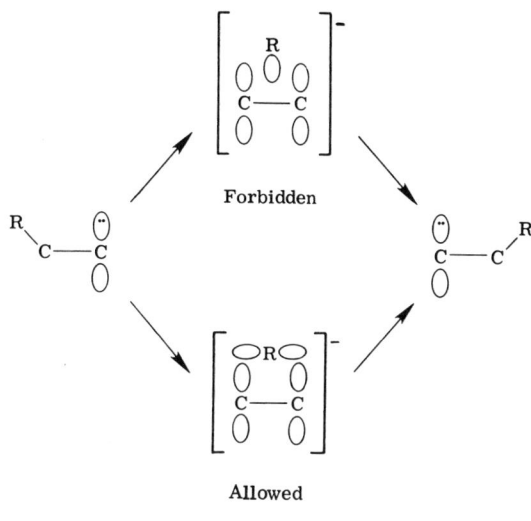

Forbidden

Allowed

Scheme 3

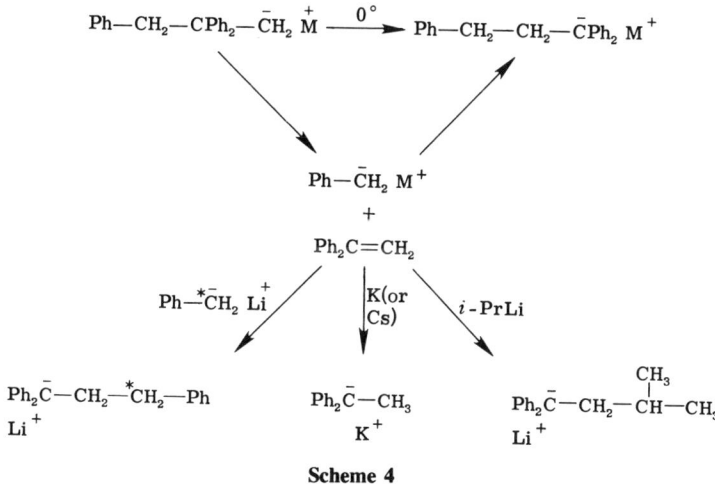

Scheme 4

9   10

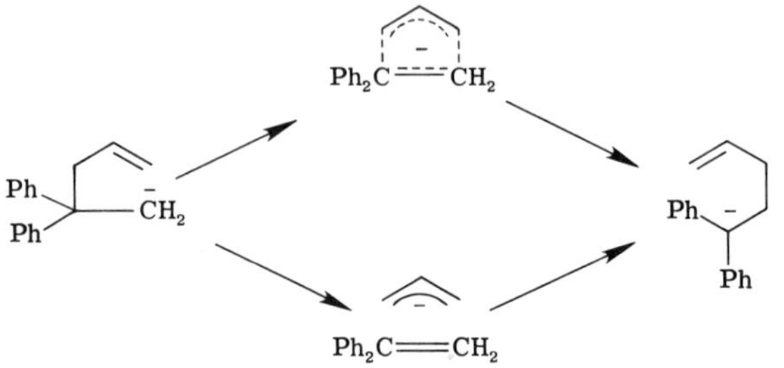

Scheme 5

cyclobutane (**10**) was described, but it was pointed out that this reaction is probably best considered a [1,8] sigmatropic migration rather than a 1,2 migration (*15*).

There is an example (*16*) of a 1,2 shift of an allyl group (Scheme 5), and it has been demonstrated that this can occur by either an elimination–readdition mechanism or by the allowed [2,3] sigmatropic process. An analogous rearrangement had been reported earlier (*17*) for the substituted fluorene **11** and was interpreted in similar terms.

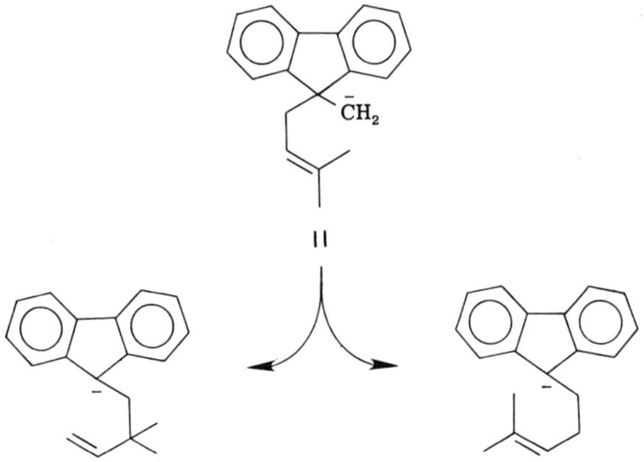

The tricyclic diketone **12** was reported (*18*) to undergo a 1,2-allyl shift, for which three mechanistic proposals were presented: an elimination–readdition process, a concerted 1,2 shift allowed by a breakdown in the simple symmetry selection rules, or successive [2,3] and [1,3] sigmatropic shifts. The actual mechanistic pathway remains uncertain.

Thus, at present, there are no clear-cut examples of concerted 1,2-alkyl shifts. Furthermore, there are no proved examples of intramolecular 1,n-alkyl shifts or of the corresponding intermolecular process. However,

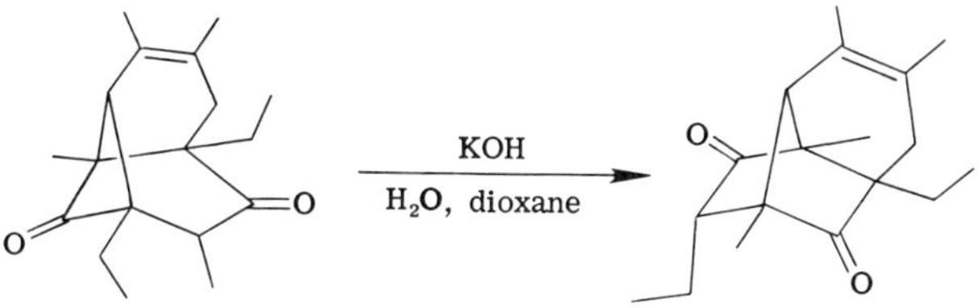

6. REARRANGEMENT IN CARBANIONS

[Structure 12: bicyclic diketone rearrangement with KOH, H₂O, dioxane]

12

it should be realized that the S_N2 reaction is but a conceptual extension of this process.

C. Heteroatoms

The 1,2 migration of the first row analogues of CH_3 (BH_3, NH_2, OH, F) all suffer the same orbital symmetry constraints of allowed suprafacial migration with inversion, but the nature of the heteroatom should affect the energetics. In fact, a very simplistic molecular orbital analysis* leads one to predict that in the series F, OH, NH_2, CH_3, BH_3 the symmetric species **13** will be more stable for boron than for fluorine. However, boron is probably less favored than carbon since the open-chain form is already negatively charged. If this simple analysis is valid, the prospects of finding examples of such 1,2 migrations seem very slim; examples, even involving carbon, appear to be unknown.

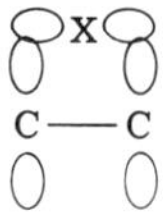

13

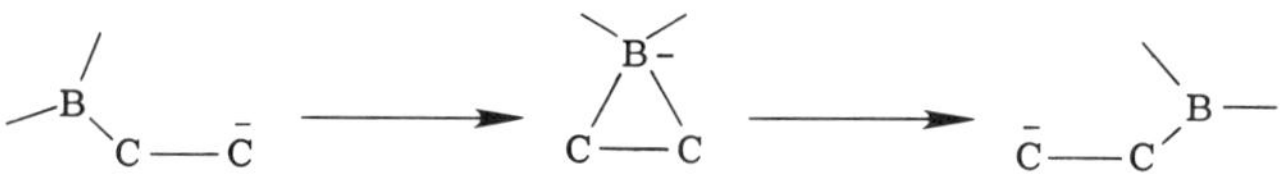

14

However, three-coordinate boron would seem to provide a lower-energy pathway for 1,n migration via a four-coordinate boron intermediate (**14**). This addition–elimination process avoids orbital symmetry

---

* If only the filled p orbital of the heteroatom X and the π orbitals of the double bond are considered, it is the π^* orbital that has appropriate symmetry for interaction with X. The amount of stabilization that this interaction will provide decreases as the heteroatom becomes more electronegative and as the p orbital of X becomes lower in energy relative to the π^* orbital.

problems. There do not seem to be examples of such 1,2 migration of boron, presumably because of the difficulty of generating carbanions in the presence of three-coordinate boron. There are, however, examples of each step of 1,n migrations. Cyclic, four-coordinate boron compounds analogous to **14** have been characterized [**15** (*19*) and **16** (*20*)], and the opening **16** → **17** has been observed (*21*). Reversible closure to form a three-membered ring (**18**) has been postulated, but in a system where net rearrangement does not result. For a discussion of rearrangements in boron compounds, see Essay 7.

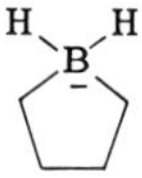

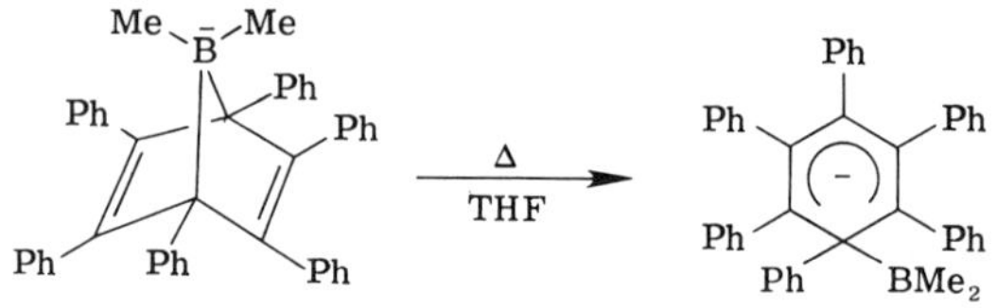

The main group elements other than those in the first long row of the periodic table have vacant orbitals in the valence shell, as does three-coordinate boron. Consequently, formation of the valence-expanded cyclic species required for 1,n migration should be energetically more facile. At the same time, substitution reactions with retention are well known for many of these elements. However, very few 1,n migrations have been observed, with silicon providing the only examples to date.

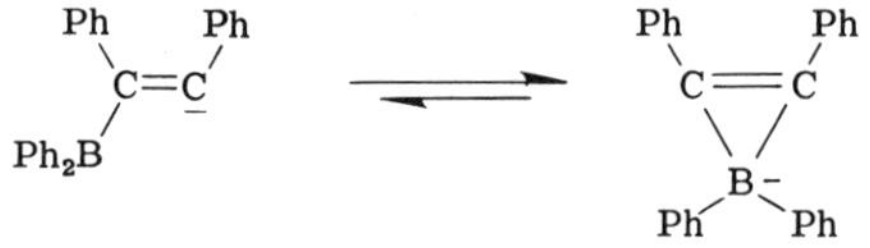

The lack of 1,n migrations may well be a result of more facile elimination and nucleophilic substitution reactions for many of these elements. Instead of 1,2 migrations, elimination reactions predominate for Group VI and VII compounds, and, instead of 1,n migrations, ring formation or elimination reactions predominate (Scheme 6). It is only from Group IV

6. REARRANGEMENT IN CARBANIONS

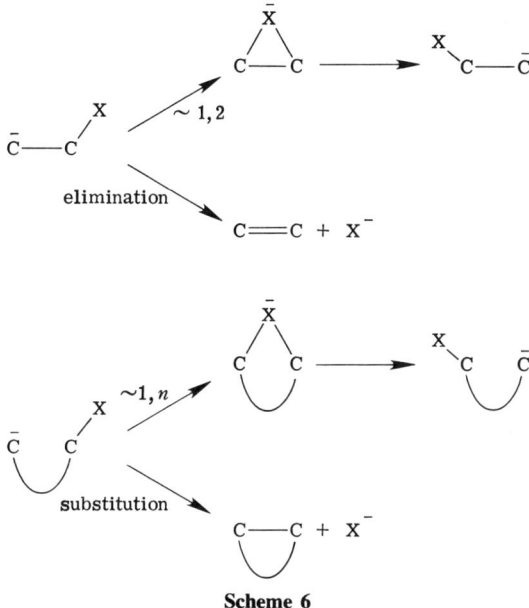

Scheme 6

that examples of 1,n migrations are found, Group V being in a borderline position. Although we are not aware of any intramolecular substitution reactions at phosphorus of the type needed for 1,n migration along a carbon chain, phosphonium salts would seem to be likely candidates.

The one example of an apparent concerted 1,2 migration of silicon involves a trimethylsilyl group (22). In **19**, the trimethylsilyl group was

$$
\begin{array}{c}
\text{SiMe}_3 \\
| \\
\text{Me}_3\text{Si}-\overset{-}{\text{C}}-\bar{\text{C}}\text{H}-\text{Ph} \\
| \\
\text{Ph}
\end{array}
\xrightarrow{\sim 1,2}
\begin{array}{c}
\text{SiMe}_3 \\
| \\
\text{Me}_3\text{Si}-\bar{\text{C}}-\text{CH}-\text{Ph} \\
| \\
\text{Ph}
\end{array}
$$

19 $\Big\downarrow n\text{-BuLi TMED}$

$$
\begin{array}{c}
\text{Me}_2\text{SiCH}_2\text{SiMe}_3 \\
| \\
\bar{\text{C}}-\bar{\text{C}}-\text{Ph} \\
| \quad | \\
\text{Ph} \ \text{H}
\end{array}
\xleftarrow{\sim 1,4}
\begin{array}{c}
\text{Me}_2\text{Si}\bar{\text{C}}\text{H}_2 \quad \text{SiMe}_3 \\
| \qquad\qquad | \\
\bar{\text{C}}\text{------}\text{C}-\text{Ph} \\
| \qquad\qquad | \\
\text{Ph} \qquad\quad \text{H}
\end{array}
$$

found to migrate rather than phenyl and under mild conditions (25°C). In fact, the major product was consistent with a subsequent 1,4 migration of trimethylsilyl. Although this is the only report of migration from carbon to carbon, there are enough analogous reactions (23) involving nitrogen, oxygen, and sulfur for one to expect that this is a general phenomenon (see also Essay 8). The Group III elements should have the same capabilities for migration as shown by boron, but as yet no example is available.

II. MIGRATION OF UNSATURATED CARBON

The 1,n migration of unsaturated carbon (C=C, C=O, aryl) through an addition–elimination pathway avoids the prohibitive energetic and stereochemical restrictions imposed by orbital symmetry. This then raises the question of whether the addition–elimination pathway, itself, is energetically viable in these three systems. On the basis of the accompanying free-energy diagram (Scheme 7)*, an attempt is made to provide answers to some aspects of this question. The only pair of isomeric compounds for which a $\Delta G°_{298}$ is available (24) is methylcyclopropane and 1-butene, with the cyclic isomer 7.6 kcal/mol less stable. No direct data are available for cyclopropanol and propanal, but comparisons of larger-ring alcohols and aldehydes (25) suggest that the $\Delta G°_{298}$ might be near 5 kcal/mol, with an uncertainty of about 5 kcal/mol. The next step in completing the free-energy diagram requires assignment of pK_a values to the hydrogens of interest.

The MSAD scale of Cram (2) has been used and pK_a values assigned assuming no long-range interaction. Thus, cyclopropanol is taken as a typical alcohol (pK_a = 16), and butene, methylcyclopropane, and propanal have been assigned the pK_a of ethane, 42. Although these are rather crude assumptions, the differences are large enough that the trends are evident.

These data indicate that the homoenolate anion should exist in the closed form, whereas the homoallylic anion should prefer the open form. The latter is observed experimentally (Section III,A), but there appear to have been no direct observations of homoenolate anions. Another difference of note concerns the transition state for deprotonation of the open-chain forms. Whereas it is probable that the olefin would proceed directly to the open-chain anion with little or no assistance from the double bond, the carbonyl compound should proceed directly to the cyclopropoxide through a transition state very much like the closed form, provided that the geometry is appropriate. Support for the latter view comes from the enhanced kinetic acidity of homoenolizable protons (see Section III,A), but no kinetic evidence concerning homoallylic systems appears to be available. In principle, it would be possible to do a similar analysis of the

---

* A free energy diagram has recently been published for 1-phenylcyclopropanol, using a different procedure for estimating relative free energies and resulting in a similar but not identical diagram (24a).

6. REARRANGEMENT IN CARBANIONS

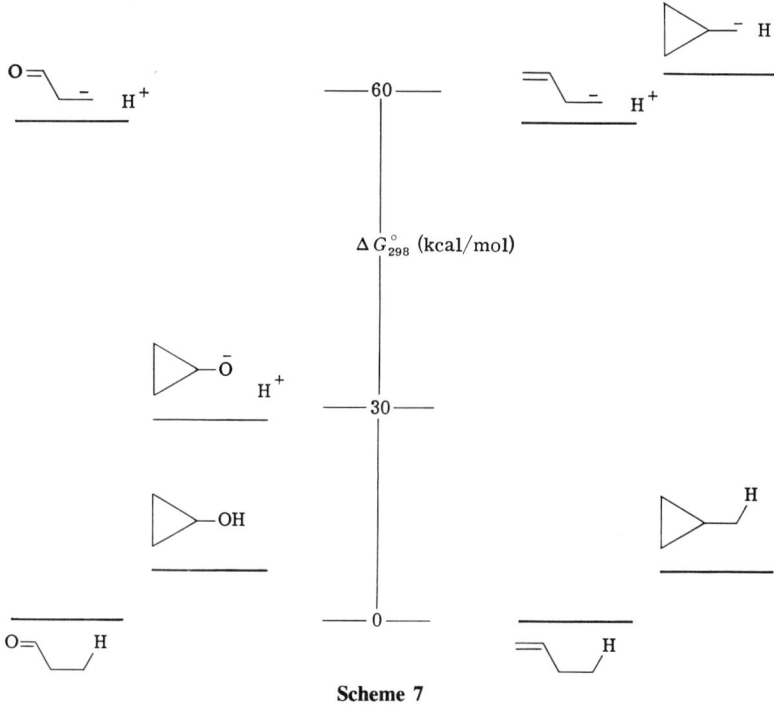

Scheme 7

homobenzyl system, but the relevant thermodynamic data are not available.

A. Olefins and Acetylenes

In the reviews mentioned (7, 8), a comprehensive survey of the earlier literature on organomagnesium rearrangements was provided. Although 1,2 shifts and allylic rearrangements were discussed, the major emphasis was on the cyclization and cleavage reactions necessary for migration of C=C groups. In addition, effects of ring size were summarized and quantitative information collected; a significant portion was also devoted to reaction mechanism. In the same year, 1,2 migration of C=C in organomagnesiums and organoalkalies was reviewed (4) under the heading of homoallylic anions, but in less detail. The aim of this essay is to provide a short summary of the types of 1,n migrations that are observed. The reader is directed to the above-mentioned review articles for more detailed discussions.

Since organoalkalies and organomagnesiums have seemed quite reluctant to add to simple alkenes, it was surprising to find examples of 1,n

migrations via addition–elimination. Nonetheless, examples of rearrangements remain plentiful, involving three-, four-, five-, and six-membered rings in both acyclic and bicyclic systems. Only recently it has been shown that Grignard and dialkylmagnesium compounds do add to simple alkenes under fairly mild conditions (26, 27). The wide variety of results in this area is summarized in three parts: Table 1 (27–31) presents some examples of 1,2 and 1,3 rearrangements; Table 2 (32–39) contains examples in which only ring closure has been observed for three-, four-, and five-membered rings; and Table 3 (7, 28, 40–43) includes ring-opening reactions of three-, four-, five-, and six-membered rings.

As the examples in Table 1 reflect, most of the studies of rearrangements have involved 1,2 migrations proceeding through three-membered rings, with some 1,3 migrations involving four-membered ring intermediates. Except for the deuterium-labeled symmetric intermediates **20**, **22**, and **24**, the rearrangements involve conversion of a secondary to a primary organomagnesium and occur under mild conditions. Although most studies have involved organomagnesiums, it seems likely that organoalkalies may also rearrange with ease, and this is illustrated by the organolithium intermediate **22**. Acetylenes also undergo a 1,2 migration (**23**), although it has been reported that the analogous terminal acetylene remains in the ring-closed form.

Although it is generally true that the three-membered rings are found as intermediates (**20–23**) and are less stable than the open-chain forms, there are several exceptions worth noting (Table 2). The equilibrium can favor the cyclic form because of the geometric constraints of a bicyclic system (**27**), by formation of a more stable vinyl organomagnesium (**28**), and by methyl substitution (**29**). For **29** the equilibrium apparently favors a primary rather than tertiary organomagnesium in spite of the three-membered ring. An interesting solvent dependence has been observed for the organolithium case (**30**). In tetrahydrofuran the red solution of the ring-closed, charge-delocalized form (**31**) is favored, whereas in diethyl ether the colorless, open-chain, primary lithium compound (**30**) predominates.

Again, in the case of 1,3 migrations, four-membered rings are found as unstable intermediates (**24–26**), but there are exceptions. For example, for **32**, a primary Grignard compound seems favored over a secondary or tertiary species. As illustrated by **33** and **34**, both olefins and acetylenes undergo cyclization reactions to form five-membered rings. Compound **35** is an example of formation of a six-membered ring, although it also involves conversion of a secondary Grignard compound to a primary one. Examples of 1,4 and 1,5 migrations, through five- and six-membered rings, are unknown.

6. REARRANGEMENT IN CARBANIONS

One feature of note in most of the ring-closure reactions is preferential formation of the smaller ring size. In general, this is also coincidental with formation of a primary Grignard reagent, although this is not always the case (**28, 30** and **34**).

The reactions listed in Table 3 help to clarify the ring-opening aspect of 1,n migrations of double bonds. The three-membered ring species **36** and **37** were found to be quite reactive and opened readily at low temperatures, which is consistent with the behavior proposed for these as intermediates in 1,2 migrations. There are also a number of ring-opening reactions involving bicyclic systems. Thus, the [3,1,0] system **38** opens to provide products from cleavage of both possible bonds in the three-membered ring, but at equilibrium the primary Grignard reagent is definitely preferred. Four-membered rings also cleave, with **39** converting to a primary Grignard reagent and **40** converting from a primary to a benzyl Grignard compound. Even five-membered rings cleave, as illustrated in the case of **41**, by the conversion of a secondary Grignard compound to an allyl system.

The examples presented here are to be regarded merely as illustrative. The number of examples of 1,n migrations and ring-closure and ring-cleavage reactions has expanded greatly in recent years, and further expansion should be expected with the continuing activity and interest in the area.

B. Aromatic Compounds

The 1,2 migration of the phenyl group from carbon to carbon (Grovenstein–Zimmerman rearrangement) was first reviewed in 1963 (*1*) and most recently in 1978 (*6*). In the early article, it was pointed out that the energetics of 1,2-phenyl migration were favorable if the reaction were treated as an addition–elimination to the migrating phenyl ring. The later article presented the accumulated evidence that strongly supports the intermediacy of a cyclic anion (**42**) in these migrations; this includes

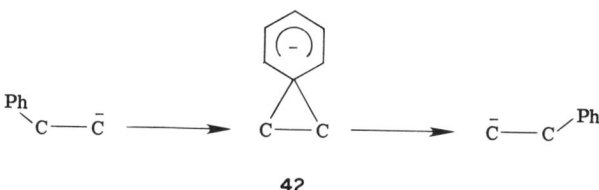

42

trapping of the ion by carbonation. Although this authoritative presentation (*6*) is available, a brief summary of observed rearrangements may be of introductory value.

Selected examples of 1,2-aryl migration are presented in Table 4 (*44*–

TABLE 1
Examples of 1,2 and 1,3 Migrations Involving Double and Triple Bonds

Substrate	Intermediate	Products	Reference
CH₂=CH–CHD–CH₂–MgCl	cyclopropane with MgCl and D (20)	CH₂=CH–CH₂–CHD–MgCl and CH₂=CH(D)–CH₂–CH₂–MgCl	28
CH₂=CH–CH₂–CH(CH₃)–MgCl	methylcyclopropane with MgCl (21)	CH₂=CH–CH₂–CH(CH₃)–MgCl and CH₂=CH–CH(CH₃)–CH₂–MgCl (Major)	28
CH₂=C(CH₃)–CD₂–CH₂–Li	cyclopropane with Li, D, D (22)	CH₂=C(CH₃)–CD₂–CH₂–Li (with D rearrangement) and CD(Li)=C(CH₃)–CH₂–CHD–	29
HC≡C–CH₂–CH(CH₃)–MgBr	methylcyclopropylidene with MgBr (23)	HC≡C–CH₂–CH(CH₃)–CH₂–MgBr	30

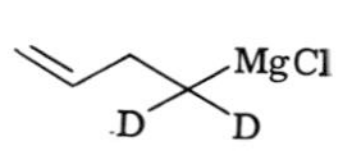

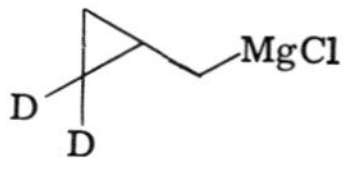

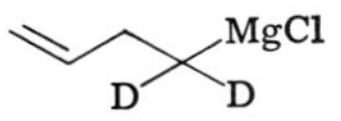

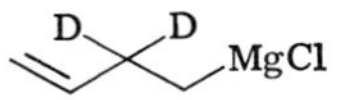

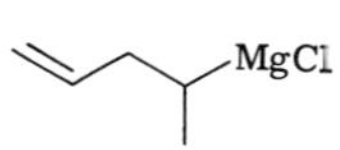

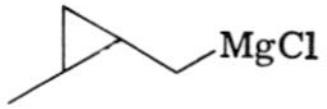

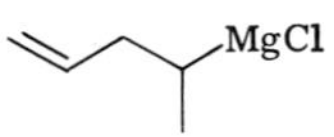

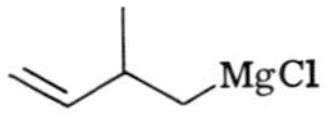

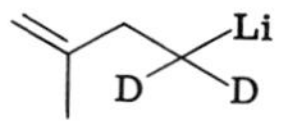

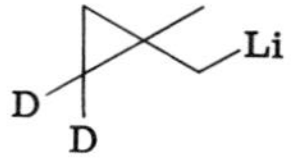

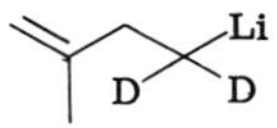

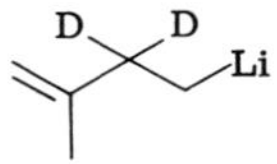

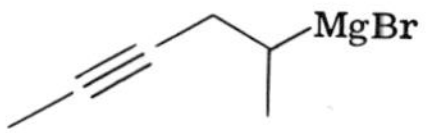

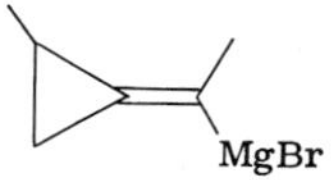

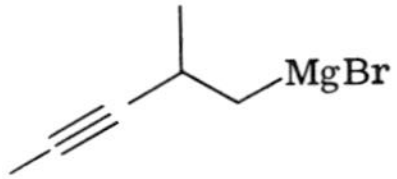

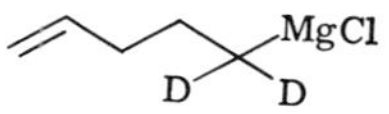

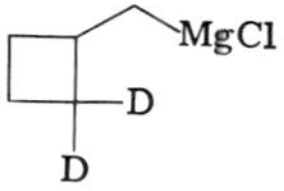

24

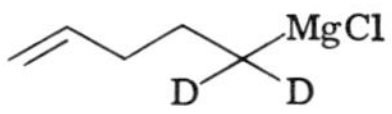

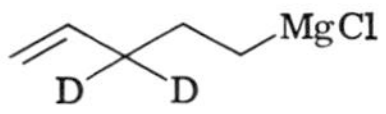

25

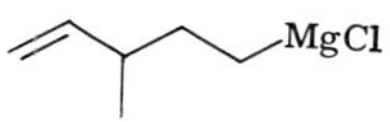

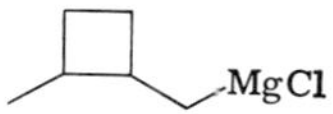

26

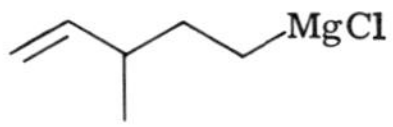

27

27

31

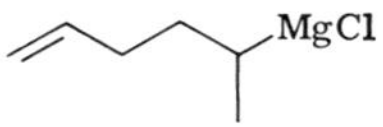

Major

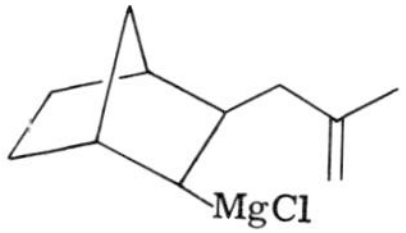

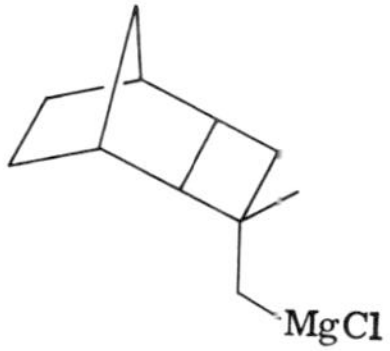

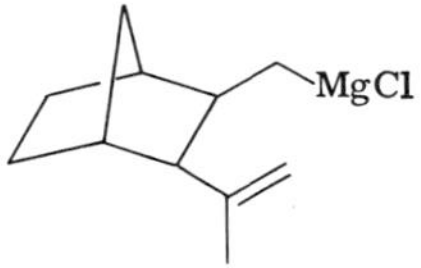

TABLE 2
Examples of Ring-Closure Reactions

Substrate	Product	Reference
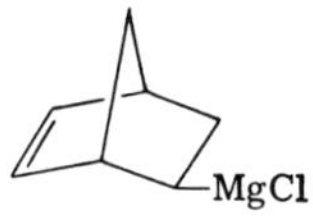 27	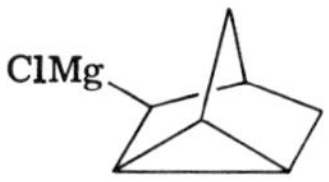	32
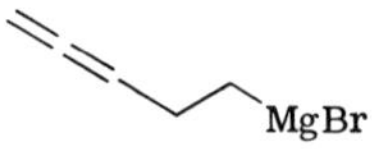 28	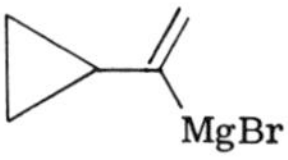	33
 29	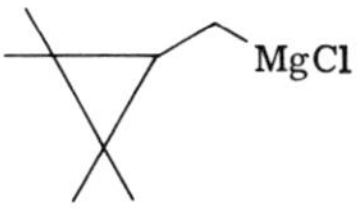	34
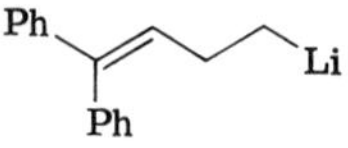 30 (THF ⇌ Et$_2$O)	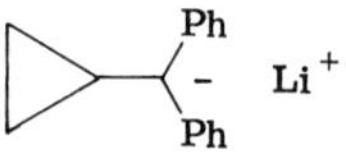 31	35
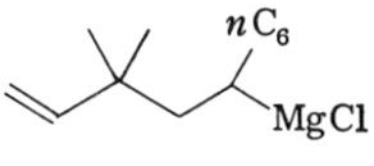 32	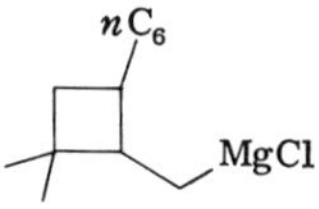	36
 33	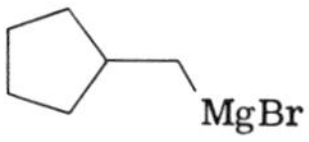	37
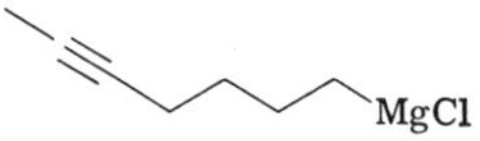 34	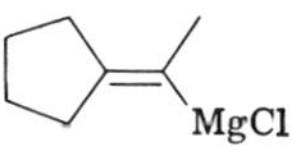	38
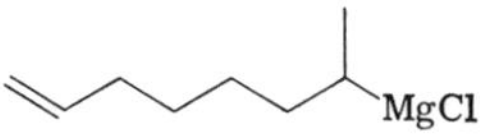 35	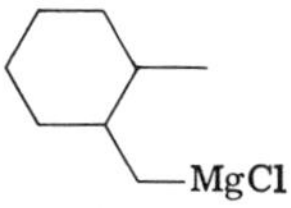	39

6. REARRANGEMENT IN CARBANIONS

TABLE 3
Examples of Ring-Opening Reactions

Substrate	Product	Reference
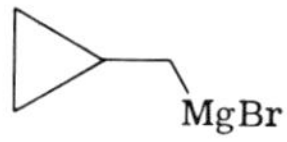 36 (cyclopropylmethyl-MgBr)	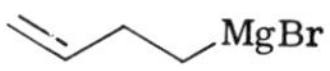 CH$_2$=CH-CH$_2$-CH$_2$-MgBr	28
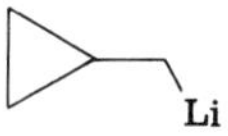 37 (cyclopropylmethyl-Li)	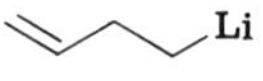 CH$_2$=CH-CH$_2$-CH$_2$-Li	40
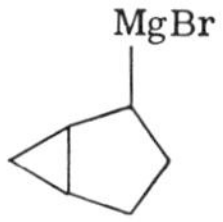 38 (bicyclo[3.1.0] with MgBr)	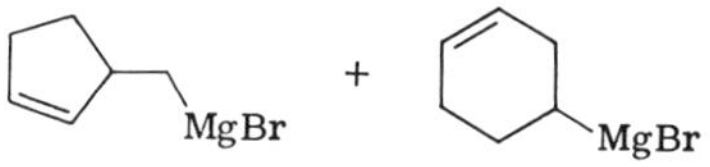 cyclopentenyl-CH$_2$-MgBr + cyclohexenyl-MgBr	41
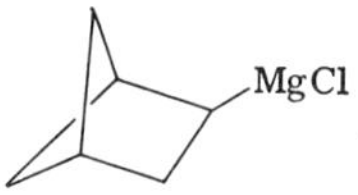 39 (norbornyl-MgCl)	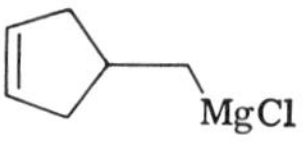 cyclopentenyl-CH$_2$-MgCl	42
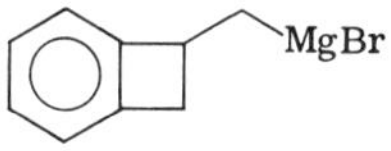 40 (benzocyclobutene-CH$_2$MgBr)	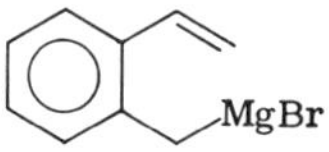 o-vinylbenzyl-MgBr	43
 41 (norbornenyl-MgCl)	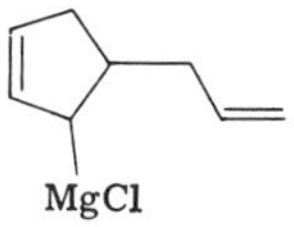 cyclopentenyl with allyl and MgCl	7

TABLE 4
Examples of 1,2-Aryl Migrations

Substrate		Product	Reference
Ph$_3$C—CH$_2$Li **43**	$\xrightarrow[\text{THF}]{0°C}$ · KOtBu $\xrightarrow[\text{THF, -65°C}]{}$	Ph$_2$C(Li)—CH$_2$Ph	44, 45
Ph$_2$C(CH$_3$)—CH$_2$Li **44**	$\xrightarrow[\text{Et}_2\text{O}]{\text{reflux}}$	Ph—C(Li)(CH$_3$)—CH$_2$Ph	46
Ph—C$_6$H$_4$—CH$_2$CD$_2$Li **45**	$\xrightarrow[\text{Et}_2\text{O}]{0°C}$	no rearrangement	47
Ph—C$_6$H$_4$—CH$_2$CD$_2$K **46**	$\xrightarrow[\text{THF}]{65°C}$	Ph—Ph—CH$_2$CD$_2$H + Ph—Ph—CD$_2$CH$_3$ 50% 50%	47
[cyclohexadienyl anion with Ph, CH$_3$] **47**	→	[cyclohexadienyl anion with Ph, H, CH$_3$] → products	48
H$_3$C—C$_6$H$_4$—C(Ph)(CH$_3$)—CH$_2$Li **48**	$\xrightarrow[\text{Et}_2\text{O}]{\text{reflux}}$	CH$_3$—Ph—C(Li)(CH$_3$)—CH$_2$Ph + Ph—C(Li)(CH$_3$)—CH$_2$—Ph—CH$_3$ 11 : 1	46
Ph—C$_6$H$_4$—C(3-Ph-C$_6$H$_4$)(4-Ph-C$_6$H$_4$)—CH$_2$Li **49**	$\xrightarrow[\text{THF}]{0°C}$	Ph—C$_6$H$_4$—C(3-Ph-C$_6$H$_4$)(CH$_2$Li)—C$_6$H$_4$—Ph >98%	49

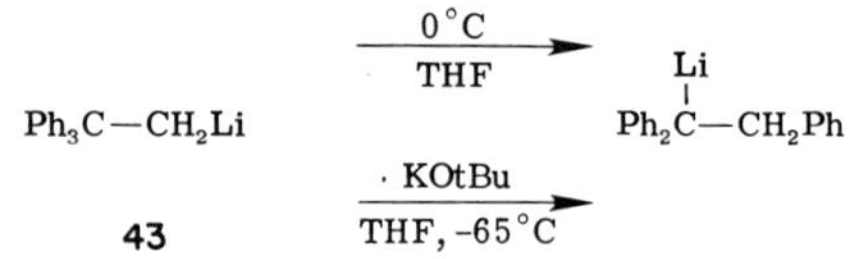

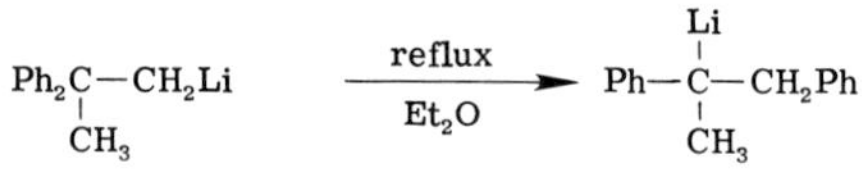

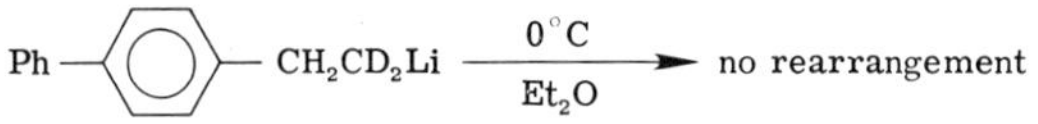

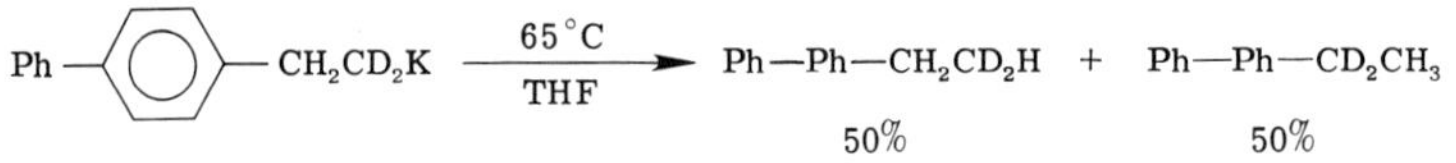

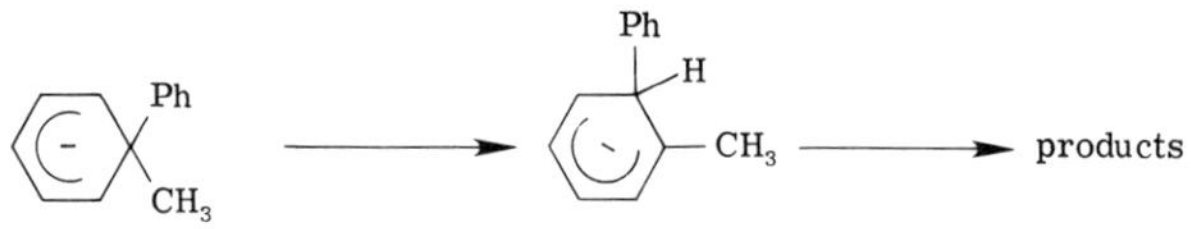

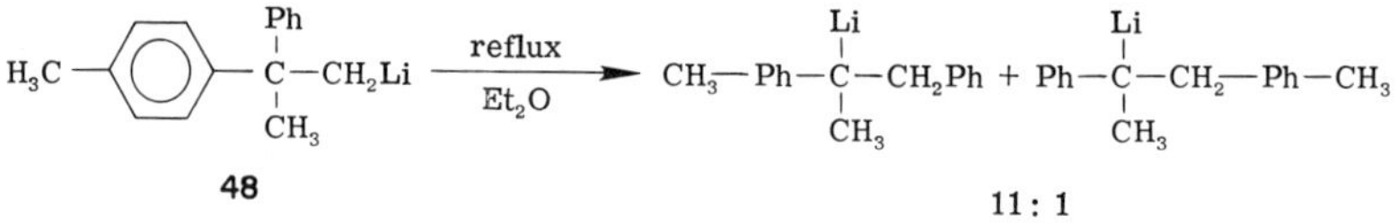

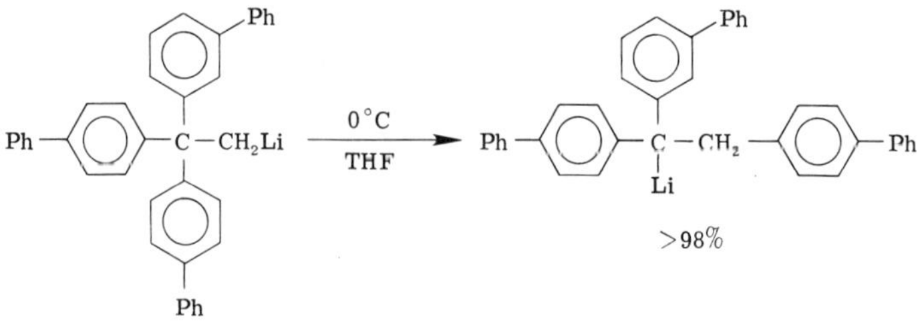

6. REARRANGEMENT IN CARBANIONS

49) and illustrate some of the features of these rearrangements. The 1,1,1-triphenylethyl system was the first, and probably the most, studied. The sensitivity of rate to the nature of the cation is shown with **43**, where the potassium salt is much more reactive than that of lithium; this appears to be a general phenomenon. A comparison of **43–45** shows that reducing the number of phenyl rings decreases the rate of rearrangement, and a comparison of **45** and **46** shows again the importance of the effect of cation on rate. A 1,2-phenyl migration in a more delocalized anion (**47**) has also been postulated to account for the observed reaction products.

Reaction of **43** in the presence of benzyllithium and of ^{14}C-labeled phenyllithium has established the intramolecular nature of the rearrangement (*50*). The relative migratory aptitudes shown in the rearrangements of **48** and **49** seem to be consistent with a spiranic intermediate similar to **42**. Furthermore, it has been possible to capture such a spiranic species (**51**) by carbonation at low temperatures after short reaction times in the Cs—K—Na reduction of **50** (*51*). It was shown earlier that an analogous

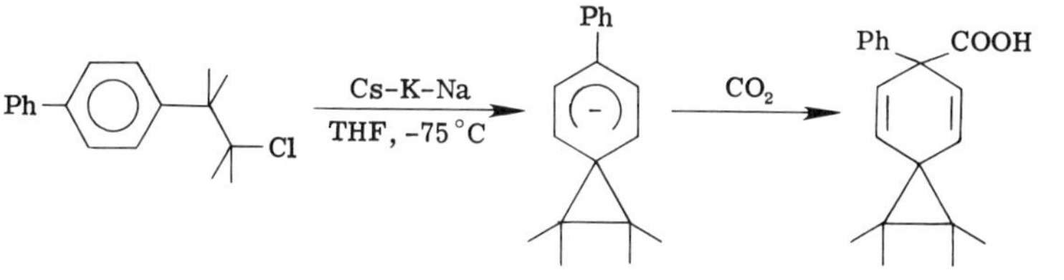

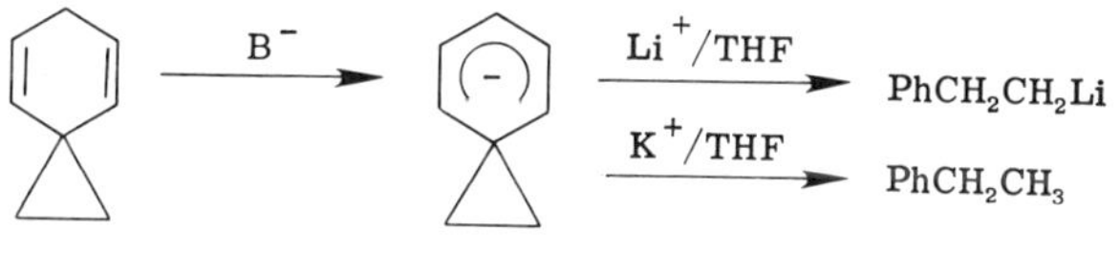

compound (**52**) also opens when converted to the lithium or potassium salt (*15*). Thus, at present, the mechanism of 1,2-aryl migrations seems most compatible with an addition–elimination process through a spiranic intermediate.

There have been systematic searches for other 1,*n* migrations, and selected examples are shown in Table 5 (*51–54*). Although there are no examples of 1,3 migrations because of side reactions (e.g., **53**), it has been possible to trap a spiranic species (**54**) by carbonation at low temperature, using a poorer leaving group and a better migrating group. There is an example of 1,4 migration (**55**), which is facilitated by the use of the biphenylyl group (**56**). In the case of **57**, a candidate for 1,5 migration,

TABLE 5
Examples of Attempts at 1,3-, 1,4-, and 1,5-Aryl Migrations

Substrate		Product	Reference
$Ph_2C(Ph)-CH_2CH_2Li$ (**53**)	$\xrightarrow{-40°C}$	$Ph_3\bar{C}$ + $CH_2=CH_2$	52
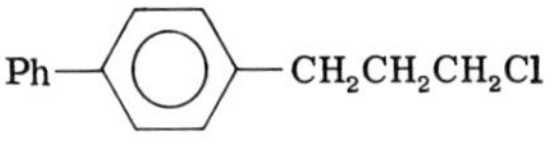 $Ph-C_6H_4-CH_2CH_2CH_2Cl$ (**54**)	$\xrightarrow[\text{2. CO}_2]{\text{1. Cs-K-Na/THF, }-75°C}$	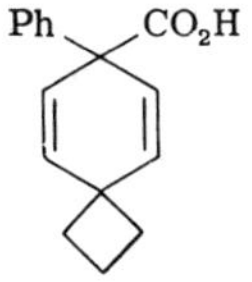 (cyclohexadiene with Ph, CO$_2$H, and cyclobutane spiro)	51
$Ph_2C(Ph)-CH_2CH_2CH_2Cl$ (**55**)	$\xrightarrow[\text{K/THF}]{65°C}$	$Ph_3CCH_2CH_2CH_3$ + 14% $PhCH_2CH_2CH_2CHPh_2$ 0.5% 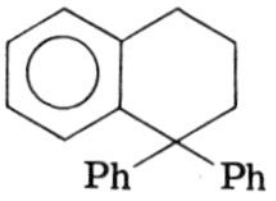 + (1,1-diphenyl tetralin) 84%	53
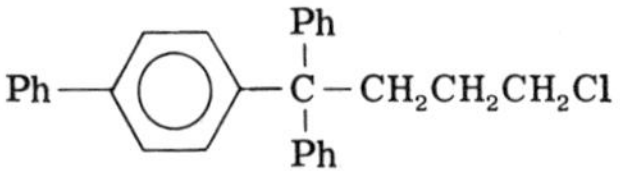 $Ph-C_6H_4-C(Ph)_2-CH_2CH_2CH_2Cl$ (**56**)	$\xrightarrow[\text{K/THF}]{65°C}$	$Ph_2\overset{K}{C}-CH_2CH_2CH_2-C_6H_4-Ph$	54
$Ph_3C-CH_2CH_2CH_2CH_2Cl$ (**57**)	$\xrightarrow[\text{K/THF}]{65°C}$	$Ph_3CCH_2CH_2CH_2CH_3$ >98%	53

rearrangement does not appear to be facile, and only unrearranged products are obtained, although the biphenylyl equivalent has yet to be attempted.

III. MIGRATIONS IN DOUBLY BONDED OXYGEN COMPOUNDS

A. *Homoenolization*

Proton loss from a carbon adjacent to a carbonyl group occurs readily because of the stability of the resultant enolate anion. The carbonyl group

6. REARRANGEMENT IN CARBANIONS

stabilizes the carbanionic center through conjugation and by inductive effects. Since the pK_a values for methyl protons in propane and acetone are estimated to be 42 and 20, respectively (2), the stabilizing effect is of the order of 20 pK_a units. The magnitude of this effect suggests that homoconjugated enolates may be generated in systems having a carbonyl group appropriately oriented for interaction with more remote centers. In such cases the homoenolate anion could be described in terms of species **58a–58c** arising by proton abstraction from positions β or γ, or even further removed from the carbonyl group, which we term β-enolates, γ-enolates, etc. Protonation of **58** could furnish the initial ketone (**59**), an isomeric ketone (**60**), and a cyclic alcohol (**61**), and even a second rearrangement or cleavage to **62** could occur. Of the three species, **58a** and

58b can be viewed as having nearly the same shape, whereas **58c** represents a form having a different geometry, with a smaller separation between the indicated carbon atoms. In principle, protonation to regenerate **59** could occur with inversion or retention of configuration. Similarly, there could be a preferred stereochemistry for the rearrangement to **60** or the cleavage to **62**. It should be noted that the ring opening of **61** constitutes an alternative route to the generation of **58** and hence the formation of **59** and/or **60**. This process has been termed "homoketonization." This process, as well as the rearrangement **59**→**60** (62), would appear to demand the intervention of **58c**. However, remote proton abstraction from **59** to form **58a,b** need not lead to rearrangement, and may result only in exchange at the carbanionic center. The precise nature of **58** for a given system undoubtedly depends on the structure of the initial ketone and, in any event, would be difficult to define other than by suggestion as to whether **58c** is required. A number of the features of the "homoenolization" and "homoketonization" processes have received attention since the discovery of the former by Nickon and Lambert (55). In this section we survey these studies to illustrate the current state of knowledge in this area.

1. Discovery

The existence of homoenolate anions was first proposed to account for the fact that base-catalyzed racemization and deuterium exchange occurred at the same rates when (+)-camphenilone (**63**) was treated with potassium *t*-butoxide in *t*-butyl alcohol-O-*d* at 185°C for varying periods of time (*55*). The findings were readily accommodated by proposing a

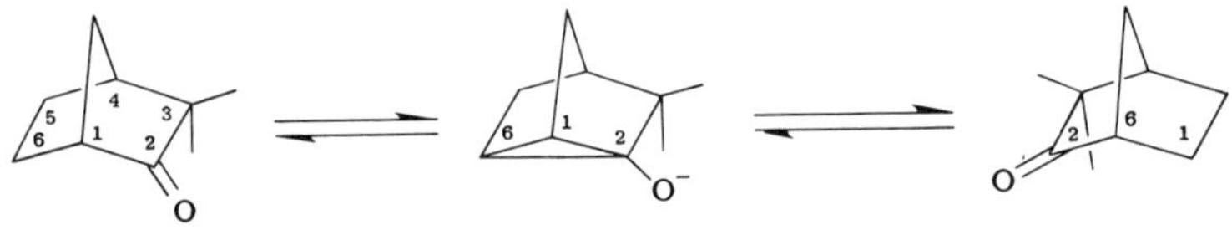

symmetric intermediate (**64**, analogous to **58c**) in which C-1 and C-6 become equivalent, and hence scrambled, and which led to the observed incorporation of up to three deuterium atoms per molecule. Although a symmetric species must be involved for the racemization, **64** may be a representation of the transition state between equivalent anions **65a,b** rather than a true intermediate. Also, **65a** and **65b** may be contributing

forms for the more delocalized species **66**, rather than separate entities. As mentioned above, it is difficult to describe homoenolate anions precisely, and throughout the following discussion the use of cyclic forms analogous to **58c** in the structural formulas is not intended to imply that these are necessarily the preferred forms in a given case, or even that these are the true intermediates; their use, however, is a convenient symbolism for **58** and is intended solely for this purpose.

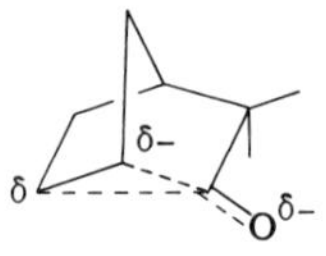

A variety of bases were examined (*55*) for comparison with the *t*-BuO⁻/*t*-BuOH system, but these were found to be much less effective and/or led to undesired side reactions. The systems of bases studied included ethylene glycoxide/ethylene glycol; potassium hydroxide/*t*-butyl alcohol; potassium triphenylmethylate/dioxane; and potassium *t*-butoxide/dimethyl sulfoxide. Consequently, nearly all subsequent examinations of homoenolization have utilized the *t*-BuO⁻/*t*-BuOH system at elevated temperatures (155°–275°C). Unless otherwise specified, it may be assumed that this is the base employed for each of the examples discussed.

In the original study (55), other possible modes of reaction were considered and rejected as highly unlikely. These included a reversible opening of the ketonic ring in a manner analogous to Haller–Bauer cleavage of nonenolizable ketones (56). Unlikely though this pathway appeared, it was tested experimentally by treatment of **67** (R = Et) under the homoenolization conditions, and *no* **63** was detected in the neutral fraction.

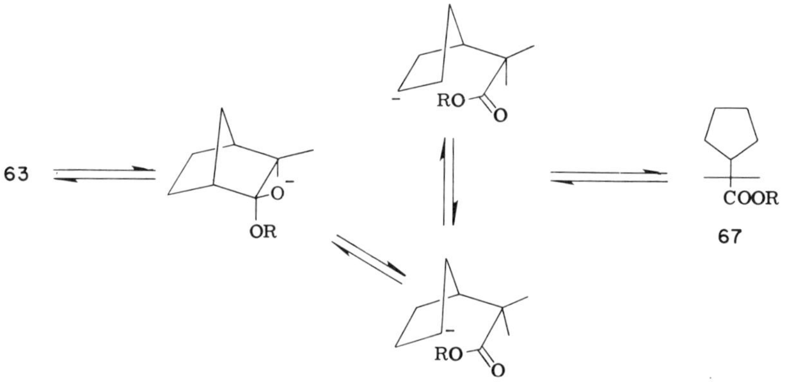

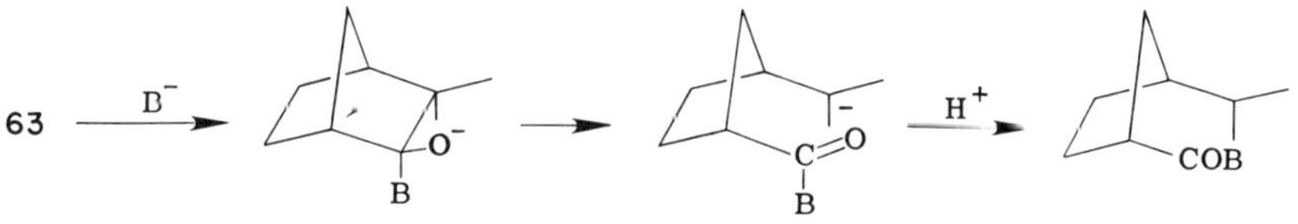

The authors pointed out that any **67** (R = *t*-Bu) that formed by transesterification underwent elimination to the corresponding carboxylic acid. In fact, it has been found that *t*-butyl esters from ketonic cleavage in other systems do not survive the homoenolization conditions but are isolated as the carboxylic acids. As an additional point, the cleavage of **63** with sodium amide was investigated to confirm an earlier report that the C-2—C-3 bond suffers cleavage (**68**) rather than the C-1—C-2 linkage. This was established, and it was shown that no significant racemization preceded the Haller–Bauer cleavage. In the homoenolization reactions the recovery of **63** was ~80% indicating the absence of substantial side reactions. However, cases were found subsequently that exhibited significant irreversible ketonic cleavage.

2. Methodology

Before turning to the variety of systems that undergo homoenolization, it seems appropriate to comment on the methodology of these studies. It has been found that the base is most readily prepared *in situ* by adding potassium metal to anhydrous *t*-butyl alcohol under a nitrogen atmosphere. It is important to minimize the water content so that potassium *t*-butoxide is the active species rather than potassium hydroxide, the catalytic activity of which is much lower (*57*). This may account for the reported failure of homoenolate formation in some compounds that would be expected, by analogy with closely related cases, to be reactive. For experiments in deuteriated media, *t*-butyl alcohol-O-*d* is readily prepared (*58*) via the borate ester. After the initial hydrolysis, a small amount of potassium metal is added to the alcohol before distillation; this may be repeated, and a third distillation from molecular sieves yields deuteriated alcohol, which is <0.005 M in water by Karl Fischer titration.

As already noted, it can be particularly informative to examine systems capable of homoenolization in deuteriated media since hydrogen–deuterium exchange can signal the occurrence of homoenolate anions for systems in which their presence is not manifest by rearrangement. In the earlier work, the location of sites of deuterium incorporation was established by direct comparison with specially prepared, selectively deuteriated derivatives. This involved rather tedious, albeit elegant, syntheses. The use of detailed comparisons of proton NMR specta became popular, especially with the advent of shift reagents to assist identification of specific absorption patterns through the increased shift dispersion. By careful integration the extent of deuterium incorporation can be assessed. The same information can be obtained directly by ^{13}C NMR since carbons bearing deuterons show distinctive patterns in the spectra. The inherent weakness of each of these approaches, however, is their limited sensitivity and lower precision in quantitative assessment. The direct observation of deuterium nuclei by ^{2}H NMR spectra has striking advantages for the examination of materials containing deuterium at levels as low as 1%. Since the natural abundance level is 0.015%, signals from centers containing 1% ^{2}H will be approximately two orders of magnitude more intense than the natural abundance signals. The magnetic relaxation of deuterium nuclei is induced entirely by an intramolecular quadrupole mechanism, and hence there is no Overhauser enhancement accompanying proton decoupling. Variations in the Overhauser enhancement of different carbons constitute one of the potential inconveniences of ^{13}C spectra in quantitative work. Proton decoupling is employed, in observing ^{2}H spectra, to collapse each absorption to a singlet (in the absence of other magnetic nuclei), and these signals typically exhibit

6. REARRANGEMENT IN CARBANIONS

line widths of a few Hz. The assignments for each signal follow from an analysis of the proton spectrum since ^{1}H and ^{2}H in the same environment have the same shielding. The shift range of ^{2}H in terms of frequency, however, is reduced by a factor of ~6.5 because of the lower magnetogyric ratio. Consequently, the ^{2}H spectra of compounds containing similar, but nonequivalent, ^{2}H nuclei may have heavily overlapping signals. In most cases, other than that of hydrocarbons, this inconvenience may be reduced through the use of lanthanide shift reagents to gain greater shift dispersion. Even with overlapping signals, precise integrated line intensities may be obtained by computer lineshape fitting; experience indicates that precisions of the order of 1–2% are readily attained. Each of these NMR techniques requires a measure of the total deuterium content by mass spectrometry, but the combination can provide precise data for several sites in the same molecule in a direct manner. The use of ^{2}H NMR techniques for mechanistic studies in a variety of systems has been reviewed (59).

3. Survey of homoenolization

a. Polycyclic Systems. The generation of β-enolates from **63** having been established, the effect of prolonged treatment with *t*-BuO$^-$/*t*-BuOD at 185° and 250°C was examined (60). At the higher temperature, **63** containing up to *nine* deuterium atoms was produced, and, with IR and NMR techniques, it was shown that C-1, C-6, and the methyl groups acquired deuterium (Fig. 1). It was also established that exchange did not occur at C-4, C-5, and C-7. For example, **63**-4-*d* and **63**-7-*d*, separately synthesized, were shown not to lose any deuterium under the reaction conditions. This clearly indicated that β-proton abstraction occurred only from sites that can interact with the carbonyl group, presumably via orbital overlap, without introducing excessive strain.

In addition to the intrinsic interest created by the discovery of homoenolization, there is the potential synthetic utility of the process since homoenolate anions could afford direct routes to systems that are otherwise difficult to obtain. For example, it was shown (61) that brexan-2-one (**69**) is transformed to brendan-2-one (**70**) in 60% yield under homoenolization conditions for 150 hr. Another example is the equilibra-

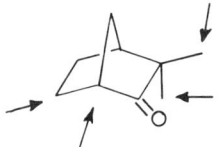

Fig. 1 Sites of H/D exchange in **63**.

tion of longicamphenilone (**71**) with the isomeric ketone **72**, as a 7:1 mixture, after 14 hr at 275°C (*62*).

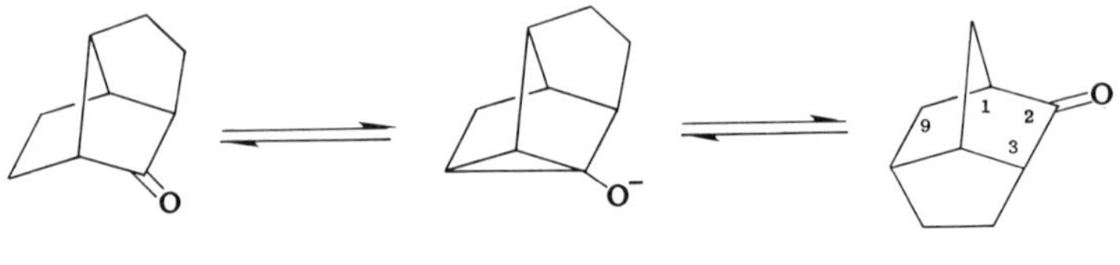

In the initial examinations of the homoenolization process, the behavior of fenchone (**73**) was studied briefly (*63*). Prolonged treatment was found to give species containing up to six atoms of deuterium per molecule, and there was some evidence for the formation of a second ketone in *ca.* 5% yield. These results were confirmed in an independent study (*64*) in which it was established that deuterium incorporation occurred only at C-6 and the methyl sites in **73**. In these experiments a second fraction, ~5% of the total neutral product, was isolated and shown to contain **74** and **75** in a 3:1 ratio. A sample of this mixture was treated under the reaction conditions, and the neutral fraction, after 200 hr, contained **73**, **74**, and **75** in a ratio of 75:5:20, indicating that **74** is the more reactive of the rearranged ketones. This finding was expected on the basis of earlier

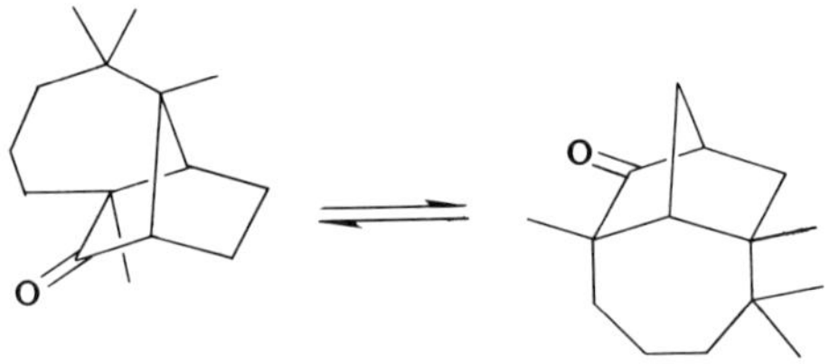

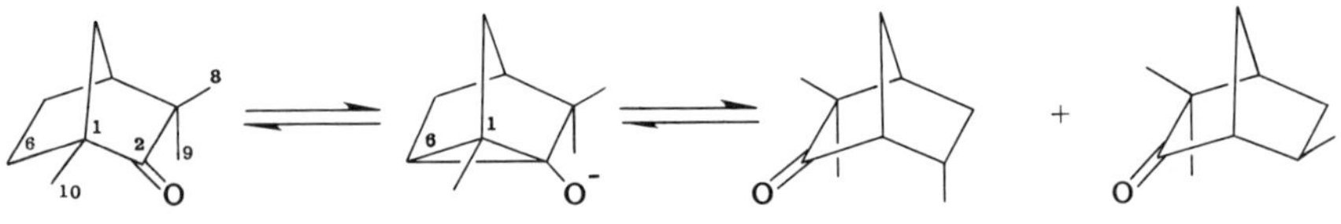

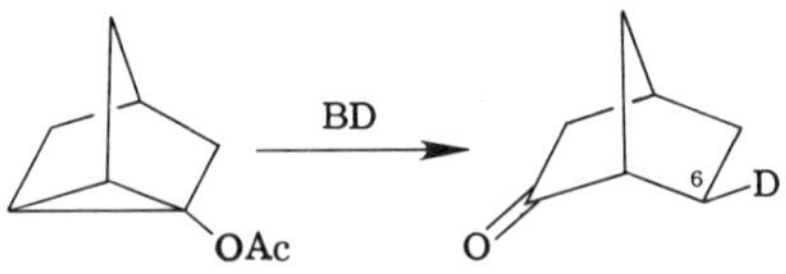

6. REARRANGEMENT IN CARBANIONS

experiments in the camphenilone (**63**) system in which 1-acetoxynortricyclene (**76**) was employed to generate the corresponding homoenolate under mild conditions (*65, 66*). In deuteriated base, homoketonization was observed to yield norcamphor-6-*d* (**77**), which acquired additional deuterium at C-3 by enolization after its formation. It was established that the cleavage of **76** produced an exo C—D bond with high stereoselectivity (95–98%). Thus, the major process involves inversion of configuration at C-6. It follows that protonation (deuteriation) of the homoenolate from **73** would exhibit similar behavior as observed in the preferential formation of **74** rather than **75**. It should be noted that the exo/endo ratio of deuterium uptake at C-6 in **73** was found to favor exo in a 3:1 ratio (*64*); this was readily monitored with ^{2}H NMR spectroscopy. Another feature emerging from the fenchone study was the observation of deuterium at each of the methyl sites. Again, by analogy with the results for **63**, this was expected, at least for the C-8 and C-9 positions, perhaps via a β-enolate such as **78**, but the appearance of deuterium at C-10 would require the highly strained species **79** if indeed a cyclic intermediate is

involved. Bridgehead methyl exchange has also been observed for camphor (*67*). In fenchone, however, an interesting regioselectivity was apparent in that deuterium was incorporated most readily at C-8 (the *exo* methyl) and least readily at C-10; the ratio of rates was roughly 20:5:2 for C-8, C-9, and C-10, respectively. In contrast to **63**, the homoenolate from **73** cannot be symmetric, and its partitioning between **73** and **74** (**75**) in a ratio of *ca.* 20:1 reflects this, although ring opening of the cyclic anion involves formation of secondary and tertiary carbanionic centers at C-6 and C-1, respectively, which could bias the process. As noted above, however, Haller–Bauer cleavage of **63** produces the tertiary center exclusively rather than a secondary carbanion. Clearly, rather subtle features play a role in the homoenolization process.

In each of the preceding examples, either the α position is blocked by alkyl substitution, or enolization is prohibited by Bredt's rule. Since it was of interest to test for the compatibility of α- and β-enolization in the same system, i.e., to determine whether the abstraction of α protons precluded the generation of higher-energy β-enolates, the isocamphanones **80** and **81** were examined. Epimerization could occur via a

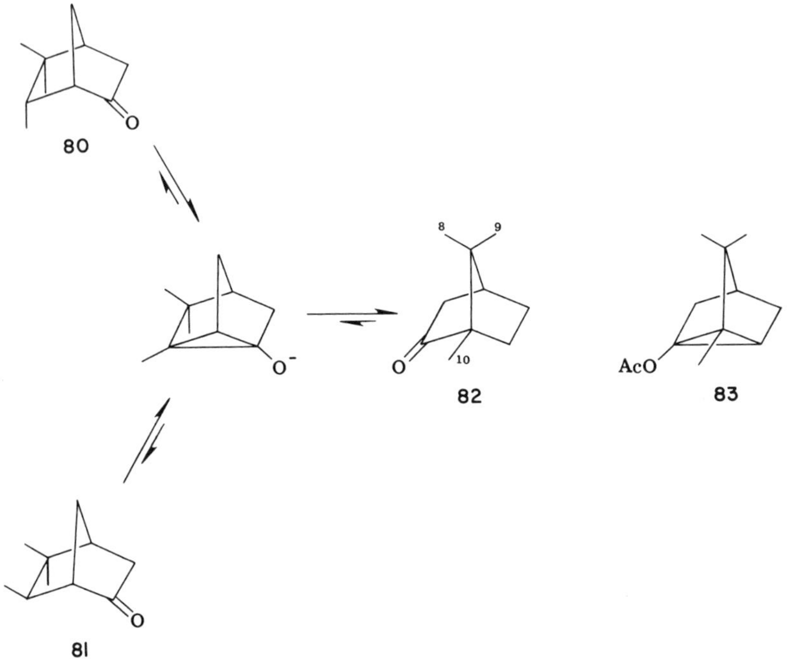

80

81

82

83

β-enolate from which camphor (**82**) might also form. Each of the ketones **80–82** was treated with t-BuO$^-$/t-BuOH at temperatures ranging from 185° to 250°C (67, 68). Both **80** and **81** exhibited ~60% β-enolization at C-6. In each case the major product was **82**, and it was found that **81** → **80** was favored over the **80** → **81** interconversion, in agreement with the expected preference for exo protonation. Even on prolonged treatment **82** gave only small amounts (2.5%) of **80** and **81**, presumably reflecting the greater stability of the camphor skeleton. Interestingly, the irreversible cleavage of **83**, the 2,6-homoenol acetate of **82**, which could be expected to generate the same homoenolate, gave 100% **82** at room temperature in t-BuO$^-$/t-BuOH. If, indeed, the same homoenolate is involved in both processes, this would represent a significant temperature effect on the partitioning of the anion, but the same homoenolate may not be involved with different reaction conditions. It may be noted that the cleavage of **83** in KOD/MeOD gave mainly **82**-exo-6-d (69), as expected from the earlier results. Camphor (**82**) that was recovered from the experiments over prolonged periods in deuteriated media contained species with as many as six deuterons per molecule and it was (67) established that incorporation occurred at C-8 and C-10 as well as at C-6 and C-3. Bridgehead methyl exchange (C-10) has analogy with the fenchone system, as already noted, but incorporation at C-8 is indicative of γ-enolate formation,* a process for which there are only a few examples. These are discussed later.

In the experiments with **80–82**, the recovery of ketonic products was of the order of 80%, showing that these reactions are relatively uncomplicated by side reactions. It was found, however, that ketones **84** ($n = 1$) and **85** suffered extensive cleavage under these reaction conditions, but the recovered ketones contained up to two deuterium atoms per molecule. Mass spectrometric evidence indicated that **84** ($n = 1$) undergoes methylene and methyl exchange, whereas in **85** exchange occurred only at the bridgehead. Treatment of other members of the monocyclic series **84** ($n = 2$–5) under comparable conditions (70) revealed that (a) these ketones are much more stable (~90% recoveries), (b) exchange is restricted almost entirely to the methyl sites, and (c) rearrangement is a very minor process. The ketonic products from the reactions of the six-, seven- and eight-membered ring ketones for long reaction times contained ~1% of a new component(s) in each case. There was no evidence of rearrangement for tetramethylcyclopentanone. The ^{2}H spectra of the recovered ketones showed that only in the cases of the six- and seven-membered rings were there indications of methylene exchange. It is interesting that the rate of methyl hydrogen exchange in the cyclopentanone was 10 times that for each of the larger rings. The behavior of **86** (R = H) and **87** (R = H) was examined (67), but it did not reveal exchange at sites other than the α-enolic positions; in the cases of the methylated analogues **86** (R =

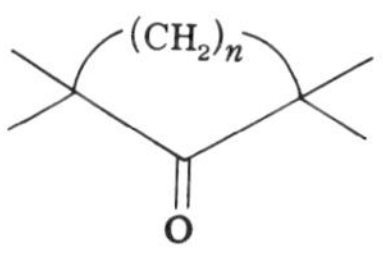

84

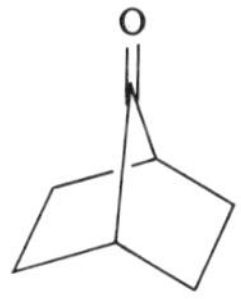

85

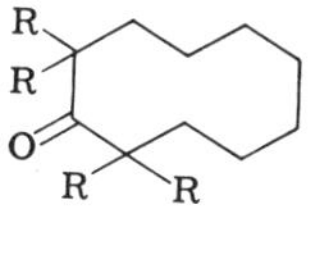

86

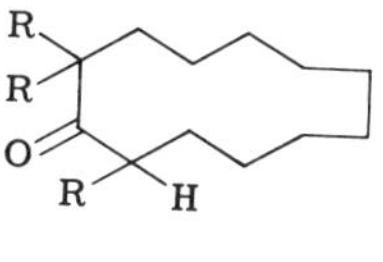

87

Me) and **87** (R = Me), there was evidence of ^{2}H incorporation at the methyl sites. In none of these was there any indication of rearrangement.

---

* The corresponding homoenol has recently been synthesized (69a) from 8-bromocamphor upon treatment with Mg/THF or Ca/NH$_3$(l). Reaction of this homoenol with t-BuO⁻/t-BuOD gives 8-deuteriocamphor exclusively.

Two additional examples having the basic [2.2.1] skeleton were examined in an attempt to obtain remote deuteriation (71). Copacamphor (**88**) and longicamphor (**89**) were treated under the standard conditions and, upon recovery, were found to contain one and up to three deuterons per molecule, respectively, with no evidence of rearrangement. In the light of the **71** ⇌ **72** isomerization mentioned earlier, it seems curious that **88** underwent only α-exchange and that in neither case was there any rearrangement. No attempt was made to identify the sites of exchange in **89** other than to show that the α position acquired deuterium. It would be interesting to reexamine these compounds.

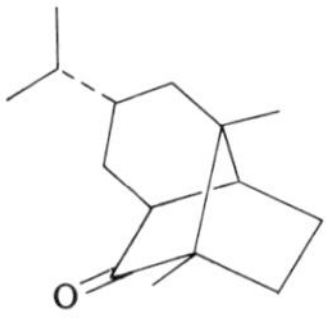

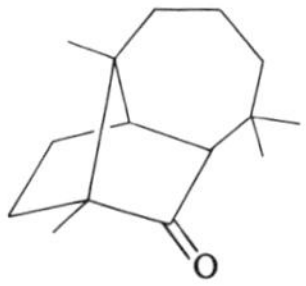

It was reported in the literature (72), and informally to the present authors by other groups, that adamantanone failed to exhibit β-enolate formation under the usual conditions. However, it has since been established that deuterium exchange indeed occurs at the α and β positions, albeit considerably slower than in most of the preceding examples (73). The order of reactivity was found to be exo β > α > endo β with the exo/endo ratio for β exchange being ~15 (see **90**), or significantly larger than that found for **73**.

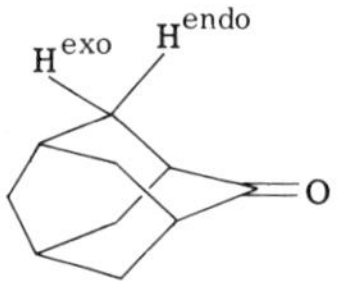

In an effort to assess the synthetic potential of the homoenolization process, other bicyclic skeletons have been examined, among which is the bicyclo[2.2.2]octanone system. In principle, ketones **91** and **92** should be interconvertible via a common homoenolate anion. Preliminary experiments (74) revealed that both ketones exhibited ^{2}H exchange, consonant with β-enolate formation, and that each was converted to the other at different and very slow rates. From a more detailed study (70) it was

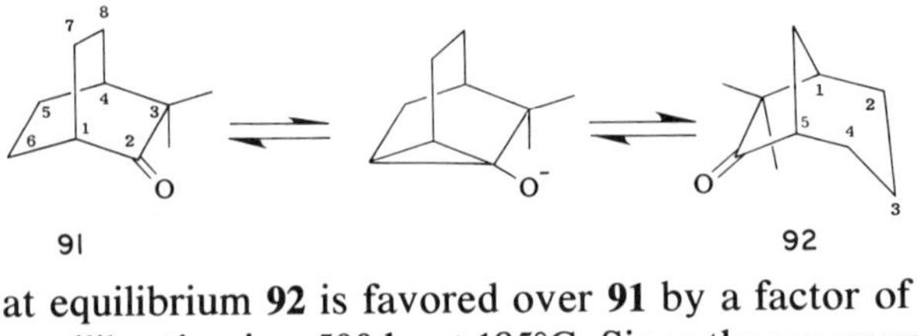

found that at equilibrium **92** is favored over **91** by a factor of 4, and the half-life for equilibration is ~500 hr at 185°C. Since the recovery of ketone is ~75%, side reactions are relatively unimportant.

The recovered ketones were examined by ^{2}H NMR to establish that

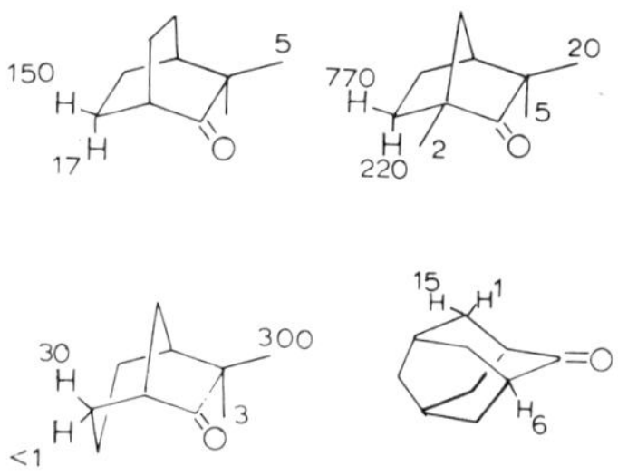

6. REARRANGEMENT IN CARBANIONS

Fig. 2 Estimated rates ($k \times 10^8$ sec^{-1}) of H/D exchange via β-enolates at 185°C.

exchange occurred only at C-6(7) and the methyl positions in **91**. Again, for the β-methylene site, exo deuteriation was favored over endo deuteriation by a factor of 10, which is intermediate between the corresponding ratios for fenchone and adamantanone. For **92**, deuterium could not be detected at the endo-4 site, indicating an even greater stereoselectivity for β-methylene exchange than in any of the other systems. In addition, there was deuterium uptake at both methyl positions, but exo exchange is 100 times faster than *endo*-methyl exchange, whereas the latter is comparable to that found for **91**. Although regioselectivity for methyl exchange in **73** was also observed, *exo*-methyl exchange was only *ca.* 3.5 times the rate found for *endo*-methyl exchange in **73**. Some estimated relative rates of H/D exchange in **73** and **90–92** are listed in Fig. 2. Samples of **92**-d_x isolated from prolonged treatment of **91** were found to contain deuterium at the exo-3 position as well. This almost undoubtedly arises by incorporation at C-7 in **91** before rearrangement to **92** through the β-enolate formed by proton abstraction from C-6. There is a slight possibility that incorporation could occur directly in **92** through γ-enolate formation by analogy with cases discussed later, but the relative amounts found at this position in the **91** → **92** interconversion, coupled with the lack of detectable ^{2}H in samples of **92** treated directly, render this possibility remote. Deuterium was found to be incorporated most efficiently at C-5, the bridgehead position, a finding similar to that noted above for **88** and **89** and reported subsequently for **93** (75). As pointed out (76), a bridgehead

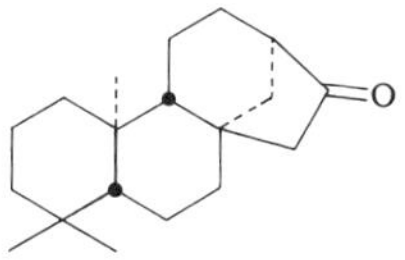

double bond is endocyclic to two rings and, therefore, transoid in one. In these systems the bigger ring is seven-membered or larger, which is evidently sufficient to accommodate the double-bond character of the bridgehead enolate through which exchange occurs. A remarkable example of unusually facile bridgehead enolization has been found for **70**. Upon treatment with MeO⁻/MeOD at 25°C, **70** undergoes essentially complete exchange at the 3 position (*77*).

Since an olefinic bond in a β-enolate could bias its partitioning to isomeric ketones, the behavior of **94** under the usual reaction conditions was examined (*74, 78*). In contrast to **91** ⇌ **92**, equilibrium was attained rapidly, with a half-life of approximately 7 hr at 185°C. At equilibrium, **94–96** are present as a 8:44:48 mixture. With either **95** or **96** as starting

material, the product is nearly a 1:1 mixture of the two within several minutes at 185°C. Clearly, these are readily interconverted via a common allylic anion. Equilibration of **94–96** was also attained at 155°C, with a half-life of *ca.* 80 hr. This indicates an activation energy of approximately 30 kcal/mol for skeletal isomerization. Deuterium exchange studies were carried out at both temperatures since incorporation at most of the sites was too rapid at 185°C to allow extraction of even crude rate data (*78*). The rate of rearrangement of **94** is *ca.* 100 times faster than that for its saturated analogue **91**, and the double bond appears to bias the equilibrium more toward the [3.2.1] skeleton. After separation of the ketones recovered in the neutral fraction (~80% yields), each was examined using ²H NMR techniques to assess the ²H incorporation. From the reactions at 185°C, each of **94–96** was found to contain deuterium at six sites. In **94**, ²H was found at the C-1, C-5, C-6, exo-7, endo-7, and *exo*-methyl positions, whereas, in **95** and **96**, ²H incorporation occurred at the C-2, C-3, C-4, C-5 and *exo*-methyl positions; exo and endo ²H were found at C-4 and C-2, respectively. As expected from the results for **92**, bridgehead exchange is facile in **95** and **96**, but the appearance of deuterium at the bridgehead in **94** contrasts with the results for **91**. However, ²H uptake at C-1 in **94** almost undoubtedly results from exchange in **95** (**96**) before conversion to **94**. Deuterium incorporation at the *exo*-methyl site in **94** almost undoubtedly occurs in the same way. Both processes are very slow in the saturated analogue **91**, and, at 155°C, no ²H was observed at these sites upon treatment and recovery of **94** (Fig. 3). The appearance of deuterium at C-6 in **94** and C-3 in **95** (**96**) cannot be ascribed to β-enolization, and it was shown that an addition–elimination process involving *t*-BuOD was

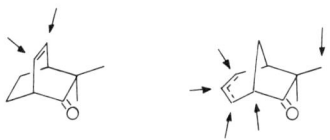

Fig. 3 Sites of H/D exchange in **94–96** at 155°C.

unlikely since the 2-*t*-butoxybicyclo[3.2.1]octanes are stable under the reaction conditions at 185°C (*79*). It can be concluded that ^{2}H incorporation at these sites occurs via direct vinyl exchange. As a measure of the activating effect of the carbonyl group on the vinyl exchange process, the corresponding bicyclic olefins **97** and **98** were subjected to homoenolization conditions in deuteriated media (*79*). The results show that exchange at C-5 and C-6 in **94** proceeds about 50 and 10^3 times faster, respectively, than vinyl exchange in **97**. For the [3.2.1] systems, the data indicate that the rate of vinyl exchange is increased by a factor of 40 in the ketones.

As in the saturated analogue **92**, the *exo*-methyl group is preferentially deuteriated in **95** (**96**), and the rates are comparable in the three ketones. Stereoselectivity was observed for deuterium exchange at C-4 in **95**, with the exo favored by a factor of 2. This is essentially the same as that found for **98** but, with allowance for the temperature difference of the two series of experiments, the process is *ca*. 250 times faster for the ketones than for the olefin. In contrast, exo exchange at C-2 in **96** is about six times faster than endo incorporation; this could, in part, be due to a steric effect of the *endo*-methyl at C-7 hindering approach of the deuterium donor to the endo site at C-2.

Apart from the interesting exchange data observed for the **94–96** system, it can be noted that the reaction provides an efficient route to the unsaturated [3.2.1] system from the more readily available bicyclo[2.2.2]octane derivative. This process would be more attractive if it were not necessary to block the α-methylene position, and consequently some experiments with the parent ketone **99** were carried out. As already noted, aldol condensation is not an important side reaction in the reactions with camphor derivatives. It was expected that even if this were to occur for **99** all steps would be reversible and β-enolization should be observable as a competitive process. However, no evidence for the latter process was found, and all of the starting material was consumed (*80*). In contrast to the [2.2.1] system, the [2.2.2] ketone readily condenses, and at 185°C the resulting adduct undergoes a retro-Diels–Alder loss of ethylene to give the 3-phenyl derivative of **99**. This product then is destroyed by both loss of a second ethylene moiety, to form 2-phenylphenol, and through a Haller–Bauer type of cleavage, to form a mixture of 4-benzylcyclohexenecarboxylic acids (Scheme 8). At 100°C, the retro-Diels–Alder reaction is suppressed, but there was no evidence of rearrangement of **99**; the product appeared to be a mixture of isomeric condensation products (*81*).

Although **92** undergoes β-enolization sluggishly, the isomeric ketone **100** is appreciably more reactive (*82*). Of the three β-methylene positions at which β-enolization could conceivably occur, reaction occurs at two of these sites. Proton abstraction from C-7 generates a β-enolate through which optically active **100** could racemize. Deuterium exchange occurs at C-7, with exo incorporation as the faster process. Deuterium also appears

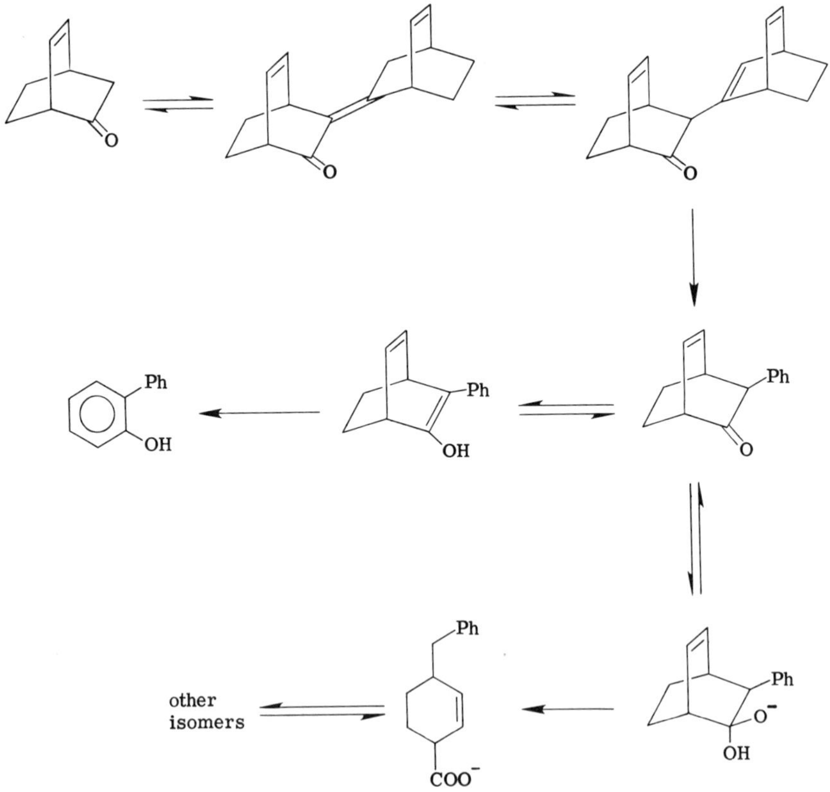

Scheme 8

6. REARRANGEMENT IN CARBANIONS

at both methyl sites, with the *exo*-methyl more heavily labeled, which shows that proton abstraction from the *exo*-methyl group is faster than that from C-7 since the latter process generates a β-enolate which equilibrates the methyl groups. No deuterium is detected at the 8 position

in **100**, although rearrangement to **102** through β-enolate **101** is observed. Clearly, the generation of **102** is essentially irreversible and at 185°C the half-life for the rearrangement is *ca.* 60 hr. Bridgehead exchange for **100** is >90% in 1 hr, as expected from the data for the systems discussed above. Incorporation of ²H in **102** arises from exchange in **100** before its rearrangement and directly after **102** is formed. To examine the latter separately, samples of **102** were treated at 185°C. The ²H spectra of recovered **102**-d_x revealed that at least *eight* sites undergo exchange (Fig.

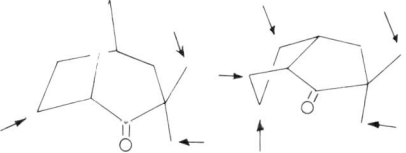

Fig. 4 Sites of direct H/D exchange in **100** and **102** via β- and γ-enolates at 185°C.

4); these include C-1, C-6, C-7, and C-8 as well as both methyl groups. For the latter, the *exo*-methyl position is *ca*. 10 times more reactive than the *endo*-methyl, whereas at C-8 exo and endo exchange occur at comparable rates, albeit slower than *exo*-methyl exchange. Unfortunately, it was not possible to assess separately the deuterium incorporation at the exo orientations and C-6 and C-7, but there was more deuterium at endo-6 than at the endo-7 location. Although the stereoselectivity of exchange at C-6 and C-7 could not be determined, these processes are clearly not stereospecific. Nevertheless, observations of exchange at these positions strongly suggest the intervention of γ-enolates to account for deuterium incorporation. These represent the first examples of γ-proton abstraction in conformationally mobile systems. In **102**, conformations in which the 6- and 7-methylene groups closely approach the carbonyl group can be readily envisaged, and the resulting tricyclic γ-enolates seem reasonable intermediates from examination of molecular models. The corresponding process in camphor, noted above, involves the *syn*-methyl group, which is constrained to a location near the carbonyl group. In **102**, however, the molecule is required to adopt a particular conformation to permit interaction.

It is appropriate to cite here the first reported example (*83*) of γ-enolization; this was found for the half-cage ketone **103**. Upon treatment of **103** under the usual conditions, isomerization to **105** proceeded with very high yields (~96%), and **104** was established as the likely intermediate by treatment of the corresponding alcohol with base, which gave the same 4:96 mixture of **103** and **105**. More recently the stereochemistry of the ring opening of **104** has been investigated (*84*) and found to lead to endo deuteriation with high stereoselectivity (90 ± 3%). The high stereoselectivity found for this process contrasts with the course of γ-exchange in **102**, which perhaps indicates a significant difference in the nature of the γ-enolates in the two systems. Of course, the difference in stereoselectivity could also arise simply from the temperature dependence of the rates of the two processes, opening with inversion or with retention of configuration. Interestingly, under the conditions employed, t-BuO$^-$/t-BuOD at 100°C, deuterium also appeared at that α bridgehead indicated in **105**.

6. REARRANGEMENT IN CARBANIONS

One other rearrangement that appears to involve γ-enolization is the conversion of **106** to **108** (*85*). The generation of γ-enolate **107** was found

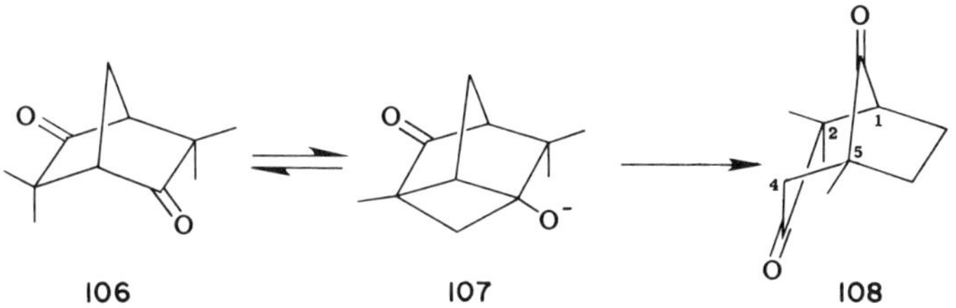

to be essentially irreversible, and the recovery of ketones was quantitative. Upon its formation, **108** undergoes facile exchange at one of the methyl sites, which was subsequently shown to be the bridgehead methyl on C-5 (*86*). In addition to rearrangement, **106** acquired deuterium at the bridgehead *and* methylene (C-7) sites. Exchange at these positions was not observed in any of the simpler norcamphor derivatives discussed above. The enhanced acidity at C-7 could result from cyclic delocalization of the negative charge through the carbonyl groups conjugated either through three σ bonds or through space (*85*).

The intermediate γ-enolate (**107**) most readily envisaged for the **106** → **108** interconversion suggests that other structures having a methylene group endo in the [2.2.1] skeleton could be reactive under homoenolization conditions. Preliminary experiments indicate that **109**

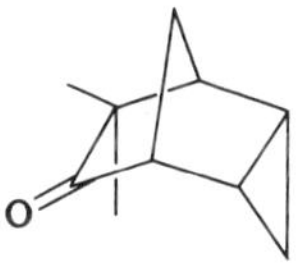

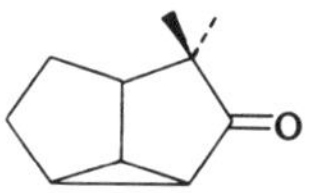

undergoes rearrangement with **110** as the initial product, which is subsequently consumed in a second reaction; under the same conditions for comparable periods of time, isomer **111** appears to be stable (*87*). In light of the results for **106**, it is apparent that a detailed examination of the behavior of **109** and **111** is warranted; this work is currently in progress in these laboratories.

b. Acyclic Systems. Although all of the preceding examples involve cyclic systems, evidence for homoenolate intermediates has also been found in a few acyclic cases. An investigation of the hydrolysis of some 3-acetoxy-Δ^1-pyrazolines (*88*) revealed that a skeletal rearrangement, presumably involving a β-enolate, occurred for the derivatives **112** [R = R' = Me; R = CH=C(CH$_3$)$_2$, R' = Me; and R = R' = Et]. For each of these, **113** was the major product formed upon alkaline hydrolysis together with small amounts of **114**, except in the case of the isobutenyl derivative, for which the product was entirely **113**. Hydrolysis of 1,2,2-trimethylcyclopropyl acetate (**115**) afforded a 90:10 mixture of **113** and **114** (R = R' = Me) under similar conditions. The results for these pyrazolines, therefore, are most readily accommodated in terms of β-enolate formation.

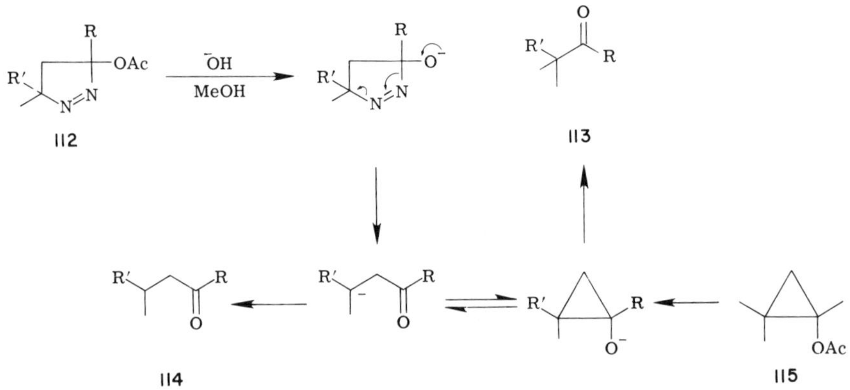

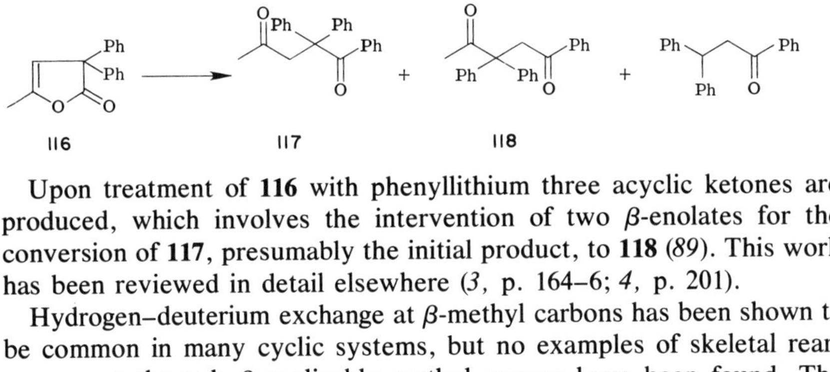

Upon treatment of **116** with phenyllithium three acyclic ketones are produced, which involves the intervention of two β-enolates for the conversion of **117**, presumably the initial product, to **118** (*89*). This work has been reviewed in detail elsewhere (*3*, p. 164–6; *4*, p. 201).

Hydrogen–deuterium exchange at β-methyl carbons has been shown to be common in many cyclic systems, but no examples of skeletal rearrangement through β-enolizable methyl groups have been found. The absence of such rearrangement could be taken to indicate that cyclic β-enolates are not involved in the exchange process, but the fact that these exchanges are regioselective shows that factors other than the inductive effect of the carbonyl group govern the reactivity of β-methyl groups. Clear evidence of the existence of cyclic β-enolates, therefore, would be welcomed, and the isolation of rearranged product(s) would provide such evidence. Deuterium incorporation was observed in di-*t*-butyl ketone (**119**, R = Me), presumably via β-enolate **120**, but no other products were reported in the original work (*2*, p. 65). A reexamination (*90*) of this ketone, however, revealed that 2,2,5-trimethylhexan-3-one (**121**, R = Me) was indeed produced slowly under the usual conditions

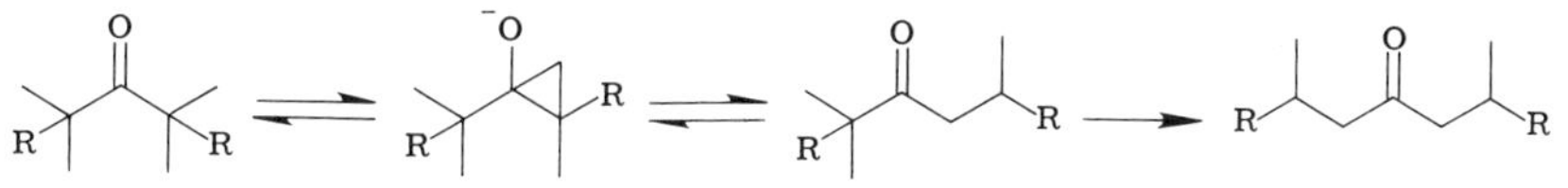

6. REARRANGEMENT IN CARBANIONS

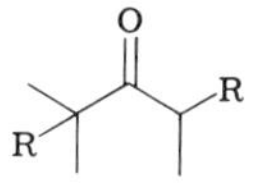

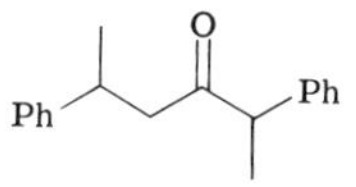

(~20% yield, 150 hr). Treatment of **121** (R = Me) under the same conditions gave, after very long reaction times, trace quantities of a third ketone identified as **122** (R = Me) by GC–MS techniques. Similar experiments with **119** (R = n-Bu) also gave rearranged ketone **121** (R = n-Bu), with no evidence of further rearrangement. Interestingly, recovered **119** (R = n-Bu) from reactions in a deuteriated medium contained deuterium at the methyl positions only; no exchange occurred at any site in the alkyl chain, even after prolonged treatment. It would appear from these results that the β-enolate produced by proton abstraction from a β-methyl group leads to a cyclic intermediate such as **120**. In these experiments the yields of recovered ketones exceeded 90%, indicating that other reactions are unimportant. In contrast, the related ketone **119** (R = Ph) undergoes significant cleavage to acidic products (*91*) in addition to rearrangement to **121** (R = Ph) and, ultimately, **122** (R = Ph). In this system, cleavage of β-enolate **120** (R = Ph) in the appropriate sense for rearrangement generates a benzylic anion, and this presumably accounts for the observation of more extensive rearrangement to **121** (R = Ph) and **122** (R = Ph). In the aliphatic examples (R = Me or n-Bu), the generation of **120** is reversible, as evidenced by deuterium acquisition by the methyl groups. The generation of **120** (R = Ph), however, is unidirectional since recovered **119** (R = Ph) contains no detectable ^{2}H at the methyl sites, although deuterium is present on the aryl rings. The occurrence of aryl exchange is hardly surprising because the aryl C—H bonds must be appreciably more acidic than alkyl C—H bonds (*2*). It may be noted that in the less highly substituted analogues of **119**, ketones **123** (R = n-Bu or Ph) show different reactivities. With R = n-Bu, **123** is recovered unchanged except for deuterium exchange, whereas **123** (R = Ph) rearranges (*92*) smoothly to **124** (R = Ph). Again, the formation of a benzylic anion upon cleavage of the requisite cyclic β-enolate must account for the tendency of the latter to undergo rearrangement.

Examination of the acidic products from **119** (R = Ph), which can constitute more than 50% of the product, depending on the reaction time, revealed the presence of three major components: **125–127** (*92*). From **123** (R = Ph), acid **128** is present in the mixture as well. Clearly, **125, 126,** and

Ph\C(CH₃)₂COOH Ph\CH(CH₃)COOH o-(iPr)C₆H₄COOH
125 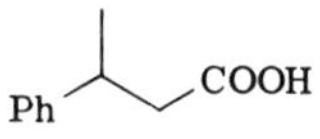   **126** 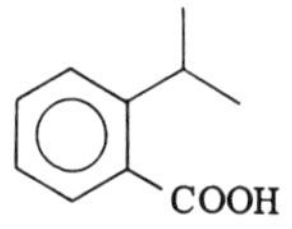   **127**

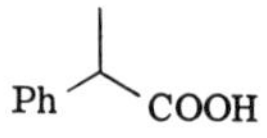

Ph\CH(CH₃)COOH   (benzocyclobutenolate with R, Ph)   o-(iPr)C₆H₄C(O)CR(Ph)
128 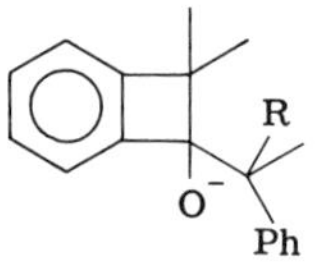   **129** 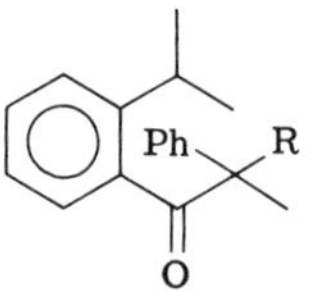   **130**

128 arise from Haller–Bauer type of cleavage of the initial ketone and the first rearrangement product, but **127** requires a different rearrangement to occur before cleavage to 2-isopropylbenzoic acid. Since reversible formation of phenyl anions proceeds throughout these reactions, it seems reasonable to propose the generation of a γ-enolate (**129**) produced by interaction of an *o*-phenyl anion with the carbonyl group. Ring opening could then lead to **130**, subsequent cleavage of which would produce **127**. Although there was evidence of **130** (R = H or Me) in the neutral fractions, these ketones could not be isolated in pure form. Both were synthesized subsequently and were found to cleave readily to **127** (*92a*); **130** could be expected to be more prone to Haller–Bauer cleavage than the others because of strain. The isolation and identification of **127**, however, establish an unprecedented 1,3-acyl shift from alkyl to aryl carbon and can be viewed as an example of the relatively rare γ-enolization process. The systems that exhibit γ-enolization are shown in Fig. 5.

4. Survey of homoketonization

In a number of instances already described, the fate of homoenolates generated from appropriate cyclopropanol and cyclobutanol derivatives has been investigated to obtain confirmatory evidence of their intermediacy in homoenolization processes. These homoketonizations have also attracted attention on their own merits, generally in a quest for stereochemical detail. Although both acid- and base-catalyzed processes have been examined (*65, 66, 93*), attention will be restricted to the latter reactions in this discussion.

Two aspects of homoenolate cleavage that require consideration are the regioselectivity of ring opening of unsymmetric homoenolates and, in all

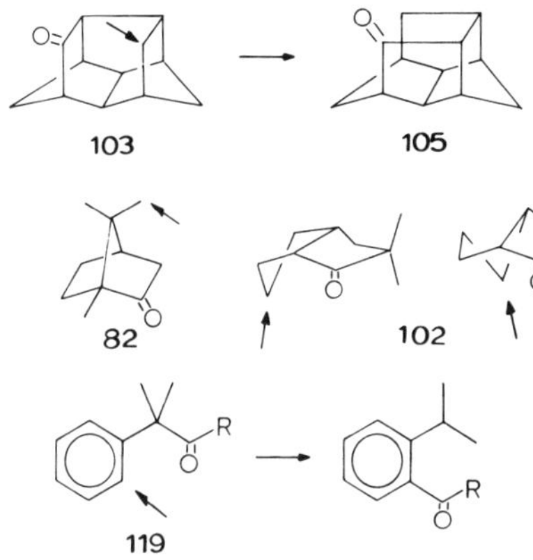

6. REARRANGEMENT IN CARBANIONS

Fig. 5 Systems and the sites exhibiting γ-enolization.

cases, the stereoselectivity of protonation (deuteriation) at the carbanionic site. In each case in the previous section for which the requisite experiments were carried out, it was found that β-enolates tend to open with predominant inversion of configuration. The examples that shed light on the regioselectivity of cleavage are listed in Table 6. In each case the partitioning between the two possible modes of ring opening is indicated as the ratio for a:b bond fission, and the types of carbanionic centers generated in the open-chain forms are given. Processes that are apparently unidirectional have been assigned a:b ratios of >100:1 as a lower limit since ≳1% of a second product or of exchange could be expected to be observable. The first four ratios show that there is a general tendency favoring generation of a primary over a tertiary carbanion and a secondary over a tertiary anion. This is in agreement with findings for base-catalyzed opening of cyclopropoxides, which favors formation of the less highly substituted carbanion (*94*). The β-enolate from **94** illustrates a preference for generation of an allylic carbanion, whereas that from **100** presumably reflects the effect of opening to form the more stable of the two possible products. The final entry shows that the preference for a primary over a tertiary carbanion can be reversed if the latter is stabilized, in this case, as a benzylic center. From these results, it is apparent that the regioselectivity of β-enolate cleavage is governed by the balance between carbanion and product stability; thus, the preference between two modes of cleavage in a given system may be difficult to predict. With these generalizations from the homoenolization experiments in mind, we now survey the results of studies of homoketonization.

The initial example, included in the pioneering study of homoenolization in camphenilone (**63**), was **76**, which upon treatment with *t*-BuO⁻/*t*-

TABLE 6
Regioselectivity of β-Enolate Cleavage in Homoenolization

Origin	β-Enolate			Regioselectivity a : b	Anion types
63, 73 92, 95, 96 91, 94	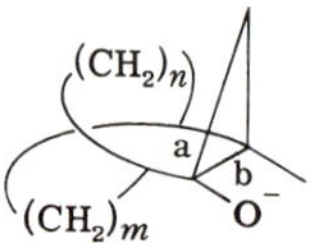 $(CH_2)_n$ ... $(CH_2)_m$ with a, b, O⁻	n: 1, 1, 2	m: 2, 3, 2	>100 : 1	1° : 3°
119	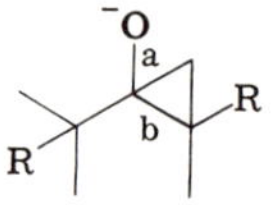 R-C(O⁻)(a)-C(b)-R			3 : 1	1° : 3°
80–82	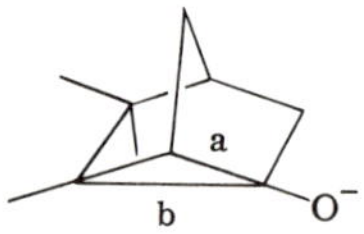 norbornyl skeleton, a, b, O⁻			~50 : 1	2° : 3°
73	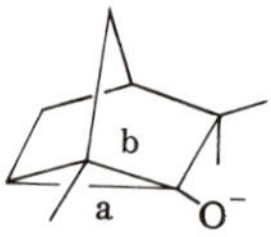 norbornyl skeleton, b, a, O⁻			20 : 1	2° : 3°
94	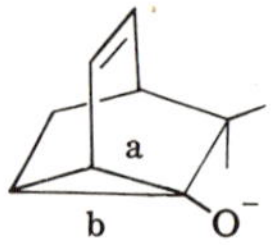 bicyclic skeleton, a, b, O⁻			10 : 1	2° : 2°
100	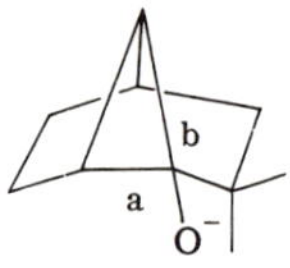 bicyclic skeleton, b, a, O⁻			>100 : 1	2° : 2°
110	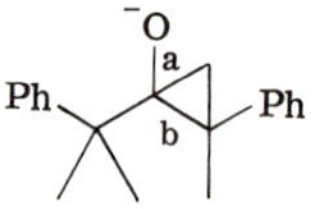 Ph-C(O⁻)(a)-C(b)-Ph			1 : >100	1° : 3° benzylic

BuOD or MeO⁻/MeOD at room temperature gave norcamphor-*exo*-6-*d* as the major product (>94.5% stereoselectivity), i.e., primarily with inversion of configuration (65, 66). Similar results (95) were found for 2-acetoxytriaxane (**131**), albeit with somewhat lower stereoselectivity (75 ± 3 and 83 ± 2%, respectively, in the two bases); the major product was **132** with an equatorial deuterium at C-4. In the same study, it was found (95) that treatment of **133** with MeO⁻/MeOD gave exclusively brendan-2-one-

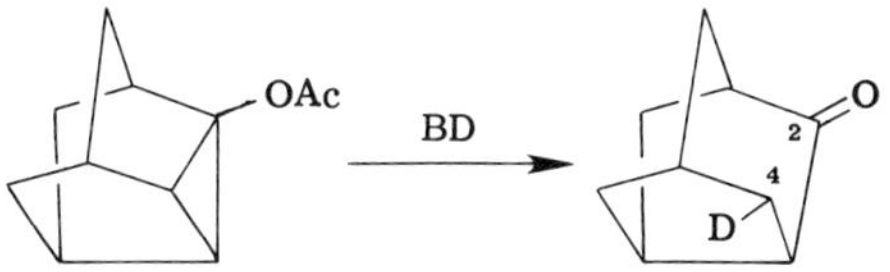

6. REARRANGEMENT IN CARBANIONS

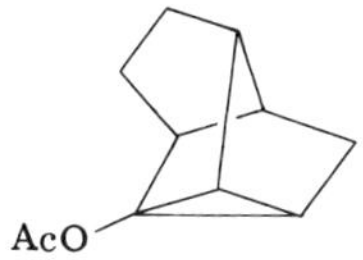

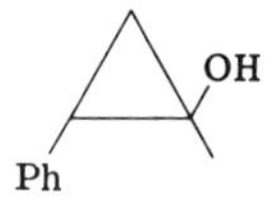

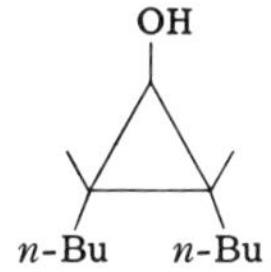

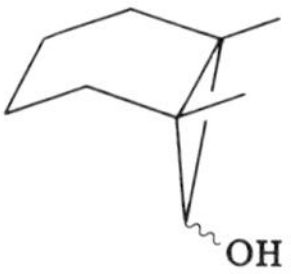

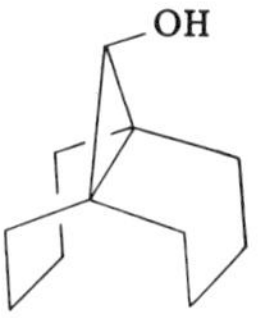

131 (structure with OAc) →(BD) **132** (structure with 2, 4, D, =O) **133** (structure with AcO)

134 Ph, OH (cyclopropane) **135** OH, n-Bu, n-Bu **136** (structure with OH) **137** (structure with OH)

exo-9-d (97 ± 2% stereoselectivity). The 100% regioselectivity, opening entirely by cleavage of bond *a* in **133**, leading to **70** with no detectable brexan-2-one (**69**), is particularly interesting, since a 60% yield of **70** was obtained from **69** after 150 hr (*8*). If the intermediate for both transformations is the same species, a complete **69** → **70** conversion should be possible. The exclusive formation of **70** from **133** has been attributed to the greater stability of the brendane skeleton relative to that of the brexane system, as indicated by molecular mechanics calculations. This implies a thermodynamic influence on the course of homoenolizations, as noted above for **100**, and has been suggested to account for the rearrangements of **119** (R = Me, *n*-Bu) and the lack of such for **123** (R = *n*-Bu) (*90*). The observation of preferred inversion in these homoketonizations is in agreement with the findings for **134** (*96*) but is in direct contrast with those for **135–137** (*97*), for which retention of configuration attends base-catalyzed ring opening. However, homoketonization proceeds under much milder conditions, and the stereochemical results need not necessarily reflect the behavior of similar species under more vigorous conditions. It can be noted that the stereoselectivity for β exchange in adamantanone is comparable to that found for **76** and **133** and distinctly higher than that for **131**. The somewhat lower stereoselectivity for a number of the other homoenolates cited earlier may merely reflect different temperature effects on the modes of homoenolate cleavage. Differences may also be caused, in part by differing degrees of strain in the variety of polycyclic species that have been found to undergo homoenolate formation. Since at room temperature a 90:10 selectivity between two modes of reaction represents a barrier difference of only 1.3 kcal/mol, it should be emphasized that the observed differences in the stereochemical course of homoketonization can depend on rather subtle features.

An example of homoketonization in a cyclopentanol derivative was shown to proceed with >98% stereoselectivity favoring retention. It was found that **138** is cleaved to **139** quantitatively by treatment with *t*-BuO⁻/*t*-BuOD at 75°C or $DO(CH_2)_2O^-(CH_2OD)_2$ at 175°C for 5 hr (*98*).

138 139

140 141

142

143

144

145

146

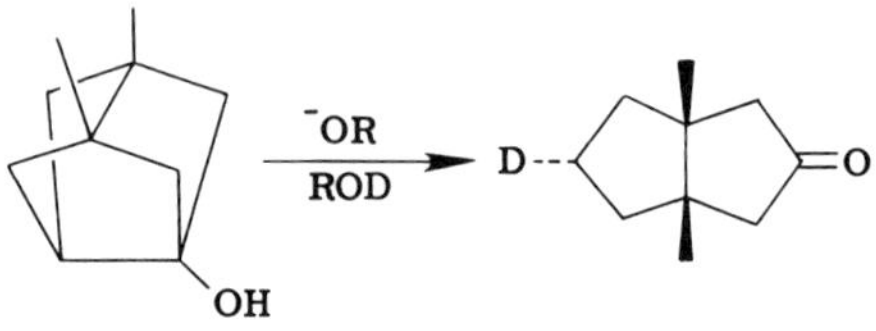

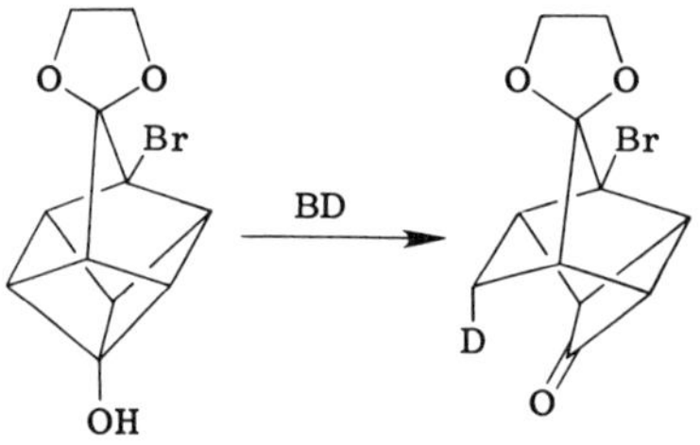

A number of more highly strained systems have been investigated. The homocubane alcohol **140** and its acetate were readily converted to the half-cage ketone **141** with MeO⁻/MeOH at room temperature. There was no evidence of the presence of a second ketone. With MeOD as solvent, a single deuterium atom was incorporated at the endo-3 position, as shown in **141**, with 96% stereoselectivity (*99*). To test for a possible influence of the ethylene ketal group on the stereochemical course of this cleavage the simpler homocubane acetate **142** was treated under similar conditions (*100*). Again, a quantitative conversion producing **143** with >96% retention of configuration at C-7 was observed. The cubane alcohol **144** and its

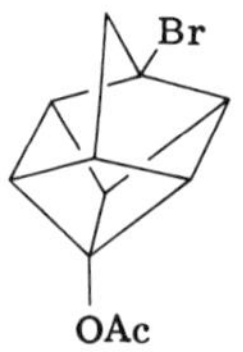

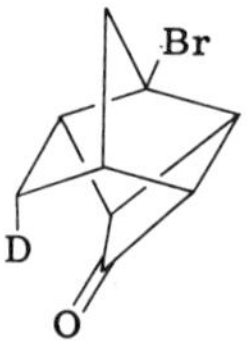

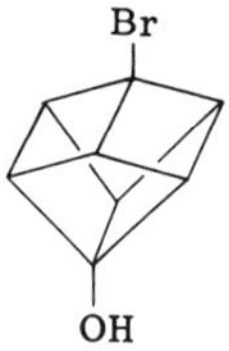

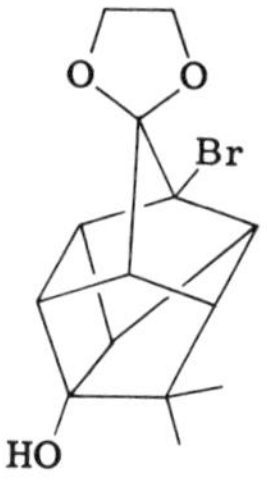

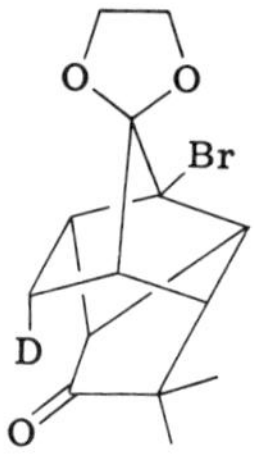

6. REARRANGEMENT IN CARBANIONS

acetate were more sensitive toward base and reacted instantly to produce a mixture of degradation products; it can be presumed that the initial step was homoketonization to the corresponding half-cage ketone, which readily fragmented, but no products were isolated or identified. This study also included the bishomocubane alcohol **145** and its formate ester as well as the corresponding alcohol lacking the ethylene ketal function. For both skeletons, the cleavage proceeded smoothly as **145 → 146** with a highly stereoselective acquisition of deuterium at the endo-2 position as indicated. Again, the reaction proceeded with 100% regioselectivity.

The homoketonization of the 4-acetoxyhomocuneane **147** was found to produce a mixture of isomeric ketones (**148** and **149**) in almost quantitative

yield (*101*). In striking contrast to the stereochemical results for the polycyclic cyclopropanol derivatives cited earlier, the cleavage of **147** proceeds with >96% retention of configuration. Clearly, inversion is not a general feature for base-catalyzed ring opening of cyclopropanols constrained in polycyclic structures.

Another homocubane system was investigated (*102*), and it was found that diol **150** undergoes bishomoketonization to yield brendandione **151** upon treatment with MeO⁻/MeOH(D) at 25°C. The same transformation was more conveniently accomplished by treatment of the bistrimethylsiloxy derivative of **150** at 0°C. In deuteriated media, two deuterium atoms were incorporated as indicated, one each into an exo and endo location. It was proposed that the process followed through the series **152–154** to account for the stereochemical result. The cleavage of **152** is analogous to the results for other homocubane derivatives (*99, 100*), and the cleavage of **154 → 151** has analogy with data for triaxane **131** (*95*). Further examination of the reactivity of the bistrimethylsiloxy derivative of **150** revealed that the corresponding derivative of **153** was formed quantitatively upon treatment of the former with MeLi at −15°C followed by quenching with cold saturated NH₄Cl solution (*103*). Attempts to purify this product showed it to be thermally unstable but smoothly converted to the trimethylsiloxy derivative of **154**, either by heating the neat material at 75°C for 40 min, or by refluxing a solution of **153** (OSiMe₃) in CCl₄ for 1.5

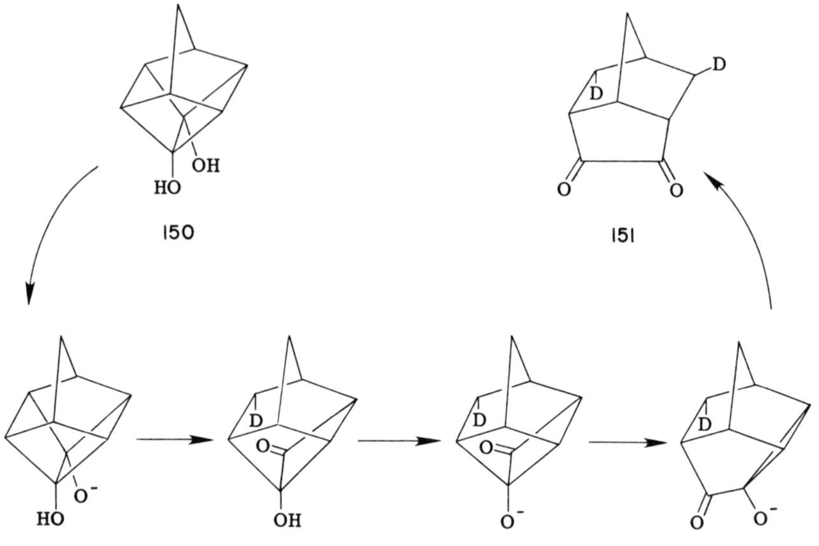

hr. The homoketonization of this product with MeO⁻/MeOH was readily accomplished. Repetition of this sequence **150 → 154** using the bistrimethylsiloxy derivative as starting material and quenching the MeLi reaction product with ND_4Cl–D_2O solution afforded the trimethylsiloxy derivative of **154** having almost exclusively endo deuterium incorporation. Treatment of a nondeuteriated sample of **154** (OSiMe₃) with MeO⁻/MeOD afforded **151**-*exo*-2-*d* almost exclusively. Thus, the suggested sequence **152 → 154** neatly accounts for the bishomoketonization of **150 → 151**, and the stereochemical results are consistent with a number of related systems.

The results for these homocubane derivatives suggested that bishomoketonization of the homocuneane derivatives **155** would lead to **151** with two exo deuterium atoms. However, treatment of the diol **155** (R = H) or its trimethylsiloxyl derivative **155** (R = TMS) with MeO⁻/MeOD gave **151**, which was *identical* with that produced from **150**, having one exo and one endo deuterium label (*104*). Apparently, the first cleavage of

155 proceeds with retention, as found for 147, and the second with inversion. Clearly, the ethylene ketal substituent in 147 does not influence the course of the ring-opening reaction. The stereochemical course of homoketonizations of cyclopropanol derivatives must be determined by a delicate balance of factors dependent on the structure and geometry of the reactant. On the basis of the available evidence it may be difficult to predict the favored stereochemical pathway in new systems. The regioselectivity of these processes, however, is very high, leading in each case to the more stable product.

B. Favorskii Rearrangement

Another type of carbanionic rearrangement in ketones, the Favorskii rearrangement, does not involve migration of the carbonyl group, but rather the formal 1,2 shift of an alkyl group from the carbonyl carbon to its adjacent α-carbon with displacement of a leaving group (156 → 161). In addition to an older general review (*105*) of this reaction, there are now several more recent ones on particular aspects (*106–110a*), and therefore this account will emphasize work reported since 1973.

The generally accepted mechanism in hydroxylic solvents (Scheme 9), for which there is now much evidence, involves halide ionization from the α'-enolate ion 157, disrotatory ring closure of the resulting oxyallyl (158) to a cyclopropanone (159), and subsequent nucleophilic cleavage of 159 to

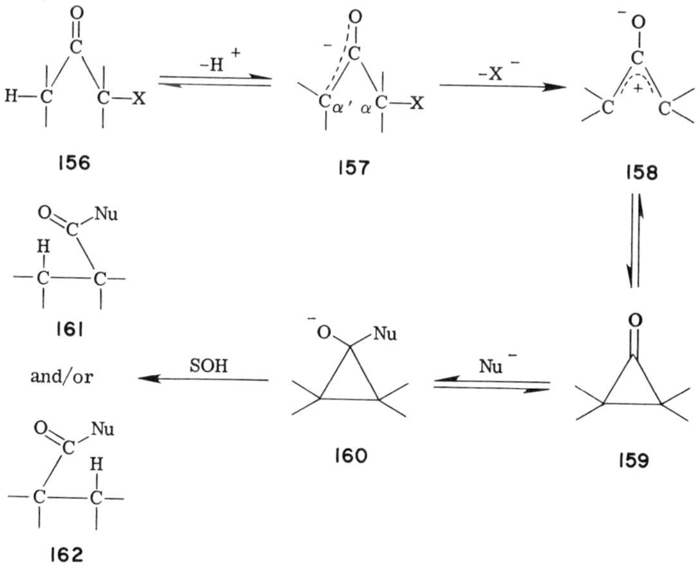

Scheme 9

an acid derivative (**161** and/or **162**). Recent work on the reaction has focused on details of this moderately well-established mechanism: Is an oxyallyl always the precursor of the cyclopropanone? Is the acid product always formed by cleavage of a cyclopropanone? Are allene oxides ever involved in the reaction sequence? Is there a preferred stereochemistry for the leaving group in **157**? Which steps in the mechanism are reversible under the reaction conditions? What factors control the direction of cyclopropanone cleavage? Can the cyclopropanone open to give α-substituted ketone side products under the reaction conditions? How can the striking solvent effects be explained? In addition to these questions of mechanism, other new facets of this complex reaction continue to be discovered, such as the involvement of one-electron transfer as an additional complication, the substitution of N for O in Scheme 9, and the homologous and vinylogous Favorskii rearrangements. These various points and others are the basis for this selective review. First, each step in the mechanistic sequence will be considered in the light of the most pertinent and recent evidence.

There is ample evidence that the first step in Scheme 9 may be reversible or not, depending on the substrate structure. For compounds **163–169**

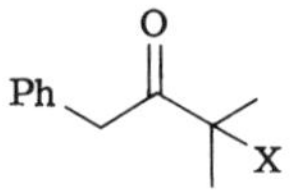

[Structures 163–169 shown]

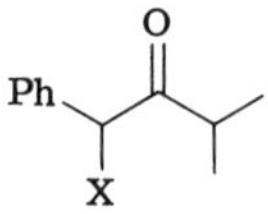

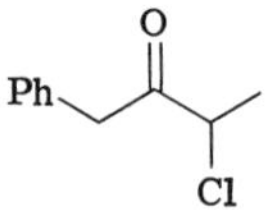

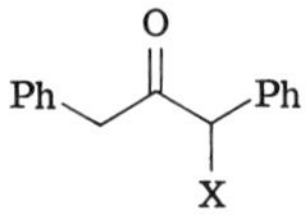

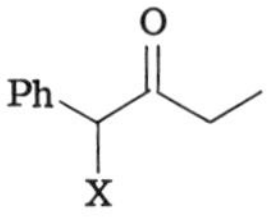

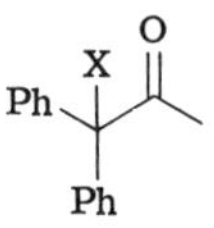

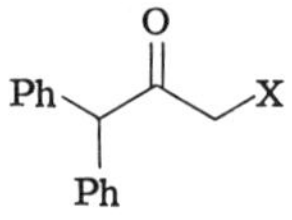

X = Cl, Br

(*111–113*) enolate formation leads irreversibly to Favorskii product, the evidence being absence of deuterium exchange before rearrangement, a low $k_{Br/Cl}$ rate ratio* near unity, and a ρ value of +1.4 for **165** (*111*). In these compounds the α and α′ substituents stabilize the transition state for halide loss so effectively that enolization becomes rate limiting. For other, less substituted α-halo ketones such as **170–176** (*113–115*) carbanion formation is reversible (i.e., not rate limiting for loss of halide) in view of deuterium exchange before rearrangement, $k_{Br/Cl}$ rate ratios of 36–116,

---

* The ratio of the rate constants for the reactions of **156** (X = Br) and **156** (X = Cl).

6. REARRANGEMENT IN CARBANIONS

and large negative ρ values for phenyl-substituted cases (e.g., -5.0 for 175) (*115*).

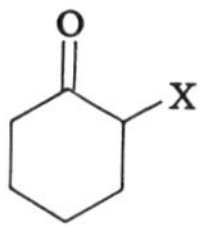

170, 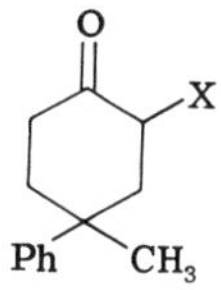 **171**, 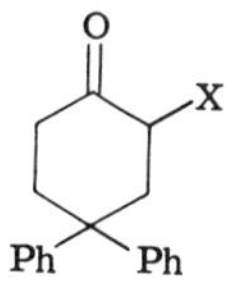 **172**, 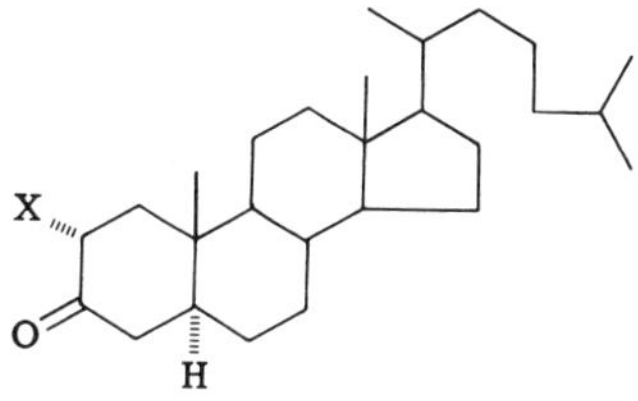 **173**, 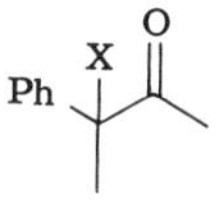 **174**, 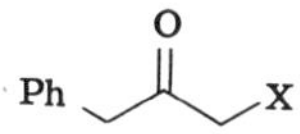 **175**, 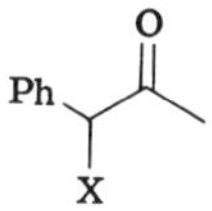 **176**

X = Cl, Br

The second step of the rearrangement is clearly *ionization* of the leaving group from the enolate **157**. Large $k_{Br/Cl}$ rate ratios and the negative ρ for **175** and **176** support this interpretation. Another piece of evidence is the strong rate acceleration of halide loss caused by increasing the solvent ionizing power or by alkyl substitution on the carbon bearing the leaving group (*111, 115*). There is no evidence that the loss of halide from **157** is reversible. The nature of the intermediate produced by halide loss from **157** is the least satisfactorily established feature of the mechanism.

At this point exact mechanistic details of the carbanionic process are intimately linked with the

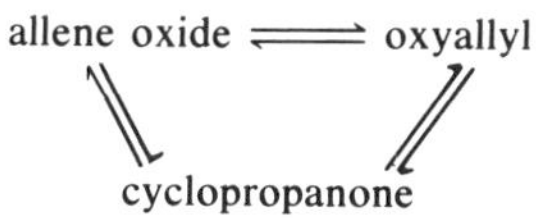

allene oxide ⇌ oxyallyl
  ↘   ↙
  cyclopropanone

interconversion question, and a short digression is in order. The results of both bond additivity (*116*) and *ab initio* calculations (*117*) agree that cyclopropanone is more stable than allene oxide by about 23 kcal/mol. Depending on whether it is a biradical (singlet or triplet) or not, oxyallyl has been calculated to be 45–232 kcal/mol higher in energy than cyclopropanone (*116–120*). Allene oxides such as **177** (*121*), **178** (*122*), and **179** (*123*) have been thermally isomerized to the corresponding cyclopropanone, but calculations (*116, 124*) indicate that oxyallyl may not be on the path for this transformation. Instead, an adjustment of C—O and

C-1–C-3 bond distances with rotation of C-3 may occur (*124, 125*). In any event, whereas both oxyallyl and allene oxide would isomerize to cyclopropanone, the reverse of either of these reactions is unlikely. However, since alkyl substitution decreases the energy difference between an oxyallyl and the corresponding cyclopropanone (*116*), the **159** → **158** process becomes less unlikely with increasing substitution.

177   **178**   **179**   **180**

The currently favored structure for the "thing" that is formed directly by halide loss from enolate **157** is oxyallyl **158** rather than cyclopropanone **159** or allene oxide **180**. The main evidence against direct, synchronous formation of a cyclopropanone in an S_N2-like process is (a) that the implied stereospecific inversion at the carbon bearing the leaving group is frequently not observed, and (b) that the leaving group is "pulled off" by ionization instead of being pushed off by an internal nucleophile (C—Lv bond breaking leads); both of these facts are accommodated by formation of oxyallyl. Moreover, direct S_N2 reaction to a cyclopropanone from **157** would involve serious distortion of its π system in violation of Baldwin's rules for ring closure (*126*). The formation of the same indanone (**184**) by cyclization during Favorskii rearrangement of **181** and **182** (X = Br and Cl) in NaOMe/MeOH was taken as strong evidence for the intermediacy of oxyallyl **183** because the ratio of Favorskii ester **185** to indanone **184** from all three starting materials was a constant 1.66 ± 0.02 under the same reaction conditions (*112*). The enols **186** and **187** were excluded as indanone precursors because **184** was not formed under less basic buffered conditions which would have favored **186** at the expense of enolate. Also, from data on the solvolysis of **186** and **187** the ratio **185**:**184** would not have been the same from the isomeric ketones **181** and **182** (X = Cl) if the indanone arose from **186** and **187**. The argument is rendered less satisfying by the fact that indanone **184** is also formed in up to 48% yield during acetolysis of **182** (X = Cl) (*127*). In this case the cyclization apparently proceeds through the hydroxyallylic cation **188** because indanone formation is suppressed by the presence of acetate ion.

The evidence for the reality of oxyallyl is stronger than the foregoing would suggest. Entities with the properties of oxyallyl have been prepared in various ways. Electrolysis of dibromo ketone **189** in a two-electron reduction gave an intermediate that reacted with HOAc and EtOH to

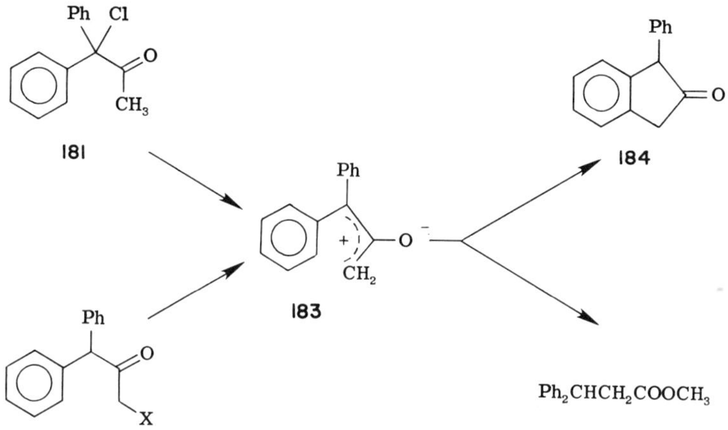

6. REARRANGEMENT IN CARBANIONS

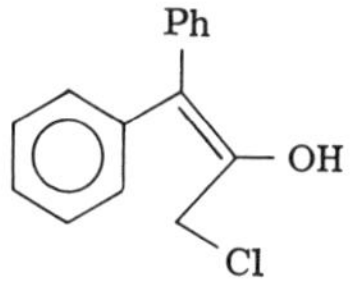

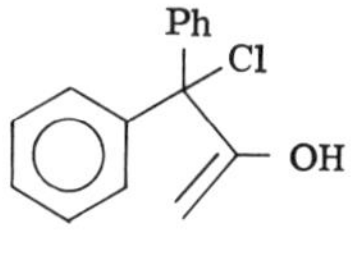

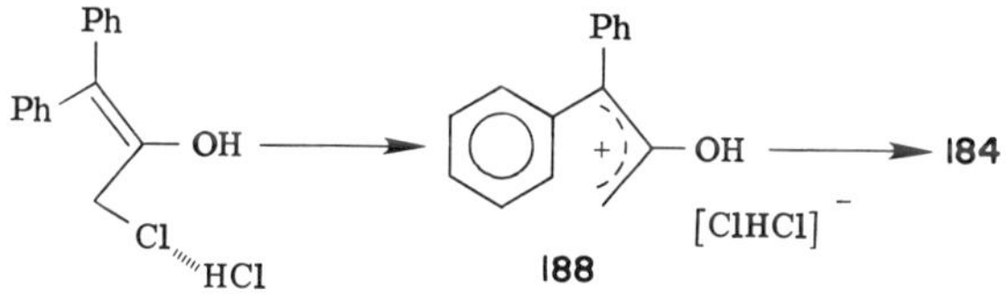

produce α-substituted ketones and with furan to yield adduct **190** (*128*). However, no cyclopropanone hemiketal or Favorskii product was isolated. In this connection it is interesting that Zn—Cu couple reduction of **191** in MeOH did give 2% of Favorskii esters **192** (*113*). Even more significant, the Zn—Cu couple reduction of **193** in MeOH gave up to 80% of the Favorskii ester **185** (*129*). The fact that no indanone **184** was produced in this reaction does not necessarily exclude oxyallyl, because a Zn salt-complexed oxyallyl might differ in reactivity from an uncomplexed oxyallyl (*130*). The enol sulfite ester **194** on reflux in MeOH afforded both epimeric α-methoxy ketones **195**, but on reflux with NaOMe in ether the product was a mixture of Favorskii esters, mostly **196** (*131*). These reactions are consistent with the intermediacy of oxyallyl

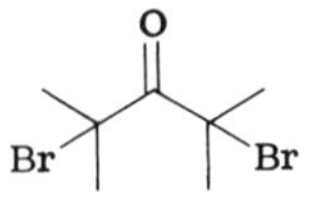

197. An intermediate reasonably assigned the conjugated oxyallyl structure **199** has been produced both by the irradiation of the bicyclo[3.1.0]-hexenone **198** (see Essay 18) and by t-BuOK/t-BuOH treatment of bromo ketone **200** (*132*). Aromatization by phenyl migration is apparently faster than cyclopropanone formation, although it was reported that a cyclopropanone was detected on irradiation of **198** at low temperature (*133*). *trans*-Di-*t*-butylcyclopropanone (**201**) undergoes thermal racemization at 79.6°C without enolization. Since the racemization is faster than the isomerization of the one known isomer of 1,3-di-*t*-butylallene oxide (**177**), and since it shows a small increase with increasing solvent polarity, the most attractive process is disrotatory ring opening to the oxyallyl **202** and disrotatory reclosure (*134*). The extreme reactivity of dibenzo[*c*,*e*]-tropone (**204**) formed *in situ* by reaction of the bromo ketone **203** with triethylamine suggests a high oxyallyl-like contribution (**205**) to the resonance hybrid (*135*). Bromo ketone **203** also undergoes Favorskii rearrangement to **206** in NaOMe/MeOH. 1,1-Di-*t*-butylallene on reaction with ozone at −78°C in CH_2Cl_2 yields as one product 1,1-di-*t*-butylcyclopropanone (**209**), most plausibly via the oxyallyl **208** and not by way of the allene oxide **178**, which, being moderately stable, would have been detected (*136*). *In toto*, the case for the existence of oxyallyls is quite persuasive, and attempts are in progress to prepare persistent examples (*136a, 136b*).

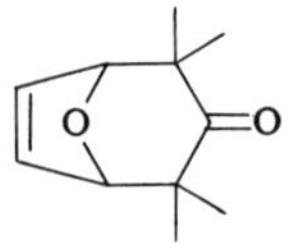

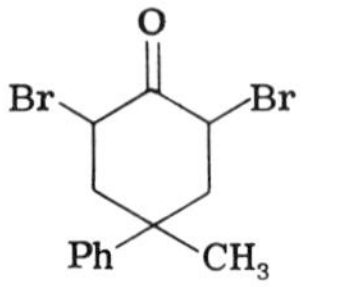

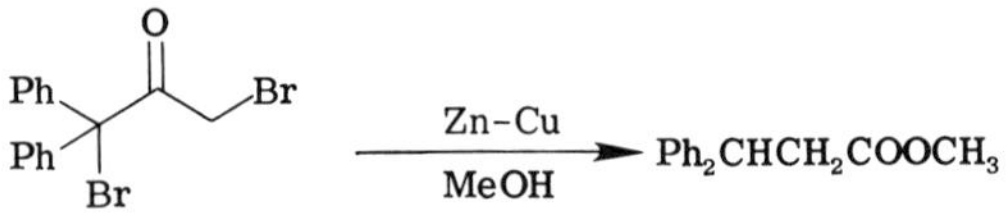

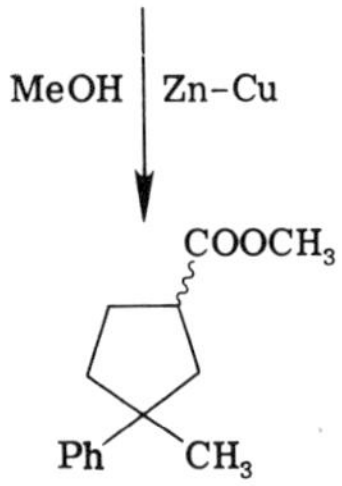

The ionization of the leaving group from the enolate to give oxyallyl should exhibit the stereochemical preference shown in **210** or **211** (*136c*). In a cyclohexenolate ion a quasi-axial halide should be lost preferentially.

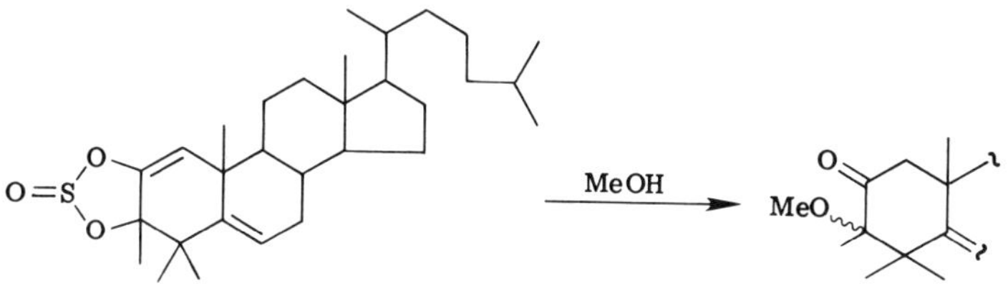

6. REARRANGEMENT IN CARBANIONS

194 → (MeOH) → **195**

↓ NaOMe / ether

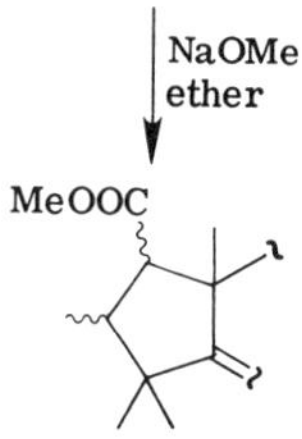

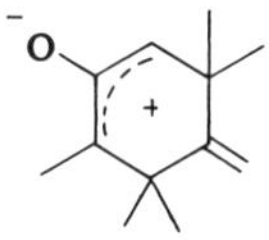

196   **197**

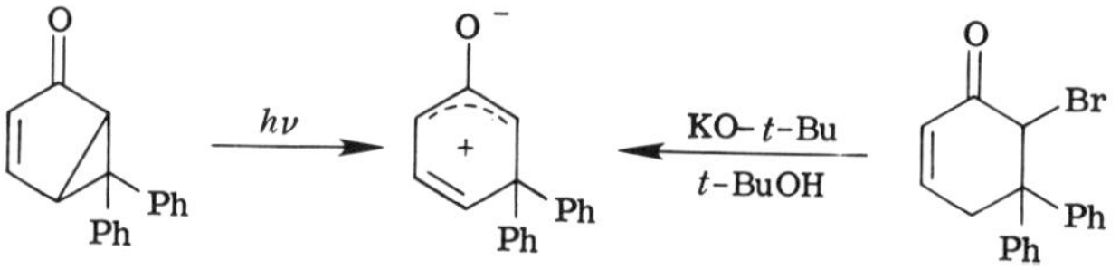

198 → (hν) → **199** ← (KO-t-Bu, t-BuOH) ← **200**

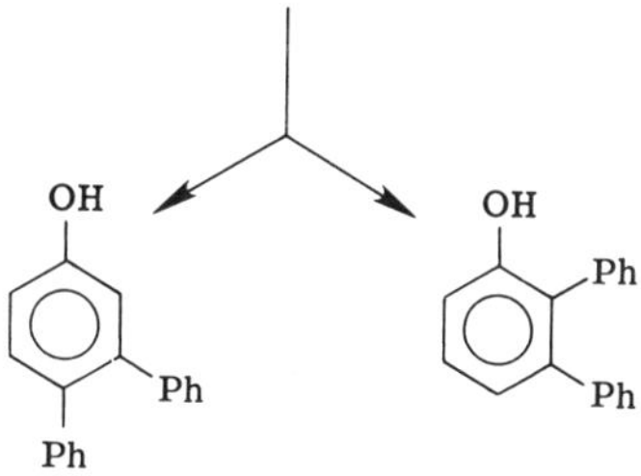

However, there is remarkably little evidence on the point. Experimental testing is complicated by the fact that an axial leaving group also favors the major side reaction of epoxy ether formation (**212**). Earlier studies with both epimers of **213** and **214** in NaOMe/MeOH were inconclusive for this reason (*137, 138*). In NaOMe/DME both epimers of **214** gave high yields of Favorskii esters at about the same rate (*138*). A more recent study with the two epimers of **215** did find that the axial Cl isomer gave

201 202

203 204 205

206

207 208 209

210

211

212

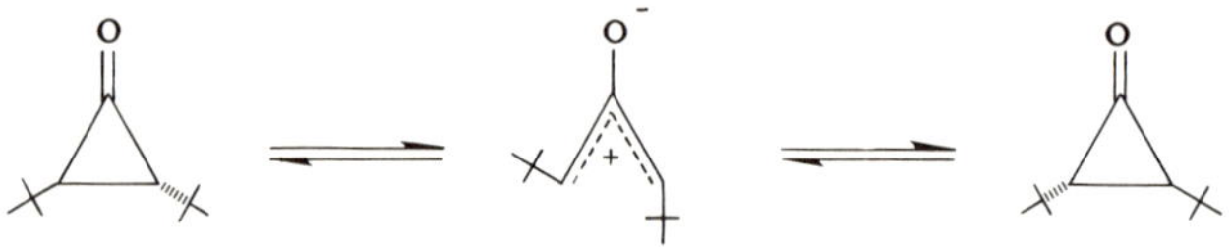

Favorskii ester, whereas the equatorial isomer did not, but the yield of Favorskii ester (2%) was so low that no valid conclusion is possible (*139*).

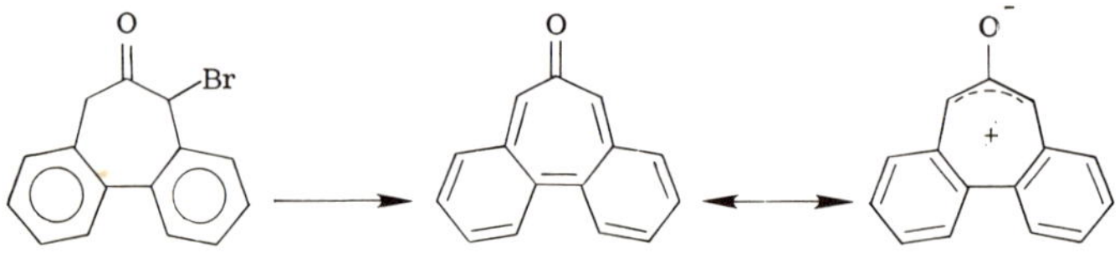

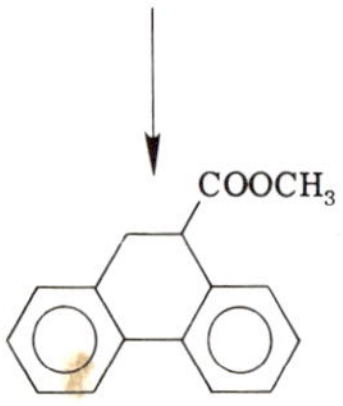

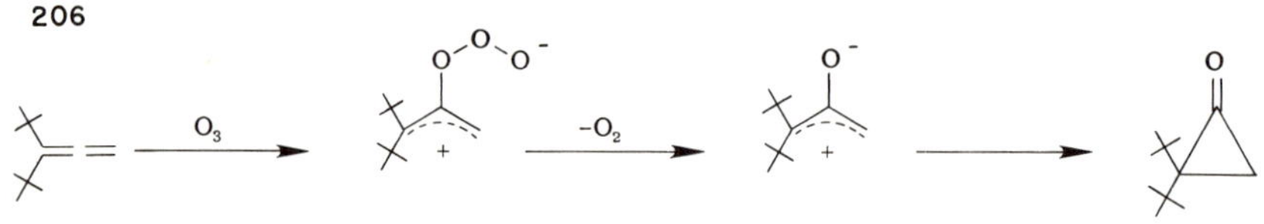

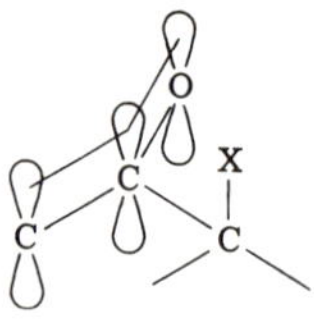

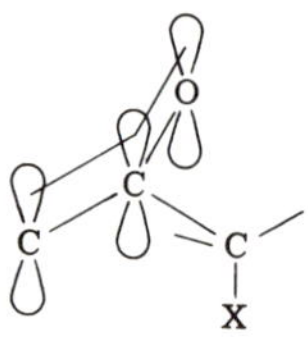

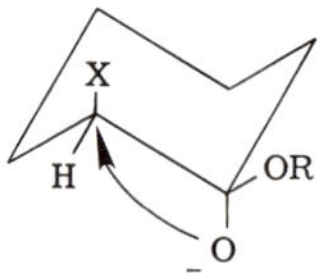

The third possibility for halide ion loss from **157**, direct formation of an allene oxide (**180**), has been less widely considered. The possibility that the Favorskii rearrangement might be initiated by S_N2 attack of an oxygen nonbonding sp^2 orbital on the α-carbon in **157** only appears stereoelectronically reasonable for noncyclic ketones, and it would not be in accord with the previously mentioned evidence that halide loss is an ionization.

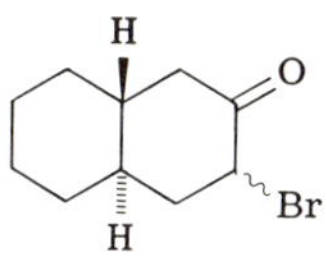

6. REARRANGEMENT IN CARBANIONS

213

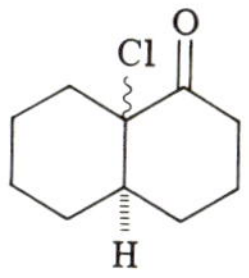

214

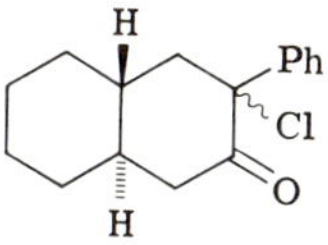

215

However, an allene oxide could conceivably be formed after ionization of halide from the enolate has proceeded beyond the transition state. It has already been noted that some allene epoxides (e.g., **177, 178,** and **179**) isomerize thermally to cyclopropanones. There is also experimental evi-

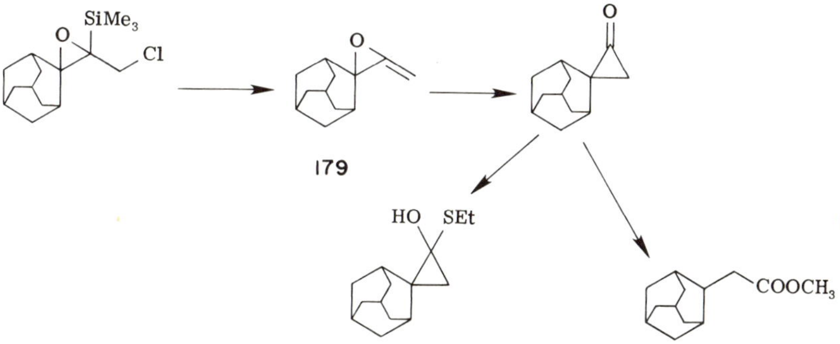

179

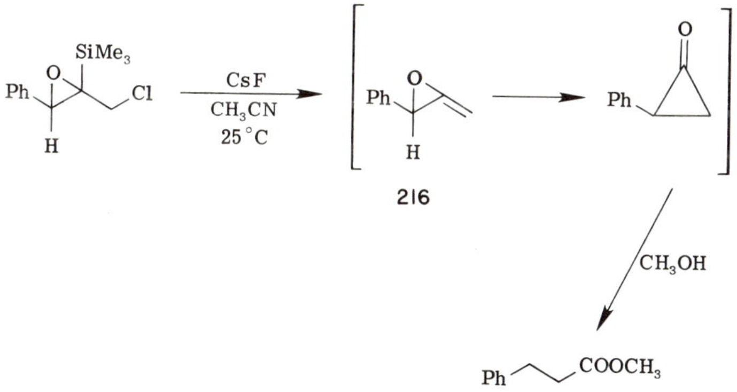

216

217

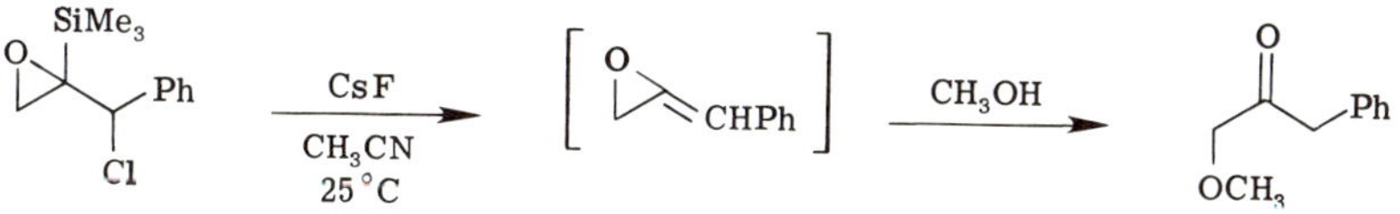

218

219

dence that some allene oxides such as **179** and **216,** derived from epoxysilanes, can yield Favorskii products *(123, 140),* presumably by first rearranging to the cyclopropanone, which is cleaved by a nucleophile. A recent intramolecular instance of this is the preparation of γ-lactones by the epoxidation of allenic alcohols (**220** → **221**) *(141).* However, the fact that, under the same conditions affording Favorskii ester **217** from **216,** the isomeric allene oxide **218** gives only substitution product **219** argues

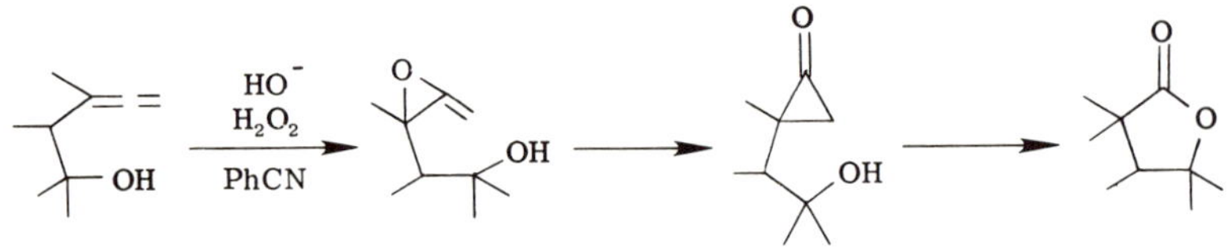

against rearrangement to a cyclopropanone in this case. Since both chloro ketones **175** and **176** do give **217** with base, allene epoxides appear unlikely as intermediates in these Favorskii reactions. Moreover, not all of the many allene oxides that have been prepared undergo ready isomerization to cyclopropanones *(142);* many, such as **222** and **223,** react directly with nucleophiles to yield α-substituted ketones *(123, 142).* Therefore, allene oxides are not likely intermediates in Favorskii reactions.

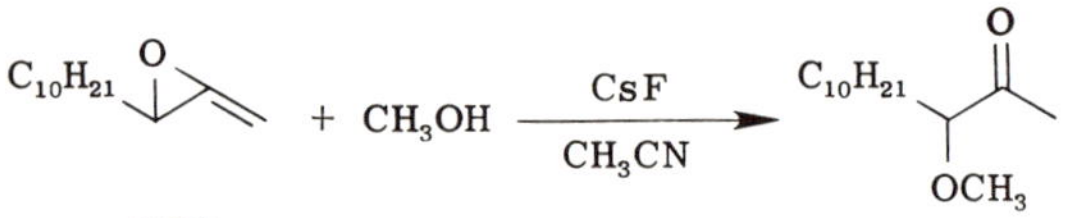

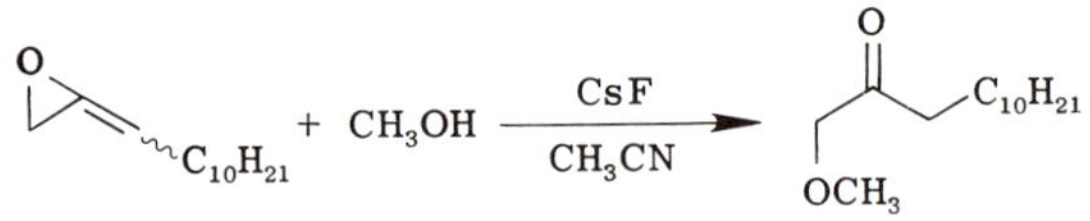

The evidence for the intermediacy of cyclopropanones is by now overwhelming. In addition to the finding that several cyclopropanones give the same Favorskii product distribution as the two corresponding α-halo ketone precursors *(143), trans*-di-*t*-butylcyclopropanone (**201**) has actually been isolated *(144)* from the reaction of α-bromoneopentyl ketone with *t*-BuOK/*t*-BuOH. However, if there is even a slight excess of base over bromo ketone, complete conversion to *t*-butyl Favorskii ester results.

Does cyclopropanone formation always proceed by way of an oxyallyl? There are certainly exceptions, such as the base-catalyzed conversion of propargyl diazotates to Favorskii products in high yields *(145).* The reaction of **224** can not proceed via a planar oxyallyl since the configuration of the chiral atom is 88% inverted in the product **225**. There are also several

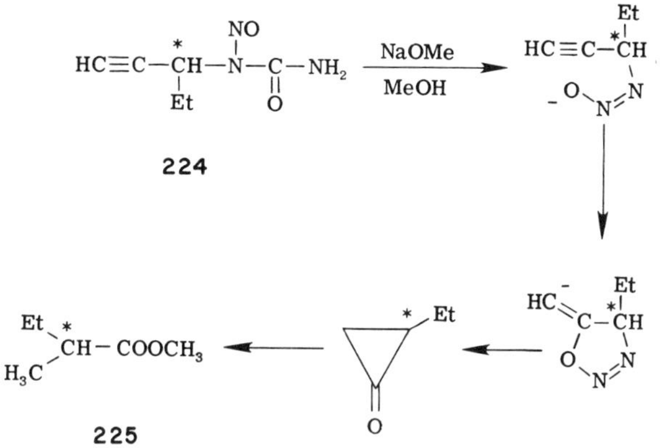

6. REARRANGEMENT IN CARBANIONS

224

225

Favorskii rearrangements (e.g., **226–233**) in nonpolar aprotic solvents (DME, ether) that proceed with exclusive or predominant inversion at the carbon bearing the leaving group (*138, 146–151*). As mentioned earlier, an S_N2 displacement of the usual aliphatic type by the enolate carbon is stereoelectronically and energetically unreasonable, but inversion can be

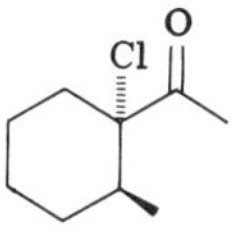

226 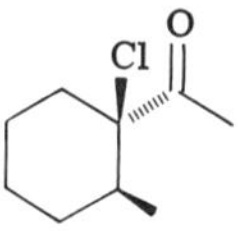   **227** 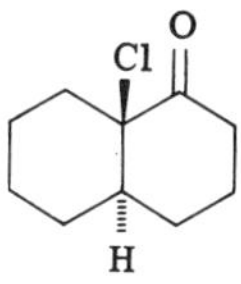   **228** 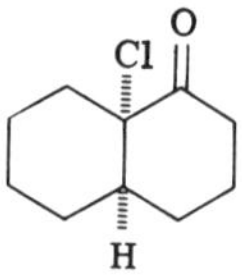   **229**

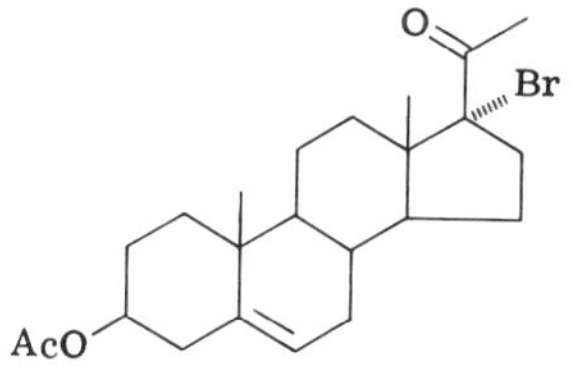

230 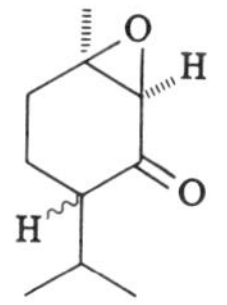   **231** 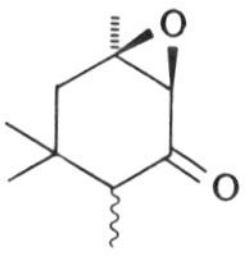   **232**

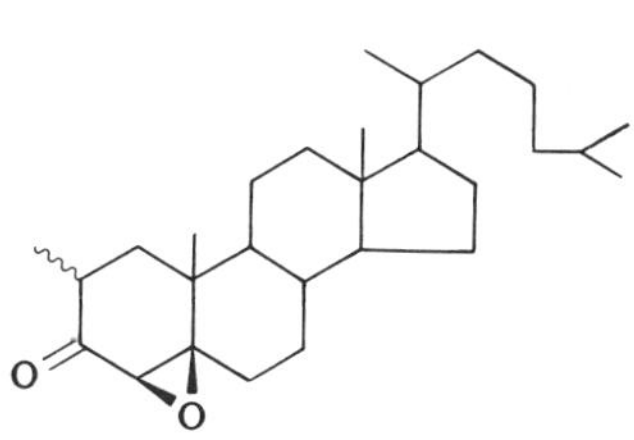

233 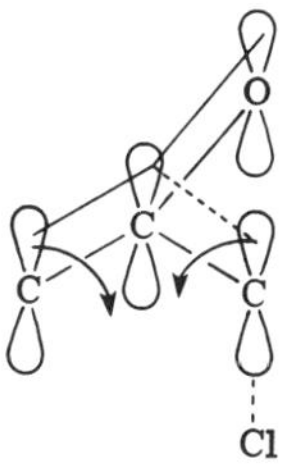   **234**

accommodated by halide ionization from the enolate anion if in nonpolar solvents the disrotatory closure of the developing oxyallyl commences before the bond to the leaving group is completely broken (bond breaking leads bond making). In this case, concerted disrotatory closure presumably would occur only in the sense of inversion depicted in **234**. Such a possibility also allows for **210** and **211** as the preferred leaving group geometries (axial preference in cyclohexanones). In more polar solvents ionization to a free oxyallyl could be complete, and disrotatory ring closure could occur in either sense to give both possible α-carbon configurations, or even mostly retention if steric factors so dictate, as with **230** (*148*). With poorer leaving groups, such as epoxide, disrotatory closure might start before oxyallyl generation is complete, even in polar solvents (MeOH) (*151*). The fact that each of the chloromethyl epimers **235** and **236** gives both Favorskii esters **237** and **238** with MeO⁻/DME (*152*) is not surprising since the C—Cl bond can assume either conformation **210** or **211** and thus give both cyclopropanones by concerted ionization–disrotatory closure.

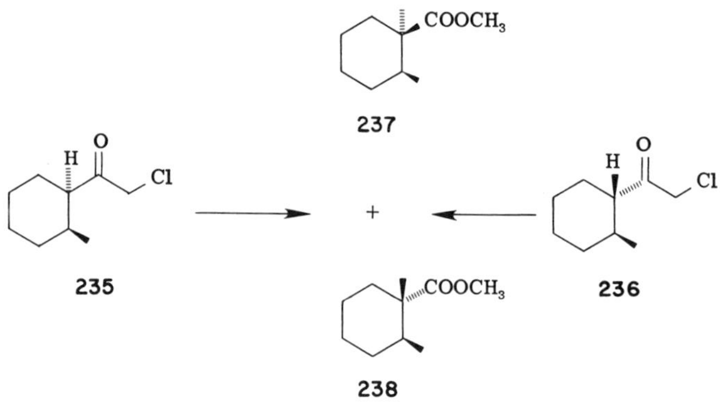

An interesting point on which there is as yet little information is whether a cyclopropanone can reopen to oxyallyl during Favorskii rearrangement. What circumstantial evidence is available suggests that reopening can occur under certain conditions. The racemization of (+)-*trans*-di-*t*-butylcyclopropanone at 79.6°C has already been mentioned (*134*). The variation in product stereochemistry (**240:241**) from the reaction of **239** in different NaOMe/MeOH concentrations is most simply explained by oxyallyl–cyclopropanone reversibility. At low methoxide concentration the ratio of **240** to **241** was 3:1, but in 2 *M* NaOMe this ratio was reversed to 1:3 (carbomethoxyl epimerization did not occur under the reaction conditions) (*153*). In a similar way, the variation in the ratio of Favorskii ester **185** to indanone **184** in the reaction of **181** was also interpreted in terms of reversible cyclopropanone formation. In 0.05 M NaOMe/MeOH the product was 98% Favorskii ester **185**, but in very dilute (~10^{-5} M) solution only 48% of **185** was formed along with 22% of indanone **184** (*112*). On the condition that the Favorskii product arise only from a cyclopropanone and that the indanone arise from oxyallyl **183**, then reversibility is required.

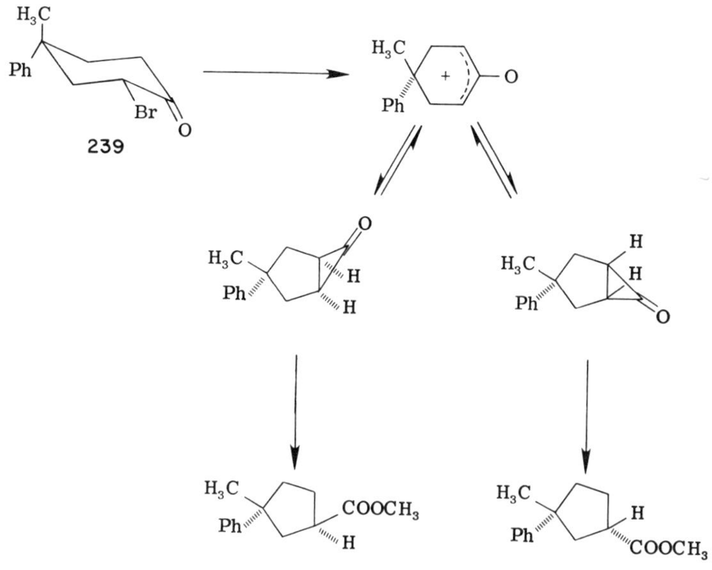

A hemiketal and/or its anion (**160**) is undoubtedly the next intermediate in the mechanistic sequence since such addition products form so readily from cyclopropanones and many nucleophiles (*144, 154, 155*). Moreover, 2,2-di-*t*-butylcyclopropanone (**209**), which does not give a hemiketal readily, requires refluxing for 2 days with NaOMe/MeOH for complete conversion to Favorskii esters (*122*). The rate of addition of alcohols to *trans*-di-*t*-butylcyclopropanone is in the expected steric order MeOH> EtOH>*i*-PrOH>*t*-BuOH (*144*). Although with less hindered cyclopropanones the equilibrium strongly favors the adduct, nevertheless the addition is readily reversible, as shown by facile exchange of addends in the reactions below.

Cyclopropanones and their hydrates and hemiketals exhibit quite different behavior depending on pH. In a neutral medium, carbonyl adducts of cyclopropanones are often stable for long periods. If the medium is capable of protonating the oxygen of a cyclopropanone or its adduct, then ring cleavage of the cation **242** formed on dissociation coupled with

6. REARRANGEMENT IN CARBANIONS

nucleophilic attack furnishes an α-substituted ketone, but, if the medium is basic enough to form the cyclopropoxide ion **243**, then ring cleavage to a Favorskii product is the exclusive reaction (*122, 144*). The only apparent exception is a report that 3% of α-methoxyketone **245** accompanied the 97% of Favorskii ester **244** on cleavage of tetramethylcyclopropanone with NaOMe/MeOH or NaOMe/DME (*156*). There is no evidence that Favorskii acid derivatives can be formed directly from oxyallyls.

The cleavage of **160** to **161** and/or **162** has been referred to in terms of "incipient carbanions," "transient carbanions" (*105, 109*), and just plain "carbanions" (*97, 155*), accompanied by the structural formula of the supposed alkyl carbanion. Actually, it is clear that simple alkyl carbanions are not usually intermediates. The energies of the transition states for cleavage reflect the energies of the potential carbanions, but carbon protonation accompanies hemiketal cleavage. It has been estimated that the $pK_a^{H_2O}$ of the β-CH bond in propionic acid is 57 ± 10 (*157*). A $pK_a^{H_2O}$ of 47 for these β-hydrogens would be just on the edge of permitting a primary carbanion in the cleavage of cyclopropanone (*157*). Therefore, although phenyl-substituted carbanions might be intermediates in cyclopropanone scission, less stable alkyl-substituted carbanions will not be intermediates; i.e., the protonation step is S_E2 and not S_E1 [see also (*24a*)]. The necessity that protonation be concerted with hemiketal cleavage may enter into the explanation of why cyclopropanone hemiketal cleavage occurs with retention. There are now a number of cyclopropanone cleavages, and Favorskii reactions proceeding through cyclopropanones, which have been unambiguously demonstrated to occur with retention of configuration at both secondary and tertiary carbon, e.g., **201, 246–251**. In fact, the authors are unaware of any examples of base-catalyzed cyclopropanone fission that do not proceed with retention, in contrast to the situation with cyclopropanol cleavage, in which both retention and inversion are found (page 433). In these hindered secondary and tertiary cases, with S_E2 protonation occurring as the carbanion begins to form, the backside of the incipient carbanion is shielded from close solvation. Consequently, protonation should take place from the solvent shell around the two oxygen atoms at the site of cleavage, leading to retention. Such an interpretation would appear to fit most cyclopropanol cleavages as well when the steric requirements of both substrate and base are taken into account.

The general rule that the direction of hemiketal cleavage can be predicted by considering the relative stabilities of the two potential carbanions breaks down if steric crowding becomes too severe. In this situation, the less sterically encumbered transition state (better solvated developing COOR?) leading to abnormal product becomes more favored, as can be seen from the examples of **209** and **251–254**. The abnormal cleavage of

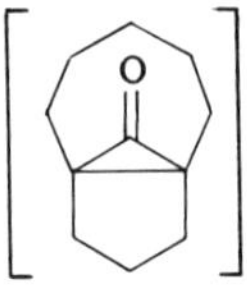

(158) 246

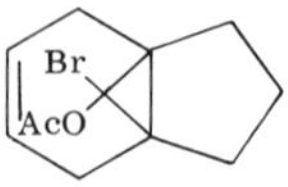

(159) 247

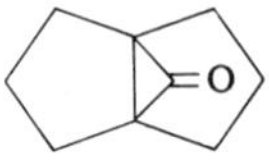

(160) 248

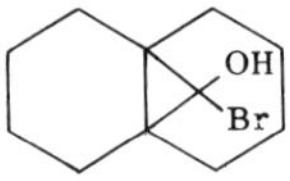

(161) 249

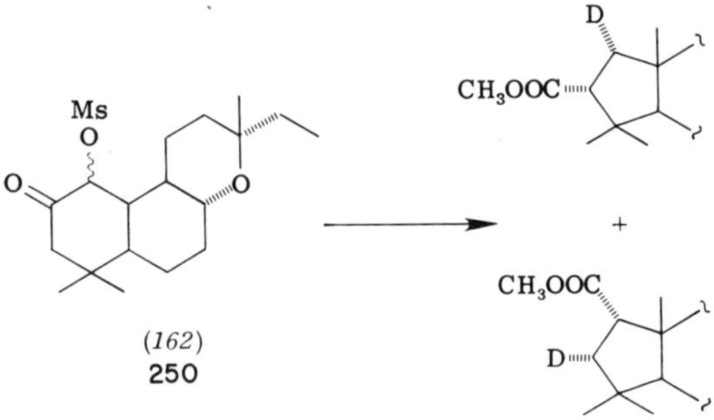

(162) 250

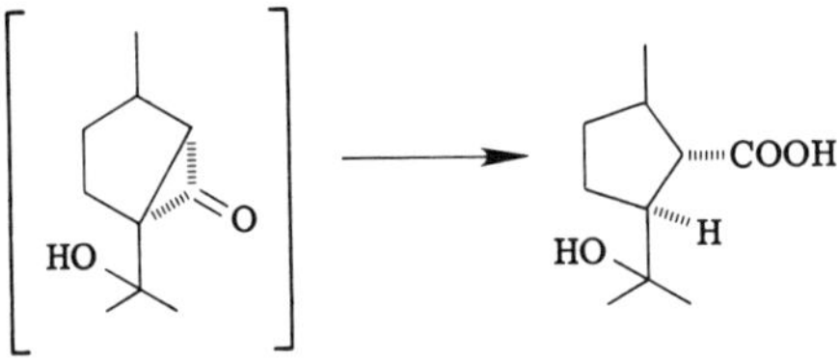

(163) 251

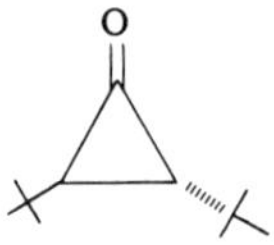

(97) 201

255 is probably due to the concomitant loss of bromide [compare to a similar case in (*164*)].

There are a number of side reactions that can accompany, or even preclude, the Favorskii rearrangement and which have made its study difficult. In the typical alcoholic alkoxide medium these reactions lead to α-alkoxy ketones, epoxy ethers, α-hydroxy ketones, α-hydroxy ketals, and alkenes, and the question arises as to which of the intermediates gives the various side products.

From product distribution studies at various base concentrations there is strong evidence that the major path for α-alkoxy ketone formation is the reaction of α'-enol allylic halide with solvent or base (*165*). However, this is not the only path because α-alkoxy ketones are also formed in nonpolar aprotic solvents and from compounds without α'-H. Direct S_N2 substitution must also occur in analogy with the many known cases of this reaction (*167, 168*). Although there is no direct evidence about the possi-

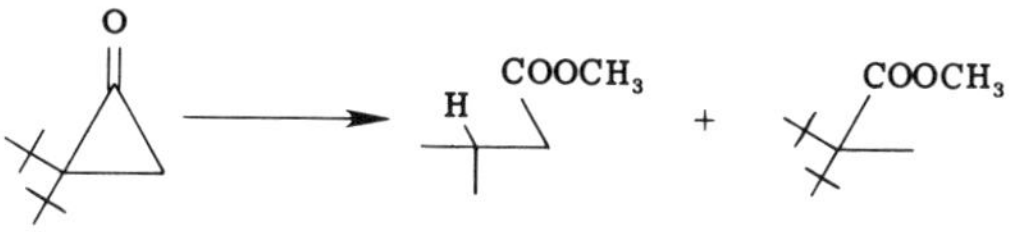

6. REARRANGEMENT IN CARBANIONS

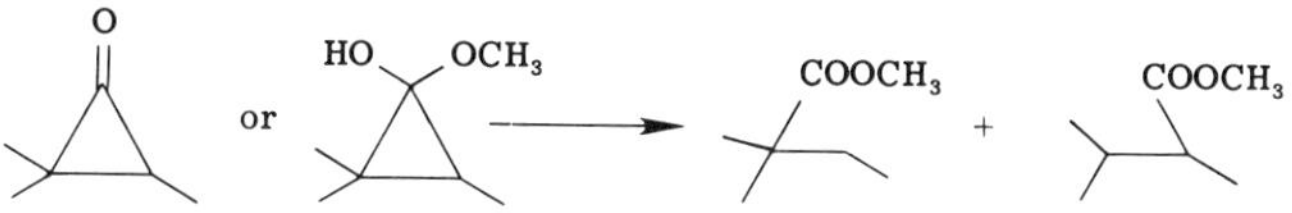

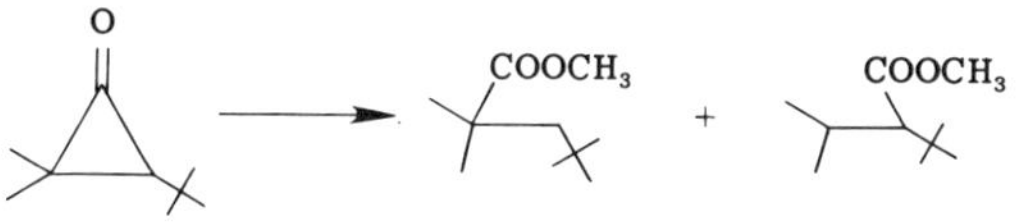

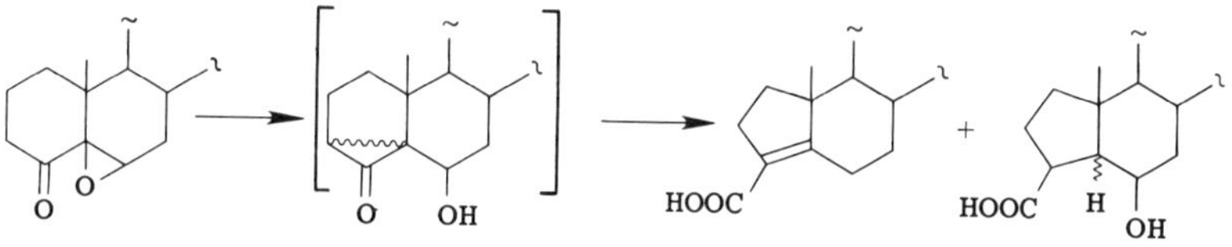

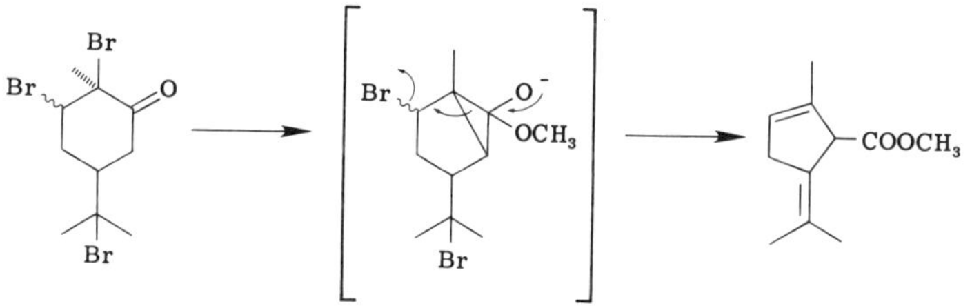

$$\text{(schemes 209, 252, 253, 254, 255)}\quad (122), (143), (143), (163), (166)$$

ble reaction of oxyallyl with nucleophiles, such as alkoxides, to give α-alkoxy ketones, the implication is that this does not occur, because in several cases increasing base concentration favors the formation of a Favorskii product at the expense of α-alkoxy ketones (113, 153).* Once the stage of the cyclopropanone or of its carbonyl adduct is reached, alkoxy ketone is not formed in *basic* medium, in spite of the fact that a number of cyclopropanone hydrates and hemiketals have been shown to

---

* The formation of α-phenoxycyclohexanone from α-chlorocyclohexanone and sodium phenoxide might appear to be an exception (168a), but the reaction may be proceeding through a symmetric ion pair from the enol allylic chloride.

open cleanly to α-substituted ketones in *acidic* medium *(144, 155)*. However, as noted earlier, α-alkoxy ketones can be produced from allene oxides should any of the latter be formed *(123, 142)*.

Epoxy ethers **257**, which can be isolated under controlled conditions *(169)*, are too reactive under the usual Favorskii rearrangement and work-up conditions [however, see *(166)*], and they give rise to α-hydroxy ketals **258**, which on aqueous acidic work-up yield α-hydroxy ketones **259**. Alternatively, α-hydroxy ketones can be formed by direct S_N2 reaction of hydroxide with **256**, by carbonyl addition of hydroxide to **256**, or by hydrolysis of α'-enol allylic halides in aqueous solution. Thus, except for the latter reaction, all of these side products result from nucleophilic attack on the carbonyl or α-carbon of starting material **256** rather than from any of the intermediates in Scheme 9. The amounts of these by-products will depend on the size of the nucleophile, its carbonyl nucleophilicity, steric access to the carbonyl carbon, and the orientation of the α leaving group for S_N2 displacement.

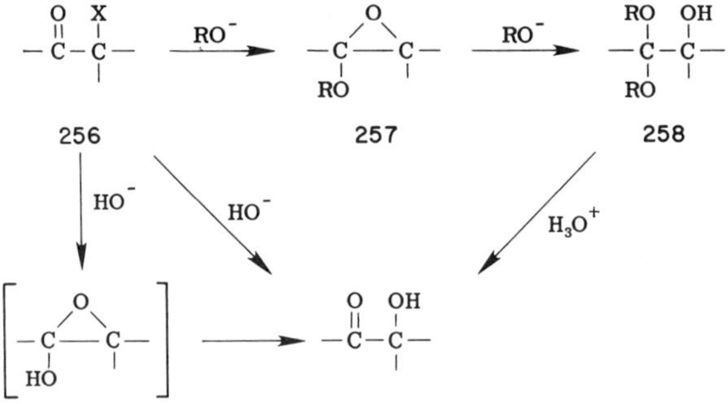

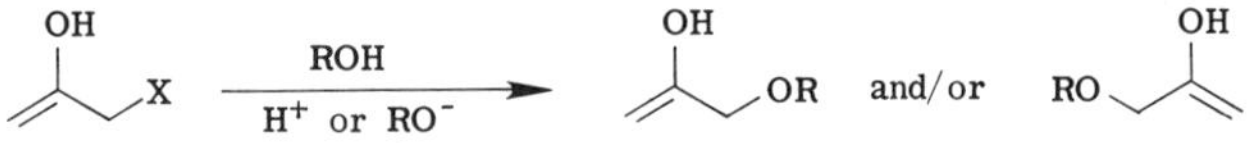

It has been found that alkene by-products can be formed in detrimental yields (up to 31%) by reaction of oxygen with an α'-enolate ion, e.g., **260** and **262 → 261** *(162)*. Since the same change can be brought about in better yield by the use of alkaline hydrogen peroxide, the oxygen reaction is

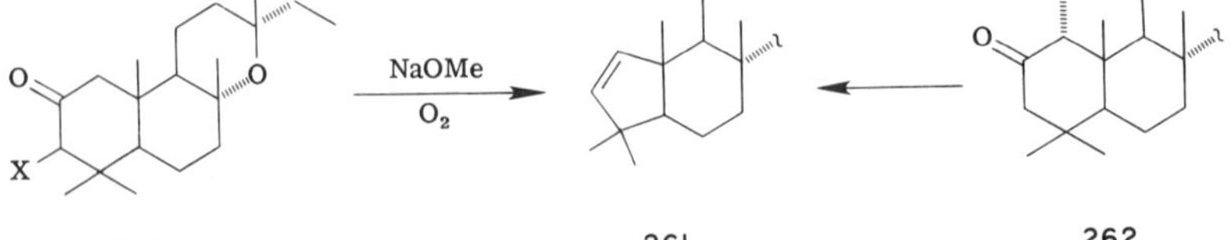

6. REARRANGEMENT IN CARBANIONS

closely related to the oxidative decarbonylation of cyclopropanones (preformed or formed *in situ* from α-halo ketones) by this reagent (*170*).

A number of other α-halo ketones that have been rearranged recently also fit the general carbanionic mechanism outlined in this essay (*166, 171–175*).

As is true for many carbanion reactions, the α'-carbanion in the Favorskii rearrangement can be replaced by an enamine in the case of some α-halo ketones. Although it has long been known that some Favorskii rearrangements could be catalyzed by secondary amines (*105*), more recently it has been noted that cyclic α-chloro ketones react at room temperature with piperidine and pyrrolidine to yield aminals 263 (*176, 177*; also see *154, 178*). These aminals will cleave to give Favorskii amides. A particularly striking example is *cis*-6-phenyl-2-chlorocyclohexanone (264), which gives only α-methoxy ketones and no Favorskii product, even with high (2 M) concentrations of NaOMe/MeOH, but with piperidine/MeOH at 0°C the trans Favorskii amide 265 was produced in 85% yield (*179*). Insight into the mechanism was provided when 175 (X = Cl) was treated with a combined excess of both NaOMe and piperidine in methanol; ester and piperidide were both major products (*179*). Since methoxide is a much more effective carbonyl nucleophile than piperidine, the Favorskii amide was not merely formed by reaction of the cyclopropanone with piperidine. Consequently, the reaction leading to amide must proceed via the enamine, as illustrated for 264→265. Note that the aminal cleavage proceeds with retention of configuration. The reaction of 2,6-dibromocyclohexanone (266) with several secondary amines to give the Favorskii amide 267 (*180, 181*) also probably follows the enamine path.

An extension of the carbanionic Favorskii mechanism of Scheme 9 can be imagined in which the leaving group is on the β- instead of the α-carbon. If β elimination were prevented by the absence of hydrogen on the α-carbon, then, in this case, there would be no ionization of the leaving group or formation of oxyallyl, but rather an S_N2 displacement by the enolate carbon, leading to a cyclobutanone. Such a reaction has been observed (*182, 183*), although it is an exception to Baldwin's rules (*126*). Furthermore, it has been proved that the cyclobutanone does not result from enolate fragmentation to a ketene + double bond and recombination 268 → 269 (*184, 185*). Normally, cyclobutanones are stable to the basic conditions of the Favorskii rearrangement, but if an α substituent capable of stabilizing an incipient carbanion is present, cleavage occurs to yield an acid derivative in what is formally a homologous Favorskii rearrangement. A phenyl group or double bond is sufficiently stabilizing to allow the reaction. Thus, although 270 gives only the cyclobutanone 271 and acids from Haller–Bauer cleavage of **270**, compounds **272–276** all give homo-Favorskii rearrangement acids as major products (*183*).

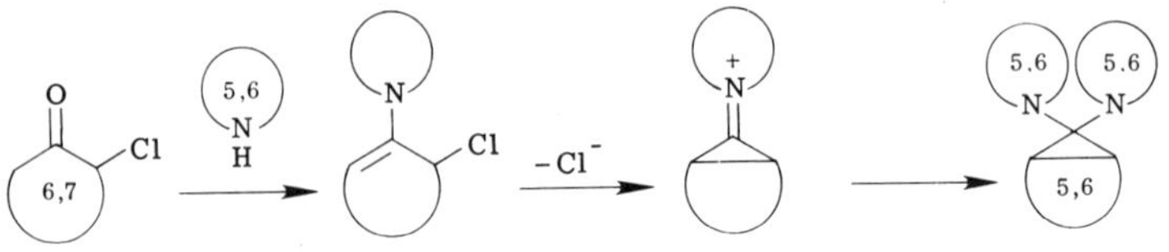

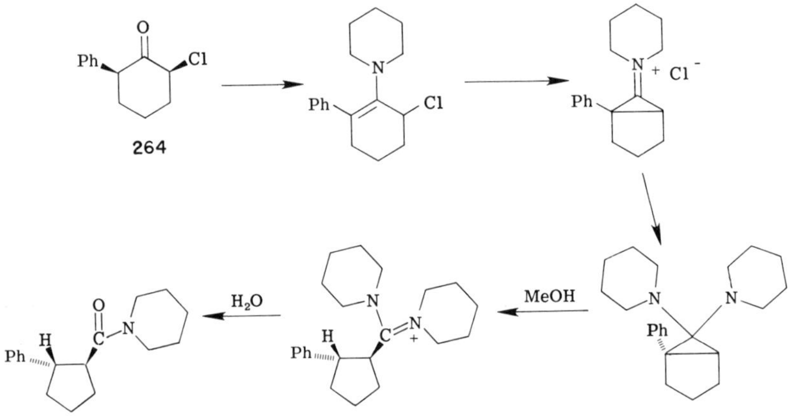

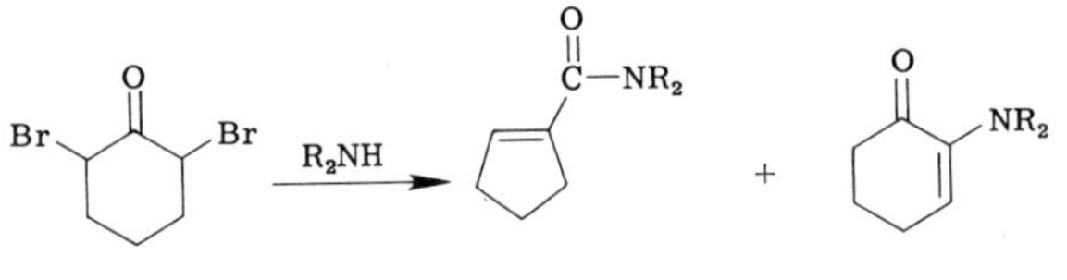

The leaving group may be even farther away at the γ- or δ-carbon if there is an α,β double bond. In this case treatment with base leads to vinylogous Favorskii reactions, as in the three examples below (**277–279**).

Finally, although the Favorskii rearrangement usually proceeds through carbanion intermediates, it has become increasingly clear that the alternative noncarbanionic semibenzilic pathway (Scheme 10) for the

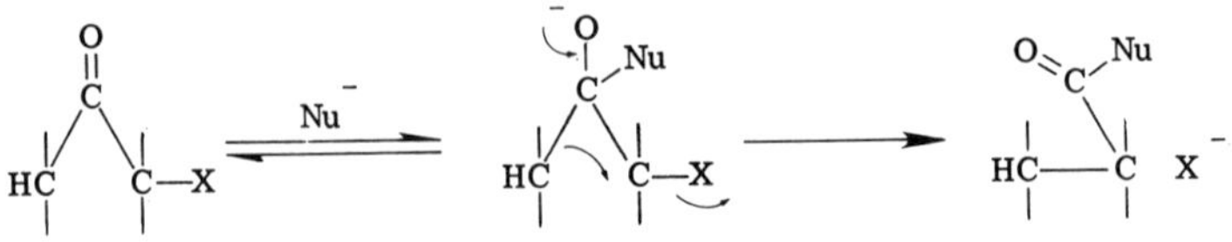

Scheme 10

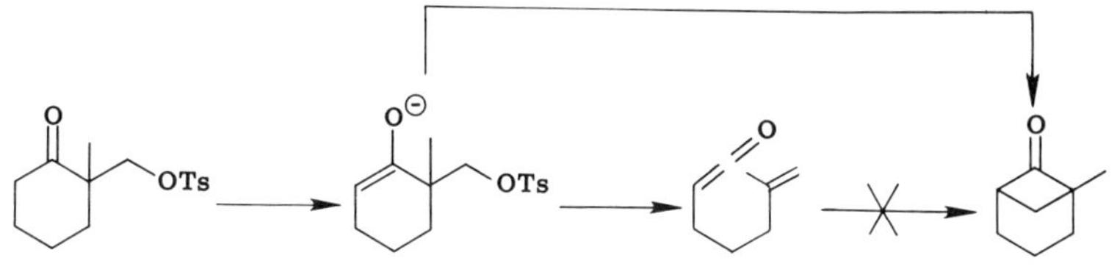

6. REARRANGEMENT IN CARBANIONS

268 → **269**

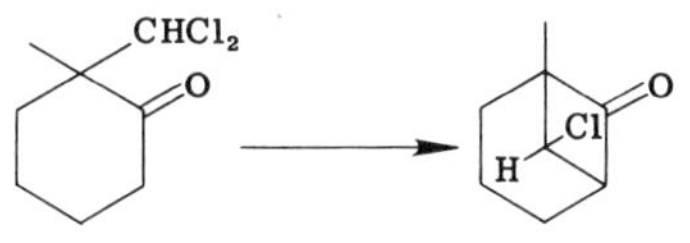

270 → **271**

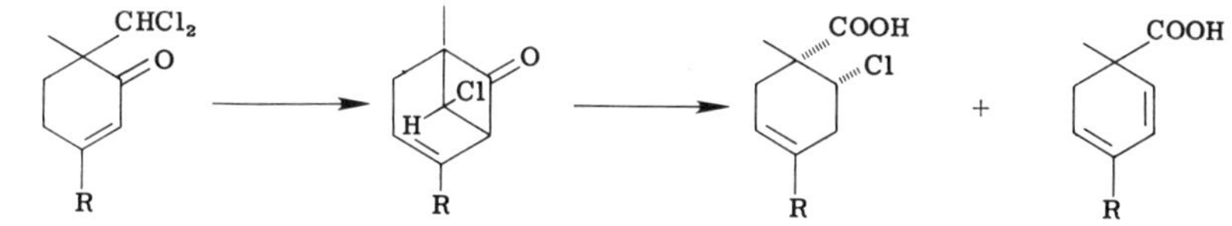

272 R = H
273 R = Me

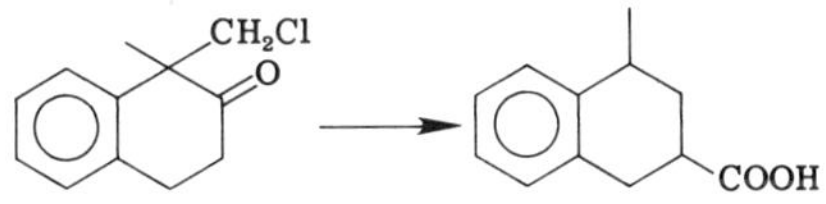

274, 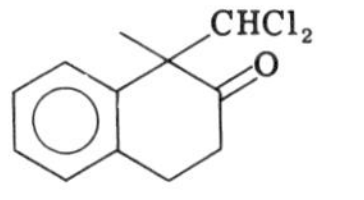 **275**, 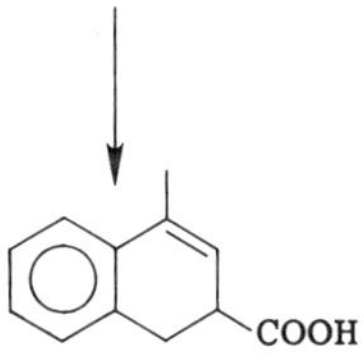 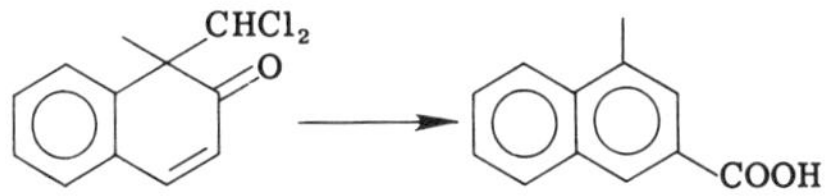 **276**

reaction may be more common than previously supposed. This carbonyl addition followed by a concerted 1,2 shift can yield only a single Favorskii acid, and it requires inversion at the carbon bearing the leaving group. It has been demonstrated a number of times that the required stereospecific inversion does occur—for example, with **280–286** (*158, 190–196*). Furthermore, there is a strong preference for the leaving group orientation shown in **287** (*197, 198*). Most of the examples of semibenzilic rearrangement have been with compounds without α'-hydrogen, or at least not bearing enolizable α'-hydrogen, and the reaction has been especially useful for preparing polycyclic fused small-ring systems as in **288** and **289**

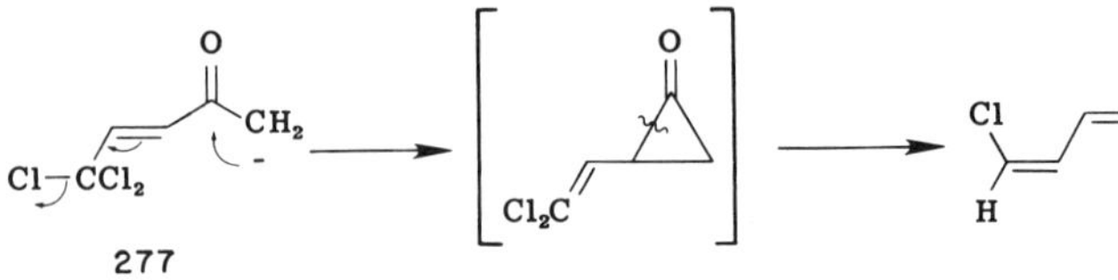

(186, 187)

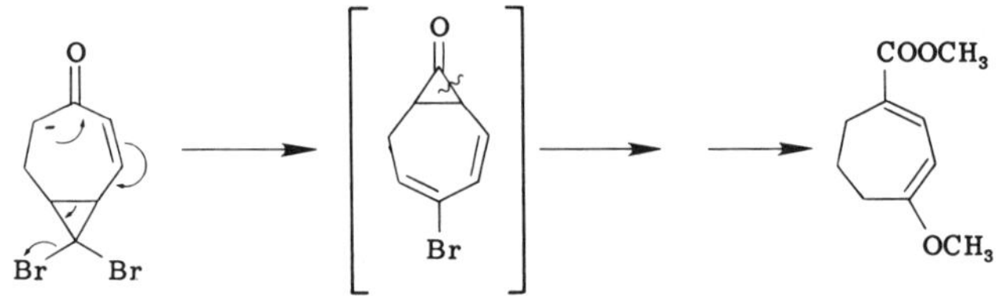

(188)

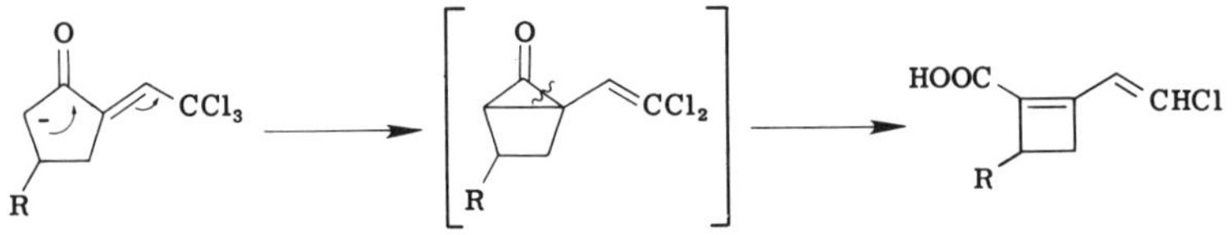

(189)

(190, 191)

R = H, Me

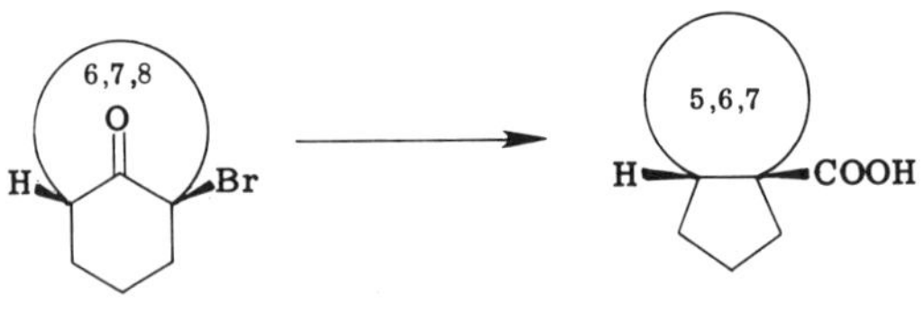

(158, 192)

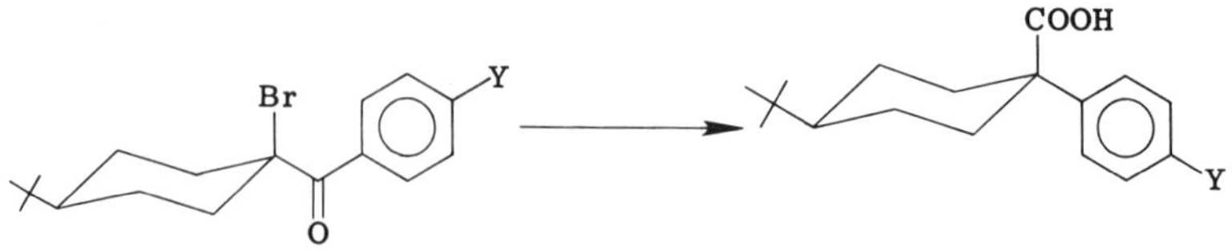

(193)

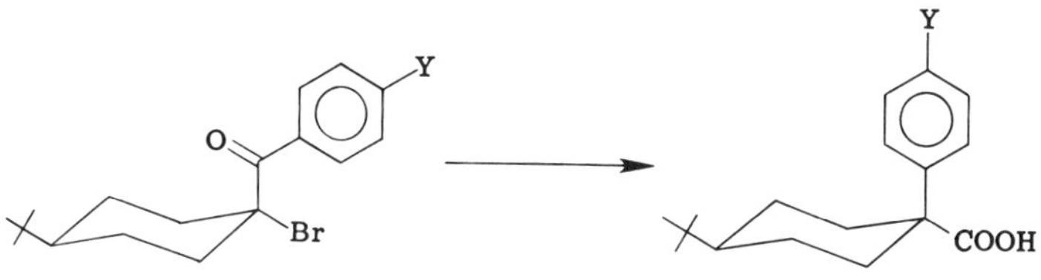

6. REARRANGEMENT IN CARBANIONS

(193)

283
Y = H, OCH$_3$, N(CH$_3$)$_2$

(194)

284

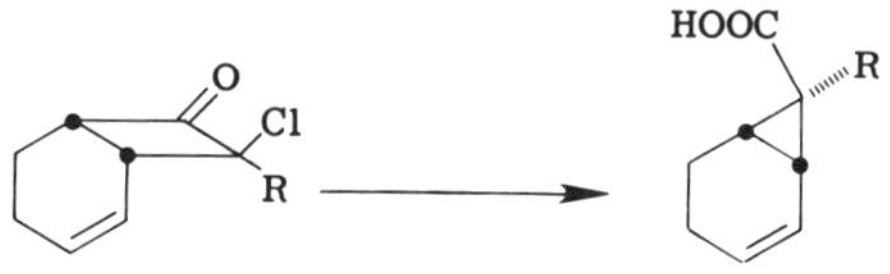

(195, 196)

285

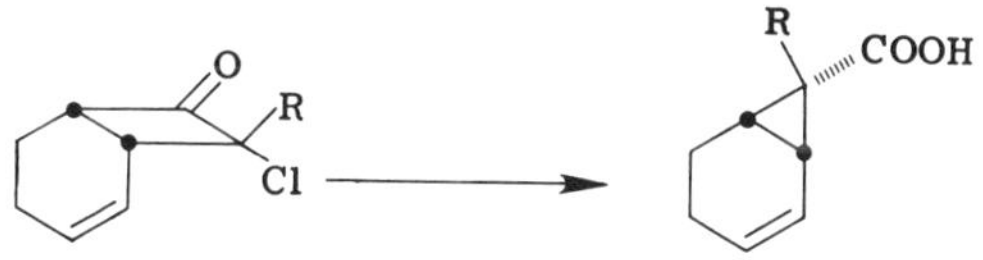

(195, 196)

286
R = Me, Et, i-Pr

[199–201; also the earlier examples tabulated in (110)]. There have even been double semibenzilic Favorskii rearrangements, such as **290** and **291** (202–204). On the other hand, some semibenzilic or probable semibenzilic rearrangements have involved α-halo ketones with enolizable α'-hydrogen atoms, e.g., **292** (205) and the many α-halocyclobutanones that have been studied (195–197, 206). The [4.3.1] bromo ketone **281** rearranges by the semibenzilic route with hydroxide because of the relatively low acidity of the bridgehead α'-hydrogen. However, if the base is the stronger and bulkier t-butoxide, the rearrangement follows the carbanion–cyclopropanone path (158). If the α'-hydrogen is now made almost normal in acidity by homologation, optically active [5.3.1] bromo

287

Y = H or OCH_3

288

289

290

291

292

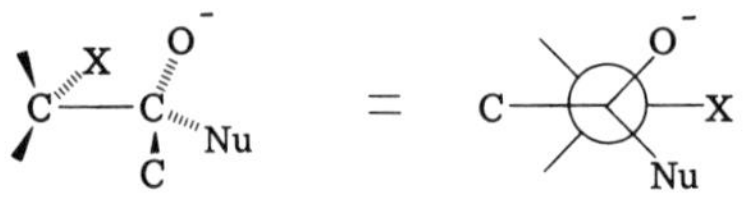

ketone **281** can be caused to rearrange by either mechanism depending on the basicity and carbonyl nucleophilicity of the reagent chosen. With hydroxide ion the mechanism is via a cyclopropanone (racemic product); with hydroperoxide ion the same acid is formed only by the semibenzilic mechanism (complete retention of optical purity); and with cyanide, hy-

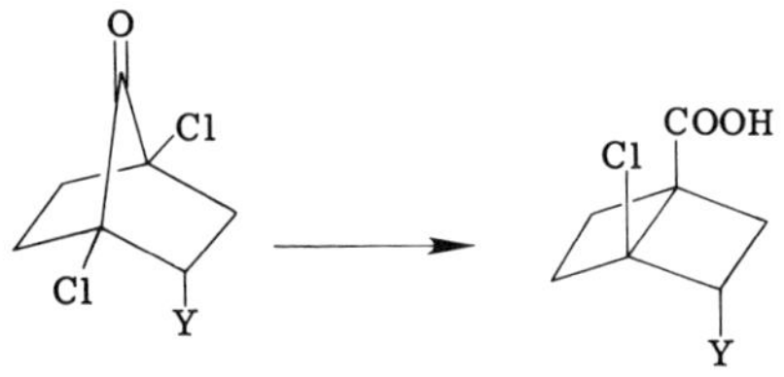

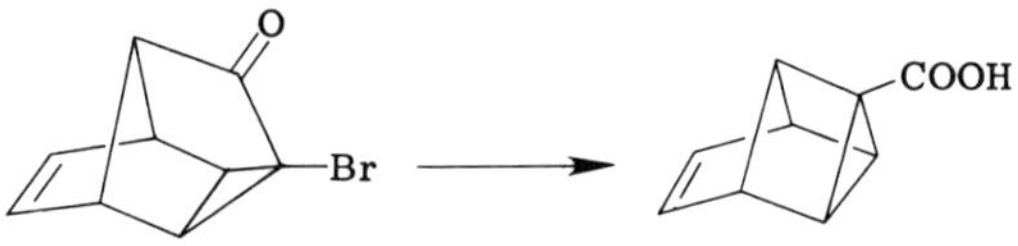

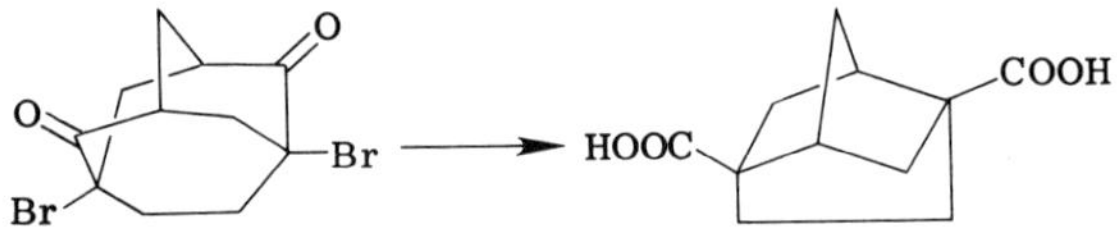

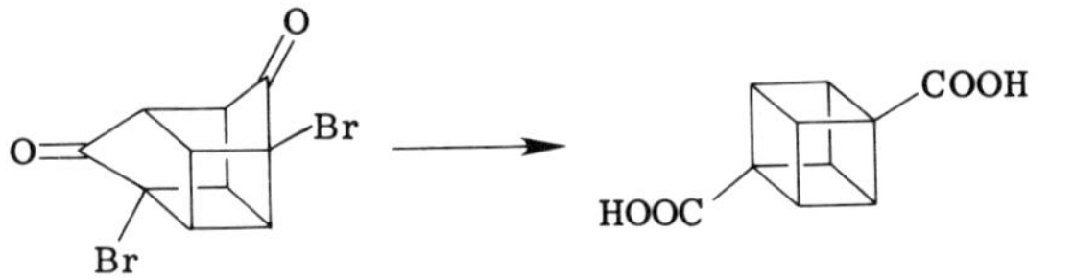

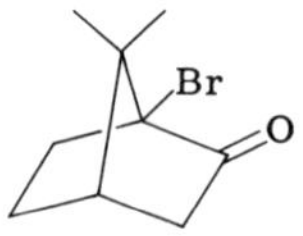

drosulfide, and bicarbonate ions both mechanisms operate at the same time (partial racemization) *(207)*. Thus, the two different mechanisms for the Favorskii rearrangement are not that unevenly matched, and it may eventually prove possible to choose either at will in many cases.

C. Ramberg–Bäcklund Reaction

The Ramberg–Bäcklund reaction *(208)* provides an illuminating contrast to the Favorskii rearrangement. Since it has been reviewed recently *(209–211)*, only the major features relevant to a comparison with the Favorskii reaction will be mentioned. Although the two reactions are formally analogous in that an α-halo sulfone **293** is converted in basic medium into an episulfone **294** (the sulfonyl equivalent of a cyclopropanone), they differ in that the episulfone loses SO_2 in alkali to give an alkene instead of ring opening to a sulfonic acid [cf. the oxidative loss of CO from some cyclopropanones *(126, 162)*]. The differences in the nature of the sulfonyl vs carbonyl groups also lead to significant differences in the mechanisms of the two reactions.

Even though sulfones are weaker acids by 5–6 pK_a units than the corresponding ketones, the Ramberg–Bäcklund reaction proceeds via a discrete α'-sulfonyl carbanion intermediate in a two-stage mechanism *(212–214)*. Unlike many Favorskii rearrangements, carbanion formation has not so far been found to be rate-determining (prior exchange of α'-H and $k_{\alpha'\text{-H}}/k_{\alpha'\text{-D}} \sim 1$) *(213)*. While the configuration of the α'-carbon of a ketone is lost in the change to the planar 3-atom 2p–π enolate, the change of the α'-carbon of a sulfone to the planar or nearly planar anion *(216)* does not necessarily lead to immediate loss of configuration since deuterium exchange is 50–100x faster than epimerization in **296** and **299** *(215)*. Initially, the chiral anion **295** is formed by deprotonation from a conformation in which the H removed is flanked by the sulfonyl oxygen atoms. The sulfonyl anion does not lose its α-halogen by ionization to generate the sulfonyl analogue of oxyallyl **158**, in contrast to the Favorskii rearrangement. Instead, even in solvents as polar as MeOH, the sulfonyl α'-carbanion displaces the halide from the α-carbon in an intramolecular S_N2 substitution to produce an episulfone. For most Ramberg–Bäcklund reactions it is this step which is rate-determining as can be seen from the large $k_{Br/Cl}$ rate ratios (88–620, depending on compound and temperature) and the lack of episulfone build-up. That the α-halogen is not lost by ionization is clear because, in contrast to the Favorskii rearrangement, additional methyl substitution on the α-carbon atom has very little effect on reaction rate; and the ρ values for halide loss (after adjustment for the ρ values of the carbanion pre-equilibrium) for several sulfone types, ArCHBr—SO$_2$—CH$_3$, ArCHBr—SO$_2$—CH$_2$Ph, and ArCHCl—SO$_2$—Ch$_2$Ph, are slightly positive, rather than the large negative values expected for ionization (213).

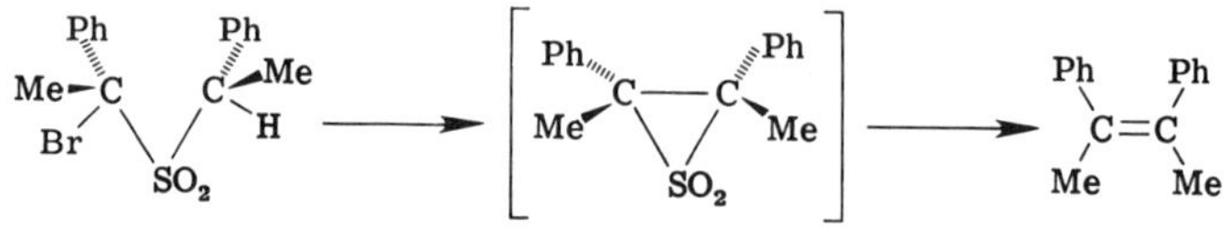

Ph   Ph   [Ph   Ph]   Ph Ph
Me—C C—Me   Me—C—C—Me   C=C
Br   H   SO$_2$ Me Me
SO$_2$
296   **297**   **298**

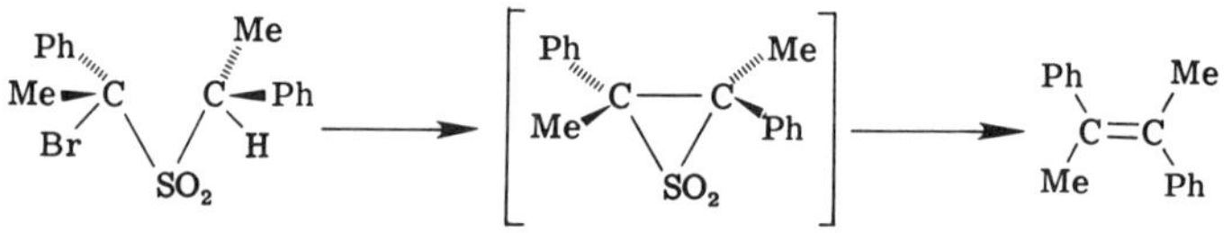

Ph Me   [Ph   Me]   Ph Me
Me—C C—Ph   Me—C—C—Ph   C=C
Br   H   SO$_2$ Me Ph
SO$_2$
299   **300**   **301**

Evidence that episulfones are actually formed in the Ramberg–Bäcklund reaction is convincing. Aside from the conclusion that halide loss by intramolecular displacement should produce an episulfone, some independently prepared episulfones have been found to give the same products as the Ramberg–Bäcklund reactions at faster rates (217, 218). More directly, in the rearrangement of *threo*-bromo sulfone **299** there is a build-up of an intermediate (presumably **300**) which breaks down to *trans*-dimethylstilbene **301** (215). Also, small amounts of 1- and 2-propane sulfonic acids have been observed in the reaction of chloromethyl ethyl sulfone with hydroxide ion (217). Finally, in a reaction analogous to the facile preparation of cyclopropenones, the Ramberg–Bäcklund reaction of α,α'-dibromodibenzyl sulfone with triethylamine allowed isolation of the more stable unsaturated episulfone **302** in 70% yield (219).

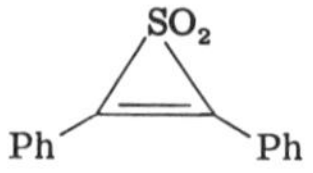

SO$_2$
△
Ph   Ph
302

Stereochemical evidence is also in agreement with an S_N2 ring closure. The acquisition of such evidence was complicated by two facts: Firstly, episulfones bearing an α- or α'-H can be epimerized via carbanion formation (217) [but apparently not by ring opening and reclosure (216); cf. the contrasting behavior of *trans*-2,3-di-*t*-butylcyclopropanone mentioned earlier], and secondly both C—S bonds are broken in the rapid loss of SO_2 from the episulfone. Fortunately, independently prepared episulfones have been found to break down stereospecifically to alkene of the same configuration on solvolysis (217). Therefore, provided that the intermediate episulfone either does not undergo epimerization faster than alkaline breakdown or else does not have an α-H, the stereochemistry of the ring closure can be determined from that of the halo sulfone and alkene product.

The constrained chloro sulfone **303** which yields the bicyclic alkene **305** must clearly do so by inversion at both α-carbon atoms to produce the

[structures: **303** → **304** → **305**]

intermediate episulfone **304**. The unconstrained (±)*erythro*-bromo sulfone **296**, whose stereochemistry was determined by single crystal X-ray analysis, produced the cis-alkene **298** in 93% diastereomeric purity, while the (±)*threo* isomer **299** gave the trans-alkene **301** in >95% diastereomeric purity (215). Since the episulfone intermediates in these examples are not subject to epimerization, the ring closures have either occurred with double retention or double inversion at both α-carbon atoms. Until it proves possible to isolate an episulfone from a Ramberg–Bäcklund reaction or else an appropriate sulfonic acid cleavage product, it is not possible to decide unambiguously between the two alternatives. However, with the constrained bicyclic example **303** → **305** and the so far invariable rule of inversion in S_N2 substitutions, buttressed by the failure to react of halo sulfones which can not undergo inversion at the halo carbon (220), a decision can be made in favor of double inversion. A similar double inversion conclusion follows from a variant of the Ramberg–Bäcklund reaction, the Ph_3P debromination of the meso and racemic dibromo sulfones **306** (220).

$$\underset{\underset{Br}{|}}{PhCH} \overset{SO_2}{\diagup\diagdown} \underset{\underset{Br}{|}}{CHPh}$$

306

The unexpected finding that the α'-carbanion undergoes inversion on ring closure, although before ring closure it has a strong tendency to retain the configuration of the initial sulfone, has been neatly rationalized as follows (215, 220). The planar chiral carbanion **295** results from the proven kinetic preference for removal of a hydrogen flanked by the two sulfonyl oxygen atoms (*216, 221*), perhaps because the base is generated from solvent hydrogen-bonded to these sulfonyl oxygen atoms (*221*), or perhaps because base is cation-bonded to them (*216*). Reprotonation of the anion occurs by reversal of this process resulting in retention of the original α'-configuration. On the other hand, the planar carbanion **295** is correctly aligned for nucleophilic attack on the halide-bearing α-carbon by merely becoming tetrahedral in the inverted sense from the original α'-configuration, without the need for any rotation about the α'—C—S bond.

The stereochemistry found for the reaction of **303, 296,** and **299** would also be consistent with episulfone formation from a W-conformation of reacting centers in a concerted 1,3-elimination not involving carbanions. Therefore, the possibility that carbanion formation was irrelevant to the Ramberg–Bäcklund reaction had to be tested. The best evidence comes from a comparison of the reactions of **307** and **308** (*214*). Since **307** is conformationally mobile while **308** is locked into a favorable arrangement for concerted elimination, a difference in the activation parameters of the two compounds would be expected if both reactions were concerted (a more positive ΔS^* for **308**), or if **308** were concerted and **307** were stepwise via a carbanion (different ΔH^* and ΔS^*). The activation parame-

$$\underset{\underset{Br}{|}}{PhCH}\overset{SO_2}{\diagup}\overset{}{\diagdown}CH_2Ph$$

307

308

ters are the same within experimental error, and therefore there is no support for a concerted 1,3-elimination with either compound.

The opening of the episulfone ring is apparently mechanistically different from the Favorskii opening of cyclopropanones discussed earlier.* The cleavage of the two C—S bonds, which is nonconcerted (Woodward–Hoffmann restriction; near identical rates for 2-phenyl- and 2,3-diphenyl episulfone), exhibits the marked sensitivity to solvent ionizing power expected for charge-separated species (*217*). However, in basic solution the almost invariable absence of expected products from the alkaline scission of a single C—S bond is not consonant with an incipient carbanion analogous to that in cyclopropanone cleavage, or with nucleophilic attack on carbon to displace a sulfinate ion. Therefore, it has been suggested that, in basic solution, nucleophilic addition to sulfur occurs,

---

* Very recent work requires modification of this statement. J. F. King and J. H. Hillhouse (Univ. West. Ont.) found that the major reaction product of episulfone with aqueous Ba(OH)$_2$ is barium ethanesulfonate. When considered with the earlier mentioned fact that methyl episulfone yields small amounts of 1- and 2-propanesulfonic acids, it appears that minimally substituted episulfones do ring open, after attack of base on sulfur, in a manner analogous to cyclopropanone cleavage. The interpretation expressed in the text may apply only to more substituted episulfones.

6. REARRANGEMENT IN CARBANIONS 465

$$\underset{309}{\underset{\underset{Nu}{|}}{\overset{O}{\underset{\diagdown}{\overset{\diagup}{S}}}\overset{O^-}{\diagup}}\underset{}{\overset{}{\diagdown C — C \diagup}}}$$

followed by cleavage of a C—S bond to a singlet diradical anion **309** from which loss of the sulfur-containing group happens faster than C—C rotation can scramble the episulfone geometry (*217*). The sensitivity to solvent ionizing power can be understood if the transition state leading to the diradical anion **309** is appreciably dipolar (*217*).

Further work will be needed to test a recent explanation (*215*) of why, when two α'-sulfonyl carbanions are possible in a Ramberg–Bäcklund reaction, the more sterically crowded episulfone is frequently formed preferentially.

REFERENCES

1. H. E. Zimmerman, *in* "Molecular Rearrangements" (P. de Mayo, ed.) Vol. 1, p. 345. Wiley (Interscience), New York, 1963.
2. D. J. Cram, "Fundamentals of Carbanion Chemistry." Academic Press, New York, 1965.
3. E. Buncell, "Carbanions: Mechanistic and Isotopic Aspects." Elsevier, Amsterdam, 1975.
4. D. H. Hunter, *in* "Isotopes in Organic Chemistry" (E. Buncel and C. C. Lee, eds.), Vol. 1, p. 135. Elsevier, Amsterdam.
5. S. W. Staley, *in* "Pericyclic Reactions" (A. P. Marchand and R. E. Lehr, eds.), Vol. 1, p. 199. Academic Press, New York, 1977.
6. E. Grovenstein, *Adv. Organomet. Chem.* **16**, 167 (1977).
6a. E. Grovenstein, *Angew. Chem. Int. Ed. Engl.* **17**, 313 (1978).
7. E. A. Hill, *Adv. Organomet. Chem.* **16**, 131 (1977).
8. E. A. Hill, *J. Organomet. Chem.* **91**, 123 (1975).
9. J. G. Carpenter, A. G. Evans, and N. H. Rees, *J. Chem. Soc., Perkin Trans. 2* p. 1598 (1972).
10. H. Pines, *Acc. Chem. Res.* **7**, 155 (1974).
11. R. M. Magid and S. E. Wilson, *Tetrahedron Lett.* p. 19 (1971).
12. J. Klein and S. Brenner, *Tetrahedron* **26**, 2345 (1970).
13. R. B. Bates, S. Brenner, W. H. Dienes, D. A. McCombs, and D. E. Potter, *J. Am. Chem. Soc.* **92**, 6345 (1970).
14. J. A. Berson, see Essay 5 of this volume.
15. S. W. Staley, G. M. Cramer, and W. G. Kingsley, *J. Am. Chem. Soc.* **95** 5052 (1973).
16. E. Grovenstein and A. B. Cottingham, *J. Am. Chem. Soc.* **99**, 1881 (1977).
17. J. E. Baldwin and F. J. Urban, *J. Chem. Soc., Chem. Commun.* p. 165 (1970).
18. J. R. Scheffer, R. E. Gayler, T. Zakouras, and A. A. Dzakpasu, *J. Am. Chem. Soc.* **99**, 7726 (1977).
18a. T. J. Greenhough, J. R. Scheffer, J. Trotter and Y.-F. Wong, *J. Chem. Soc., Chem. Commun.* p. 933 (1979).
19. H. C. Brown and E. Nigishi, *J. Am. Chem. Soc.* **93**, 6682 (1971).
20. J. J. Eisch and J. E. Galle, *J. Organomet. Chem.* **127**, C9 (1977).

21. J. J. Eisch, K. Tamao, and R. J. Wilcsek, *J. Am. Chem. Soc.* **97,** 895 (1975).
22a. J. J. Eisch and M.-R. Tsai, *J. Am. Chem. Soc.* **95,** 4065 (1973).
22b. M. Daney, R. Lapouyade, B. Labrande, and H. Bouas-Laurent, *Tetrahedron Lett.* p. 153 (1980); a recent example of a 1,4 migration.
23. R. West, *Adv. Organomet. Chem.* **16,** 1 (1977).
24. S. W. Benson and H. E. O'Neal, *Nat. Stand. Ref. Data Ser., Nat. Bur. Stand.* **21,** 225 (1970).
24a. A. Thibblin and W. P. Jencks, *J. Am. Chem. Soc.* **101,** 4963 (1979).
25. J. D. Cox and G. Pilcher, "Thermochemistry of Organic and Organometallic Compounds." Academic Press, New York, 1970.
26. H. Lehmkuhl, D. Reinehr, D. Henneberg, G. Schomburg, and G. Schroth, *Justus Liebigs Ann. Chem.* p. 108 (1975).
27. H. Lehmkuhl, D. Reinehr, G. Schomburg, D. Henneberg, H. Damen, and G. Schroth, *Justus Liebigs Ann. Chem.* p. 119 (1975).
28. M. S. Silver, P. R. Shafer, J. E. Nordlander, C. Ruchardt, and J. D. Roberts, *J. Am. Chem. Soc.* **82,** 2646 (1960).
29. A. Maercker and K. Weber, *Justus Liebigs Ann. Chem.* **43,** 756 (1972).
30. A. M. Rothman, Ph.D. Thesis, Pennsylvania State University, University Park (1969).
31. E. A. Hill and H. R. Ni, *J. Org. Chem.* **36,** 4133 (1971).
32. D. O. Cowan, N. G. Krieghoff, J. E. Nordlander, and J. D. Roberts, *J. Org. Chem* **32,** 2639 (1967).
33. H. G. Richey and W. C. Kossa, *Tetrahedron Lett.* p. 2313 (1969).
34. A. Maercker, P. Guthlein, and H. Wittmayr, *Angew. Chem. Int. Ed. Engl.* **12,** 774 (1973).
35. D. J. Patel, C. L. Hamilton, and J. D. Roberts, *J. Am. Chem. Soc.* **82,** 2646 (1960).
36. H. Lehmkuhl, D. Reinehr, D. Henneberg, and G. Schroth, *J. Organomet. Chem.* **57,** 49 (1973).
37. H. G. Richey and H. S. Veale, *J. Am. Chem. Soc.* **96,** 2641 (1974).
38. H. G. Richey and A. M. Rothman, *Tetrahedron Lett.* p. 1457 (1968).
39. W. G. Kossa, T. C. Rees, and H. G. Richey, *Tetrahedron Lett.* p. 3455 (1971).
40. P. T. Lansbury, V. A. Pattison, W. A. Clement, and J. Sidler, *J. Am. Chem. Soc.* **86,** 2247 (1964).
41. A. Maercker and R. Geuss, *Angew. Chem. Int. Ed. Engl.* **9,** 909 (1970).
42. E. A. Hill, R. J. Theissen, and K. Taucher, *J. Org. Chem.* **34,** 3061 (1969).
43. L. Horner, P. V. Subramanian, and K. Eiben, *Justus Liebigs Ann. Chem.* **714,** 91 (1968).
44. E. Grovenstein and L. P. Williams, *J. Am. Chem. Soc.* **83,** 412 (1961).
45. E. Grovenstein, *Adv. Organomet. Chem.* **16,** 172 (1977).
46. H. E. Zimmerman and A. Zweig, *J. Am. Chem. Soc.* **83,** 1196 (1961).
47. E. Grovenstein and Y.-M. Cheng, *J. Am. Chem. Soc.* **94,** 4971 (1972).
48. S. W. Staley and J. P. Erdman, *J. Am. Chem. Soc.* **92,** 3832 (1970).
49. E. Grovenstein and G. Wentworth, *J. Am. Chem. Soc.* **89,** 2348 (1967).
50. E. Grovenstein and G. Wentworth, *J. Am. Chem. Soc.* **89,** 1852 (1967).
51. J. A. Bertrand, E. Grovenstein, P.-C. Lu, and D. Vanderveer, *J. Am. Chem. Soc.* **98,** 7835 (1976).
52. E. Grovenstein, *Adv. Organomet. Chem.* **16,** 181 (1977).
53. E. Grovenstein, J. A. Beres, Y.-M. Cheng, and J. A. Pegolotti, *J. Org. Chem.* **37,** 1281 (1972).
54. E. Grovenstein, and J.-U. Rhee, *J. Am. Chem. Soc.* **97,** 769 (1975).
55. A. Nickon and J. L. Lambert, *J. Am. Chem. Soc.* **84,** 4604 (1962); **88,** 1905 (1966).
56. K. E. Hamlin and A. W. Weston, *Org. React.* **9,** 1 (1957).
57. D. J. Cram, B. Rickborn, C. A. Kingsbury, and P. Haberfield, *J. Am. Chem. Soc.* **83,** 3678 (1961).
58. A. T. Young and R. D. Guthrie, *J. Org. Chem.* **35,** 853 (1970).

6. REARRANGEMENT IN CARBANIONS

59. H. H. Mantsch, H. Saito, and I. C. P. Smith, *Prog. Nucl. Magn. Reson. Spectrosc.* **11**, 211 (1977).
60. A. Nickon, J. L. Lambert, and J. E. Oliver, *J. Am. Chem. Soc.* **88**, 2787 (1966).
61. A. Nickon, H. Kwasnik, T. Swartz, R. O. Williams, and J. B. DiGiorgio, *J. Am. Chem. Soc.* **87**, 1615 (1955).
62. R. M. Coates and J. P. Chen, *J. Chem. Soc., Chem. Commun.* p. 1481 (1970).
63. J. L. Lambert, Ph.D. Thesis, Johns Hopkins University, Baltimore, Maryland (1963).
64. A. L. Johnson, J. B. Stothers, and C. T. Tan, *Can. J. Chem.* **53**, 212 (1975).
65. A. Nickon, J. H. Hammons, J. L. Lambert, and R. O. Williams, *J. Am. Chem. Soc.* **85**, 3713 (1963).
66. A. Nickon, J. L. Lambert, R. O. Williams, and N. H. Werstiuk, *J. Am. Chem. Soc.* **88**, 3354 (1966).
67. A. Nickon, J. L. Lambert, J. E. Oliver, D. F. Covey, and J. Morgan, *J. Am. Chem. Soc.* **98**, 2593 (1976).
68. D. H. Hunter, A. L. Johnson, J. B. Stothers, A. Nickon, J. L. Lambert, and D. F. Covey, *J. Am. Chem. Soc.* **94**, 8582 (1972).
69. G. C. Joshi and E. W. Warnhoff, *J. Org. Chem.* **37**, 2383 (1972).
69a. T. Money, personal communication.
70. A. K. Cheng, J. B. Stothers, and C. T. Tan, *Can. J. Chem.* **55**, 447 (1977).
71. K. W. Turnbull, S. J. Gould, and D. Arigoni, *J. Chem. Soc., Chem. Commun.* p. 597 (1972).
72. J. E. Nordlander, S. P. Jindal, and D. J. Kitko, *J. Chem. Soc., Chem. Commun.* p. 1136 (1969).
73. J. B. Stothers and C. T. Tan, *J. Chem. Soc., Chem. Commun.* p. 738 (1974).
74. D. M. Hudyma, J. B. Stothers, and C. T. Tan, *Org. Magn. Reson.* **6**, 614 (1974).
75. D. H. Bowen and J. MacMillan, *J. Chem. Soc., Chem. Commun.* p. 4111 (1972); D. H. Bowen, C. Cloke, and J. MacMillan, *J. Chem. Soc., Perkin Trans. I* p. 378 (1975).
76. J. R. Wiseman, *J. Am. Chem. Soc.* **89**, 5966 (1967); J. R. Wiseman and W. A. Pletcher, *ibid.* **92**, 956 (1970).
77. A. Nickon, D. F. Covey, F. Huang, and Y.-N. Kuo, *J. Am. Chem. Soc.* **97**, 904 (1975).
78. A. K. Cheng and J. B. Stothers, *Can. J. Chem.* **56**, 1342 (1978).
79. A. K. Cheng and J. B. Stothers, *Can. J. Chem.* **55**, 50 (1977).
80. A. K. Cheng and J. B. Stothers, *Can. J. Chem.* **55**, 4184 (1977).
81. A. K. Cheng, V. Patel, and J. B. Stothers, unpublished results.
82. A. L. Johnson, N. O. Petersen, M. B. Rampersad, and J. B. Stothers, *Can. J. Chem.* **52**, 4143 (1974).
83. R. Howe and S. Winstein, *J. Am. Chem. Soc.* **87**, 915 (1965); T. Fukunaga, *ibid.* p. 916; T. Fukunaga and R. A. Clement, *J. Org. Chem.* **42**, 270 (1977).
84. A. B. Crow and W. T. Borden, *Tetrahedron Lett.* p. 1967 (1976).
85. N. H. Werstiuk, *Can. J. Chem.* **53**, 2211 (1975).
86. N. H. Werstiuk, personal communication.
87. A. K. Cheng and J. B. Stothers, unpublished results.
88. J. P. Freeman and J. H. Plonka, *J. Am. Chem. Soc.* **88**, 3662 (1966).
89. P. Yates, G. D. Abrams, and S. Goldstein, *J. Am. Chem. Soc.* **91**, 6868 (1969); M. J. Betts and P. Yates, *ibid.* **92**, 6982 (1970); P. Yates, G. D. Abrams, M. J. Betts, and S. Goldstein, *Can. J. Chem.* **49**, 2850 (1971); P. Yates and M. J. Betts, *J. Am. Chem. Soc.* **94**, 1965 (1972).
90. M. B. Rampersad and J. B. Stothers, *J. Chem. Soc., Chem. Commun.* p. 709 (1976).
91. P. C. Coleman, R. A. Dyllick-Brenzinger, and J. B. Stothers, unpublished work.
92. R. A. Dyllick-Brenzinger and J. B. Stothers, *J. Chem. Soc., Chem. Commun.* p. 108 (1979).
92a. V. Patel and J. B. Stothers, unpublished work.
93. A. Nickon, J. J. Frank, D. F. Covey, and Y.-i. Lin, *J. Am. Chem. Soc.* **96**, 7574 (1974).

94. C. H. DePuy, *Acc. Chem. Res.* **1**, 33 (1968).
95. A. Nickon, D. F. Covey, G. D. Pandit, and J. J. Frank, *Tetrahedron Lett.* p. 3681 (1975).
96. C. H. DePuy, F. W. Breitbeil, and K. R. DeBruin, *J. Am. Chem. Soc.* **88**, 3347 (1966).
97. P. S. Wharton and A. R. Fritzberg, *J. Org. Chem.* **37**, 1899 (1972).
98. W. T. Borden, V. Varma, M. Cabell, and T. Ravindranathan, *J. Am. Chem. Soc.* **93**, 3800 (1971).
99. A. J. H. Klunder and B. Zwanenburg, *Tetrahedron Lett.* p. 1721 (1971).
100. A. J. H. Klunder and B. Zwanenburg, *Tetrahedron* **29**, 1683 (1973).
101. N. B. M. Arts, A. J. H. Klunder, and B. Zwanenburg, *Tetrahedron Lett.* p. 2359 (1976).
102. R. D. Miller and D. L. Dolce, *Tetrahedron Lett.* p. 1151 (1973).
103. R. D. Miller and D. L. Dolce, *Tetrahedron Lett.* p. 5217 (1973).
104. R. D. Miller and D. L. Dolce, *Tetrahedron Lett.* p. 1023 (1977).
105. A. S. Kende, *Org. React.* **11**, 261 (1960).
106. A. A. Akhrem, T. K. Vstynyuk, and Y. A. Titov, *Russ. Chem. Rev. (Engl. Transl.)* **39**, 732 (1970).
107. F. G. Bordwell, *Acc. Chem. Res.* **3**, 286 (1970).
108. D. Redmore and C. D. Gutsche, *Adv. Alicyclic Chem.* **3**, 46 (1971).
109. E. Buncel, "Carbanions: Mechanistic and Isotopic Aspects," p. 143. Elsevier, Amsterdam, 1975.
110. P. J. Chenier, *J. Chem. Educ.* **55**, 286 (1978).
110a. V. P. Andreev, *Sovrem. Probl. Organ. Khim.* **6**, 21 (1978).
111. F. G. Bordwell and M. W. Carlson, *J. Am. Chem. Soc.* **92**, 3370 (1970).
112. F. G. Bordwell and R. G. Scamehorn, *J. Am. Chem. Soc.* **93**, 3410 (1971).
113. F. G. Bordwell and J. Almy, *J. Org. Chem.* **38**, 575 (1973).
114. F. G. Bordwell, R. R. Frame, R. G. Scamehorn, J. G. Strong, and S. Meyerson, *J. Am. Chem. Soc.* **89**, 6704 (1967).
115. F. G. Bordwell, R. G. Scamehorn, and W. R. Springer, *J. Am. Chem. Soc.* **91**, 2087 (1969).
116. J. F. Liebman and A. Greenberg, *J. Org. Chem.* **39**, 123 (1974).
117. A. Liberles, A. Greenberg, and A. Lesk, *J. Am. Chem. Soc.* **94**, 8685 (1972).
118. N. Bodor, M. J. S. Dewar, A. Harget, and E. Haselbach, *J. Am. Chem. Soc.* **92**, 3854 (1970).
119. J. F. Olsen, S. Kang, and L. Burnelle, *J. Mol. Struct.* **9**, 305 (1971).
120. R. C. Bingham, M. J. S. Dewar, and D. H. Ho, *J. Am. Chem. Soc.* **94**, 8685 (1972).
121. R. L. Camp and F. D. Greene, *J. Am. Chem. Soc.* **90**, 7349 (1968).
122. J. K. Crandall and W. H. Machleder, *J. Am. Chem. Soc.* **90**, 7347 (1968).
123. T. H. Chan and B. S. Ong, *J. Org. Chem.* **43**, 2994 (1978).
124. M. E. Zandler, C. E. Choc, and C. K. Johnson, *J. Am. Chem. Soc.* **96**, 3317 (1974).
125. P. C. Martino, P. B. Shevlin, and S. D. Worley, *J. Am. Chem. Soc.* **99**, 8003 (1977).
126. J. E. Baldwin and L. I. Kruse, *J. Chem. Soc., Chem. Commun.* p. 235 (1977).
127. P. Beltrame, V. Rosnati, and F. Sannicolo, *Tetrahedron Lett.* p. 4219 (1970).
128. J. P. Dirlam, L. Eberson, and J. Casanova, *J. Am. Chem. Soc.* **94**, 240 (1972).
129. H. M. R. Hoffmann and T. A. Nour, *J. Chem. Soc., Chem. Commun.* p. 37 (1975).
130. H. M. R. Hoffmann, *Angew. Chem. Int. Ed. Engl.* **12**, 819 (1973).
131. J. Levisalles, E. Rose, and I. Tkatchenko, *J. Chem. Soc., Chem. Commun.* p. 445 (1969).
132. H. E. Zimmerman and G. A. Epling, *J. Am. Chem. Soc.* **94**, 7806 (1972).
133. O. L. Chapman, as quoted in Zimmerman and Epling (*132*).
134. D. B. Sclove, J. F. Pazos, R. L. Camp, and F. D. Greene, *J. Am. Chem. Soc.* **92**, 7488 (1970).
135. C. E. Hudson and N. L. Bauld, *J. Am. Chem. Soc.* **95**, 3822 (1973).

136. J. K. Crandall and W. W. Conover, *J. Chem. Soc., Chem. Commun.* p. 340 (1973).
136a. D. Mitchell, J. H. Eilert, and N. L. Bauld, *Tetrahedron Lett.* p. 2865 (1979).
136b. A. S. Kende, R. Greenhouse, and J. A. Hill, *Tetrahedron Lett.* p. 2867 (1979).
136c. A. J. Fry, W. A. Donaldson, and G. S. Ginsburg, *J. Org. Chem.* **44,** 349 (1979).
137. E. E. Smissman, T. L. Lemke, and O. Kristiansen, *J. Am. Chem. Soc.* **88,** 334 (1966).
138. H. O. House and G. A. Frank, *J. Org. Chem.* **30,** 2948 (1965).
139. S. Vickers and E. E. Smissman, *J. Org. Chem.* **40,** 749 (1975).
140. B. S. Ong and T. H. Chan, *Tetrahedron Lett.* p. 3257 (1976).
141. M. Bertrand, J. P. Dulcere, G. Gil, J. Grimaldi, and P. Sylvestre-Panthet, *Tetrahedron Lett.* p. 3305 (1976).
142. T. H. Chan, B. S. Ong, and W. Mychajlowskij, *Tetrahedron Lett.* p. 3253 (1976).
143. C. Rappe, L. Knutson, N. J. Turro, and R. B. Gagosian, *J. Am. Chem. Soc.* **92,** 2032 (1970).
144. J. F. Pazos, J. G. Pacifici, G. O. Pierson, D. B. Sclove, and F. D. Greene, *J. Org. Chem.* **39,** 1990 (1974).
145. W. Kirmse, A. Engelmann, and J. Heese, *J. Am. Chem. Soc.* **95,** 625 (1973).
146. G. Stork and I. J. Borowitz, *J. Am. Chem. Soc.* **82,** 4307 (1960).
147. H. O. House and W. F. Gilmore, *J. Am. Chem. Soc.* **83,** 3980 (1961).
148. C. H. Engel, S. K. Roy, J. Capitaine, J. Bilodeau, C. McPherson-Foucar, and P. Lachance, *Can. J. Chem.* **48,** 361 (1970).
149. A. Skrobek and B. Tchoubar, *C. R. Hebd. Seances Acad. Sci. Ser. C* **263,** 80 (1966).
150. H. O. House and W. F. Gilmore, *J. Am. Chem. Soc.* **83,** 3972 (1961).
151. R. W. Mouk, K. M. Patel, and W. Reusch, *Tetrahedron* **31,** 13 (1975).
152. H. O. House and F. A. Richey, *J. Org. Chem.* **32,** 2151 (1967).
153. F. G. Bordwell and J. G. Strong, *J. Org. Chem.* **38,** 579 (1973).
154. W. J. M. Van Tilborg, S. E. Schaafsman, H. Steinberg, and T. J. DeBoer, *Recl. Trav. Chim. Pays-Bas* **86,** 417 (1967).
155. N. J. Turro, *Acc. Chem. Res.* **2,** 25 (1969).
156. N. J. Turro and W. B. Hammond, *J. Am. Chem. Soc.* **87,** 3258 (1965).
157. J. P. Guthrie, personal communication.
158. E. W. Warnhoff, C. M. Wong, and W. T. Tai, *J. Am. Chem. Soc.* **90,** 514 (1968).
159. P. M. Warner, S. L. Lu, E. Myers, P. W. DeHaven, and R. A. Jacobson, *J. Am. Chem. Soc.* **99,** 5102 (1977).
160. P. M. Warner and S. L. Lu, *J. Am. Chem. Soc.* **98,** 6752 (1976).
161. P. M. Warner, R. F. Palmer, and S. L. Lu, *J. Am. Chem. Soc.* **99,** 3773 (1977).
162. M. J. A. McGrath, *Tetrahdron* **32,** 377 (1976).
163. W. Reusch and P. Mattison, *Tetrahedron* **23,** 1953 (1967).
164. R. Verhé, N. Schamp, L. DeBuyck, and R. Van locke, *Bull. Soc. Chim. Belg.* **84,** 381 (1975).
165. F. G. Bordwell and M. W. Carlson, *J. Am. Chem. Soc.* **92,** 3377 (1970).
166. J. Wolinsky and R. O. Hutchins, *J. Org. Chem.* **37,** 3294 (1972).
167. F. G. Bordwell and W. T. Brannen, *J. Am. Chem. Soc.* **86,** 4645 (1964).
168. A. Halvorsen and J. Songstad, *J. Chem. Soc., Chem. Commun.* p. 327 (1978).
168a. W. B. Smith and C. Gonzalez, *Tetrahedron Lett.* p. 5751 (1966).
169. C. L. Stevens and E. Farkas, *J. Am. Chem. Soc.* **74,** 618 (1952).
170. J. E. Baldwin and J. H. I. Cardellina, *J. Chem. Soc.* p. 558 (1968).
171. J. W. Wilt and R. R. Rasmussen, *J. Org. Chem.* **40,** 1031 (1975).
172. P. J. Chenier and J. C. Kao, *J. Org. Chem.* **41,** 3730 (1976).
173. A. Baretta and B. Waegell, *Tetrahedron Lett.* p. 753 (1976).
174. A. Kurek, L. Kohout, J. Fajkos, and F.Šorm, *Collect. Czech. Chem. Commun.* **38,** 279 (1973).
175. N. Schamp, N. De Kimpe, and W. Coppens, *Tetrahedron* **31,** 2081 (1975).

176. J. Szmuszkovicz, E. Cerda, M. F. Grostic, and J. F. Zieserl, *Tetrahedron Lett.* p. 3969 (1967).
177. J. Szmuszkovicz, D. J. Duchamp, E. Cerda, and C. G. Chidester, *Tetrahedron Lett.* p. 1309 (1969).
178. J. C. Blazejewski, D. Cantecuzene, and C. Wakselman, *Tetrahedron* **29**, 4233 (1973).
179. F. G. Bordwell and J. Almy, *J. Org. Chem.* **38**, 571 (1973).
180. K. Sato, S. Inoue, M. Ohashi, and S. I. Kuranami, *Chem. Lett.* p. 405 (1975).
181. K. Sato, S. Inoue, and S. I. Kuranami, *J. Chem. Soc. Perkin Trans.* 1 p. 1666 (1977).
182. K. B. Wiberg and G. W. Klein, *Tetrahedron Lett.* p. 1043 (1963).
183. E. Wenkert, P. Bakuzis, R. J. Baumgarten, C. L. Leicht, and H. P. Schenk, *J. Am. Chem. Soc.* **93**, 3208 (1971).
184. R. H. Bisceglia and C. J. Cheer, *J. Chem. Soc., Chem. Commun.* p. 165 (1973).
185. S. Wolff and W. C. Agosta, *J. Chem. Soc., Chem. Commun.* p. 771 (1973).
186. A. Takeda and S. Tsuboi, *J. Org. Chem.* **38**, 1709 (1973).
187. S. Acevado, K. Bowden, and M. P. Henry, *Tetrahedron Lett.* p. 4837 (1976).
188. G. M. Iskander and F. Stansfield, *J. Chem. Soc. C* p. 669 (1969).
189. A. Takeda, S. Tsuboi, F. Sakai, and M. Tanabe, *Tetrahedron Lett.* p. 4961 (1973).
190. J. M. Conia and J. L. Ripoll, *Bull. Soc. Chim. Fr.* p. 773 (1963).
191. J. M. Conia and J. Salaun, *Bull. Soc. Chim. Fr.* p. 1957 (1964).
192. E. W. Warnhoff, W. T. Tai, and Y. C. Toong, *Can. J. Chem.* **56**, 93 (1978).
193. D. Baudry, J. P. Bégué, and M. Charpentier-Morize, *Bull. Soc. Chim. Fr.* p. 1416 (1971).
194. C. L. Stevens, P. M. Pillai, and K. G. Taylor, *J. Org. Chem.* **39**, 3158 (1974).
195. W. T. Brady and J. P. Hieble, *J. Am. Chem. Soc.* **94**, 4278 (1972).
196. W. T. Brady and P. T. Ling, *J. Chem. Soc. Perkin Trans.* 1 p. 456 (1975).
197. J. M. Conia and J. Salaun, *Acc. Chem. Res.* **5**, 33 (1972).
198. P. R. Brook and J. M. Harrison, *J. Chem. Soc., Chem. Commun.* p. 997 (1972).
199. K. V. Scherer, *Tetrahedron Lett.* p. 5685 (1966).
200. R. N. McDonald and C. A. Curi, *Tetrahedron Lett.* p. 1423 (1976).
201. G. W. Klump, W. G. J. Rietman, and J. J. Vrielink, *J. Am. Chem. Soc.* **92**, 5266 (1970).
202. B. R. Vogt, *Tetrahedron Lett.* p. 1579 (1968).
203. M. Nakazaki and K. Naemura, *J. Org. Chem.* **42**, 4108 (1977).
204. T. Y. Luh and L. M. Stock, *J. Org. Chem.* **37**, 338 (1972).
205. W. C. Fong, R. Thomas, and K. V. Scherer, *Tetrahedron Lett.* p. 3789 (1971).
206. V. R. Fletcher and A. Hassner, *Tetrahedron Lett.* p. 1071 (1970).
207. E. W. Warnhoff, C. Bonnice, and Y. C. Toong, unpublished results.
208. L. Ramberg and B. Bäcklund, *Ark. Kemi Mineral. Geol.* **13A**, No. 27 (1940); CA 34:4725 (1940).
209. L. A. Paquette, *Org. React.* **25**, 1 (1977).
210. F. G. Bordwell, *Acc. Chem. Res.* **3**, 281 (1970).
211. S. W. Schneller, *Int. J. Sulfur Chem.* **8**, 491 (1973); also reprinted without updating in **8**, 585 (1976).
212. F. G. Bordwell and J. B. O'Dwyer, *J. Org. Chem.* **39**, 2519 (1974).
213. F. G. Bordwell and M. D. Wolfinger, *J. Org. Chem.* **39**, 2521 (1974).
214. F. G. Bordwell and E. Doomes, *J. Org. Chem.* **39**, 2531 (1974).
215. F. G. Bordwell and E. Doomes, *J. Org. Chem.* **39**, 2526 (1974).
216. F. G. Bordwell, J. C. Branca, C. R. Johnson, and N. R. Vanier, *J. Am. Chem. Soc.* (1980).
217. F. G. Bordwell, J. M. Williams, E. B. Hoyt, and B. B. Jarvis, *J. Am. Chem. Soc.* **90**, 429 (1968).
218. F. G. Bordwell and J. M. Williams, *J. Am. Chem. Soc.* **90**, 435 (1968).
219. L. A. Carpino and L. V. McAdams, *J. Am. Chem. Soc.* **87**, 5804 (1965).
220. F. G. Bordwell and B. B. Jarvis, *J. Am. Chem. Soc.* **95**, 3585 (1973).
221. E. J. Corey and T. H. Lowry, *Tetrahedron Lett.* p. 793 (1965).

INDEX

A

Acetals, cyclic, ring-opening reactions, 224–226
Acetone, protonated, dissociation, 86–90
4-Acetoxyhomocuneane, homoketonization, 435
Acetylenes, rearrangement reactions with organomagnesiums, 401–407
Acetyl migration, 181
β-Acyloxyalkyl radicals, rearrangement, 242–244
Adamantane, by tetrahydrodicyclopentadiene isomerization, 25
Adamantane matrix technique, kinetic studies, 167
Adamantanone, homoenolization, 420, 433
1-Adamantyl cation
  1,2-hydride shift, 42
  stability, 19
1-Adamantylethyne, retro 1,2 rearrangements, 118
Alkanoylalkyl radicals, rotational isomerization, 272, 273
Alkenaminyl radical, ring-closure reactions, 203, 205–207
Alkenes, retro 1,2 rearrangements, 116–119
Alkenoxy radical, ring-closure reactions, 203
Alkenylaryl radicals, cyclization, 209–211
Alkenylperoxy radical, cyclization, 205
Alkenylthiyl radicals, cyclization, 207, 208
Alkoxy migration, 1,2 rearrangement, 98, 99
Alkylaminyl radical, 1,2 hydrogen migration, 264
Alkyl migration, 98, 99, 394–397
Alkynes
  retro 1,2 rearrangements, 118
  thermal rearrangements, 313–315, 318
Alkynyl radicals, cyclization, 202, 203
Allene oxide, in Favorskii rearrangement, 438–446

Allylcarbenes, intramolecular reactions, 108–113
Allyl cation
  rearrangement, 3
  solvent effect, 18
Allylic radicals, rotational cis–trans isomerization, 270–272
Allyl migration, 396
Allyloxyprop-2-yl radical, cyclization, 191, 192
Allyl resonance stabilization energy, 270
Aminium radical, hydrogen migration, 264
Amino migration, 1,2 rearrangement, 98, 99
Aminyl radicals, cyclization of unsaturated, 205–207
t-Amyl cation, solvent effect on rearrangement, 15, 16
Antimony pentafluoride, for solvation of carbocations, 14
Apollan-11-oxy radical, 1,5-hydrogen migration, 261
Arylcarbenes
  interconversion, 128–136
  rearrangements, 123–125
Aryl migration, 403, 408–410
  1,2 rearrangement, 99
  1,2 shift, kinetic data, 170–175
  1,3, 1,4, and 1,5 shifts, 175–178
Arylnitrenes, interconversion, 128–136
1-Aza-1,2,4,6-cycloheptatetraene, formation from phenylnitrene, 134, 135
Azepine, formation from phenylnitrene, 134
Azetidin-2-yl radical, ring-opening reaction, 226
Azibenzil-carbonyl-^{13}CO, photolysis, 322
Azides
  photolysis, 101, 133
  thermal reactions, 101, 133
Aziridinyl radicals
  inversion, 279
  ring-opening reactions, 223

Azirine, formation from phenylnitrene, 134, 135
Azo compounds, ring-closure reactions, 218

B

Benzanilide, ring-closure reactions, 211–213
Benzannelation, effect on carbene rearrangements, 128–131
Benzazulene, formation mechanism, 314
Benzenium ions, rearrangements, 44, 45
Benzotropone tosylhydrazone, rearrangement, 129
Benzyl cation, detection, 79
Benzyl migration, 394, 395
Bicyclic radicals, ring-opening reactions, 222
Bicyclo[2.2.1]heptanedione, homoenolization, 427
Bicyclo[4.1.0]heptatriene
 rearrangement, 132
 theory, 123–125
Bicyclo[3.2.0]hept-2-ene, rearrangement, 375
Bicyclo[2.2.0]hexane-2,3,5,6-d_4, pyrolysis, 369–371
Bicyclo[3.1.0]hex-3-en-2-yl cation, rearrangement, 45, 46
Bicyclo[2.1.1]hexyl cation, deuterium effect, 9
Bicyclo[2.2.2]octanone, homoenolization, 420–422
Bicyclo[3.2.1]octanone, homoenolization, 420–422
Bicyclo[2.2.2]octenone, homoenolization, 422–424
Bicyclopropenyl, thermal rearrangements, 320–322
Biradical hypothesis, thermal unimolecular rearrangement pathways, 311–390
Biradicals
 cleavage-cyclization mechanism, 360–368
 cyclization-cleavage mechanism, 368–372
 internal rotational rates, 356–358
 1,4, from olefin dimerizations, cyclobutane reversions and diazene decompositions, 353–372
 in sigmatropic rearrangements, 372–383
 1,3, trimethylene, 324–352

Bishomocubane alcohol, homoketonization, 435
1,2-Bismethylenecyclobutane
 formation and thermal rearrangement, 351, 352
 pyrolysis, Cope rearrangement, 372
1,3-Bismethylenecyclobutane, formation, 351, 352
1,4-Bismethylenecyclohexane, rearrangement, 367
3,4-Bismethylenecyclopentanone, photolysis, 351
Born equation, 19
Boron migration, 397, 398
Brexan-2-one, homoenolization, 415, 433
Brexanylidene, rearrangement, 107
Bridged radicals, rotational isomerization, 273–275
1-Bromo-2-methylbutane, photobromination, 273, 274
Bullvalene, biradical intermediates, 359
1-Butene
 rearrangement, 400
 retro 1,2 rearrangement, 116
Butenyl radical, rotational isomerization, 270
3-Butenyl radical, cyclization, 198, 199
t-Butyl bromide, reaction with t-butylhypochlorite, 250
t-Butyl cation
 solvent effect, 15
 stability, 19
t-Butyl chloride, reaction with t-butyl hypobromite, 248

C

Camphenilone, homoenolization, 412, 417, 431, 432
Camphor, rearrangements, 31, 32
Carbanion
 definition, 392
 rearrangements, 391–470
Carbene
 "foiled," 108–113
 intermediate, in acetylene pyrolysis, 313, 314
 in cyclopropene pyrolysis, 315–322
Carbene–carbene rearrangement, 96
 in gas phase, 137–143

INDEX 473

in solution, 126–136
theory, 122–126
type I and type II, 119–122
type II, 149–153
Carbene–nitrene rearrangement, 96, 100
in gas phase, 137, 143–149
theory, 122, 125, 133–136
type, I, 120–122
Carbene rearrangement, 95–119
1,2, 97–101
stereochemistry, 105–108
theory and mechanism, 102–104
definition, 95–97
retro 1,2, 114–119
Carbocations
migratory aptitudes, 34–41
observable, 3–7
rearrangements, 1–53
synthetic potentials, 47, 48
solvation, 14–21
1-Carbomethoxy-2,4-dimethylspiropentane, thermal rearrangement, 350
Carbon, migration of unsaturated, 400–410
Carbonium ion
rearrangements, 1–3
solvent effects on kinetics and equilibria, 18
Carbonylcarbenes, rearrangement, theory, 122, 126–136
Carbonyl group migration, 391, 392, 400
Cephalotaxine, synthesis, 210
N-Chloroazacyclononane, hydrogen migration isomerization, 264
N-Chloroazacyclooctane, hydrogen migration isomerization, 264
N-Chloro-5-deuterio-2-hexylamine, Hoffmann–Löffler reaction, 265, 266
Collisional activation, ion detection, 77–79
Copacamphor, homoenolization, 420
Cope rearrangement
1,4 biradical intermediates, 358–372
epimerically unfavorable, 360–364
geometrically unfavorable, 364–368
Cubane alcohol, homoketonization, 434, 435
Cyano migration, 182
Cycloalkenyl cations, rearrangements, 44–47
Cycloalkylcarbinyl radicals, ring-opening reactions, 227–237

Cycloalkylimino radicals, ring-opening reactions, 241
Cycloalkyl radicals, ring-opening reactions, 220–227
Cyclobutanes, fragmentation/internal rotation rates, 357, 358
Cyclobutanones, retro 1,2 rearrangements, 115, 116
Cyclobutenylcarbinyl radical, rearrangement, 237
Cyclobutenyl cation, rearrangement, 45, 46
Cyclobutylcarbinyl radicals, rearrangements, 236, 237
Cyclobutyloxy radical, rearrangement, 238
1,5-Cyclodecadiene
cyclization, 196
rearrangement, 368
Cycloheptatrienylidene, rearrangements, 128–132
Cyclohexa-1,4-diyl biradical, formation, 368–371
Cyclohexanol, synthesis by ring closure, 216
Cyclohexanylidene, rearrangement, 103, 105, 106
Cyclohexenylbutyl radical, cyclization, 192
Cyclohexyl cation
deuterium effect, 8, 9
rearrangement, 4
1,5-Cyclooctadiene, cyclization, 195
Cyclopentanones, retro 1,2 rearrangements, 115
2-Cyclopentenecarbinyl radical, vinyl migration, 180
Cyclopentenyl cation
deuterium effect, 8, 9
hydride shifts, 41–43
rearrangement, 38
rearrangement mechanism, graph theory, 24
Cyclopentyloxy radical, rearrangement, 239
Cyclopropane
pyrolysis, moninteracting biradical model, 324–326
quantum mechanical model, 326–329
stereomutations, 328–334
trimethylene biradicals, 324–352
Cyclopropanone, formation and cleavage, in Favorskii rearrangement, 437–461
Cyclopropene
pyrolysis, 325

thermal rearrangements, 315
Cyclopropylallyl radicals, rearrangement, 231, 232
Cyclopropyl–allyl rearrangement 221
Cyclopropyl-1,1-biscarbinyl biradical, 349
Cyclopropylcarbinyl radical, ring-opening reactions, 227–235
Cyclopropyldiphenylcarbinyl radical rearrangement, 178, 179
1-Cyclopropyl-1-hydroxyalkyl radical, ring opening by 1,5 hydrogen migration, 267
Cyclopropylidenes, ring opening, 96
Cyclopropyloxy radical, rearrangement, 238
Cyclopropyl radicals, inversion, 276–278
Cycloreversion, thermal, 353

D

9-Decalyl radical
   formation, 192
   inversion, 280
Degenerate rearrangement, 5–11, 24
Deuterium, effect on degenerate 1,2-shifts, 8–11
Diamantane, formation graph, 29–31
Diazenes
   fragmentation/internal rotation rate, 357, 358
   thermal decomposition, 354
Diazo ketones, photochemical Wolff rearrangements, 322, 323
Dibenzo[c,e]tropone, formation and reactivity, 442, 444
Di-t-butylcyclopropanone, thermal rearrangement, 442, 444, 449, 451, 453
1,2-Dideuterocycloheptene, retro 1,2 rearrangement, 117
Dienes, cyclopolymerization, 195
1,3-Diethylcyclopropene, thermal rearrangements, 316–320
Diethynylcyclopropane, rearrangement, in gas phase, 142
1,1-Dimethylallyl radical, rotational isomerization, 271
3,3′-Dimethylbicyclopropenyl, thermal rearrangements, 321
1,2-Dimethylcyclobutane, thermal cycloreversion and stereomutation, 353, 354

1,6-Dimethylcyclodecyl cation, structure, 10, 11
2,7-Dimethylcycloheptatrienylidene, rearrangement, 128
1,3-Dimethylcyclohexyl cation, 1,3-hydride shift, 12, 13
1,4-Dimethylcyclohexyl cation, hydride shift, 13
1,2-Dimethylcyclopentyl cation, deuterium effect on hydride shift, 8, 10
3,3-Dimethylcyclopropene, thermal rearrangement, 316
2,6-Dimethylheptyl cation, 1,5-hydride shift, 14
3,4-Dimethylhexa-1,5-diene, stereospecific conversions, 371
2,5-Dimethyl-2-hexyl cation, 1,4-hydride shift, 12, 13
2,3-Dimethylmethylenecyclopropane, pyrolysis, 336
Dimethyl-2-norbornyl cation
   deuterium effect, 10
   rearrangement graph, 32–34
3,5-Dimethylpyrazoline, pyrolysis, stereochemical crossover, 326
3,4-Dimethyltetrahydropyridazine, thermal decomposition, 354
2,2-Dimethyltetrahydrothiophene, formation by ring closure, 219
1,3-Dioxanyl radical, ring-opening reactions, 225
γ,γ-Diphenylallylcarbinyl radical, rearrangement, 178, 179
1,3-Diphenylallyl radical, rotational isomerization, 272
6,6-Diphenylbicyclo[3.1.0]hexenone, irradiation, 442, 443
2,7-Diphenylcycloheptatrienylidene, rearrangement, 128
Diphenylcyclopropyl radical, ring-opening reaction, 221, 222
Diphenylethylenes, retro 1,2 rearrangements, 117
1,2-Dipropenylcyclobutane, thermal rearrangement rate constants, 362–364
Dithioacyloxy radicals, ring-closure reactions, 218
1,3-Dithiolanes, synthesis by ring closure, 218
1,1-Divinyl-3-methylenecyclobutane, thermal rearrangement, 364–366

E

Electrolysis, free radical generation, 212
Electron spin resonance spectroscopy, determination of unimolecular radical reaction rate, 165–168
Energy profile, *see* Potential energy profile
Episulfones, in Ramberg–Bäcklund reaction, 461–465
Ethanonoradamantane, isomers, 27
Ethyl cation, 3, 4
  rearrangement mechanism, graph theory, 24, 25
Ethylidene, rearrangement, 102, 103
1-Ethylidene-2-cyano-3-methylcyclopropane, thermal rearrangement, 341, 342
2-Ethynylbiphenyl, thermal rearrangement, 314
1-Ethynyl-1-methylcyclohexane, thermal rearrangement, 314
Ethynylpyrroles, formation, 147
o-Ethynyltoluene, thermal rearrangement, 314

F

Favorskii amides, 455, 456
Favorskii rearrangement, 437–461
Fenchone, homoenolization, 416
Fenchyl cation, rearrangement graph, 32, 33
Fluorobenzenium ions, rearrangements, 44, 45
α-Fluorocyclobutyl radical, inversion, 279
α-Fluorocyclopropyl radicals, inversion, 277
Free-radical generation, 212
Free-radical rearrangement, 161–310, *see also* Isomerization
  by carbon-centered group transfer, 168–182
  carbon to carbon migration, 1,2-acyloxy, 242–244
    organothiyl, 244–246
  group mobilities, 168–170, 241, 242
  migration to or from heteroatoms, 247, 248
  by transfer of heteroatom-centered group, 241–248
Fulveneallene, formation, 140–142

G

Gas-phase ions, 2, 19
  rearrangements, 55–93
  versus solution experiments, 56, 57
Germyl radicals, inversion, 281, 282
Grignard compounds, rearrangement reactions with alkenes and acetylenes, 402–407
Grovenstein–Zimmerman rearrangement, phenyl migration, 403

H

Halogen transfer, isomerization, 248–251
Heat of formation, of cations in gas phase, 67–69
1,6-Heptadiene, cyclization, 195
1,6-Heptadiyne, cyclization, 203
Heptafulvalene, formation, 137, 140, 141, 142
Heptamethylbenzenonium ion, solvent effect on rearrangement, 16
6-Heptenyl radical, cyclization, 198, 201
6-Hepten-2-yl radical, cyclization, 190, 191
Heptynyl radical, cyclization, 202
Heteroatom migration, 163, 397–400
Heterocycles, synthesis, 212
5-Hexenyl nitrite, photolysis, 204
5-Hexenyl radical
  cyclization, of substituted, 189–198
  generation and cyclization, 185–189
5-Hexynyl radical, cyclization, 202
Hoffmann–Löffler reaction, hydrogen migration from carbon to nitrogen, 265, 266
Homoadamantyl cation, rearrangements, 34
Homoallylic anion, open form, 400, 401
Homocubane alcohol, homoketonization, 434
Homocubyl cation, rearrangement, 38
Homoenolate anion, closed form, 400
Homoenolization, 410, 411
  acyclic systems, 427–430
  discovery, 412, 413
  methodology, 414, 415
  polycyclic systems, 415–427
Homoketonization, 411, 417, 430–437
Homovinylcyclopropylidene, rearrangement, 150

Hydrogen migration; Hydrogen shift; Hydrogen transfer, 98, 102, 163, 392–394
1,2-Hydrogen migration
   carbon to carbon, free-radical rearrangement, 253
   carbon to oxygen, 258, 259
   degenerate, 6
   effect of ring size, 41–44
1,3-Hydrogen migration
   carbon to carbon, free-radical rearrangement, 253, 254
   degenerate, 11
   oxygen to carbon, 266
1,4-Hydrogen migration, 12, 13
   carbon to carbon, free-radical rearrangement, 255, 256
   from carbon to nitrogen, 264
1,5-Hydrogen migration, 13, 14, 140
   carbon to carbon, free-radical rearrangement, 256–258
   from carbon to nitrogen, 264–266
   carbon to oxygen, 260–264
   oxygen to carbon, 267
   oxygen to oxygen, 268
1,6-Hydrogen migration, carbon to carbon, free-radical rearrangement, 256–258
1,7-Hydrogen migration, carbon to carbon, free-radical rearrangement, 257

I

Iminocarbenes, rearrangement, theory, 122, 126–128
Iminocyclohexadienylidene, rearrangement, in gas phase, 146
Iminoxy radicals, inversion, 282, 283
2-Indanone, 1-phenyl, formation, 440, 441
Indanones, synthesis by ring closure, 214
Indene, formation mechanism, 314
Inversion, *see also* Isomerization
   carbon-centered radicals, aziridinyls, 279
   cycloprophyls, 276–278
   9-decalyl, 280
   α-fluorocyclobutyl, 279
   2-methyl-1,3-dioxolan-2-yl, 279
   oxiranyls, 277–279
   vinyl radicals, 280, 281
   definition, 164
   heteroatom-centered radicals, iminoxy radicals, 282, 283

phosphoranyl radicals, 283
silyl, germyl, and stannyl radicals, 281, 282
Ion cyclotron resonance, ion detection, 77–79
Isocamphanone, homoenolization, 417–419
Isomerization, *see also* Inversion
   by bromine 1,2 migration, 250
   by chlorine 1,2 migration, 248–250
   definition, 162, 163
   by halogen transfer to more remote radical center, 251
   by hydrogen atom transfer, atom mobilities, 251–253
   carbon to carbon migration, 253–258
   carbon to nitrogen, 264–266
   carbon to oxygen, 258–264
   oxygen to carbon, 266–268
   oxygen to oxygen, 268–269
   rotational, 164
   for alkanoylalkyl radicals, 272, 273
   of allylic radicals, 270–272
   for bridged radicals, 273–275
Isopropyl cation, rearrangement, 4
Isotope labeling, in potential energy profile approach to ion rearrangement, 64–67

K

Ketocarbene–oxirene system, rearrangements, 322–324
Ketones
   carbanionic rearrangements, 437–461
   cyclic, synthesis by ring closure, 210, 217
   retro 1,2 rearrangements, 115, 116

L

Lactones, synthesis by ring closure, 216, 217
Lewis acids, in solvation of carbocations, 15
Longicamphenilone, homoenolization, 416
Longicamphor, homoenolization, 420

M

Mass spectrometry, historical development, 55, 56
Meldrum's acid, pyrolysis of benzylidene derivative, 313

Metastable ions
  dissociation, 56
  metastable peak, broad gaussian, 73–77
  channeling, 58, 59
  flat-topped or dished, 71–73
  kinetic energy release, 59–62, 70–77
  narrow and gaussian, 70, 71
  slow reactions, 57
1,6-Methano[10]annulene-11-yl radical rearrangement, 222
Methanoannulenylcarbene, rearrangement, 132, 133
Methyl cation, 4
Methylcyclopentyl cation, 5
Methylcyclopropane, rearrangement, 400
2-Methyl-2,3-dioxolan-2-yl radical, inversion, 279
6-Methylenebicyclo[3.2.1]oct-2-ene, rearrangement, 367, 368
3-Methylenebicyclo[3.2.1]oct-6-ene, rearrangement, 367
Methylenecyclobutane, rearrangement, 348, 349, 375
Methylenecyclohexadienylidene, rearrangement in gas phase, 142
Methylenecyclopropane
  pyrolysis, concerted reactions, 341, 342
  mechanism role of trimethylenemethanes, 334–341
  scope of reaction, 334
  structures and energies of hypothetical intermediates, 342–348
Methylenespiropentane, thermal rearrangements, 348, 351, 352
2-Methyl-6-ethylidenebicyclo[3.1.0]hexane, pyrolysis of stereoisomers, 340, 341
3-Methylpenta-1,4-diyl biradical, 354–356
3-Methyl-4-pentenal, decarbonylation, 178, 179
p-Methylphenylcarbene, rearrangement, in gas phase, 139
Methyltropone tosylhydrazone, rearrangement in gas phase, 139
Migratory aptitude, 34–41, 98–101
Monte-Carlo method, analysis, rearrangement, 23, 24

N

Naphthylbutyl radical, cyclization, 214, 215
Naphthylcarbene, rearrangement, 131, 132

Neophyl rearrangement, 170–175
Nesmeyanov rearrangement, 1,2 chlorine shift, 248
Nitrene–nitrene rearrangements, in gas phase, 137, 148
Nitrene rearrangement, 95–119
  definition, 95–97
Nitrobenzene, rearrangement to nitrene, 133
Nitro compounds, ring-closure reactions, 218
Nitrosobenzene, rearrangement to nitrene, 133
7-Norbornadienyl acetate, pyrolysis, 141
7-Norbornadienylidene, rearrangement, 150
Norbornane, rearrangement mechanisms, graph theory, 30, 31
Norbornene, hydride shifts, 44
Norbornenyl radical
  cyclization, 199
  rearrangement, 199, 233
anti-7-Norbornenyl tosylate, solvolysis, 7
Norbornyl cation
  deuterium effect, 9, 10
  rearrangement pathways, 26
  rearrangements, 31, 43
  solvation energy, 20
Norbornyl tosylate
  hydride shift rates, 21
  solvolysis, 7
Norcaradiene, sigmatropic rearrangements, 376–383

O

7-Octenyl radical, cyclization, 198, 201
Olefins, see also Alkenes
  biradicals by dimerization, 353
  rearrangement reactions with organomagnesiums, 401–407
Oxepane, ring-opening reactions, 224
2-Oxetanyl radical, ring-opening reactions, 224
Oxiranyl radicals
  inversion, 277–279
  ring-opening reactions, 223, 224
Oxirenes, rearrangements, 322, 323
Oxyallyl anion, formation in Favorskii rearrangement, 437–442

P

Paracyclophanes, synthesis, 97
Pentacyclotetradecane, rearrangement pathways, 29–31
Pentamethylcyclobutenyl cation, rearrangement of deuteriated, 45, 46
4-Pentenoxy radical, ring-closure reactions, 203
4-Pentenyl radical, cyclization, 198, 200
4-Pentenylsilyl radicals, cyclization, 208
4-Pentynoxy radical, cyclization, 204
Phenanthrene, synthesis by ring closure, 211
Phenylacetylene, formation mechanism, 313–315
Phenylacetyl migration, 182
4-Phenylbutyl radical, 177
  ring closure, 215
Phenylcarbene
  rearrangement in gas phase, 137–143
  in solution, 133
cis-6-Phenyl-2-chlorocyclohexanone, reaction with piperidine, 455, 456
3-Phenylcycloheptene, retro 1,2-rearrangement, 117
2-Phenylethyl radical, 170, 172
7-Phenyl-6-heptynyl radical, cyclization, 202
3-Phenyl-3-methylhepta-2,5-diene, stereospecific conversions, 371
Phenyl migration, 403, 408–410
Phenylnitrene
  rearrangement, 125, 133
  in gas phase, 143–149
  in solution, 133–136
N-Phenyloxazolidine, rearrangement to nitrene, 133
3-Phenylpropyl radical, ring closure, 214
Phosphoranyl radicals
  cyclization of unsaturated, 208, 209
  inversion, 283
Phosphorus compounds, cyclic, synthesis by ring closure, 219
Photoionization, heat of formation analysis, 67
Photolysis, free radical generation, 212
Potential energy profile
  collisional activation and ion cyclotron resonance, 77–79
  complete equilibration of isomers before unimolecular dissociation, 80–82
  energetics, 67–69
  isotope labeling, 64–67, 86
  kinetic energy release, 70–77
  metastable abundance data, 63, 64
  no equilibration of isomers before unimolecular dissociation, 82–85
  predictive capacity, 80
  rate-determining isomerization before unimolecular dissociation, 85–90
  of slow reactions and rearrangements of ions, 62, 63
Propionaldehyde, protonated, dissociation, 86–90
Propylene sulfide, ring-opening reactions, 226
Prostaglandins, biosynthesis, 205
Proton affinity, heat of formation analysis, 67
Pschorr reaction, 211, 212
Pyrazoline, pyrolysis of 3,5-disubstituted, 326–329
Pyridazines, pyrolysis, 354, 355
Pyridylcarbene
  rearrangements, 125
  in gas phase, 143–149
Pyridyl radical, photolytic ring opening, 226, 227

R

Radical reactions, rate measurement of unimolecular, 165–168
Radicals, generation of free, 212
Ramberg–Bäcklund reaction, 461–465
Rearrangement, see also Cope rearrangement; Degenerate rearrangement; Free-radical rearrangement; Neophyl rearrangement; Sigmatropic rearrangement; Wagner–Meerwein rearrangement; Wolff rearrangement
  carbanion, 391–470
  carbenes and nitrenes, 95–160
  carbocations, 1–53
  definition, 312, 391, 392
  gas-phase ion, 55–93
  least-motion pathway, 376
  mechanism, graph theory, 24–34
  migratory aptitudes, 34–41
  Monte-Carlo method of analysis, 23, 24
  multiple, 22–34
  potential energy profile, 62–80
  radical, definitions, 162–165
  retro 1,2, of carbenes, 114–119, 141

INDEX

thermal unimolecular, hypothetical biradical pathways, 311–390
Regioselectivity
  cyclization, 201, 213, 266, 267
  free-radical cyclization, 198, 203
  1,5 hydrogen migrations, 261
  ring opening, 430, 432
Ring-closure reactions, 163, 172, 179, 182–210
  comparison with analogous intermolecular processes, 182–185
  5-hexenyl system, 185–198
  intramolecular aromatic homolytic substitution, 211–216
  rearrangement, alkenes and acetylenes with organomagnesiums, 402, 403, 406
  organomagnesium compounds, 402, 403, 407
  regioselectivity, 201, 213, 266, 267
Ring-opening reactions, 163
  cycloalkylcarbinyl radicals, 227–237
  cycloalkyloxy and similar heteroradicals, 238–241
  cycloalkyl radicals and related species, 220–227
Ring size
  effect on hydride 1,2-shifts, 41–44
  on migratory aptitude, 37
Rotation, see Isomerization
Rotation rate, of biradicals, 356–358

S

Selenocarbonylcarbenes, interconversion, theory, 126–128
Sigmatropic rearrangement, 341, 362
  benzyl migration, 396
  biradicals, 372–383
  hydrogen, 393
Silicon migration, 399
Silyl radicals
  cyclization of unsaturated, 208
  inversion, 281, 282
Skattebøl rearrangement, mechanism, 150–153
Solvation, of carbocations, 14–21
Solvents, effect on properties of ions, 2
Solvolysis rate, 2
Spiro compounds, synthesis, 212
Spirocyclopropanepyrazoline, thermal rearrangement, deazetation, 351

Spirodecyl radical, formation, 192
Spiro[2.5]octadienyl radical, 170, 172
Spiropentane, thermal rearrangements, 348–352
Stannyl radicals, inversion, 281, 282
Stereochemistry, in carbene 1,2 rearrangements, 105–108
Stereomutation
  cyclopropane thermolysis, 328–334
  and cycloreversions, 353
  spiropentane thermolysis, 348, 349
Stereospecificity, 266, 353, 356, 371
Substituents, effect on migratory aptitude, 39–41
  on ring-closure reactions, 184, 185, 189–198
  on rotational rates of biradicals, 356
α'-Sulfonyl carbanion, 461–465
Sulfuranyl radicals, migration, 245
Sulfur compounds, cyclic, synthesis by ring closure, 219

T

Tetracycloundecane, rearrangement pathways, 27, 28
Tetrahydrodicyclopentadiene, rearrangement pathways to adamantane, 25
3,3'-Tetramethylenebicyclopropenyl, thermal rearrangement, 320, 321
Tetramethylene biradical, properties, 353, 354
Tetraphenylcyclopropene, thermal rearrangement, 316
Thermal rearrangement, see Rearrangement
Thermochemistry, ring-closure reactions, 183, 184
Thietane, ring-opening reactions, 226
Thioalkoxy migration, 1,2 rearrangement, 98, 99
Thiocarbonylcarbenes, rearrangement, theory, 122, 126–128
Thiyl radicals, migration of organic, 244–246
Tricyclodecane, rearrangement pathways, 25–27
Tricyclononyl cation, deuterium effect, 9
Tricycloundecane, rearrangement mechanism and graph, 27, 28
Tricyclyl radical, rearrangement, 199, 233
Trimethylene biradicals
  in cyclopropane pyrolysis, 324–352

properties, 327, 328, 353
Trimethylenemethane, in pyrolysis of methylenecyclopropanes, 334–348
Trimethylsilyl migration, 399
1,2,4-Trimethylspiropentane, stereomutations, kinetic analysis, 350
1,2,3-Triphenyl-3-vinylcyclopropene, thermal rearrangement, 316
Tropilidene, pyrolysis, sigmatropic rearrangement, 376–383
Tropylium cation, detection, 79

V

Vinylcarbene, ring closure, 316, 317
anti-Vinylcyclopropane-2,3-*cis*-d_2, thermolysis, stereomutation, 330
Vinylcyclopropylidene, rearrangement, 150–153
Vinylidenes, 1,2 rearrangements, 113, 114
Vinylmethylenecyclopropane, thermal rearrangement, 366
Vinyl migration, 108–113, 178–181
Vinylnitrenes, rearrangement, 101
Vinyl radicals, inversion, 280

W

Wagner–Meerwein rearrangement, 9, 30–32
Wolff rearrangement
  1,2, of carbenes, 97, 136
  of α-diazo ketones, 322

ORGANIC CHEMISTRY
A SERIES OF MONOGRAPHS

EDITOR

HARRY H. WASSERMAN

Department of Chemistry
Yale University
New Haven, Connecticut

1. Wolfgang Kirmse. CARBENE CHEMISTRY, 1964; 2nd Edition, 1971

2. Brandes H. Smith. BRIDGED AROMATIC COMPOUNDS, 1964

3. Michael Hanack. CONFORMATION THEORY, 1965

4. Donald J. Cram. FUNDAMENTALS OF CARBANION CHEMISTRY, 1965

5. Kenneth B. Wiberg (Editor). OXIDATION IN ORGANIC CHEMISTRY, PART A, 1965; Walter S. Trahanovsky (Editor). OXIDATION IN ORGANIC CHEMISTRY, PART B, 1973; PART C, 1978

6. R. F. Hudson. STRUCTURE AND MECHANISM IN ORGANO-PHOSPHORUS CHEMISTRY, 1965

7. A. William Johnson. YLID CHEMISTRY, 1966

8. Jan Hamer (Editor). 1,4-CYCLOADDITION REACTIONS, 1967

9. Henri Ulrich. CYCLOADDITION REACTIONS OF HETEROCUMULENES, 1967

10. M. P. Cava and M. J. Mitchell. CYCLOBUTADIENE AND RELATED COMPOUNDS, 1967

11. Reinhard W. Hoffmann. DEHYDROBENZENE AND CYCLOALKYNES, 1967

12. Stanley R. Sandler and Wolf Karo. ORGANIC FUNCTIONAL GROUP PREPARATIONS, VOLUME I, 1968; VOLUME II, 1971; VOLUME III, 1972

13. Robert J. Cotter and Markus Matzner. RING-FORMING POLYMERIZATIONS, PART A, 1969; PART B, 1; B, 2, 1972

14. R. H. DeWolfe, CARBOXYLIC ORTHO ACID DERIVATIVES, 1970

15. R. Foster. ORGANIC CHARGE-TRANSFER COMPLEXES, 1969

16. James P. Snyder (Editor). NONBENZENOID AROMATICS, VOLUME I, 1969; VOLUME II, 1971

17. C. H. Rochester. ACIDITY FUNCTIONS, 1970

18. Richard J. Sundberg. THE CHEMISTRY OF INDOLES, 1970

19. A. R. Katritzky and J. M. Lagowski. CHEMISTRY OF THE HETEROCYCLIC N-OXIDES, 1970

20. Ivar Ugi (Editor). ISONITRILE CHEMISTRY, 1971

21. G. Chiurdoglu (Editor). CONFORMATIONAL ANALYSIS, 1971

22. Gottfried Schill. CATENANES, ROTAXANES, AND KNOTS, 1971

23. M. Liler. REACTION MECHANISMS IN SULPHURIC ACID AND OTHER STRONG ACID SOLUTIONS, 1971

24. J. B. Stothers. CARBON-13 NMR SPECTROSCOPY, 1972

25. Maurice Shamma. THE ISOQUINOLINE ALKALOIDS: CHEMISTRY AND PHARMACOLOGY, 1972

26. Samuel P. McManus (Editor). ORGANIC REACTIVE INTERMEDIATES, 1973

27. H. C. Van der Plas. RING TRANSFORMATIONS OF HETEROCYCLES, VOLUMES 1 AND 2, 1973

28. Paul N. Rylander. ORGANIC SYNTHESES WITH NOBLE CATALYSTS, 1973

29. Stanley R. Sandler and Wolf Karo. POLYMER SYNTHESES, VOLUME I, 1974; VOLUME II, 1977; VOLUME III, 1980

30. Robert T. Blickenstaff, Anil C. Ghosh, and Gordon C. Wolf. TOTAL SYNTHESIS OF STEROIDS, 1974

31. Barry M. Trost and Lawrence S. Melvin, Jr. SULFUR YLIDES: EMERGING SYNTHETIC INTERMEDIATES, 1975

32. Sidney D. Ross, Manuel Finkelstein, and Eric J. Rudd. ANODIC OXIDATION, 1975

33. Howard Alper (Editor). TRANSITION METAL ORGANOMETALLICS IN ORGANIC SYNTHESIS, VOLUME I, 1976; VOLUME II, 1978

34. R. A. Jones and G. P. Bean. THE CHEMISTRY OF PYRROLES, 1976

35. Alan P. Marchand and Roland E. Lehr (Editors). PERICYCLIC REACTIONS, VOLUME I, 1977; VOLUME II, 1977

36. Pierre Crabbé (Editor). PROSTAGLANDIN RESEARCH, 1977

37. Eric Block. REACTIONS OF ORGANOSULFUR COMPOUNDS, 1978

38. Arthur Greenberg and Joel F. Liebman, STRAINED ORGANIC MOLECULES, 1978

39. Philip S. Bailey. OZONATION IN ORGANIC CHEMISTRY, VOL. I, 1978

40. Harry H. Wasserman and Robert W. Murray (Editors). SINGLET OXYGEN, 1979

41. Roger F. C. Brown. PYROLYTIC METHODS IN ORGANIC CHEMISTRY: APPLICATIONS OF FLOW AND FLASH VACUUM PYROLYTIC TECHNIQUES, 1980

42. Paul de Mayo (Editor). REARRANGEMENTS IN GROUND AND EXCITED STATES, VOLUME I, 1980; VOLUME II, 1980; VOLUME III, 1980

43. Elliot N. Marvell. THERMAL ELECTROCYCLIC REACTIONS